T0375900
9781848820722-2

Sources and Studies in the History of Mathematics and Physical Sciences

For other titles published in this series, go to
http://www.springer.com/series/4142

Gaṇita-Yukti-Bhāṣā
(Rationales in Mathematical Astronomy)
of Jyeṣṭhadeva

Volume II: Astronomy

A Critical Translation of the Original Malayalam Text by
K.V. Sarma

With Explanatory Notes by
K. Ramasubramanian
M. D. Srinivas
M. S. Sriram

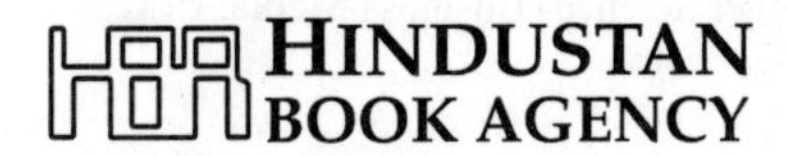

Sources Editor:
J. L. Berggren
Department of Mathematics
Simon Fraser University
Burnaby
BC Canadaa
V5A 1S6

British Library Cataloguing in Publication Data
A catalogue record for this book is available from the British Library

Library of Congress Control Number: 2008937472

ISBN: 978-1-84882-072-2 Printed on acid-free paper

A co-publication with the Hindustan Book Agency, New Delhi, licensed for sale in all countries outside of India. Sold and distributed within India by Hindustan Book Agency, P 19 Green Park Extn., New Delhi 110 016, India.

The work leading to this publication was undertaken under the auspices and with the financial support of the Indian Institute of Advanced Study, Simla.

9 8 7 6 5 4 3 2 1

Springer Science+Business Media, LLC
springer.com

Table of Contents

Gaṇita-Yukti-Bhāṣā

English Translation

Chapters 8 – 15

Chapter 8

Computation of Planets

8.1 Planetary motion

Now, all planets move in circular orbits. The number of degrees which each planet moves in its orbit in the course of a day is fixed. There again, the number of *yojana*-s moved per day is the same for all planets. For planets which move along smaller orbits, the circle would be completed in a shorter time. For those which move along larger orbits, the circle would be completed only in a longer period. For instance, the Moon would have completely moved through the twelve signs in 28 days, while Saturn will complete it only in 30 years. The length of time taken is proportional to the size of the orbit. The completion of the motion of a planet once in its orbit is called a *bhagaṇa* of that planet. The number of times that a planet completes its orbit during a *catur-yuga* is called its *yuga-bhagaṇa* (revolutions per aeon).

Now, if the Moon is seen with an asterism on a particular day, it will be seen the next day with the asterism to the east of it. From this, it might be understood that the Moon has proper motion (relative to the stars), and that the motion is eastwards. The sequence of the signs can also be understood to be eastwards. For all these orbits, a particular point is taken as the commencing point. This point is termed as the first point of Aries (*Meṣādi*). All the circles considered in a sphere are divided into 21,600 equal parts. Each part is a minute (*ili*). They are larger in bigger circles and smaller in smaller circles, the number of parts being the same in all. The number of minutes that a planet will move along its orbit during the course of a day is fixed. If one observes the said motion placing himself at the centre of the orbit of a planet, then the motion of the planet would appear equal every day. The centre of the planetary orbit is slightly above the centre of the Earth. The observer is, however, situated on the Earth. Conceive a circle touching the planet and with the observer at its centre. The observer would find the planet that much advanced from the first point of Aries as it has advanced in the said circle. The method by which this is ascertained is called the 'computation of the true planet' (*sphuṭa-kriyā*). We state it here, deferring the specialties to later sections.

8.2 Celestial Sphere (*Bhagola*)

Now, there is what is called *bhagola-madhya* (centre of the celestial sphere). That is a point from where the stars in general are all taken to be at the same distance. There, it would seem that the centre of the Earth and the *bhagola-madhya* are one and the same. Whatever difference there might be, will be dealt with later.

8.3 Motion of planets: Conception I

First is stated the computation of the true positions of the Sun and the Moon, for the reason that it is simple. Now, consider a circle with its centre at the centre of the celestial sphere. This circle is much smaller than the orbital circle of the planet. The centre of the orbital circle of the planet (*graha-bhramaṇa-vṛtta*) will be on the circumference of this (smaller) circle. This smaller circle is called *mandocca-nīca-vṛtta* (or *manda-nīca-vṛtta*, *manda*-circle). The orbital circle of the planet is called *pratimaṇḍala* (eccentric circle). The centre of the *pratimaṇḍala* will move on (the circumference of) the *ucca-nīca-vṛtta*. The rate of motion of this circle (*pratimaṇḍala*) will be the rate of motion of the *mandocca*. The rate of motion of the planet on the circumference of the *pratimaṇḍala* will be the same as the mean motion (*madhya-gati*) of the planet. The circles should be so constructed that there is no gap between the centre (of the *pratimaṇḍala*) and the circumference (of the *manda-vṛtta*), both touching each other.

In the methodologies of computation of true planets, where the centre of a circle is assumed to be moving on the circumference of another circle, the east-west line of the moving circle should always be conceived to be along the east-west direction. The transverse of this line, viz., the north-south line, is the same as the the up-down line (i.e., the *ūrdhvādho-rekhā*). That line should always be positioned the same way. There should not be any change in their directions. It is in this way that the motion should be conceived. This being the case, when the centre of this circle moves a certain extent on the circumference of a circle of a certain size, it would be that all the parts of that moving circle would be moving together on the circle of that size. When the centre of the moving circle has completed one cycle, it would be that all the parts of the moving circle have also completed one cycle. Here, for a planet situated on the circumference of a circle, even if it (the planet) does not have a motion on its own, it would ultimately result that the planet would be executing a motion along the same (similar) circle, which the centre of the circle supporting the planet is executing; the rate of motion being the same as that of the *mandocca*. (This motion is) similar to the motion of persons travelling in a vehicle. Thus, this motion of the planet is due to the motion of the centre of the *pratimaṇḍala*.

Now, the Sun and the Moon have *manda-nīcocca-vṛtta*-s, with their centres at the centre of the *bhagola* (celestial sphere). Further, they have a planetary orbital circle (*graha-bhramaṇa-vṛtta*) with their centres on the circumference of these (*manda-nīcocca-vṛtta*-s). The centre of the planetary orbital circle will move on the circumference of this *mandocca-vṛtta* with a rate of motion equal to that of the *mandocca*. Besides (this motion), the planets will also move on their orbital circles with their own mean rates of motion. Thus, the motion of the planets and the *graha-bhramaṇa-vṛtta*-s have to be conceived. This is the actual situation.

8.4 Motion of planets: Conception II

The same result can be achieved also through another conception. Construct a circle which is similar to the *graha-bhramana-vṛtta*, with its centre at the centre of the celestial sphere. This is called *kakṣyā-vṛtta* (orbital circle). Construct an *ucca-nīca-vṛtta* (or *nīcocca-vṛtta*) with its centre on the circumference of the above-said (orbital) circle. The size of the *ucca-nīca-vṛtta* will be the same as stated earlier (in conception I). The centre of the *ucca-nīca-vṛtta* will move on the circumference of the orbital circle at the rate of the mean planet. And, along the circumference of the *ucca-nīca-vṛtta*, the planet will move with the speed of the *mandocca*. Here the *ucca-nīca-vṛtta* is the support for the motion of the planet. Then, conceive that the centre of the *ucca-nīca-vṛtta* has the same rate of motion as had been previously proposed for the planet on the *pratimaṇḍala* (eccentric circle). Also, suppose the rate of motion originally proposed for the centre of the *pratimaṇḍala* (eccentric circle) earlier, to be the rate of motion of the planet moving on the *ucca-nīca-vṛtta*, whose centre is now supposed to move on the circumference of the *kakṣyā-vṛtta* (orbital circle). Even in this conception, the result will be the same. In this case, when the centre of the *ucca-nīca-vṛtta* moves on an orbital circle equal in size to the eccentric circle, every part of this *ucca-nīca-vṛtta* will move on a circle with the same size as the orbital circle. Hence, the planet moving on the circumference of the *ucca-nīca-vṛtta*, on account of its support on the (orbital) circle, will consequently be moving on an eccentric circle of the same size. Here, for the motion of the centre of *ucca-nīca-vṛtta*, the support is the *kakṣyā-maṇḍala*: note its centre; the centre of the eccentric circle, which is the support of the motion of the circumference of the *ucca-nīca-vṛtta*, will be removed from the (previously mentioned) centre by the radius of the *ucca-nīca-vṛtta*.

In the present section on (the computation of) true planets, the motion of the orbital and other circles must be conceived in such a way, that the (north-south and east-west) direction lines marked (on them) remain unchanged in all cases. Then, it is also to be noted that the measure of the circle, on which the centre of a circle moves, will be same as the measure of the circle on which all its parts move. Therefore, the mean motion of a planet can be conceived for the centre of the *nīcocca-vṛtta*, which lies on the circumference of the orbital circle (*kakṣyā-vṛtta*), or for the planet on the *pratimaṇḍala*, which has its support on the circumference of this (i.e., *nīcocca-vṛtta* with centre at the centre of the celestial sphere), since in both cases the result is the same. In other words, for computing the true planet it is sufficient to have the two circles, the *kakṣyā-maṇḍala* (orbital circle) with its centre at the centre of the *bhagola*, and the *ucca-nīca-vṛtta* with its centre on the circumference of the *kakṣyā-maṇḍala*; or the *ucca-nīca-vṛtta* with its centre at the centre of the *bhagola* and the *pratimaṇḍala* (eccentric circle) with its centre on the circumference of the *ucca-nīca-vṛtta*. One can also have all the four circles.

8.5 The position of *Ucca*

Now, find the deviation from the first point of Aries of the apogee of the Moon (*candra-tuṅga*) as calculated by the rule of three. Mark that point on the *ucca-nīca-vṛtta*, whose centre is at the centre of the *bhagola*, and with that as the centre, construct the eccentric circle. The location of the mean planet must be marked on the circumference of the eccentric circle by finding the mean position using the rule of three. Let the centre of the *ucca-nīca-vṛtta* be marked on the circumference of the *kakṣyā-vṛtta* at the point where the mean planet should be. Then, place the planet on the circumference of the *ucca-nīca-vṛtta* where the apogee (*tuṅga*) should be. In this model, the planet will be located at that point of intersection of the circumferences of the *ucca-nīca-vṛtta* on the *kakṣyā-vṛtta* and the *pratimaṇḍala*, which is close to the location of the *ucca*. (In fact) the circumferences of these circles intersect at two places. The planet will be at that point of intersection of the circumferences which happens to be in the region of the *ucca* (*ucca-pradeśa*).

8.6 *Ucca*, *Madhyama* and *Sphuṭa*

When the *ucca* and *madhya* as derived using the rule of three coincide, then the centres of all the four circles will be on the same line. Assuming this (phenomenon) to occur on the east-west line (*pūrva-sūtra*), herein below is described how to ascertain the difference between the *ucca* and the *madhya* and the motion of the circles and of the planet.

There, the centre of the *kakṣyā-maṇḍala* and that of the *ucca-nīca-vṛtta* have been presumed to be at the centre of the *bhagola*. The centre of the *pratimaṇḍala* had been presumed to be at the eastern point on the *ucca-nīca-vṛtta*. Now, presume another *ucca-nīca-vṛtta* on the east-west line itself with its centre on the circumference of the *kakṣyā-vṛtta*. The tip of the east-west line of this (second) *ucca-nīca-vṛtta* and the tip of the east-west line of the *pratimaṇḍala* will touch one another. Since the intersection of the circumferences of the *ucca-nīca-vṛtta* and of the *pratimaṇḍala* is at the tip of the east-west line, the planet will also be at the tip of the east-west line. At this moment, since the line from the centre of the *kakṣyā-maṇḍala* and that from the centre of the *pratimaṇḍala* touching the planet are the same, there is no difference between the true and mean planets. Now, the difference between the true and mean (positions of the planet) commences from this situation where the mean meets the *ucca*.

8.7 Computation of true Sun

First, the procedure for obtaining true Sun is being explained. Since, the motion of the centre of the *pratimaṇḍala* in this case is so small, it might be considered as if the motion does not exist. The advantage in this presumption is that it would then be sufficient to consider the motion of the planet alone. This is so in the first conception. In the second conception, we suppose that the centre of the *ucca-nīca-vṛtta* alone moves on the circumference of the *kakṣyā-vṛtta;* then also, the result will be the same. It will also be advantageous to explain the two types of motion considering them simultaneously.

Now, when the *madhya* has moved three signs from its *ucca*, the centre of the *ucca-nīca-vṛtta* will be (at the north-point) on the circumference of the *kakṣyā-vṛtta*. Also, the east-point of the *ucca-nīca-vṛtta* would be touching the north-point of the *pratimaṇḍala*. The planet will be at that point at that time. Here, the motion of the planet on the circumference of the *pratimaṇḍala* and the motion of the centre of the *ucca-nīca-vṛtta* on the circumference of the *kakṣyā-vṛtta* would be the same.

Now, two bodies, starting from the same point and at the same time and moving at the same rate on circles having the same dimension, actually move through the same degrees in their respective circles. Hence, when the planet and the centre of the *ucca-nīca-vṛtta* have travelled through one-fourth the circumference in their respective circles, they will be at the north-point of their circles. Here, the east-west line (*pūrva-sūtra*) common to the *kakṣyā-vṛtta* and the *pratimaṇḍala* is called *ucca-nīca-sūtra*. (This is called so) because it touches the points on the circumference of the *pratimaṇḍala* farthest from (i.e., *ucca*), and nearest to (i.e., *nīca*), the centre of the *bhagola*. Here, the *madhyama* would be on the *pratimaṇḍala* at a distance of three signs from the east-point.

Now, the *sphuṭa* (true longitude of the planet) is equal to the distance moved on that circle, whose centre is the centre of the *bhagola* and whose radius is equal to the line joining the said centre and the planet. Here, when the (mean) planet is on the circumference of the *kakṣyā-vṛtta* at the north point, the *sphuṭa* would have moved three signs from the *ucca*. Therefore, when the *madhyama* has moved three signs, the planet would be towards the east of the north-point of the *kakṣyā-vṛtta*, at a distance separated from it by the radius of the *ucca-nīca-vṛtta*. Hence, at that moment, the difference between the *sphuṭa* and *madhyama* will be equal to the radius of the *ucca-nīca-vṛtta*. In other words, the *sphuṭa* will be less than the *madhyama* by a measure equal to the radius of the *ucca-nīca-vṛtta* when it (the *madhyama*) has moved by three signs.

Now, the circle that is constructed with its centre at the centre of the *bhagola* and with radius equal to the distance therefrom to the planet, would be called *karṇa-vṛtta* (hypotenuse-circle). Since this circle and the *kakṣyā-vṛtta* have their centres at one place, the number of minutes of arc (*ili*) in both are the same. Hence, the mean planet, which has been presumed (above) to be at the centre of the *ucca-nīca-vṛtta* on the circumference of the *kakṣyā-vṛtta* at the tip of its north-line, can be assumed to be at the north-point of the *karṇa-vṛtta*. Now, the difference from that point (the mean planet on the *karṇa-vṛtta*) to the point where the planet lies, will be the *sputa-madhyāntarāla-cāpa* (arc of the difference between the true and mean). Hence, this *sphuṭa-madhyāntarāla-cāpa* would be got by taking the radius of the *ucca-nīca-vṛtta* as the *jyā* (Rsine) in the *karṇa-vṛtta* and finding its arc. Now, since the *madhyama* has moved by three signs from the east-point which is on the *ucca-sūtra*, if this *sphuṭa-madhyāntarāla-cāpa* is subtracted from three signs, the remaining part will be the difference between the planet and the *ucca-sūtra* on the circumference of the *karṇa-vṛtta*. When the *ucca* (the longitude of apogee) is added to this, the true position of the planet from the first point of Aries will result. The above result for the true planet, i.e., how far has the planet moved in the *karṇa-vṛtta*, can be obtained even by subtracting, from the *madhyama*, that portion of the arc in the hypotenuse circle which is equal to the difference between *sphuṭa* and *madhyama*.

Now, when it has been conceived that the *ucca* is on the east-line and the *madhya* is at the *ucca*, it has also been conceived that the planet is at the east-point of the *pratimaṇḍala*, and that the centre of the *ucca-nīca-vṛtta* is at the east-point of the *kakṣyā-vṛtta*. In both these conceptions, the east-line is the same for both the *kakṣyā-vṛtta* and the *pratimaṇḍala*. Thus, since the minutes of motion is the same (in both these conceptions) at that instant, the mean and the true are also the same.

Now, when the planet and the centre of the *ucca-nīca-vṛtta*, both of which have the same rate of motion, move by three signs, the planet will reach the north-point on the *pratimaṇḍala* and the centre of the *ucca-nīca-vṛtta* will be at the north-point of the *kakṣyā-vṛtta*. While the centre of the *ucca-nīca-vṛtta* is conceived to move in such a manner that there is also no change in the direction lines (drawn on these circles), the planet will not deviate from the east-point of the *ucca-nīca-vṛtta*. Therefore, at that time, the difference between the true and mean planet (*sphuṭa-madhyāntarāla*) will be equal to the radius of the *ucca-nīca-vṛtta*.

Then, when it moves by another quarter of a circle, (i.e., three signs), the planet will be at the west-point on the *pratimaṇḍala*, and the centre of the *ucca-nīca-vṛtta* will be on the west-point of the *kakṣyā-vṛtta*. Then also it will be the case that the planet is at the east-point of the *ucca-nīca-vṛtta*. Since, the west-line of the *kakṣyā-vṛtta* is the same as that of the *pratimaṇḍala*, and the minutes of arc at that place is also the same, the true and the mean planets are the same even at that situation. Thus, even when the *madhyama* and the *nīca* are the same, there will be no difference between the true (*sphuṭa*) and the mean (*madhyama*).

Now, when the two move by still another quarter of a circle, they will be at the south-point. Here also, since the planet is at the east-point of the *ucca-nīca-vṛtta*, the centre of the *ucca-nīca-vṛtta* is to the west of the planet, by a measure equal to the radius of the *ucca-nīca-vṛtta*. Hence, here, the arc of the radius of the *ucca-nīca-vṛtta* should be added to the *madhyama*. That will be the true planet (*sphuṭa*). Again, moving three signs further, when (the planet) reaches the *ucca*, there will be no difference between the *sphuṭa* and the *madhyama*.

Thus, the increase and decrease in the difference between *sphuṭa* and *madhyama* occur, starting from the conjunction of the (*madhyama* and) *ucca*, in accordance with the quarter of the circle (*vṛtta-pāda*) occupied by the *madhyama*. It is significant to note that if the *jyā* (Rsine) of the difference between the *ucca* and the *madhyama* on the *pratimaṇḍala* is converted by the rule of three to the *ucca-nīca-vṛtta*, then it will be the *jyā* (Rsine) of the difference between the true and the mean planet.

If it is asked, how it is so (here is the explanation): Now, consider the line drawn from the centre of the *kakṣyā-vṛtta* passing through the centre of the *ucca-nīca-vṛtta*, which is on the circumference of the former, and meeting the circumference on the other side (outer side of the *ucca-nīca-vṛtta*). This line, will represent the minutes of the *madhyama-graha*. Now, the planet is situated at the point where the east-line of the *ucca-nīca-vṛtta* and its circumference meet the circumference of the *pratimaṇḍala*. The difference between the planet at that point and the minutes of arc of the *madhyama* is the *madhyama-sphuṭāntara*. Now, that line is the *madhya-sūtra* which cuts the two points of the *ucca-nīca-vṛtta* that are farthest from, and nearest to, the centre of the *kakṣyā-vṛtta*. Therefore, the tip of this line on *ucca-nīca-vṛtta* is the *ucca*. Hence, the Rsine of the arc in the *ucca-nīca-vṛtta*, which is the traversed portion between the *ucca* on the *ucca-nīca-vṛtta* and (the tip of) the east-line (on it), would be the interstice between the true and mean planets.

Now, if the centre of the *ucca-nīca-vṛtta* is at the east-point on the circumference of the *kakṣyā-vṛtta*, then the tip of east-line on the *ucca-nīca-vṛtta* will be the location of the *ucca*-point. If, however, the centre of the *ucca-nīca-vṛtta* is at the north-east corner in the *kakṣyā-vṛtta*, the north-east point of the *ucca-nīca-vṛtta* would be the *ucca*-point. If the centre is at the northpoint, the *ucca* will be at that point. Thus, the difference between the east-line (of the *kakṣyā-vṛtta*) which has been conceived as the *ucca*-line, and the centre of the *ucca-nīca-vṛtta* whose centre is on the *kakṣyā-vṛtta*, will be equal to the interstice between the east-line and the *ucca*-point on the *ucca-nīca-vṛtta*, in its own measure. Now, calculate the Rsine of the arc on the *kakṣyā-vṛtta*, between *ucca* and *madhyama*. Convert this, by the rule of three, into Rsine on the *ucca-nīca-vṛtta*. This Rsine will be the Rsine of the difference between the *ucca* (*sphuṭa*?) and *madhyama*. The Rsine of the difference between the *sphuṭa* and the *madhyama* will be obtained, even if the rule of three is applied using the Rsine of the arc between the planet and the *ucca*-point on the circumference of the *pratimaṇḍala*. Moreover, the separation between the *ucca-sūtra* and the planet in the *pratimaṇḍala* is the same as the separation between *ucca-sūtra* and the planet on the *ucca-nīca-vṛtta*.

Here, when the planet makes one revolution starting from the *ucca*-point which has been conceived as the east-point on the *pratimaṇḍala*, the *ucca*-point on the *ucca-nīca-vṛtta* will also complete one revolution. Hence, the difference between the *ucca-sūtra* and the planet in the *pratimaṇḍala*, and the difference between the *ucca-sūtra* and the planet on the *ucca-nīca-vṛtta* are equal in degrees. Therefore, when the Rsine of the difference between the *ucca* and *madhyama* is multiplied by the radius of the *ucca-nīca-vṛtta* and divided by Rsine of three signs (*trijyā* or radius), we will get the Rsine of the difference between the *sphuṭa* and *madhyama.* If this Rsine is taken as the Rsine on the *karṇa-vṛtta* and converted to arc, and applied to the mean planet, the true planet will be obtained.

8.8 Computation of the *Karṇa*

Now is stated the method of the computation of the Rsines in the *karṇa-vṛtta.* Here, the *ucca-sūtra* is the line drawn from the centre of the *kakṣyā-vṛtta* and passing through the centre of the *pratimaṇḍala* and touching the circumference of the *pratimaṇḍala* (on the other side). As stated above, it has been taken as the east-line. The (extended) part of that line towards the west is the *nīca-sūtra.* And the entire line is termed *ucca-nīca-sūtra.* Consider the Rsine of the arc on the *pratimaṇḍala* from the planet to this *sūtra*; that Rsine will be the Rsine of the portion corresponding to the *madhyama*–minus–*ucca.* This segment has its tip at the planet and the base on the *ucca-nīca-sūtra.* This will be the *bhujā* (lateral) for deriving the radius of the *karṇa-vṛtta* (hypotenuse-circle). The *koṭi* (upright) is the distance from the base of the Rsine to the centre of the *kakṣyā-vṛtta.* And the *karṇa* (hypotenuse) is the distance from the centre of the *kakṣyā-vṛtta* to the planet.

Now, when the planet is at the *ucca* (on the circumference) of the *pratimaṇḍala*, the *koṭi* would be the difference (sum?) of the Rcosine of the *ucca*–minus–*madhyama* and the radius of the *ucca-nīca-vṛtta.* When, however, the planet is at the *nīca* on the *pratimaṇḍala*, the *koṭi* would be the difference of the Rcosine of the *ucca*–minus–*madhyama* and the radius of the *ucca-nīca-vṛtta.* The Rcosine of the *ucca*–minus–*madhyama* is the distance from the centre of the *pratimaṇḍala* to the base of the Rsine. The radius of the *ucca-nīca-vṛtta* would be the distance between the centre of the *pratimaṇḍala* and the centre of the *kakṣyā-maṇḍala.*

When, the planet is to the east of the north-south line passing through the centre of the *pratimaṇḍala*, the *koṭi* of the hypotenuse circle (*karṇa-vṛtta-koṭi*) would be the sum of the Rcosine (found) with respect to the centre (of the *pratimaṇḍala*) and the radius of the *ucca-nīca-vṛtta*. If, however, the planet is to the west of the north-south line passing through the centre of the *pratimaṇḍala* (and lies just below it), then the base of the Rsine would fall inside the *ucca-nīca-vṛtta* (drawn at the centre of *bhagola*). Now, if the base of the Rsine is to the east of the north-south line of the *ucca-nīca-vṛtta*, situated at the centre of the *kakṣyā-vṛtta*, the Rcosine would be the distance from the base of the Rsine to the circumference of the *ucca-nīca-vṛtta*. When this (segment) lying inside the *ucca-nīca-vṛtta*, is subtracted from the radius of the *ucca-nīca-vṛtta*, the remainder, which is the distance between the centre of the *kakṣyā-vṛtta* and the base of the Rsine, would be the *koṭi* for the hypotenuse circle. However, if the base of the Rsine is to the west of the north-south line of the *ucca-nīca-vṛtta* situated at the centre of the *kakṣya-vṛtta*, the *koṭi* of hypotenuse circle would be the Rcosine from the centre (of the *pratimaṇḍala*) minus the radius of the *ucca-nīca-vṛtta*.

Now, *madhya*–minus–*ucca* is stated to be the *kendra*. When the *bhujā* and *koṭi* thus obtained and related to the *karṇa-vṛtta*, are squared added together and the square root of the sum is calculated, the result obtained will be the distance between the centre of the *kakṣyā-vṛtta* and the planet, which is equal to the radius of the the *karṇa-vṛtta* in terms of the minutes of arc of the *pratimaṇḍala*. When this itself is measured in terms of the minutes of arc of the hypotenuse circle, it would be equal to *trijyā* (Rsine of three signs). Now, if (the circumference of) any circle is divided by 21,600, each part would be equal to one minute in that circle. And the radius of the circle would be equal to Rsine of three signs (*trijyā*) in its own measure. Hence, it was said that (the radius of the hypotenuse circle) measured in terms of the minutes of arc of the hypotenuse circle, would be equal to *trijyā*. Since there would be increase and decrease in the (dimension of) *mandocca-nīca-vṛtta* on account of (the increase and decrease of) the *manda-karṇa* (the hypotenuse), it (i.e., the dimension of the *mandocca-nīca-vṛtta*) is always measured in terms of the minutes of arc of the hypotenuse circle. Only when this hypotenuse is calculated by *aviśeṣa* (iteration) will it be converted to minutes of arc of the *pratimaṇḍala*. Thus (has been explained) the method of knowing the measure on the *karṇa-vṛtta* from the minutes of arc of the *pratimaṇḍala*.

8.9 Alternative method for finding the *Karṇa*

Here is an alternative method of deriving (the hypotenuse). Now, the line starting from the centre of the *kakṣyā-vṛtta* passing through the centre of the *ucca-nīca-vṛtta* (which is) on the circumference of the *kakṣyā-vṛtta*, and meeting its circumference (i.e., the circumference of the *ucca-nīca-vṛtta*), is called *madhyama-sūtra*, as mentioned earlier. The (perpendicular) distance from this line to the planet is the difference between the *madhyama* and *sphuṭa*. This is called *bhujā-phala* (or *doḥ-phala*). Take this as having its tip at the planet and base on the *madhyama-sūtra*. The *koṭi-phala* would be the distance from the foot of the *bhujā* and the centre of the *ucca-nīca-vṛtta* which is on the circumference of the *kakṣyā-vṛtta*.

If the planet happens to be situated on (the portion of the circumference of) the *pratimaṇḍala* which is outside the circumference of the *kakṣyā-vṛtta*, the foot of the *bhujā-phala* would also be outside the circumference of the *kakṣyā-vṛtta*. In that case, if the *koṭi-phala* is added to the radius of the *kakṣyā-vṛtta*, the result will be the interstice between the foot of the *bhujā-phala* and the centre of the *kakṣyā-vṛtta*. If, on the other hand, the planet is on the (portion of the) circumference of the *pratimaṇḍala* which happens to be inside the circumference of the *kakṣyā-vṛtta*, the foot of the *bhujā-phala* would be inside the circumference of the *kakṣyā-vṛtta*. Then, the *koṭi-phala* subtracted from the radius of the *kakṣyā-vṛtta* will give the interstice between the base of the *bhujā-phala* and the centre of the *kakṣyā-vṛtta*. Then, take the difference between the foot of the *bhujā-phala* and the centre of the *kakṣyā-vṛtta* as the *koṭi* and the *bhujā-phala* as the *bhujā*. Square the two, add together (the results) and find the root; the result would be the distance between the planet and the centre of the *kakṣyā-vṛtta*, in terms of the minutes of arc of the *pratimaṇḍala*, which is the same as the *karṇa* obtained earlier. Thus the radius of the *karṇa-vṛtta* can be derived in two ways.

Now, it is learnt from the *madhyama* as to how far the planet has moved in the *pratimaṇḍala*. The *karṇa* has been derived above, using which the distance through which the planet has moved on the *karṇa-vṛtta* can be found.

8.10 *Viparīta-karṇa* (Inverse hypotenuse)

Now, is explained the method to derive the radius of the *kakṣyā-vṛtta* and the *pratimaṇḍala* from the minutes of arc of hypotenuse-circle (*karṇa-vṛtta*). Since the working here is just the opposite of the derivation of the hypotenuse, this is called the (method for) inverse hypotenuse (*viparīta-karṇa*). Here, in the computation of the *manda-sphuṭa*, the difference between the *madhya* and *sphuṭa* is measured in terms of the minutes of arc of the *manda-karṇa-vṛtta*. When the *bhujā-phala*, which is Rsine of the difference between *madhya* and *sphuṭa*, is squared and subtracted from the square of the radius (*trijyā*) and the square root found, the result will be the interstice between the base of the *bhujā-phala* and the centre of the *kakṣyā-vṛtta*. Subtract the *koṭi-phala* from this, if the base of the *bhujā-phala* is outside the circumference of the *kakṣyā-vṛtta*, and add otherwise. The result will be the radius of the *kakṣyā-vṛtta* in terms of the minutes of arc of the *karṇa-vṛtta*.

8.11 Another method for *Viparīta-karṇa*

Here is another method to derive the radius of the *pratimaṇḍala* in terms of the minutes of arc of the *karṇa-vṛtta*. Now, the *bhujā-jyā* or the Rsine of the difference between the *ucca* and *sphuṭa* is in terms of the minutes of arc of the *karṇa-vṛtta*. As is well known, *sphuṭa* is the distance moved by the planet on the *karṇa-vṛtta*. The *bhujā-jyā* referred to above has its foot in the *nīcocca* line and its tip at the planet. The Rcosine of the difference between the *sphuṭa* and *ucca* is the distance from the foot of the *bhujā* and the centre of the *kakṣyā-vṛtta*. Subtract from this the radius of the *ucca-nīca-vṛtta* which is the distance between the centres of the *kakṣyā-vṛtta* and the *pratimaṇḍala*, in case the foot of the *bhujā-jyā* is outside the circumference of the *ucca-nīca-vṛtta*, otherwise add. Square the *koṭi* which is left, and the *bhujā-jyā*, add them and find the root. The result will be the distance from the centre of the *pratimaṇḍala* to the planet, which is the radius of the *pratimaṇḍala* in terms of the minutes of arc of the *karṇa-vṛtta*.

8.12 Still another method for *Viparīta-karṇa*

Now, a method is given to derive the radius (of the *pratimaṇḍala*) using the *bhujā-phala* and the *koṭi-phala* corresponding to the difference between the *sphuṭa* and the *ucca*. Now, what is called the Rsine of the difference between *sphuṭa* and *ucca* is the Rsine of the arc between the line passing through the planet and the *ucca-nīca* line on the *karṇa-vṛtta*. When this Rsine of the arc between the two lines (*sūtra*-s) is conceived with respect to the *ucca-nīca-vṛtta* at the centre of the *karṇa-vṛtta*, it will be the Rsine of the difference between the *sphuṭa* and the *ucca* (i.e., *doḥ-phala*). This *doḥ-phala* should be conceived to have its tip at the centre of the *pratimaṇḍala* and its foot on the planet-line. The interstice between the foot of this *doḥ-phala* and the centre of the *karṇa-vṛtta* on the planet-line will be the *koṭi-phala* here. And the *karṇa* minus this *koṭi-phala* will be the *koṭi*. Here, the *bhujā* is the *doḥ-phala*. When the two are squared, added together and the root calculated, the result will be the distance between the centre of the *pratimaṇḍala* and planet, which is the radius of the *pratimaṇḍala* in terms of the minutes of arc of the *karṇa-vṛtta*.

This will be the case when the planet is in the *ucca* region (eastern half of the *pratimaṇḍala*). If it is in the *nīca* region (western half of the *pratimaṇḍala*) there is a distinction. Here, the interstice between the *nīca*-line and the planet-line on the *karṇa-vṛtta* is the Rsine of difference between the *sphuṭa* and the *ucca*. This interstice in the *nīcocca-vṛtta* will be the *doḥ-phala*.

Now, there is, on the other side of the *nīca-sūtra*, the remaining portion of the *ucca-nīca-sūtra*; extend the planet-line also to that side through the centre of the *karṇa-vṛtta*. Now, in this case also, the Rsine of the arc between the extension of the planet-line and the *ucca*-line is indeed the above-said *bhujā-phala*. Here also conceive the *doḥ-phala* with its tip at the centre of the *pratimaṇḍala* and with its foot on the extended tail of the planet-line. The *koṭi-phala* is the distance between the foot of the *doḥ-phala* and the centre of the *karṇa-vṛtta* along the extension of the planet-line. When this *koṭi-phala* is added to the (portion of the) planet-line, which is the radius of the *karṇa-vṛtta*, the result will be the distance from the planet to the foot of the said *doḥ-phala*. If the square of this is added to the square of the *doḥ-phala* and the square root is taken, the result will be the distance from the planet to the centre of the *pratimaṇḍala* which is the radius of the *pratimaṇḍala* in terms of the minutes of arc of the *karṇa-vṛtta*. It is to be noted that taking the intervening Rsines in reverse direction does not cause any change in their measures.

Thus has been stated the methods for deriving the radius of the *kakṣyā-vṛtta* and the *pratimaṇḍala* in terms of the minutes of arc of the *karṇa-vṛtta*. The result so obtained, is called the *viparīta-karṇa* (reverse-hypotenuse). Now, the *karṇa* is nothing but the radius of the *karṇa-vṛtta* measured in terms of the minutes of arc of the *pratimaṇḍala*. Since, instead, we employed the reverse process what we obtained is the *viparīta-karna* (inverse hypotenuse). If the square of the radius is divided by this *viparīta-karṇa*, the result would be the *karṇa* (hypotenuse) which is the radius of the *karṇa-vṛtta* measured in terms of the minutes of the *pratimaṇḍala*. Here, the radius of the *pratimaṇḍala* specified in its (*pratimaṇḍala-vṛtta*'s) own measure in minutes (*anantapurāṃśam* or 21,600 equal parts), is equal to *trijyā* (Rsine of three signs) and it is equal to the *viparīta-karṇa* when measured in terms of the minutes of arc of the *karṇa-vṛtta*. The radius of the *karṇa-vṛtta* is equal to the *trijyā* when specified in its own measure. Now, what would it be if measured in terms of the minutes of arc of the *pratimaṇḍala*, has to be calculated by the rule of three. The result would be the radius of the *karṇa-vṛtta* measured in terms of the minutes of arc of the *pratimaṇḍala*.

8.13 *Manda-sphuṭa* from the *Madhyama*

Now, find the Rsine of the arc *madhyama*–minus–*ucca*. That will be the Rsine of the portion of the *pratimaṇḍala* lying between the planet and the *ucca-nīca-sūtra*. If this Rsine is measured in terms of the minutes of arc of the *karṇa-vṛtta* and converted into arc, the result will be the portion of the *karṇa-vṛtta* lying between the planet and the *ucca-nīca-sūtra*. When this arc is applied to the *ucca* or *nīca*, the angle covered by the planet along the *karṇa-vṛtta* is obtained. And this will be the *sphuṭa* (true position of the planet).

We have here the rule of three: The radius of the *karṇa-vṛtta* in terms of the minutes of arc of the *pratimaṇḍala* is equal to the *karṇa*. This is the *pramāṇa*. When the said radius is in terms of the minutes of arc of the *karṇa-vṛtta*, it is equal to *trijyā*. This is the *pramāṇa-phala*. The *icchā* is the Rsine of (the portion of) the *pratimaṇḍala* lying between the planet and the *ucca-nīca-sūtra*. And, that itself, when converted in terms of the minutes of arc and treated as a Rsine of the *karṇa-vṛtta*, would be the *icchā-phala*. When this is applied to the *ucca* or to the *nīca* in accordance to its nearness to either, it is the *sphuṭa* (true planet). This process of obtaining *sphuṭa* is called *pratimaṇḍala-sphuṭa*.

Now, the Rsine of *sphuṭa*–minus–*ucca* will be the *icchā-phala* which has been mentioned above. The Rsine of *madhya*–minus–*ucca* will be the *icchā-rāśi*. Therefore, when the *icchā-phala* is considered as *pramāṇa*, the *icchā-rāśi* is taken as *pramāṇa-phala* and the radius of the *karṇa-vṛtta* which is equal to *trijyā* taken as *icchā-rāśi* (and the rule of three applied), the *icchā-phala* got would be the *karṇa* mentioned above. Here, since the Rsine of *madhya*–minus–*ucca* is the Rsine of a portion of the *pratimaṇḍala*, it is in terms of the minutes of arc of the *pratimaṇḍala*. This is the very Rsine of *sphuṭa*–minus–*ucca* also. Further, the two Rsines are equal, because this segment is perpendicular to the *ucca-sūtra* and represents the distance between the planet and the *ucca-sūtra*. The difference is only because of employing different units for measurement. Since the arc of the *sphuṭa-kendra* (*sphuṭa*–

minus–*ucca*) is a portion of the *karṇa-vṛtta*, the Rsine of *sphuṭa-kendra* is in terms of the minutes of arc of the *karṇa-vṛtta.* That is, this Rsine is nothing but the Rsine of *sphuṭa-kendra* in terms of minutes of the *karṇa-vṛtta.* When this itself is measured in terms of the arc of the *pratimaṇḍala*, it is Rsine of *madhya-kendra* (*madhya*–minus–*ucca*). If this is equal to *trijyā* when measured in terms of the arc of the *karṇa-vṛtta*, by finding what it will be in terms of the minutes of arc of the *pratimaṇḍala* (using rule of three), we obtain the *karṇa* that was stated earlier.

The *karṇa* may also be obtained thus. In this connection, a doubt might arise as to how the *bhujā-phala* of the *madhya-kendra* would be in terms of the minutes of arc of the *karṇa-vṛtta*, when Rsine of *madhya-kendra* is in terms of the minutes of arc of the *pratimaṇḍala.* Here is the answer: When the *karṇa* is large, the *mandocca-nīca-vṛtta* would also be correspondingly large. When the *karṇa* is smaller than *trijyā*, the *mandocca-nīca-vṛtta* would also be correspondingly smaller. Hence, the Rsine in this circle would always be in terms of the minutes of arc of the *karṇa-vṛtta.* Hence it is that the *manda-karṇa* can be derived in this manner. It is again the reason why it is not necessary to resort to the rule of three to convert the *madhya-kendra-bhujā-phala* to minutes of arc of the *karṇa-vṛtta*, when it has to be applied to the *madhyama.* In this manner, since there is an increase and decrease of the dimension of the *mandocca-nīca-vṛtta* in accordance with the *manda-karṇa*, there is this distinction for the *manda-karṇa* and for the *sphuṭa* derived from the *manda-bhujā-phala.* (On the contrary), in the *śīghra* (*phala*) there is no increase or decrease in the dimension of *śīghrocca-nīca-vṛtta* with reference to its *karṇa.* Thus (has been stated) the derivation of *manda-sphuṭa.*

8.14 *Śighra-sphuṭa* (True planets): General

Next is stated the process of *śīghra-sphuṭa.* Since the centre of the *manda-nīcocca-vṛtta* of the Sun and the Moon is at the centre of the *bhagola*, for the Sun and the Moon the *manda-sphuṭa* as computed will give their (true) motion in the *bhagola.* For Mars and other planets, if we presume a circle with its centre as the centre of the *bhagola* and joining the planet, the (true) motion in the *bhagola* would be equal to the measure by which it (the planet) has moved in that circle. The speciality of (Mars and other planets) is this: There is a *śīghra-nīcocca-vṛtta* with its centre at the centre of the *bhagola*. The *manda-nīcocca-vṛtta* moves on the circumference of that (*śīghra-nīcocca-vṛtta*) at the rate of the *śīghrocca*. Hence, at a particular moment, the centre of the *manda-nīcocca-vṛtta* is that point on the circumference of the *śīghra-nīcocca-vṛtta*, where the *śīghrocca* would lie. The *mandocca* moves on this circle (*manda-nīcocca-vṛtta*). Now, presume a *pratimaṇḍala* circle with its centre on the *manda-nīcocca-vṛtta*, at that point where the *mandocca* is located on the *manda-nīcocca-vṛtta*. Presume also that the planet (*graha-bimba*) moves on the circumference of this *pratimaṇḍala*. Then, the extent of motion of the planet at any time along the circumference of the *pratimaṇḍala* as measured from *Meṣādi*, is known by the *madhyama* or the mean planet.

Now, presume another circle with its centre at the centre of the *manda-nīcocca-vṛtta* and touching the planet. This circle is called *manda-karṇa-vṛtta*. *Manda-sphuṭa* is ascertained by calculating how much the planet has moved from *Meṣādi* on this *manda-karṇa-vṛtta* by taking it as the *prati-maṇḍala*. Now, presume a circle with its centre at the centre of the *śīghra-nīcocca-vṛtta* and touching (i.e., having at its circumference at) the planet. This (circle) is called *śīghra-karṇa-vṛtta*. The *śīghra-sphuṭa* (true planet) is known by ascertaining the the number of signs etc., through which the planet has moved in this circle from *Meṣādi*.

Śīghra-sphuṭa can be ascertained by presuming the *manda-karṇa-vṛtta* as the *pratimaṇḍala* and the *manda-sphuṭa-graha* as the mean planet (*madhyama*) and carrying out computations in a manner similar to that (followed in the case) of *manda-sphuṭa*, and thus the number of signs etc. traversed by the planet from *Meṣādi*, in the *śīghra-karṇa-vṛtta*, would be obtained.

There is a special feature in the case *śīghra-sphuṭa*. Here, if the *śīghra-bhujā-phala* is calculated and measured in terms of the minutes of arc of the *śīghra-karṇa* , it will become a *jyā* (Rsine) of the *śīghra-karṇa-vṛtta*. If this is converted into arc and applied, the result would be the distance traversed by the planet in the *śīghra-karṇa-vṛtta.* For this purpose, the *śīghra-bhujā-phala* should be multiplied by *trijyā* and divided by the *śīghra-karṇa.* Since the *śīghra-bhujā-phala* is obtained in terms of the minutes of *manda-karṇa-vṛtta*, it should be multiplied by *trijyā* and divided by *śīghra-karṇa.* In this way, the *śīghra-bhujā-phala* is in terms of the minutes of arc of the *śīghra-karṇa-vṛtta.* Here, in order to get the *manda-bhujā-phala* in terms of *manda-karṇa-vṛtta*, it is not necessary to do such an application of the rule of three. If the *manda-kendra-jyā*-s are multiplied by the radius of the *mandocca-nīca-vṛtta* and divided by *trijyā*, the result will be in terms of the minutes of arc of the *manda-karṇa-vṛtta.* The reason for this is this: when the *manda-karṇa* becomes large, the *mandocca-nīca-vṛtta* will also become large; when it becomes small, the other will also become small. Thus, the *manda-bhujā-phala* and *koṭi-phala* are always measured in terms of the degrees of the *manda-karṇa-vṛtta.* On the other hand, there is no increase or decrease for the *śīghrocca-nīca-vṛtta* in relation to the *śīghra-karṇa-vṛtta.* Hence, the *śīghra-koṭi-phala* and *śīghra-bhujā-phala* will be only in terms of the *pratimaṇḍala.* So, in order to reduce them in terms of the *śīghra-karṇa-vṛtta*, another application of the rule of three is required.

When the dimensions of the *manda-nīcocca-vṛtta* and *śīghra-nīcocca-vṛtta* were given earlier, it was in terms of the dimensions of their own *pratimaṇḍala.* Hence, they have to be first determined in terms of the minutes of arc of the *pratimaṇḍala.* But, there is a distinction: the *manda-nīcocca-vṛtta* has increase and decrease, but the *śīghra-nīcocca-vṛtta* has no increase and decrease.

Now, the *jyā*-s for the differences between the *manda-sphuṭa-graha* and its *śīghrocca* are called *śīghra-kendra-jyā*-s . Since these *jyā*-s are measured in the *manda-karṇa-vṛtta* they are in terms of the minutes of arc of the *manda-karṇa.* Since the *śīghra-vṛtta* is measured in terms of the *pratimaṇḍala*, if the *śīghrocca-nīca-vṛtta* and its radius, which is the *śīghrāntya-phala*, are multiplied by *trijyā* and divided by the *manda-karṇa*, the results will be the *śīghrocca-nīca-vṛtta* and its radius in terms of the *manda-karṇa-vṛtta.* If these are considered as *pramāṇa* and *phala*, and also the *manda-karṇa-vṛtta*, measured in terms of the minutes of arc of itself, and its radius are considered as *pramāṇa* and *phala* and the *śīghra-kendra-bhujā* and *koṭi* as *icchā-rāśi*, (and the rule of three applied), the *icchā-phala*-s thereby obtained would be the *śīghra-bhujā-phala* and *śīghra-koṭi-phala* in terms of the minutes of arc of the *manda-karṇa-vṛtta*. Then, apply this *koṭi-phala* to the radius of the *manda-karṇa-vṛtta* which is equal to *trijyā* as measured by itself; add its square to the square of the *bhujā-phala*, and find the square root. The result would be the distance, in terms of the *manda-karṇa-vṛtta*, from the planet to the centre of the *śīghrocca-nīca-vṛtta* which is also the centre of the *bhagola*. This will be the *śīghra-karṇa*. This (*śīghra-karṇa*) can be computed in different ways.

The *śīghra-antya-phala* which has been measured in terms of the *manda-karṇa-vṛtta*, has to be added or subtracted, depending on whether it is *Makarādi* or *Karkyādi*, to the *śīghra-koṭi-jyā*; the result thus obtained and the *śīghra-kendra-bhujā-jyā* have to be squared, added together and the square root found; this will be the *śīghra-karṇa*, the one which is stated earlier.

Here, if the *manda-sphuṭa-graha* and the *śīghrocca*, which is the *āditya-madhyama* (mean Sun), are subtracted from each other, the result will be *śīghra-kendra*. The *bhujā* and *koṭi-jyā*-s of this are measured in terms of the minutes of arc in the *manda-karṇa-vṛtta*. Since these are the *jyā*-s measured in this circle, if they are multiplied by the *manda-karṇa* and divided by *trijyā*, the result will be *jyā*-s of the *manda-karṇa-vṛtta* measured in terms of the minutes of arc of the *pratimaṇḍala*. Now, if these are multiplied by the *śīghrāntya-phala* (radius of the *śīghrocca-nīca-vṛtta*) stated earlier in terms of the dimensions of the *pratimaṇḍala* and divided by the *manda-karṇa*, we get the *śīghra-bhujā-phala* and *śīghra-koṭi-phala* in terms of the minutes of arc of the *pratimaṇḍala*. Now, if the *koṭi-phala* thus obtained is applied to the *manda-karṇa*, and its square and the square of this *bhujā-phala* are added together and the root found, the result will be *śīghra-karṇa* in terms of the degrees of the *pratimaṇḍala*.

Again, if the *śīghra-kendra-koṭi-jyā* and the *antya-phala* which are measured in terms of the minutes of arc of the *pratimaṇḍala* are added to or subtracted from each other, as the case may be, and the square of the result and the square of the *bhujā-jyā* are added together and the square root found, then also will be obtained the *śīghra-karṇa* in terms of the degrees of the *pratimaṇḍala*.

Now, multiply the *śīghra-kendra-bhujā-jyā* by the *trijyā* and divide by the (*śīghra*) *karṇa*. The result will be the *jyā* of the interstice between the planet and the *śīghrocca-nīca-sūtra*, in terms of the minutes of arc of the *śīghra-karṇa-vṛtta*. If this is converted to arc and applied to the *śīghrocca*, one can find the position of the planet in the *śīghra-karṇa-vṛtta* which has its centre at the *bhagola-madhya*. Now, multiply the *bhujā-phala* by *trijyā* and divide by the (*śīghra*) *karṇa* and find the arc. Apply it to the *manda-sphuṭa-graha*, and this will be the *sphuṭa* as above. Here, if either the *bhujā-jyā* or the *bhūjā-phala*, measured in terms of the *manda-karṇa*, is multiplied by *trijyā*, it has to be divided by the *śīghra-karṇa* which is in terms of the minutes of arc of the *manda-karṇa*. (On the other hand) if it is measured in terms of the minutes of arc of the *pratimaṇḍala*, then the division has to be made by the *śīghra-karṇa* measured in terms of the *pratimaṇḍala*. This is the only distinction.

Jñāta-bhoga-graha-vṛtta is that circle on whose circumference the planet's motion is known. This is taken to be the *pratimaṇḍala*. Then, we have the circle passing through the planet with an appropriate centre (*bhagola-madhya*), on whose circumference the portion traversed is desired to be found. Such a circle is called *jñeya-bhoga-graha-vṛtta*. This circle is taken to be the *karṇa-vṛtta*. Then we construct a circle whose centre is the same as that of the *jñeya-bhoga-graha-vṛtta* and whose circumference passes through the centre of the *jñāta-bhoga-graha-vṛtta*. Such a circle is called *ucca-kendra-vṛtta*. Constituting the three circles as above, derive the *karṇa* according to the *śīghra-nyāya*, and find the *sphuṭa* as instructed above. If this is done, we can ascertain the motion of the planet on a circle with the desired centre, and on whose circumference the planet moves. Thus has been explained the general procedure for finding true planets.

Now, we describe the construction of the *kakṣyā-vṛtta*, when the motion of the planet is conceived of in terms of the *kakṣyā-vṛtta* and the *ucca-nīca-vṛtta* on the circumference of it. With the centre of the *jñeya-bhoga-graha-vṛtta* as centre, construct another circle (whose radius is) equal to the *jñāta-bhoga-graha-vṛtta*. This is the *kakṣyā-vṛtta*. On the circumference of this construct the *ucca-nīca-vṛtta*, with a radius equal to the distance between the centres of the *jñāta* and *jñeya-bhoga-vṛtta*-s. Here, the centre of the *ucca-nīca-vṛtta* has to be fixed at that point on the *kakṣyā-vṛtta* by considering the measure of arc traversed by the planet in the *jñāta-bhoga-graha-vṛtta*. In this manner, the rationale behind the *śīghra-sphuṭa* can be explained by constructing five circles. The above is the method for ascertaining true Mars, Jupiter and Saturn.

8.15 True Mercury and Venus

There is a distinction (in the method to be adopted) for Mercury and Venus. There too the computation of *manda-sphuṭa* is as above. In the case of *śīghra-sphuṭa*, the *śīghrocca-nīca-vṛtta* is large and the *manda-karṇa-vṛtta* is small. Therefore, the centre of the *śīghrocca-nīca-vṛtta* will fall outside the circumference of the *manda-karṇa-vṛtta*. In such cases where the radius of the *ucca-nīca-vṛtta*, which is the distance between the centres of the *jñāta* and *jñeya-bhoga* circles, is larger than the radius of the *jñāta-bhoga-graha* circle, the circle which stands for the *ucca-nīca-vṛtta* is to be considered as the *kakṣyā-vṛtta*, and the *jñata-bhoga-graha-vṛtta* which stands for the *pratimaṇḍala* is to be considered as the *ucca-nīca-vṛtta*, lying on the circumference of the *kakṣyā-vṛtta*. Construct the *karṇa-vṛtta*, which is said to be the *jñeya-bhoga-graha-vṛtta*, in such a way that its centre is the centre of the *kakṣyā-vṛtta* itself and the planet is on its circumference. The *sphuṭa-kriyā* has to be done with the above as the basis.

Now, if two more circles have to be constructed, construct one circle with its centre at the centre of the *kakṣyā-vṛtta* and of size equal to that of the *jñāta-bhoga-graha-vṛtta*, which has its centre on the circumference of the *kakṣyā-vṛtta*. Since this new circle is equal in size to *jñāta-bhoga-graha-vṛtta* and has the same centre as the *jñeya-bhoga-graha-vṛtta*, this should be considered as the *kakṣyā-vṛtta* in accordance with the arguments stated earlier. Still, since it does not touch the *jñāta-bhoga-graha-vṛtta*, consider it as the *ucca-nīca-vṛtta*. Now, construct a second circle, a *pratimaṇḍala* equal in size to the *kakṣyā-vṛtta* with its centre on this *ucca-nīca-vṛtta* at the point, which corresponds in minutes to the distance traversed by the *manda-sphuṭa* on the *jñāta-bhoga-graha-vṛtta*. Thus, in this set up, it would be as if the *kakṣyā* and *pratimaṇḍala* would have been taken as the *ucca-nīca-vṛtta*-s and the *ucca-nīca-vṛtta*-s as the *kakṣyā-pratimaṇḍala*-s. The *jñeya-bhoga-vṛtta* would have been taken as the *karṇa-vṛtta*. Therefore, it would result that the centre of the assumed *pratimaṇḍala* will be moving with the velocity of the planet. Still its motion should be considered as the *ucca-gati* and its centre should be considered as *ucca*. Though the motion of the centre (*kendra-gati*) of the *jñāta-bhoga-graha-vṛtta* is to be taken as the the *ucca-gati*, since the *jñāta-bhoga-graha-vṛtta* has been considered as the *ucca-nīca-vṛtta*, we should take it as the *graha-gati*.

Now, take the centre of the *jñāta-bhoga-graha-vṛtta* on the circumference of the *kakṣyā-vṛtta* at the point which has the same measure in minutes as the *madhyama-graha*. Then the planet will move along the *pratimaṇḍala* which has been conceived in accordance with the proposed picture. Thus, in this case, the planet will lie where the circumferences of the *jñāta-bhoga-graha* and the assumed *pratimaṇḍala* intersect each other, and this will always be the intersection near the *ucca*. Now assume the *graha-gati* on the *jñāta-bhoga-graha-vṛtta* to be the same as the *graha-gati* on the circumference of the *ucca-nīca-vṛtta*, which has its centre on the circumference of the *kakṣyā-vṛtta*. In this set up, it would be as if the *graha* is taken as *ucca*, and the *ucca* is taken as the *graha*. Hence, the *śīghra-bhujā-phala* which is to be applied to *manda-sphuṭa* is applied to *śīghrocca*, and the *śīghra-kendra-bhujā-jyā* which has been measured by the minutes of *śīghra-karṇa* is applied to *manda-sphuṭa-graha*. Thus the true motion of Mercury and Venus will be obtained. Since it is necessary, we have shown here the motions of the *grahocca*-s and the rationale of true planets in terms of the scheme of five circles discussed earlier. Thus when carefully set forth, these concepts will become clear.

Here, the measures of the *manda-vṛtta*-s and *śīghra-vṛtta*-s for Mars etc. (i.e., Mars, Jupiter and Saturn) have been set out in tables in terms of the minutes of arc of the *pratimaṇḍala.* For Mercury and Venus, however, since the *śīghra-vṛtta*-s are large, it is the *pratimaṇḍala* which is measured in terms of the minutes of arc of this (i.e., *śīghra-vṛtta*) and set out as the *śīghra-vṛtta* in the text *Tantrasaṅgraha.* In other texts, the *manda-vṛtta*-s of Mercury and Venus are also measured by the measure of the *śīghra-vṛtta* and set out (in tables). In *Tantrasaṅgraha*, the *manda-nīcocca-vṛtta*-s have been measured by the minutes of arc of the *pratimaṇḍala* and set out. For this reason, the *mandocca* is subtracted from the *madhyama* and the *manda-phala* is calculated according to the *manda-sphuṭa-nyāya.* Applying this result to the *madhyama* the *manda-sphuṭa* is derived. This (*manda-sphuṭa*) is taken as the *śīghrocca* and the *āditya-madhyama* (the mean Sun) is taken as *graha-madhyama* and the *śīghra-sphuṭa* is calculated. Since the *mandocca-nīca-vṛtta* is smaller than the *pratimaṇḍala* for these two, calculating the *manda-sphuṭa* for Mercury and Venus is similar to that for the other planets. Only, in the *śīghra-sphuṭa*, it is necessary to reverse their *grahocca*-s, their *gati*-s and *vṛtta*-s. There, if the *manda-karṇa* is multiplied by the *śīghra-antya-phala*, and divided by *trijyā*, we get the radius of the *manda-karṇa-vṛtta* in terms of the minutes of arc of the *śīghra-vṛtta.* The reason is that the *pratimaṇḍala* has been taken as *śīghra-karṇa-vṛtta* and therefore the *manda-karṇa-vṛtta* has to be taken as *śīghra-karṇa-vṛtta.* This is all the distinction in the case of Mercury and Venus. Thus has been stated the derivation of true planets when there is no *vikṣepa* (latitude).

8.16 *Śīghra* correction when there is latitude

Now, for the situation when there is *vikṣepa* (latitude), there is a difference. That is stated here. Now, at the centre of the *bhagola* (with its centre as the centre), there is a circle called *apakrama* (ecliptic). For the present calculations, a consideration of its change of position with reference to place and time is not required and hence it might (simply) be taken as an exact vertical circle, situated east-west. Mark off on its circumference twelve (equal) divisions, then construct six circles, passing through those two division-marks which are diametrically opposite. These (circles) will meet at the north and south directions of the *apakrama* as seen from its centre. These two meeting points (of the circles) are called *rāśi-kūṭa*-s (poles of the ecliptic). There will result twelve interstices due to the six circles. The interstices between two circles will make the twelve *rāśi*-s (signs). The middle of these signs will be in the *apakrama* circle and the two meeting points at the two *rāśi-kūṭa*-s. These signs will be such that the middle portions are broad and the ends are pointed. These signs have then to be divided into minutes, seconds, etc.

In the above set up, the *śīghra-vṛtta* is presumed with its centre at the centre of the *apakrama* circle and its circumference along the *mārga* (in the plane) of the *apakrama* circle. It may be recalled that the *apakrama* circle near the centre is called *śīghrocca-nīca-vṛtta*. The size of the *śīghra-vṛtta*-s will be different for the different planets. That is all the difference (between the *śīghra-vṛtta*-s) and there is no difference in their placement as they are located the same way (i.e., with their centre at the centre of the *apakrama* circle and also lying in the same plane).

Now, the *manda-nīcocca-vṛtta* is a circle having its centre on the circumference of the *śīghra-vṛtta* at the point where the mean Sun is. This is the case for all (the planets). The ascending node (*pāta*) has its motion along the circumference of the *manda-nīcocca-vṛtta* in the retrograde manner. The point in the *manda-nīcocca-vṛtta* where the *pāta* is, will touch the *apakrama-maṇḍala*. One half of the *manda-nīcocca-vṛtta*, commencing from the *pāta* will lie on the northern side of the *apakrama-maṇḍala*. Again, the point which is six signs away from the *pāta* will touch the *apakrama-maṇḍala*. The other half (of the *manda-nīcocca-vṛtta*) will lie on the southern side of the *apakrama-maṇḍala*. Here, that point, which is displaced maximum from the (plane of the) *apakrama-maṇḍala*, will indicate the maximum *vikṣepa* (*parama-vikṣepa*) of the planets in terms of the minutes of arc of their respective *mandocca-vṛtta*-s. Further, the plane of this *nīcocca-vṛtta* itself will be the plane of the *pratimaṇḍala*. Hence, the *pratimaṇḍala* too will be inclined towards the north and south from the plane of *apakrama-maṇḍala* in accordance with the *nīcocca-vṛtta*. The *manda-karṇa-vṛtta* will also be inclined accordingly. Now, the *vikṣepa* has to be obtained from the *manda-sphuṭa*.

Here, since the centre of the *manda-karṇa-vṛtta* is the same as the centre of the *mandocca-vṛtta* and since it will be inclined to the plane of *apakrama-maṇḍala*, south and north, accordingly as the *mandocca-vṛtta*, the maximum divergence of the circumference of the *manda-karṇa-vṛtta* from the plane of the *apakrama-maṇḍala* will be the maximum *vikṣepa* in the measure of the *manda-karṇa-vṛtta*. Hence, if the Rsine of the *manda-sphuṭa* minus *pāta* is multiplied by the maximum *vikṣepa* and divided by *trijyā*, the result will be the *iṣṭa-vikṣepa* of the planet on the *manda-karṇa-vṛtta*. This inclination (deflection from the ecliptic) is called *vikṣepa*.

This being the situation, when the position of the planet is displaced from the (plane of the) *apakrama-maṇḍala*, since the *dik* (direction or plane) of the *śīghrocca-nīca-vṛtta* is not the same as that of the *manda-karṇa-vṛtta*, it would not be proper to consider the *manda-karṇa-vṛtta* as the *pratimaṇḍala* in (evaluating) the *śīghra-sphuṭa*. However, when the *pāta* and *manda-sphuṭa* occupy the same position (i.e., they have the same longitude), the *manda-karṇa-vṛtta* can be taken to be in the plane of the *śīghrocca-nīca-vṛtta*. (In other words) when the planet has no *vikṣepa*, this *manda-karṇa-vṛtta* need not be conceived to be inclined. However, if the planet in the *manda-karṇa-vṛtta* is assumed to be removed maximum from plane of the *apakrama-maṇḍala*, then by moving a quarter of a circle it will be in the plane of the *apakrama-maṇḍala*, and from the *vikṣepa* of the planet the inclination (of the planetary orbit) can be obtained.

(We shall consider the case) when there is no *vikṣepa* for *manda-karṇa-vṛtta* (*śīghrocca-nīca-vṛtta* ?). Now, calculate the *vikṣepa-koṭi* by subtracting the square of *vikṣepa* from the square of *manda-karṇa-vyāsārdha* (radius of the *manda-karṇa-vṛtta*) and taking the root (of the difference). This *vikṣepa-koṭi* would be (the base of a triangle) with its tip at the planet and having its base along the line from the centre of the *manda-karṇa-vṛtta* to the *vikṣepa* (foot of the perpedicular from the planet on the *apakrama-maṇḍala*). Construct a circle with its radius parallel to this *vikṣepa-koṭi*, with the *vikṣepa-koṭi* as radius. This *vikṣepa-koṭi-vṛtta* would have all its parts (i.e., centre and the circumference) equally away (i.e., parallel) from the *apakrama-maṇḍala* just as the *ahorātra-vṛtta* would be from the *ghaṭikā-maṇḍala*. Construct the *śīghra-nīcocca-vṛtta* parallel and away from it.

Since the *vikṣepa-koṭi-vṛtta* is now parallel to the *śīghra-nīcocca-vṛtta*, it (the *vikṣepa-koṭi-vṛtta*) will be the *pratimaṇḍala* for the (calculation of) the *śīghra-sphuṭa*. Subtract the square of the *vikṣepa* in the measure of the *pratimaṇḍala* from the square of the *manda-karṇa* (in the same measure) and find the square-root. This is the *vikṣepa-koṭi* in the measure of the *pratimaṇḍala* and the *śīghra-phala* shall have to be calculated with this *vikṣepa-koṭi*. Taking the *vikṣepa-koṭi* mentioned above as the semi-diameter and taking it as the *manda-karṇa*, calculate the *śīghra-sphuṭa* as directed above. The result will be the *graha-sphuṭa* (true planet) on the *śīghra-karṇa-vṛtta* which has its circumference touching the planet and its centre at a place removed from the centre of the *apakrama-maṇḍala* to the south or north by the extent of the *vikṣepa* . This itself will be the *sphuṭa* on the *apakrama-maṇḍala*. The minutes *(kalā)* in the (*vikṣepa*) *koṭi-vṛtta* on either side of the *apakrama-maṇḍala* will be the same as in the *apakrama-maṇḍala* itself. In the *koṭi-vṛtta* the *kalā*-s will be smaller (in length) but there is equality in number. Just as the measures in the *svāhorātra-vṛtta*-s will be the same in number as in the bigger *ghaṭikā-maṇḍala*, so also the *kalā*-s in the *vikṣepa-koṭi-vṛtta*. This will be clear later.

Now, when the square of *vikṣepa* is added to the square of the *śīghra-karṇa* and the root calculated, the result will be the distance from the centre of the *apakrama-maṇḍala* to the planet. This is called the *bhū-tārāgraha-vivara* (the distance between the Earth and the planet). Now, the *vikṣepa* got by multiplying the previously stated *vikṣepa* by *trijyā* (radius) and dividing by *bhū-tārāgraha-vivara* will be the *bhagola-vikṣepa*. *Bhagola-vikṣepa* is the extent by which the circumference of the *bhū-tārāgraha-vivara-vṛtta*, which has its centre at the centre of the *apakrama-maṇḍala*, is inclined from the plane of the latter. For computing the true planet, the *bhū-tārāgraha-vivara* is not needed. Here, the minutes (*ili*-s) are small in accordance with the nearness of the *rāśi-kūṭa*-s, since the number of *rāśi*-s etc., in the (*vikṣepa*) *koṭi-vṛtta* and the *apakrama-vṛtta* are same. This case is similar to the case of the *svāhorātra-vṛtta*-s and *ghaṭikā-maṇḍala*, with reference to the *prāṇa*-s. Hence, there is no need for the *bhū-tārāgraha-vivara* for calculating the *śīghra-bhujā-phala*. Thus has been stated the calculation of *sphuṭa*.

Now, when the *śīghrocca-nīca-vṛtta* itself has a *vikṣepa* from the plane of the *apakrama-maṇḍala*, and that *vikṣepa* is not along the path of the *manda-karṇa-vṛtta*: If the *manda-karṇa-vṛtta* has a different *vikṣepa* than the *śīghra-vṛtta*, it is shown below how to know the *sphuṭa* and *vikṣepa* .

For this, first ascertain the position of the *pāta* in the *śīghrocca-nīca-vṛtta* and the maximum *vikṣepa* therefor. Then ascertain the *vikṣepa* at that moment for the centre of the corresponding *manda-karṇa-vṛtta*. For this, subtract the *pāta* of the *śīghra-vṛtta* from the *śīghrocca*; find the Rsine of the difference, multiply by its maximum *vikṣepa* and divide by *trijyā*; the result will give the *vikṣepa* of the centre of the *manda-karṇa-vṛtta* on the circumference of the *śīghrocca-nīca-vṛtta* from the plane of the *apakrama-maṇḍala*. Like in the case of the desired *apakrama*, find its square and subtract it from the square of *trijyā*. The square root of the result will be *vikṣepa-koṭi*. Then with this *vikṣepa-koṭi* as radius, draw a circle parallel to the plane of *apakrama-maṇḍala*. Then that (*vikṣepa-koṭi-vṛtta*) will be removed from the plane of the *apakrama-maṇḍala* by the extent of the *vikṣepa*. Now, if the *vikṣepa-koṭi* is multiplied by *śīghra-antya-phala* and divided by *trijyā*, the radius of the *vikṣepa-koṭi-vṛtta* in terms of the minutes of arc of the *pratimaṇḍala* would result. Now, taking this *vikṣepa-koṭi-vṛtta* as the *śīghra-nīcocca-vṛtta* and the earlier stated *vikṣepa-koṭi-vṛtta* of the *manda-karṇa-vṛtta* as the *pratimaṇḍala*, the *śīghra-bhujā-phala* has to be derived. This has to be applied to the *manda-sphuṭa* (to find the *sphuṭa*).

Thus, if the *śīghra-nīcocca-vṛtta* has a deflection in some other direction, then the measure by which the *manda-karṇa-vṛtta* will be deflected from the *śīghra-nīcocca-vṛtta*, which itself is deflected (from the *apakrama-maṇḍala*) as described above, will be known. Then, if the planet has a *vikṣepa* directly to the south along the *manda-karṇa-vṛtta* whose centre lies on the circumference of the *śīghra-vṛtta*, which has a northerly *vikṣepa* from the path of the *apakrama-maṇḍala*, then the *vikṣepa* of the planet at that moment would be the difference between the *vikṣepa* of the *śīghra-nīcocca-vṛtta* and the *manda-karṇa-vṛtta*. If both the *vikṣepa*-s are either to the north and or to the south, then the *vikṣepa* of the planet would be the sum of the two. This would be the *vikṣepa* from the plane of the *apakrama-maṇḍala*.

Thus have been specified the method for the derivation of the *sphuṭa* and the *vikṣepa* when there is *vikṣepa* for the *jñāta-bhoga-graha* (the true planet) and the *ucca-nīca-vṛtta* which is (i.e., whose radius is) the difference between the *jñāta-*(*bhoga-graha-vṛtta*) and the *jñeya-*(*bhoga-graha-vṛtta*). Here, the method for *sphuṭa* has been stated to show the procedure for all possible situations that can occur, not that it has actually occurred here.

If it is desired to compute how much Mars has travelled in the circle with its centre at the centre of the lunar sphere, when the measure by which it has travelled in a circle with its centre at the centre of the *bhagola* is known, then the *kakṣyā-vṛtta* of the Moon has to be taken as the *ucca-nīca-vṛtta* for finding the *sphuṭa*. In such a case, the above situation may occur. This is also the case (that is, the above procedure has to be adopted) even when the computations (of the planetary motion) are known for the centre of Moon, and they need to be converted in terms of the circle with centre at the centre of the *bhagola* (celestial sphere).

8.17 Calculation of the mean from true Sun and Moon

Now is described the method of calculating the mean (planet) from the true (planet). Here, for the Sun and Moon, the Rsine of the distance between the planet and the *ucca-nīca-sūtra* is the *bhujā-jyā* of *sphuṭa*–minus–*ucca*. If this is multiplied by the *karṇa* and divided by *trijyā* and the quotient reduced in terms of the minutes of arc of *pratimaṇḍala*, the result will be the Rsine of the relevant portion of the *pratimaṇḍala*. If the arc of this is found, and is applied to the *ucca* or the *nīca*, the extent to which the planet has travelled along the *pratimaṇḍala* is known. If the *doḥ-phala* (*bhujā-phala*) is similarly reduced in terms of the minutes of arc of the *pratimaṇḍala*, converted to arc, and applied to the *sphuṭa* reversely (when it lies between) *Meṣa* and *Tulā*, then also the mean planet would result. The rationale here is as follows: The difference (ratio) between Rsine of the *ucca*–minus–*sphuṭa* and the *ucca*–minus–*madhyama* will be similar to the difference (ratio) between the *trijyā* and the *karṇa*; also the relation between *pramāṇa* and its *phala* and *icchā* and its *phala* are similar.

8.18 Another method for the mean from true Sun and Moon

Now, even by a successive iteration process (*aviśeṣa-karma*) involving the *doḥ-phala*, the mean planet can be obtained from the true planet. Here is the method therefor: The *ucca* is subtracted from the *sphuṭa* (true planet) and *doḥ-phala* is found. If that is applied inversely to the *sphuṭa*, according to *Meṣa-Tulādi*, the approximate mean planet is obtained. Subtract the *ucca* from this *madhyama*, find the *doḥ-phala* and apply it to the *sphuṭa*. Again, from this mean, subtract the *ucca*, find the *doḥ-phala* and apply it to the original *sphuṭa* itself. When these (successive approximations), lead to indistinguishable results (*aviśeṣa*), the *madhyama* will be exact. (In this method) the *karṇa* need not be found at all for deriving the *manda-sphuṭa*.

Now, instead of the *doḥ-phala* of the *sphuṭa*–minus–*ucca*, being multiplied by the *karṇa* and divided by *trijyā*, if the *trijyā*–minus–*karṇa* is multiplied by *sphuṭa-doḥ-phala* and divided by *trijyā*, the *phalāntara* (difference between the *phala*-s) will result. Add this to the *sphuṭa-doḥ-phala* if *Makarādi*, and subtract if *Karkyādi*. The result will be the *doḥ-phala* of *madhya*–minus–

ucca. Here *trijyā*–minus–*karṇa* is practically the *koṭi-phala*, since the contribution (to the *karṇa*) due to the square of the *doḥ-phala* would be very little (negligible).

Here, if the *doḥ-phala* is multiplied by the *koṭi-phala* and divided by *trijyā*, the result will be the difference between the *sphuṭa-doḥ-phala* and the *madhya-kendra-doḥ-phala*. And these will practically be the current *khaṇḍa-jyā*-s of the *sphuṭa-doḥ-phala*. Since, it is common knowledge that (the value of) the *bhujā-khaṇḍa* is according to the *koṭi-jyā*, the *bhujā-phala-khaṇḍa* will be according to the *koṭi-phala*. Take the *bhujā-phala-cāpa* as the *jyā* (*manda-jyā*), multiply it by the *koṭi-phala* and divide by *trijyā*; the *bhujā-phala-khaṇḍa* would be obtained. Here, multiply the *bhujā-phala* by the *khaṇḍa-jyā* and divide by its *cāpa*. Then also we will get the *bhujā-phala-khaṇḍa* of this *bhujā-phala*. Here, the *bhujā-phala-khaṇḍa* of this *bhujā-phala* might be greater or less than the *bhujā-phala* calculated from the *kendra* to which has been applied the *bhujā-phala* derived from itself. Thus, the *madhya-kendra-bhujā-phala* can be obtained by applying reversely the *sphuṭa-kendra-bhujā-phala* successively. When this is applied to the true planet the mean planet is obtained. Through the above methods, the mean of Sun and Moon can be derived from their true positions.

8.19 Calculation of the mean from true planet

In the same manner, the mean of the other planets can be derived from their *manda-sphuṭa*. The method of deriving the *manda-sphuṭa* from the *bhujā-phala* of the *śīghra-sphuṭa-kendra* is also similar. But there is a difference that the *aviśeṣa* (successive iteration to near equality) need not be done. Multiplication by the *karṇa* and division by *trijyā* too are not necessary. The *manda-sphuṭa* can be got thus: Multiply the *kendra-bhujā-jyā* of the *śīghra-sphuṭa* by the *vṛtta* (360) and divide by 80, and convert this Rsine (*jyā*) in the *śīghra-nīcocca-vṛtta* to arc and apply the result to the *śīghra-sphuṭa* inversely for *Meṣa* and *Tulā*, then the *manda-sphuṭa* is obtained.

Here, it may be noted that when the *bhujā-phala* is calculated for deriving the (*manda*) *sphuṭa* from the *madhyama*, the rule of three using the *karṇa* should not be resorted to. On the other hand, there is a need for iteratively finding the *bhujā-phala* (doing *aviśeṣa)*, when the *madhyama* is calculated using the said *manda-sphuṭa*. The rationale for this has been stated. It will be clear from this that, because *karṇa* is required for calculating *śīghra-sphuṭa* from the *manda-sphuṭa*, it (the *karṇa*) is not required for calculating *manda-sphuṭa* from *śīghra-sphuṭa*. Therefore, it is not necessary to iterate the *bhujā-phala* till *aviśeṣa*, since the rationale is the same.

That being the case, when *śīghra-sphuṭa* is calculated from the *manda-sphuṭa* without using the *karṇa*, if the *śīghra-bhujā-phala* is iterated till *aviśeṣa* and applied, the *śīghra-sphuṭa* would result. On the other hand, if the *bhujā-phala* is calculated by the rule of three using the *karṇa*, even if successive iteration is done without the use of the *karṇa*, the *bhujā-phala* will be the same. Here, in the rule of three using *koṭi-phala* and *trijyā*, the *icchā-phala* should be obtained using the sum of the *bhujā-phala-khaṇḍa* and the *cāpa-khaṇḍa*. This has been stated elaborately in the section on Rsines (*jyā-prakaraṇa*) and so might be referred to there.

8.20 Computation of true planets without using *Manda-karṇa*

By using the same reasoning, it would be possible to obtain the difference which arises in the *śīghra* circumference due to the *manda-karṇa*, and the consequent difference which occurs in the *śīghra-bhujā-phala* may be obtained as *manda-phala-khaṇḍa*. To derive this, first calculate the *śīghra-bhujā-phala* from *madhyama*–minus–*śīghrocca*; apply this to the *madhyama* and subtract from it the *mandocca* and get the *manda-phala*. In that *manda-phala*, the *manda-phala-khaṇḍa-jyā*-s of the *śīghra-bhujā-phala-bhāga* might be increasing or decreasing. Now, when this (*manda*) *phala* is derived in this manner, the difference that occurs in the *śīghra-phala* due to the *manda-karṇa*, would have been included also. Now, when this *manda-phala* is applied to the *madhyama*, it would be that the difference in *phala* that occurs in the *śīghra-bhujā-phala* due to *manda-karṇa* too would have been (automatically) applied.

Here, when it is intended to separately obtain (in a different manner) the difference that occurs in the *śīghra-bhujā-phala* due to the *manda-karṇa*, two *trairāśika*-s shall have to be used. The first is to multiply the *śīghra-bhujā-phala* by *trijyā* and divide by *manda-karṇa.* The second is to multiply the result by *trijyā* and divide by *śīghra-karṇa.* Then apply the result according to the *śīghra-kendra.*

Now is set out as to how these three, viz., the two *trairāśika*-s and the third being the condition for their positive or negative nature, arise when we calculate *manda-phala* after first applying *śīghra-doḥ-phala.* There (in the first *trairāśika*), the *śīghra-doḥ-phala* is multiplied by *trijyā* and divided by *manda-karṇa.* The difference, between the result obtained and the original *śīghra-doḥ-phala*, is the difference between the *icchā* and its *phala* of the first *trairāśika.* This result will practically be the same if the first *guṇya* is multiplied by the difference of the multiplier and the divisor (*guṇa-hārāntara*) and divided by the divisor. This is practically the same as multiplying by the *manda-koṭi-phala* and dividing by the *trijyā.*

Here, if the *manda-doḥ-phala* is read off after applying the *śīghra-doḥ-phala* there-through, there also the *manda-khaṇḍa-jyā*-s related to the *śīghra-doḥ-phala* are obtained. And this will be the distinction in the *śīghra-doḥ-phala* due to the *manda-karṇa.* Hence, the *phala* of the first *trairāśika* in the *śīghra-doḥ-phala* can be derived by applying it to the *manda-doḥ-phala.* Here again, the difference in the *śīghra-doḥ-phala* due to the *manda-karṇa* will be the difference between *śīghra-doḥ-phala* and the *manda-doḥ-phala* calculated from the basic *madhyama*, and that obtained after applying to the basic *madhyama* the *manda-doḥ-phala* and *śīghra-doḥ-phala.* This will be the result of the first *trairāśika.*

The result of the second *trairāśika* is derived thus: Find the *śīghra-doḥ-phala* calculated from the *madhyama* to which has been applied the *manda-phala* which latter has been derived from the basic *madhyama*; find also the *śīghra-doḥ-phala* calculated from that *madhyama*, which is obtained by applying *śīghra-doḥ-phala* to the *madhyama* which has been obtained by applying the *manda-phala*, which latter has been derived from the basic *madhyama*. The difference between the two is the required result. (It might be noted that) the difference arising from the *śīghra-karṇa* will result in the *śīghra-karṇa-bhujā-khaṇḍa*-s, and that from the *manda-karṇa* will result in the *manda-bhujā-khaṇḍa*-s.

Now, we consider the *karṇa* as the *trijyā*, the difference between the *trijyā* and *karṇa* as the *koṭi-phala*, the arc of the *doḥ-phala* as the full chord (i.e., double the Rsine), and the *koṭi-phala* of the *cāpa-khaṇḍāgra* as a part of the *madhyama*, and we also ignore the grossness (*sthaulya*) in the calculations mentioned above. Then, just as the difference that occurs in the *śīghra-doḥ-phala* due to the *manda-karṇa* is added to the *manda-doḥ-phala*, the correction need not be carried out for the *manda-kendra*, but has to be appropriately carried out for the *śīghra-kendra*.

Now, it will be shown that even if the correction is made in terms of the *manda-kendra*, the result will be same. Here, as regards the increase and decrease of the *śīghra-doḥ-phala* due to the *manda-karṇa*, the increase will be when the *trijyā* becomes greater than the *manda-karṇa*, and the decrease will be when it is less. This will be according to whether the *manda-kendra* is *Karkyādi* or *Makarādi*. This result would be reduced in the same manner, as was the case when earlier, the *śīghra-phala* was corrected by the *manda-phala*. Now, when first the *śīghra-phala* is positive and the *manda-kendra* is within the three signs from *Meṣa*, then since the *manda-karṇa* is large, the corresponding *śīghra-phala* would be small. In this case, the *śīghra-phala* corresponding to the *manda-koṭi* should be subtracted. The *manda-phala* should also be subtracted. So the two can be subtracted together. However, when the *śīghra-phala* is positive and the *manda-kendra* is in the three signs beginning with *Karki*, then the contribution to the *śīghra-phala* due to the *manda-karṇa* will be positive. Then the *śīghra-bhujā-phala*–plus–*manda-kendra* would be greater than the basic *manda-kendra*. When, however, it is in the even quarters, the further away it moves, the *bhujā-phala* will be less. When this *bhujā-phala* becomes negative, it is so small that in effect the *śīghra* degrees will be positive. Then, when the *śīghra-phala* is positive and the *manda-kendra* is within the three signs beginning with *Tulā*, the *manda-karṇa* will be less than *trijyā*, and the *śīghra-phala* derived from it will be more. And, since the *manda-phala* is *Tulādi*, it is positive. When however, the *manda-kendra* is in the odd quandrants, the *manda-phala*, calculated from the *madhyama* to which *śīghra-phala* had been applied, would be large. Since this *phala* is *Tulādi*, it is positive. Here also it would be proper to apply the *śīghra* degrees in accordance with the *manda-kendra*.

When the *manda-kendra* is in the three signs beginning with *Makara*, and the *śīghra-phala* is positive, then the *manda-kendra* with the *śīghra-phala* applied to it will be greater than the basic *manda-kendra*. Since this is an even quadrant, and the part passed over is more, the part to be passed over, which is the *bhujā-cāpa* is smaller. Therefore its *manda-phala* will be less than the *manda-phala* of (i.e., computed from) the basic *madhyama*. When this is added to the *madhyama* and (the *manda-phala*) is slightly increased, the correction due to the *śīghra* degrees which is negative would also be effected herein, since the negativity is due to the *manda-karṇa* being larger than *trijyā*.

Thus, it is seen that when the *śīghra-phala* is positive in all the four quadrants of the *manda-kendra*, it would be appropriate if the correction due to *śīghra* is done in accordance with the *manda-kendra*. In the same manner, the positive and negative nature of the *śīghra* derived in accordance with the *manda-kendra* is to be inferred even when the *śīgrhra-phala* is negative. Thus, though the correction to the *śīghra-phala* due to the *manda-karṇa* is normally to be applied in accordance with the *śīghra-kendra*, if that is added to the *manda-phala* and applied according to the *manda-kendra*, there will not be any appreciable difference in the result. This being the case, there is no necessity of the *manda-karṇa* for (the derivation of) the *śīghra-phala*.

Therefore, for ease in the computation of *sphuṭa*, the *śīghra-phala* can be computed and listed (in a table) for making calculations and so also (a table can be made) for the *manda-phala*. Here, by obtaining the three *bhujā-phala*-s and applying two of them to the *madhyama* (mean), the *sphuṭa* (true planet) is obtained. This is one School (of explanation).

There is another School which explains that the *śīghrocca-nīca-vṛtta* increases and decreases in accordance with half the difference between the *manda-karṇa* and the *trijyā*. In that School, the *śīghra-doḥ-phala* has to be multiplied by *trijyā* and divided by half the sum of *manda-karṇa* and *trijya*. The result in degrees has to be added to the *manda-phala*; for this, the *manda-phala* has to be derived from the *madhyama* to which has been applied half the *śīghra-phala*. This is the only difference (in this School). Other things are as stated earlier. This is the idea behind the *sphuṭa* correction that is stated in the *Parahita* (School) for Mercury and Venus.

The author of *Laghumānasa* (i.e., Muñjāla) follows the School, which states that the *manda-nīcocca-vṛtta* also increases and decreases in accordance with the half the difference between *manda-karṇa* and *trijyā*. According to that School, the *manda-phala* and *śīghra-phala* should be multiplied by *trijyā* and divided by half the sum of the *manda-karṇa* and *trijyā*. The *manda-phala* should be corrected having obtained the result thus. The *śīghra-phala* should be multiplied by this difference between the multiplier and divisor (*guṇa-hārāntara*) and divided by the divisor. The result should again be multiplied by *trijyā* and divided by the *śīghra-karṇa*, and the correction applied. Thus is explained the computation of the *sphuṭa* in that School. Therefore, it was directed in the *Laghumānasa* to correct the *manda-phala* and the *śīghra-phala* by the *mandaccheda*, which has been obtained by applying half-*koṭi*. According to this School, if the *manda-phala* is to be obtained without the use of *manda-karṇa*, the *manda-phala* and the *śīghra-phala* have to be halved and applied to the *madhyama*. Then, the *manda-phala* thus derived is applied to the basic *madhyama* (to get the *manda-sphuṭa*). The *śīghra-phala* derived from this is now applied to the *manda-sphuṭa*. The result will give the *sphuṭa*. The computation, as described in this school, is set down as four *sphuṭa*-s in several places. In case the *manda-karṇa* is not used, the *śīghra-karṇa-bhujā-phala* may be set out in a table. Here, since both the *bhujā-phala*-s are to be multiplied by half the *manda-koṭi-phala*, and *manda-phala* has to be derived for both the halves of *manda* and *śīghra-phala*-s, the *manda-phala* is calculated after first applying half of both the *bhujā-phala*-s. This is the reason for the above-said calculation. Thus has been stated the 'computation of true planets'.

Now, for Mercury and Venus, the true planet is to be found using the *manda-nīcocca-vṛtta* and *pratimaṇḍala*, which are tabulated in terms of their *śīghrocca-vṛtta*. Here, after mutually interchanging the *śīghra-nīcocca-vṛtta* and *pratimaṇḍala*, their *manda-sphuṭa* and *śīghra-sphuṭa* can be computed in the same manner as in the case of Mars etc (i.e., Mars, Jupiter and Saturn). Their *manda-sphuṭa* could be supposed to be obtained by applying the *manda-phala* to the mean Sun which is conceived as the *madhyama*. The *manda-karṇa-vṛtta* would be that circle whose circumference meets the centre of the *pratimaṇḍala*, which is taken as the *śīghra-nīcocca-vṛtta* which in turn is constructed at the centre of the *bhagola*. The centre of the *manda-pratimaṇḍala* will be on the circumference of the *manda-nīcocca-vṛtta*. (In this set up) it would be as if the planet is at the circumference of the *manda-nīcocca-vṛtta* (whose centre is) on the circumference of the *kakṣyā-vṛtta*. Hence, the *śīghra-phala* derived from the *manda-sphuṭa* is multiplied by *trijyā* and divided by the *manda-karṇa*, so as to convert it to minutes of arc of the *manda-karṇa*. If it is desired to derive this without the use of the *manda-karṇa*, (the method is this): Now, the *manda-sphuṭa* is obtained by applying the *manda-phala* on the mean Sun. Apply the *śīghra-phala* calculated from that *manda-sphuṭa* to the basic *madhyama*. Since that has to be applied to the *manda-sphuṭa*, apply on itself the *manda-phala* obtained from that *śīghra-sphuṭa*. The *sphuṭa* (true planet) will be the result. It has to be remembered here that the difference arising due to the *śīghra-karṇa* has been incorporated in the table. Hence (the computation) has to be done as above. Thus (has been stated) the computation of true planets.

Chapter 9

Earth and Celestial Spheres

9.1 *Bhūgola* : Earth sphere

Now is demonstrated the situation and motion of the *bhūgola*, *vāyugola* and *bhagola*. The Earth is a sphere supporting on its entire surface all things, moving and non-moving, maintaining itself (suspended) in the sky at the centre of the celestial sphere (*nakṣatra-gola*) by its own power, and not depending on any other support. Now, it is the nature of all heavy things to fall on the Earth from all regions of the sky all around. Hence, the Earth, everywhere, is below the sky. Similarly, from all locations on the Earth, the sky is above. Now, the southern half of the Earth-sphere is abundant with regions of water. And, in the northern half, the land region is in profusion and watery region less. Then, with the land of India (*Bhārata-khaṇḍa*) appearing to be in the upward (northern) direction, at the confluence of the landed and watery division (of the Earth), there is a city known as *Laṅkā*. Conceive a circular line (*vṛttākāra-rekhā*) from that place, east-west, cycling round the Earth. On this line are situated four cities (including *Laṅkā*), to the west *Romakapurī*, to the other (diametrically opposite) side *Siddhapura*, and to the east *Yavakoṭi*.

Similarly from *Laṅkā*, conceive another circle round the Earth, which is north-south across and passing through the upper and lower halves of the Earth. On this line (are situated), *Mahāmeru* to the north, *Baḍavāmukha* to the south, *Siddhapura* on the opposite side. This line is the *samarekhā* (north-south standard meridian line). In this line is a city called *Ujjayinī*.

Now, the places lying on the east-west line mentioned above are called *nirakṣa-deśa* (equatorial places having no latitude).

From all places on that (equatorial) line, can be seen two *nakṣatra*-s (stars) called *Dhruva*-s (pole stars), one in the north and the other in the south, which have no rising or setting. If one moves towards the north from this line one can see only the northern *Dhruva*. This *Dhruva* would have as much altitude as one moves towards the north. This altitude of the *Dhruva* is called *akṣa* (the terrestrial latitude). From this point on the surface, the southern *Dhruva* cannot be seen, since it has gone down (i.e., lies below the horizon). Where the *Dhruva* is seen at a particular altitude, there will be seen near the *Dhruva* certain stars, some below and moving towards the east and some above and moving towards the west, but without rising or setting. On the other hand, similar stars around the southern *Dhruva* can never be observed as they are moving below the horizon. However, from the *nirakṣa-deśa* (the equator), it would be possible to see the rising and setting of all the stars, in regular order. There again, for an observer on the equator, the measure by which a star is removed at its rise from the east towards the north or south, is the same as the measure by which it is removed from the zenith of the observer at the meridian transit. It will also set in the west, at a point which is exactly opposite to the (rising point on the) east. Rising and setting (of stars) take place in this manner at the equator (for an observer on the equator). Even for a place having latitude, the phenomenon is similar. But the meridian transit would be shifted a little towards the south, if one's place has a northern latitude.

9.2 *Vāyugola:* Equatorial celestial sphere

Now, (for an observer) on the equator, for a particular star at a particular place, the vertical circle from the east to the west passing through it would seem to be the rising-setting path (diurnal circle). Here again, for a star rising exactly in the east (point on the horizon), the diurnal path would be the biggest circle. The path of the stars on its either side would be smaller circles. These circles would gradually become smaller, and the diurnal path of the stars very close to the *Dhruva* will be the smallest of all. This being the situation, it would seem that this celestial sphere is like a sphere with an axis fixed at two posts at its two ends, here the posts being the two *Dhruva*-s. Now, (for an observer) at the equator, the circle passing through the east-west points and touching the top and the bottom, right above the top (of the observer) is known as *ghaṭikā-vṛtta* (celestial equator). The several smaller circles on the two sides of the (*ghaṭikā-vṛtta*) are known as *svāhorātra-vṛtta*-s (diurnal circles).

Now, from *Laṅkā*, there is another (great) circle rising right above (and below) touching the two *Dhruva*-s. This is known as *dakṣiṇottara-vṛtta* (prime meridian). Then there is another (great) circle around the Earth, passing through the east and west points, and touching the two *Dhruva*-s (the north and the south poles). This is *Laṅkā-kṣitija* (horizon at *Laṅkā*). The stars are said to rise (at the equator) when they touch this *Laṅkā-kṣitija* along that half of it which lies to the east of the *dakṣiṇottara-vṛtta*, and are said to set when they touch its western half. And, when they touch the *dakṣiṇottara-vṛtta*, the stars have their meridian transit.

Thus, the three (great) circles, *ghaṭikā*, *dakṣiṇottara*, and *Laṅkā-kṣitija* are mutually perpendicular to each other. The points where they meet each other are known as *svastika*-s (cardinal points). There are six of them: i.e., along the horizon on the four directions, and at the top and at the bottom. Between the interstices of all these *svastika*-s, one-fourth of a (great) circle will be contained. Therefore, there will be formed eight divisions of a sphere of equal sizes, cut off by these three circles, four being below the horizon (at the equator) and four above.

9.3 *Bhagola*: Zodiacal celestial sphere

Now, the path traced by the Sun in its eastward (annual) motion is known as *apakrama-maṇḍala* (ecliptic). This will intersect the *ghaṭika-maṇḍala* (celestial equator) at two points. From these (two points), at the distance of one-fourth of a circle (*vṛtta-pāda)*, the *apakrama-maṇḍala* will be removed from the *ghaṭikā-maṇḍala* by 24 degrees towards north and south. (These points) will move further westwards along with the *ghaṭikā-maṇḍala.* The first point of contact between the *ghaṭikā-maṇḍala* with the *apakrama-maṇḍala* is near the first point of Aries *(Meṣādi)*. Then it (*apakrama-maṇḍala*) will gradually separate away from it (*ghaṭikā-maṇḍala*) towards the north. When a semi-circle has been completed, the second contact occurs, near *Tulādi* (beginning point of Libra). From there, it (*apakrama-maṇḍala*) will again be oriented towards the south. Again, when half the circle has been completed they will meet each other. These two points of contact are respectively called *pūrva-viṣuvat* (vernal equinox) and *uttara-viṣuvat* (autumnal equinox). Now, the points that are at the middle of these two contacts (equinoxes), where the circles are separated away the most, are called *ayana-sandhi*-s (solstices).

Now, when on account of the motion caused by the *Pravaha* wind, the *Meṣādi* (first point of Aries) rises, at that time *Tulādi* (first point of Libra) sets, *Makarādi* (first point of Capricorn) will touch the *dakṣinottara* towards the south from the zenith (*kha-madhya*) and *Karkyādi* (first point of Cancer) will touch the *dakṣiṇottara-vṛtta* towards the north from the *ghaṭikā-maṇḍala* right below. There, the difference between the *apakrama-maṇḍala* and the *ghaṭikā-maṇḍala* along the *dakṣiṇottara-vṛtta* will be 24 degrees, as it is the place of maximum divergence (between them). Now, this (*apakrama-maṇḍala*) will rotate according to the *ghaṭikā-maṇḍala.* Thus when *Meṣādi* (first point of Aries) is at the peak (on the prime meridian), *Tulādi* (first point of Libra) is at the bottom, *Makarādi* (first point of Capricorn) would touch the *kṣitija* (horizon) at a point which is shifted towards the south from the west point by 24 degrees, and *Karkyādi* (first point of Cancer) would touch the *kṣitija* (horizon) at a point which is shifted towards the north from the east point by that much (i.e., 24 degrees). The *apakrama-maṇḍala* would be a vertical circle at that time. When, however, the *Meṣādi* is at the west point, *Tulādi* will be at the east point, *Karkyādi* would be on the prime meridian, separated away from the *kha-madhya* (zenith) towards the north by 24 degrees, and *Makarādi* would be on the prime meridian, separated from the bottom-most point (nadir), towards the south (by 24 degrees). When *Tulādi* is at the peak (on the prime meridian), *Meṣādi* would be at the bottom, *Makarādi* would touch the *kṣitija* towards the south of east point and *Karkyādi* would touch the *kṣitija*, towards the north of the west point. At this time also, the *apakrama-maṇḍala* would be vertical (i.e., a vertical circle). Thus, the situation of the *apakrama-maṇḍala* changes according to the rotation of the *ghaṭikā-maṇḍala*, the reason being that the two are bound together, in a specific way.

Then again, just as the *ghaṭikā-maṇḍala* is the central (great) circle of the *Pravaha-vāyugola* (equatorial celestial sphere), the *apakrama-maṇḍala* will be the central (great) circle of the *bhagola* (zodiacal celestial sphere). Just as the two *Dhruva*-s are situated on the two sides of the *ghaṭikā-maṇḍala*, on the two sides of the *apakrama-maṇḍala* are situated the two *rāśi-kūṭa*-s (poles of the ecliptic). At one of the poles (*rāśi-kūṭā*-s) the southern heads (ends) of all the *rāśi*-s (signs) would have gathered together, and at the other, all the northern heads (i.e., ends) meet. The points where the ends of *rāśi*-s meet are called the *rāśi-kūṭa*-s.

Herein below is described the situation of the *rāśi-kūṭa*-s when the *pūrva-viṣuvad* (vernal equinox) is at the centre of the sky (zenith). At that time, the *apakrama-maṇḍala* would be a vertical (circle). The *ayanānta*-s (solstices) of the *apakrama-maṇḍala* would touch the *kṣitija* (horizon) north of the eastern cardinal point and south of the western cardinal point. Between the *ayanānta*-s and the eastern and the western cardinal points, there would be a difference of 24 degrees.

Again at that time, the *rāśi-kūṭa*-s (poles of the ecliptic) would be on the horizon (*kṣitija)*, 24 degrees west of the north *Dhruva* and that much to the east of the south *Dhruva.* Conceive a (great) circle touching the two *rāśi-kūṭa*-s and *kha-madhya* (zenith). This will be a *raśi-kūṭa-vṛtta.* Now conceive of another *rāśi-kūṭa-vṛtta* towards the east of *Meṣādi* at a distance equal to one-twelfth of the *apakrama-maṇḍala* and passing through the *rāśi-kūṭa*-s. That (*rāśi-kūṭa-vṛtta*) would touch a point that much to the west from *Tulādi* at the bottom. The distance would be 30 degrees. This will be the second *rāśi-kūṭa-vṛtta*. Now, the interstice stretching between these two *rāśi-kūṭa-vṛtta*-s, east of *kha-madhya*, is the *rāśi* of *Meṣa* (Aries). Down below, the interstitial stretch between these two *rāśi-kūṭa-vṛtta*-s would be the *rāśi* of *Tulā* (Libra).

Now, construct another *rāśi-kūṭa-vṛtta* from the second *rāśi-kūṭa-vṛtta* this much degrees (i.e., 30 degrees) to the east and down below that much to the west. The intersticial stretch between the second *rāśi-kūṭa-vṛtta* and the third one is the *rāśi* of *Vṛṣabha* (Taurus); down below it is *Vṛścika* (Scorpio). Now, the intersticial stretch between the third (*rāśi-kūṭa-vṛtta*) and the *kṣitija* is the *rāśi* of *Mithuna* (Gemini), and down below their intersticial stretch is *Dhanus* (Sagittarius). Thus are the six *rāśi*-s.

Then, from the zenith (*kha-madhya*) towards the western side of the *apakrama-maṇḍala*, conceive of two *vṛtta*-s of equal interstice as above. Then the other six *rāśi*-s can be identified, as was done with the first *rāśi-kūṭa-vṛtta* and *kṣitija*. Now, inside the different *rāśi*-s, conceive various circles to represent the divisions of the *rāśi*, viz., degrees, minutes, and seconds. It is to be noted that here, in the case of the horizon (*kṣitija*) and *dakṣiṇottara-vṛtta*, there is no rotation due to the *Pravaha-vāyu* as in the case of the *ghaṭikā-vṛtta* and *apakrama-maṇḍala*. Therefore, conceive of another *rāśi-kūṭa-vṛtta* similar to (i.e., along) the *kṣitija* for conceiving its rotation. Thus, the entire celestial globe (*jyotir-gola*) is completely filled by the twelve *rāśi*-s. When this celestial globe is conceived with the *apakrama-maṇḍala* as the centre and the *rāśi-kūṭa*-s on the sides (*pārśva*), it is known as *bhagola* (zodiacal celestial sphere). When the *ghaṭikā-maṇḍala* is conceived as the centre with the *Dhruva*-s on the sides, it is known as *vāyu-gola* (equatorial celestial sphere).

When the point of intersection of *ghaṭikā-maṇḍala* and *apakrama-maṇḍala*, at *Meṣādi*, is at the zenith, the solstice which is at the end of Gemini (*Mithuna*) and the southern pole of the ecliptic (*rāśi-kūṭa*) will rise (in the east). Similarly, (the solstice at) the end of Sagittarius (*Cāpa* or *Dhanus*) and the northern *rāśi-kūṭa* will set (in the west). Then, on account of the rotation caused by the *Pravaha*-wind, those that have risen, reach up to the prime meridian, in other words, they touch *dakṣiṇottara-vṛtta* (in the visible hemisphere), and those that had set will touch the *dakṣiṇottara* down below. Then, when the end of Gemini and the southern *rāśi-kūṭa* set, the end of Sagittarius and northern *rāśi-kūṭa* will rise. Thus, the southern *rāśi-kūṭa* will revolve in consonance with the end of Gemini, and the northern *rāśi-kūṭa* in consonance with the end of Sagittarius. Now, on both sides of the *ghaṭikā-maṇḍala*, at 24 degrees (from it), there are two solsticial diurnal circles. Again, from the two *Dhruva*-s, at a distance of 24 degrees, there are two diurnal circles corresponding to the two *rāśi-kūṭa*-s. They (the two solstices and the *rāśi-kūṭa*-s) have constant motion along these diurnal circles.

9.4 *Ayana-calana*: Motion of the equinoxes

Now, on a day when there is no motion of the equinoxes (*ayana-calana*), the ends of Virgo and Pisces will be the meeting points of the (great circles of the) spheres (i.e., the equinoxes); and the ends of Gemini and Sagittarius will be the meeting points of the *ayana*-s (solstices). And, on a day when precession of equinoxes is to be added, the said four points will be at places removed from the aforesaid ends along the earlier *rāśi* by a measure equal to the degrees of the precession of the equinoxes. Again, on a day when precession of the equinoxes is to be deducted, the four points will be at places removed from the aforesaid ends along the next *rāśi* by a measure equal to the degrees of the motion of the equinoxes. These four points are the points where the *ghaṭikā-maṇḍala* and *apakrama-maṇḍala* meet and where they are most apart respectively. The distance of separation is of course equal to 24 degrees. It is to be noted that what moves would only be the points of contact of the *ghaṭikā-maṇḍala* and *apakrama-maṇḍala*.

9.5 The manner of *Ayana-calana*

Now the manner of the motion. Ascertain the point on the *apakrama-maṇḍala* which the *ghaṭikā-maṇḍala* cuts on the day when there is no motion.

Then, for any day for which motion has to be added, the intersection of these two circles will be at a point behind the first mentioned point by the measure of the motion (of the equinoxes) for that day. In the same manner, for a day for which the motion has to be deducted, the intersection of the two circles will take place at a point in advance of the first mentioned point. (Actually), the *ghaṭikā-maṇḍala* will not be moving, and the movement therein would only be for the point of intersection. (On the other hand), the *apakrama-vṛtta* will be moving. On account of this, the *rāśi-kūṭa*-s will also have a motion. But they will not move away from their *svāhorātra-vṛtta*-s. The motion is only backward and forward in the *rāśi-kūṭa-svāhorātra-vṛtta*. Again the deviation of the *rāśi-kūṭa*-s from the *Dhruva*-s and that of the *ayanānta*-s (solstices) in the *apakrama-vṛtta* from the *ghaṭikā-maṇḍala* is always 24 degrees. All these four deviations, (two above and two below), can be demonstrated on an *ayanānta-rāśi-kūṭa-vṛtta*. Thus, when one leg of a pair of compasses is fixed at a point and the other leg is turned to make a circle, the centre of the circle would be at the point of the fixed leg. That centre is called *nābhi* and also '*kendra*'. The line around (traced by the moving leg) is called '*nemi*' (circumference).

Now, when considering the great celestial circles, it is always taken that the centre of all of them is the centre of the *bhagola* which is (practically) the same as the centre of the Earth, and that the magnitude of all these circles is the same. This is the general conception except in the case of the diurnal circles (*svāhorātra-vṛtta*) and the (*ucca-nīca-vṛtta*) circles conceived in the computation of true positions of the planets. Now, the two great circles, *ghaṭikā-maṇḍala* and *apakrama-maṇḍala*, which have a common centre, intersect each other at two points. But the diameter of the two circles that passes through the centre and touches the two points of intersection is the same. But the two diameters which touch the points of maximum divergence (*paramāntarāla*) are different for the two circles. The term *paramāntarāla* means the place of (or the extent of) the maximum separation (i.e., solstices). The diameters at the points of maximum divergence of the two circles would be at right angles to the diameter passing through the intersection of the *ghaṭikā-vṛtta* and *apakrama-vṛtta*. Hence the *ayanānta-rāśi-kūṭa-vṛtta* which touches the points of maximum divergence, will be perpendicular to the two circles (*ghaṭikā-vṛtta* and *apakrama-vṛtta*). It is always the case that this perpendicular circle touches the poles (*pārśva*) of both circles; (and conversely) touching the poles would imply that the circles are mutually perpendicular. Here, the *ayanānta-rāśi-kūṭa-vṛtta* is at right angles, both to the *ghaṭikā-vṛtta* and *apakrama-vṛtta*. Hence, it will touch the two *Dhruva*-s and the *rāśi-kūṭa*-s which are the poles of the two (circles namely *ghaṭikā-vṛtta* and *apakrama-vṛtta* respectively). Thus, it is definitely the case that they, the poles of the two circles *ghaṭikā* and the *apakrama*, lie in the same circle, and the distance between the poles and the maximum divergences between the two circles are equal.

Taking account of the fact that, when on account of motion of the equinoxes the *ayanānta* (solstice) moves, the circle which passes through the *ayanānta* will also pass through the *rāśi-kūṭa*, it follows that the *ayanānta-raśi-kūṭa* (poles of the ecliptic) too would have moved in the direction in which the *apakramāyanānta* (solstices) has moved. Since it is also the rule that the *ayanānta* (solsticial point) on the *apakrama-maṇḍala* would on all days (i.e., always) be removed from the *ghaṭikā-maṇḍala* by 24 degrees, it follows that the *rāśi-kūṭa*-s on the two sides (*pārśva*) of the *apakrama-maṇḍala* would be removed by the same extent, on all days (i.e., always) from the two *Dhruva*-s on the sides of the *ghaṭikā-maṇḍala*. Hence, the *svāhorātra-vṛtta* of the two *rāśi-kūṭa*-s will be the same always. Thus, it has to be understood that the two *rāśi-kūṭa*-s will swing to the east and the west, on account of the motion of equinoxes, in their own *svāhorātra-vṛtta*-s. Then, the distance which a planet has moved from *Meṣādi* can be ascertained through computing the true planet.

And in order to learn how much it has moved from the point of contact of the *ghaṭikā-maṇḍala* and *apakrama-maṇḍala*, the amount of motion of the equinoxes (*ayana-calana*) has to be applied to it (i.e., to the true planet). Then it (i.e., the corrected true planet) is said to be *golādi*. Thus (has been stated) the mode of the motion of the equinoxes.

9.6 Changes in placement due to terrestrial latitude

What is explained above is applicable when one considers the celestial sphere for an observer having zero-latitude (i.e., on the equator). For him it would appear that it (the celestial sphere) is rotating towards west (from the east) due to the (motion of the) *vāyu-gola*. It has been stated that, because of this, it would appear that all the diurnal circles beginning with the circle at the centre of the *vāyu-gola*, namely the *ghaṭikā-vṛtta*, would appear as vertical circles. It has also been stated that the *bhagola* is inclined to the *vāyu-gola* and that it has a slow motion. Now, when the celestial sphere is considered from a place having a latitude, it would appear that the *vāyu-gola* itself has an inclination, and that the *bhagola* too has an inclination in accordance with the former. This is being explained below.

9.7 Zenith and horizon at different locations on the surface of the Earth

Now, what is perfectly spherical is called a *gola* (sphere). The Earth is in the form of a sphere. On the Earth, which is of this shape, there are people all over its surface. The feeling that anybody has at any place would be that the place that he is standing on is the top (of the Earth), that the surface of the Earth (over which he stands) is flat (horizontal), and that he is standing perpendicular (to the Earth's surface). Consider the spherical Earth, which is suspended in the centre of the sky, as having two halves, the upper half and the lower half. Then, for the upper half, the centre seems to be the place where one stands. Then, the sky below the horizon around on the sides (*pārśva*) would be hidden by the Earth. This being the case, when celestial bodies enter the horizon (*bhū-pārśva*), their rising and setting take place. The sky will be visible above this (horizon). The centre of the (visible portion) is the *kha-madhya* (zenith). It will be right above the head of the observer. Here, what has been stated as *ghaṭikā-maṇḍala* is the east-west vertical circle at the place with no latitude. Its centre will be exactly at the centre of the Earth. At the two far sides of this will be the two *Dhruva*-s (poles). In this configuration, consider a north-south axis passing through the centre of the Earth and extending to the two poles. Let it be called the *akṣa-daṇḍa* (polar axis). This would be like an axle. Consider the celestial sphere to be attached to it so that when it spins the celestial sphere will also spin according to it. If so, it is easy to conceive of the variation in the inclination of the *vāyu-gola* in accordance with the difference in the locations on the Earth.

Here, in the region of no latitude, the *ghaṭikā-maṇḍala* is a circle which is exactly east-west, and passes through the zenith. It has been stated earlier that at the place of no latitude, the horizon (horizontal circle) passing through the poles at the two sides of the Earth, is the 'equatorial horizon' (*nirakṣa-kṣitija*). Now, if looked at from the *Meru* in the North (pole) of the Earth, the *Dhruva* will appear at the zenith. Then the equatorial horizon would be vertical and the *ghaṭikā-maṇḍala* will appear as the horizon. There, everybody will have the feeling that the place they are located is one of uniform motion around (*sama-tiryak-gata*); and there too, they will feel that their posture is vertical. This accounts for the difference between the zenith (*kha-madhya*) and the horizon (*bhū-pārśva*) at each place (on the surface of the Earth). This being the case, as one moves from the equator northwards, the pole will be seen higher and higher up from the horizon. And, as one moves from the *Meru* (north pole) southwards, it (*Dhruva*) will be seen lower and lower with respect to the zenith, up to the equator. Thus, for each (observer) in different parts of the Earth, the zenith and the horizon are different.

Now, conceive one's place to be on the meridian (*sama-rekhā*) right northwards of *Laṅkā*. Then, conceive of a (great) circle passing through the zenith, which is a point lying towards the north of the point of intersection of the *ghaṭikā-vṛtta* and the *dakṣiṇottara-vṛtta* on the *dakṣiṇottara-vṛtta*, and passing through the previously mentioned east and west cardinal points. This circle is called *sama-maṇḍala* (prime vertical). Ascertain on the *dakṣiṇottara-*

vṛtta, the distance between the *ghaṭikā-maṇḍala* and the *sama-maṇḍala*. Conceive of a circle passing through the east and west cardinal points and the two points on the *dakṣiṇottara-vṛtta*, one below the north pole by the abovesaid difference, and another by the same measure above the south pole. That circle is called the *svadeśa-kṣitija* (local horizon). The portion lying to the north of the east and the west cardinal points of the equatorial horizon described above, will be above the local horizon, and the portion lying to the south of it (i.e., the east and the west cardinal points) will be below (the local horizon). Now, when such a local horizon is conceived of separately, the equatorial horizon is called *unmaṇḍala*. Now, just as the six equidistant cardinal points generated by the three (circles), viz., *dakṣiṇottara*, *ghaṭikā* and equatorial horizon circles (*unmaṇḍala*) gave rise to eight equal spherical sections, in a similar manner, eight equal divisions of the sphere can be conceived of by (the set of three circles) *dakṣiṇottara*, *sama-maṇḍala* (prime vertical) and the local horizon (*svadeśa-kṣitija*). In this manner, six equidistant cardinal points and eight equal spherical divisions are formed whenever we have three mutually perpendicular great circles.

Now conceive of a fourth circle. Let it be constructed such that, it passes through the cardinal points (*svastika*-s) formed by two of the said three circles. Then, by means of this circle, it would seem as if four of the eight spherical sections (mentioned above) are divided apart. This circle is called *valita-vṛtta* (deflected circle). The computation of the distance of the other two circles from this *valita-vṛtta* is carried out using the rule of three pertaining to the difference in the circles, and this will be explained later, in detail. Thus has been explained the nature of *vāyu-gola*. The locational distinction between the *vāyu-gola* and the *bhagola* has already been stated.

Since the Earth is spherical, for observers on different locations of the surface of the Earth, the altitude of the pole (*Dhruva*) that is along the tip of the *akṣa-daṇḍa* (polar axis) will appear different. Therefore, the *vāyu-gola* that rotates in accordance with the spin of the said axis will appear to rotate with different inclinations, as has been mentioned earlier. Then, it is to be noted that the nature of the *vāyu-gola*, the difference in the location (*saṃsthāna*-

bheda) of the *bhagola* from that of the *vāyu-gola*, and the spherical shape of the Earth – these three provide the basis for those calculations pertaining to the planets, which are to be carried out after the computation of true planets (*graha-sphuṭa*). Hence, their nature has been stated here in advance.

9.8 Construction of the armillary sphere

Now, in case a clear mental conception of the the circles mentioned above and their rotation has not been achieved, then construct an armillary sphere with the necessary circular rings tied appropriately (rotating around the polar axis) and having a spherical object representing the Earth fixed to the middle of the axis, and perceive the rotation of the sphere. In this construction, the prime vertical, north-south circle, the local horizon and the equatorial horizon need not have to revolve. So, to keep them fixed, employ a few larger circles and tie them up from outside. The other circles have to revolve. Hence, tie them up inside by choosing them to be smaller circles. Represent the *jyā*-s by means of strings. Thus (experimenting with this), clearly understand the situation of (the circles making up) the armillary sphere and their revolutions.

9.9 Distance from a *Valita-vṛtta* to two perpendicular circles

Now, let there be certain (say, three) great circles with same dimension and with a common centre. Herein below is described a method to ascertain the distance from one circle, namely the *valita-vṛtta* to the other two. This is first illustrated by the derivation of the *apakrama-jyā* and its *koṭi*. For this, suppose the vernal equinox to be coinciding with the zenith for an equatorial observer. There, the *viṣuvad-viparīta-vṛtta* (the circle passing through the vernal equinox and the north-south poles) which is perpendicular (*viparīta*) to the *ghaṭikā-maṇḍala* at the vernal equinox would also (incidentally) coincide with the *dakṣinottara-vṛtta*. The *ayanānta-viparīta-vṛtta* (the perpendicular circle passing through the solstice) will coincide with the equatorial horizon. When this is the situation of the celestial sphere, construct the *apakrama-maṇḍala*, which is nothing but the locus (*mārga*) of the eastward motion of the Sun, such that it passes through (1) the *svastika*-s (cardinal points) at the top and bottom (zenith and the nadir), (2) the point on the horizon which is 24 degrees away from the eastern *svastika* towards the north and (3) the point on the horizon which is 24 degrees away from the western *svastika* towards the south. Then, conceive of a desired Rsine (*iṣṭa-jyā*) with its foot at the place which forms the *śara* with its beginning at the equinox, and with the tip at the desired point on the *apakrama-maṇḍala* which lies to the east of the zenith. This will be the Rsine of the desired part of the arc of the *apakrama-maṇḍala*. Now, first it has to be ascertained as to what would be the distance between the tip of the desired Rsine and the *ghaṭikā-maṇḍala* along the north-south direction, and secondly it has also to be ascertained as to what would be the distance between the tip of the Rsine and the *dakṣiṇottara-maṇḍala* along the east-west direction. Herein below (is given) a method to ascertain the above.

Now, the maximum divergence between the *apakrama-maṇḍala* and *ghaṭikā-maṇḍala* can be found on the horizon which is the same as the *ayanānta-viparīta-vṛtta*. Here, the maximum divergence is the Rsine of 24 degrees and this is the Rsine of the maximum declination (*paramāpakrama*). Then again, the maximum divergence between the *apakrama-vṛtta* and the *dakṣiṇottara-vṛtta* can also be seen from the *ayanānta-viparīta-vṛtta* itself. Here, the Rcosine of the maximum declination would be the divergence of the *apakrama-maṇḍala* from the pole (i.e., the line joining the north and the south pole). This is called *parama-svāhorātra*.

Now, conceive as the *pramāṇa* the hypotenuse (*karṇa*), the radius of the *apakrama-maṇḍala*, which is the distance between the centre and the circumference of the *ayanānta-viparīta-vṛtta*. Conceive the two maximum divergences, the *bhujā* and *koṭi* of this hypotenuse, as the respective *pramāṇa-phala*-s. Then conceive as *icchā*, the desired Rsine (*iṣṭa-dorjyā*) which has its tip at the desired place on the *apakrama-maṇḍala.* Apply the rule of three. Then the *bhujā* and *koṭi* of the *iṣṭa-dorjyā* will be got as *icchā-phala*-s, being respectively the distances from the tip of the *dorjyā* to the *ghaṭikā-maṇḍala* and to the *dakṣiṇottara-vṛtta.* These two are called *iṣṭāpakrama* (Rsine of the desired declination) and *iṣṭāpakrama-koṭi.* This is the rationale of the rule of three for finding the distances between great circles having the same dimension and a common centre.

9.10 Some *Viparīta* and *Nata-vṛtta*-s

Herein below (is stated) the method to arrive at the above in an easy manner. There, we have the *ghaṭikā-maṇḍala, viṣuvad-viparīta-vṛtta* and *ayanānta-viparīta-vṛtta,* being three circles mutually perpendicular (*tiryak-gata*) to each other. Construct an *apakrama-vṛtta,* a little inclined to the *ghaṭikā-maṇḍala.* Then, conceive of three more circles besides these four circles (as follows). First, a circle which passes through the two poles and the desired place in the *apakrama-vṛtta* is constructed. This (circle) is called *ghaṭikā-nata-vṛtta.* The maximum divergence from this circle to the *viṣuvad-viparīta-vṛtta* and the *ayanānta-viparīta-vṛtta* can be seen on the *ghaṭikā-maṇḍala.*

Construct (the second) circle touching the point of intersection of the *ghaṭikā-vṛtta* and the *ayanānta-viparīta-vṛtta,* and the desired point on the *apakrama-maṇḍala.* This is called *viṣuvad-viparīta-nata-vṛtta* , and since the *viṣuvad-viparīta* is the same as the *dakṣiṇottara-vṛtta,* it is (also) called *dakṣiṇottara-nata-vṛtta.* The maximum divergence between this circle and (i) the *ayanānta-viparīta-vṛtta* and (ii) the *ghaṭikā-vṛtta* can be seen along the *viṣuvad-viparīta-vṛtta.*

It might be noted that in the above-said situation of the *apakrama-maṇḍala,* the two *rāśi-kūṭa*-s (poles of the ecliptic) would be situated on the horizon, which is the *ayanānta-viparīta-vṛtta,* at 24 degrees towards the east from the south pole and by the same amount towards the west from the northern pole. Conceive of still another circle, which passes through the two *rāśi-kūṭa*-s and a point on the *apakrama-maṇḍala*, which is one-fourth of the circumference (90 degrees) away from the desired point on the *apakrama-maṇḍala* and lies to the west of the zenith. This is called the *rāśi-kūṭa-vṛtta*. The maximum divergence between the *rāśi-kūṭa-vṛtta* and the *ghaṭikā-maṇḍala* occurs at a point which is one-fourth of the circumference (90 degrees) removed from the place where these two circles intersect. This (maximum) divergence will occur on the *ghaṭikā-nata-vṛtta*.

Since, however, this *ghaṭikā-nata-vṛtta* passes through the two poles, it is *viparīta* (perpendicular) to the *ghaṭikā-maṇḍala*. Then again, the tip of the desired Rsine of the declination (*krāntīṣṭa-jyāgrā*) on the ecliptic forms the pole (*pārśva*) of the *rāśi-kūṭa-vṛtta*. Since it passes through that point, the *ghaṭikā-nata-vṛtta* is perpendicular also to the *rāśi-kūṭa-vṛtta*. Since, as indicated here, the *ghaṭikā-nata-vṛtta* is perpendicular both to the *ghaṭikā* and the *rāśi-kūṭa-vṛtta*-s, the maximum divergence of the latter two will occur on this *ghaṭikā-nata-vṛtta*. And, that will be equal to the desired *dyujyā*. Thus, the maximum divergence between the *rāśi-kūṭa-vṛtta* and the *dakṣiṇottara-vṛtta*, which itself is perpendicular to the *viṣuvad-vṛtta* (celestial equator), would occur on the *yāmyottara* (i.e., *dakṣiṇottara-nata-vṛtta*) which is perpendicular to both these circles.

Since the *yāmyottara-nata-vṛtta* touches the east and west cardinal points and the tip of the desired Rsine, the *yāmyottara-nata* circle is *viparīta* (perpendicular) to both (the above circles). As is known, when two (equal) circles (inclined to each other) intersect at two points, a third (equal) circle passing through the points which are at one-fourth the circumference (90 degrees) away from these two intersecting points, happens to be a *viparīta-vṛtta* (perpendicular circle). Thus it is appropriate that the maximum divergence that occurs between the first mentioned two circles is on this (perpendicular) circle.

Here, the circles, viz., the *dakṣiṇottara-vṛtta* which is the same as the *viṣuvad-viparīta-vṛtta*, and the *ayanānta-viparīta-vṛtta* which is the same as the horizon, and the *ghaṭikā-maṇḍala* are mutually perpendicular. Here is a set up where the division into quadrants (*pāda-vyavasthā*) and division of the sphere (*gola-vibhāga*) have been determined by the said three circles. In this set-up, the divergence amongst the circles is determined by the two *nata-vṛtta*-s, the *apakrama-vṛtta* and the *rāśi-kūṭa-vṛtta.* There, the maximum divergence between the *ghaṭikā* and *apakrama* circles is equal to the maximum declination and occurs in the horizon. The desired Rsine (*dorjyā*) is the distance (*agra*) from the *viṣuvad* (equinox) to the desired point on the *apakrama-maṇḍala* . Rcosine thereof (*dorjyā-koṭi*) is the distance from the *ayanānta-viparīta-vṛtta* to the point of desired declination. The declination at the desired place is the Rsine on the *ghaṭikā-nata* circle from the point of contact of the *nata* (*ghaṭikā-nata*) and *apakrama* circles to the *ghaṭikā-maṇḍala.* And the *iṣṭa-dyujyā* (day radius) is the Rsine on the *nata-vṛtta* from the pole to the desired point on the *apakrama-vṛtta.*

9.11 Declination of a planet with latitude

It might be noted that the *iṣṭāpakrama* (declination at a desired point) is also to be found in the above-said circle. Now, the *iṣṭāpakrama-koṭi* is the Rsine on the *dakṣiṇottara-nata-vṛtta* which is from the *dakṣiṇottara-vṛtta* to the desired point which is the tip of the (previously stated) *dorjyā.* The Rsine on the *dakṣiṇottara-vṛtta* from the east-west cardinal points to the tip of the *dorjyā* is the *koṭi* of this *iṣṭāpakrama-koṭi. Laṅkodaya-jyā*, which is nothing but the *kāla-jyā*, is the Rsine from the equinox to the point of contact of the *ghaṭikā* and *nata-vṛtta*-s. *Laṅkodaya-jyā-koṭi* is the one which has its tip at the tip of the above *jyā* and extends up to the east-west cardinal points. *Kāla-koṭi-jyā* is that which starts from the zenith and with its tip at the point of contact of the *rāśi-kūṭa* and *ghaṭikā-vṛtta* on the *ghaṭikā-vṛtta.* The *kāla-koṭyapakrama* (declination of the *kāla-koṭi* on the *rāśi-kūṭa-vṛtta*) is that which has its tip at the tip of the *kāla-koṭi* and commences from the point of contact of the *rāśi-kūṭa-vṛtta* and the *apakrama-maṇḍala.* This has to be derived from the maximum declination which has been specified as the hypotenuse in the *ghaṭikā-maṇḍala.*

If a planet has a latitudinal deflection from the point of contact of the *rāśi-kūṭa* and the *krānti-vṛtta* (ecliptic), then the deflection would be along the *rāśi-kūṭa-vṛtta* and so the latitude arc (*vikṣepa-cāpa*) would be a remainder (i.e., extension) of the arc of the *kāla-koṭyapakrama*. The sum or difference of these arcs would be the distance between the latitudinally deflected planet and the point of contact of the *ghaṭikā* and *rāśi-kūṭa-vṛtta*-s. Now, the maximum divergence between the *ghaṭikā* and *rāśi-kūṭa-vṛtta*-s is seen in the *ghaṭikā-nata-vṛtta*. And, that is equal to the desired *dyujyā*.

Now, the poles (*pārśva*) of the *rāśi-kūṭa-vṛtta* are the points of contact of the *nata* and *apakrama* circles. Since it is a fact that from the poles (*samapārśva*) all its (i.e., the circle's) parts are away by one-fourth of a circle (90 degrees), the distance between the (point of intersection of) *rāśi-kūṭa* and *apakrama* is one-fourth of a circle (90 degrees) away from the (point of intersection of) *ghaṭikā-nata* and the *dakṣiṇottara-nata*. These quadrants would have been divided into two by the *ghaṭikā-maṇḍala* and the *yāmyottara* (north-south circle). Here, the northern part of the *ghaṭikā-maṇḍala* would be the desired declination. But the southern part would be the *dyujyā*. This would be the maximum divergence between the *ghaṭikā-vṛtta* and the *rāśi-kūṭa-vṛtta*.

Now, the *ghaṭikā-nata* passes through the poles (*pārśva*) of *ghaṭikā* and the *rāśi-kūṭa-vṛtta*-s. Since the *ghaṭikā* and *rāśi-kūṭa-vṛtta*-s are passing through the poles of the *ghaṭikā-nata-vṛtta*, (we have the following): (Consider) the radius hypotenuse (*trijyā-karṇa*) of the *rāśi-kūṭa-vṛtta* which commences from the point of intersection of the *ghaṭikā* and *rāśi-kūṭa-vṛtta*-s, and having its tip at its contact with the *nata-vṛtta*. For this *karṇa*, the maximum divergence stated above, viz., the desired *dyujyā*, would be *koṭi*. If the above is the case, how much will be the *koṭi* of the *jyā* which is the hypotenuse on the *rāśi-kūṭa-vṛtta* stretching from the point of intersection of the *ghaṭikā-vṛtta*, and having its tip at the planet with latitude (*vikṣipta-graha*). (This *koṭi* will be) the distance between the planet-with-latitude and the *ghaṭikā-vṛtta*. This will be the declination of the planet that has a latitudinal deflection.

This is the method of deriving the declination of the planet-with-latitude by applying the rule of three using the Rsine of the arc got by finding the sum or difference of the arc of the *kāla-koṭi-krānti* and the arc of the *vikṣepa.* This *icchā-phala* and the *pramāṇa-phala* might be taken as triangles. Instead of adding the arcs, the Rsines might be added. There again, by mutually multiplying the *koṭi*-s with the *jyā*-s and dividing the product by *trijyā* and adding or subtracting the results appropriately, and again multiplying by the *iṣṭa-dyujyā* and dividing by *trijya*, the declination of the planet-with-latitude is obtained. Here, the *vikṣepa-koṭi* and *iṣṭa-dyujyā* are the *guṇakāra*-s (multipliers) for *kāla-koṭi-krānti.* Multiply first the *kāla-koṭi-krānti* by *iṣṭa-dyujyā* and divide by *trijyā.* The result will be the distance from the point of contact of the *rāśi-kūṭa-vṛtta* and the *krānti-vṛtta* to the *ghaṭikā-vṛtta.* This will be the *jyā* on the *apakrama-maṇḍala*, from that point on it from which the planet has a latitudinal deflection, if it is presumed that it has no latitude. Here, instead of multiplying the latitude by *dyujyā*, one might multiply the *kāla-koṭi-krānti-koṭi*, which is the multiplier of the latitude, by the *iṣṭa-dyujyā* and divide by *trijyā* since, both ways, the result will be the same. Then it would be as if the *kāla-koṭi-krānti* and its *koṭi* had been multiplied by *iṣṭa-dyujyā* and divided by *trijyā.* The results obtained will then be the *bhujā* and *koṭi* of a circle having its radius equal to that of the *iṣṭa-dyujyā.* So, multiply the *kāla-koṭi-krānti* and its *koṭi* by the *iṣṭa-dyujyā-vyāsārdha*, and respectively by the *vikṣepa-koṭi* and the *vikṣepa.* It has already been stated that if *kāla-koṭi-krānti* is converted in terms of *iṣṭa-dyujyā-vṛtta*, the result will be the declination of the planet with latitude.

Now, when the square of the declination of the planet without latitude (*avikṣipta-graha*) is subtracted from the square of the *iṣṭa-dyujyā*, the result will be the square of the *ādyanta-dyujyā* (*antya-dyujyā*). Its root is the *kālakoṭi-krānti-koṭi* on the *dyujyā-vṛtta.* That will also be the *koṭi* of the maximum declination. Now, when the square of the *iṣṭa-dorjyā-krānti* is subtracted from the square of *trijyā*, the result will be the square of *iṣṭa-dyujyā.* Here, suppose the planet without latitude is at the tip of the *dorjyā.* Then, its *krānti-koṭi* will be its *krānti* (declination). When the square of this declination is also subtracted, it would be as if the square of the *koṭi-krānti* and the square of the *bhujā-krānti* have also been subtracted. When the square of the *koṭi-krānti* and the square of the *bhujā-krānti* are added together, the sum would be the square of the *parama-krānti* (maximum declination). When it is subtracted from the square of *trijyā* the result would be the square of the *parama-krānti-koṭi* (Rcosine maximum declination). The square root thereof is the *parama-krānti-koṭi.* Hence multiply the *vikṣepa* by *parama-krānti-koṭi.* Multiply also the *krānti-jyā* of the planet-without-latitude (*avikṣipta-graha*) by the *vikṣepa-koṭi.* These two added together or subtracted from one another appropriately, and divided by *trijyā* will result in the declination of the planet with latitude. Thus has been explained the method of arriving at the declination of a planet with latitude.

9.12 *Apakrama-koṭi*

Now is explained the method of ascertaining the *apakrama-koṭi* of a planet-with-declination, extending east-west, being the distance between the planet and the north-south circle which is the same as the *viṣuvad-viparīta-vṛtta.* The east and west cardinal points are the poles of the north-south circle. The tip of the *jyā* of the desired declination is the pole of the *rāśi-kūṭa-vṛtta.*

Now, consider the location on the *dakṣiṇottara-nata-vṛtta* passing through the poles of these two circles, which touches the tip of the *jyā* of the desired declination. A point that is one-fourth circumference (90 degrees) away from this will touch the *rāśi-kūṭa-vṛtta*, since all the points in a circle from its pole are at a distance of one quadrant. Divide this quadrant into two parts by the *dakṣiṇottara-vṛtta.* The distance between the tip of the *iṣṭa-kranti-dorjyā* and the north-south circle is the *iṣṭā-kranti-koṭi.* This remainder of *koṭi* extends from the north-south circle to the *rāśi-kūṭa-vṛtta* and is the *koṭi* of the *iṣṭāpakrama-koṭi.* In all circles, quadrants divided into two will have complementary *bhujā* and *koṭi.* Thus, it follows that the *koṭi* of the *iṣṭāpakrama-koṭi* is the maximum divergence between the *rāśi-kūṭa* and *dakṣiṇottara-vrtta*-s. Here the *pramāṇa* is the *trijyā-karṇa* which extends from the point of intersection of the north-south and *rāśi-kūṭa* circles to the *dakṣiṇottara-nata-vṛtta* along the *rāśi-kūṭa-vṛtta.* The Rsine of this maximum divergence is the *pramāṇa-phala.* The distance from the point of contact of the north-south circle up to the planet with latitude on the *rāśi-kūṭa-vṛtta* could be taken as the *icchā.* From this the distance between the planet and the north-south circle can be got as the *icchā-phala.*

Here is the method for the derivation of the *icchā-rāśi.* Now, for a circle with its tip at the southern cardinal point and having a radius equal to the radius of the north-south circle, its divergence from the *apakrama-vṛtta* that occurs on the horizon and is equal to the maximum *dyujyā* (i.e., *antya-dyujyā*) is the *pramāṇa-phala.* How much is the distance between the *apakrama-vṛtta* and the *yāmyottara-vṛtta-jyā*, the Rsine which starts from the zenith and has its tip on the contact with the *rāśi-kūṭa-vṛtta.* The *icchā-phala* would be the divergence between the north-south and *apakrama* circles which will be seen on the *rāśi-kūṭa-vṛtta.* Then to this *jyā* add or subtract, appropriately, the *vikṣepa-jyā* (Rsine latitude). The result would be the *jyā* on the *rāśi-kūṭa-vṛtta* being the *jyā* which commences from the point of contact of the north-south circle, and having its tip on the planet-with-latitude. Multiply this by the maximum divergence between the *rāśi-kūṭa* and north-south circles, and divide by *trijyā.* The result is the distance from the deflected planet to the north-south circle.

Here, when for the purpose of deriving the *icchā-rāśi*, addition and subtraction of *vikṣepa-jyā* is carried out, mutual multiplication of the *koṭi*-s and division by *trijyā* are required. Multiplication by maximum divergence is also needed. For this the following order might be employed: that is, first multiply by the maximum divergence and then multiply by *vikṣepa-koṭi*, since there will be no difference in the result. Here, when the *jyā* on that part of the *rāśi-kūṭa-vṛtta* which lies on the divergence between the north-south and *apakrama* circles, is multiplied by the *jyā* of maximum divergence between the *rāśi-kūṭa* and north-south circles, and divided by *trijyā*, the result will be the distance between the point of contact from the *rāśi-kūṭa* and *apakrama* circles to the north-south circle. And that will be the *koṭi* of the declination which is the hypotenuse to the *jyā* of the planet-without-latitude.

Now, multiply the *koṭi* of the *jyā* of the divergence between the north-south and *apakrama-vṛtta*-s by the maximum divergence between the north-south circle and the *rāśi-kūṭa-vṛtta* and divide the product by *trijyā*. The result will be the square root of the difference between the square of the Rcosine of the declination of the planet-with-latitude, and the square of the above-derived maximum divergence. The rationale here is that in a circle which has *trijyā* (Rsine 90 degrees) as radius, if any Rsine and Rcosine therein are multiplied by the same multiplier and divided by *trijyā*, they will be converted respectively into the Rsine and Rcosine of a circle having the said multiplier as radius.

In the above case, the Rcosine in the circle with *trijyā* as radius, the maximum divergence will be the maximum declination. Then again, when the square of the desired declination is subtracted from the square of the *iṣṭa-dorjyā*, the remainder is the square of the Rcosine of the desired declination. If this is subtracted from the square of *trijyā*, the remainder will be the square of the maximum divergence between the north-south circle and the *rāśi-kūṭa-vṛtta*. Subtract from this the square of the Rcosine of the declination of the planet-without-latitude. The result will be the desired square of the Rcosine. That will be the square of maximum declination.

Now, when the square of the *bhujāpakrama-koṭi* and that of the *koṭyapakrama-koṭi* are added, the sum will be the square of the *antyāpakrama-koṭi*. When that is subtracted from the square of *trijyā*, the result will be the square of *antyāpakrama*. The root thereof is the *antyāpakrama*. Now, the *vikṣepa* (Rsine of the latitude) is multiplied by the maximum declination and *vikṣepa-koṭi* (Rcosine of the latitude) by the Rcosine of the declination of the planet-with-latitude. When these two results are added or subtracted appropriately and divided by *trijyā*, the result obtained would be the distance from the planet-with-latitude to the north-south circle.

Suppose, however, that it is not divided by *trijyā* and the square of the declination of the planet with latitude is subtracted from the square of *trijyā*. The root thereof would be the *dyujyā* of the planet. By this, divide the earlier result and the result will be the *kāla-dorguṇa* (*kāla-jyā*) of the planet-with-latitude. This *kāla-dorguṇa* has been explained earlier.

Now, besides (the above circles), conceive of still another circle passing through the planet with latitude, and the two poles. (Ascertain) the Rsine starting from the point where it intersects the *ghaṭikā-vṛtta* up to *viṣuvad* (the intersection point of the ecliptic and the celestial equator) along the *ghaṭikā-maṇḍala*. This Rsine is called *kāla-dorguṇa* (*kāla-jyā*). The arc of this Rsine is measured in *prāṇa*-s (units of time equal to one-sixth of a *vināḍī*).

It is the case that the portion between the planet-with-latitude and the equinox will revolve during this specified time (i.e., above said *prāṇa*-s). The Rsine of this, which is in time units, is *kāla-jyā*. The number of *prāṇa*-s in the *ghaṭikā-vṛtta* are equal to the number of minutes (*ili*) in the twelve *rāśi*-s. This will revolve once in 21,600 (*anantapura* in *kaṭapayādi* notation) *prāṇa*-s or minutes. Hence, the identity in the number of *prāṇa*-s and time (*kāla*). This being the case, just as in the *ghaṭikā-vṛtta*, in all the *svāhorātra-vṛtta*-s (diurnal circles) also, there will be a revolution of one minute (*anantapurāṃśa* = 1/21,600) in one *prāṇa*. Hence, all the *svāhorātra-vṛtta*-s have to be divided into the number of minutes in a circle (i.e., 21,600), when the measure of time is required. Then, the distance between the planet-with-latitude and the north-south circle would be as derived above. Since, that (21,600) is the number (of *kalā*-s) when measured on the *svāhorātra-vṛtta* of the planet-with-latitude, *kāla-dorguṇa* can be taken also as the *jyā* on that *svāhorātra-vṛtta*. The arc thereof can also be the *kāla*-arc, being the measure of difference between the north-south circle and the *rāśi-kūṭa-vṛtta*, as seen on the *svāhorātra-vṛtta* of the planet with latitude. Thus has been stated the method to ascertain the distance of the *ghaṭikā-vṛtta* and of the *viṣuvad-viparīta-vṛtta* from the planet with latitude. The *Ācārya* (Nīlakaṇṭha Somayājī) has stated so in his *Siddhānta-darpaṇa* .

antyadyujyeṣṭabhakrāntyoḥ kṣepakoṭighnayoryutiḥ |
viyutirvā grahakrāntistrijyāptā kāladorguṇaḥ ||

antyakrāntīṣṭatatkoṭyoḥ svadyujyāptāpi pūrvavat |

[*Siddhānta-darpaṇa*, 28-29]

'Mutiply the Rcosine of the maximum declination (24 degrees) and the declination of the (*avikṣipta*) planet, separately, by the Rsine and Rcosine of the latitude, and add or subtract the products (as the case may be). The result is the declination of the planet (with latitude).

'Multiply (separately) the maximum declination (24 degrees) and the *koṭi* of the *iṣṭāpakrama* of the (*avikṣipta*) planet as before (by the Rcosine and Rsine of the latitude); add or subtract the products (as the case may be), and divide by the *dyujyā*. The result will be the *kāla-jyā*.

Thus have been described the derivation of the declination of a planet-with latitude, and also the *kāla-jyā*. And therethrough have been described also the complete details of the rule of three (for calculating) the divergences between the (great) circles.

Chapter 10

The Fifteen Problems

10.1 The fifteen problems

Now, towards demonstrating in detail, the above-stated principles, fifteen problems are posed in relation to the divergence between the said seven (great) circles.

Now, there are the six items to be known: the maximum declination (*antya-krānti*), the desired declination *(iṣṭa-krānti)*, the *iṣṭa-krānti-koṭi*, the Rsine of the desired longitude (*dorjyā*), Rsine of the right ascension (*kāla-jyā*) and the *nata-jyā*. When two of these (six) are known, herein below (are) the methods to derive the other four. This can occur in fifteen ways. When one item is known, mostly, its *koṭi* can be found by subtracting its square from the square of *trijyā* and finding the root of the result.

Now, the (portions of the) *ghaṭikā*, *apakrama* and *viṣuvat-viparīta-nata* circles lying between the *ghaṭikā-nata-vṛtta* and the *rāśi-kūṭa-vṛtta* are separated by a quadrant of the circle (90 degrees). The quadrants of these circles are bifurcated by the *viṣuvad-viparīta-vṛtta*. And (the portions of) *viṣuvad-viparīta* and *ghaṭikā-nata* circles lying between the *viṣuvad-viparīta-nata* and *rāśi-kūṭa-vṛtta* are separated by the quadrant of a circle (90 degrees). All these quadrants are bifurcated by the *ghaṭikā-vṛtta*. These bifurcated parts will mutually be *bhujā* and *koṭi*, for it follows that when quadrants are bifurcated there will be (mutual) *bhujā* and *koṭi*.

10.2 Problem one

Now, when the maximum declination and actual declination are known, here is the method to find the other four. For the maximum declination, *trijyā* is the hypotenuse. By finding how much is it for the desired declination, the *dorjyā* can be found. Applying the rule of three: If the divergence between *ghaṭikā* and *apakrama* is the *antyāpakrama* (maximum declination) and the divergence between the north-south circle and *apakrama-vṛtta* is the *antya-dyujyā*, what will it (i.e., divergence between the north-south circle and *apakrama-vṛtta*) be when it (i.e., the divergence between *ghaṭikā* and *apakrama*) is the desired declination (*iṣṭāpakrama*): we get this divergence (*iṣṭāpakrama-koṭi*) from the tip of the *dorjyā* to the north-south circle. For all these three, the *koṭi*-s can be got by subtracting their squares from the square of the *trijyā* and calculating the roots. By the rule of three, the *iṣṭāpakrama* (declination) is the divergence between *ghaṭikā* and *yāmyottara-nata* while going from the east-west cardinal points to the tip of the *dorjyā* along the *yāmyottara-nata-vṛtta*. Then, by finding the maximum extent of these along the north-south circle, the *yāmyottara-nata-jyā* is got. Again by the rule of three: The *iṣṭāpakrama-koṭi* is the distance from the north pole to the tip of the *dorjyā* and is the divergence between the *yāmyottara* and (*ghaṭikā*) *nata*. Then, by finding the maximum divergence in the *ghaṭikā-vṛtta*, the *laṅkodaya-jyā* will be obtained. Such being the case, for the *pramāṇa-phala*-s in the form of the *iṣṭāpakrama-koṭi*, the mutually corresponding *koṭi*-s form the *pramāṇa*-s. Since, for these *pramāṇa*-s, *dorjyā-koṭi* is the *pramāṇa-phala* and *trijyā* forms the *icchā*, we get the divergence between (*yāmyottara* or *ghaṭikā*) *nata* and *kṣitija* as *nata-koṭi* and *laṅkodayajyā-koṭi*. Thus is the solution to the first problem.

10.3 Problem two

The second (problem) is when the maximum declination and *iṣṭa-krānti-koṭi* are known. If for the Rcosine of maximum declination (*paramāpakrama-koṭi*), *trijyā* is the hypotenuse, then what would be the hypotenuse for *iṣṭāpakrama-koṭi*. From this the *dorjyā* can be got. Then, calculate (the other quantities) as in the previous case (i.e., the first problem).

10.4 Problem three

The third (problem) is when the maximum declination and *dorjyā* (are known). Here, (it is to be noted that) the maximum declination is the distance of the *ghaṭikā-vṛtta* from the point of contact of the *apakrama-vṛtta* and the *ayanānta-viparīta-vṛtta*, and maximum *dyujyā* is the distance from the same point to the *viṣuvad-viparīta-vṛtta*. Taking these as *pramāṇa-phala*-s and taking the *dorjyā* as *icchā*, the actual *apakrama* and its *koṭi* can be obtained. The rest is as before.

10.5 Problem four

The fourth problem is when the maximum declination and *kāla-jyā* are known. Now, the *kāla-jyā* is that portion of the *ghaṭikā-vṛtta* from the *viṣuvat* to the (*ghaṭikā*) *nata-vṛtta*. *Kāla-koṭi* is the portion of the *ghaṭikā-vṛtta* from the *viṣuvat* to the *rāśi-kūṭa-vṛtta*. By finding the divergence for this in the *apakrama-vṛtta*, we get the divergence of the *ghaṭikā* and *apakrama-vṛtta*-s on the *rāśi-kūṭa-vṛtta*. This would be *kāla-koṭyapakrama*. Now, construct a *rāśi-kūṭa-vṛtta* touching the zenith which is the *viṣuvat*. The point of contact of this (*rāśi-kūṭa-vṛtta*) and the earlier (referred) *rāśi-kūṭa-vṛtta* will be on the *rāśi-kūṭa-vṛtta* at the horizon (i.e., point of contact of the earlier *rāśi-kūṭa-vṛtta* and the horizon). That will be towards the west from the north pole by a measure equal to the maximum declination, and as much to the east from the south pole. Subtract the square of the *kāla-koṭyapakrama* from the square of *kāla-koṭi* and find the root. The result will be the distance from the point of intersection of the *ghaṭikā* and *rāśi-kūṭa-vṛtta*-s to the second *rāśi-kūṭa-vṛtta*. Now, if the square of the *kāla-koṭyapakrama* is subtracted from the square of *trijyā* and the root extracted, the result will be the *jyā* of the arc of the *rāśi-kūṭa-vṛtta* from the point of contact on the horizon to the point of contact on the *ghaṭikā-vṛtta*. When this *jyā* is taken as the hypotenuse and considered as the *pramāṇa*, the root derived above, being the divergence between the *rāśi-kūṭa-vṛtta*-s, can be taken as the *bhujā* and considered as the *pramāṇa-phala*. In that situation, the *trijyā* would be the *icchā*. The maximum distance between the two *rāśi-kūṭa-vṛtta*-s, which is the *jyā* of the distance between the zenith to the *rāśi-kuṭa-vṛtta* on the *apakrama-maṇḍala* would be the *icchā-phala*. The *koṭi* of this (*jyā*) would be the *dorjyā* of the distance on the *apakrama-maṇḍala* from the zenith to the *nata-vṛtta*. The rest (of the calculation) is as done earlier.

10.6 Problem five

Now, the fifth problem relates to knowing the *nata-jyā* and maximum declination. Now, *nata* is that portion of the north-south circle from the zenith to the (*ghaṭikā*) *nata-vṛtta*. And *nata-koṭi* is that part of the north-south circle from the zenith to the *rāśi-kūṭa-vṛtta*. From the consideration that if the *antya-dyujyā* is the distance on the horizon from the north-south circle to the *apakrama-vṛtta*, what would be the distance from the tip of the *nata-koṭi*, one would get the distance between the north-south circle and the *apakrama-vṛtta* on the *rāśi-kūṭa-vṛtta*. Square this and subtract it separately from: (1) the square of the *nata-koṭi*, and (2) the square of *trijyā*. When the roots of the two remainders are extracted they would be: (1) the distance from the point of contact of the north-south circle and the first *rāśi-kūṭa-vṛtta* to the second *rāśi-kūṭa-vṛtta*, which is to be taken as the *pramāṇa-phala*, and (2) the *jyā* on the *rāśi-kūṭa-vṛtta* from the point of contact of the north-south circle to the horizon, which is to be taken as the *pramāṇa* and the hypotenuse. When *trijyā* is the *icchā*, the maximum divergence of the two *rāśi-kūṭa-vṛtta*-s is the *icchā-phala* and that is the earlier-said maximum declination. The *koṭi* of this is the *dorjyā*. The rest (of the calculation) is as before. Thus (have been explained), the five problems involving maximum declination.

10.7 Problems six to nine

Then the sixth problem is the one not involving maximum declination but involving actual declination and *iṣṭa-krānti-koṭi*. The root of the sum of the squares of the above (two) would be the *dorjyā*, which is to be taken as the hypotenuse.

The seventh problem involves the knowledge of the actual declination and *dorjyā*, and calculations are as before.

The eighth problem is when the actual declination and *kāla-jyā* are known. Find the squares of these two and subtract them from the square of *trijyā*. Find the roots thereof. The results will be the actual *dyujyā* and *kāla-koṭi-jyā*, which also happens to be the maximum divergence of the *nata-vṛtta* and the horizon. Here *trijyā* will be the *pramāṇa*, *kāla-koṭi-jyā* is the *pramāṇa-phala*, and actual *dyujyā* is the *icchā*. The resultant *icchā-phala* will be *dorjyā-koṭi*. The rest (of the calculation) is as before.

Then, the ninth problem is where the actual declination and the *nata-jyā* are known. While *nata-jyā* is the distance between the *yāmyottara-nata-vṛtta* and the *ghaṭikā-vṛtta*, the difference between *nata-vṛtta* and the horizon is the *nata-koṭi-jyā*. From the consideration: when the actual declination is the first difference, what will be the second difference, the result obtained is *dorjyā-koṭi*; the earlier is the distance from the tip of the *dorjyā* to the horizon. These are the four problems involving actual declination.

10.8 Problems ten to twelve

Then, leaving the above, there is the tenth problem when the *iṣṭa-krānti-koṭi* and the *dorjyā* (are known). The root of the difference of the squares of these (two) is the actual declination. The rest is as before.

The eleventh problem is when the *kāla-jyā* and the *iṣṭāpakrama-koṭi* are known. From the consideration: If *trijyā* is the hypotenuse for *kāla-jyā* what is it for *iṣṭa-krānti-koṭi*, the result would be *dyujyā*. Now, when *dyujyā* is multiplied by *kāla-koṭi* and divided by *trijyā*, the result will be *dorjyā-koṭi*. The first *trairāśika* is done by the divergence between *nata-vṛtta* and north-south circle. And the second *trairāśika* is done by the distance between *nata-vṛtta* and horizon.

The twelfth problem involves the knowledge of the *iṣṭa-krānti-koṭi* and *nata-jyā*. When the squares of these two are (separately) subtracted from the square of *trijyā* and roots extracted, the two results will respectively be the *jyā* of the portion of *yāmyottara-nata-vṛtta* from eastern cardinal point to the tip of the *dorjyā*, and the maximum divergence between the *yāmyottara-nata* and the horizon. When these two are multiplied together and divided by *trijyā*, the result will be *dorjyā-koṭi*.

10.9 Problems thirteen and fourteen

Then the thirteenth problem is when the *dorjyā* and *kāla-jyā* are known. When these two are squared separately, and each subtracted from the square of *trijyā* and the roots found, their *koṭi*-s will be got. Then from the consideration: If *trijyā* is the hypotenuse for the *kāla-koṭi*, what is the hypotenuse for *dorjyā-koṭi*, we get the *dyujyā*.

The fourteenth problem is where the *dorjyā* and *nata-jyā* are known. By the consideration, if *trijyā* is the hypotenuse for *nata-koṭi*, what will be the hypotenuse for the *dorjyā-koṭi*, will be obtained the *jyā* on the (*yāmyottara*) *nata-vṛtta* which is the line from the tip of the *dorjyā* to eastern cardinal point. The *koṭi* of this is *iṣṭa-krānti-koṭi*.

10.10 Problem fifteen

Now, knowing the *kāla-jyā* and the *nata-jyā*, to derive the other (four items) is the fifteenth (problem): Here, the distance from the east-west cardinal points to the point of contact of the *rāśi-kūṭa-vṛtta* and the *ghaṭikā-vṛtta* is the *kāla-jyā*. Then the distance between the east-west cardinal points to the point of contact of the *rāśi-kūṭa-vṛtta* and the *yāmyottara-nata-vṛtta* is *krānti-koṭi*. Now, the remaining portion of the *kāla-koṭi* from the zenith is the portion between *kāla-bhujā* and the horizon. It is also to be noted that the extent from the *ghaṭikā-nata-vṛtta* to the *rāśi-kūṭa-vṛtta* is a *pāda* (quadrant) on the *yāmyottara-nata-vṛtta*. From this, (it follows that) from the point of contact in the *yāmyottara-vṛtta* the horizon is also a *pāda* (on the *yāmyottara-nata-vṛtta*). Such are the modalities here.

In the same manner, the distance between the *yāmyottara-svastika* and the *rāśi-kūṭa-vṛtta* on the *yāmyottara-nata-vṛtta* will be the *nata-jyā*. Here the declination is the divergence between the *rāśi-kūṭa-vṛtta* and horizon on the *ghaṭikā-nata-vṛtta*. Here too, the extent from the point of contact of the other *nata* (i.e., point of contact of *yāmyottara-nata* and *ghaṭikā-nata*) to the *rāśi-kūṭa* is a *pāda*; the extent from the southern cardinal point to the *ghaṭikā-vṛtta* is also a *pāda*. This will be the situation.

Here, the application of the rule of three is thus: For the hypotenuse which is equal to the radius of the *ghaṭikā-vṛtta* which extends from the western cardinal point to zenith, the *nata-jyā* is the maximum divergence of the *yāmyottara-nata-vṛtta*; and the *kāla-jyā* is the *jyā* for the portion of the circle from the western cardinal point to the end of the *rāśi-kūṭa-vṛtta*. (The consideration is): When that (i.e., the *kāla-jyā*) is taken as the hypotenuse, what will be the difference between the *yāmyottara* and *nata-vṛtta*-s. The result will be the *ghaṭikā-natāntarāla* (the divergence between the *ghaṭikā* and *yāmyottara-nata*) on the *rāśi-kūṭa-vṛtta*. In the same manner, the *nata-jyā* from the southern cardinal point to the *rāśi-kūṭa-vṛtta* on the north-south circle is the *icchā*. The *pramāṇa-phala* is the *kāla-jyā* which is the maximum distance from the zenith to the (*ghaṭikā*) *nata-vṛtta*. The *icchā-phala* is the distance from the north-south circle to the end of the (*ghaṭikā*) *nata-vṛtta* on the *rāśi-kūṭa-vṛtta*. This will be equal to the *icchā-phala* derived earlier. The square of this distance, when subtracted from the square of *trijyā* and the root extracted, would be the divergence between the *ghaṭikā-vṛtta* and north-south circle, a portion of the *rāśi-kūṭa-vṛtta.* When the *jyā* of this divergence is taken as the hypotenuse and as *pramāṇa*, there will arise two divergences in the circles as *pramāṇa-phala*-s. Now, the maximum divergences, being the *icchā-phala*-s of these two, will be *nata-jyā*-s from the points of contact between the *nata* and *rāśi-kūṭa* to the (two) *svastika*-s. Here, that on the *yāmyottara-nata* is the *iṣṭāpakrama-koṭi*, and that on the *ghaṭikā-nata* is the *iṣṭāpakrama* (actual declination).

The method of deriving the *pramāṇa-phala* of them is as follows: First (we derive), the divergence between the *ghaṭikā-vṛtta* and (*yāmyottara*) *nata-vṛtta.* Subtract the square of the *jyā* on the (relevant) section of the *rāśi-kūṭa-vṛtta* from the square of *kāla-jya* and find the root. The result would be the distance of the first *tiryag-vṛtta* (transverse circle described below), and when subtracted from the square of the *nata-jyā* and the root found, the result would be the distance of the second *tiryag-vṛtta.*

Now, to the depiction of the *tiryag-vṛtta*-s. The first *tiryag-vṛtta* is to be constructed so as to pass through the points of contact of the *rāśi-kūṭa* and north-south circles, which happen to be the poles of the *yāmyottara-nata* circle, and also through the east-west cardinal points. The second (*tiryag-vṛtta*) is to pass through the points of contact of the *ghaṭikā* and *rāśi-kūṭa-vṛtta*-s, which happen to be the poles of the *ghaṭikā-nata-vṛtta*, and also through the north-south cardinal points. The maximum divergences of these two circles with the *rāśi-kūṭa-vṛtta*-s will be at the two *nata-vṛtta*-s. These will be the actual declination (*iṣṭāpakrama*) and its *koṭi*.

Thus have been stated the fifteen problems. And thus are the methods of extension of the rule of three in the case of divergences of circles.

Chapter 11

Gnomonic Shadow

11.1 Fixing directions

Now, the method to identify the (four) directions. First prepare a level surface. It should be such that if water falls at its centre, the water should spread in a circle and flow forth on all the sides uniformly. That is the indication for a level surface. On this surface draw a circle (in the following manner): Take a rod slightly bent at both ends and, with one end of the rod fixed at the centre, rotate the other end on all sides (so that a circle will result). The point where the end (of the rod) is fixed is known by the terms *kendra* and *nābhi* (centre). The line resulting from the rotation of the other end is called *nemi* (circumference). Fix (vertically) at the centre a uniformly rounded gnomon (*śaṅku*). On any morning, observe the point on the circumference where the tip of the shadow of the gnomon graces and enters into the circle and, in the same manner, also the point where the tip of the shadow graces the circumference and goes out of the circle in the afternoon. Mark these two points on the circle with dots. These two points, between themselves, will be almost along the east-west. For this reason, these are termed east and west points. These would have been the exact east and west points if they were the shadow-points of the stars which do not have any north-south motion. The Sun has a north-south motion on account of (its motion between) the solstices, and during the interval from the moment, when the western shadow-point gets marked, to the moment when the eastern shadow-point is formed, if the Sun has moved north due to the change in its declination, then to that extent the tip of the shadow would have moved to the south. If (the correction were to be) done on the eastern shadow-point, it has to be moved to the north, in order that (the line connecting the two shadow-points) is along the true east-west. The east-point shall have to be shifted south appropriately if the Sun is moving towards south (*dakṣiṇāyana*). This shifting would be (measured by) the difference in the amplitude of the Sun in inches (*arkāgrāṅgula*) which corresponds to the difference in the declinations at the two instants (of time at which the shadow-points were marked). Multiply the difference in *apakrama* (Rsine of declination) by the inches of the shadow-hypotenuse (*chāyā-karṇāṅgula*) of that moment and divide by the local co-latitude (*svadeśa-lambaka*). The result is the *arkāgrāṅgula* in the shadow-circle (*chāyā-vṛtta*). Then shift by this measure, the east shadow-point (towards north or south), in accordance to the *ayana* (northward or southward motion of the Sun). If a line is drawn connecting the shifted point and the western shadow-point that will be the correct east-west line. Had the above correction been done on the western shadow-point, the shifting would have to be done in the reverse direction of the *ayana*. Then, by constructing intersecting fish-figures (*matsya*) with this line, obtain the north-south line. The rising and setting of stars would be exactly east and west. From this also the directions can be identified.

11.2 Latitude (*Akṣa*) and co-latitude (*Lamba*)

Now, that day when the declinations at sunrise and sunset, which are usually different, are equal, that would be the day of equinox when the Sun will be at the zenith at noon. The 12-inch gnomonic shadow at that time would be the equinoctial shadow (*viṣuvacchāyā*). Take the measure of this shadow as the *bhujā,* and the 12-inch gnomon as *koṭi,* square them, add the squares and find the square root thereof and thus derive the hypotenuse (*karṇa*). This hypotenuse (should be taken as) the *pramāṇa* and the gnomon and the shadow as the *pramāṇa-phala*-s. *Trijyā* is the *icchā*. The two *icchā-phala*-s are the latitude (*akṣa*) and the co-latitude (*avalamba* or *lambana*). Corrections enunciated for the *viparītacchāyā* (to be discussed below) have to be applied to these. They will then be more exact. Here, the latitude is the distance between the zenith and the *ghaṭikā-maṇḍala* (celestial equator). It is the same as the distance between the *Dhruva* (pole star) and horizon (*kṣitija*), measured on the north-south circle. And, the co-latitude is the distance between the *ghaṭikā-maṇḍala* and the horizon measured on the north-south circle. It is also equal to the distance between the zenith and the *Dhruva*.

11.3 Time after sunrise or before sunset

Now, the shadow. Here, for the Sun which moves eastwards on the ecliptic, there will be a shift north and south, in accordance with the inclination of the ecliptic. (Now, picture the following): Let the Sun (whose motion is as stated above) be at a certain point (on the ecliptic) at a desired moment. Then, construct a circle passing through the Sun on the ecliptic at the given moment, with its centre on the axis which passes through the two poles and the centre of the celestial sphere, in such a way that all its parts are equally removed from the celestial equator (i.e., parallel to the *ghaṭikā-maṇḍala*) by the measure of the declination of the Sun at that moment. This circle is the diurnal circle (*svāhorātra-vṛtta*) at that moment. Its radius would be the *iṣṭa-dyujyā* (day-radius). Its quadrants shall have to be demarcated through the six-o'clock circle (*unmaṇḍala*) and the north-south circle. On account of the motion along the diurnal circle, which occurs due to the *Pravaha-vāyu*, the sunrise and sunset occur. Here, the rate of motion of the *Pravaha-vāyu* is constant and so it is possible to ascertain in a definitive manner by how much the diurnal circle will move in a specific time. Hence it is possible to calculate correctly the position of a planet on the diurnal circle, i.e., as to how much it has risen from the horizon on the diurnal circle at a specific time after rising, or how much it has to go before it sets.

11.4 *Unnata-jyā*

Now, the *Pravaha-vāyu* revolves once in 21,600 *prāṇa*-units of time. The diurnal circle will also complete one revolution during this period. So, demarcate each diurnal circle into 21,600 *kalā* divisions. Hence, one division will rotate (by one minute of arc) in one *prāṇa.* Therefore, in ordinary parlance, that portion of the diurnal circle that moves in one *prāṇa*, is also called a *prāṇa* by secondary extension of meaning (*lakṣaṇā*). Thus, the *prāṇa*-s elapsed after sunrise and the *prāṇa*-s yet to elapse before sunset are spoken of (in ordinary speech) as *gata* (past) *prāṇa*-s and *gantavya* (to-go) *prāṇa*-s.

These *gata-prāṇa*-s or *gantavya-prāṇa*-s would be equal to the difference between the horizon and the position of the Sun on the diurnal circle of measure 21,600. Since this is an arc, its Rsine has to be calculated to get the actual (distance). Now, (it is known that) when Rsines are conceived north-south, the limit is the east-west line at (i.e., passing through) the centre of the circle. So also, when the Rsines are conceived east-west, the limit is the north-south line at the centre of the circle. In the same manner, in the conception of up and down Rsines in the diurnal circle, the limits would be the lines at right angles to it passing through the centre of the diurnal circle. (This is so for the following reason): There will be a total chord (*samasta-jyā*) passing through the points of contact of the *unmaṇḍala* and the diurnal circle and the axis (*akṣa-daṇḍa*). A Rsine has to be constructed with the above as the limit from the point of sunrise on the horizon. Since it is on the horizon that the Sun rises, the *gantavya-prāṇa*-s are reckoned from the horizon. Now, the portion of the diurnal circle lying between the horizon and the *unmaṇḍala* forms the ascensional difference (*cara*) in *prāṇa*-s. This has to be subtracted from the *gata-prāṇa*-s and *gantavya-prāṇa*-s in the case of the northern hemisphere, since the horizon is to the north of the east-west *svastika*-s and is below the *unmaṇḍala* (six-o'clock circle). In the southern hemisphere, however, the ascensional difference in *prāṇa*-s has to be added to the *gata* and *gantavya-prāṇa*-s, since there, the horizon is above. The result will be the *unnata-prāṇa,* i.e., the time in *prāṇa*-s elapsed from the *unmaṇḍala* to the position of the Sun as indicated in the diurnal circle. Calculate the Rsine for this (arc). Then, apply to this the Rsine of the ascensional difference inversely, i.e., by adding it in the northern hemisphere and subtracting in the southern hemisphere. The result will be the *unnata-jyā* (i.e., Rsine of the *unnata-prāṇa*) from the horizon. This is the full Rsine pertaining to the two quadrants, of this *svāhorātra-vṛtta,* and so it will not be just a half-sine. Therefore for addition and subtraction, multiplication of the *koṭi* is not required. Since it itself is the remainder of a Rsine, mere addition and subtraction could be done. Thus shall be derived the Rsine of the portion of the diurnal circle for the portion between the Sun and the horizon. Since the (measure in) seconds (*ili*) is small, it should be multiplied by the *dyujyā* and divided by *trijyā.* The result would be the *unnata-jyā* which would be in terms of seconds of *trijyā-vṛtta.*

11.5 *Mahā-śaṅku* and *Mahācchāyā*: Great gnomon and great shadow

[Here, it might be noted that] the diurnal circle is inclined to the south exactly as the celestial equator (*ghaṭikā-vṛtta*). Hence, when the *unnata-jyā* which forms as it were the hypotenuse, is multiplied by the *laṃbaka* and divided by the *trijyā*, the result will be the interstice between the Sun and the horizon. This is called the *mahā-śaṅku* (great gnomon, celestial gnomon). The *koṭi* of this is the distance between the zenith and the planet. This is termed *mahācchāyā* (great shadow, celestial shadow).

11.6 *Dṛṅmaṇḍala*

Now, construct a circle passing through the zenith and the planet. This (circle) is termed *dṛṅmaṇḍala.* The Rsine and Rcosine in this circle are the *mahā-śaṅku* and *mahācchāyā* which have their tips at the location of the planet. Since the horizon is on the sides (centered around) the centre of the Earth (*ghana-bhū-madhya*) and the foot of the *mahā-śaṅku* is on the plane of the horizon, the *dṛṅmaṇḍala* has its centre at the centre of the Earth.

11.7 *Dṛggolacchāyā*

People (residing) on the surface of the Earth see a planet only by how much it has risen from, or is lower than, their horizon at the level of their heads. Therefore, the great gnomon and great shadow which people on the surface of the Earth (actually) see are the ones on that *dṛṅmaṇḍala* which has its centre at the location (*dṛṅmadhya*) of the observer on the surface of the Earth and its circumference passing through the planet and the zenith. So, construct a (observer-centric) horizon, tangential to the surface of the Earth, having all its parts equally raised by a measure equal to the radius of the Earth from the horizon through the centre of the Earth. The altitude from this (horizon) is the gnomon for those on the surface of the Earth. This is called *dṛggola-śaṅku*. What has been stated earlier is the *bhagola-śaṅku*. Subtracting the radius of the Earth from the *bhagola-śaṅku*, the *dṛggola-śaṅku* results. Therefore, the difference between the bases of the two gnomons is equal to the radius of the Earth on account of the difference between the two horizons. Now, for the shadow, the base is the vertical line. Since this (vertical) drawn from the centre of the solid-Earth-sphere and that drawn from (the observer on) the surface of the Earth are the same, the base of the shadow will be at the same point. Hence, there is no difference in the (length of the) shadow. In all cases, the tips of the shadows and the gnomons are at the centre of the planet.

Now, by squaring and adding the two, viz., the gnomon with its base on the observer-centric horizon tangential to the surface of the Earth, and the (related) shadow, and finding the square root, a hypotenuse will be obtained with respect to (the observer on) the surface as the centre. That is called *dṛkkarṇa*. This hypotenuse is in fact derived by the *pratimaṇḍala-nyāya* (rule of calculating the *karṇa* in eccentric circle). Here the *pratimaṇḍala* has its centre at the centre of the Earth, whereas the *karṇa-vṛtta* has its centre on the surface of the Earth. The distance between the centres of these two circles, viz., the radius of the Earth, corresponds to the *ucca-nīca-vyāsārdha*. Since, the *nīca*-point is the zenith, the minutes of the *karṇa-vṛtta* would naturally be small. Therefore the (length of) the shadow that is measured in the units of the *karṇa-vṛtta,* when converted into those of the *trijyā-vṛtta,* will undergo an increase in its magnitude. To that extent the drop from the zenith will appear to be large. When the shadow in the celestial circle is multiplied by *trijyā* and divided by *dṛkkarṇa,* the result will be the shadow in the *dṛggola* (*dṛggolacchāyā*). Thus (is explained) the method of deriving the shadow using the principle of the *pratimaṇḍala-sphuṭa.*

11.8 *Chāyā-lambana*

Now, multiply the celestial shadow (*bhagolacchāyā*) by the *yojana*-s of the Earth's radius and divide by the *yojana*-s of the hypotenuse (*sphuṭa-yojana-karṇa*), because we want it in terms of the *yojana*-s of *dṛkkarṇa.* The result will take the place of *bhujā-phala.* This will be the *chāyā-lambana* in terms of minutes. Add this to the celestial shadow. And the result will be the shadow in the *dṛggola.* Thus has been stated the method to derive the *chāyā-lambana* in minutes by the principle of the *ucca-nīca-sphuṭa.*

11.9 Earth's radius

Now, the radius of the *ucca-nīca-vṛtta* measured in terms of the minutes of the *pratimaṇḍala* is called *antya-phala.* Here, since the *sphuṭa-yojana-karṇa* is the radius of the *pratimaṇḍala* measured in its units, the radius of the *ucca-nīca-vṛtta* is equal to the *yojana*-s of the Earth's radius.

Now is stated the method to derive the minutes of the radius of the Earth in terms of the *sphuṭa-kakṣyā* of the respective planets, when the said *sphuṭa-yojana-karṇa* is taken as *trijyā.* Now, when *trijyā* is taken as the shadow, the minutes of the *yojana*-s of the Earth radius will be the *lambana* (vertical). In the calculation of what would be the minutes of *lambana* for a particular shadow, since the *trijyā* is the shadow and is both a multiplier and a divisor, it can be dropped. So, multiply the desired shadow by the *yojana*-s of the radius of the Earth and divide by the *sphuṭa-yojana-karṇa.* The result will be the minutes of the *lambana* of the shadow. Here, between the *sphuṭa-yojana-karṇa* and the *madhya-yojana-karṇa* there is not much difference. Hence one can divide the *yojana* of the Earth's radius by the *madhya-yojana-karṇa.* The result in the case of the Sun would be 863. With this divide the desired shadow. The result will be the *chāyā-lambana* (*lambana* of the shadow) in terms of minutes. Now, when the *chāyā-lambana* of the *dṛṅmaṇḍala* is taken as the hypotenuse, its *bhujā* and *koṭi* will be the *nati* and *lambana* which will be stated below. How it is to be done is also being stated later. There has to be such a correction for the shadow.

11.10 Corrected shadow of the 12-inch gnomon

These shadows and gnomons have their tip at the centre of the sphere of Sun. Now, the rays of the Sun emanate from all over its surface. The shadow of a gnomon should be taken to extend to the point upto which the rays from the uppermost part of Sun's circumference is obstructed by the gnomon. The shadows of all 12-inch gnomons are not merely formed by the rays emanating from the centre of the solar sphere. Hence the gnomon should (be made to) extend up to the upper part of the circumference of the solar orb. The distance of separation between that point and the zenith will be the shadow. Now, the measure of half the orb is the distance from the centre of the solar sphere to its upper circumference. This will be a full *jyā* in the *dṛṅmaṇḍala.* Hence, if the radius of the orb is multiplied, respectively, by the gnomon and the shadow and divided by *trijyā,* the results will be the Rsine-differences (*khaṇḍa-jyā*). Now, add to the gnomon the result got from the shadow, and subtract from the shadow the result got from the gnomon. Thus can be derived the gnomon and the shadow relating to the top of the Sun's orb. These form the useful tools for the *dṛgviṣaya* (values related to the observer). Though the Rsine-differences have actually to be derived from the *bhujā-jyā* and *koṭi-jyā,* which have their tips at the 'centre' of the full *jyā,* there would be little difference even if they are derived from the tip of the full *jyā.* Hence it was directed above to make use of them.

In the same manner, if the *lambana* and the Rsine-differences relating to the radius of the orbit are corrected, the corrected gnomon and shadow which have their tips at the top circumference of the solar orbit in the *dṛggola* will be obtained. This shadow is multiplied by 12 and divided by the gnomon calculated as above. The result will be the (correct) shadow of the 12-inch gnomon.

11.11 *Viparītacchāyā* : Reverse shadow

Now, the reverse shadow. The method (of the reverse shadow) is applied for the problem: When the shadow of the 12- inch gnomon is known, how to find the time in *prāṇa*-s, elapsed or yet to elapse. Now, if the 12-inch gnomon and the shadow are (separately) squared, added together and the root found, the result will be the *chāyā-karṇa* (hypotenuse of the shadow) in inches (*aṅgula*-s). Then, multiply the above-said shadow and the gnomon by *trijyā* and divide by the above-said *chāyā-karṇa* in inches. The results got will be the *mahā-śaṅku* (great gnomon) and *mahācchāyā* (great shadow). Since they have been derived through the shadow corresponding to the observer (*dṛgviṣaya*), they will have their tips at the top circumference of the orb. Hence, when these gnomon and the shadow are multiplied separately by the radius of the orb and divided by *trijyā* and the results obtained are, respectively, added to the shadow and subtracted from the gnomon, they would have been reduced to what they would have been if their tips were at the centre of the object (*bimba*, i.e., Sun). Then divide the shadow by *gatija* (863) and subtract the result from the shadow. Add to the gnomon the radius of the Earth in minutes (*liptā*). The calculations up to this should be carried out on the latitude and the co-latitude as well.

Then multiply this gnomon by the square of *trijyā* and divide by the product of the *dyujyā* (radius of the diurnal circle) and *lambaka* (co-latitude). The result will be the distance from the centre of the solar sphere to the horizon. The Rsine on the diurnal circle is 21,600 in its own measure. Then apply to this the *cara-jyā* (Rsine of the ascensional difference), positively or negatively in accordance as it is *Meṣādi* or *Tulādi,* then convert it to arc and apply the *cara-prāṇa*-s positively or negatively, as the case may be. The result would be the *prāṇa*-s elapsed or yet to elapse. Thus (has been stated) the method to derive the *prāṇa*-s elapsed or yet to elapse through the reverse process from the *kramacchāyā*, which in turn is obtained from *karṇa* and the shadow of the 12 inch gnomon observed at a desired time.

11.12 Noon-time shadow

Now (is stated) the derivation of noon-day shadow. Now, the noon-day shadow is the distance between a planet and the zenith (measured) in the north-south circle, when the planet comes into contact with the north-south circle. The angular separation between the zenith and the celestial equator is the latitude. The separation between the Sun and the celestial equator is the declination (*apakrama*). The *ghaṭikā-maṇḍala* (celestial equator) is always inclined to the south of the zenith. The Sun shifts south or north of the *ghaṭikā-maṇḍala* in accordance with (northern or southern) hemisphere. Hence, the sum or difference between the celestial latitude and the declination, depending on the hemisphere (in which the Sun lies), is the noon shadow. Hence, the declination is the sum or difference between the noon-time shadow and the latitude. Hence, the latitude is the sum or difference (as the case may be) of the noon-time shadow and the declination. Thus, if two among these three are known, the third can be found.

11.13 *Chāyā-bhujā, Arkāgrā* and *Śaṅkvagrā*

Now, the *chāyā-bhujā*. *Chāyā-bhujā* is the distance from the tip of the shadow in the *dṛṅmaṇḍala* (vertical at the given time) up to the *sama-maṇḍala* (prime vertical). *Chāyā-koṭichāyā-bhujā* is the distance between the tip of the shadow to the north-south circle.

Arkāgrā is the distance from the point of contact of the horizon and the relevant diurnal circle to the east or the west point along the horizon. The Sun rises at that point (the point of contact of the diurnal circle and the horizon). Then, on account of the (effect of the) *Pravaha-vāyu*, while intersecting the north-south circle, it would have shifted towards the south from the rising point. This shift is called *śaṅkvagrā*. Now, draw a line connecting the points of the rising and the setting (of the Sun). The distance of the base of the gnomon from this line is the *śaṅkvagrā*. (It is to be noted that) the tip of the gnomon would also have shifted that much. Hence it has got the name *śaṅkvagrā*.

11.14 Some allied correlations

Now, the Rsine of the *arkāgrā* is along the horizon and the Rsine of the *apakrama* (declination) is along the *unmaṇḍala* (six-o' clock circle). These two (Rsines) will be respectively equal to the distance from the east-west cardinal points to the diurnal circle (along the respective circles). Now, *kṣiti-jyā* (Earth-sine) is the Rsine of that part of the diurnal circle intercepted between the horizon and the *unmaṇḍala*. This could be taken as the *bhujā*. Declination would be the *koṭi*. The hypotenuse is the *arkāgrā*. Thus is formed a triangle. This has been formed due to the latitude (of the place). This triangle has been formed because the horizon and the *unmaṇḍala* are different circles, (which again is) due to the latitude. Hence if the desired *apakrama* (Rsine of declination) is multiplied by *trijyā* and divided by *lambaka*, the result will be *arkāgrā*.

Now, there is another triangle made up of the *unnata-jyā* (Rsine of the hour angle) on the diurnal circle, the gnomon and the *śaṅkvagrā*. This is also latitudinal. This triangle has also been formed since the *unnata-jyā* has an inclination, because of the latitude. Here, the hypotenuse is made up by the *unnata-jyā* on the diurnal circle, the gnomon is the *koṭi* and the distance between the base of the *unnata-jyā* and the base of the gnomon is the *bhujā*. This *bhujā* is the *śaṅkvagrā*. This is directly north-south. Since, at the equator, the diurnal circle is vertical, there the *unnata-jyā* is also vertical.

However, the inclination (of the *unnata-jyā*) due to latitude is towards the south. For this reason, the line (distance) between the base of the gnomon and the base of the *unnata-jyā* is also exactly north-south. The *arkāgrā* is also exactly north-south. Since, at that time, the direction of both are the same, it is enough to add them both or subtract one from the other, according to the northern or southern hemisphere. Mutual multiplication by the *koṭi* is not needed here. This addition or subtraction will give the *chāyā-bhujā,* which is the distance between the east-west line and the base of the gnomon on the horizon. This is also the distance between the planet on the *dṛṅmaṇḍala* and the *sama-maṇḍala* (prime vertical). When this is considered as the *bhujā* and the shadow as the hypotenuse, the *koṭi* will be the *chāyā-koṭi,* which is the distance between the planet and the north-south circle.

The above-said (*chāyā-koṭi*) is the Rsine on the diurnal circle also. When this is measured by its own 21,600 minute measure, the *nata-prāṇa*-s will be obtained. Since this pertains also to the 12-inch gnomon, the directions can be determined therefrom as well. For obtaining this, the *arkāgrā* is multiplied by the hypotenuse of the shadow and divided by *trijyā.* The result is called *agrāṅgula.* Here, *śaṅkvagrā* will always be the *viṣuvacchāyā* (the equinoctial shadow) of the 12-inch gnomon. Hence, when the *viṣuvacchāyā* and *agrāṅgula* are added together or subtracted from one another, the result will be the *bhujā* of the shadow (*chāyā-bhujā*) of the 12-inch gnomon. Its direction will be opposite to the *bhujā* of the *mahācchāyā* (great shadow), since the tip of the direction of the shadow has to be opposite to the direction in which the Sun is.

11.15 Determination of the directions

Now, when the shadow, and the corresponding *bhujā* and *koṭi* at a desired (place and time) for a 12-inch gnomon have been derived, construct a circle with the shadow as radius and fix the gnomon at its centre. Mark with a dot, the point on the circumference, where the tip of the shadow of the gnomon falls. Touching the said point, place two rods, one being twice the length of the *chāyā-bhujā* laying it north-south, and the other being double the length of the *chāyā-koṭi,* laying it east-west, in such a manner that the other ends (of these two rods) also touch the circumference (of the circle drawn). The directions having been known roughly, the *koṭi*-rod will be along the east-west and the *bhujā*-rod along the north-south. This is another method to ascertain the direction.

11.16 *Sama-śaṅku* : Great gnomon at the prime vertical

Now *sama-śaṅku* is explained. Now, the *sama-maṇḍala* (prime vertical) is a great circle which passes through the east and the west cardinal points, and the zenith (and nadir). The *ghaṭikā-vṛtta* (celestial equator) is a great circle which passes through the east and the west cardinal points touching the north-south circle at a place removed from the zenith towards the south by (the extent of) the latitude (of the desired place). The descent of the *ghaṭikā-maṇḍala* from the zenith would be equal to the ascent of the north pole from the horizon. The day on which the diurnal circle (of the Sun) becomes identical with the *ghaṭikā-maṇḍala,* on that day the rising and the setting take place at the east and west cardinal points. Midday occurs at a place removed south from the zenith by (the extent of) the latitude. All the diurnal circles will be inclined southwards. Hence the midday will occur to the south, from where the rising had taken place. However, on the day when the northern declination is smaller than the latitude, the rising and setting will be to the north of the east and west cardinal points, and midday will be to the south of the zenith. Since there is the meeting point (of the diurnal circle) with the north-south circle, the planet will cross the *sama-maṇḍala* (prime vertical) once between its rising and noon. In the same manner, in the afternoon, it will cross the *sama-maṇḍala* once before setting. The gnomon at that time would be *sama-śaṅku.* (Again), at that time the shadow will be exactly east-west.

Now, on the day when the northern declination is equal to the latitude, the planet will meet the *sama-maṇḍala* at the zenith. When the northern declination becomes greater than the latitude the diurnal circle does not intersect with the *sama-maṇḍala*. Hence, on that day, the *sama-śaṅku* does not occur. Also, during southern declination, the *sama-maṇḍala* and the diurnal circle do not intersect and hence on that day (or during that period) also *sama-śaṅku* does not occur. Here, when the northern declination becomes equal to the latitude, the planet meets the *sama-maṇḍala* at the zenith and the *sama-śaṅk* is equal to *trijyā*. Then, applying the same argument for a given northern declination less than the latitude as to what the *sama-śaṅku* would be, the *sama-śaṅku* can be calculated. By a calculation reverse to this, the northern declination can be obtained from the *sama-śaṅku*. And, from that the *bhujā-jyā* of the planet can also be derived. This is one way of obtaining the *sama-śaṅku*.

11.17 *Samacchāyā*

Now is (stated) the method to derive the hypotenuse of the 12-inch gnomon corresponding to the *sama-śaṅku*. Here, it has been stated above that the *sama-śaṅku* is got by multiplying the *trijyā* by the (Rsine of the) northern declination, which is less than the latitude, and dividing the product by Rsine of the latitude (of the place). By the proportion: If for this *sama-śaṅku* the hypotenuse is the *trijyā*, what will be the hypotenuse for the 12-inch gnomon, the hypotenuse for the *sama-śaṅku* will be got. Then, the hypotenuse of the *samacchāyā* in (terms of) *aṅgula*-s is got by multiplying *trijyā* by 12 and dividing by *sama-śaṅku*. Here, since the *mahā-śaṅku* is the divisor and that is got from the product of the *trijyā* and the *apakrama* (Rsine of declination), the divisor would be the product of *trijyā* and *apakrama*, and the dividend is the product of *trijyā* and 12. Then, since *trijyā* occurs both in the divisor and in the dividend, *trijyā* can be left out (in the calculation). Since the *akṣa* (Rsine of latitude) is the divisor-of-the-divisor it will form a multiplier to the dividend. Hence, (ultimately) when *akṣa* is multiplied by 12 and divided by the northern declination which is less than the latitude, the result will be the hypotenuse of the *samacchāyā*.

Here (it might be noted that) the product of the *akṣa* and 12 would be equal to the product of the equinoctial shadow and the *lambaka*, for the reason that the product of *icchā* and *pramāṇa-phala* is equal to the product of the *pramāṇa* and *icchā-phala*. Hence, this (product of equinoctial shadow and the *lambaka*) might be divided by the *apakrama* to derive the hypotenuse of the *samacchāyā*.

Now, the 12-inch gnomon shadow of a planet at noon on the equinoctial day is its equinoctial shadow (*viṣuvacchāyā*). When the *krānti* (declination) of the Sun is to the north, *samacchāyā* will occur only when the noon-time shadow is less than equinoctial shadow. The difference between this noon-time shadow and the equinoctial shadow is the noon-time *agrā* in *aṅgula*-s (*madhyāhnāgrāṅgula*). This is equal to the equinoctial shadow on the day when midday occurs at the zenith. On that day, the hypotenuse of the midday shadow will be equal to the hypotenuse of the *samacchāyā*. On a day when the *agrāṅgula* is very small, the hypotenuse of the *samacchāyā* is very much longer than the hypotenuse of the midday shadow. In proportion to the increase of the *agrāṅgula*, the difference between the hypotenuse of the midday shadow and the hypotenuse of the *samacchāyā* will become lesser and lesser. Hence, inverse proportion is to be applied here. Therefore, when the equinoctial shadow is multiplied by the midday hypotenuse and the product divided by the *agrāṅgula* at midday (*madhyāhnāgrāṅgula*), the result will be the hypotenuse of the *samacchāyā*.

11.18 The *Sama-śaṅku*-related triangles

Now are explained the characteristics of certain planar figures (*kṣetra-viśeṣa*) which arise in places with latitude, on account of the latitude. Now, on the day when the diurnal circle meets the horizon to the north of the east and west cardinal points and similarly meets the north-south circle towards the south of the *sama-maṇḍala*, a triangle can be conceived wherein the hypotenuse is that portion of the diurnal circle between the horizon and the *sama-maṇḍala,* the *koṭi* is the *sama-śaṅku* and the *bhujā* is the *arkāgrā.* At places where there is no latitude, since there will be no inclination of the diurnal circle (from the vertical), the above-said triangle will not occur. Now, consider the three, viz., (1) the distance between the east and west cardinal points and the point of contact with the diurnal circle, occurring on the horizon, which is the *arkāgrā*; (2) the declination on the *unmaṇḍala* (equinoctial colure or six-o' clock circle); and (3) the portion of the diurnal circle between the horizon and the *unmaṇḍala,* which is the *kṣiti-jyā.* These three make a triangle arising due to the latitude, with the above three taking the place of *bhujā, koṭi* and *karṇa.* Then, the portion of the diurnal circle above the *unmaṇḍala* (may be taken as) the *koṭi,* the portion of *apakrama* along the *unmaṇḍala* would be the *bhujā* and *sama-śaṅku* would be the hypotenuse. Thus will be formed another triangle. All the above three triangles are as if made up of the latitude, co-latitude and *trijyā.* Thus, if one of them is known, the others can be derived using *trairāśika.*

11.19 The ten problems

Now, let there be two equal circles, their centres being at the same place cutting each other. It might be necessary to know as to what would be the distance of separation between the circumferences when we proceed by a given distance from the point of contact of their circumferences, and also what would be the distance from the meeting point of the circumferences at a place where their circumferences are at a given distance. Herein below is given in detail as to which *trairāśika*-s have to be used to know the above, and as an illustration of their application, the ten problems are discussed.

Now, there are five entities, viz., the *śaṅku* (gnomon), the *nata-jyā* (Rsine hour angle), *apakrama* (Rsine of declination), desired *āśāgrā* (Rsine of amplitude) and *akṣa-jyā* (Rsine of latitude). When three of the above are known, here are stated the methods to derive the other two. This can happen in ten ways and so it is called 'The Ten Problems'.

11.20 Problem one: To derive *Śaṅku* and *Nata*

11.20.1 Shadow and gnomon at a desired place

First is stated the method to derive the *śaṅku* and the *nata-jyā* when the declination, amplitude (*āśāgrā* or *digagrā*) and latitude are known. Now conceive of a circle passing through the zenith and the planet. Such a circle is called *iṣṭa-digvṛtta*, or *dṛṅmaṇḍala* (also *diṅmaṇḍala*). Now, the Rsine of the distance from the meeting point of the *dṛnmaṇḍala* and the horizon to the east-west cardinal points along the horizon is the *iṣṭāśāgrā* (the desired sine amplitude). Now, construct a circle passing through the zenith and that part of the horizon whose separation from the south-north *svastika* (cardinal point) is equal in measure to the *iṣṭa-āśāgrā*. This circle is called *viparīta-digvṛtta*. Construct another circle passing through the point of intersection of the *viparīta-digvṛtta* and the horizon, and also passing through the two poles. This is called a *tiryag-vṛtta,* for the reason that it is at right angles to the *iṣṭa-digvṛtta* and the *ghaṭikā-vṛtta*. In this circle occurs the maximum divergence between the *iṣṭa-digvṛtta* and the *ghaṭikā-vṛtta*. The distance on the horizon, between the north-south circle and the *viparīta-digvṛtta* would be their maximum divergence. This maximum divergence occurs on the horizon since their point of contact is at the zenith. This would be equal to the *iṣṭa-āśāgrā*. Taking this as the *pramāṇa-phala,* the Rsine of the arc-bit on the north-south circle between the zenith and the *Dhruva* will be the *lambaka* (co-latitude). This is the *icchā*. The *icchā-phala* would be the Rsine of the distance between the *Dhruva* and the *viparīta-digvṛtta*. Take this as the *koṭi,* the Rsine latitude as *bhujā*; derive the *karṇa* (hypotenuse) by finding the square root of the sum of the squares (of the two). The result will be the Rsine of the distance between the horizon and the *tiryag-vṛtta,* with its tip at the *Dhruva*. This will also be the maximum divergence between the *iṣṭa-dṛṅmaṇḍala* and the *ghaṭikā-maṇḍala*. The poles (*pārśva*) of the *digvṛtta* are at the points of contact of the *viparīta-digvṛtta* and the horizon. The poles of the *ghaṭikā-vṛtta* are the *Dhruva*-s. (These) will be touching the four sides of the *iṣṭa-digvṛtta* and the *ghaṭikā-vṛtta*. Since the distance between the poles on the *tirvag-vṛtta* is equal to the maximum divergence amongst these, the hypotenuse derived as above would be the maximum divergence between the *iṣṭa-digvṛtta* and the *ghaṭikā-vṛtta.* Here, since it is difficult to grasp the geometrical situation when we deal with all the directions (and possibilites), a specific direction (and situation) should be considered.

Specified below is the case, when the *śaṅku* is in the south-west direction in the southern hemisphere. In this case, the *digvṛtta* would be passing through the horizon at the south-west and the north-east directions. And, the *viparīta-digvṛtta* would be passing through the south-east and north-west directions. The Rsine in the *tiryag-vṛtta,* which is the distance between the north-west corner and the northern *Dhruva*, would serve as the divisor. Take this divisor as the *pramāṇa* and the latitude which is the height of the *Dhruva* as the *pramāṇa-phala.* Then the *icchā-phala* will be obtained, as it is the maximum divergence between the *tiryag-vṛtta* and the horizon on the *digvṛtta* in the north-east.

Here, whatever be the extent of the altitude of the point of intersection of the *tiryag-vṛtta* from the horizon on the *digvṛtta* on the north-east, that much will be depression of the point of intersection of the *ghaṭikā-vṛtta* and *digvṛtta* from the zenith in the south-west on the *digvṛtta.* Now, it is known that when there are two circles (with a common centre and inclined to one another) having a common circle (*tiryag-vṛtta*) at right angles to them, the two circles meet at points a quarter of the circumference (*vṛtta-pāda*) away from where they touch the *tiryag-vṛtta.* Hence, in the instance discussed earlier, the divergence on the *tiryag-vṛtta* from the point where it meets the *digvṛtta* in the north-west, to the *ghaṭikā-vṛtta*, is equal to the divisor mentioned above. Take this as the *bhujā* and as the *pramāṇa.* Now, at the north-east consider the arc from the point of contact of the *digvṛtta* to the *ghaṭikā-vṛtta* in the south-west, which is a quarter of the circumference (*vṛtta-pāda*) along the *digvṛtta,* the Rsine of this (arc) is the radius. Take this as the hypotenuse and *pramāṇa.* The divergence from the zenith to the *ghaṭikā-vṛtta* on the north-south circle is the latitude and this would be the *icchā.* The divergence from the zenith to the *ghaṭikā-vṛtta* on the *dṛṅmaṇḍala* is the *icchā-phala*. Here is one set up where *icchā* is the *bhujā* and *icchā-phala* is the hypotenuse.

Now, from the point on the *dṛṅmaṇḍala* where it meets the *tiryag-vṛtta,* at a distance of a quarter of the circumference (*vṛtta-pāda*), would be the meeting point of the *ghaṭikā-vṛtta* on the *diṅmaṇḍala.* Hence, the ascent of the *tiryag-vṛtta* from the horizon and the descent of the *ghaṭikā-vṛtta* from the zenith on the *digvṛtta* are equal. Thus, this can be considered in two ways. There will be no difference in the derivation of the *icchā-phala.* In the derivation of the gnomon at a desired place, the above would represent the latitude. The *koṭi* of this is the distance from the *ghaṭikā-maṇḍala* to the horizon. On the *diṅmaṇḍala,* this would represent the *lambana.* Then the divergence between the desired diurnal circle and the *ghaṭikā-vṛtta* on the north-south circle, is the desired declination. Take this as the *bhujā* and *icchā* and take the divergence of the *ghaṭikā-vṛtta* and the diurnal circle on the *diṅmaṇḍala* as the hypotenuse, and calculate the *icchā-phala.* This would represent the declination. Here, since the mere (i.e., actual) latitudes and the declination on the north-south circle are representatives (*sthānīya*) of the latitudes and the declination on the *diṅmaṇḍala,* the differences/divergences are equal.

For the above reason, the same *pramāṇa-phala* which is representative of the latitude (*akṣa-sthānīya*) can be used to derive the representatives of the declination. Here, the distance from the zenith to the *ghaṭikā-vṛtta* on the *diṅmaṇḍala* is the representative of the latitude. Again, the divergence of the *ghaṭikā-vṛtta* from the diurnal circle on the *diṅmaṇḍala* is the representative of the declination. If these are added together or subtracted from each other, the result will be the distance from the zenith to the diurnal circle on the *diṅmaṇḍala.* And, that would be the shadow at the desired place (*iṣṭadik-chāyā*). Then, the divergence between the horizon and the *ghaṭikā-vṛtta* on the *diṅmaṇḍala* represents the *lambana.* The divergence of the *ghaṭikā-maṇḍala* and the diurnal circle on the *diṅmaṇḍala* represents the declination. When these have been added to or subtracted from in accordance with their hemisphere, the result would be the gnomon at the desired place (*iṣṭa-dikchaṅku*). Now, in any place, it is possible to derive the midday shadow and gnomon by the addition or subtraction of the latitude and declination or of the *lambaka* and declination on the north-south circle. In the same manner, the shadow and gnomon at a desired place can be derived applying them on the *digvṛtta* at the desired place. Here, after the addition or subtraction of the desired arcs, their Rsines can be derived. Or, the Rsines themselves can be added or subtracted amongst themselves.

Now, square the representatives of the latitude and the declination, subtract from the square of *trijyā* and find the square roots; thus, the respective *koṭi*-s would be obtained. Then, multiply the representatives of the latitude and the declination by the *koṭi*-s of each other, add them together or subtract one from the other and divide by the *trijyā.* The result will be the shadow at the desired place. Again, when the representatives of the *lambaka* (co-latitude) and the declination are also cross-multiplied by the *koṭi-s,* and the result divided by *trijyā,* then also the result will be the shadow of the desired place.

Then, take the actual latitude and declination, add together or subtract one from the other, and derive the midday shadow. Multiply it by *trijyā* and divide by the divisor obtained earlier, which is the maximum divergence of the *digvṛtta* at the desired place and the *ghaṭikā-vṛtta,* and thus obtain the shadow at the desired place. Here, the multiplication by *trijyā* and division by the divisor can be done either before or after the addition or subtraction of the latitude and declination, since there will be no difference in the final result. Since in such cases, multiplication has to be done by the latitude, co-latitude and declination, and division by the divisor, we might consider the divisor as the *pramāṇa,* the actual latitude and declination, as the *pramāṇa-phala, trijyā* as *icchā,* and the representatives of the latitude and declination as *icchā-phala.* In this, the place occupied by the actual latitude and declination in a circle having the divisor as radius, will be the same as that occupied by the representatives of the latitude and declination in the circle which has *trijyā* as radius. Therefore, if the actual latitude and declination are squared and subtracted from the square of the divisor and the roots calculated, the results will be the *koṭi*-s of the latitude and declination in the circle with the divisor as radius. The same *koṭi*-s will be obtained also when the *dyujyā* (radius of the diurnal circle) and co-latitude are multiplied by the divisor and divided by *trijyā*. Multiply the *koṭi* of the latitude by the *koṭi* of the declination, simiiarly multiply the declination by the latitude and divide both by the divisor. The two results obtained shall be added together or subtracted from one another. The result would be the gnomon at the desired direction on the circle of which the divisor is the radius. When this is multiplied by the *trijyā* and divided by the divisor, the gnomon in the required direction is obtained. (It is to be noted that) the gnomon in the southern direction is obtained in the southern hemisphere by the difference of the co-latitude and the declination, and in the northern hemisphere, by the sum of the co-latitude and the declination.

When the declination is larger than the *koṭi* of the latitude, the point of intersection of the diurnal circle with the desired *digvṛtta* would be below the horizon. Therefore, when subtraction is done, there will be no gnomon in the desired direction. When the northern declination is greater than the latitude, the midday would be to the north of the zenith. On that day too, there will be no gnomon in the southern direction. When, however, the *āśāgrā* is north, the gnomon will occur. When the sum of the arcs of the representatives of the co-latitude and declination is greater than *trijyā,* the *koṭi-jyā* thereof would be the gnomon in the northern direction. If the sum of the *jyā*-s exceeds a quarter of a circle, the result will be *koṭi-jyā.*

Now, in the northern hemisphere, when the declination is greater than the latitude, the gnomon with northern *āśāgrā* will result. When the northern declination is less than the latitude, in certain cases depending on the *āśāgrā,* the gnomon with northern *āśāgrā* and the gnomon with the southern *āśāgrā* might occur on the same day. Here, by the sum of the co-latitude and declination and by their difference, gnomons will occur with equal amounts of southern *āśāgrā* and the northern *āśāgrā,* respectively.

Then again, when the desired declination is greater than the divisor, the representative of declination will become greater than *trijyā*. Since there can be no such *jyā*, no gnomon will be there for that *āśāgrā.* Thus has been explained the methods of deriving the desired gnomon.

11.20.2 Corner shadow

Now, herein below is stated the equivalence of the above procedure (*nyāya-sāmya*) for the case of *koṇa-śaṅku* (corner shadow) with that stated in the *Sūrya-siddhānta.* Since here, the desired *diṅmaṇḍala* is facing the corner, the *āśāgrā* is the Rsine of one-and-a-half *rāśi*-s. And, this will be half of the total chord (*samasta-jyā*) of three *rāśi*-s, for the reason that the *diṅmaṇḍala* touches the horizon at the middle of the interstice between the east-west cardinal points and the north-south cardinal points. When the Rsine (*ardha-jyā*) and Rversine are squared, added and the root of the sum found, the result would be the total chord. In a quadrant both the Rsine and the Rversine are equal to *trijyā.* Therefore the sum of their squares is twice the square of *trijyā.* And one-fourth of that is the square of (Rsine) of one and a half *rāśi*-s. Now, when half the square of *trijyā,* which is the same as the square of above mentioned *āśāgrā,* is taken in the circle of the co-latitude (*lamba*), it will be half the square of the co-latitude. When the square of Rsine of latitude is added to half the square of Rsine of co-latitude and the root found, it will be the divisor here also. When the product of declination and latitude and the product of their *koṭi*-s in the *hāraka*-circle are added together or subtracted from one another, as the case may be, and the result divided by the *hāraka* (divisor), then the corner-shadow on this *hāraka*-circle will be obtained. Then again, when the product of the squares of these arcs are added together or subtracted from one another, and the result divided by the square of the divisor, the square of the gnomon is obtained. When the root of this is found and is multiplied by *trijyā* and divided by the divisor, the result will be the gnomon on the *trijyā*-circle.

Here, the product of the squares of the *koṭi*-s of the declination and latitude is a divisor. Now, the squares of the (two) *koṭi*-s are the remainders obtained by subtracting from the square of the divisor, the squares of the latitude and the declination, respectively. Now, consider the square of the *koṭi* of the latitude as the multiplicand and the square of the *koṭi* of declination as the multiplier. Then, the square of the declination will be the difference between the multiplier and the divisor. Now, consider the calculation: Multiply the square of the co-latitude by the square of the declination. Divide by the square of the divisor. Subtract the result from half the square of the co-latitude. The result obtained will be equal to the product of the squares of the Rcosine of the latitude and declination, divided by the square of the divisor. This is the method of calculation when one takes the multiplier as simply half of the square of the co-latitude. There is the rule:

> *iṣṭonayuktena guṇena nighno'bhīṣṭaghnaguṇyānvitavarjito vā.*
>
> (Bhāskara's *Līlāvatī,* 16)
>
> (Multiplication can be done also by) deducting or adding a desired number to the multiplier and multiplying the multiplicand, and adding or deducting from that, the product of the said number and the multiplicand.

As stated above, the multiplicand, which is half the square of the co-latitude, is added to the required number represented by the square of the latitude. Take the multiplicand as equal to the square of the divisor. Then, the square of the declination which is the difference between the multiplier and the divisor should be subtracted from half the square of the co-latitude which is the multiplicand. There is a distinction here, viz., that a correction is to be made to the square of the declination which is to be subtracted. The said correction is as follows: Here the simple multiplicand is half the square of the co-latitude. To this has been added, (as stated earlier), the square of the latitude as the desired additive number. Hence that square of the latitude should be multiplied by the square of the declination which is the difference between the multiplier and the divisor. This, divided by the square of the divisor, is the correction. This correction has to be subtracted from the square of the declination, for the reason that the desired number had been added to the multiplicand. On the other hand, in case the square of the declination had been deducted from half the square of the co-latitude, that correction should have been added. The square root of the remainder (after the abovesaid deduction) is one part of the gnomon. The other part is got by multiplying the latitude and the declination and dividing the result by the root of the divisor, which is the square mentioned above. When this result is squared, it will be the above-mentioned correction to the square of the declination. Then, these (two) parts of the gnomon should respectively be multiplied by the *trijyā* and divided by the divisor.

Now, substract the square of *arkāgrā* from the square of *trijyā*. Multiply the remainder by the square of the co-latitude and divide by the square of *trijyā*. The result obtained would be equal to half the square of the co-latitude minus the square of the declination. Multiply this by the square of the *trijyā* and divide by the square of the divisor. Since this result has to be converted to the *trijyā-vṛtta* (circle with *trijyā* as the radius), the multiplication and division by the square of the *trijyā* can be dropped. Now, from half the square of *trijyā*, subtract the square of the *arkāgrā;* multiply the remainder by the square of co-latitude and divide by the square of the divisor; the result will be on the *trijyā*-circle. In the same manner, when the declination has to be multiplied by the latitude, if instead the *arkāgrā* is multiplied and the product is multiplied by the co-latitude and divided by the divisor, the result will be a part of the gnomon on the *trijyā*-circle, the reason being that the relation between *trijyā* and co-latitude is the same as that between *arkāgrā* and declination. Then, the (two) parts of the gnomon have to be added or subtracted, depending on whether the hemisphere is south or north; the results would be the southern and northern corner-gnomons.

Here, in place of latitude and co-latitude, the equinoctial shadow and 12-inch gnomon can be used. There, the only distinction is that the square of the equinoctial shadow is added to half the square of 12, being 72, to get the square of the divisor. Thus has been explained the method of deriving the desired gnomon under problem one. It has also been indicated that there are the above-mentioned easier methods for the case of the *koṇa-śaṅku* (corner-shadow).

11.20.3 Derivation of *Nata-jyā* (Rsine hour angle)

Now, *nata-jyā* (Rsine of hour angle) has to be obtained. Conceive of the maximum divergence between the desired *diṅmaṇḍala* and the north-south circle as the Rcosine of the desired *āśāgrā* (*iṣṭāśāgrā-koṭi*). A consideration of what it would be on the tip of the shadow, would lead to the *chāyā-koṭi*. This *chāyā-koṭi* is the *nata-jyā*. The distinction is that, in order to convert it in terms of its own minutes of arc in the diurnal circle, the above should be multiplied by *trijyā* and divided by *dyujyā*. Now, the products of Rcosine *āśāgrā* and shadow, of *chāyā-koṭi* and *trijyā*, and of *nata-jyā* and *dyujyā* would all be numerically the same. Hence, when a factor of one of these is used to divide the product of the other two, the second factor would be got. This fact might be kept in mind in the present problem, as also in all the other problems. Thus ends problem one.

11.21 Problem two: *Śaṅku* and *Apakrama*

11.21.1 Derivation of the gnomon

Now, the second problem. Here, using the *nata-jyā*, *āśāgrā* and *akṣa,* the other two, viz., *śaṅku* and *krānti* (or *apakrama*) are to be derived. The geometrical construction (*kṣetra-kalpana*) here is as follows: Construct a circle touching the two poles and the planet. This is called *nata-vṛtta.* The maximum divergence between the *nata-vṛtta* and the north-south circle is on the *ghaṭikā-maṇḍala.* Construct another great circle passing through the zenith and the point where the *nata-vṛtta* meets the horizon. This circle is called *nata-sama-maṇḍala.* Mark the point on the horizon a quarter of the circumference away along the horizon from the point where the *nata-sama-maṇḍala* and the horizon meet each other. Construct another circle passing through this point and the zenith. This circle is called *nata-dṛkkṣepa-vṛtta.* It is on this circle that the maximum divergence between the *nata-vṛtta* and the *nata-sama-maṇḍala* occurs. It is again on this circle that the maximum divergence between the *nata-vṛtta* and the horizon occurs. The above-said two maximum divergences are called, respectively, *svadeśa-nata* and *svadeśa-nata-koṭi.* Now, construct the *iṣṭa-digvṛtta* and *vyasta-digvṛtta* as instructed earlier (in connection with problem one). Now, the extent by which the *nata-vṛtta* is below the zenith on the *nata-dṛkkṣepa-maṇḍala,* to that extent the pole of the *nata-maṇḍala* would be higher than the horizon on the same *vṛtta* (*nata-dṛkkṣepa-maṇḍala*). The point of intersection of the *nata-dṛkkṣepa-maṇḍala* and the *ghaṭikā-vṛtta* is also the pole of the *nata-vṛtta.*

Now, construct another circle passing through the point of intersection of the horizon and the *vyasta-digvṛtta* and the pole of the *nata-vṛtta* on the *nata-dṛkkṣepa-maṇḍala.* This will be a *tiryag-vṛtta* which is perpendicular (?) both to the *nata-vṛtta* and *iṣṭa-digvṛtta.* Note how much this *tiryag-vṛtta* is above the horizon on the *iṣṭa-digvṛtta;* to that extent will the point of intersection of the *digvṛtta* and the *nata-vṛtta* be lower from the zenith. The distance between the zenith and the *nata-vṛtta* on the *digvṛtta* will be the shadow. Its *koṭi* will be the gnomon.

Now, moving from the northern *Dhruva* to the *ghaṭikā-vṛtta* on the north-south circle, the divergence between the *nata-vṛtta* and the north-south circle is the *nata-jyā.* The co-latitude thereof is the distance between the *Dhruva* and the zenith. The distance from this to the *nata-vṛtta* will be the *svadeśa-nata.* Put one end of this at the point of intersection of the *nata-vṛtta* and the *svadeśa-nata-vṛtta.* The *koṭi* of this *svadeśa-nata-jyā* would be the distance from this point to the horizon. Now, presume that the *iṣṭāśāgrā* meets the horizon a little to the south of the east cardinal point. In that, the gnomon shall also be in the same way. In this situation, the *nata-vṛtta* will meet the horizon a little to the west of the north *svastika* and a little to the east of the south *svastika.* The *svadesa-nata-vṛtta* will then touch the horizon that much to the north from the east *svastika* and that much south from the west *svastika.*

The *vidiṅ-maṇḍala* (*vidig-vṛtta*) will meet the horizon at a point west of the south *svastika* by a length equal to the *iṣṭāśāgrā.* From that point, the *tiryag-vṛtta* will begin to rise. When it reaches the *svadeśa-nata-vṛtta,* it would touch the pole (of the *nata-vṛtta*) which is above the horizon by the extent of Rsine of *svadeśa-nata.* The distance between the pole of the *nata-vṛtta* and the horizon is the divisor here. When this *tiryag-vṛtta* reaches the *digvṛtta,* it would have traversed one quadrant from its point of contact with the horizon. Hence, in this *digvṛtta,* the *tiryag-vṛtta* and the horizon would have their maximum divergence. And that (divergence) is equal to the shadow. Its *koṭi,* which is equal to the gnomon, would be the distance from the zenith on the *digvṛtta* to the point of intersection of the *digvṛtta* and the *tiryag-vṛtta.* This will be the maximum divergence between the *tiryag-vṛtta* and the *vidig-vṛtta.* When the difference between the *nata-vṛtta* and the horizon becomes equal to *svadeśa-nata-koṭi,* its hypotenuse is *trijyā.* This being so, by applying the rule of three to find what it will be for *dhruvonnati,* we will obtain the distance between the north pole and the horizon on the *nata-vṛtta.* Now, the maximum difference between the *nata-vṛtta* and the north-south circle is the *nata-jyā.* Then, consider the proportion: When so much is the divergence in the north-south circle for the *dhruva-kṣitijāntarāla-jyā* on the *nata-vṛtta,* what will be the divergence in the north-south circle for the *dhruva-kṣitijāntarāla-jyā*; thus the divergence between the *nata-vṛtta* and the north-south circle on the horizon would be obtained. The same will be the divergence of the point of intersection of the *nata-dṛkkṣepa* and the horizon, to the south of the western *svastika.* Subtract this from the *koṭi* of *āśāgrā.* The remainder will be the divergence of the *svadeśa-nata-vṛtta* and the *vidig-vṛtta* on the horizon.

Here, when Rsines are added or subtracted, they should mutually be multiplied by their *koṭi*-s and the results added or subtracted and then divided by *trijyā.* [According to the above rule, the following is to be done]: The *nata-jyā* is multiplied by the latitude and divided by the *svadeśa-nata-koṭi.* The square of this is subtracted from the square of *trijyā* and the square root found. This root is multiplied, respectively, by *āśāgrā* and *āśāgrā-koṭi.* Find the difference between the products found, if the gnomon is in the south, and add them if the gnomon is in the north. If this is divided by *trijyā,* the result would be the divergence between the *svadeśa-nata-vṛtta* and the *vidig-vṛtta,* on the horizon. Now if the *āśāgrā* is to the north in the forenoon, the point of contact of the *vidig-vṛtta* and the horizon would be away from the north *svastika* towards the west at a distance equal to the *āśāgrā*. From this point the *trijyā-vṛtta* begins to rise. And, from this point, the distance up to the west *svastika* is the *koṭi* of the *āśāgrā*. Now, the meeting point of the *svadeśa-nāta-vṛtta* and the horizon is to the south of the west *svastika*; therefore, add this distance to the *koṭi* of *āśāgrā*. The result will be the distance from the *vidig-vṛtta* to the *svadeśa-nata-vṛtta*. This would be the maximum divergence between the *digvṛtta* and the *svadeśa-nata-vṛtta* on the horizon. Now, when one proceeds on the *svadeśa-nata-vṛtta* from the zenith up to the pole of the *nata-vṛtta* (*nata-vṛtta-pārśva*), the divergence would be equal to the *svadeśa-nata-koṭi*. Derive the *vidig-vṛttāntara* for this, it being the *vidig-vṛttāntara* from the pole (*nata-pārśva*) of the *nata-vṛtta* . Square this. Square also the *svadeśa-nata-jyā,* it being the altitude (*nata-pārśvonnati*) of the pole of the *nata-vṛtta.* Add the two, and find the root. The result will be the divergence between the pole of the *nata-vṛtta* and the horizon on the *tiryag-vṛtta*. Take this as the *pramāṇa*. The *pramāṇa-phala*-s are the altitude of the pole of the *nata-vṛtta* from the horizon and the divergence between the pole of the *nata-vṛtta* and the *vidig-vṛtta*. For this *pramāṇa* the above two are the shadow and gnomon. *Trijyā* is the *icchā* here. The *icchā-phala*-s are *iṣṭa-dikchāyā* and gnomon.

11.21.2 Derivation of the declination

Now, when the shadow and the *koṭi* of *āśāgrā* are multiplied together and divided by *nata-jyā,* the result would be *iṣṭa-dyujyā*. Square this, subtract from the square of *trijyā* and find the root. The result will be the desired declination. Thus has been stated the second problem.

11.22 Problem three: *Śaṅku* and *Āśāgrā*

11.22.1 Derivation of *Śaṅku*

In the third problem, given *nata, apakrama* and *akṣa, śaṅku* and *āśāgrā* are to be found. Here, when the *nata* and *trijyā* are squared, subtracted from each other, and the root found, it will be the distance of the planet on the diurnal circle (*svāhorātra-vṛtta*). This is what it would be if the radius of the *dyu-vṛtta* (diurnal circle) is taken as *trijyā*. Then multiply this *koṭi* of the *nata* by *dyujyā* and divide by *trijyā*. The result will be the Rsine of the *dyu-vṛtta* in terms of the measure of *trijyā*. Subtract *kṣiti-jyā* from this in the southern hemisphere and add in the northern hemisphere. Multiply the result by the co-latitude and divide by *trijyā*. *Śaṅku* (gnomon) will result.

11.22.2 Derivation of *Āśāgrā*

The *koṭi* of the *śaṅku* derived above is the shadow. Multiply the *nata-jyā* and *dyujyā* and divide by the gnomon. The result would be *āśāgrā-koṭi*.

11.23 Problem four: *Śaṅku* and *Akṣa*

Next, given *nata, krānti* and *āśāgrā,* to derive the *śaṅku* (gnomon) and *akṣa* (latitude).

11.23.1 Derivation of *Śaṅku* (gnomon)

Now, when *nata-jyā* and *dyujyā* are multiplied together, place the product at two places. Divide one by the *koṭi* of *āśāgrā* and the other by *trijyā*. The results will be the shadow and the *chāyā-koṭi*. By squaring the shadow and the *trijyā,* subtracting them from one another and finding the root thereof, the gnomon is got.

11.23.2 Derivation of *Akṣa* (latitude)

The following is the geometrical construction (*kṣetra-kalpana*) for the derivation of *akṣa*. Construct a (smaller) north-south circle parallel to the north-south circle at a distance equal to the *chāyā-koṭi*. This will be like the diurnal circle with respect to the *ghaṭikā-maṇḍala*. In relation to the (normal) north-south circle, this (new circle) would be the one on which the true planet is situated. Construct another circle touching the planet and the east-west *svastika*-s. In this situation, the distance from the planet to the north-south circle is the *chāyā-koṭi*. The distance from the planet to the east-west-*svastika* is the *koṭi* of the *chāyā-koṭi*. This latter *koṭi* would be the radius of the *koṭi*-circle conceived here. The *chāyā-bhujā* is the Rsine in this *koṭi*-radius (circle). And, that would be the distance from the planet to the *sama-maṇḍala*. The *koṭi* of this (*chāyā-bhujā*) is the gnomon.

Now, the distance between the planet to the *ghaṭikā-vṛtta* is the declination. The *koṭi* of this (declination) will be square root of the difference between the squares of the *chāyā-koṭi* and *dyujyā*. This will be the interstice between the planet and the *unmaṇḍala* on this *koṭi*-circle. Now, multiply the *chāyā-bhujā* by the *koṭi* of *apakrama*; add them or find the difference between them as the case may be. Divide the result by the radius of the *koṭi*-circle, which is nothing but the square root of the difference of the squares of *trijyā* and *chāyā-koṭi*. The result will be the *akṣa* on this *koṭi*-circle. Now, multiply this *akṣa* by *trijyā* and divide by the radius of the *koṭi*-circle. The result will be the latitude of the place (*svadeśa-akṣa*). Here, if the *krānti* (declination) and *āśāgrā* are in opposite directions, the two should be added, and if in the same direction, they are to be subtracted from each other. Again, there will be addition if the planet is between the *unmaṇḍala* and the horizon. (It is also to be noted that) the root of the difference of the squares of the *chāyā* and *chāyā-koṭi* is the *chāyā-bāhu*.

Thus the four problems involving *śaṅku* (gnomon) have been discussed.

11.24 Problem five: *Nata* and *Krānti*

Now is discussed the derivation of the *nata* (hour angle) and *krānti* (declination) (when the other three are known).

Construct a circle with its radius as the Rsine of the arc extending from the east or west *svastika* to the planet. The Rsine on this circle is the *chāyā*-

bhujā. It was stated earlier that *chayā-bhujā* is the sum or difference of the *akṣa* on the *koṭi*-circle and the declination on the *trijyā-vṛtta*. Hence, when the *akṣa* on the *koṭi*-circle and the *chāyā-bhuja* are added together or mutually subtracted, the result will be declination on the *trijyā*-circle. Now, since the co-latitude and latitude have to be converted to the *koṭi*-circle, multiplication by the radius of the *koṭi*-circle and division by *trijyā* are needed. By the co-latitude and latitude so obtained, multiply, respectively, the *chāyā-bhujā* and *śaṅku*. Add or subtract the results as the case may be and divide by the radius of the *koṭi*-circle. The result will be the desired declination. Here, multiplication or division by the *koṭi*-circle is not required. Now, when simply the co-latitude and latitude are multiplied, respectively, by the *chāyā-bhujā* and *śaṅku,* and the results added or subtracted as the case may be and divided by *trijyā,* the result will be the desired *apakrama* (declination). The *koṭi* of this would be *iṣṭa-dyujyā*. Divide with this the product of *chāyā-koṭi* and *trijyā*. The result will be *nata-jyā* (Rsine of hour angle).

11.25 Problem six: *Nata* and *Āśāgrā*

Now are derived *nata* and *āśāgrā* (from the other three). For this, first, the *chāyā-bāhu* is obtained. And that is got by the addition or subtraction of *arkāgrā* and *śaṅkvagrā*. Now, *arkāgrā* is the divergence between the east and west *svastika*-s and the rising and setting points of the Sun on the horizon. And, *śaṅkvagrā* is the distance by which the planet has moved south at the desired time, in accordance with the slant of the diurnal circle from the place of its rising. Since it moves only to the south it (i.e., *śaṅkvagrā*) is said to be ever to the south (*nitya-dakṣiṇa*). Then (the planet) will rise in the northern hemisphere towards the north of the east-west *svastika*. Hence, on that day the *arkāgrā* is 'northern', and in the southern hemisphere the *arkāgrā* is 'southern'. Hence, the sum of the *arkāgrā* and the *śaṅkvagrā* in case both are in the same direction, or their difference if they are in opposite directions, will give the distance of the planet from the *sama-maṇḍala*. And, that is the *chāyā-bhujā*.

Now, here the *arkāgrā* and declination bear the relationship obtaining between *trijyā* and *lambaka*, and the *saṅku* and *saṅkvagrā* bear the relationship between *lambaka* and *akṣa.* Hence, if the declination is multiplied by *trijyā* and the *śaṅku* is multiplied by *akṣa* and added to or subtracted from, in accordance with the hemisphere in which they are, and divided by the *lambaka*, the result will be *chāyā-bhujā.* When this *chāyā-bhujā* is multiplied by *trijyā* and divided by the *chāyā,* the result will be *āśāgrā.* And, when the product of *chāyā* and *āśāgrā-koṭi* is divided by *dyujyā* the result will be *nata-jyā.*

11.26 Problem seven: *Nata* and *Akṣa*

Next is derived *nata* and *akṣa* (when the other three are known). Here, the *nata-jyā* is to be derived in the manner explained above. Now, the root of the difference between the squares of *chāyā-koṭi* and *dyujyā* is the Rsine of the distance of separation between the planet and the *unmaṇḍala* on the diurnal circle. This Rsine which rises from the horizon is termed *unnata-jyā.* The Rsine on that part of the diurnal circle situated between the horizon and the *unmaṇḍala* is called *kṣitija-jyā* (*kṣiti-jyā*). Since, however, in the southern hemisphere the *unmaṇḍala* is below the horizon, the *unnata-jyā*-plus-*kṣitija-jyā* is the root of the difference between the squares of *chāyā-koṭi* and *dyujyā.* In the northern hemisphere however, it would be equal to the *unnata-jyā*-minus-*kṣitija-jyā.*

Now, the *unnata-jyā* is the hypotenuse in the triangle whose sides are *śaṅku* and *śaṅkvagrā.* The *kṣitija-jyā* is the *bhujā* of a triangle which is similar to this triangle. In the southern hemisphere, this is the sum of the *bhujā* and the hypotenuse of two triangles. (Again), in the southern hemisphere, the *chāyā-bhujā* is the sum of *arkāgrā* and *saṅkvagrā.* Add this *chāyā-bhujā* to the *unnata-jyā* to which *kṣiti-jyā* has been added. This will then be the sum of the *bhujā* and *karṇa* of two triangles. In the northern hemisphere, however, it would be the difference between the *bhujā* and *karṇa.* This would, again be the sum of the *chāyā-bhujā* and the root of the sum of the squares of the *chāyā-koṭi* and *dyujyā.*

Here, there is a triangle with *śaṅku, śaṅkvagrā* and *unnata-jyā* as its three sides. There is another triangle with *apakrama, kṣiti-jyā* and *arkāgrā* as its sides. In the southern hemisphere there will be, the addition of the *bhujā*-s and of the *karṇa*-s in these two triangles. In the northern hemisphere, however, there will be the subtraction of the sum of the two *bhujā*-s from the sum of the two *karṇa*-s. Since these two triangles are similar *(tulya-svabhāva),* even when addition or subtraction is made, it will be as if the sum and difference of the *bhujā* and *karṇa* has been done in the same triangle. By its nature, the sum of the *śaṅku* and *apakrama* will be the *koṭi* of the triangle. Therefore, in the southern hemisphere, divide the square of the sum of this *śaṅku* and *apakrama* by the sum of the *bhujā* and *karṇa.* The result will be their difference. In the northern hemisphere, however, divide by the difference between the *bhujā* and *karṇa.* The result will be their sum. When the sum and difference of the *bhujā* and *karṇa* are found thus, half their sum will be the *karṇa,* and half their difference will be *bhujā.* Now, the *bhujā* is multiplied by *trijyā* and divided by the *karṇa.* The result will be the *akṣa,* since the above said two triangles are similar to the triangle formed by the *lamba, akṣa* and *trijyā.*

11.27 Problem eight: *Apakrama* and *Āśāgrā*

Next is stated the derivation of the *apakrama* and *āśāgrā.* Now, the maximum divergence between the *nata-vṛtta* and the horizon is the *svadeśa-nata-koṭi.* Take this as the *pramāṇa.* The divergence between the *svadeśa-nata-vṛtta* and the horizon on the *nata-vṛtta,* is a quarter of the circumference (90 degrees). The *jyā* thereof is the radius, and is the *pramāṇa-phala.* The *śaṅku* is the *icchā.* The distance between the planet and the horizon on the *nata-vṛtta* will be the *icchā-phala.* For these *pramāṇa-phala*-s, when the altitude of *Dhruva* is the *icchā,* there will be the interstice between *Dhruva* and the horizon on the *nata-vṛtta.*

Now, when the planet is north of the point of intersection of the *svadeśa-nata-vṛtta* and the *nata-vṛtta,* then subtract from one another the arcs of the (Rsines of the) *icchā* and *phala.* The result will be the arc between the north pole and the planet, on the *nata-vṛtta.* When, however, the planet is to the south of the abovesaid point of intersection, add together the arcs of the Rsines of the *icchā* and *phalā.* The result will be the arc between the south pole and the planet on the *nata-vṛtta.* The Rsine of this is the *dyujyā.* The *koṭi* thereof is the *apakrama.* The *āśāgrā* can be derived as discussed before.

11.28 Problem nine: *Krānti* and *Akṣa*

Next, are *krānti* (declination) and *akṣa* (latitude). First, derive the *dyujyā* and (using that) derive the *krānti.* The *akṣa* shall be derived by one of the (two) methods described earlier (in the fourth and the seventh problems).

11.29 Problem ten: *Āśāgrā* and *Akṣa*

Next are, derived *digagrā* and *akṣa.* Multiply *dyujyā* and *nata-jyā* or, *chāyā-koṭi* and *trijyā.* Either of the two shall be divided by the *chāyā.* The result of the division would be *āśāgrā-koṭi.* The *akṣa* can be derived as stated earlier.

Thus have been stated the answers for all the ten problems.

11.30 *Iṣṭa-dik-chāyā* : Another method

Now is stated, a method for the derivation of the *iṣṭadik-chāyā* (shadow in any desired direction). Now, consider the shadow of a 12-inch gnomon when the planet is at the point of intersection of the *ghaṭikā-maṇḍala* and the *iṣṭa-diṅmaṇḍala* (i.e., the vertical passing through the planet at the desired location). When the planet is at the equinox, the *chāyā-bhujā* of gnomonic shadow will be equal to the equinoctial shadow. The *āśāgrā* on the *trijyā* circle will be the *chāyā-bhujā* in the circle whose diameter is the shadow of the 12-inch gnomon. In order to find the *chāyā-koṭi* when it (i.e., the *āśāgrā*) becomes equal to the equinoctial shadow, multiply *āśāgrā-koṭi* and equinoctial shadow and divide by the *āśāgrā*. The result will be the *chāyā-koṭi*. Square this and the equinoctial shadow, add and find the root. The result will be the shadow of the 12 inch gnomon when the planet is on the *ghaṭikā-maṇḍala*. If this shadow is converted to (the shadow in) the *trijyā-vṛtta*, the result will be the distance between (the zenith and) the *ghaṭikā-maṇḍala* in the *iṣṭa-diṅmaṇḍala*. This can be conceived to be representive of the latitude (*akṣa-sthānīya*). Here, the distance between the zenith and the *ghaṭikā-maṇḍala* on the north-south circle is the latitude. The distance on the same between the *ghaṭikā-maṇḍala* and the diurnal circle is the declination. Hence the latitude and the representative of latitude will be the *pramāṇa* and *pramāṇa-phala* and the *icchā* will be the declination. For that *icchā*, the *icchā-phala* would be the distance from the *ghaṭikā-maṇḍala* to the diurnal circle on the *iṣṭa-digvṛtta*. This will be the representative of the declination (*apakrama-sthānīya*). Then, following the method by which the noon-day shadow is computed, when the arcs of the representatives of the latitude and the declination are added together or subtracted from each other and the Rsine thereof is found, the result will be the shadow in the desired direction.

11.31 *Kāla-lagna, Udaya-lagna* and *Madhya-lagna*

Now is stated the method for deriving the *kāla-lagna* (time elapsed since the rise of the first point of Aries) and *udaya-lagna* (the orient ecliptic point). Here, the ecliptic, i.e., the great circle which is the central circle of the zodiac (*rāśi-cakra*), which revolves westwards on account of the *Pravaha-vāyu*, touches the horizon at the desired time, (at two points) either towards the north or south of the east and west *svastika*-s. *Lagna* is a point of contact of the ecliptic and the horizon. Conceive of a circle touching the two *lagna*-s and the zenith. This is called *lagna-sama-maṇḍala*. Now, conceive of another circle touching the zenith and the points on the horizon, which are as much removed from the north-south *svastika*-s as the *lagna-sama-maṇḍala* is removed from the east-west *svastika*-s. This circle is termed *dṛkkṣepa-vṛtta* .

This circle and the *lagna-sama-maṇḍala* will be at right angles (*viparīta-dik*). These two circles and the horizon divide the celestial sphere into octants. Here, in the centre would be situated the *apakrama-vṛtta*. Here, the *rāśi-kūṭa* (the converging points of the sign segments of the ecliptic), which is the pole of the ecliptic, will be raised with respect to the horizon on the *dṛkkṣepa-vṛtta* by the same amount by which the point of contact of the ecliptic and the *dṛkkṣepa-vṛtta* is depressed from the zenith, which also represents the maximum divergence between the *lagna-sama-maṇḍala* and the ecliptic. This is because the zenith is removed by a quarter of the circle (from the horizon).

Now, that point on the ecliptic which is removed farthest from the *ghaṭikā-maṇḍala* is called *ayanānta*. Here, that circle which passes through the farthest points of the ecliptic and the *ghaṭikā-maṇḍala* will pass through the four poles of the *ghaṭikā-maṇḍala* and the ecliptic. Hence, the two interstices between the poles will be contained in this circle passing through the *ayanānta*-s.

Let it be conceived that the observer is on the equator, and the vernal equinox is at the zenith. Then the north solstice would be removed to the north from the eastern *svastika* by the measure of the maximum declination on the equatorial horizon, and the south solstice would be that much removed from the western *svastika* towards the south. The two *rāśi-kūṭa*-s would be on the horizon, one to the east of the south-*svastika*, and the other to the west of the north-*svastika*. When the north solstice would rise from the horizon, on account of the *Pravaha-vāyu*, the south *rāśi-kūṭa* will also rise. In the same manner, they reach the north-south circle and the horizon in the west at the same instant. Thus the rising and setting of the the southern *ayanānta* (winter solstice) and the northern *rāśi-kūṭa* occur at the same moment. Thus, the altitude of the *rāśi-kūṭa*-s is in exact accordance with the altitude of the solstices. Hence the Rsine altitude of the solstice is equal to the Rsine altitude of the *rāśi-kūṭa*-s.

Now, the gnomon for the *rāśi-kūṭa* has to be derived. That will be the altitude of the *rāśi-kūṭa* from the horizon. Now, since the *rāśi-kūṭa* is removed from the pole as much as the maximum declination, the maximum declination would be equal to the (radius of the) diurnal circle of the *rāśi-kūṭa*. Therefore multiply the Rsine altitude of the solstice (*ayanāntonnata-jyā*) with the maximum declination and divide by *trijyā*. The result would be the gnomon of the *rāśi-kūṭa* at the equator. At places with latitude, since it (i.e., the equator) would be inclined, this should be multiplied by the co-latitude and divided by *trijyā* and the result should be added to the portion of the gnomon (derived) above corresponding to the interstice between the horizon and the hour circle (*unmaṇḍala*), in the case of the gnomon pertaining to the northern *rāśi-kūṭa,* and subtracted from the gnomon pertaining to the southern *rāśi-kūṭa*. The result will be the gnomon of the *rāśi-kūṭa*.

Now, conceive of a circle touching the *rāśi-kūṭa*-s and the zenith. That is clearly the *dṛkkṣepa-vṛtta*. On this circle, the meeting point with the ecliptic would be located below the zenith by the same amount by which the gnomon is lower than the *rāśi-kūṭa*. That will be the *dṛkkṣepa*. Hence, it follows that the gnomon of the *rāśi-kūṭa* will itself be the *dṛkkṣepa*. Now, arises the proportion: If the latitude is the gnomon for the distance between the horizon and the hour circle (*unmaṇḍala*) when (a point) moves from east-west *svastika* to the pole on the hour circle (*unmaṇḍala*), then what will be the gnomon for the *antya-dyujyā* when it moves by the distance between the diurnal circle and the *rāśi-kūṭa*. The result will be the portion of the gnomon for the interstice between the equator and the hour circle at the desired place.

Then the Rsine of 90 degrees-minus-*kāla-lagna* will be equal to the Rsine of the altitude (*unnata-jyā*) of the *rāśi-kūṭa* from the hour circle. If this Rsine is converted to the required diurnal circle and from it is subtracted the correction to the extent of the *lamba* due to the inclination of the latitude, the result will be the gnomon of the *rāśi-kūṭa*. *Kāla-lagna* with respect to the equinox (*golādi*) when subtracted from 90 degrees will be the *kāla-lagna* with respect to the solstice (*ayanādi*). Hence it is enough to take the Rsine of the *kāla-lagna-koṭi*. The *koṭi* of *dṛkkṣepa-jyā* thus derived, will be the greatest distance between the horizon and the ecliptic. Take this as the *pramāṇa,* and *trijyā* as the *pramāṇa-phala.* Now, the gnomon of the planet at the required time is the distance between the horizon and the point on the ecliptic where the planet is. That will be the *icchā-rāśi.* The *icchā-phala* would be the interstice between the horizon and the planet along the ecliptic. Convert it into arc and add or subtract the arc (i.e., the longitude) of the planet; the result would be the portion of the ecliptic between the equinox and the point of contact with the horizon. That will be the *lagna* at the time of the setting of the planet (*asta-lagna*) in the west, and in the east, it will be the time of the rising of the planet (*udaya-lagna*).

Next, the method to derive the gnomon at this place. Now the *nata* (hour angle) is the interstice, on the diurnal circle, between the planet and the north-south circle. Now, all diurnal circles would have revolved once during a day and night. In a day-night, the number of *prāṇa*-s would be equal to 21,600 (*cakra-kalā-tulya*) in number. Hence, when all the diurnal circles are conceived as divided into *prāṇa*-s, they would be 21,600 in number. Hence, the *nata-prāṇa*-s are only part of the minutes in the diurnal circle. Therefore, the Rsine of 90 degrees less the Rversine of the hour angle, would be equal to the Rsine of the portion of the diurnal circle in the interstice between the planet and the hour-circle. If to this, the correction due to the *cara-jyā* is applied, the result will be the Rsine of the altitude from the horizon (*unnata-jyā*). In order to convert it to the circle with *trijyā* as radius, it should be multiplied by *dyujyā,* and to correct for the inclination on account of the latitude, it should be multiplied by the co-latitude (*lambaka*) and divided by the square of *trijyā.* The result got would be the distance on the required *diṅmaṇḍala* (vertical circle), from the horizon to the place in the diurnal circle where the planet is situated. This will be the (required) gnomon.

Since the point where the Sun is on the diurnal circle would touch the ecliptic, the distance between the horizon and that point on the ecliptic would be this very same gnomon. Hence, this gnomon will be the *icchā.* Hence, that portion of the ecliptic which is between the planet and the horizon of which the gnomon is the *icchā,* even during night time, the gnomon derived as above would be the distance between the specified point on the ecliptic and the horizon. Hence, the gnomon would be the *icchā-rāśi* during night time also. Here, the difference between half the measure of the night and the portion of the night which has either gone (*gata*) or is yet to go (*eṣya*), would be the *nata-prāṇa*-s (*prāṇa*-s of the hour angle). This is for the reason that this is the portion of the diurnal circle corresponding to the difference between the planet and the north-south circle below. When the Rversine of this is subtracted from *trijyā*, the result will be the Rsine of that portion of the diurnal circle between the planet and the hour circle. To convert this to the difference between the horizon and the planet, subtract from it the *cara* in the northern hemisphere, and add the *cara* in the southern hemisphere. Then derive the gnomon as before. Then multiply that by *trijyā* and divide by the *dṛkkṣepa-koṭi*. The result will be the Rsine of the interstice between the planet and the horizon on the ecliptic. Find the arc of this Rsine and add to the (longitude of the) planet in the eastern hemisphere and subtract from the planet if the gnomon is directed below. The result will be the *lagna* at the time of rising. In the western hemisphere if this is applied to the planet in the reverse order, the *lagna* at setting would be obtained. The *lagna* exactly at the middle of the rising and setting is the *dṛkkṣepa-lagna*. And, that will fall at the point of intersection of the *dṛkkṣepa* circle with the ecliptic.

Now, *madhya-lagna* is the point of intersection of the north-south circle and the ecliptic. This can be obtained using the method of fifteen questions (*pañcadaśa-praśna-nyāya*) dealt with earlier. The *madhya-kāla* is the point of intersection of the north-south circle and the celestial equator (*ghaṭikā-maṇḍala*). This can be obtained from the method of *madhya-lagna*.

11.32 *Kāla-lagna* corresponding to sunrise

Kāla-lagna is *madhya-kāla* plus three *rāśi*-s (90 degrees), which will fall at the point where the celestial equator meets the eastern *svastika*. The method to derive this is stated now. If the *sāyana* Sun is in the first quadrant, then derive its *bhujā-prāṇa*-s as stated earlier. Construct a transverse circle (*tiryag-vṛtta*) passing through the Sun on the ecliptic and the two poles (of the equator). Note the point where it touches the celestial equator. The distance from that point to the point of equinox on the celestial equator would be the measure of the *bhujā-prāṇa*-s. If the Sun is supposed to be at the horizon, then the meeting point of the celestial equator and the *tiryag-vṛtta* will be a little below the eastern *svastika*, with the distance being equal to the *cara*. Hence when the required *cara* is subtracted from the *bhujā-prāṇa,* the result will be the distance from the eastern *svastika* to the equinox on the celestial equator. This will be the *kāla-lagna* when the *sāyana* Sun is in the first quadrant.

Similarly, in the second quadrant, when the Sun rises calculate the *bhujā-prāṇa*-s of the Sun. Construct also the transverse circle (*tiryag-vṛtta*) as before. Here too, as stated earlier, the distance between this transverse circle and the equinox to the north would be the *bhujā-prāṇa*-s. Here, the *bhujā-prāṇa*-s would occur below the horizon and the intersection with the transverse circle would occur below the eastern *svastika*. Therefore, add the *cara* to the *bhujā-prāṇas*-s and subtract (the sum) from 6 *rāśi*-s (180 degrees). The remainder would be the difference from the eastern *svastika* to the equinox in the east, on the celestial equator. That will be the *kāla-lagna* at the time of sunrise.

In the third quadrant, the sunrise is to the south of the eastern *svastika*. There, the horizon would be above the hour circle and hence the transverse circle constructed would be above the eastern *svastika*. Therefore, to reach upto the *svastika*, the *prāṇa*-s of *cara* have to be added to the *bhujā-prāṇa*-s. And, that has its beginning in the equinox to the north. Hence six *rāśi*-s (180 degrees) are to be added. The result will be the *kāla-lagna*.

In the fourth quadrant also, as in the second quadrant, there is a portion of arc yet to be traversed (*eṣya*), the *bhujā-prāṇa*-s are below the horizon. Since the transverse circle is above the eastern *svastika*, to reach up to the horizon, the *prāṇa*-s of *cara* are to be subtracted from the *bhujā-prāṇa*-s. Thus the subtraction has to be done from all the twelve *rāśi-s,* since it is yet to be traversed. This is the *kāla-lagna* for sunrise. In this manner, the *kāla-*

lagna for all the twelve *rāśi*-s have to be calculated and the earlier ones have to be subtracted from the subsequent ones, successively. The differences would be, in order, the *rāśi-pramāṇa*-s (appropriate time measures for the *rāśi*-s). Now, the twelve *rāśi*-s are formed by dividing the ecliptic equally, commencing from equinox in the east. When, as a result of the motion of the *Pravaha-vāyu*, when the forefront of a *rāśi* meets the horizon, that *rāśi* is said to commence. When its hind end leaves the horizon, that *rāśi* is said to end. The time interval between these two events are said to be the time measure of the *rāśi-prāṇa*-s. Thus, incidentally, the *rāśi*-s and time measure of *rāśi*-s have also been stated.

11.33 *Madhya-lagnānayana*: Calculating the meridian ecliptic point

In this manner, calculate the *kāla-lagna* for (the required) sunrise and add to it the time elapsed (since sunrise) in terms of *prāṇa*-s. That will be the *kāla-lagna* of the desired time. When three *rāśi*-s are subtracted from it the result will be the point of contact of the celestial equator and the north-south circle. This would be the *madhya-kāla*.

The *koṭi* of this would be that portion of the celestial equator lying between the equinox and the east-west cardinal points. Calculate the Rsine declination (*apakrama-jyā*) corresponding to this *koṭi*. That will be the distance between the celestial equator and the ecliptic on the *rāśi-kūṭa-vṛtta* which touches the east and west cardinal points. Then derive the *koṭi-jyā* and *dyujyā* for this and obtain the *bhujā-prāṇa*-s. That will be the *koṭi* of the interstice on the ecliptic between the equinox and the point of intersection of the *rāśi-kūṭa-vṛtta* and ecliptic mentioned above. Then, consider the portion of the ecliptic contained in the interstice between the equinox and the north-south circle. This will be *madhya-bhujā*. The rest (of the work) is as per the different quadrants, as has already been stated for *kāla-lagna*. The only distinction here is that the ecliptic is considered as the celestial equator and the celestial equator is considered as the ecliptic. Thus has been stated the method of deriving the *madhya-lagna.*

11.34 *Dṛkkṣepa-jyā* and *Koṭi*

Now is stated the method for deriving *dṛkkṣepa-jyā* using the *udaya-lagna* and *madhya-lagna.* First conceive the ecliptic and *dṛkkṣepa-maṇḍala* as directed above. Then, that point of the ecliptic which meets the horizon to the east of the north-south circle is called *udaya-lagna* (rising or orient ecliptic point), and that point which touches (the horizon) at the west is called *asta-lagna* (setting or occident ecliptic point). That point (of the ecliptic) which touches the north-south circle is called *madhya-lagna.* The method to ascertain these has been stated earlier.

Now, *udaya-jyā* would be the distance from the east and west *svastika*-s of the point of contact of the ecliptic with the horizon. The *udaya-jyā* should be derived in the same manner as the *arkāgrā*, (with the difference) that the *udaya-lagna* is (here) taken as the Sun. Now, *madhya-jyā* would be the distance from the zenith to the point of intersection of the ecliptic and the north-south circle. This has to be derived in the same manner as the *madhyāhnacchāyā*, with the difference that *madhya-lagna* is taken as the Sun.

Now, the prime vertical and *dṛkkṣepa-sama-maṇḍala* meet at the zenith, and have their maximum divergence on the horizon. This maximum divergence is the *udaya-jyā.* Now, the *dṛkkṣepa-vṛtta* is perpendicular (*viparīta*) to the *dṛkkṣepa-sama-maṇḍala.* Hence the maximum divergence between the *dṛkkṣepa-vṛtta* and the north-south circle on the horizon will be equal to the *udaya-jyā.* Such being the case, if the *madhyama-jyā* is towards the south, take as *pramāṇa* the *karṇa* (i.e., the radius) of the north-south circle which has its end in the southern *svastika.* If however, the *madhyama-jyā* is towards the north, take as *pramāṇa* the radius which has its end in the northern *svastika.* Then, the *pramāṇa-phala* would be the distance between the (south or north) *svastika* to the *dṛkkṣepa-vṛtta* on the horizon, which is the same as the *udaya-jyā*. *Icchā* will be the *madhya-jyā*. And, *icchā-phala* would be the interstice between the end of the *madhya-jyā* to the *dṛkkṣepa-vṛtta* on the ecliptic. This is taken as the *bhujā*. This would be the Rsine of that portion of the ecliptic lying between *madhya-lagna* and *dṛkkṣepa-lagna*. The square of this is subtracted from the square of *trijyā* and the root of this would be its *koṭi*. This will be the Rsine of the portion of the ecliptic which is between the north-south circle and the horizon. When the square of the *bhujā* is subtracted from the square of the *madhya-jyā* and the root extracted, the result would be the distance between *madhya-lagna* and *dṛkkṣepa-sama-maṇḍala*. Take this as the *pramāṇa-phala* and the *koṭi* mentioned above as the *pramāṇa*. Then take the radius of the ecliptic which has its end at the *dṛkkṣepa-lagna* as *icchā* and derive the *icchā-phala* by the rule of three. The result would be *dṛkkṣepa-jyā*. It is the *icchā-phala* since it is the maximum divergence between the ecliptic and the *dṛkkṣepa-maṇḍala*.

Now take as *pramāṇa* what has been stated as *pramāṇa* above, the distance between the *madhya-lagna* and the horizon, which is an Rsine on the north-south circle. Consider the *madhyama-jyā-koṭi* as the *pramāṇa-phala*, and the *trijyā* as *icchā*. The resultant *icchā-phala* would be the maximum divergence between the ecliptic and the horizon, since the divergence between the *dṛkkṣepa-lagna* and the horizon is a Rsine on the *dṛkkṣepa-vṛtta*. This is called *dṛkkṣepa-śaṅku* or *para-śaṅku* and *dṛkkṣepa-koṭi*. Thus has been stated the methods for deriving *dṛkkṣepa-jyā* and *dṛkkṣepa-koṭi*.

11.35 Parallax in latitude and longitude (*Nati* and *Lambana*)

Next are stated the methods for deriving *nati* (parallax in latitude) and *lambana* (parallax in longitude) which are used in computations relating to the Moon's shadow, eclipses and the like. Here, *lambana* is the amount by which the shadow on the *dṛṅmaṇḍala* which has its centre at *dṛṅmadhya* (the location of the observer), is more than the shadow on the *dṛṅmaṇḍala* whose centre is *bhagola-madhya* (the centre of the celestial sphere). This has been stated in *chāyā-prakaraṇa.* Conceiving this *lambana* as the *karṇa,* a method to derive the actual *nati* and *lambana* is going to be stated presently. Conceive, as before, the *dṛkkṣepa* circle (vertical circle through the central ecliptic point), ecliptic and *dṛṅmaṇḍala* (vertical circle). Conceive also another *rāśi-kūṭa-vṛtta* passing through the two *rāśi-kūṭa*-s (poles of the ecliptic) and the planet. Such being the case, the planet will be at the meeting place of *rāśi-kūṭa-vṛtta* passing through the planet and *dṛṅmaṇḍala* and the ecliptic.

Now, consider these three circles as having not been shifted by parallax and the planet to have a parallactic shift. When the planet is shifted by parallax, it is shifted downwards along the *dṛṅmaṇḍala.* Here, the interstice along the *dṛṅmaṇḍala* between the planet shifted by parallax and the (location of the unshifted planet at the) meeting point of the circles is the *chāyā-lambana* (parallax of the shadow). Then, the distance from this parallactically shifted planet to the ecliptic is its *nati* (parallax in latitude). And, the interstice between this shifted planet to the *rāśi-kūṭa* circle passing through the (unshifted) planet will be its *lambana* (parallax in longitude). These parallaxes in latitude and in longitude form the *bhujā* and *koṭi,* and the *chāyā-lambana* will be the hypotenuse.

11.36 Second correction for the Moon

Now the method to compute the *chāyā-lambana.* This can be done as stated earlier by computing the *dṛkkarṇa* which is in terms of minutes; it can also be obtained (in terms of *yojana*-s) by converting the *dṛkkarṇa* into *yojana*-s. The method for this is stated herein below.

Now, the *manda-karṇa* of the Sun and the Moon is the distance between the planet – (here Sun and Moon) – and the centre of the *bhagola* (sphere of the asterisms). For them, the second-true-hypotenuse (*dvitīya-sphuṭa-karṇa*) is the interstice between the planet and the centre of the Earth. Here, the interstice between the centre of the *bhagola* and the centre of the Earth, will vary in accordance with that between the *candrocca* and the Sun.

Conceive that (later) interstice as the radius of the *ucca-nīca* circle. Now, the centre of the Earth and the centre of the *bhagola* are displaced along the line connecting the centre of the orb of the Sun and the centre of the shadow of the Earth. Hence, that line is the *ucca-nīca* line. Since the Sun is always on this *ucca-nīca* line, in both the circles, the circle with its centre at the centre of the *bhagola*, and the circle with centre of the Earth as its centre, its *sphuṭa-kalā* (true longitude in minutes) is (the same); there would be difference only in the hypotenuse (*karṇa*). For the Moon, however, there is motion from the *ucca-nīca* line. And that will be the motion away from the Sun. Hence, the *tithi*-s (lunar days) of specific lengths, commencing from *pratipad* would be the *kendra*, being the planet-minus-*ucca*. Therefore, calculate the *bhujā-phala* and *koṭi-phala* using the radius of the *ucca-nīca* circle, which is the distance from the centre of the *bhagola* and the centre of the Earth, and the Rsines and Rcosines of the required *tithi*. Then, using these (*bhujā* and *koṭi-phala*-s) and the *manda-karṇa* compute the *dvitīya-sphuṭa-karṇa*, either in terms of minutes or in terms of *yojana*-s. Then, using this *karṇa*, correct the *bhujā-phala* and apply that *bhujā-phala* to the Moon. The result will be the *candra-sphuṭa* on the circle with the centre of the Earth as the centre. Thus is computed the *dvitīya-sphuṭa* by the principle of *śīghra-sphuṭa*.

Now, the radius of the *ucca-nīca* circle is variable. Here is the rule relating to it. Now, conceive of a line passing through the centre of the *bhagola* and at right angles to the *ucca-nīca* line passing through (the orbs of) the Sun and the shadow of the Earth. If the *candrocca* lies on that side of the line where the Sun is, then the centre of the *bhagola* will move to that side from the centre of the Earth. The Sun will be at its *ucca* (apogee) at that time.

If, however, the *candrocca* is on that side of the transverse line where the Earth's shadow is, then the centre of the *bhagola* will move from the centre of the Earth towards the Earth's shadow. Since, at that time, the apogee is at the Earth's shadow, here also the lengthening and shortening of the radius of the *ucca-nīca-vṛtta* will be according to the Rcosine of Sun-minus-*candrocca*. Here also, if the quadrants (beginning with) *Mṛga* and *Karki* are the same for both Rcosine Sun-minus-*candrocca* and Moon-minus-Sun, then the Rcosine of Moon-minus-Sun has to be applied positively in the *manda-karṇa*, otherwise negatively. When there is *vikṣepa* (latitude) the above-said *koṭi-phala* shall have to be applied to the *vikṣepa-koṭi.*

Just as the *vikṣepa* derived from the *manda-sphuṭa* is squared and it is subtracted from the square of the *manda-karṇa,* if it is measured in terms of the minutes of the *pratimaṇḍala,* and from the square of *trijyā* being the radius of the *manda-karṇa* circle, if it is measured in terms of the minutes of *manda-karṇa* circle, and the root found and the resulting *vikṣepa-koṭi* is corrected by *koṭi-phala* measured in similar units (angular or in *yojana*-s): In the same manner, here also, square the latitude derived from the Moon's first *sphuṭa,* and subtract it from the square of the first hypotenuse or the square of the *trijyā* and extract the root. The result would be *vikṣepa-koṭi.* To this should be applied the second *koṭi-phala* (*dvitīya-sphuṭa-koṭi-phala*). Here, the *antya-phala* in the case of *dvitīya-sphuṭa* is half the Rcosine of Sun-minus-Moon. Since this would be in terms of *yojana*-s, the *bhujā-phala* and *koṭi-phala* of *dvitīya-sphuṭa* which are derived by multiplying the Rsine and Rcosine of Sun-minus-Moon by the above-said (*antya-phala*) and dividing by *trijyā,* would also be in terms of *yojana*-s. Therefore, the *vikṣepa-koṭi* should also be converted into *yojana*-s and the *koṭi-phala* should be applied to it. To the square of this add the square of the *bhujā-phala,* and extract the root. The result will be the *yojana*-s between the centre of the Earth and the centre of the Moon. Then, multiply the *bhujā-phala* by *trijyā* and divide by this *karṇa* and apply the result to the *sphuṭa* of the Moon. The method of this correction will be stated later. If the Rcosine of Sun-minus-*candrocca* is in the *Makarādi* quadrant, subtract the *bhujā-phala* from the Moon in the bright fortnight, add in the dark fortnight. If it is the *Karkyādi* quadrant, add in the bright fortnight and subtract in the dark fortnight.

Then, multiply the mean motion (of the Moon) by ten and by *trijyā* and divide by the second *sphuṭa-karṇa.* The result will be the (mean) *dvitīya-sphuṭa-gati.* Thus is the method of *dvitīya-sphuṭa.* With this, the true Moon on the circle with its centre at the centre of the Earth, and having at its circumference the centre of the Moon's orb can be derived. From this, the *sphuṭa* on the circle with its centre at the observer standing on the Earth's surface (*bhū-pṛṣṭha*) can be derived.

11.37 *Chāyā-lambana:* Parallax of the gnomon

The method for this (derivation) using the correction for parallaxes in latitude and longitude (*nata-lambana-saṃskāra*) is stated here. This method is only slightly different from the method stated for the *chāyā-lambana.* The *chāyā-lambana* in this case is conceived of in two parts. By how much has the planet, which is shifted along the path of the shadow, been deflected along the ecliptic, and secondly, by how much it has been deflected along the *rāśi-kūṭa* circle passing through the planet. The first is called *lambana* (parallax of longitude) which will be the difference between the *sphuṭa*-s. The latter is called *nati* (parallax in latitude). This will be in the form of latitude. Now, consider a situation when a planet without latitude and hence located on the ecliptic itself, happens to be passing through the zenith in the course of its motion caused by the *Pravaha-vāyu.* At that time, the ecliptic itself will be the *dṛṅmaṇḍala* (vertical circle). Hence, the *chāyā-lambana* will be the apparent depression towards the horizon along the ecliptic. Then, when the planet is on the *dṛkkṣepa-maṇḍala,* since at that time both the *dṛṅmaṇḍala* and the *rāśi-kūṭa* circle passing through the planet are identical, the *chāyā-lambana* which will be along the *dṛṅmaṇḍala* will be at right angles to the ecliptic. Hence, the *chāyā-lambana* will be wholly in latitude and there will be no difference between the *sphuṭa*-s. On the other hand, when the *rāśi-kūṭa* circle passing through the planet, the ecliptic and the *dṛṅmaṇḍala,* are all different, then the planet which is deflected from the meeting point of the three circles due to parallax along the *dṛṅmaṇḍala,* will deviate from both the *rāśi-kūṭa* circle and the ecliptic. There, the deflection from the *rāśi-kūṭa* circle will be the difference between the *sphuṭa*-s and the deflection from the ecliptic is the latitude. If there is already a latitudinal deflection, then this will be the difference between the latitudes.

Here the division into quarters is through the *rāśi-kūṭa* circle and the ecliptic. Conceive the *dṛṅmaṇḍala* as the *valita* (inclined) circle to these. Conceive also the *chāyā*, which is the distance of separation between the planet and the zenith along the *dṛṅmaṇḍala,* to have its foot at the meeting point of the three circles and its tip at the zenith. Then, ascertain at what distance are the ecliptic and the *rāśi-kūṭa* circle, passing through the planet, from the tip of the *chāyā.*

Now, the distance from the zenith to the ecliptic is the *dṛkkṣepa-jyā.* The *dṛkkṣepa* circle and the *rāśi-kūṭa* circle passing through the planet, have their meeting point at the *rāśi-kūṭa*-s. Their maximum divergence is on the ecliptic and is the distance between *dṛkkṣepa-lagna* and the planet. It is to be *pramāṇa-phala* here. The interstice between the *rāśi-kūṭa-vṛtta* and the *dṛkkṣepa-lagna*, being the *trijyā* on the relevant section of the *dṛkkṣepa*, is the *pramāṇa.* In this *dṛkkṣepa* circle itself, the interstice between the zenith and the *rāśi-kūṭa* is *dṛkkṣepa-koṭi.* This is the *icchā.* The *icchā-phala* is the interstice between the zenith and the *rāśi-kūṭa* circle passing through the planet. This is called *dṛggati-jyā.* These two, the *dṛggati* and the *dṛkkṣepa,* would be the *bhujā* and *koṭi* for the *chāyā*, and the *chāyā* itself would be the hypotenuse. In the same way, on the other side of the meeting point of the three circles, the differences between the latitudes and of the *sphuṭa*-s will be the *bhujā* and *koṭi* for the hypotenuse formed by the portion forming the *chāyā-lambana* in the *dṛṅmaṇḍala.* Here, *chāyā* is the *pramāṇa, dṛkkṣepa* and *dṛggati* are the *pramāṇa-phala*-s, the *chāyā-lambana* is the *icchā* and *nati* and *lambana* are the *icchā-phala*-s.

Therefore, the *nati* and the *lambana* might be derived using the *dṛkkṣepa* and *dṛggati.* There, using the proportion: when the *chāyā* becomes equal to *trijyā,* the *chāyā-lambana* would be equal to the radius of the Earth, then how much it will be for the desired *chāyā.* Similarly, when the *dṛkkṣepa* and *dṛg-gati* become (separately) equal to *trijyā,* the *nati* and the *lambana* will each be equal to the number of minutes in the radius of the Earth, then what will the *nati* and *lambana* be for the desired *dṛkkṣepa* and *dṛggati.*

Now, multiply the *dṛkkṣepa* and *dṛggati* by the *yojana*-s of the radius of the Earth and divide by the *yojana*-s of *dṛkkarṇa.* Here, we can avoid multiplying and dividing by *trijyā* since there will be no difference in the result. Here, it is to be noted that if the planet is on the east of the *dṛkkṣepa-lagna,* it will be (seen) depressed towards the east, and the *sphuṭa* related to the observer on the surface of the Earth would be greater than the *sphuṭa* related to the centre of the Earth. If, however, the planet is on the west of the *dṛkkṣepa-lagna,* it will be less. Similarly, If the *vikṣepa* is to the south, there will be a depression to the south and hence the *nati* will be towards the south, and, if it is the other way, (the *nati*) will be towards the north. All these follow logically (*yukti-siddha*). Thus has been stated the method for *nati* and *lambana.*

11.38 *Dṛkkarṇa* when the Moon has no latitude

Now is stated the specialities in the matter of deriving the shadow and the gnomon from which the *dṛkkarṇa* can be calculated when the Moon has latitude. It is always the case that, when there is no latitude for the shadow, the root of the sum of the squares of *dṛkksepa-jyā* and *dṛggati-jyā* will be the *koṭi-śaṅku* of the shadow. When the shadow and gnomon are calculated in this manner and multiplied, individually, by the *yojana*-s of the radius of the Earth and divided by *trijyā,* the *bhujā* and *koṭi-phala*-s in the calculation of *dṛkkarṇa* are in terms of *yojana*-s. Now, the *koṭi-phala* is subtracted from the hypotenuse of the second *sphuṭa* in terms of *yojana*-s (*dvitīya-sphuṭa-yojana-karṇa*). The remainder is squared and added to the square of the *bhujā-phala* and the root found. The result will be the *dṛkkarṇa* in terms of *yojana*-s.

11.39 Shadow and gnomon when the Moon has latitude

Now is stated the method of deriving the shadow and gnomon for the Moon when it has latitude. Conceive a circle as much removed (in all its parts) from the ecliptic as the latitude of the planet. The centre of this circle will also be removed from the centre of the ecliptic by the measure of the latitude. These two circles will be like the *ghaṭikā-maṇḍala* (celestial equator) and the *ahorātra-vṛtta* (diurnal circle). This circle is called *vikṣepa-koṭi-vṛtta.* The planet will be situated in this circle at the point where the *rāśi-kūṭa-vṛtta* passing through the planet meets it. Here, the rising and setting *lagna* (of the planet) are the two points where the ecliptic and the horizon meet. The *lagna-sama-maṇḍala* passes through these two points and the zenith. Conceive the division of the sphere (into equal parts) made by the said *lagna-sama-maṇḍala, dṛkkṣepa-maṇḍala* and the horizon. Conceive the ecliptic as the *valita-vṛtta* for these. Then the maximum divergence of the ecliptic to the *lagna-sama-maṇḍala* would be *dṛkkṣepa-jyā. dṛkkṣepa-koṭi* would be the maximum divergence between the horizon and the ecliptic. This would be the *pramāṇa-phala.* The *pramāṇa* is *trijyā. Icchā* is the interstice between the planet and the point of intersection of the horizon and the ecliptic on the ecliptic. *Icchā-phala* is the interstice between the planet and the horizon. This will be the gnomon of the planet which has latitude and the *chāyā* is the root of the sum of the squares of the *dṛkkṣepa-jyā* and *dṛggati-jyā.*

Now to the speciality of the gnomon and shadow of the planet on the *vikṣepa-koṭi-vṛtta.* Here *dṛkkṣepa* is that part of the *dṛkkṣepa-vṛtta* forming the interstice between the zenith and the *dṛkkṣepa-lagna,* which in turn is equal to the maximum divergence between the *lagna-sama-maṇḍala* and the ecliptic. The *vikṣepa* (latitude) is the interstice between the *dṛkkṣepa-lagna* and the *vikṣepa-koṭi-vṛtta* along the *dṛkkṣepa-vṛtta.* Now, add together or subtract one from the other, the *vikṣepa* and the *dṛkkṣepa.* This will be the interstice between the zenith and the *vikṣepa-koṭi-vṛtta* along the *dṛkkṣepa-vṛtta.* This is called *nati* (parallax in latitude). The *koṭi* of this is the maximum distance between the horizon and the *vikṣepa-koṭi-vṛtta* being a portion of the *dṛkkṣepa-vṛtta.* This is called *parama-śaṅku* (maximum gnomon). Now, the noon shadow and the noon gnomon on the north-south circle are derived by the sum or difference of the latitude and declination and the sum or difference of the co-latitude (*lambaka*) and the declination, and by taking the interstices. In the same manner the *nati* and *parama-śaṅku* on the *dṛkkṣepa-vṛtta* can be derived from the sum or difference of the *dṛkkṣepa* and *vikṣepa* and the *dṛkkṣepa-koṭi* and *vikṣepa.* Here, the Rsine of the interstice between the *lagna* and planet, which has been taken as the *icchā,* should be subtracted from *trijyā* which is taken as the *pramāṇa.* The remainder shall then be considered as the *icchā.* Then the *icchā-phala* would be the difference between the *pramāṇa-phala* and the *icchā-phala.*

The *śara* (celestial latitude) of the portion of the ecliptic that lies between the *rāśi-kūṭa-vṛtta* touching the planet and the *dṛkkṣepa-vṛtta* has to be derived first. Then, this *śara* should be multiplied by *vikṣepa-koṭi* and divided by *trijyā.* The result will be the *śara* in the *vikṣepa-koṭi* circle that lies between the *rāśi-kūṭa-vṛtta* touching the planet and the *dṛkkṣepa-vṛtta.* Now, multiply the *śara* of the *vikṣepa-koṭi* circle by *dṛkkṣepa-koṭi* and divide by *trijyā.* Subtract the result from the *parama-śaṅku.* The remainder will be the required *śaṅku* of the planet on the *vikṣepa-koṭi* circle.

It is to be noted that if the multiplication is done by *parama-śaṅku,* it would not be correct to divide by *trijyā.* The division should be made by the *vikṣepa-koṭi* which has been corrected by the difference between the horizon and the *unmaṇḍala.* The reason for this is as follows: When the *unnata-jyā* of the diurnal circle or the *śara* of the *nata-jyā* is multiplied by *lambaka* (Rcosine of the latitude) and the result is divided by *trijyā,* the result will be the desired *śaṅku* or the difference between the noon-day *śaṅku* and the desired *śaṅku.* Here, if the multiplication has to be done by the noon-day *śaṅku,* then the division is not to be done by *trijyā.* On the other hand, the division should be made by that part of the diurnal circle which is above the horizon and which is the radius of the diurnal circle as corrected by the *kṣiti-jyā.* In the same way, here, it is the *dṛkkṣepa-koṭi* which is in place of the *lambaka,* and the *parama-śaṅku* which is in place of the noon-day *śaṅku.*

Now, the slant of the diurnal circles is in the same way as (i.e., parallel to) the slant of the celestial equator. And, the slant of the *vikṣepa-koṭi* circle is in the same way as the slant of the ecliptic. Since the two are alike in nature, there would be similarity in methodology (*nyāya-sāmya*) also. Hence, the *śaṅku* is derived thus.

Now, the shadow. Here, the sum or difference of the *vikṣepa* and the *dṛkkṣepa* is the distance between the *vikṣepa-koṭi* circle and the *lagna-sama-maṇḍala,* on the *dṛkkṣepa* circle. This is called *nati.*

Now, take the interstice between the planet and the *dṛkkṣepa* circle on the *vikṣepa-koṭi* circle, and derive the Rsine and Rversine. These will result when the Rsine and Rversine of the interstice between the *dṛkkṣepa-lagna* and Moon are, respectively, multiplied by *vikṣepa-koṭi* and divided by *trijyā.* Here too, calculate first Rversine, and then multiply this Rversine by *dṛkkṣepa-jyā* and divide by *trijyā.* The result will give the slant of the tip of the Rversine from its foot. Add to or subtract this from the above stated *nati*, derived earlier, in accordance with its direction. The result obtained will be the distance between the foot of the Rversine to zenith. This will also be the same as that obtained from the foot of the Rsine stated here. From this argument it will be clear that this is the distance between the planet at the tip of the Rsine and the *lagna-sama-maṇḍala.* This is called *bāhu.* When this and the *bhujā* mentioned earlier are squared, added together and the root calculated, the result will be the shadow. Thus have been stated the methods to derive the gnomon and the shadow. Now, it is also possible to calculate one of these two by the methods enunciated above, and calculate the other by squaring it and subtracting it from the square of *trijyā* and finding the root of the difference.

Chapter 12

Eclipse

12.1 Eclipsed portion at required time

Calculate the gnomon and shadow of the Moon in the above manner. From these calculate the *dṛkkarṇa* in terms of *yojana*-s. Using the *dṛkkarṇa-yojana* calculate the minutes of the corresponding *lambana*. The minutes of *lambana* of the Sun and Moon are to be applied, respectively, to the (true longitudes of) Sun and Moon. When the resulting true longitudes of the two are the same, that will indicate the time of the middle of the eclipse.

Or the time of the *lambana* can be calculated from *dṛggati*. Here, when the *dṛggati* is equal to *trijyā,* the *lambana* will be four *nāḍikā*-s. Then using the rule of three, find out what will be the *lambana* for the desired *dṛggati.* It is known that when the *dṛkkṣepa* and *dṛggati* are equal to *trijyā* then the *yojana*-s of *nati* and of *lambana* are equal to the radius of the Earth. It is also known that the minutes of *madhya-yojana-karṇa* are equal to *trijyā.* Multiply the minutes of *lambana* thus obtained by the true motion and divide by mean motion. Then the *lambana* will be obtained in terms of the minutes of *bhagola.* Therefore, multiply the *madhya-yojana-karṇa* and the mean motion and divide by the *yojana*-s of the Earth's radius. The result will be 51,770 (*asau sakāmaḥ*). Now multiply *dṛkkṣepa* and *dṛggati* by true motion and divide by 51,770. The results, *dṛkkṣepa* and *dṛggati,* can be derived for everyday. Derive the time of *lambana* in this manner and apply it to the syzygy (*parvānta*). Then calculate the *dṛkkṣepa-lagna* and the planet for the required time and (from them) find the *lambana* in time units and apply it to the *parvānta.* In this manner do the *aviśeṣa-karma* (iteration or repetition of the calculation till the results do not vary). Here, only by knowing the correct *lambana*, the *sama-liptā-kāla* (the *parvānta*, which represents the time of equality in minutes of true Sun and Moon) can be ascertained. And, only by knowing the *sama-liptā-kāla* can the *lambana*-minutes be ascertained. Hence, the necessity of *aviśeṣa-karma*.

Since, at this moment, there is no difference in the true longitudes for the Sun and the Moon, there will be no east-west divergence. Their divergence will only be north-south, on account of *nati* and *vikṣepa* . These two have to be ascertained and shall have to be subtracted from the sum of the halves of the orbs (*bimbārdha*-s of the Sun and Moon). The remainder will give the extent of the eclipsed portion (of the orbs).

Now, when the distance between the spheres (*bimba-ghana-madhyāntara* of the Sun and Moon) is equal to half the sum of the orbs (*bimba-yogārdha*), the circumferences of the two orbs will be touching each other. The commencement or end of the eclipse will occur at that time. When, however, the distance between the spheres (*bimbāntara*) is greater, there will be no eclipse, since the circumferences will not touch.

Now, at the desired time apply the *lambana* to the true longitudes of the Sun and the Moon. Find the square of their difference. To it add the square of the true *vikṣepa* and find the root. The result will be the distance between the spheres at that time. Subtract this from the sum of the minutes of half the sum of the two orbs. The remainder will be the extent of the eclipsed portion at that time. This is the method to ascertain the eclipsed portion at any required time.

12.2 Time for a given extent of eclipse

Here is the method to calculate the moment of time when a specified portion (of the orbs) have been eclipsed. Now, the eclipsed portion subtracted from half the sum of the orbs will give the distance between the centres of the two spheres (*bimba-ghana-madhyāntarāla*). This is called *bimbāntara* (difference between the spheres). Using this, the desired time is to be calculated. When the square of true *vikṣepa* is subtracted from the square of the difference of the *bimbāntara,* the root of the remainder will be the difference between the true longitudes (*sphuṭāntara*). Then, calculate the time using the proportion: If 60 *nāḍikā*-s pertain to the difference between daily motions, how many *nāḍikā*-s would it be for the given *sphuṭāntara.* The time got thus is to be applied to the time of the *parvānta.* Calculate the true *vikṣepa* for that time, square it and subtract from the square of half the sum of the orbs. The root thereof would again be the *sphuṭāntara.* In this manner, the result obtained by *aviśeṣa-karma* will be the true time of the required (extent of the) eclipse. Then calculate in this manner, the times, before and after mid-eclipse, which are required for the *bimbāntara* to be equal to half the sum of the orbs, after finding the *nati* and *vikṣepa* by *aviśeṣa-karma.* The results will be the times of the commencement and end of the eclipse.

In computing eclipses, it is necessary to know, first, the actual (moment in) time when the longitudes of the planets are identical. Now, when the Moon is exactly six *rāśi*-s (180 degrees) away from the Sun, it is the end of the full-Moon. When that Moon is hidden by the Earth's shadow it is lunar eclipse.

When at the end of new-Moon, the Moon hides the Sun, then it is solar eclipse. Now, when either of the (two) eclipses (their times, as stated above) occurs near sunset, then calculate the longitudes of the Sun and the Moon for that time. If such times occur at near sunrise, then also calculate the Sun and Moon for that time. There, if the (longitude of) Moon is more, the distance will keep increasing. If the *candra-sphuṭa* is lesser, there will be further and further decrease. Then use the difference in daily motion to find the time of conjunction.

12.3 Computation of *Bimbāntara*

Here is the method for computing the *bimbāntara.* Now, the orbs of the Sun, Moon and the Earth's shadow will appear to be large when they are close to the Earth, and appear to be small when they are far from the Earth. The dimension of their (the Sun and the Moon) orbs is dependent on the magnitude of *sva-bhūmyantara-karṇa* (the distance from the Earth). When they move away from the Earth, they (the orbs) look small. Therefore, when the *bimba* is derived using the *karṇa*, the reverse rule of three is to be employed. Now, the minutes of the orb would be changing every moment. But the *yojana* measure of the orb always remains the same. Now, the rule of three here is: If the minutes of the *sphuṭa-yojana-karṇa* is that of *trijyā*, how much it would be for the *yojana* measure of the orbs. So, the *yojana* measure of the diameters of the orbs of the Sun and the Moon are multiplied by *trijyā* and divided by the *yojana* measure of the *sva-bhūmyantara-karṇa* (the distance between the planet and the Earth). The result will be the diameter of the orb (of the planet) in minutes. Here the division should be made by *dṛkkarṇa* since it is a case of the use of the reverse rule of three. As is well known, the rule is:

> *vyastatrairāśikaphalam icchābhaktapramāṇaphalaghātaḥ.*
>
> [*Brāhmasphuṭa-siddhānta*, *Gaṇita,* 11]
>
> The result of the reverse rule of three is the product of *pramaṇa* and (*pramāṇa*)*-phala* divided by the *icchā*.

12.4 Orb measure of the planets

Now, the orb of the Sun is a large sphere of effulgence. Somewhat much smaller is the sphere of the orb of the Earth. That half of the (Earth's orb) which is facing the Sun will be illuminated. The other half will be dark. And that is the shadow of the Earth. Of this, the base will be large and the tip pointed. Here, since the orb of the Sun is large, the rays that go beyond the Earth's (circumference) will be those emanating from the circumference of the Sun. These rays will converge. At that point will be the tip of the Earth's shadow; its radius at its base will be that of the Earth. From then on, being in the form of a circle (based cone), it tapers to a point. The Sun's rays emanating from its circumference pass over the circumference of the

Earth and would converge to a point on the other side of the Earth. Now, the distance of the Earth from the Sun is equal to the *sva-bhūmyantara-karṇa* in *yojana*-s. For this distance, the rays of the Sun emanating from the circumference of the Sun come up to the Earth according to the Earth's diameter. Thus, for the rays to taper by an amount of measure equal to the difference between the diameters of the Sun and the Earth, the distance required is the above *karṇa* in *yojana* measures. Then, what that distance would be for the tapering by an amount equal to the diameter of the Earth, that would give the length of the shadow. Now, for the shadow of the Earth, for the distance from the tip of the shadow to the base of the shadow (cone), the diameter is equal to the diameter of the Earth in *yojana*-s: then, for the distance from the tip of the shadow to the point where the Moon's path cuts it, what is the diameter of the Earth's shadow at that place. To know this, subtract the *candra-karṇa* from the length of the Earth's shadow, multiply it by the diameter of the Earth and divide by the length of the Earth's shadow. The result will be the *yojana* measure of the diameter of Earth's shadow along the path of the Moon. For this (*yojana* measure) derive the diameter in terms of minutes.

Thus has been stated the method for obtaining the orb-measures of the eclipsed and eclipsing planets. From these (measures of the) orbs of the planets, the times of begining, middle, and that of any desired extent of eclipse can be calculated as explained earlier.

12.5 Direction of the eclipses and their commencement

Here is stated how to know the direction where the eclipse commences and what its configuration (*samsthāna*) would be at any desired time. Now, when the solar eclipse commences, the Moon which is in the west, moves a little towards the east, and a little of the Sun's orb at its circumference in the west will begin to be hidden. It is now intended to identify that portion thereof. Now, the ecliptic is a circle which touches the centre of the Sun's sphere and the centre of the Moon's sphere when there is no *vikṣepa* for the Moon. At that time, it is that portion in the west of the Sun's sphere from where the ecliptic passes through, that will be the portion that gets hidden first by the Moon when it has no *vikṣepa.* The diurnal circle of the Sun at that moment will be touching the centre of the solar sphere. Since that is exactly east-west in places of zero latitude, there the diurnal circle emerges exactly to the west.

12.6 *Āyana-valana*

Since, however, the ecliptic deviates from the diurnal circle, the emergence of the ecliptic will be a little to the north or south of the western direction. Hence the beginning of the eclipse which occurs on the solar orb will be deflected from the west by a certain amount at that time. This deflection is called *Āyana-valana.*

Now, it is necessary to know how much this would be. Conceive of the following set up: Let the winter solstice on the ecliptic touch the north-south circle at the meridian ecliptic point. Let the equinox be at the eastern rising point of the ecliptic and the Sun be one *rāśi* (30 degrees) from the winter solstice in the eastern hemisphere. There, the intersection of the ecliptic and the diurnal circle would be at the centre of the Sun's sphere. The emergence of the diurnal circle would be exactly west thereof and the emergence of the ecliptic would be deflected a little to the south. It is to be known what this divergence is. Now, Rcosines on the ecliptic are the Rversines on the radius which has its tip at the point of intersection of the ecliptic and the north-south circle drawn from the centre of the Earth. Hence the bases of the Rcosines are on that line.

Now, conceive of a Rsine corresponding to the Rcosine (*koṭi-cāpa*) which has its tip at the centre of the Sun and its foot at the point of intersection of the north-south circle and the ecliptic. Then, conceive of a Rcosine having its tip at the point where the ecliptic emerges through from the western side of the Sun's orb. Then the feet of both of them, will touch the diameter which has its tip at the meridian ecliptic point. There, that which has its tip at the centre of the planetary sphere (*bimba-ghana-madhya*) will touch the bottom circumference, and the interstice between the feet of the Rcosines on the diameter, having its tip on the circumference, will touch the top (of the circumference). It is well known that the *koṭi-khaṇḍa* is equal to the distance between the feet of the *koṭi-jyā*-s. Therefore, conceive of a vertical line from the centre of the Earth's sphere and with its tip at the zenith. Now, when the *ayanānta* touches the north-south circle, the maximum distance between the *ayanānta* and the vertical line will be the maximum declination.

Then consider the point where the base of the Rcosine which has its tip at the centre of the planet's sphere touches the *ayanānta-sūtra*. The distance from that point to the vertical line would be equal to the required declination (*iṣṭāpakrama*). Now, the distance from the base of the Rcosine with its tip at the circumference of the orb to the vertical line will be greater than the required declination. This excess will be the declination pertaining to the *bhujā-khaṇḍa*. The *āyana-valana* will be equal to the said (excess) declination associated with the *bhujā-khaṇḍa*.

This declination of the said *bhujā-khaṇḍa* will be the distance between the points of emergence of the points of intersection of the diurnal circle and of the ecliptic on the circumference at the west end of the orb. Here, the *bhujā-khaṇḍa* is to be derived using the Rcosine with its tip at the middle of the arc. The Rcosine with its tip at the centre of the sphere is *sphuṭa-koṭi*. Since the distance between the centre of the sphere and the circumference is the *cāpa-khaṇḍa*, the *bhujā-jyā-khaṇḍa* is derived by multiplying the *sphuṭa-koṭi-cāpa-jyā*–minus–one-fourth-the-orb with the *icchā-rāśi* formed by half the orb which is the full chord, and dividing by the radius. When this is multiplied by maximum declination and divided by the radius, the result is the *āyana-valana*. There, it should be possible to derive the declination of the *koṭi-jyā* by first multiplying the *koṭi-jyā* by the maximum declination and dividing by the radius. In the result, there will be no difference. Thus the derivation of *āyana-valana*.

12.7 *Ākṣa-valana*

In a place having *akṣa* (terrestial latitude), (besides the above) the diurnal circle is also inclined and it is necessary to find this inclination. For this conceive of an east-west circle. The centre of this circle, its circumference and all its parts should be removed from the prime vertical by an amount equal to *chāyā-bhujā* at the desired time. This will be related to the prime vertical even as the *ghaṭikā-maṇḍala* is to the *svāhorātra-vṛtta.* This circle is called *chāyā-koṭi-vṛtta.* It is to be noted that the *chāyā-koṭi-vṛtta, apakrama-vṛtta* and *svāhorātra-vṛtta* cut one another at the centre of the planetary sphere (*bimba-ghana-madhya*). And the circumferences of the three will emerge in three different ways. The *chāyā-koṭi-vṛtta* will go straight westwards from the centre of the solar sphere. The *svāhorātra-vṛtta* will be inclined southwards from this. Therefore, when the Sun is in the eastern side of the north-south circle, the *svāhorātra-vṛtta* will emerge deflected to the south from the west. When, however, the planet is on the western side of the north-south circle, the *svāhorātra-vṛtta* (diurnal circle) is deflected to the north. This removal is called *ākṣa-valana.*

12.8 Combined *valana*

When the two *valana*-s arrived at as above are added together when their directions are similar, and subtracted from each other when their directions are dissimilar, the distance between the *icchā-koṭi-vṛtta* and the ecliptic is obtained. At the periphery of the orb, this will be the *valana* in a place with latitude, when there is no *vikṣepa* . When, however, there is *vikṣepa* there is a shift of it by the measure of the *vikṣepa* and along the direction of the *vikṣepa.* There the *vikṣepa* which had been derived earlier applying the rule of three will be that relating to the *bimbāntara* (difference of the orb). Therefore, multiply that *vikṣepa* by half the diameter of the Sun's orb and divide by *bimbāntara.* The result will be the *vikṣepa-valana* at the circumference of the solar orb. In this way, the point of contact and release will also deflect according to this, on the circumference of the orb. Again, on the eastern side of the circumference of the orb, the directions of the *valana* will be correspondingly reversed. This alone is the speciality here.

Here, since the Sun is being eclipsed, the Sun is called *grāhya-graha* (the planet that is eclipsed). Thus has been stated the derivation of the *āyana-valana*-s by the use of the rule of three. The same principle is applicable to the derivation (also of the) *ākṣa-valana*.

Thus it is seen that the *chāyā-koṭi-vṛtta* and *svāhorātra-vṛtta* meet at the centre of the planetary sphere (*bimba-ghana-madhya*) and have the maximum divergence on the north-south circle. The last will be a section of the latitude pertaining to the *natotkrama-jyā* at that time. *Nata* is the difference between the planet and the north-south circle along the *svāhorātra-vṛtta*. Here, the *nata-jyā* is the Rcosine.

Since the *akṣa* represents the maximum declination, we now have the proportion; If *nata-jyā* multiplied by *akṣa-jyā* and divided by *trijyā* is the *ākṣa-valana* for the *trijyā-vṛtta,* then what would it be for the radius of the orb eclipsed (*grāhya-bimbārdha*); this would give the *valana* for the circumference of the eclipsed orb.

12.9 Graphical chart of the eclipse

Calculate the *valana* for the desired time and the times of commencement and release of the eclipse and draw the eclipsed orb. Further, mark on it the east-west line and the north-south line. Then identify a point removed from the east-west line by a measure equal to the *valana.* Now, construct a *valana*-line passing through the above-said point and the centre of the eclipsed orb. Then draw the orb of the eclipsing planet with its centre on the said *valana*-line at a point which is removed from the centre of the eclipsed body by a distance equal to the difference between the orbs at that time. Then, that portion of the eclipsed orb which falls outside the eclipsing orb would be bright. And, that portion of the eclipsed body which falls inside the eclipsing body would be hidden. The setup of the eclipse has to be understood in this manner. Here, it is not essential to make the *valana* for the eclipsed body. It can as well be made for a circle of desired radius. In that case, care should be taken to move by the needed *valana* from the *dṛk-sūtra* of that circle. This is the speciality. Thus has been stated the computation of the solar eclipse.

12.10 Lunar eclipse

What is of note in the lunar eclipse is that the Moon's orb is the orb that is eclipsed, the Earth's shadow is the eclipser. Here, the (circular) extent of the Earth's shadow along the path of the Moon is called *tamo-bimba* (orb of darkness). Since here, both the eclipsed and the eclipser are at the same distance from the observer, the *nati* and *lambana* are the same for both, and hence both (*nati* and *lambana*) might be ignored in this case (of lunar eclipse). All the other rules are the same here too (as in the solar eclipse). Thus have been stated the procedures for the computation of eclipses.

It is to be noted however, that for the *kendra-bhujā-phala* of both the Sun and the Moon there is a correction called *ahardala-paridhi-sphuṭa* (half day-true-circumference). There will occur difference of true longitude on account of this. And, for that reason, there will occur some difference in the time of equality of the longitudes (of the Sun and the Moon). There is a view (*pakṣa*) that, on account of this, there will be a difference also in the time of the eclipse.

Chapter 13

Vyatīpāta

13.1 *Vyatīpāta*

Next is stated *vyatīpātā*. Now, if the declinations of the two, Sun and Moon, become equal at some time, when one of them is in an odd quadrant with the declination increasing and the other in an even quadrant with declination decreasing, then at that moment *vyatīpāta* is said to occur.

13.2 Derivation of declination

A method of computing the declination of the Sun and the Moon has been stated earlier. Now, another method of computing the declination of the Moon is stated here. Now, when a set up is conceived where there are several circles of equal measure and have a common centre but with their circumferences diverging, it will be that the circumferences of all circles (considering them in pairs) will intersect with all other circles (again considering them in pairs) at two places, and will have maximum divergence at two places. Now, we know where the ecliptic and the celestial equator meet and where they have maximum divergence. Now, if it is known that the ecliptic and the *vikṣepa-vṛtta* meet at this place and that this much is the maximum divergence and that from their point of intersection the Moon has moved this much on the *vikṣepa-vṛtta,* then how far the celestial equator is from the Moon can be computed as in the case of the declination of the Sun.

13.3 *Vikṣepa*

Here is stated a method to know at which place the *ghaṭikā-vṛtta* and the *vikṣepa-vṛtta* meet and what is their maximum divergence. Now, on a particular day towards the middle of the *Mīna,* the meeting of the *ghaṭikā* and *apakrama-vṛtta*-s will occur. From that meeting point, the *apakrama-vṛtta* will diverge northwards. From the same day it will diverge southwards from the middle of *Kanyā.* When it has fully diverged, it would have diverged by 24 degrees. The *vikṣepa-vṛtta* will touch the *apakrama-vṛtta* at the point where *Rāhu* (the ascending node of the Moon) is situated. It will then diverge northwards from the point (*Rāhu*) and from *Ketu* (the descending node of the Moon), it will diverge southwards. Conceive that *Rāhu* is situated at the point of contact of the *apakrama-maṇḍala* and the *ghaṭikā-maṇḍala*, and that this point is rising at the equator. Then, the maximum declination and maximum *vikṣepa* are on the north-south circle. There, from the *ghaṭikā-vṛtta*, the *apakrama-vṛtta,* and from that, the *vikṣepa-vṛtta,* will both diverge in the same direction. For this reason, the *ayanānta-pradeśa* (the solsticial points on Moon's orbit) of the *vikṣepa-vṛtta* is removed from the *ghaṭikā-vṛtta* by the sum of the maximum declination and maximum *vikṣepa.* Hence, that will be the maximum declination of the Moon on that day. Therefore, taking it as the *pramāṇa-phala,* it should be possible to derive the declination of the Moon at that time from the equinox.

This being the case, the northern *rāśi-kūṭa* (pole of the ecliptic) is on the north-south circle, raised from the nothern *Dhruva* (north pole) by a measure equal to the maximum declination. Since the northern *vikṣepa-pārśva* (pole of the *vikṣepa-vṛtta*) is raised above this by the measure of the maximum *vikṣepa,* the distance of the *vikṣepa-pārśva* to the (north) pole is equal to the sum of the maximum *vikṣepa* and the maximum *apakrama.* The relation of the *vikṣepa-vṛtta* to the *vikṣepa-pārśva* is the same as that between the *Dhruva* (north pole) and the *ghaṭikā-vṛtta* and that between the *rāśi-kūṭa* and the *apakrama-vṛtta.* Therefore, the distance between the *Dhruva* and the *vikṣepa-pārśva* will be equal to the maximum divergence between the *ghaṭikā* and *vikṣepa-vṛtta*-s. Now, conceive of a circle touching the *Dhruva* and the *vikṣepa-parśva*. In this circle will occur the maximum divergence between the *ghaṭikā* and *vikṣepa-vṛtta*-s.

Now, the distance between the *Dhruva* and the *vikṣepa-pārśva* has to be computed. Conceive the set-up as above and consider *Rāhu* to be at the *ayanānta* in the middle of the arc. Then the *vikṣepa-vṛtta* would be deflected towards the north by the measure of the maximum *vikṣepa* from the vernal equinox along the *rāśi-kūṭa-vṛtta* which touches the equinox. Therefore, the *vikṣepa-pārśva* would be shifted to the west by the above-said measure from the *uttara-rāśi-kūṭa* (north pole of the ecliptic). Since in this set-up, the maximum *vikṣepa* and maximum declination form the *bhujā* and *koṭi*, the distance between the pole and the *vikṣepa-pārśva* will be the *karṇa*. Consider the circle which passes through the *vikṣepa-pārśva* and the poles. The maximum divergence of *ghaṭika* and *apakrama-vṛtta*-s on this is the *vikṣepāyanānta*. Hence this circle is called *vikṣepāyanānta-vṛtta*. The points of intersection of this with the north-south circle are the poles. Starting from here, as we traverse a measure of the maximum declination, the *vikṣepāyana-vṛtta* would have moved towards the west by the measure of the maximum *vikṣepa*. When we traverse a quadrant it would have inclined towards the west from the north-south circle and will have its maximum divergence in the *ghaṭikā-vṛtta*. Therefore, the *vikṣepāyanānta* would shift to the west from the north-south circle by the above said measure to touch the *ghaṭikā-vṛtta*. Therefore, the *vikṣepa-viṣuvat* would be on the *ghaṭikā-maṇḍala* raised by the above measure from the vernal equinox on the horizon. The reason for what is said above is that the meeting point and maximum divergence between two circles would occur at a distance equal to the quadrant of the circle. This shift is called *vikṣepa-calana*. Now, when this correction (of *vikṣepa-calana*) is applied to the commencing point of *apakrama-viṣuvat* the result will be the commencement of the *vikṣepa-viṣuvat*.

For this reason, when *Rāhu* arrives at the *viṣuvat* in the middle of sign *Kanyā* (Virgo), the *vikṣepa-vṛtta* would have shifted towards the north of the *ayanānta* at the centre of the arc by a measure equal to the maximum declination. The *vikṣepa-pārśva* would have been depressed to that extent from the northern *rāśi-kūṭa.* The distance of separation between the poles at that time would be the maximum declination of the Moon. That will be half a degree less than twenty (i.e., 19.5 degrees). Since the difference between *vikṣepa-pārśva* and *Dhruva* is on the north-south circle, the *vikṣepāyanānta* will lie on the *apakramāyanānta* only. The *vikṣepa-viṣuvat* and the *apakrama-viṣuvat* will be at the same point. At that time there will be no *vikṣepa-calana.*

When, however, *Rāhu* reaches the *ayanānta* at the middle of the *Mithuna-rāśi* (Gemini), the *vikṣepa-vṛtta* will touch the *rāśi-kūṭa-vṛtta* which passes through the middle of *Kanyā-rāśi* northwards at a distance equal to the maximum declination. Hence, the *vikṣepa-pārśva* will be shifted to the east from the northern *rāśi-kūṭa* by a measure equal to the maximum *vikṣepa.* There again, the distance between the *vikṣepa-pārśva* and the *Dhruva* will represent the hypotenuse. The *vikṣepa-pārśva* will be above the northern *rāśi-kūṭa* just as in the case when the *Rāhu* is on the vernal equinox. In this way, the location of southern *vikṣepa-pārśva* will be on the southern *rāśi-kūṭa-vṛtta.* Thus the *vikṣepa-pārśva* is going around, according to the motion of *Rāhu*, at a place which is removed from the *rāśi-kūṭa* by a distance of the maximum *vikṣepa.*

13.4 *Vikṣepa-calana*

Now, conceive of a circle with radius equal to the maximum *vikṣepa.* The centre of this circle should be at a place on the line from the *rāśi-kūṭa* to the centre of the celestial sphere at a distance of the Rversine of the maximum *vikṣepa.* Conceive of another circle with its circumference passing through the centre of the above-mentioned circle and having its centre on the polar axis (*akṣa-daṇḍa*). These two circles will then be mutually like the *kakṣyā-vṛtta* and *ucca-nīca-vṛtta.* Here the ascent of the *kṣepa-pārśva* (*vikṣepa-pārśva*) from the polar axis would represent *trijyā.* Now, note the point where the *kṣepa-pārśva* falls on the *vikṣepa-pārśva* (*vṛtta*); from that point draw a vertical line to its own centre; that line will represent the *koṭi-phala.* The east-west distance on the north-south circle represents the *bhujā-phala*. When the *vikṣepa-pārśva,* which is revolving, happens to be above the *rāśi-kūṭa*, add the *koṭi-phala* to the ascent of the *kṣepa-pārśva* (*kṣepa-pārśvonnati*); if it is below, subtract it. When *Rāhu* happens to be at the middle of *Mīna-rāśi*, the height from the pole will be maximum. And, when *Rāhu* happens to be in the middle of *Kanyā-rāśi*, it will be lowest. Hence it turns out that the updown distance is *koṭi-phala*, and that it is positive in the (six) *rāśi*-s commencing from *Makara-rāśi* and negative in the (six) *rāśi*-s commencing from *Karki*.

At the *ayanānta*, the *bhujā* is full (i.e., 90 degrees). When *Rāhu* is situated there, there will be a east-west shift from the *kṣepa-pārśva*, and the *bhujā-phala* is also east-west. The day when *Rāhu* is situated at the beginning of *Tulā* (Libra) the *kṣepa-pārśva* is to the west of the north-south circle and so the *vikṣepa-calana* is to be added to the beginning of Libra. At the beginning of Aries (*Meṣādi*), the *vikṣepa-pārśva* is to the east of the north- south circle, and hence (the *vikṣepa-calana*) is to be subtracted. Now, multiply the Rsine and Rcosine of *Rāhu* at the beginning of *viṣuvat* by maximum *vikṣepa* and divide by *trijyā*. The results will be the *bhujā-phala* and *koṭi-phala*.

13.5 *Karṇānayana*

Now is given the method to derive the *karṇa* (hypotenuse) from the above. The *karṇa* is the Rsine of the distance between the pole and *vikṣepa-pārśva*, at the time when the above *kṣepāyanānta-vṛtta* passes through the *rāśi-kūṭa*. If the maximum declination and maximum *vikṣepa* have to be added to or subtracted from each other, then mutual multiplication by the Rcosines and division by *trijyā* are necessary.

When the maximum declination and the *koṭi-phala* are to be added to or subtracted from each other, the multipliers would be *antya-kṣepa-koṭi* and *antyāpakrama-koṭi*. Now, the maximum *vikṣepa*, which is the Rsine of a portion of the north-south circle, is the line from the centre of the *kṣepa-pārśva-vṛtta* to the circumference of this (north-south circle). A portion of this (line) is *koṭi-phala*. This is all the difference, and there is no difference in placement. Therefore, in the addition or subtraction there will be no difference in the multipliers; the difference is only in the multiplicands. Now when the maximum declination is multiplied by the Rcosine of the maximum *vikṣepa* and divided by *trijyā*, the result will be the distance from the centre of the *kṣepā-parśva* to the polar axis (*akṣa-daṇḍa*). The *koṭi-phala* will be the remnant of this. When multiplication is made by *paramāpakrama-koṭi* and division by *trijyā*, the result obtained, being the distance from the *koṭi-phalāgrā* to the *vikṣepa-pārṣva*, is the *bhujā-phala*. Then, find the square of the sum or the difference of these two, add it to the square of the *bhujā-phala* and find the square root. The result will be the Rsine of the arc forming the distance between the pole and the *vikṣepa-pārśva*. This is also the same as the maximum declination, which is the maximum divergence between the *ghaṭikā-vṛtta* and the *vikṣepa-vṛtta*.

13.6 Determination of *Vikṣepa-calana*

Now, the maximum divergence between the *kṣepāyanānta-vṛtta* and the north-south circle on the *ghaṭikā-vṛtta* is got by applying the rule of three: If the divergence in the north-south is equal to the *bhujā-phala* when one moves along the *kṣepayananta-vṛtta* from the pole to the *kṣepa-pārśva*, then what will it be if one moves through quarter of a circle. This will be the distance between the two *ayanānta*-s, since this is the distance between the *viṣuvat*-s. This is called *vikṣepa-calana*. When *vikṣepa-calana* is applied to *sāyana-candra* (i.e., Moon to which *āyana-calana* has been applied) the result will be the distance between the point of contact of the *ghaṭikā-vṛtta* and *vikṣepa-vṛtta* to the Moon, on the *vikṣepa-vṛtta*.

13.7 Time of *Vyatīpāta*

Now, the auspicious time of *vyatīpāta* occurs when the declination of the Moon to which *vikṣepa-calana* and *ayana-calana* have been applied, and that of the Sun to which *ayana-calana* has been applied, become identical, when one of them is in an odd quadrant and the other in an even quadrant.

13.8 Derivation of *Vyatīpāta*

Now is explained the procedure for finding the time at which the declinations of the two become equal. First estimate an approximate time when there is equality of the longitudes (*bhujā-sāmya*) for the Sun and the Moon when one is in an odd quadrant and the other in an even quadrant. Using the *bhujā-jyā* of the Sun find out its declination at that time. Then, using the rule of three, ascertain what the *bhujā-jyā* of the Moon should be, for it to have the same declination as the Sun. Now, the maximum declination of the sun is 1398 ('*dugdhaloka*'). Here, the rule of three would be as follows: If this (i.e., 1398) is the *bhujā-jyā* for the Sun, then what would be the *bhujā-jyā* for the Moon which has a given maximum declination at the moment, to become equal in declination to the Sun. This is the rule of three to be applied. Here, the Sun's maximum declination is the *pramāṇa*, its *bhujā-jyā* is *pramāṇa-phala*, the Moon's *antyāpakrama* is the *icchā* and the Moon's *bhujā-jyā* is *icchā-phala.*

Now, if the *antyāpakrama* is large, the *bhujā-jyā* will be small; for small *antyāpakrama*, the other will be big. Then, at that time the declinations would become equal. Hence the inverse rule of three should be applied. For this, multiply the *bhujā-jyā* and *antyāpakrama* of the Sun and divide by the *antyāpakrama* of the Moon. The result will be the Moon's *bhujā-jyā.* Compute its arc and apply it to the *ayana-sandhi* or *gola-sandhi* according to the quadrant and compute the Moon.

Then subtract the Moon computed (as above) from the Sun, and the Moon which has been computed (independently) for the given time. Place the result in two places and multiply by the daily motions of the Sun and the Moon, respectively, and divide by the sum of the daily motions. This correction is to be applied to the two separately. It has to be subtracted if the *vyatīpāta* is past and to be added if the *vyatīpāta* is yet to occur. In the case of the node, the application should be made the other way. In this way do *aviśeṣa-karma* (repeating the process till results do not vary) till the Moon's longitude-arc (*bhujā-dhanus*) derived from the Sun and that of the Moon computed for the desired time become equal. There, in the odd quadrants, if the longitude-arc of the Moon calculated for the desired time is larger, the *vyatīpāta* has already occured; if it is smaller, then the *vyatīpāta* is yet to occur. In the even quadrants it is the other way round. Here, when, for the Sun and the Moon, and for the Earth's shadow and the Moon, the diurnal circle is the same, *vyatīpāta* occurs. When however, even if parts of the orbs do not have identical diurnal circles, there will be no *vyatīpāta*. Hence a *vyatīpāta* will last for about four *nāḍika*-s.

Chapter 14

Mauḍhya and Visibility Corrections of Planets

14.1 Computation of visibility correction

Next is stated *darśana-saṃskāra*. This is indicated by that part of the ecliptic which touches the horizon when a planet having *vikṣepa* rises above the horizon. Consider a set up in which the northern *rāśi-kūṭa* is raised and the planet is in one of the first three *rāśi*-s beginning from *Meṣa*; let the point of contact of the ecliptic and the *rāśi-kūṭa-vṛtta* passing through the planet be rising on the horizon. Further, suppose that the planet has *vikṣepa* towards the northern *rāśi-kūṭa*. Then, the planet will be raised above the horizon. Therefore, the gnomon of the planet at that time is computed first. When this gnomon is taken as Rcosine, its hypotenuse will be the distance between the planet and the horizon on the *vikṣepa-koṭi-vṛtta*. Now, the *dṛkkṣepa-vṛtta* meets the *apakrama-vṛtta* towards the south at a distance equal to the distance between the zenith and the *dṛkkṣepa*. In the *dṛkkṣpepa-vṛtta* itself, at a place north of the horizon, at a height equal to the distance between the horizon and the *dṛkkṣepa*, is the northern *rāśi-kūṭa*. The northern *vikṣepa* is that which moves towards the northern *rāśi-kūṭa*. Applying the rule of three: If the maximum distance between the horizon and the *rāśi-kūṭa* (*vṛtta*) touching the planet is the *dṛkkṣepa*, how much will be the distance from the horizon to the planet with *vikṣepa*; the result would give the gnomon of the planet with *vikṣepa*. Then, the proportion: If for the *dṛkkṣepa-koṭi* the hypotenuse is *trijyā*, then what will be the hypotenuse for this gnomon, will give as result the portion of the arc of the ecliptic between the planet and the horizon. In the same manner, the portion of the *vikṣepa-koṭi-vṛtta* for the distance between the horizon and the planet with latitude is also obtained.

14.2 Rising and setting of planets

Now, conceive of a *rāśi-kūṭa-vṛtta* touching the meeting point of *vikṣepa-koṭi-vṛtta* and the horizon. This circle will intersect the *rāśi-kūṭa* circle passing through the planet and the *rāśi-kūṭa*. The planet is situated at a distance of the *vikṣepa-koṭi* from the said point of contact. At that place, the divergence between the two *rāśi-kūṭa-vṛtta*-s is equal to the hypotenuse of the *śaṅku* of a planet with *vikṣepa* which has been obtained. In this set up, the maximum divergence between the two *rāśi-kūṭa-vṛtta*-s would be on the ecliptic. Here, when the true planet is the *lagna*, for the reason that the planet would be raised by that number of minutes at the time, the distance between the true planet and the *lagna* when the planet rises will be the maximum distance between *rāśi-kūṭa-vṛtta*-s. Since the rising has taken place earlier here, this difference is subtracted from the true planet to get the *lagna* at the time of the rising of the planet. This is the case when the *vikṣepa* is north.

In the case of the south *vikṣepa* when the same set up is conceived, the planet will be below the horizon, since due to latitude it is deflected from the point of contact of the horizon and the ecliptic, above and southwards on the *rāśi-kūṭa-vṛtta*. When this is the case, just as the rising and setting *lagna* were directed earlier to be computed using the downward gnomon (*adho-mukha-śaṅku*), (working in the same manner), the minutes of the distance between the true planet and the *lagna* at the time of the rising of the planet which is at the tip of its *vikṣepa* would be obtained. Since the planet will rise only after that much time, these minutes of the difference should be added to the true planet to derive the *lagna* at the rising of the planet.

In the same manner, derive the setting *lagna* at the time of the setting of the planet. Now, if it is the downward gnomon, the planet will set earlier, if it is the upward gnomon, the planet will set later than the setting *lagna* of the true planet. Hence, there is an inversion in the addition and subtraction. This is all the difference (for this case).

When the southern *rāśi-kūṭa* is raised from the horizon, the planet will be raised when the *vikṣepa* is to the south, and will be lowered when the *vikṣepa* is to the north. Hence, in this case, the nature of addition and subtraction will be opposite to that stated when the northern *rāśi-kūṭa* is raised. This is the only difference (for this case).

Now, when the *dṛkkṣepa* is south, the northern *rāśi-kūṭa* would be raised, and, when it is north, the southern (*rāśi-kūṭa* will be raised). Hence, if the direction of the *vikṣepa* and the *dṛkkṣepa* happen to be the same, the *darśana-saṃskāra-phala* should be added to the planet when it rises. If the directions are different, it is to be subtracted. At setting, (all this) is in the reverse.

14.3 Planetary visibility

Now, we find difference in the *kāla-lagna*-s corresponding to the rising of the planet and of the Sun in minutes. It is (empirically) found that there is a critical value exceeding which the planet would be visible and below which the planet is not visible. The position and the rising of the planet based on this will be stated later. The method for obtaining the *madhya-lagna* of the planet with *vikṣepa* at the noon is also similar. Here, the difference is that computations have to be done with that *dṛkkṣepa* which is derived without taking the latitude into account. The reason for this is that the north-south circle is the same for both places with latitude and without latitude. Thus has been stated the visibility correction.

Chapter 15

Elevation of the Moon's Cusps

15.1 The second true hypotenuse of the Sun and the Moon

Now is stated the (computation of) the elevation of the cusps of the Moon. For this, first compute the second true hypotenuse (*dvitīya-sphuṭa-karṇa*) of the Sun and the Moon. Apply also the second true correction (*dvitīya-sphuṭa-saṃskāra*) for the Moon. Here, the view of (Śrīpati, author of) *Siddhāntaśekhara* is that when the radius of the *ucca* and *nīca* circles have been ascertained, a correction has to be applied to them. The view of Muñjāla, author of *Laghumānasa* is that the *antya-phala* of the Moon is to be multiplied by Moon's *manda-karṇa* and five and divided by *trijyā*. These two views are worth consideration. Then, (for the Moon), compute the *dṛkkarṇa* and apply the corrections of *bhū-pṛṣṭha* and *nati*. Then compute the *nati* for the Sun. Compute and apply the correction of *lambana* for both the Sun and the Moon. Ascertain also the distance, at the required time, between the centres of the solar and lunar spheres.

15.2 Distance between the orbs of the Sun and Moon

Now, at a time when there is no *nati* or *vikṣepa*, compute the Rsine and Rversine of the difference in the *sphuṭa*-s; square them, add them together and derive the root of the sum. The result will be the *samasta-jyā* (complete chord of the arc) on the circle which has the observer as the centre and whose circumference passes through (the centres of) the two orbs.

Here, in the matter of ascertaining the distance between the two orbs: For the sake of convenience, conceive the ecliptic as the prime vertical of the observer, touching the zenith and lying east-west. Conceive of the Sun at the zenith. Conceive the *rāśi-kūṭa-vṛtta* passing through the Sun as the north-south circle. A little away, place the Moon and passing through the Moon conceive of a *rāśi-kūṭa-vṛtta.* Conceive also of two lines from the centre of the circle, one passing through the Sun, and the other passing through the Moon. It will be seen that the line drawn through the Sun is vertical and that passing through the Moon will be a little inclined to it. Here, consider that (segment) which has its tip at the meeting place of the *candra-rāśi-kūṭa-vṛtta* and *apakrama-vṛtta* and the foot on the vertical line. This would be the *bhujā-jyā*, the half chord of the part of the arc on the *apakrama-vṛtta* cut-off by the two *rāśi-kūṭa-vṛtta*-s. The Rversine (*śara*) would be the distance from the foot of the above to the location on the vertical circle where the Sun is situated. The root of the sum of the squares of these two is the full chord of the distance between the two orbs. When this is halved and the arc thereof is doubled, the result will be the arc of the difference between the two orbs, when there is no *nati* or *vikṣepa.*

When, however, the Moon has a *vikṣepa* on the *rāśi-kūṭa-vṛtta*, then the base of the *vikṣepa-jyā* will meet the *candra-sūtra* at a point lower to the Moon by the measure of the *vikṣepa-śara.* Then apply the rule of three: If the, *bhujā-jyā* is the difference between the tip of the *candra-sūtra* and the vertical line, then what would be the distance between the base of the *vikṣepa-jyā* and the vertical line. This rule of three would be: If the *bhujā-jyā* is the distance between the tip of the *candra-sūtra* to the vertical line, then what will it be for a distance less by the Rversine of the *vikṣepa.* Or, one might do the rule of three using the Rversine of the *vikṣepa* and subtract the result from the *bhujā-jyā.* Now, the *bhujā-jyā*-s derived by subtracting the square of the *vikṣepa-śara-phala* from the square of the *kṣepa-śara* and finding the root would be equal to the vertical distance between the *bhujā-jyā* which touches the tip of the *candra-sūtra* and the *bhujā-jyā* which touches the foot of the *vikṣepa*. Add this distance to the Rversine of the difference between the true positions. The result will be the distance between the Sun and the foot of the *bhujā-jyā* which touches the foot of the *vikṣepa*.

However, now, the *śara* would be a little longer, and the *bhujā-jyā* would be a bit shorter. The root of the sums of the squares of these two will be the line from the Sun to the foot of the *vikṣepa-jyā*. If to the square of this, the square of the *vikṣepa* is added and the root of the sum found, it will be the full chord of the difference between the orbs. When there is *nati* for the Sun, then assume that it has deflected from the zenith along the north-south circle. There, from the *śara* of the difference of the true longitudes, the *nati-śara* of the Sun has to be subtracted. The remainder would be the portion of the vertical line between the foot of the *nati-jyā* of the Sun and the foot of the *bhujā-jyā*–less–*vikṣepa-śara*. This would also be the *śara* of the difference between the true positions less the *nati-śara* of the Sun plus the *koṭi-phala* of the Moon's *kṣepa-śara*. This would be one *rāśi* (the first quantity). The *bhujā-jyā* of the difference between the true positions will also be one *rāśi*.

If the Sun and the Moon move on the same side of the ecliptic the difference in their *nati*-s is to be taken and if (they move) on the two sides, the sum of the *nati*-s should be taken. This will be one *rāśi*. The only distinction is that, here, the *bhujā-jyā* and *śara* of the difference in the *sphuṭa*-s should be conceived straight from that planet which has the smaller *nati*. By adding up the squares of all these three (quantities) and finding the root of the sum, the full chord of the difference in the orbs would be obtaind. This is the case when the difference of the true planets is less than three *rāśi*-s.

When it is more (than three *rāśi*-s) also, the ecliptic is to be conceived as follows: The Sun and the Moon are to be conceived on the two sides of the zenith, equally removed from it, since at that time there is no *nati* for both. The distance between the orbs would be double the half-*jyā* of half the difference between the true longitudes when there is no *nati* for both.

Now, the Rsines of half the difference in the true longitudes are the distances from the points of contact of the *rāśi-kūṭa-vṛtta* and the ecliptic to the vertical line. Here, subtract the *bhujā-jyā-phala* derived from the respective *nati-śara*-s from the respective halves. The results would be the respective distances from the foot of the *nati-jyā* to the vertical line. Here, too, they would have touched the vertical line along its verticality according to the magnitude of the *nati*.

Now, calculate the distance between the feet of the *bhujā-jyā* in the vertical line. That will be the *koṭi-phala* of the *śara* of the *nati*, which is, the vertical length of this *śara*. But the difference between the *koṭi-phala*-s of the two *śara*-s is the vertical distance between the feet of the two *bhujā-jyā*-s. This is a *rāśi*. When the *bhujā-phala* of the respective *nati-śara*-s are subtracted from the *jyā*-s of half the difference of the true longitudes: the remainders would be the *bhujā-jyā*-s of the difference of the true longitudes. The sum of these two is the second *rāśi*. The difference between the *nati*-s or their sums forms the third *rāśi*. The root of the sum of squares of these is the full chord of the difference between the orbs. The sum or difference of the *nati*-s is the north-south distance between the Sun and the Moon. The *nati-phala* subtracted from the sum of the *antarārdha-jyā*-s will be the distance in the east-west. The sum of half of the Rcosines of the *nati* and *śara* is the up-down difference. The root of the sum of the squares of these three is the full chord of the difference between the orbs. Thus has been stated the difference between the orbs when the difference between the true longitudes is more than three *rāśi*-s. The same procedure will apply also for the derivation of the difference of the orbs in computation of eclipses.

Gaṇita-Yukti-Bhāṣā

Explanatory Notes

Chapters 8–15

Chapter 8

Computation of Planets

8.1 Planetary motion

In Indian astronomical texts, as a first approximation, the planets are taken to move uniformly along different circular orbits; the linear velocity of all the planets is taken to be a constant. In other words, if R_p be the radius of the planetary orbit (usually given in *yojanā*-s), and T_p be the sidereal time period, then

$$\frac{R_p}{T_p} = C, \tag{8.1}$$

where C is a constant. Given C, the radius of the planetary orbit is determined, if the time period of a planet is known. The term *yuga-bhagaṇa* refers to the number of complete revolutions made by the planet in a *catur-yuga* consisting of 43,20,000 years. This period is also called a *Mahā-yuga* and consists of four parts namely *Kṛta-yuga*, *Tretā-yuga*, *Dvāpara-yuga* and *Kali-yuga*.

The centre of the planetary orbit is not the centre of the Earth. As seen by an observer on the surface of the Earth, there are two types of motion for the planets: (i) the proper motion, which is eastward due to the motion of the planet in its own orbit with respect to the stars, and (ii) the diurnal motion, the uniform westward motion of all celestial objects, as seen from the Earth. The proper motion is discussed in this chapter, whereas the diurnal motion is considered in a later chapter[1]. The 'true planet' should be computed with respect to the observer with the first point of Aries (*Meṣādi*) as the reference point.

[1]Chapter 11 primarily deals with the diurnal problems.

8.2 Zodiacal celestial sphere

The terms *bha* and *gola* mean stars and sphere respectively. Hence, *bhagola* refers to the sphere dotted with stars. In modern terminology, it is called the celestial sphere. At this stage, the centre of the zodiacal celestial sphere is stated to be the centre of the Earth. Any finer distinction will be dealt with later.

Following this, two different conceptions are proposed for perceiving the motion of the planets. In modern terminology they are known as the eccentric and the epicycle models.

8.3 Motion of planets: Eccentric model

To start with, the computation of true positions of the Sun and Moon, which involve just the *manda-saṃskāra* (equation of centre) is discussed. In Figure 8.1a, the planet at P is conceived to be moving on an eccentric circle (*pratimaṇḍala*). The centre of the *pratimaṇḍala* is O', and the centre of the zodiacal sphere (*bhagola-madhya*) is O. The point O' is located from O along the direction of the *mandocca*, which is the apogee (for Sun and Moon) or aphelion (for other planets) in modern terminology. O' is moving on a circle called *manda-vṛtta* which is a small circle centred around O.[2]

It is further conceived that as O' moves on the circle around, it carries the *pratimaṇḍala* along with it also. In other words, even if the planet does not move on its own on the *pratimaṇḍala* it has motion with respect to the *bhagola-madhya* due to the motion of the *mandocca*. As the Text notes, this is like the motion of persons travelling in a vehicle.

In Figure 8.1a, Γ represents the direction of the fixed star which is taken as the reference point for the measurement of the *nirayaṇa* longitude of the planet. O and O' are the centres of the *manda-vṛtta* and *pratimaṇḍala* respectively. The two lines $NO'S$ and $EO'W$ passing through O', and per-

[2]The word *vṛtta* means circle. With the adjective *manda* added to it, the word *manda-vṛtta* suggests that this circle plays a key role in the *manda-saṃskāra*. The same circle is also called *manda-nīcocca-vṛtta* for reasons explained in the next section.

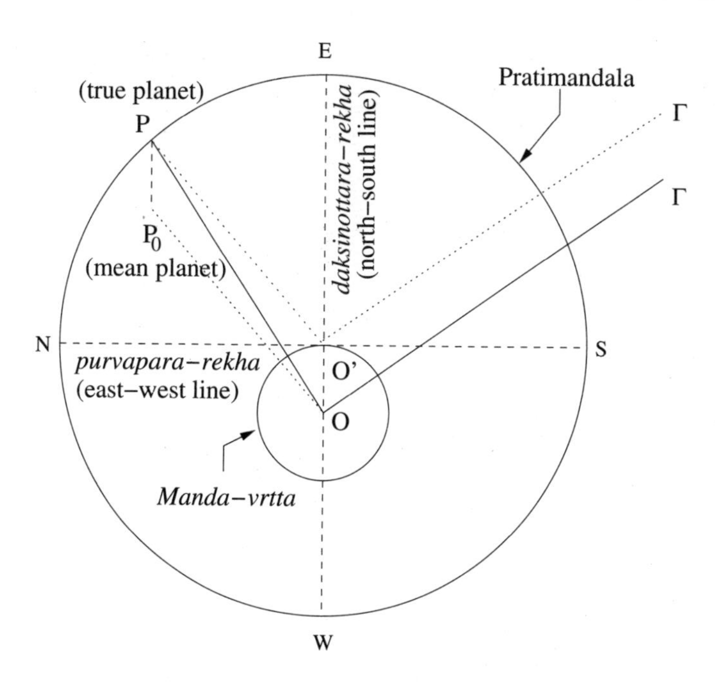

Figure 8.1*a*: The eccentric model of planetary motion.

pendicular to each other, represent the *dakṣiṇottara-rekhā* (north-south line) and *pūrvāpara-rekhā* (east-west line). It is further mentioned that even as O' moves at the rate of *mandocca*, the directions of the east-west line and north-south line remain unchanged. In the figure,

$$\begin{aligned} \Gamma\hat{O}O' &= \text{longitude of } mandocca, \\ \Gamma\hat{O}'P &= \Gamma\hat{O}P_0 = \text{longitude of } madhyama\text{-}graha, \\ \text{and} \quad \Gamma\hat{O}P &= \text{longitude of } sphuṭa\text{-}graha \text{ (the true planet).} \qquad (8.2) \end{aligned}$$

8.4 Motion of planets: Epicyclic model

As suggested by the title, this model explains the irregularities in the planetary motion by considering an epicycle instead of an eccentric circle discussed in the previous section. Apart from explaining the epicyclic model, the Text also establishes the equivalence of the two models.

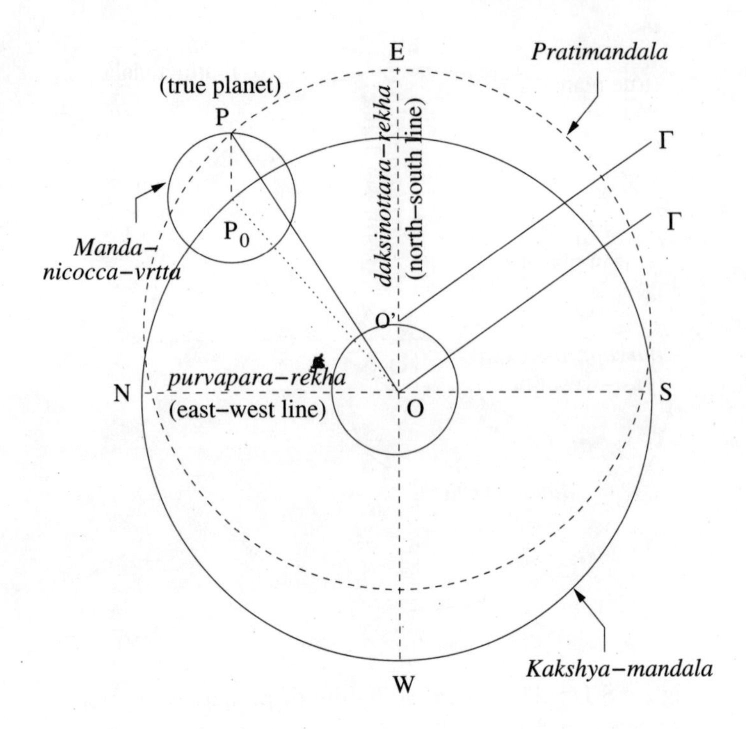

Figure 8.1*b*: The epicyclic model of planetary motion.

In Figure 8.1b, the deferent circle called *kakṣyā-vṛtta* is the circle centred around O, with a radius equal to the radius of the *pratimaṇḍala* described earlier. The mean planet P_0 moves on this circle with mean uniform velocity. Around P_0 we draw a circle whose radius is the same as the radius of the *manda-vṛtta* described earlier. Here it is called the *manda-nīcocca-vṛtta.*[3] At any given instant of time, the actual planet P is located on the *manda-nīcocca-vṛtta* by drawing a line from P_0 along the direction of *mandocca.* The point of intersection of this line with the *manda-nīcocca-vṛtta* is the true position of the planet. In fact, it can be easily seen that this point happens to be the point of intersection of the *pratimaṇḍala* and the *manda-nīcocca-vṛtta* centered around the mean planet on the *kakṣyā-vṛtta.* Thus we see the equivalence of the two models.

[3]The adjective *nīcocca* is given to this *vṛtta* because, in this conception, it moves from *ucca* to *nīca* on the deferent circle along with the mean planet P_0. The other adjective *manda* is to suggest that this circle plays a crucial role in the explanation of the *manda-saṃskāra.*

8.5 The position of *Ucca*

The term *ucca* or *tuṅga* means 'peak'. With reference to planetary motion, this refers to the direction of apogee/aphelion of the planet. This is because it is along this direction that the distance of the true planet from the centre of the *kakṣyā-maṇḍala* becomes maximum.

The direction of *ucca* varies from planet to planet. It may be noted that the true planet P is always at the intersection of *manda-nīcocca-vṛtta* and *pratimaṇḍala*, in fact at that intersection which is close to the *ucca* or in the *ucca* region. The portion above the north-south line of the *pratimaṇḍala* (see Figure 8.2) is the *ucca* region.

8.6 *Ucca*, *Madhyama* and *Sphuṭa*

When the *ucca* and *madhya* coincide, that is, the longitude of *madhya* is the same as that of the *ucca*, the centres of *kakṣyā-maṇḍala*, *pratimaṇḍala*, and the two *ucca-nīca-vṛtta*-s are on the same straight line, namely *pūrvāpara-rekhā* (east-west line). This is depicted in Figure 8.2.

Then the *sphuṭa-graha* (true planet) is the same as the *madhyama-graha* (mean planet). $\Gamma\hat{O}P = \Gamma\hat{O}P_0$. When the *madhya* moves away from the *ucca*, the true planet begins to differ from the mean planet.

8.7 Computation of true Sun

In the case of the Sun, it is noted that the *mandocca* moves so slowly (actually a few seconds of arc per century) that its motion can be neglected. The Text then gives a detailed description of how to find the difference between the true planet (*sphuṭa*) and the mean planet (*madhyama*).

In Figure 8.3, when the *madhyama* is at the east point E or the west point W, the true planet is at E' or W' and there is no difference between *madhyama* and *sphuṭa* (mean and the true longitudes). When the mean planet is at

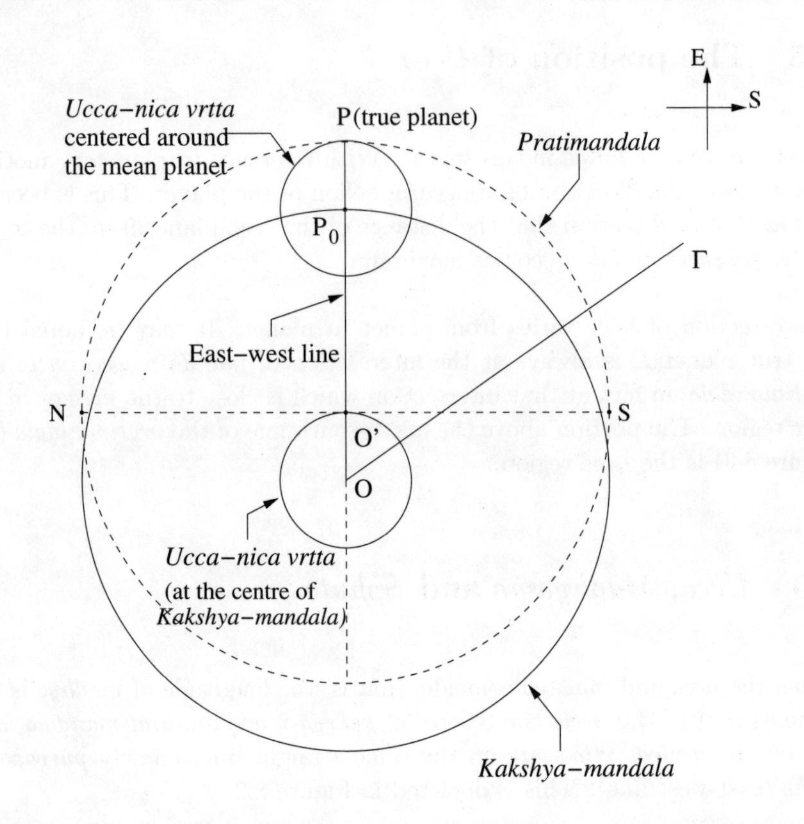

Figure 8.2: The four circles when the *madhya* coincides with the *ucca*.

N, the north point of the *kakṣyā-vṛtta*, the true planet is at N', the north point of *pratimaṇḍala*. Draw a circle with *bhūmadhya* O as the centre, with $ON' = K_N$ as the radius. This is the *karṇa-vṛtta* at this point. The arc $N'N'' = \Delta\theta_N$ on the *karṇa-vṛtta* is the difference between the *sphuṭa* N' and the *madhyama* N. Clearly, at this point,

$$K_N \sin \Delta\theta_N = r, \tag{8.3}$$

where r is the radius of the *ucca-nīca-vṛtta* (epicycle). The *sphuṭa* is less than the *madhyama* in this position. Similarly, when the planet is at S', the difference between *sphuṭa* and *madhyama* is given by the same relation. However, the *sphuṭa* will now be more than *madhyama*. When the planet is at E' and W', the *sphuṭa* and *madhyama* coincide. The procedure to find the radius of the *karṇa-vṛtta* is given in the next section.

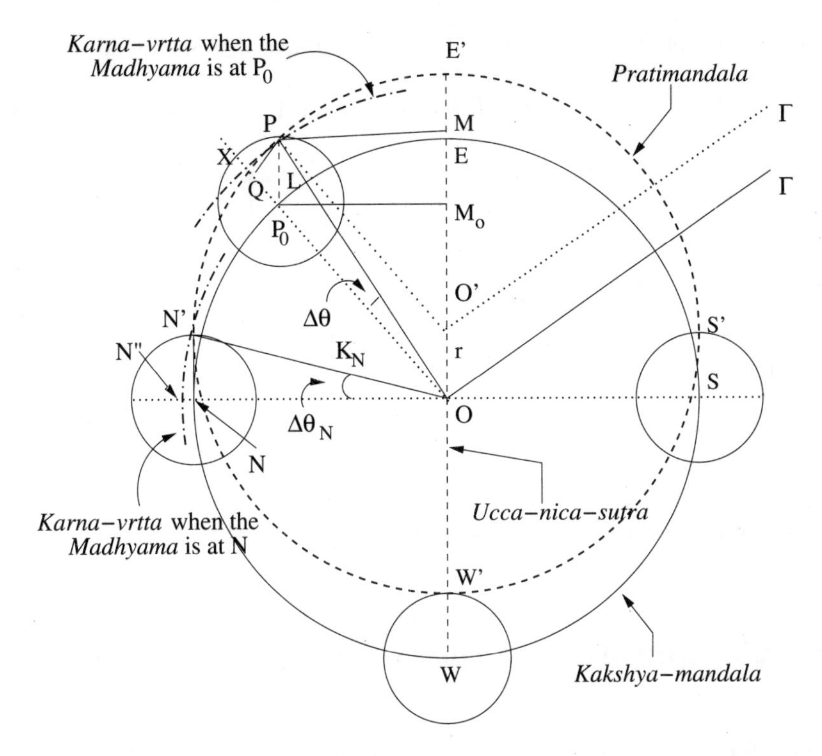

Figure 8.3: The *karṇa-vṛtta* of the planet.

For an arbitrary position P_0 of the mean planet on the *kakṣyā-vṛtta*,, the true planet is at P as shown in Figure 8.3. The line joining the centre of the *kakṣyā-maṇḍala* and the mean planet, OP_0 when extended cuts the *ucca-nīca-vṛtta* at X. Then,

$$\begin{aligned} E\hat{O}P_0 = E'\hat{O}'P &= \textit{madhyama} - \textit{ucca} \\ &= P\hat{P_0}X \\ &= A, \end{aligned} \tag{8.4}$$

where A is what is called the anomaly in modern astronomy. Now, from the figure it may be seen that

$$\begin{aligned} \Delta\theta &= \textit{madhyama} - \textit{sphuṭa} \\ &= \Gamma\hat{O}P_0 - \Gamma\hat{O}P \\ &= P\hat{O}P_0. \end{aligned} \tag{8.5}$$

Draw the perpendicular PQ from P to OX. Then, it is seen that

$$PQ = K \sin \Delta\theta = r \sin A, \tag{8.6}$$

where $K = OP$ is the radius of the *karṇa-vṛtta* when *madhya* is at P_0. Similarly draw the perpendiculars P_0M_0 and PM from P_0 and P on OE. Then, it is seen that

$$PM = K \sin(P\hat{O}E) = P_0M_0 = R \sin(P_0\hat{O}E). \tag{8.7}$$

In other words,

$$K \sin(\textit{sphuṭa} - \textit{ucca}) = R \sin(\textit{madhyama} - \textit{ucca}). \tag{8.8}$$

Both these prescriptions (8.6) and (8.8) are given in the text.

8.8 Computation of the *Karṇa*

Karṇa refers to the hypotenuse drawn from the centre of the *kakṣyā-maṇḍala* to the planet on the *pratimaṇḍala* (OP, in Figure 8.3). Let the radius of the *kakṣyā-vṛtta* (which is also the radius of the *pratimaṇḍala*) be R, and the radius of the *ucca-nīca-vṛtta* be r. The radius of the *karṇa-vṛtta* denoted by K is to be determined. In Figure 8.4, OP_i $(i = 1 \ldots 4)$ are the radii of the *karṇa-vṛtta*-s corresponding to the positions of the planet at P_i. From the planet at P_i $(i = 1 \ldots 4)$ on the *pratimaṇḍala*, we drop the perpendicular P_iB_i on the *ucca-nīca-sūtra*. Measuring with respect to O', $\Gamma\hat{O}'P_i$ represent the longitude of the *madhya* (M) and $\Gamma\hat{O}'B_1$ that of the *ucca* (U). Then,

$$\begin{aligned} P_i\hat{O}'B_i &= \textit{madhya} \sim \textit{ucca} \\ &= M \sim U, \end{aligned} \tag{8.9}$$

when the planet is to the east (upper portion) of the north-south line of the *pratimaṇḍala*.

When the planet is to the west (lower portion) of the north-south line of the *pratimaṇḍala*, then

$$P_i\hat{O}'B_i = 180° - (M \sim U). \tag{8.10}$$

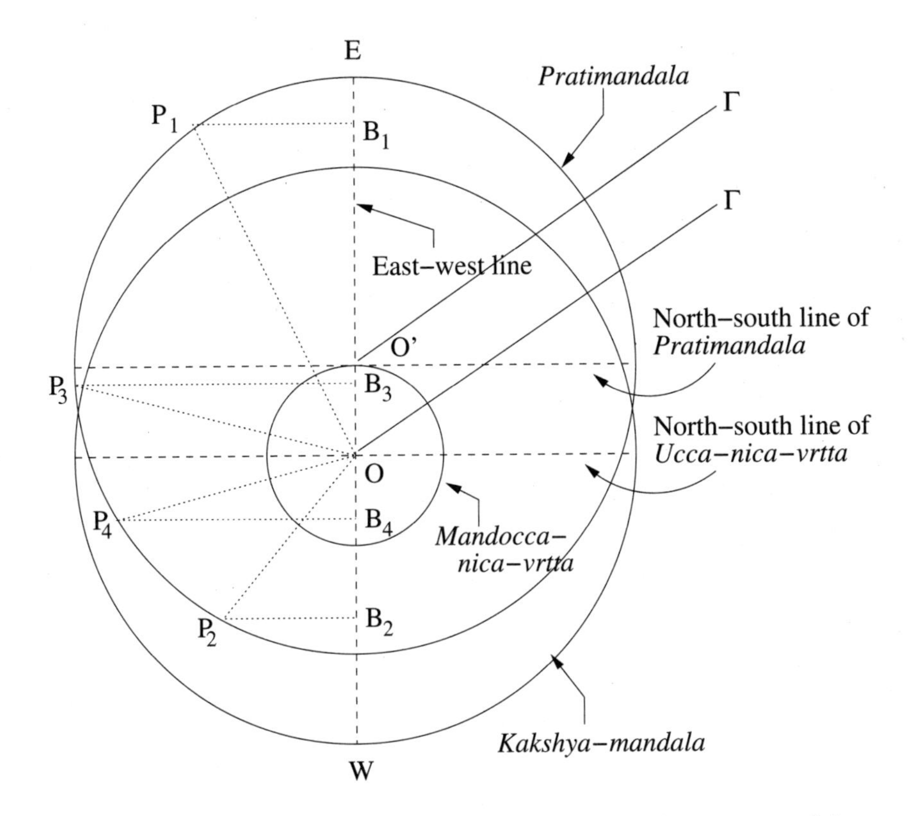

Figure 8.4: The planet P at different positions on the *pratimaṇḍala*.

Here $M \sim U$ represents the magnitude of difference between M and U. The sine and cosine of these angles called *bhujājyā* and *koṭijyā* are to be found for deriving the radius of the *karṇa-vṛtta*. The sines are given by

$$P_i B_i = R \sin(M \sim U), \tag{8.11}$$

($i = 1 \ldots 4$), as $O'P_i = R$. The cosine of the hypotenuse, $B_i O$, called the *karṇa-vṛtta-koṭi* (written as K_k henceforth) is determined thus:

1. When the planet is at P_1, above (to the east of) the *mandocca-nīca-vṛtta* (called simply as *ucca-nīca-vṛtta* for convenience hereafter),

$$\begin{aligned} K_k &= B_1 O \\ &= B_1 O' + O'O \\ &= R\cos(M - U) + r. \end{aligned} \tag{8.12}$$

2. When the planet is at P_2, below the *ucca-nīca-vṛtta*,

$$\begin{aligned} K_k &= B_2O \\ &= B_2O' - O'O \\ &= R\cos(P_2\hat{O}'B_2) - r \\ &= |R\cos(M-U)| - r. \end{aligned} \tag{8.13}$$

3. When the planet is at P_3, such that B_3, the base of the Rsine, is within the *ucca-nīca-vṛtta* and above its north-south line,

$$\begin{aligned} K_k &= B_3O \\ &= OO' - B_3O' \\ &= r - R\cos(P_3\hat{O}'B_3) \\ &= r - |R\cos(M-U)|. \end{aligned} \tag{8.14}$$

4. When the planet is at P_4, such that B_4 is within the *ucca-nīca-vṛtta* and below its north-south line,

$$\begin{aligned} K_k &= B_4O \\ &= B_4O' - O'O \\ &= R\cos(P_4\hat{O}'B_4) - r \\ &= |R\cos(M-U)| - r. \end{aligned} \tag{8.15}$$

Now, the radius of the hypotenuse circle (*karṇa-vṛtta*) K is given by

$$K = \sqrt{(bhuj\bar{a}jy\bar{a})^2 + (karṇa\text{-}vṛtta\text{-}koṭi)^2}.$$

Note: All the four cases above can be expressed in terms of a single formula by taking the signs of sine and cosine into account, and denoting the radius of the *pratimaṇḍala* by R, as follows:

$$\begin{aligned} K &= \sqrt{R^2\sin^2(M-U) + [R\cos(M-U)+r]^2} \\ &= \sqrt{R^2 + r^2 + 2rR\cos(M-U)}. \end{aligned} \tag{8.16}$$

The above expression is valid even when the planet P is to the south, that is to the right of *ucca-nīca-sūtra* in Figure 8.4.

At this point, the Text draws attention to the important feature of the *manda*-correction that the dimension of the *manda-nīcocca-vṛtta* r is assumed to increase and decrease in the same manner as *manda-karṇa* K. The mean radius of the *manda-nīcocca-vṛtta* r_o tabulated in texts corresponds to the radius R of the *pratimaṇḍala* (or the *kakṣyā-maṇḍala*), usually taken to be $3438'$. However, the mean and the actual radii are related by

$$\frac{r}{K} = \frac{r_0}{R} = C, \tag{8.17}$$

where C is a constant.[4] This will ensure that while calculating the *manda*-correction by using (8.6), we need to know only the mean values of the radius of the epicycle r_0 and the radius of the *pratimaṇḍala* $R = 3438'$, because

$$\sin \Delta\theta = \frac{r}{K}\sin(M - U) = \frac{r_0}{R}\sin(M - U). \tag{8.18}$$

Note: To obtain the actual values of r or K in terms of the minutes of arc of the *pratimaṇḍala*, usually a process of iteration *aviśeṣa-karma* is employed which is outlined in all standard texts of Indian Astronomy starting from *Mahābhāskarīya* (629 AD) to *Tantrasaṅgraha* (1500 AD).[5] Here, we shall briefly summarise this process of iteration and refer the reader to the detailed discussion in K. S. Shukla's translation of *Mahābhāskarīya*[6] for further details.

In Figure 8.5, P_0 is the mean planet moving in the *kakṣyā-maṇḍala* with O as the centre, and E' is the direction of *mandocca*. Draw a circle of radius r_0 with P_0 as centre. Let P_1 be the point on this circle such that P_0P_1 is in the direction of *mandocca* (parallel to OE'). Let O'' be a point on the line OE', such that $OO'' = r_0$. Join P_1O'' and let that line meet *kakṣyā-maṇḍala* at Q. Extend OQ and P_0P_1 so as to meet at P. The true planet is located at P. Then, it can be shown that, $OP = K$ and $P_0P = r$ are the actual *manda-karṇa* and the corresponding (true) radius of the epicycle as will result by the process of successive iteration which is described below. Since P_1O'' is parallel to P_0O, the triangles OP_0P and $QO''O$ are similar and we have

$$\frac{r}{K} = \frac{P_0P}{OP} = \frac{O''O}{QO} = \frac{r_0}{R}. \tag{8.19}$$

[4]The value of C varies from planet to planet.

[5]*Mahābhāskarīya*, IV.9-12; *Tantrasaṅgraha* II.41-42.

[6]*Mahābhāskarīya*, Ed. and Tr. by K. S. Shukla, Lucknow 1960, p.111-119.

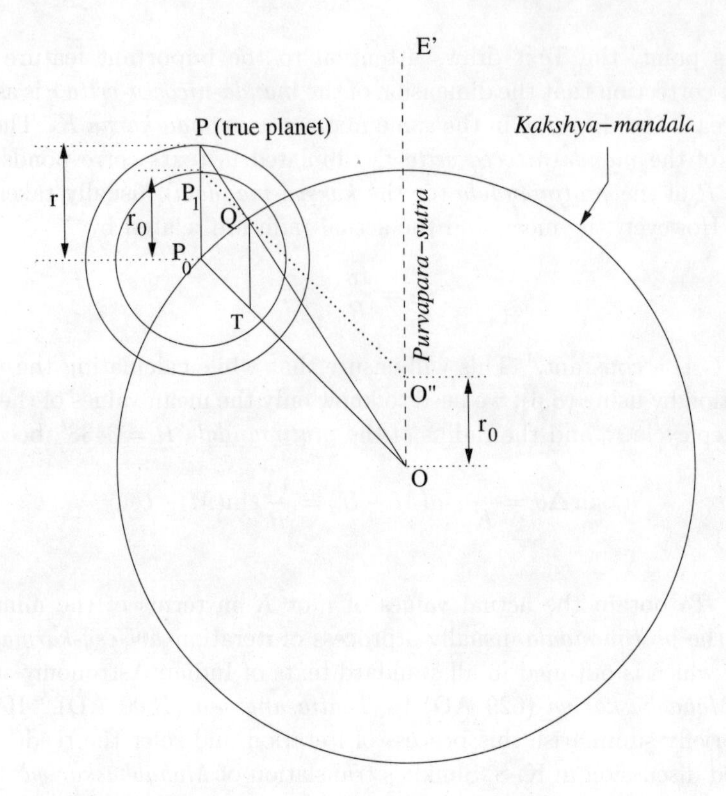

Figure 8.5: Mean and true epicycles.

The process of successive iteration to obtain K is essentially the following. In triangle OP_1P_0, with the angle $P_1\hat{P}_0O = 180° - (M - U)$, the first approximation to the *karṇa* (*sakṛt-karṇa*) $K_1 = OP_1$ and the mean epicycle radius $r_0 = P_1P_0$, are related by

$$K_1 = \sqrt{R^2 + r_0^2 + 2r_0 R\cos(M - U)}. \tag{8.20}$$

In the RHS of (8.20), we replace r_0 by the next approximation to the radius of the epicycle

$$r_1 = \frac{r_0}{R}K_1, \tag{8.21}$$

and obtain the next approximation to the *karṇa*

$$K_2 = \sqrt{R^2 + r_1^2 + 2r_1 R\cos(M - U)}, \tag{8.22}$$

and so on. This process is iterated till K_i and K_{i+1} become indistinguishable, and that will be the *aviśiṣṭa-karṇa* K,[7] which is related to the corresponding epicycle radius r as in (8.21) by

$$r = \frac{r_0}{R} K. \tag{8.23}$$

It can actually be shown[8] that the sequence $K_1, K_2, K_3, \ldots$ indeed converges and the limit is $OP = K$. Also, from the triangle OP_0P it follows that K and r are also related by

$$K = \sqrt{R^2 + r^2 + 2rR\cos(M - U)}. \tag{8.24}$$

8.9 Alternative method for the *Karṇa*

Here, another approach for the determination of *karṇa* (hypotenuse) is presented, primarily using the *ucca-nīca-vṛtta* (epicycle). This can be understood with the help of Figure 8.6.

In fact, two cases are considered here: (i) the foot of the *bhujā-phala* of the planet on the *pratimaṇḍala* lies outside the circumference of the *kakṣyā-vṛtta*, and (ii) the foot of the *bhujā-phala* is inside the circumference of the *kakṣyā-vṛtta.*[9]

1. **Case 1:** Planet at P_1

   Considering the triangle $P_1B_1M_1$, the sine and the cosine are given by

$$\begin{aligned} \textit{bhujā-phala} &= P_1B_1 = r\sin(M - U), \\ \textit{koṭi-phala} &= B_1M_1 = r\cos(M - U). \end{aligned} \tag{8.25}$$

[7] The term *viśeṣa* means 'distinction'. Hence, *aviśeṣa* is 'without distinction'. Therefore the term *aviśiṣṭa-karṇa* refers to that *karṇa* obtained after doing a series of iterations such that the successive values of the *karṇa* do not differ from each other.

[8] vide K. S. Shukla cited above in footnote 6.

[9] These also correspond to the situations when the planet is located in the *pratimaṇḍala* to the east or west of the north-south line passing through the centre of the *pratimaṇḍala.* The Text seems to wrongly suggest that these cases also correspond to the situations when the planet, located in the *pratimaṇḍala*, lies outside or inside the *kakṣyā-maṇḍala.* This error, however, is not made in the next section, 8.10, where only the location of the foot of the *bhujā-phala* is considered.

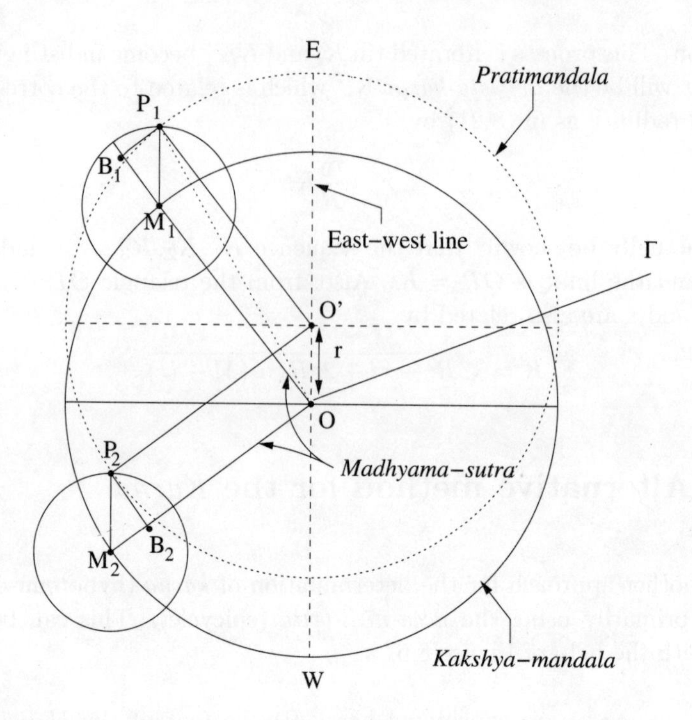

Figure 8.6: The determination of *karṇa* using epicyclic approach.

The distance between the centre O and the base of the *bhujā-phala*

$$\begin{aligned} OB_1 &= B_1M_1 + R \\ &= R + r\cos(M-U). \end{aligned} \tag{8.26}$$

Hence,

$$\begin{aligned} karṇa &= OP_1 \\ &= \sqrt{P_1B_1{}^2 + OB_1{}^2} \\ &= \sqrt{r^2\sin^2(M-U) + \{R + r\cos(M-U)\}^2}. \end{aligned} \tag{8.27}$$

2. **Case 2:** Planet at P_2

Considering the triangle $P_2B_2M_2$, the sine and the cosine are given by

$$\begin{aligned} bhujā\text{-}phala &= P_2B_2 = r\sin(M-U) \\ koṭi\text{-}phala &= B_2M_2 = |r\cos(M-U)| \\ &= -r\cos(M-U), \end{aligned} \tag{8.28}$$

as $90^o < M - U < 180^o$. Now the distance between the centre O and the base of the *bhujā-phala*

$$\begin{aligned} OB_2 &= OM_2 - B_2M_2 \\ &= R - |r\cos(M-U)|. \end{aligned} \qquad (8.29)$$

Hence,

$$\begin{aligned} karṇa &= OP_2 \\ &= \sqrt{P_2{B_2}^2 + O{B_2}^2} \\ &= \sqrt{r^2\sin^2(M-U) + \{R - |r\cos(M-U)|\}^2}. \end{aligned} \qquad (8.30)$$

In either case, (8.27) or (8.30) lead to the same expression for the *karṇa*, viz.,

$$\begin{aligned} K &= \sqrt{r^2\sin^2(M-U) + \{R + r\cos(M-U)\}^2} \\ &= \sqrt{R^2 + r^2 + 2rR\cos(M-U)}, \end{aligned} \qquad (8.31)$$

which is the same as the formula (8.24) in the last section. From K, we can find how much the planet has moved on the hypotenuse circle by (8.18).

8.10 *Viparīta-karṇa* : Inverse hypotenuse

It appears that it was the celebrated Mādhava (c.1320-1400) who gave an exact formula for evaluating the true *manda-karṇa*, without employing the iterative process. As noted in *Tantrasaṅgraha* II.44, Mādhava expressed the true *manda-karṇa* in terms of the so called *viparīta-karṇa* or inverse hypotenuse. The expression for *viparīta-karṇa* is based on the inverse relation between the *karṇa* and radius, which is being dealt with first. Mādhava's expression for the *aviśiṣṭa-manda-karṇa* will be discussed later towards the end of the section 8.12.

Here, the aim is to obtain R from K. That is the radius of the *kakṣyā-vṛtta* when the radius of the *karṇa-vṛtta* is taken to be *trijyā* ($= 3438'$). As in the previous section, we consider two cases (refer to Figure 8.6).

1. **Case 1:** Planet is at P_1 and B_1, the base of *bhujā-phala*, is outside the *kakṣyā-vṛtta*. Now, the radius of the *kakṣyā-vṛtta* is

$$\begin{aligned} OM_1 &= OB_1 - B_1M_1 \\ &= OB_1 - \textit{koṭi-phala}, \end{aligned} \tag{8.32}$$

where OB_1 is the distance between the *kakṣyā-kendra* and the base of the *bhujā-phala* and is given by

$$OB_1 = \sqrt{K^2 - \textit{bhujā-phala}^2}. \tag{8.33}$$

2. **Case 2:** Planet is at P_2 and B_2, the base of *bhujā-phala*, is inside the *kakṣyā-vṛtta*. Now, the radius of the *kakṣyā-vṛtta* is

$$\begin{aligned} OM_2 &= OB_2 + B_2M_2 \\ &= OB_2 + \textit{koṭi-phala}, \end{aligned} \tag{8.34}$$

where OB_2 is the distance between the *kakṣyā-kendra* and the base of the *bhujā-phala* and is given by

$$OB_2 = \sqrt{K^2 - \textit{bhujā-phala}^2}. \tag{8.35}$$

In both the cases we get

$$R = \sqrt{K^2 - r^2 \sin^2(M-U)} - r\cos(M-U). \tag{8.36}$$

8.11 Another method for *Viparīta-karṇa*

Here, another method for determining the *trijyā* from the *karṇa* is described. We explain this with the help of Figure 8.7. In this, and the subsequent section, we use P and U to represent the longitude of the planet and the *ucca* respectively. Consider the case when B_1, the base of the *bhujājyā*, is outside the *ucca-nīca-vṛtta* and to the east of it (as is the case when the planet is at P_1). The angle

$$\begin{aligned} U\hat{O}P_1 &= \textit{sphuṭa} - \textit{ucca} \\ &= P - U, \end{aligned} \tag{8.37}$$

is the difference between the longitudes of the *mandocca* and the planet. Also,

$$\begin{aligned} P_1B_1 &= K\sin(P-U) \\ OB_1 &= K\cos(P-U) \\ O'B_1 &= OB_1 - OO' \\ &= K\cos(P-U) - r. \end{aligned} \tag{8.38}$$

Then the radius of the *pratimaṇḍala/kakṣyā-vṛtta* is

$$O'P_1 = R = \sqrt{K^2\sin(P-U)^2 + (K\cos(P-U) - r)^2}. \tag{8.39}$$

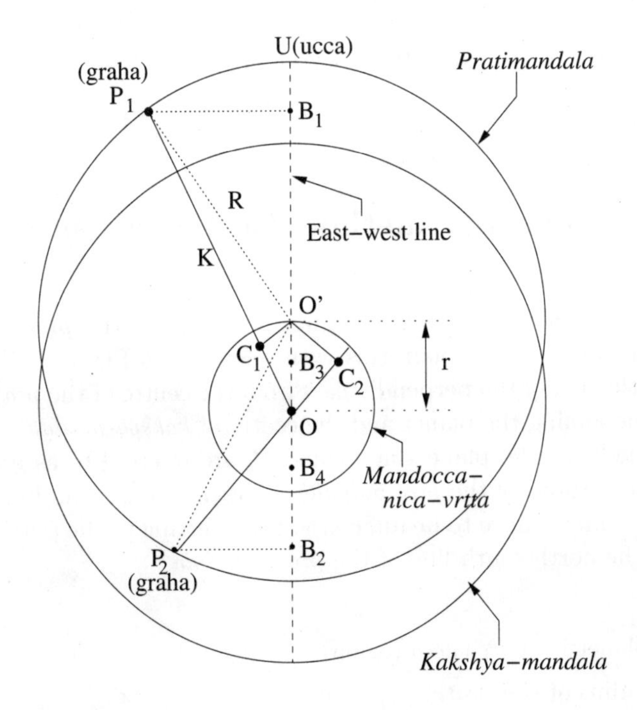

Figure 8.7: The determination *trijyā* from *karṇa* – alternate approach.

When the base of the *bhujājyā* B_3, is inside the *ucca-nīca-vṛtta* and east of O,

$$O'B_3 = r - K\cos(P-U), \tag{8.40}$$

and the expression for R is same as above. In both cases, $P - U < 90°$.

Similarly when B_2, the base of the *bhujājyā*, is outside the *ucca-nīca-vṛtta* and to the west of it (as is the case when the planet is at P_2),

$$\begin{aligned} R = O'P_2 &= \sqrt{P_2B_2{}^2 + O'B_2^2} \\ &= \sqrt{P_2B_2^2 + (O'O + OB_2)^2} \\ &= [K^2 \sin^2(P-U) + (r + |K\cos(P-U)|)^2]^{\frac{1}{2}}. \end{aligned} \quad (8.41)$$

It is easy to see that this is also the case when the base B_4 is inside the *ucca-nīca-vṛtta* and west of O. In both these cases $\cos(P - U)$ is negative, as $90^o \leq P - U < 180^o$. Hence, taking the sign of $\cos(P - U)$ into account, we get in all cases,

$$\begin{aligned} R &= [K^2 \sin^2(P-U) + (K\cos(P-U) - r)^2]^{\frac{1}{2}} \\ &= [K^2 + r^2 - 2rK\cos(P-U)]^{\frac{1}{2}}. \end{aligned} \quad (8.42)$$

8.12 Still another method for *Viparīta-karṇa*

There is yet another method for finding the radius of the *pratimaṇḍala* in terms of the *karṇa*. We explain this with reference to Figure 8.7. Here, C_1 and C_2 are the feet of the perpendiculars from the centre of the *pratimaṇḍala* O' to the line joining the planet and the centre of *kakṣyā-maṇḍala* (OP_1 and OP_2). In the Text, the planets at P_1 and P_2 are referred to as lying in the *ucca* and *nīca* regions of the *pratimaṇḍala*. The phrase *ucca* and *nīca* regions used in this context, have to be understood as referring to the portions above and below the north-south line of the *pratimaṇḍala*.

1. The planet is at P_1 (*ucca* region).

   The radius of the *pratimaṇḍala* is

   $$R = O'P_1 = \sqrt{(O'C_1)^2 + (P_1C_1)^2}. \quad (8.43)$$

   We need to calculate $O'C_1$ and P_1C_1, which are given by

   $$O'C_1 = \textit{doḥphala} = r\sin(P-U),$$

and

$$OC_1 = \textit{koṭi-phala} = r\cos(P-U).$$

In the figure,

$$\begin{aligned} P_1C_1 &= OP_1 - OC_1 \\ &= K - r\cos(P-U). \end{aligned} \qquad (8.44)$$

Hence, (8.43) reduces to

$$R = \sqrt{r^2\sin^2(P-U) + \{K - r\cos(P-U)\}^2}. \qquad (8.45)$$

2. The planet is at P_2 (*nīca* region).

   The radius of the *pratimaṇḍala* is

$$R = O'P_2 = \sqrt{(O'C_2)^2 + (P_2C_2)^2}. \qquad (8.46)$$

We need to calculate $O'C_2$ and P_2C_2, which are given by

$$\begin{aligned} O'C_2 = \textit{doḥphala} &= r\sin(O'\hat{O}C_2) \\ &= r\sin(P_2\hat{O}B_2) \\ &= r\sin(P-U), \\ \text{and} \qquad OC_2 = \textit{koṭi-phala} &= r\cos(O'\hat{O}C_2) \\ &= r\cos(P_2\hat{O}B_2) \\ &= |r\cos(P-U)|. \end{aligned} \qquad (8.47)$$

In the figure,

$$\begin{aligned} P_2C_2 &= OP_2 + OC_2 \\ &= K + |r\cos(P-U)|. \end{aligned} \qquad (8.48)$$

Hence, (8.46) reduces to

$$R = [r^2\sin^2(P-U) + \{K + |r\cos(P-U)|\}^2]^{\frac{1}{2}}. \qquad (8.49)$$

In (8.49), $\cos(P-U)$ is negative, as $90^o \leq P-U \leq 180^o$. Taking the sign of $\cos(P-U)$ into account, in all cases, the expression for *trijyā* is

$$\begin{aligned} R &= [r^2\sin^2(P-U) + \{K - r\cos(P-U)\}^2]^{\frac{1}{2}} \\ &= [K^2 + r^2 - 2rK\cos(P-U)]^{\frac{1}{2}}, \end{aligned} \qquad (8.50)$$

which is the same as (8.42).

Now, the *aviśiṣṭa-manda-karṇa* is to be determined in terms of the minutes of arc of the *pratimaṇḍala*. The radius of the *karṇa-vṛtta* is equal to the *trijyā* R, when measured in the minutes of arc of *karṇa-vṛtta* (i.e., its own measure). Then the radius of the *pratimaṇḍala* will be given (in the measure of the *karṇa-vṛtta*) by the *viparīta-karṇa* R_v, which is obtained by setting $K = R$ and $r = r_0$ in (8.36) and (8.42), where r_0 is the mean radius of the *manda-nīcocca-vṛtta* :

$$R_v = \sqrt{R^2 - r_0^2 \sin^2(M-U)} - r_0 \cos(M-U), \tag{8.51}$$

and

$$R_v = [R^2 + r_0^2 - 2r_0 R \cos(P-U)]^{\frac{1}{2}}. \tag{8.52}$$

By the rule of three: R_v is the radius of the *pratimaṇḍala* when the radius of the *karṇa-vṛtta* is *trijyā* or R. Now when the radius of the *pratimaṇḍala* is R, then the radius of the *karṇa-vṛtta* K will be given by

$$\frac{K}{R} = \frac{R}{R_v}$$

$$\text{or} \quad K = \frac{R^2}{R_v}. \tag{8.53}$$

This is the Mādhava's formula for the true or *aviśiṣṭa-manda-karṇa.*

Note: We may briefly indicate the geometrical representation of the *aviśiṣṭa-manda-karṇa* and the *viparīta-karṇa* with reference to Figure 8.5 on page 632. Here, T is a point on the line OP_0, such that the line QT is parallel to P_0P_1. Then, it can be easily seen that OT will be the *viparīta-karṇa* R_v. Now, in the triangle OQT, $OQ = R$, $QT = r_0$, $O\hat{Q}T = P\hat{O}E' = P - U$, and we have

$$OT = [R^2 + r_0^2 - 2Rr_0 \cos(P-U)]^{\frac{1}{2}}, \tag{8.54}$$

which is the same as the *viparīta-karṇa* R_v as given by (8.52). Again, the triangles OPP_0 and OQT are similar. Hence

$$\frac{OP}{OQ} = \frac{OP_0}{OT},$$

$$\text{or} \quad OP = \frac{OP_0 \times OQ}{OP}. \tag{8.55}$$

Since $OP_0 = OQ = R$ and $OT = R_v$ is the *viparīta-karṇa*, we get

$$OP = \frac{R^2}{R_v}, \tag{8.56}$$

which is the same as the *aviśiṣṭa-karṇa* K given by (8.53).

8.13 *Manda-sphuṭa* from the *Madhyama*

In Figure 8.8, O is the *bhū-madhya*, O' is the centre of the *pratimaṇḍala*, P is the planet on the *pratimaṇḍala* and U is the *ucca*. R and K are the radii of the *pratimaṇḍala* and the *karṇa-vṛtta*. In the figure,

$$\begin{aligned} \hat{A} = P\hat{O'}U &= madhya - ucca \\ &= M - U, \\ \text{and} \qquad \hat{B} = P\hat{O}U &= sphuṭa - ucca \\ &= P - U. \end{aligned} \tag{8.57}$$

It may be noted that while $\hat{A}$ corresponds to the arc measured along the *pratimaṇḍala*, $\hat{B}$ corresponds to the arc measured along the *karṇa-vṛtta*. Draw a perpendicular from P to OU, meeting it at C. Obviously,

$$PC = R\sin(M - U) = K\sin(P - U). \tag{8.58}$$

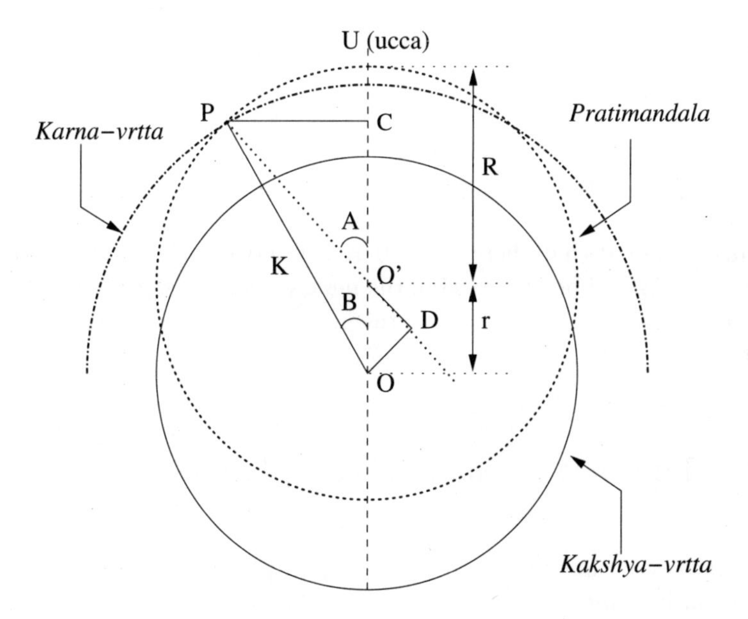

Figure 8.8: The determination *manda-sphuṭa* from *madhya*.

This means that the measures of the *pratimaṇḍala* and *karṇa-vṛtta* are different. *Jyā* of $\hat{B}$ in the measure of the *karṇa-vṛtta* is equal to the *jyā* of $\hat{A}$

in the measure of the *pratimaṇḍala*. $\hat{B}$ can be determined from the knowledge of *trijyā* R, $\hat{A}$ and the *karṇa* K which can be found using any of the methods described earlier. Adding *ucca* to $\hat{B}$, *sphuṭa* is determined. This is the *manda-sphuṭa*, that is the mean planet to which the *equation of centre* is added to get the *sphuṭa-graha*.

On the other hand, if the *sphuṭa* is known, the above relation (8.58) can be used to obtain the *manda-karṇa*; and from that the radius of the epicycle can also be determined using (8.23).

Here, it is again reiterated that the radius of the epicycle (*ucca-nīca-vṛtta*), r, increases or decreases as the *manda-karṇa* K, that is, $\frac{r}{K}$ is constant. It is noted that this simplifies the calculation of $P - M$, as it can be simply determined from the relation

$$\begin{aligned} K \sin(M - P) &= r \sin(madhya - ucca) \\ &= r \sin(M - U). \end{aligned} \tag{8.59}$$

If r_0 is the mean radius of the epicycle, or the radius of the epicycle in terms of the minutes of arc of the *pratimaṇḍala*, then

$$\sin(M - P) = \frac{r}{K} \sin(M - U) = \frac{r_0}{R} \sin(M - U). \tag{8.60}$$

Thus, in calculating the *manda*-correction, there is no need to compute the *manda-karṇa* K, or the true epicycle radius r.

It is further noted that there is a difference between the *manda* and *śīghra* procedures. As will be discussed in the next section, in *śīghra* correction, the radius of the *śīghra-nīcocca-vṛtta* is taken to be a constant, and not varying with the *śīghra-karṇa*.

8.14 The *Śīghra-sphuṭa* of the planets

Note: It may be mentioned that the revised planetary model proposed by Nīlakaṇṭha in *Tantrasaṅgraha* forms the basis of the discussion in this and the subsequent sections. An overview of the conventional planetary model employed in Indian Astronomy (at least from the time of Āryabhaṭa) and the important revision of this model by Nīlakaṇṭha Somayājī is presented in the Epilogue to this Volume.

For the Sun and the Moon, the *sphuṭa* obtained above is itself the true planet. For the planets (Mars, Jupiter, Saturn, Mercury and Venus), another correction has to be applied to find the true planet which involves the use of a *śīghrocca*. This would be equivalent to the determination of the true geocentric planet called the *śīghra-sphuṭa* from the true heliocentric planet called the *manda-sphuṭa*. We first, delineate the procedure given in the Text. In Figure 8.9, the *śīghra-nīcocca-vṛtta* is a circle with the

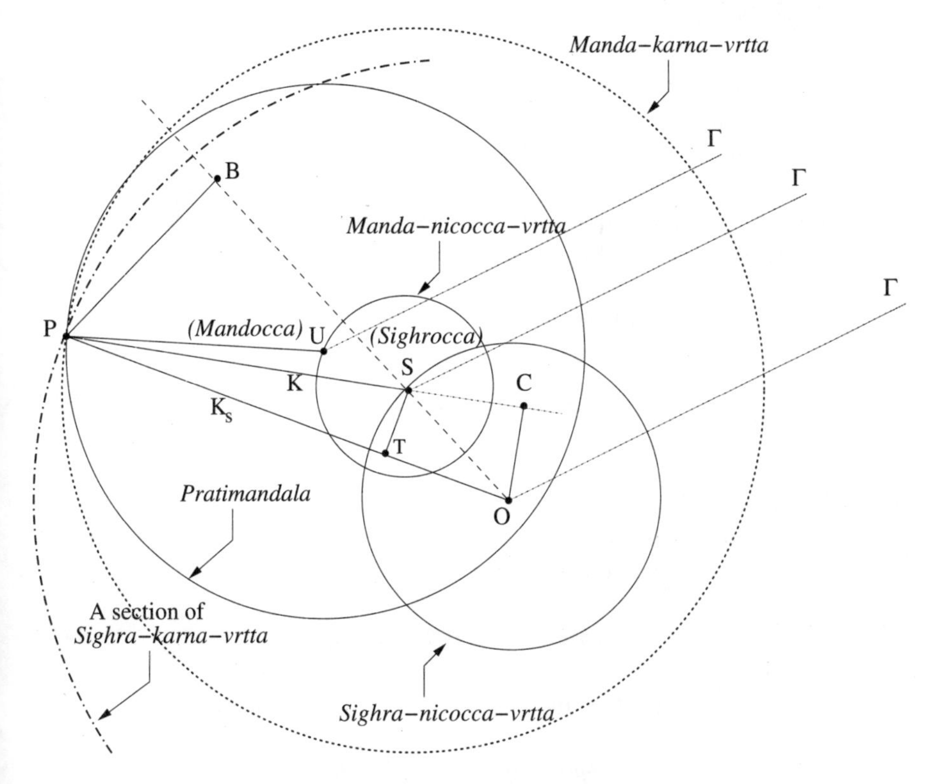

Figure 8.9: The determination of *śīghra-sphuṭa* for exterior planets.

bhagola-madhya as the centre, and whose radius is the *śīghrāntyaphala*, r_s. The *śīghrocca*, S is located on this circle. It is also stated that *śīghrocca* is the *āditya-madhyama* (the mean Sun). The *manda-nīcocca-vṛtta* is a circle with the *śīghrocca* as the centre. The *mandocca* is located on this circle. The planet P is located on the *pratimaṇḍala* which is centered at the *mandocca*. SP is the *manda-karṇa* and $P\hat{S}\Gamma$ is the *manda-sphuṭa*. $P\hat{O}\Gamma$ is the true geocentric planet known as the *śīghra-sphuṭa*. The *śīghra-sphuṭa* is found in the same manner from the *manda-sphuṭa*, as the *manda-sphuṭa* is found from the mean planet, *madhyama-graha*. Thus it may be noted that, in the computation of the *śīghra-sphuṭa*, the *śīghrocca* and the *manda-karṇa-vṛtta* play the same roles as the *mandocca* and the *pratimaṇḍala* did in the computation of the *manda-sphuṭa*. The *śīghra-karṇa* $K_s = OP$ can be determined in terms of the *manda-karṇa*, $SP = K$, or the *trijyā* R. Apart from the similarities, there is one difference, as was pointed out in the previous section. In *manda*-correction, the radius of the *manda-nīcocca-vṛtta* r varies. It increases or decreases in the same way as the *manda-karṇa* K. In the *śīghra* correction, the radius of the *śīghra-nīcocca-vṛtta* r_s, does not vary with the *śīghra-karṇa*. To start with, both the mean radius r_0 of the *manda-nīcocca-vṛtta* and the radius r_s of the *śīghra-nīcocca-vṛtta* are specified in the measure of the *pratimaṇḍala* radius, being *trijyā* or $R = 3438'$.

We first define the basic quantities/angles which are used in the later discussion, with reference to Figure 8.9:

$$\begin{aligned} \textit{madhyama-graha} &= \Gamma\hat{U}P, \\ \textit{manda-karṇa} &= SP = K, \\ \textit{manda-sphuṭa} &= \Gamma\hat{S}P = M_s. \end{aligned} \tag{8.61}$$

It is this *manda-sphuṭa*, (the last of the above), which is determined by the *manda-saṃskāra* discussed in the sections 8.7 and 8.13. The *śīghra-sphuṭa*, to be determined, is defined by

$$\textit{śīghra-sphuṭa} = \mathcal{P} = \Gamma\hat{O}P. \tag{8.62}$$

In connection with this, two more quantities are defined, namely

$$\begin{aligned} \textit{śīghrāntyaphala} &= OS = r_s, \\ \text{and} \qquad \textit{śīghrocca} &= \Gamma\hat{O}S = S. \end{aligned} \tag{8.63}$$

The former is the radius of the circle in which *śīghrocca* moves, and the latter is the longitude of *śīghrocca*. Now, the *śīghra-kendra* is given by,

$$\begin{aligned} \textit{śīghra-kendra} &= \textit{manda-sphuṭa} - \textit{śīghrocca} \\ &= \Gamma\hat{S}P - \Gamma\hat{O}S \\ &= \Gamma\hat{S}P - \Gamma\hat{S}B \\ &= P\hat{S}B. \end{aligned} \tag{8.64}$$

The Rsine of the above is called *śīghra-kendra-jyā*. It is given by

$$\begin{aligned} \textit{śīghra-kendra-jyā} = PB &= PS \sin P\hat{S}B \\ &= K \sin(M_s - S). \end{aligned} \tag{8.65}$$

Similarly, the *koṭijyā* is given by

$$\begin{aligned} \textit{śīghra-kendra-koṭijyā} = SB &= PS \cos P\hat{S}B \\ &= K \cos(M_s - S). \end{aligned} \tag{8.66}$$

The *jyā* and the *koṭi* above are defined in the measure of the *manda-karṇa* K. Now, the *bhujā-phala* and *koṭi-phala* will be defined in the measure of the *pratimaṇḍala* (i.e., taking $R = 3438'$) as follows:

$$\begin{aligned} \textit{śīghra-bhujā-phala} = OC &= OS \sin O\hat{S}C \\ &= OS \sin P\hat{S}B \\ &= r_s \sin(M_s - S), \end{aligned} \tag{8.67}$$

and

$$\begin{aligned} \textit{śīghra-koṭi-phala} = SC &= OS \cos O\hat{S}C \\ &= OS \cos P\hat{S}B \\ &= r_s \cos(M_s - S). \end{aligned} \tag{8.68}$$

Now the *śīghra-karṇa* is given by

$$\begin{aligned} K_s = OP &= \left[PC^2 + OC^2\right]^{\frac{1}{2}} \\ &= \left[(SC + PS)^2 + OC^2\right]^{\frac{1}{2}} \\ &= \left[(\textit{śīghra-koṭi-phala} + \textit{manda-karṇa})^2 + (\textit{śīghra-bhujā-phala})^2\right] \\ &= \left[(r_s \cos(M_s - S) + K)^2 + r_s^2 \sin^2(M_s - S)\right]^{\frac{1}{2}}. \end{aligned} \tag{8.69}$$

This *śīghra-karṇa* is in the measure of the *pratimaṇḍala*. That is, the expression for K_s has been obtained under the assumption that R is taken to be *trijyā* ($= 3438'$) and K is the calculated value of *manda-karṇa* (could be less than or greater than *trijyā*). However, when the *manda-karṇa* is itself taken to be *trijyā*, then the *śīghra-karṇa* will be

$$\begin{aligned} \tilde{K}_s &= \frac{R}{K} K_s \\ &= \left[\left(r_s \frac{R}{K} \cos(M_s - S) + R\right)^2 + \left(r_s \frac{R}{K}\right)^2 \sin^2(M_s - S)\right]^{\frac{1}{2}}, \end{aligned} \tag{8.70}$$

where $r_s\frac{R}{K}\cos(M_s - S)$ and $r_s\frac{R}{K}\sin(M_s - S)$ would be *śīghra-koṭi-phala* and *śīghra-bhujā-phala* in the measure of the *manda-karṇa* respectively.

The expression (8.69) for *śīghra-karṇa* was derived using the triangle OCP. Now considering the triangle OPB, we have

$$\begin{aligned} K_s = OP &= (OB^2 + PB^2)^{\frac{1}{2}} \\ &= \left[(SB + OS)^2 + PB^2\right]^{\frac{1}{2}} \\ &= \left[(\text{\textit{śīghra-koṭijyā}} + \text{\textit{śīghrāntya-phala}})^2 + (\text{\textit{śīghra-bhujājyā}})^2\right]^{\frac{1}{2}} \\ &= \left[(K\cos(M_s - S) + r_s)^2 + K^2\sin^2(M_s - S)\right]^{\frac{1}{2}}. \qquad (8.71) \end{aligned}$$

This is another expression for the *śīghra-karṇa* K_s in the measure of the *pratimaṇḍala*, and is equivalent to (8.69). However, in the measure of the *manda-karṇa*, when it is taken to be *trijyā* (that is, when $K =$ *trijyā*), it will be

$$\begin{aligned} \tilde{K}_s &= \frac{R}{K}K_s \\ &= \left[\left(R\cos(M_s - S) + r_s\frac{R}{K}\right)^2 + R^2\sin^2(M_s - S)\right]^{\frac{1}{2}}, \qquad (8.72) \end{aligned}$$

where $r_s\frac{R}{K}$ is the *śīghrāntya-phala* in the measure of the *manda-karṇa*.

Now, considering the two triangles PSB and POB, we have

$$OP\ \sin(P\hat{O}B) = PB = SP\sin(P\hat{S}B). \qquad (8.73)$$

Therefore,

$$K_s\sin(\Gamma\hat{O}P - \Gamma\hat{O}B) = K\ \sin(M_s - S). \qquad (8.74)$$

That is,

$$K_s\sin(\mathcal{P} - S) = K\ \sin(M_s - S). \qquad (8.75)$$

In other words,

$$R\ \sin(\mathcal{P} - S) = \frac{R}{K_s}K\ \sin(M_s - S), \qquad (8.76)$$

where R is the *trijyā*. From this, the arc $\mathcal{P} - S$ is found. When this is added to the *śīghrocca* S, the result will be the *śīghra-sphuṭa* $\mathcal{P}$.

Also, it can be easily seen that

$$\begin{aligned} O\hat{P}C &= \textit{manda-sphuṭa} - \textit{śīghra-sphuṭa} \\ &= M_s - \mathcal{P}, \\ OP\sin(O\hat{P}C) &= OC. \end{aligned} \tag{8.77}$$

Using (8.67) in the RHS, the above equation reduces to

$$K_s \sin(M_s - \mathcal{P}) = r_s \sin(M_s - S). \tag{8.78}$$

Multiplying by *trijyā* and dividing by *karṇa*, we get

$$R \sin(M_s - \mathcal{P}) = \frac{R\, r_s \sin(M_s - S)}{K_s}. \tag{8.79}$$

From this, the arc $M_s - \mathcal{P}$ is found. When this is subtracted from the *manda-sphuṭa* M_s, the result will be the *śīghra-sphuṭa*, $\mathcal{P}$.

It is again emphasized that one has to be careful about the measure employed. In the two alternative ways of finding the *śīghra-sphuṭa* $\mathcal{P}$, if the *śīghra-bhujā-jyā* $K \sin(M_s - S)$ and the *śīghra-bhujā-phala* $r_s \sin(M_s - S)$ are in the measure of *manda-karṇa* or *pratimaṇḍala*, the divisor K_s (*śīghra-karṇa*) should also be in the same measure.

A geometrical summary of finding the *manda-sphuṭa*/*śīghra-sphuṭa* is then provided. The motion of the planet on the *pratimaṇḍala*, whose centre is the *ucca*, is known. From this, one should determine the motion of the planet on the *karṇa-vṛtta* whose centre is the *bhagola-madhya*. Here the *pratimaṇḍala* and *karṇa-vṛtta* are called the *jñāta-bhoga-graha* and *jñeya-bhoga-graha-vṛtta*-s respectively. The terms *jñāta* and *jñeya* mean 'known' and 'to be known'. *Bhoga* in this context means the 'arc covered'. Hence, *jñāta-bhoga-graha-vṛtta* refers to the circle in which the arc covered by the planet is known, which is the *pratimaṇḍala*. Similarly, *jñeya-bhoga-graha-vṛtta* refers to the circle in which the arc covered by the planet is to be known. Obviously this is the *karṇa-vṛtta*. It could be *śīghra-karṇa-vṛtta* or *manda-karṇa-vṛtta* as the case may be. These two, along with the other three *vṛtta*-s are shown in Figure 8.10.

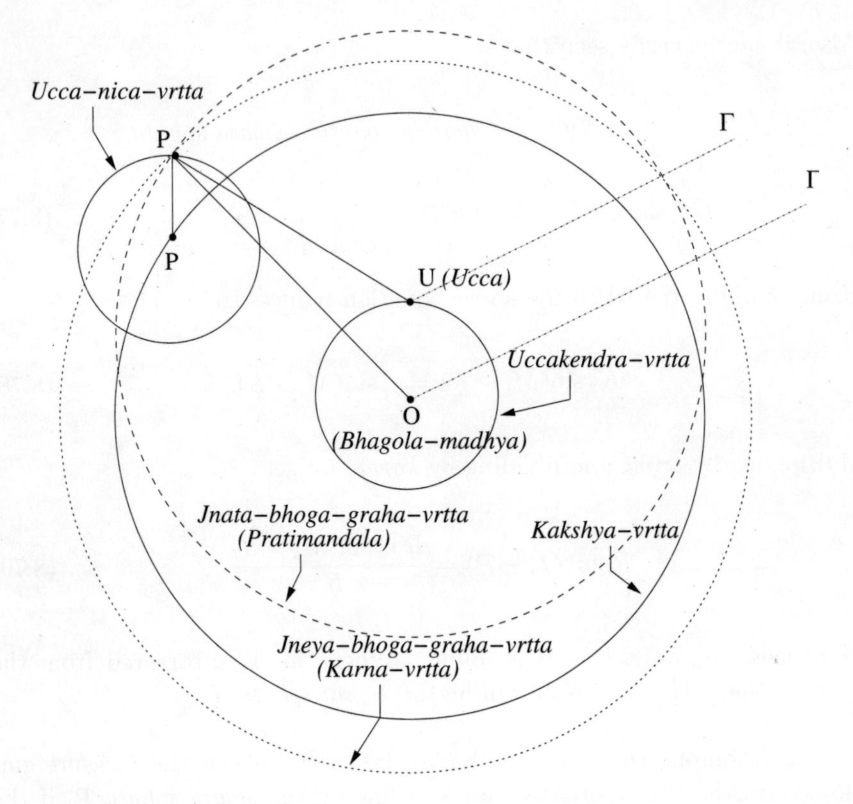

Figure 8.10: The *jñāta-bhoga-graha-vṛtta* and *jñeya-bhoga-graha-vṛtta*.

8.15 The *Śīghra-sphuṭa* of Mercury and Venus

For Mercury and Venus, *śīghra-nīcocca-vṛtta* (*śīghra-vṛtta*) is large and the *manda-karṇa-vṛtta* is small. Hence, the *śīghra-vṛtta* with its centre at the centre of *bhagola*, is taken to be the *kakṣyā-vṛtta*. On this, the *śīghrocca* (S) moves (see Figure 8.11). The *jñāta-bhoga-graha-vṛtta* is a circle with the *śīghrocca* as the centre, and on the circumference of this the planet moves with the same speed as the *manda-sphuṭa*. Here it is to be considered as the *ucca-nīca-vṛtta*. As we shall see later, this is essentially the *manda-karṇa-vṛtta* with its radius reduced from K to $\tilde{r}_s = K\frac{r_s}{R}$. Also construct another *ucca-nīca-vṛtta* /*jñāta-bhoga-vṛtta* whose centre is same as the centre of the *kakṣyā-vṛtta* (which is same as *bhagola-madhya*). On this the *manda-sphuṭa-graha* is located such that $\Gamma\hat{O}O' = \Gamma\hat{S}P$. With this ($O'$) as the centre, the *pratimaṇḍala* is constructed whose radius is the same as the *śīghra-*

vṛtta or the *kakṣyā-vṛtta*. The planet is located at the intersection of this *pratimaṇḍala* and the *jñāta-bhoga-graha-vṛtta*. The *jñeya-bhoga-graha-vṛtta* is the circle whose centre is the same as that of the *kakṣyā-vṛtta* and touches the planet. That is, it is the circle with OP as radius in Figure 8.11. We now define the following:

$$\begin{aligned} \textit{śīghrocca} &= \Gamma\hat{O}S = \Gamma\hat{O'}P = S, \\ \textit{manda-sphuṭa} &= \Gamma\hat{O}O' = \Gamma\hat{S}P = M_s. \end{aligned} \tag{8.80}$$

The *śīghra-kendra* is given by

$$\begin{aligned} \textit{śīghra-kendra} &= \textit{manda-sphuṭa} - \textit{śīghrocca} \\ &= \Gamma\hat{S}P - \Gamma\hat{O}S \\ &= \Gamma\hat{S}P - \Gamma\hat{S}B \\ &= P\hat{S}B \\ &= M_s - S. \end{aligned} \tag{8.81}$$

The sine of it, called the *śīghra-kendra-bhujājyā* is

$$\begin{aligned} R\sin(M_s - S) &= R\sin P\hat{S}B \\ &= R\sin P\hat{O'}C \\ &= PC \\ &= K_s \sin P\hat{O}C \\ &= K_s \sin(M_s - \mathcal{P}). \end{aligned} \tag{8.82}$$

Considering the triangle POB,

$$\begin{aligned} \textit{śīghra-bhujā-phala} &= PB = K_s \sin P\hat{O}B \\ \textit{śīghra-bhujā-phala-cāpa} &= P\hat{O}B = \mathcal{P} - S. \end{aligned} \tag{8.83}$$

We compare this with Figure 8.9 and determine the *śīghra-sphuṭa* in a similar manner. Here, we take the motion of O', on the *ucca-nīca-vṛtta* as the *graha-gati*, and the motion of P on the *pratimaṇḍala* (whose centre is O' and whose radius is same as the radius of the *kakṣyā-maṇḍala*) as the *śīghrocca-gati*. In this sense, the roles of *kakṣyā/pratimaṇḍala* and the *ucca-nīca-vṛtta* are reversed in this case.

The procedure for finding the *śīghra-sphuṭa* $\Gamma\hat{O}P$ is the same as that for finding the *manda-sphuṭa* of a planet, with O' which moves with *graha-gati*

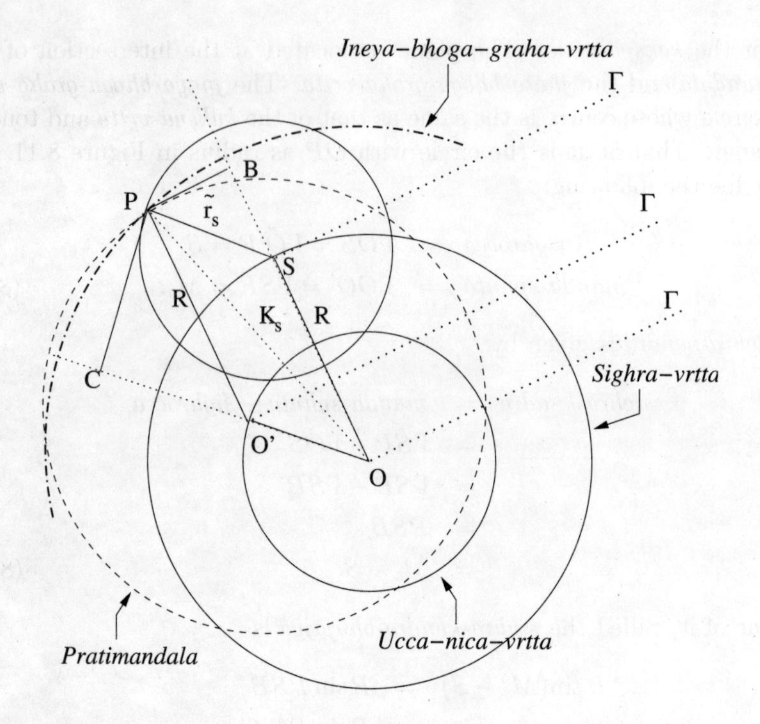

Figure 8.11: The five circles employed in elucidating the *śīghra-sphuṭa* of Mercury and Venus.

playing the role of *ucca* and P, which moves on *pratimaṇḍala* with *śīghrocca-gati*, playing the role of *madhyama-graha*. Now, the *śīghra-sphuṭa* is given by

$$\begin{aligned} \mathcal{P} &= \Gamma\hat{O}P && (8.84) \\ &= \Gamma\hat{O}S + P\hat{O}B \\ &= \textit{śīghrocca} + \textit{śīghra-phala-cāpa}, && (8.85) \end{aligned}$$

where the *śīghra-bhujā-phala-cāpa* $P\hat{O}B$ is determined from

$$\begin{aligned} K_s \sin(P\hat{O}B) &= \tilde{r}_s \sin P\hat{S}B \\ &= \tilde{r}_s \sin(M_s - S), && (8.86) \end{aligned}$$

where $K_s = OP$, is the radius of the *śīghra-karṇa-vṛtta*, and $\tilde{r}_s = SP = OO'$ is the radius of the *ucca-nīca-vṛtta*, which is the the *manda-karṇa* reduced to the scale of the *śīghra-vṛtta* and is given by

$$\tilde{r}_s = \frac{K}{R} r_s, \tag{8.87}$$

where K is the *manda-karṇa* and r_s is the radius of the *śīghra-nīcocca-vṛtta* or equivalently the *śīghra-antya-phala*. Now, for obtaining the *śīghra-sphuṭa* (ΓOP), the *śīghra-phala-cāpa* ($P\hat{O}B$) has to be applied to *śīghrocca*. Using (8.86) and (8.87), $P\hat{O}B = \mathcal{P} - S$ is given by

$$R\sin(\mathcal{P} - S) = R\frac{\tilde{r}_s}{K_s}\sin(M_s - S) = K\frac{\tilde{r}_s}{K_s}\sin(M_s - S), \tag{8.88}$$

where the *śīghra-karṇa* K_s is given by

$$\begin{aligned} K_s &= \sqrt{(PB^2 + OB^2)} \\ &= [\{\tilde{r}_s \sin(M_s - S)\}^2 + \{R + \tilde{r}_s \cos(M_s - S)\}^2]^{\frac{1}{2}}. \end{aligned} \tag{8.89}$$

Using (8.87) in the above, we get

$$K_s = [\{\frac{r_s K}{R}\sin(M_s - S)\}^2 + \{R + \frac{r_s K}{R}\cos(M_s - S)\}^2]^{\frac{1}{2}}. \tag{8.90}$$

The *śīghra-karṇa* is also given by

$$K_s = [\{R\cos(M_s - S) + \frac{r_s K}{R}\}^2 + \{R\sin(M_s - S)\}^2]^{\frac{1}{2}}. \tag{8.91}$$

Alternatively, *śīghra-sphuṭa*

$$\begin{aligned} \Gamma\hat{O}P &= \Gamma\hat{O}O' - P\hat{O}C \\ &= \textit{manda-sphuṭa} - \textit{śīghra-kendra-bhujājyā-cāpa}, \end{aligned} \tag{8.92}$$

where the *śīghra-kendra-bhujājyā-cāpa* $P\hat{O}C = M_s - \mathcal{P}$ is determined from (8.82). Since M_s and S are known, the *cāpa* $P\hat{O}C$ is known. This has to be applied to the *manda-sphuṭa* to obtain the true planet.

The Text clearly notes the difference between the exterior planets, Mars, Jupiter and Saturn and the interior planets Mercury and Venus. For the former, the stated values of *manda-vṛtta* and *śīghra-vṛtta* are in terms of their *pratimaṇḍala*-s. For Mercury and Venus, since the *śīghra-vṛtta*-s are larger, the *pratimaṇḍala* is measured in terms of the (larger) *śīghra-vṛtta* and set out as the *śīghra-vṛtta*.

Note:

The Text also notes that the procedure for finding the true planet for Mercury and Venus in *Tantrasaṅgraha* is different from that in other works. For these planets, since the *śīghra-vṛtta*-s are large, it is the *pratimaṇḍala* which has been measured in terms of the minutes of this (*śīghra-vṛtta*) and set out as *śīghra-vṛtta* in *Tantrasaṅgraha*. In the earlier texts, the *manda-vṛtta*-s of Mercury and Venus are in the measure of the *śīghra-vṛtta*. In *Tantrasaṅgraha*, the *manda-nīcocca-vṛtta*-s are in the measure of the *pratimaṇḍala*. Though it is not stated here, it is implied that the *manda-sphuṭa-nyāya* is wrong in the earlier texts, as the equation of centre is applied to the *āditya-madhyama* (mean Sun), whereas it should be applied to the mean planet (which is termed the *śīghrocca* in earlier texts).

On the other hand, according to *Tantrasaṅgraha*, the equation of centre is applied to the mean planet (termed as such – it is the mean heliocentric planet in the modern technology) to find the *manda-sphuṭa* (the true heliocentric planet). Then the *manda-karṇa* (radius of the orbit of the *manda-sphuṭa*) is reduced by a factor of $\frac{r_s}{R}$, where r_s is the *śīghra-antya-phala* and R is the *trijyā*. This reduced *manda-karṇa-vṛtta* on which the *manda-sphuṭa* moves is centered around that mean Sun (*śīghrocca*), which itself moves around the *bhagola-madhya* in an orbit of radius R. With this, the true geocentric planet is found. This is essentially the same as the standard planetary model employed in modern astronomy since Kepler, (except that here the *śīghrocca* is the mean Sun, whereas it should be the true Sun), as the stated valued of r_s/R is equal to the ratio of the planet-Sun and Earth-Sun distances in the modern picture.

It is noteworthy that the procedure for finding the true planet is essentially the same for both the exterior and the interior planets. In both the cases, the true heliocentric planet is found first from the mean heliocentric planet with the *manda-sphuṭa-nyāya*, that is, by the application of what is called the *equation of centre* in the modern terminology. Then the true geocentric planet (*śīghra-sphuṭa*) is found taking the Sun as the *śīghrocca*. The difference is that the orbit of the planet around the *śīghrocca* is larger than the orbit of the Sun around the Earth (*śīghra-vṛtta*) for exterior planets, and smaller for the interior planets. This is all as it should be.[10]

[10] For further details regarding the planetary model outlined in *Tantrasaṅgraha*, see the discussion in the Epilogue to this Volume.

8.16 *Śīghra* correction when there is latitude

In the earlier sections, while discussing the procedure for finding the true longitudes, the deflection of the planet from the ecliptic as it moves along its orbit was not considered. A detailed discussion of it is taken up in this section. Since the diurnal motion is not of any significance in this discussion, the *apakrama-maṇḍala* (ecliptic) is taken as an exact vertical circle situated east-west in the middle of the *bhagola*. This is the circle with the centre of the Earth as the centre. This is divided into 12 *rāśi*-s. Considering the two *rāśi-kūṭa*-s (poles of the ecliptic, which are the points of intersection of all the *rāśi*-s), which are diametrically opposite to each other, six circles are constructed. These are shown in Figure 8.12. It may be noted that these circles meet at the poles (*rāśi-kūṭa*-s) on the north-south line drawn through the centre of the *apakrama-maṇḍala*.

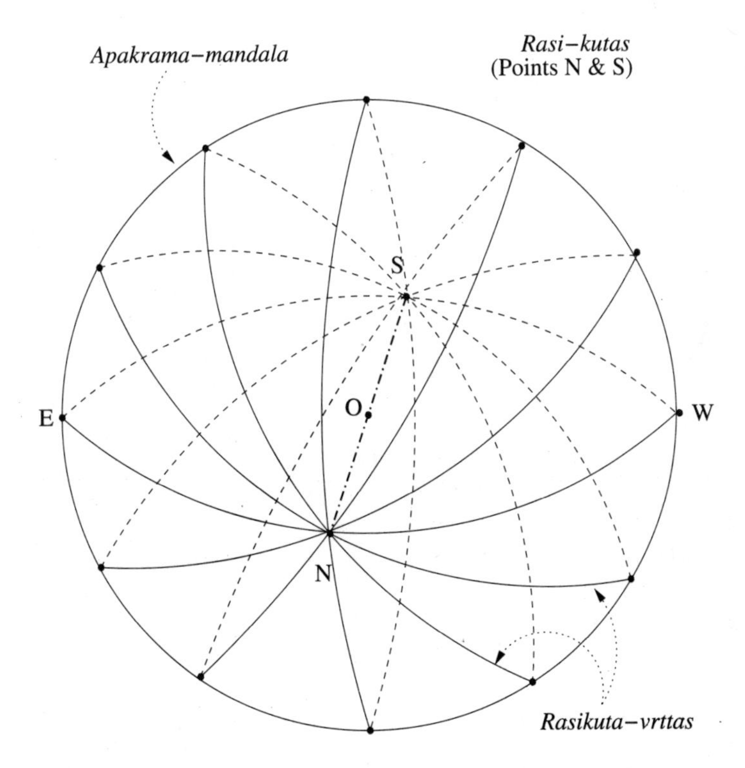

Figure 8.12: The *apakrama-maṇḍala* and the six *rāśi-kūṭa-vṛtta*-s.

By construction, the *śīghra-vṛtta* is in the plane of the *apakrama-maṇḍala*, with its centre at the centre of the Earth. The size of the *śīghra-vṛtta* will be different for different planets. The *manda-nīcocca-vṛtta* is a circle inclined to the *śīghra-vṛtta*, with its centre on the *śīghra-vṛtta*, and intersecting the *apakrama-maṇḍala* at the *pāta*-s (nodes, which have a retrograde motion). These are shown in Figure 8.13. The *pratimaṇḍala* and the *manda-karṇa-vṛtta* will be in the plane of *manda-nīcocca-vṛtta*, which is inclined to the plane of ecliptic. The planet P is on the *manda-karṇa-vṛtta* whose centre is S and will have *vikṣepa* (latitudinal deflection from the *apakrama-maṇḍala*).

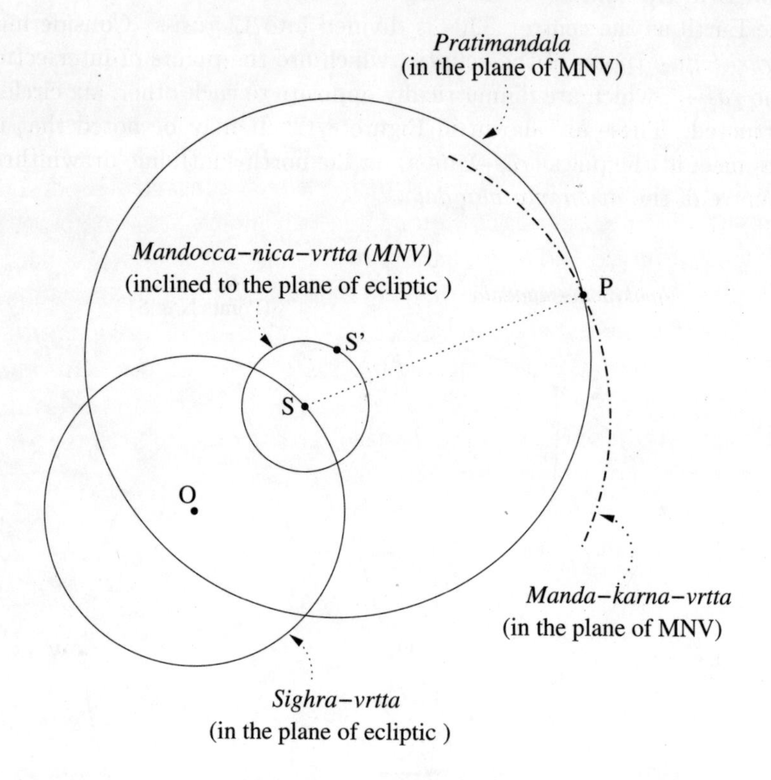

Figure 8.13: *Manda-karṇa-vṛtta* when there is latitude.

In Figure 8.14,

$$\widehat{PN} = \textit{manda-sphuṭa} - \textit{pāta}. \tag{8.93}$$

When the planet is 90° from the *pāta*, we have the maximum *vikṣepa* given by

$$v_{max} = K \sin i, \tag{8.94a}$$

where K is the radius of the *manda-karṇa-vṛtta* and i the inclination of the *manda-karṇa-vṛtta* to the *pratimaṇḍala*. At an arbitrary position P, the *vikṣepa* is given by

$$v \;= PB = K \sin\beta, \tag{8.94b}$$

where β is the latitude as observed from the *śīghrocca* S.[11] In arriving at the above result, we have used the planar triangle PSB. Considering the spherical triangle PNQ and applying the sine formula, we get

$$\frac{\sin PQ}{\sin i} = \frac{\sin(P - N)}{\sin 90}, \tag{8.95}$$

where P and N are the longitudes of the planet and the node respectively. Hence the *vikṣepa* ($K \sin PQ = v \; = K \sin\beta$) is given by

$$\begin{aligned} v &= K \sin PQ \\ &= K \sin i \; \sin(P - N) \\ &= K \sin i \; \sin(\textit{manda-sphuṭa} - \textit{pāta}) \\ &= \frac{K \sin i}{R} \times R \sin(\textit{manda-sphuṭa} - \textit{pāta}) \\ &= \frac{v_{max}}{\textit{trijyā}} \times R \sin(\textit{manda-sphuṭa} - \textit{pāta}). \end{aligned} \tag{8.96}$$

It is precisely the above equation (8.96) that is given in the Text.

In Figure 8.14, let $SS' = PB$, be perpendicular to the *apakrama-maṇḍala*. *Vikṣepa-koṭi-vṛtta* is the circle parallel to the *śīghra-vṛtta* with $S'P = SB$ as the radius. Considering the triangle SPB, since $SP = K$, the radius of *vikṣepa-koṭi-vṛtta* $S'P$ is equal to

$$\begin{aligned} SB &= K \cos\beta \\ &= \sqrt{K^2 - K^2 \sin^2\beta} \\ &= \sqrt{\textit{manda-karṇa}^2 - \textit{vikṣepa}^2}. \end{aligned} \tag{8.97}$$

This is in the measure of *pratimaṇḍala*, when the *manda-karṇa* and *vikṣepa* are in that measure.

The *vikṣepa-koṭi-vṛtta* is essentially *manda-karṇa-vṛtta* projected on to the plane parallel to the *śīghra-vṛtta* in which the planet is located. The *śīghra-sphuṭa* should be calculated taking the *vikṣepa-koṭi* as the *manda-karṇa*. The

[11] This is essentially the heliocentric latitude of the planet.

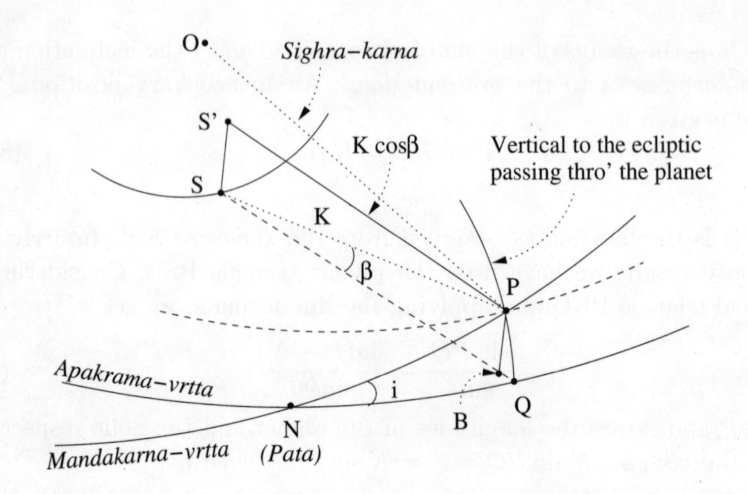

Figure 8.14: The latitudinal deflection (*vikṣepa*) of a planet.

result is the *graha-sphuṭa* (true planet) on the *śīghra-karṇa-vṛtta*, which is a circle with O' as the centre. O' is the point in the plane of *vikṣepa-koṭi-vṛtta*, at distance $OO' = PB$ from the *apakrama-maṇḍala.* The distance between the centre of the *apakrama-maṇḍala* and the planet P, represented by OP in Figure 8.15, is the *bhū-tārāgraha-vivara. Vivara* is distance of separation; *tārāgraha* is planet; and *bhū* is Earth. Hence, the term *bhū-tārāgraha-vivara* means the distance of separation between the Earth and the planet.

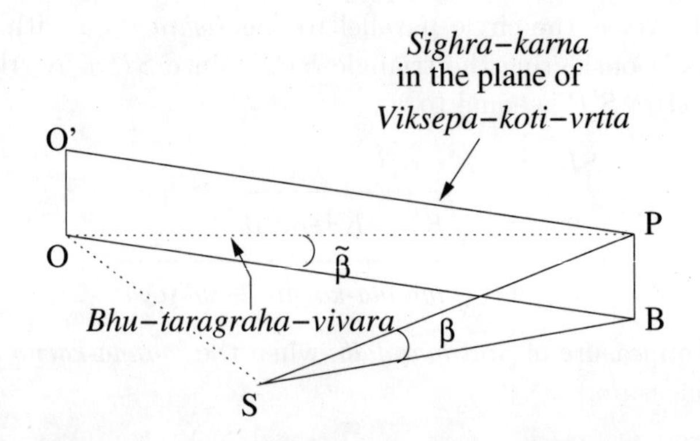

Figure 8.15: The *bhū-tārāgraha-vivara* and the *bhagola-vikṣepa.*

The angle of deflection of the planet, $\tilde{\beta}$ as seen from *bhagola-madhya*, is different from the angle of deflection β, as seen from S, which represents the *āditya-madhyama*. *Bhū-tārāgraha-vivara* is given by,

$$OP = \sqrt{O'P^2 + OO'^2} = \sqrt{(\textit{śīghra-karṇa}^2 + \textit{vikṣepa}^2)}. \tag{8.98}$$

Now the *vikṣepa* is also given by

$$v \;= OP \sin \tilde{\beta}. \tag{8.99}$$

Therefore,

$$R \sin \tilde{\beta} = \textit{vikṣepa} \times \frac{R}{OP}. \tag{8.100}$$

In the above equation, LHS is nothing but the *bhagola-vikṣepa*. That is, the latitude of the planet as seen from the Earth. The term *bhagola* is used as an adjective to *vikṣepa* to indicate the fact that the Earth is taken to be at the centre of the *bhagola* and hence the *vikṣepa* as seen from the Earth is the same as the *bhagola-vikṣepa*. Thus, we see that

$$\textit{Bhagola-vikṣepa} = \textit{vikṣepa} \times \frac{\textit{trijyā}}{\textit{bhū-tārāgraha-vivara}}. \tag{8.101}$$

Thus the angle $\tilde{\beta}$ is found from (8.96) and (8.100). This is the geocentric latitude. Though the *vikṣepa-koṭi-vṛtta* is smaller than the *apakrama-vṛtta*, the angles are the same for the both, just as the hour-angle in the diurnal circle is the same as in the equatorial circle.

The case when the *śīghra-vṛtta* itself is inclined to the *apakrama-maṇḍala*

Next, the more general case when the *śīghra-vṛtta* itself is inclined to the *apakrama-maṇḍala* is considered. This is a hypothetical case, as the *śīghra-vṛtta*, the orbit of the Sun around the Earth, is stated to be in the plane of the *apakrama-maṇḍala*. In Figure 8.16, let i and i' be the inclinations of *śīghra-vṛtta* with respect to the *apakrama-maṇḍala*, and that of the *manda-karṇa-vṛtta* (with respect to the *śīghra-vṛtta*) respectively. Let S be the *śīghrocca*, N the *pāta* of the *śīghra-vṛtta* (intersection point of *śīghra* and *apakrama-maṇḍala*-s). P is the *manda-sphuṭa-graha* in the *manda-karṇa-vṛtta* with S as the centre, and N', the *pāta* of the *manda-karṇa* (intersection point of *manda-karṇa* and the *śīghra-vṛtta* plane). Here $SC = OO'$ and $PB = SS'$.

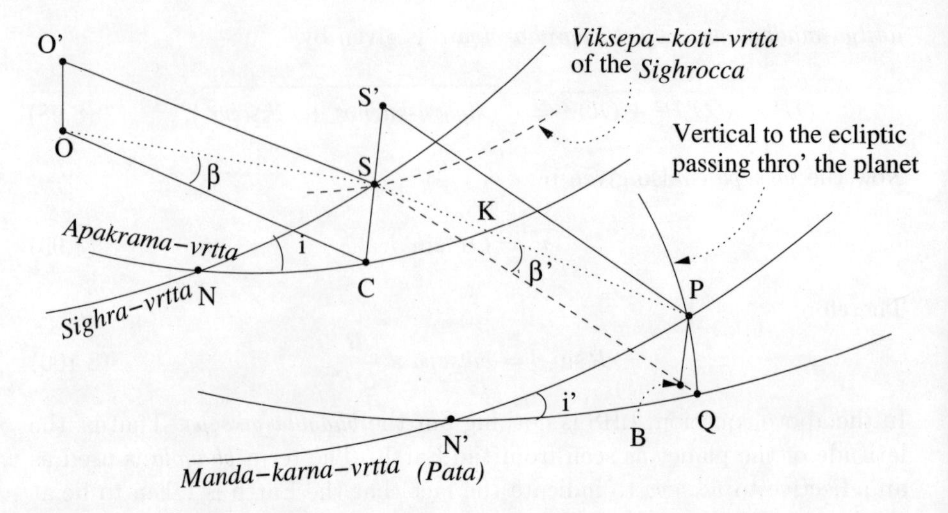

Figure 8.16: The latitude of a planet when *śīghra-vṛtta* itself is inclined to *apakrama-maṇḍala.*

Now, the *vikṣepa* of the centre of the *manda-karṇa-vṛtta* (S), which lies on the circumference of *śīghra-vṛtta*, is given by

$$\begin{aligned} SC &= OS \sin\beta \\ &= OS \sin i \sin(S-N) \\ &= r_s \sin i \sin(S-N). \end{aligned} \tag{8.102}$$

This *vikṣepa* is in the measure of the *pratimaṇḍala*, where r_s is the radius of the *śīghra-vṛtta* which is also the *śīghrāntyaphala.* Then, we need to find $\sqrt{OS^2 - SC^2}$. This gives $OC = O'S$ which is equal to *vikṣepa-koṭi* in terms of the minutes of the arc of the *pratimaṇḍala.* The latitude of the planet with reference to the *śīghrocca* (point S in the *śīghra-vṛtta*), called the *manda-karṇa-vikṣepa*, is

$$\begin{aligned} PB &= SP \sin\beta' \\ &= SP \sin i' \sin(P-N') \\ &= K \sin i' \sin(P-N'). \end{aligned} \tag{8.103}$$

This is in the measure of the *pratimaṇḍala*, where K is the *manda-karṇa.* If both CS and BP are north (of *apakrama* and *śīghra-vṛtta* respectively), or south, then the net *vikṣepa* of the planet will be

$$v_{tot} = SC + PB. \tag{8.104}$$

This case is represented in Figure 8.16. If one of them is north and other is south, then the net *vikṣepa* of the planet will be

$$v_{tot} = SC \sim PB. \tag{8.105}$$

When the *śīghra-vṛtta* is also inclined to the *apakrama-maṇḍala*, we have to find the radius of the *vikṣepa-koṭi-vṛtta* of the *śīghrocca* first, for finding the longitude of P. The radius of this is $r_s \cos\beta$. We have to find the *mandakarṇa-vikṣepa-koṭi-vṛtta* too. The *śīghra-bhujā-phala* is determined using the first circle as the *śīghra-nīcocca-vṛtta*, and the second as the *pratimaṇḍala*. This is applied to the *manda-sphuṭa* to obtain the *śīghra-sphuṭa* P.[12]

The Text mentions that it is only giving the procedure for the hypothetical situation when the *śīghra-vṛtta* happens to be inclined to the ecliptic – not that such a situation actually arises in practice. Then it gives a remarkable example where the above general discussion may find application: namely, when we seek to make computations with respect to an observer at the centre of the Moon. The Text also makes a very perceptive remark that, then, the Moon's orbit is to be considered as the *ucca-nīca-vṛtta*. The motion with respect to the *bhagola-madhya* is determined from the position with respect to the Moon's centre. The Text also notes that we can use this procedure to convert computations from a Moon-centric frame of reference to the geocentric frame.

8.17 Calculation of the mean from the true Sun and Moon

In this and the next few sections, the reverse problem of finding the mean position from the true position of the planet is considered. First, the Sun and the Moon are considered, for which only *manda-saṃskāra* is applicable. The corresponding problem for the planets is more involved as it involves two *saṃskāras*, and is considered later.

[12]Here the Text does not specify how the *manda-karṇa-vikṣepa-koṭi-vṛtta* may be found; For this, we have to find the angle between SP and the *apakrama-maṇḍala*.

In (8.58), it was shown that

$$\begin{aligned} R\ \sin(madhya - ucca) &= K\ \sin(sphuṭa - ucca) \\ R\ \sin(M - U) &= K\ \sin(P - U), \end{aligned} \tag{8.106}$$

where R is *trijyā* and K is the *karṇa.* Here, if the *manda-karṇa* K is known, then we know $(M - U)$ in terms of $(P - U)$. If we add *ucca* to this we will get the *madhya.* That is,

$$(M - U) + U = M. \tag{8.107}$$

Or, with reference to Figure 8.7 on page 637

$$\begin{aligned} O'\hat{P}_1C_1 &= B_1\hat{O}'P_1 - B_1\hat{O}P_1 \\ &= madhya - sphuṭa. \end{aligned} \tag{8.108}$$

Considering the triangle $O'OC_1$, the *bāhu-phala* is given by

$$O'C_1 = r\ \sin\ (sphuṭa - ucca),$$

where r is the radius of the *manda-nīcocca-vṛtta.* Considering the triangles $P_1O'C_1$ and $O'OC_1$, we have

$$R\ \sin(M - P) = O'C_1 = r\ \sin(P - U). \tag{8.109}$$

Here r should be in the measure of *pratimaṇḍala.* That is,

$$r = r_0 \frac{K}{R}, \tag{8.110}$$

where, r_0 is the mean epicycle radius given in the Text. As before, if K is known (*madhya* – *sphuṭa*) is determined. Adding *sphuṭa* to this, we find the *madhyama-graha* (mean planet).

Note: In both the above relations (8.106) and (8.110), the *aviśiṣṭa-manda-karṇa* K is used which itself has to be determined. The method for determining this (when the *manda-sphuṭa* is known) is given in *Tantrasaṅgraha* II. 46-47. It is based on first computing the *viparīta-karṇa*, which can be expressed in terms of the *manda-sphuṭa* P, the *ucca* U, the mean epicycle radius r_0 and *trijyā* R, by the relation given earlier (8.52):

$$R_v\ = [R^2 + r_0^2 - 2r_0 R\cos(P - U)]^{\frac{1}{2}}. \tag{8.111}$$

The *manda-karṇa* is then found from the relation $K = \frac{R^2}{R_v}$.

8.18 Another method for the mean from true Sun and Moon

Again, from (8.59), we get

$$K\ \sin(madhya - sphuṭa) = r\ \sin(madhya - ucca).$$

Therefore,

$$\begin{aligned} R\ \sin(M - P) &= \frac{rR}{K} \sin(M - U) \\ &= r_0\ \sin(M - U), \end{aligned} \qquad (8.112)$$

where r_0, the mean radius of the *manda-nīcocca-vṛtta*, is of course a known parameter. The *madhya* is obtained from this equation using an iterative procedure. First, the *sphuṭa* itself is taken to be the *madhya* in the *RHS* of (8.112), and the *madhya* – *sphuṭa* is calculated. Adding *sphuṭa* to this, we get the new *madhya*. This is approximate. This is used in the *RHS* now, and *madhya* – *sphuṭa* is again calculated. Adding *sphuṭa* to this, the next iterated value of *madhya* is found. The process is repeated. It is noteworthy that here the *aviśiṣṭa-karṇa* K does not come into the picture at all.

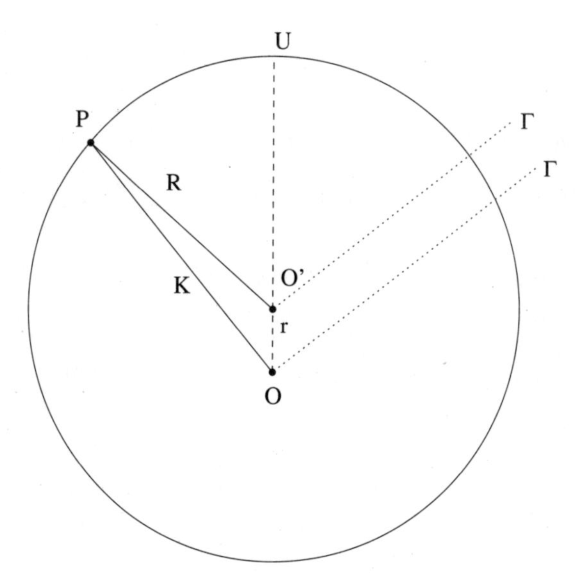

Figure 8.17: Finding the mean planet from the true planet.

Another interesting iterative procedure for determining the *madhya* from the *sphuṭa* is described next. In Figure 8.17, $O'P = R$, $OP = K$, $OO' = r$, $U\hat{O}P = P - U$ and $U\hat{O'}P = M - U$. Hence,

$$O\hat{P}O' = (M - U) - (P - U) = M - P. \tag{8.113}$$

Considering the triangle OPO' we have

$$\frac{OP}{\sin O\hat{O'}P} = \frac{OO'}{\sin O\hat{P}O'} = \frac{O'P}{\sin O'\hat{O}P}. \tag{8.114}$$

Therefore we have

$$R\sin(M - P) = r\sin(P - U), \tag{8.115}$$

$$\text{and} \quad K\sin(M - P) = r\sin(M - U). \tag{8.116}$$

From the above relations, we get

$$r\sin(M - U) = \frac{K}{R}\, r\sin(P - U). \tag{8.117}$$

Therefore,

$$r\sin(M - U) - r\sin(P - U) = \frac{(K - R)}{R} r\sin(P - U). \tag{8.118}$$

Again, in the triangle OPO', the *karṇa* can be expressed in terms of the *sphuṭa* via the relation

$$K = \{R^2 - r^2\sin^2(P - U)\}^{\frac{1}{2}} + r\cos(P - U). \tag{8.119}$$

Neglecting the term containing square of *phala-varga* ($r^2\sin^2(P - U)$), we get

$$K \approx R + r\cos(P - U). \tag{8.120}$$

Using the above in (8.118), we have

$$r\sin(M - U) - r\sin(P - U) \approx r\cos(P - U)\frac{r\sin(P - U)}{R}. \tag{8.121}$$

If the true epicycle radius r is known (it can be found by computing the *karṇa* K), then the above equation can be used to determine the *manda-kendra* $(M - U)$ and hence the *madhyama*. From (8.115) and (8.121), we also obtain

$$r\sin(M - U) - r\sin(P - U) \approx r\cos(P - U)\frac{R\sin(M - P)}{R}. \tag{8.122}$$

As $(M-P)$ is small, $R\sin(M-P) \approx (M-P)$ in minutes. Therefore the above equation reduces to

$$r\sin(M-U) - r\sin(P-U) \approx r\cos(P-U)\frac{(M-P)}{R} \text{ in minutes.} \quad (8.123)$$

In the above equation, LHS is the *bhujāphala-khaṇḍa* and $(M-P)$ is the difference in the arc (*cāpa*) between the true and the mean planet. Therefore,

$$\textit{bhujā-phala-khaṇḍa} = \textit{koṭi-phala} \times \frac{\textit{cāpa} \text{ corr. to difference}}{\textit{trijyā}}. \quad (8.124)$$

This is nothing but the relation

$$\begin{aligned} R\sin(\theta+\delta\theta) - R\sin\theta &= R\cos\theta \times \delta\theta \\ &= R\cos\theta \times \frac{R\,\delta\theta}{R}. \end{aligned} \quad (8.125)$$

It is further mentioned that *bhujā-khaṇḍa* is according to *koṭijyā*. The mean planet M is to be found iteratively from (8.123) as mentioned earlier. Equation (8.123) is an approximate relation. If the approximate value of M is found by any method, that can be used in the RHS and M can be determined iteratively from (8.123).

8.19 Calculation of the mean from true planet

The mean of all the planets can be obtained from their *manda-sphuṭa* in the same way as outlined above. The process of determining the *manda-sphuṭa* from *śīghra-sphuṭa* is indeed simpler. Considering the triangle OPS in Figure 8.9 on page 643, we have the following relation

$$r_s\sin(\mathcal{P}-S) = K\sin(M_s-\mathcal{P}). \quad (8.126)$$

Given that the longitude of *śīghrocca* is known, it follows from the above relation that if the *śīghra-sphuṭa* $\mathcal{P}$, radius of the *śīghra-nīcocca-vṛtta* r_s and the *manda-karṇa* K are known, $(M_s-\mathcal{P})$ and hence M_s can be determined.

The term in the LHS of (8.126) is *śīghra-khaṇḍa-bhujā-jyā* on *śīghra-nīcocca-vṛtta*. This equation could be written as

$$R\sin(M_s-\mathcal{P}) = R\sin(\mathcal{P}-S)\frac{r_s}{K}. \quad (8.127)$$

Here it is noted that R is taken to be 80 in *Tantrasaṅgraha* and 360 in other texts.[13] From this relation, M_s or *manda-sphuṭa* on *manda-karṇa-vṛtta* is obtained.

It is noted that while calculating *manda-sphuṭa* from *manda-kendra*, the *karṇa* K has to be found by *aviśeṣa-karma* or iteration, but the *karṇa* does not appear while calculating the *bhujā-phala*. When we want to calculate *madhyama* from *manda-sphuṭa* we do not have the simple relation as above, and we have to either evaluate the *aviśiṣta-manda-karṇa* K (in terms of the *sphuṭa*) or do iteration on the equation

$$r_0 \sin(M - U) = R \sin(M - P), \tag{8.128}$$

where the unknown *madhyama* appears on both sides of the equation.

On the other hand, when *śīghra-bhujā-phala* is calculated, we need to compute the *karṇa* K_s. But when we calculate *manda-sphuṭa* from *śīghra-sphuṭa* no iteration is required.

For deriving M_s from M, the *karṇa* is not required. We have

$$\begin{aligned} K \sin(M_s - M) &= r \sin(M - U) \\ \text{or} \quad R \sin(M_s - M) &= r_0 \sin(M - U), \end{aligned} \tag{8.129}$$

as $\frac{r}{K} = \frac{r_0}{R}$. It may be noted that though *karṇa* does not appear in the above equation, when M is to be calculated from M_s, we need a *aviśeṣa-karma* or successive iteration process.

On the other hand, when $\mathcal{P}$ (*śīghra-sphuṭa*) is calculated from M_s, *karṇa* is required. From the triangle OPS in Figure 8.9 on page 643, we have

$$K_s \sin(\mathcal{P} - S) = K \sin(M_s - S). \tag{8.130}$$

But, for deriving M_s from $\mathcal{P}$ using

$$r_s \sin(\mathcal{P} - S) = K \sin(M_s - \mathcal{P}), \tag{8.131}$$

no iteration is required. However, it is noted that using the above equation, $\mathcal{P}$ can also be found by an *aviśeṣa* process. That is, we need to take $\mathcal{P} = M_s$ in LHS, and find $M_s \sim \mathcal{P}$ and then $\mathcal{P}$. Then, put the new value of $\mathcal{P}$ in LHS, find $M_s - \mathcal{P}$, thus a new $\mathcal{P}$ and so on. Thus the *śīghra-sphuṭa* $\mathcal{P}$ can be found by an *aviśeṣa* process.

[13] There is a complication that the *manda-karṇa* varies with the *manda-kendra* – but the text seems to imply that K in the RHS is replaced by R itself.

8.20 Computation of true planets without using *Manda-karṇa*

The Text has so far clearly prescribed a two step process to compute the true planet from the mean planet – *manda-saṃskāra* (which is essentially converting the mean heliocentric planet to the true heliocentric planet) followed by *śīghra-saṃskāra* (converting heliocentric planet to geocentric planet). Here the *manda* correction can be read-off from a table as, given the mean epicycle radius, the *manda-phala* is not a function of the *manda-karṇa*. But this is not the case for *śīghra* correction, for the *śīghra-phala* depends not only on the *śīghra-kendra*, but also on the *śīghra-karṇa* which (as we see from (8.69)) depends on *manda-karṇa*, which in turn depends on *manda-kendra*. Hence, given the radius of *śīghra-nīcocca-vṛtta*, *śīghra-phala* cannot be read off from a table as a function of *śīghra-kendra* alone, as it also depends on *manda-karṇa* and hence on the *manda-kendra*.

The Text presents an elaborate derivation showing that it is possible to simulate, to some extent, the effect of *manda-karṇa* in *śīghra-phala* by doing a four-step process instead of the two-step precess discussed so far. For the exterior planets, texts of the Āryabhaṭa school from *Mahābhāskarīya* to *Tantrasaṅgraha* prescribe the following steps: (i) If M is the *madhyama*, apply half-*manda-phala* to it to obtain M'. (ii) Using M' evaluate the *śīghra-phala*, where the *śīghra-karṇa* is calculated as in (8.69), but with the *manda-karṇa* K replaced by the *trijyā* R, and apply half of this *śīghra-phala* to M' to obtain M''. (iii) Using M'' evaluate the *manda-phala* and applying that to M to obtain the *manda-sphuṭa* M_s. (iv) Use the *manda-sphuṭa* to calculate the *śīghra-phala*, where the *śīghra-karṇa* is calculated with *manda-karṇa* replaced by the *trijyā* R, to obtain the *śīghra-sphuṭa*, the true planet $\mathcal{P}$.

The Text outlines a derivation, which purports to show that under certain approximations, there is no appreciable difference between the above *śīghra-sphuṭa*, and the one obtained by calculating the *śīghra-phala* with the *manda-karṇa*-dependent *śīghra-karṇa*, as described earlier in section 8.14.

For the interior planets, Mercury and Venus, earlier texts such as *Mahābhāskarīya* prescribe a three-stage process: Application of half *manda-phala* followed by *manda-saṃskāra* and the *śīghra-saṃskāra*, where, in the latter correction, the *śīghra-karṇa* is calculated in terms of the radius R only, and not in terms of the *aviśiṣṭa-manda-karṇa*. However, *Tantrasaṅgraha* does not prescribe any three-stage process for the interior planets. Instead, it prescribes just the *manda-saṃskāra* followed by the *śīghra-saṃskāra*,[14] where the latter involves the use of *aviśiṣṭa-manda-karṇa*. Further, as was noted earlier, *Tantrasaṅgraha* also stipulates that the *manda-phala* should be applied to the mean planet and not the mean Sun as stipulated in the earlier texts.

The Text presents an elaborate justification to show how the effect of the *aviśiṣṭa-manda-karṇa* in the simple two step process of *manda-saṃskāra* followed by *śīghra-saṃskāra* can be simulated by employing a multi-stage process. It also presents a discussion of alternative models proposed by the *Parahita* School, by Muñjāla and others, who employ different rules for the variation of *manda-karṇa*. The Text also discusses the pre-*Tantrasaṅgraha* formulations for interior planets.

However, details of the argument presented in the Text are not entirely clear to us. Perhaps, a study of the discussion of the same topic as found in Śaṅkara Vāriyar's commentary *Yukti-dīpikā* on *Tantrasaṅgraha* may help in explicating all the details of the argument as presented in the Text.

[14] *Tantrasaṅgraha*, II.68–79.

Chapter 9

Earth and Celestial Spheres

The chapter commences with a discussion on the three spheres, (i) *Bhūgola* – the terrestrial sphere, (ii) *Vāyugola* – the equatorial celestial sphere (described with reference to the celestial equator which is revolving uniformly due to *Pravaha-vāyu*) and (iii) *Bhagola* – the zodiacal celestial sphere (described with reference to the ecliptic). This is followed by a discussion on the motion of equinoxes. Then, we find the description of some of the important great circles and their secondaries, which are used as the reference circles for describing the location of a celestial object using different co-ordinates. Finally, there is an elaborate discussion on the determination of the declination of a celestial object with latitude.

9.1 *Bhūgola*

Bhūgola[1] means the spherical Earth. Some of the physical properties of the Earth that are mentioned here are listed below:

- It is a sphere situated at the centre of the *Bhagola* or *Nakṣatra-gola*[2].
- It is suspended in space without any support.
- It supports all living and nonliving beings on its surface.
- It is the nature of all heavy things to fall towards the Earth from all the directions around.
- It is situated *below* when viewed from any part of the sky.

[1] *Bhū* is Earth and *gola* is sphere.

[2] The terms *bhaṃ* and *nakṣatraṃ* are synonyms and refer to a star.

- The sky is *above* from all locations on its surface.
- Its southern half is predominantly filled with water, whereas the northern half is predominantly land.
- India (*Bhārata-khaṇḍa*) is located in the northern half.

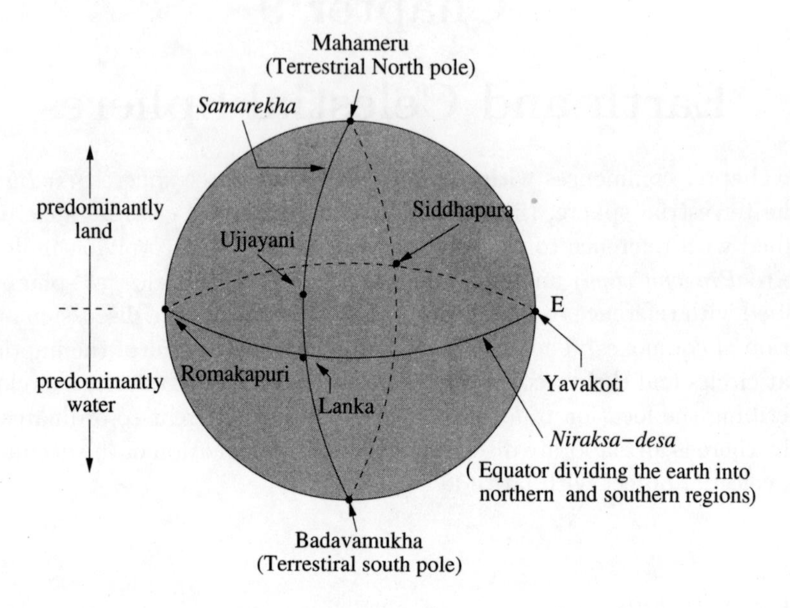

Figure 9.1: *Bhūgola* - The spherical Earth.

Continuing with the description, a few important locations on the surface of the Earth are mentioned. They are specified with reference to *nirakṣa-deśa* and *samarekhā*. *Nirakṣa-deśa* refers to the locus of points with zero latitude (the terrestrial equator). *Samarekhā* is a longitude circle (secondary to the equator). The names of the cities located at the four corners on the terrestrial equator which are ninety degrees apart are mentioned. The names of the north and the south poles are also given. Ujjayanī is situated on the *samarekhā* passing through Laṅkā, and has a northern latitude. The names of these places and their locations on the Earth are indicated in Figure 9.1.

Then we find the description of *Dhruva*-s (celestial poles) and the diurnal circles of celestial objects. For an observer having a northern latitude, the northern *Dhruva* P_1 is visible, whereas the southern *Dhruva* P_2 is not visible, as it lies below the horizon (see Figure 9.2).

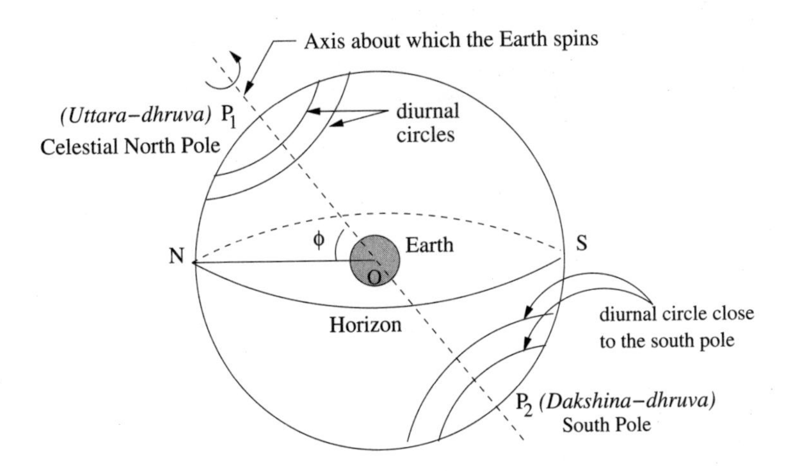

Figure 9.2: The celestial sphere for an observer having northern latitude.

On the other hand, for an observer on the equator, both the *Dhruva*-s (celestial poles) P_1 and P_2 lie on the horizon and hence both are visible. The relationship between the location of the *Dhruva* and the latitude of the place is given by:

$$N\hat{O}P_1 = \text{Altitude of the } Dhruva = \text{Latitude of the place} = \phi,$$

as in Figure 9.2. Stars near the northern *Dhruva* P_1 would be circumpolar (they never rise or set). Similarly, stars near the southern *Dhruva* P_2 would never be observed as they are always below the horizon. However at the equator, all the stars would be visible, as can be seen in Figure 9.3.

9.2 *Vāyugola*

In Figure 9.3, S_1, S_2 are the diurnal paths of the stars which are close to the *Dhruva* P_1. The horizons for an equatorial observer and an observer with a northern latitude ϕ, are also indicated. P_1, P_2 are the north and south poles. S_3 and S_4 are the diurnal circles (*svāhorātra-vṛtta*-s) of two stars which are far removed from the *Dhruva*-s. The *svāhorātra-vṛtta*-s are shown by dotted lines. As viewed from the equator, these are vertical circles parallel to the celestial equator which is called the *ghaṭikā-maṇḍala*. The radius of the *svāhorātra-vṛtta*-s keep gradually decreasing as they approach

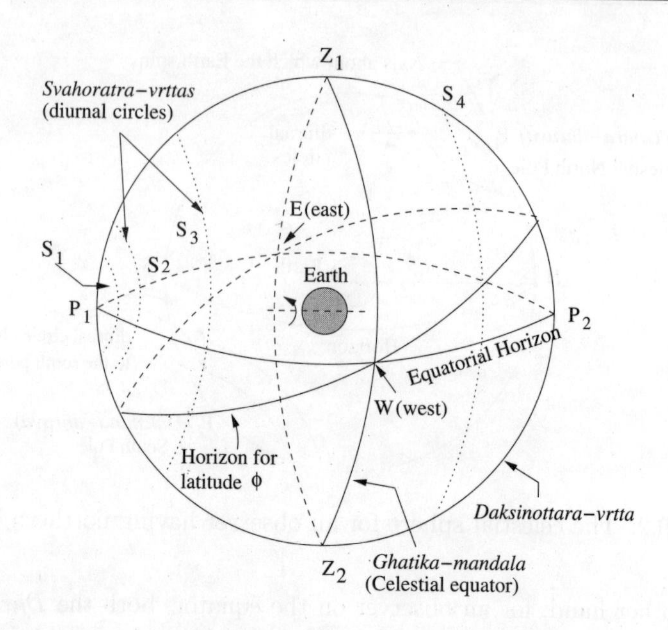

Figure 9.3: The celestial sphere for an observer on the equator.

the *Dhruva* from the equator. The axis of the celestial sphere passes through the two *Dhruva*-s, P_1 and P_2.

The *dakṣiṇottara-vṛtta* (prime meridian) is the great circle passing through the poles and the zenith. *Laṅkā-kṣitija* is the equatorial horizon. Further, it may be noted that the three great circles *ghaṭikā-maṇḍala*, *dakṣiṇottara-vṛtta* and *Laṅkā-kṣitija* are perpendicular to each other. They intersect at six points: P_1, W, P_2, E, Z_1 and Z_2. While the first four points lie on the horizon, the latter two are the poles of the horizon right above and below. These six points are called the *svastika*-s, cardinal points. The three great circles divide the celestial sphere into eight equal parts, four above the horizon, and four below.

9.3 *Bhagola*

The celestial sphere described with reference to the ecliptic as the central circle is the *Bhagola.* This may be contrasted with the *Vāyugola* described earlier, which has celestial equator as the central circle and the diurnal circles

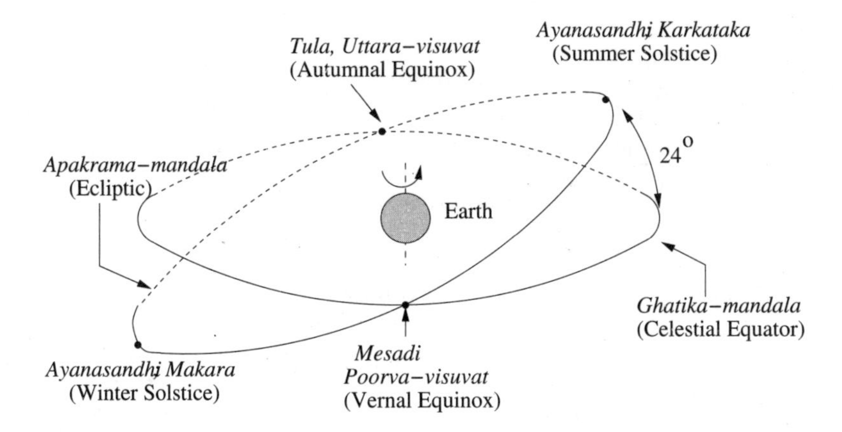

Figure 9.4: The celestial equator and the ecliptic.

on its sides. The *apakrama-maṇḍala* or the ecliptic is the path traced by the Sun in its eastward (annual) motion. In Figure 9.4, the four important points on the ecliptic and its orientation with respect to the celestial equator are indicated.

In Figure 9.5, the different orientations of the ecliptic with respect to the celestial equator at different times during the day are depicted for an equatorial observer. In Figure 9.5(a) *Meṣādi* is shown at the east point of the horizon; it is just rising. In (b) it is at the zenith. In (c) it is setting and is at the west point and in (d) it is at the nadir. In 9.5(d), the other halves of the equator and the ecliptic (which is usually shown by dashed lines) have not been shown.

Just as the celestial equator is the central great circle of the *Vāyugola*, the ecliptic is the central great circle of the *Bhagola*. The two poles of the ecliptic K_1 and K_2 are the *rāśikūṭa*-s.[3] They bear the same relation to the ecliptic, as the *Dhruva*-s P_1 and P_2 to the celestial equator. A *rāśi-kūṭa-vṛtta* (secondary to the ecliptic) is a great circle passing through K_1 and K_2.

Consider the situation when the *Meṣādi* is at the zenith. Then the ecliptic is a vertical circle. In this situation, the poles of the ecliptic, K_1, K_2 lie

[3]The word *rāśi-kūṭa* refers to a point of intersection of all the *rāśi*-s. That the poles of the ecliptic are the points where all the *rāśi*-s meet can be seen from Figure 9.6 on page 673.

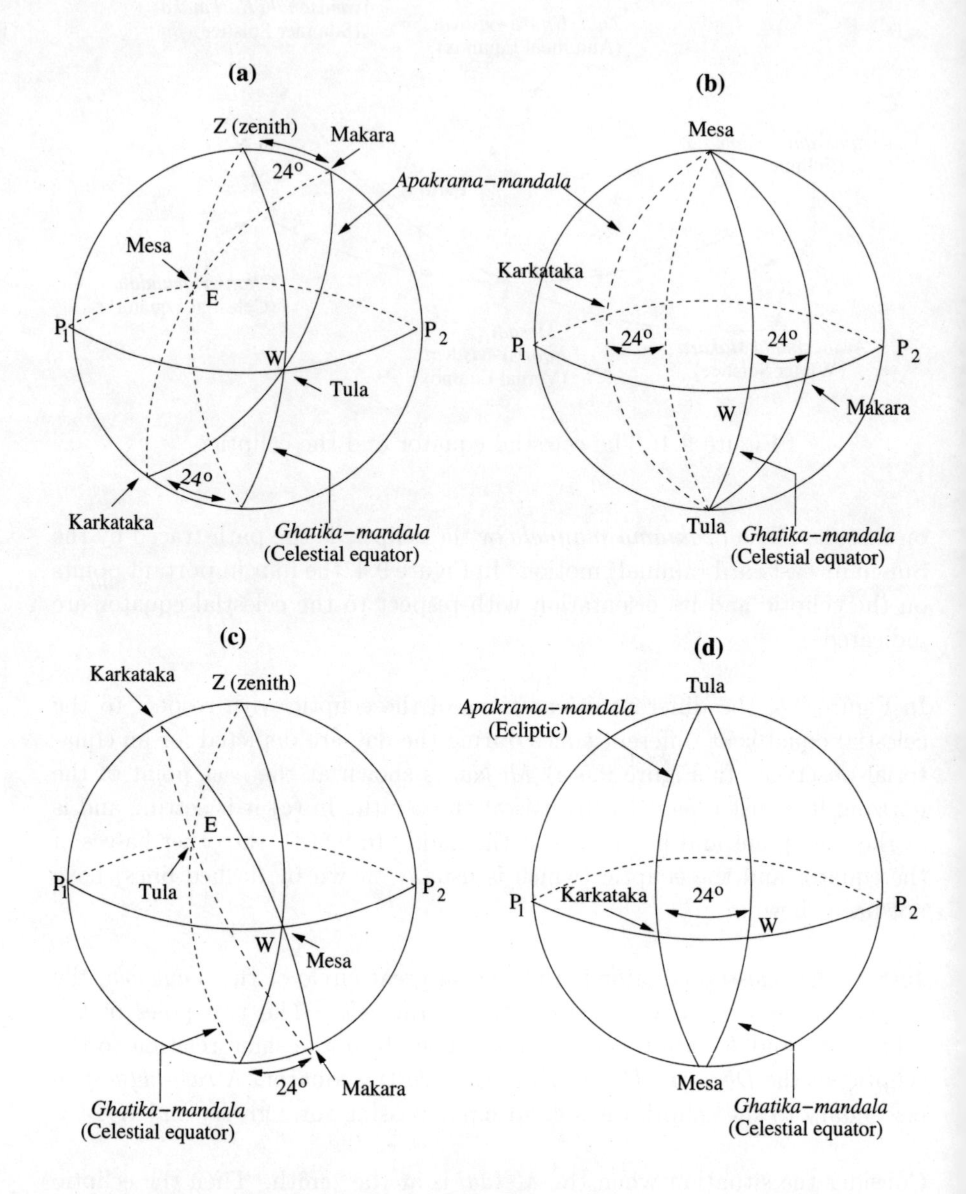

Figure 9.5: (a) *Meṣādi* is rising; (b) *Meṣādi* is at the zenith; (c) *Meṣādi* is setting; (d) *Meṣādi* is at the nadir.

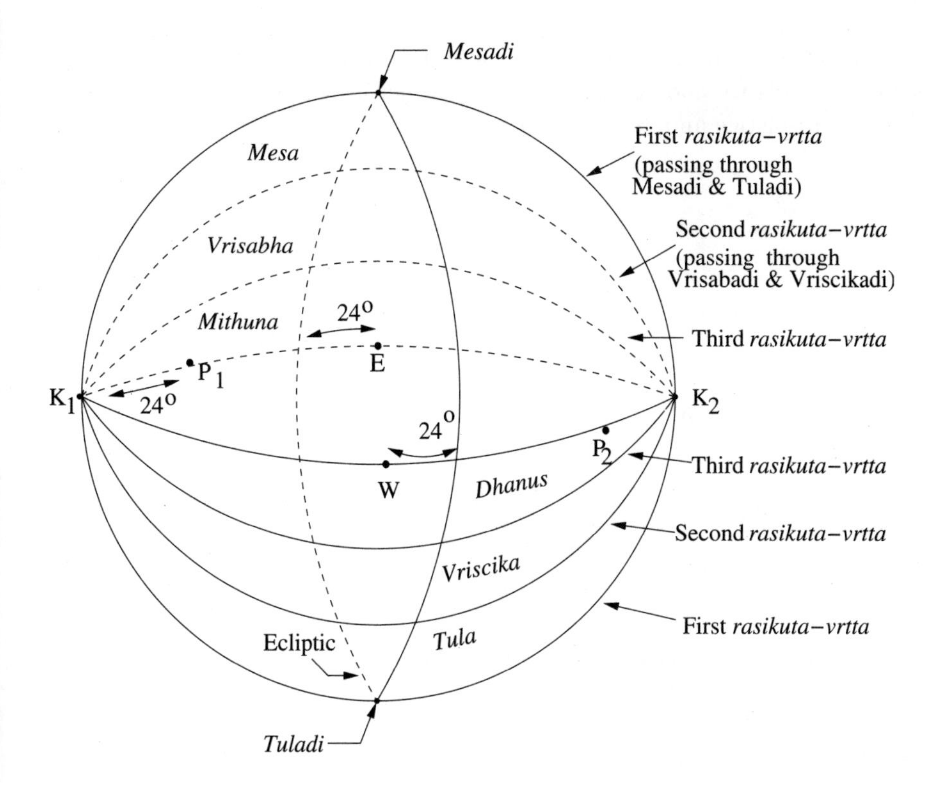

Figure 9.6: The *rāśi-kūṭa-vṛtta*-s.

on the *kṣitija* (horizon) and are 24° west of the north *Dhruva* (P_1) and 24° east of south *Dhruva* (P_2), respectively. The *rāśi-kūṭa-vṛtta* passing through *Meṣādi* and *Tulādi* is the north-south circle. Similarly we can conceive of the *rāśi-kūṭa-vṛtta* passing through *Vṛṣabhādi* and *Vṛścikādi* which is separated from the earlier one by 30° along the ecliptic; similarly the one through the *Mithunādi* and *Dhanurādi*, and so on, as shown in Figure 9.6. The *Bhagola* with the ecliptic at the centre and the *rāśi-kūṭā*-s as the poles is completely spanned by these six *rāśi-kūṭa-vṛtta*-s passing through the beginning points of the twelve *rāśi*-s. Inside each *rāśi*, we can concieve of various circles to represent the division of the *rāśi* into degrees, minutes and seconds.

In Figure 9.7, the diurnal circles of the solstices, denoted by dotted lines and marked M_1 and M_2 are 24° away for the celestial equator. Similarly, the diurnal circles of the poles of the ecliptic K_1, K_2, denoted by the solid lines and marked C_1 and C_2, are 24° away from the poles P_1 and P_2. The other halves of the diurnal circles are not shown in the figure. It is clear that the northern solstice and K_2 rise and set together at the equator. Similarly, the southern solstice and K_1 rise and set together.

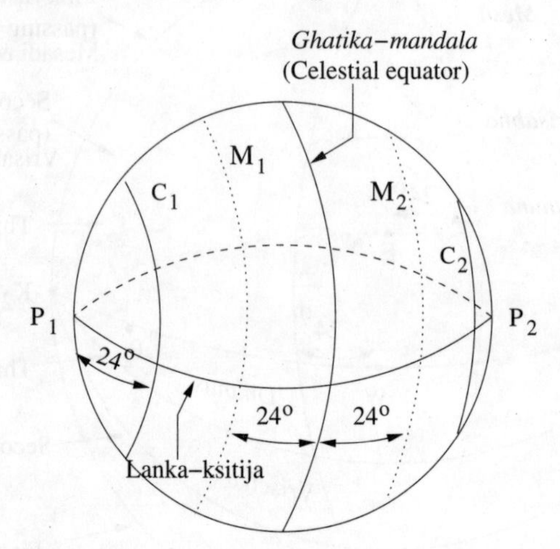

Figure 9.7: The diurnal circles of the poles of the ecliptic and the solstices.

9.4 *Ayana-calana*

The points of intersection of the celestial equator and the ecliptic, denoted by Γ and Ω, are called the equinoxes. The ends of Virgo (*Kanyā*) and Pisces (*Mīna*), or equivalently *Tulādi* and *Meṣādi*, would be the equinoxes at some epoch as shown in Figure 9.8. This would be the case when there is no *ayana-calana* and the equinoctial points are taken as the reference points for the measurement of *sāyana* or tropical longitude. But actually these points are in motion with respect to the fixed stars. The manner in which they move is described in the following section.

9.5 The nature of the motion of equinoxes

It is stated that the motion of equinoxes can be eastward or westward. These are schematically shown in Figures 9.9*a* and 9.9*b*. Actually, the motion described in the Text represents the phenomenon called *Trepidation*

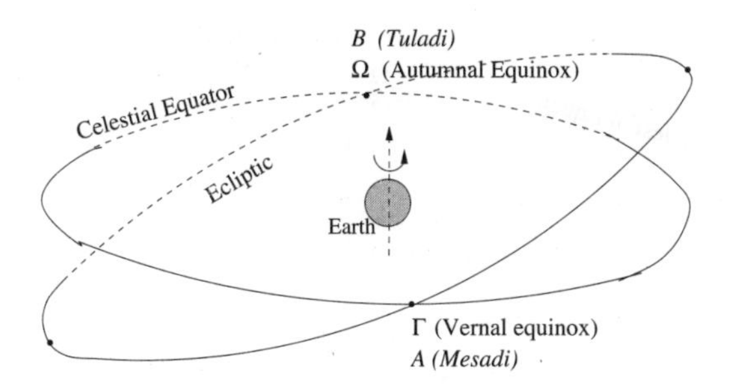

Figure 9.8: Equinoxes when there is no *ayana-calana*.

of equinoxes, where the equinox executes an oscillatory motion, going both eastwards and westwards from *Meṣādi* to a maximum extent of 24°. This is different from the continuous retrograde motion, which is usually referred to as the *Precession of the equinoxes*.

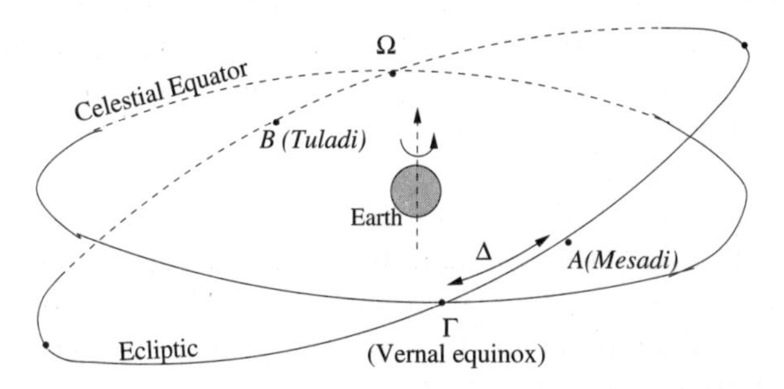

Figure 9.9*a*: The westward motion of the vernal equinox.

In Figure 9.9*a*, the motion of the equinox is shown westward (retrograde). Hence, the amount of precession/trepidation should be added to the *nirayaṇa* longitude, longitude measured from the *Meṣādi* eastwards, to obtain the tropical longitude, longitude measured from the vernal equinox eastwards.

In Figure 9.9*b*, where the motion of the equinox is shown eastward (direct), the amount of precession/trepidation should be subtracted from the *nirayaṇa* longitude to obtain the tropical longitude. The obliquity of the ecliptic remains the same at 24° even as the motion of the equinoxes takes place.

With respect to an observer on the Earth, it is the ecliptic which is moving and not the celestial equator. Because of this, the *rāśi-kūṭā*-s also have a

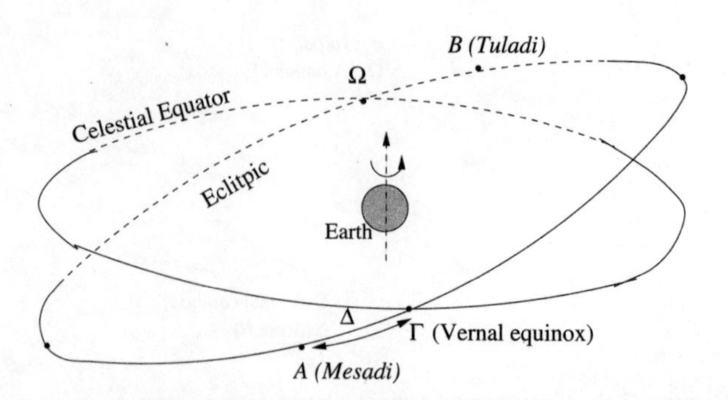

Figure 9.9*b*: The eastward motion of the vernal equinox.

motion. But their diurnal circles remain the same as the deviation of the *rāśi-kūṭā*-s from the *Dhruva*-s is always 24°. This can be explained through the *ayanānta-rāśi-kūṭa-vṛtta* which is the *rāśi-kūṭa-vṛtta* (see Figure 9.10) passing through the *ayanānta*-s (the solsticial points) and of course through the poles of the ecliptic K_1 and K_2. It is further mentioned that all these circles can be drawn with the aid of a pair of compasses (*karkaṭaka-śalākā*).

The celestial equator and the ecliptic are both great circles which intersect at two points. Consider the common diameter of these two circles, passing through the common centre and the equinoxes. The diameter joining the two solstices would be perpendicular to the common diameter. These are indicated by dotted lines in Figure 9.10.

The *ayanānta-rāśi-kūṭa-vṛtta* is perpendicular to both the celestial equator and the ecliptic. The solstices (*ayanānta*-s) will move on account of precession/trepidation. Due to this, the *ayanānta-rāśi-kūṭa-vṛtta* will also move in the same direction and so will the *rāśi-kūṭa*-s, K_1 and K_2. The latter move around the *Dhruva*-s, maintaining a distance of 24°. This implies that their diurnal circles remain the same, though they swing to the west or east on these, due to the motion of the equinoxes. The picture described here is the same as the modern geocentric picture of precession, except that the motion considered here is oscillatory (can be in either direction).

The longitude of the true planet obtained from calculations, called the *sphuṭa-graha* corresponds to the distance of the planet from *Meṣādi*. To this, the amount of motion of the equinoxes has to be added to obtain the corrected true planet which is referred to here as *golādi*.

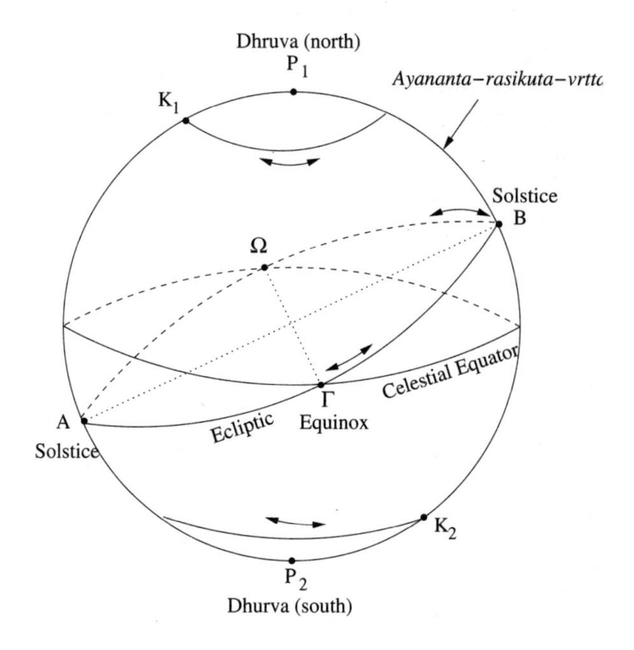

Figure 9.10: The motion of the vernal equinox.

9.6 *Vāyugola* for a non-equatorial observer

For an observer having zero latitude the central circle of the *vāyugola* (celestial equator), and the diurnal circles are all vertical circles and the *bhagola* is inclined to the *vāyugola*. For an observer having a northern or southern latitude, the *vāyugola* is not vertical but is inclined. The *bhagola* whose orientation is fixed with respect to the *vāyugola*, is also correspondingly inclined and has a slow motion (corresponding to the motion of equinoxes).

9.7 Zenith and horizon at different locations on the surface of the Earth

The Earth is a sphere. Hence, at any place on Earth, a person would feel that he is standing on top of the Earth. But the surface of the Earth (over which he stands) looks spread and so the observer feels that he is standing perpendicular to the flat Earth surface. In fact, a *kṣitija* (horizon) is conceived at every point on the surface of the Earth. This is the *svadeśa-kṣitija* or the local horizon. All the celestial bodies are rising and setting on that horizon. Only that portion of the sky which is above the horizon is visible. The centre of this visible part is the zenith called *khamadhya.* The celestial spheres for observers at different locations on the Earth are described below. These are illustrated in Figures. 9.11.

The *akṣa-daṇḍa* is the north-south axis passing through the centre of the Earth and extending to the poles. The celestial sphere is attached to it and rotates around it. The celestial equator and the equatorial horizon would have different inclinations with the local horizon at different places. For an equatorial observer, the celestial equator passes through the east(E), west(W) points and the zenith(Z); and the horizon (*nirakṣa-kṣitija*) passes through the poles (refer to Figure 9.11(a)). For an observer at the north pole, the *Dhruva* is the zenith and the celestial equator is the horizon. As one moves gradually from the equator northwards, the altitude of the north pole also increases correspondingly. The zenith, the horizon and the altitude of the pole star, are different for observers at different parts of the Earth. These are illustrated in Figure 9.11 (b) and (c).

For a place with a northern latitude, the meridian circle passing through E, W and Z is called the *sama-maṇḍala.* The local horizon which passes through the four cardinal points N, E, S, W is perpendicular to this. The *unmaṇḍala* is the equatorial horizon passing through E, W and the north pole P_1. This is called 6 o$'$ clock circle in modern astronomy. The inclination of the *unmaṇḍala* to the local horizon is the same as that of the celestial equator (*ghaṭikā-maṇḍala*) to the *sama-maṇḍala*, which is equal to the latitude of the place ϕ. Just as the three great circles, namely the celestial equator, equatorial horizon, and the north-south circle (*dakṣiṇottara-vṛtta*) are perpendicular to each other, the *sama-maṇḍala* (prime vertical), local horizon and the north-south circle are three great circles perpendicular to each other. The globe can be divided into eight equal parts even with these circles, the six *svastika*-s being N, S, E, W, Z and Z' (the nadir, opposite of zenith).

Consider a fourth circle called *valita-vṛtta*[4] passing through any pair of *svastika*-s formed by two of the three circles, and inclined to them. The

[4]The term *valita* means 'bent' or 'inclined'.

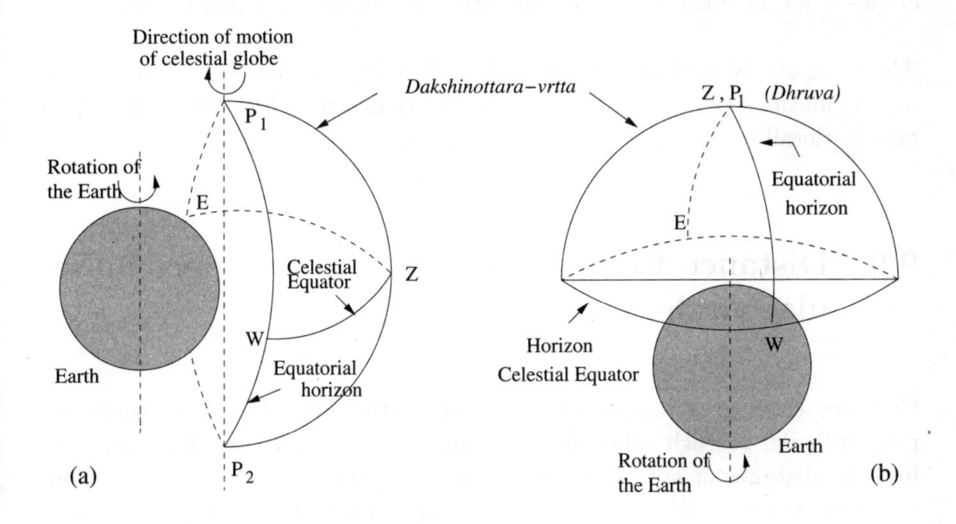

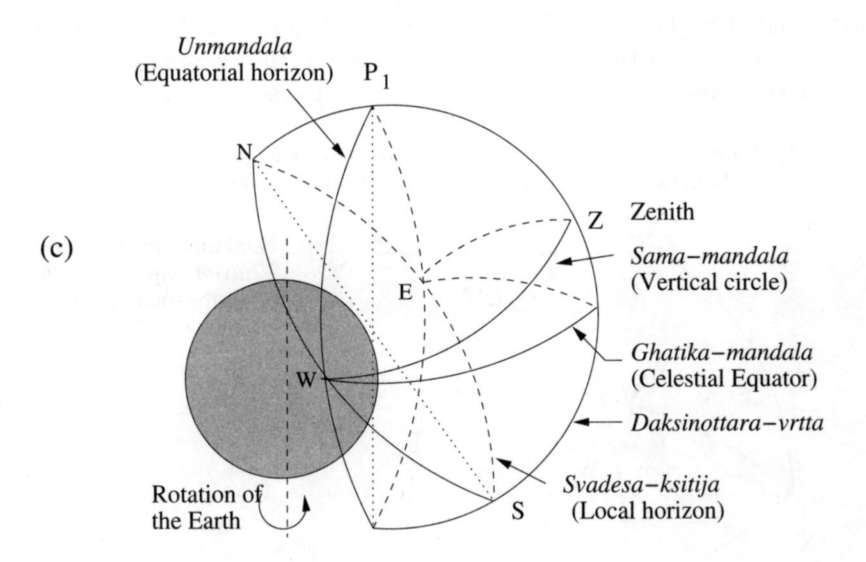

Figure 9.11: The celestial sphere for (a) an equatorial observer (b) observer at the north pole and (c) observer with a northern latitude.

distance of separation between points on this *valita-vṛtta* and the other two circles is found through the rule of three, as will be explained below.

The *vāyugola*, *bhagola* and the *bhūgola* and their interrelations are important for calculations pertaining to the planets. Hence they have been explained here in detail.

9.9 Distance from a *Valita-vṛtta* to two perpendicular circles

Consider three great circles in the sphere with radius R; two of them are perpendicular to each other and the third in between them. The aim is to find the distance of any point on the circumference of the third circle from the the other two (which are perpendicular to each other). This problem is illustrated by considering the celestial equator, the meridian (*dakṣiṇottara-vṛtta*) and the ecliptic. It may be noted from Figure 9.12, that the ecliptic is situated between the two great circles namely, the celestial equator and the *dakṣiṇottara-vṛtta* which are perpendicular to each other.

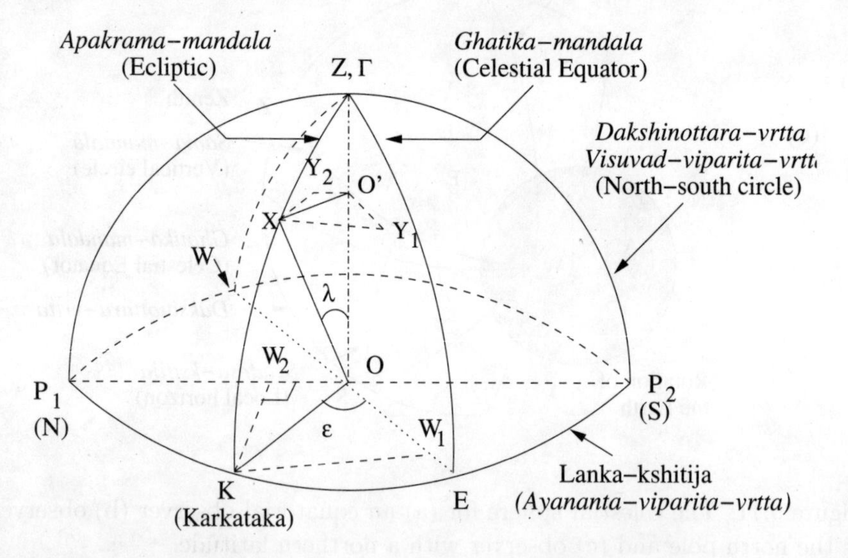

Figure 9.12: The perpendicular distance of a point on the circumference of a *valita-vṛtta* from two mutually perpendicular great circles.

In Figure 9.12, $E\Gamma W$ is the equator, ΓXK the ecliptic, and $P_1\Gamma P_2$ is the *dakṣiṇottara-vṛtta*. P_1EP_2 is the *ayanānta-viparīta-vṛtta* which is perpendicular to all the above circles. For convenience, the vernal equinox Γ is taken to be at the zenith. X is a point on the ecliptic whose *sāyana* longitude is λ. XY_1 and XY_2 are perpendiculars to the planes of the celestial equator and the *dakṣiṇottara-vṛtta* respectively. $KE = \epsilon$, is the obliquity of the ecliptic.[5] Let XO' be perpendicular to OZ $(= O\Gamma)$. $O'Y_1$ and $O'Y_2$ are in the plane of the celestial equator and *dakṣiṇottara-vṛtta* respectively. $\Gamma X = \lambda$ is the celestial longitude of X. Now, XO' is the *iṣṭa-dorjyā* given by

$$XO' = OX \sin\lambda = R\sin\lambda. \tag{9.1}$$

KW_1 and KW_2 are perpendicular to EW and NS respectively. Then,

$$KW_1 = OK \sin\epsilon = R\sin\epsilon, \tag{9.2}$$

$$\text{and} \quad KW_2 = OK \cos\epsilon = R\cos\epsilon, \tag{9.3}$$

are the *paramāpakrama* (maximum declination) and the *parama-svāhorātra* (radius of the diurnal circle at the maximum declination). The triangles $O'XY_1$ and OKW_1 are similar right angled triangles. Hence,

$$\frac{XY_1}{KW_1} = \frac{O'X}{OK} = \frac{R\sin\lambda}{R} = \sin\lambda.$$

Using (9.2) in the above, the *iṣṭāpakrama* $R\sin\delta$ is given by

$$R\sin\delta = XY_1 = KW_1 \ \sin\lambda = R\sin\epsilon\sin\lambda. \tag{9.4}$$

This is the distance between X on the ecliptic and the celestial equator. Similarly, triangles $O'XY_2$ and OKW_2 are similar right angled triangles. Therefore,

$$\frac{XY_2}{KW_2} = \frac{O'X}{OK} = \frac{R\sin\lambda}{R} = \sin\lambda.$$

Using (9.3), the *iṣṭāpakrama-koṭi* XY_2 is given by

$$XY_2 = KW_2 \ \sin\lambda = R\cos\epsilon\sin\lambda. \tag{9.5}$$

This is the distance between X on the ecliptic and the *dakṣiṇottara-vṛtta*. Thus, the use of the rule of three prescribed in the Text to find the distances, amounts to using the appropriate similar triangles.

[5] Here, and in what follows, we represent the angle corresponding to an arc by the arc itself. For instance, KE means $K\hat{O}E$.

9.10 Some *Viparīta* and *Nata-vṛtta*-s

Here, the problem of finding the distance of a point on a great circle from a set of three mutually perpendicular great circles is further elaborated geometrically.

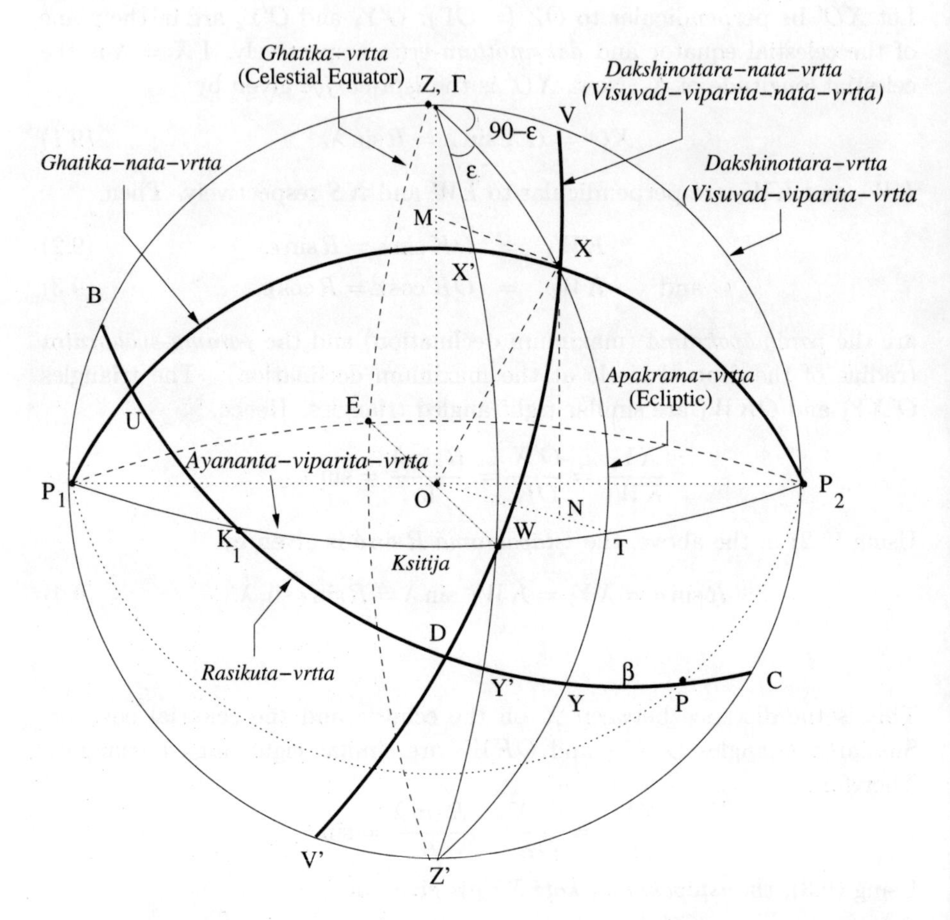

Figure 9.13*a*: The *viparīta-vṛtta*-s, *nata-vṛtta*-s and the *apakrama-maṇḍala*.

In Figure 9.13*a*, the vernal equinox Γ coincides with the zenith. P is a planet with latitude $YP = \beta$ (as will be specified in later sections) where Y is on the ecliptic. The *ghaṭikā-vṛtta*, the *dakṣiṇottara-vṛtta* and the *ayanānta-viparīta-vṛtta* are three mutually perpendicular great circles. As a fourth circle, the *apakrama-vṛtta* which is inclined to the celestial equator is considered. X is a point on it 90° away from Y. At this stage, X is referred to as the desired point on the ecliptic. Now three more circles are considered.

1. The first is the *ghaṭikā-nata-vṛtta* which passes through X and the poles P_1 and P_2. This is perpendicular to the *ghaṭikā-vṛtta* and intersects it at X'. That is, $P_1\hat{X}'W = P_2\hat{X}'W = 90°$. The maximum separation between *ghaṭikā-nata* and *dakṣiṇottara-vṛtta* which is also called the *viṣuvad-viparīta-vṛtta* is ZX', along the *ghaṭikā-vṛtta* The maximum separation between *ghaṭikā-nata* and the *ayanānta-viparīta-vṛtta* is $X'W$, which is also along the *ghaṭikā-vṛtta.*

2. The second is the *viṣūvad-viparīta-nata-vṛtta* or the *dakṣiṇottara-nata-vṛtta,* WXV passing through X and the intersection point W of *ghaṭikā-vṛtta* and *ayanānta-viparīta-vṛtta.* As W is the pole of the *viṣuvad-viparīta-vṛtta,* this *viṣuvad-viparīta-nata* is perpendicular to it. The maximum separation between *viṣuvad-viparīta-nata-vṛtta* and the *ghaṭikā-vṛtta* is ZV, along the *viṣuvad-viparīta-vṛtta.*

3. The pole of the ecliptic K_1 is on the *ayanānta-viparīta-vṛtta,* at a separation of $\epsilon = 24°$ away from the pole P_1. The *rāśi-kūṭa-vṛtta* passing through Y, P and K_1 intersects the celestial equator at Y' and the *ghaṭikā-nata* at U.

Now we show that *ghaṭikā-nata-vṛtta* is perpendicular to *rāśi-kūṭa-vṛtta.* Since K_1 is the pole of the ecliptic, $XK_1 = 90°$. By choice, the point Y on the ecliptic is such that $XY = 90°$. Therefore, any point on the great circle passing through K_1 and Y is at 90° from X. In other words, X is the pole of the *rāśi-kūṭa-vṛtta* $UK_1Y'Y$. This implies that $XY' = 90°$. But $P_1Y' = 90°$ as Y' is on the celestial equator. Therefore any point on the great circle passing through P_1 and X is at 90° from Y'. In other words, Y' is the pole of the *ghaṭikā-nata* P_1UXP_2. Hence, $Y'U = 90°$. This also implies that the *ghaṭikā-nata* is perpendicular to the *rāśi-kūṭa-vṛtta.*

The maximum divergence between the *rāśi-kūṭa-vṛtta* $UK_1Y'Y$ and the *ghaṭikā-vṛtta* is UX', along the *ghaṭikā-nata,* which is perpendicular to both the circles. X which lies on the *viṣuvad-viparīta-nata* is the pole of the *rāśi-kūṭa-vṛtta.* Hence, the *viṣuvad-viparīta-nata* is perpendicular to *rāśi-kūṭa-vṛtta.* It is also perpendicular to the *dakṣiṇottara-vṛtta.* Hence, the maximum divergence between the *rāśi-kūṭa-vṛtta* and *viṣuvad-viparīta-vṛtta* is DV', along the *viṣuvad-viparīta-nata.* The three circles (i) *ghaṭikā-nata,* (ii) *viṣuvad-viparīta-nata,* and (iii) the *rāśi-kūṭa-vṛtta* are shown by bold solid lines in Figure 9.13*a*.

Now, let the longitude of X be $\Gamma X = ZX = \lambda$. The distance between X and OZ is

$$XM = R \sin \lambda. \tag{9.6a}$$

Similarly, the distance between X and *ayanānta-viparīta-vṛtta* is

$$XN = R \cos \lambda. \tag{9.6b}$$

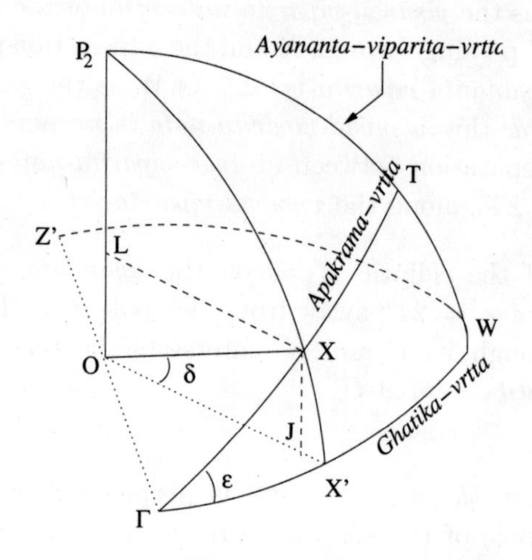

Figure 9.13*b*: A section of Figure 9.13*a*.

In Figure 9.13b, $X'X = \delta$, the declination measured along the *ghaṭikā-nata-vṛtta.* Therefore, the distance between X and the *ghaṭikā-vṛtta* is

$$XJ = R \sin \delta, \tag{9.7a}$$

and the distance between X and the polar axis P_1P_2 is

$$\begin{aligned} XL &= R \sin XP_2 \\ &= R \sin(90^\circ - \delta) \\ &= R \cos \delta. \end{aligned} \tag{9.7b}$$

Actually XL gives the radius of the diurnal circle of X, called *dyujyā*. This is also determined by considering an arc on the *ghaṭikā-nata-vṛtta*. For, in Figure 9.13a,

$$UX' = UX - XX' = 90^\circ - \delta. \tag{9.8}$$

Also,

$$P_2X = P_2X' - XX' = 90^\circ - \delta. \tag{9.9}$$

Therefore,

$$R \sin UX' = R \sin P_2X = R \cos \delta = dyujy\bar{a}. \tag{9.10}$$

Thus, *dyujyā* is the maximum separation between the *rāśi-kūṭa-vṛtta* and *ghaṭikā-vṛtta* on the *ghaṭikā-nata-vṛtta*. It is also the Rsine on the *nata-vṛtta* from the pole P_2 to the desired point on the *apakrama-vṛtta*.

9.11 Declination of a planet with latitude

Consider a planet P on the *rāśi-kūṭa-vṛtta* as in Figure 9.13a. In the following, a 'declination type' formula is employed at different stages to determine the declination of the planet with latitude. By this, we mean a formula of the form

$$\sin \delta = \sin \epsilon \ \sin \lambda, \tag{9.11}$$

where δ is the declination of the Sun whose longitude is λ. ϵ is the inclination of the ecliptic with respect to the equator (Figure 9.14(a)).

Consider any two great circles which are inclined to each other by an angle, say ρ, as in Figure 9.14(b). Then, the distance (d) of a point P on one of the circles, corresponding to an arc λ' from the point of intersection O, from the other circle is

$$d = R \sin \xi = R \sin \rho \ \sin \lambda'. \tag{9.12}$$

This can be proved along the same lines as was followed in section 9.9 for deriving (9.4). This also follows from the application of 'sine formula', to the spherical triangle OPN in Figure 9.14(b).

It may be noted that (9.12) reduces to (9.11) when the two great circles considered happen to be the celestial equator and the ecliptic.

In Figure 9.13a, it may be noted that the *apakrama-maṇḍala* and *dakṣiṇottara-vṛtta* intersect at Z and the angle of inclination is $90 - \epsilon$. Hence, the

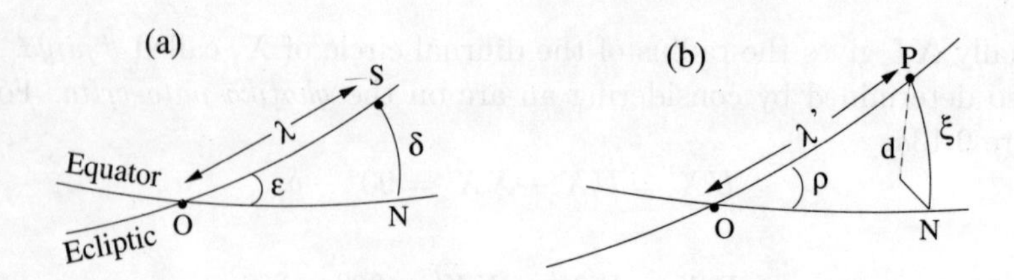

Figure 9.14: (a) Declination of the Sun; (b) Declination of a planet.

iṣṭāpakrama-koṭi is equal to the distance of X from *dakṣiṇottara-vṛtta*, which is

$$\begin{aligned} R\sin VX &= R\sin(ZX)\sin(90-\epsilon) \\ &= R\sin(ZX)\cos\epsilon. \end{aligned} \tag{9.13}$$

The Rsine of the arc from X to W on the *dakṣiṇottara-nata-vṛtta* is the *koṭi* of the above and is given by

$$\begin{aligned} R\sin XW &= R\sin(90-VX) \\ &= R\cos VX. \end{aligned} \tag{9.14}$$

Consider the arc ZX'. As it lies along the equator, it is related to time (*kāla*), and hence the Rsine of it is called *kālajyā* and is given by

$$kālajyā = R\sin(ZX'). \tag{9.15}$$

It is also called *laṅkodaya-jyā*. In the above equation, $ZX' = 360° - \alpha$ and $ZX = 360° - \lambda$, where λ and α are the longitude and Right Ascension (RA) of X. Here we have subtracted λ and α from 360°, because both the longitude and RA are measured eastwards. The *koṭi* of (9.15) is

$$Laṅkodayajyā\text{-}koṭi = R\sin X'W = R\cos ZX'. \tag{9.16}$$

Further,

$$ZY' = ZX' + X'Y' = ZX' + 90°, \tag{9.17}$$

as Y' is the pole of the *ghaṭikā-nata* and $X'Y' = 90°$. It may be noted that the *kāla-koṭi-jyā*, which is defined to be $R\sin ZY'$ is the same as the *Laṅkodayajyā-koṭi* given by (9.16).

Now, $Y'Y$ is a segment of *rāśi-kūṭa-vṛtta* which is perpendicular to the *apakrama-maṇḍala*. Now, *kāla-koṭyapakrama* also called *kālakoṭi-krānti* given by $R\sin Y'Y$, is the distance of Y' (on celestial equator) to the ecliptic. The inclination between the two being ϵ, we have

$$\begin{aligned} R\sin Y'Y &= R\sin\epsilon\sin(ZY') \\ &= R\sin\epsilon\cos(ZX'). \end{aligned} \tag{9.18}$$

Let the planet P be situated on the *rāśi-kūṭa-vṛtta* as shown in Figure 9.13*a*. YP is *vikṣepa* or the latitude of P. $Y'P$ is the arc from Y' (intersection of *rāśi-kūṭa-vṛtta* and celestial equator) to P on the *rāśi-kūṭa-vṛtta.*

The maximum separation between *rāśi-kūṭa-vṛtta* and *ghaṭikā-maṇḍala* (both of which are perpendicular to *ghaṭikā-nata*) is UX' = 90 - XX'.

This is also equal to the inclination of the *rāśi-kūṭa-vṛtta* with the *ghaṭikā-maṇḍala* ($= U\hat{Y'}X'$), since Y' is the pole of *ghaṭikā-nata* along which UX' is measured. The Rsine of the declination of P ($= R\sin\delta$) is equal to the distance of P from the celestial equator, and is given by

$$\begin{aligned} |R\sin\delta| &= R\sin(Y'P)\sin(UX') \\ &= R\sin(Y'P)\sin(90^\circ - XX') \\ &= R\sin(Y'Y+YP)\cos(XX') \\ &= R(\sin Y'Y\cos YP + \cos Y'Y\sin YP)\cos XX' \\ &= R\sin Y'Y\cos XX'\cos YP \\ &\qquad + R\cos XX'\cos Y'Y\sin YP. \end{aligned} \tag{9.19}$$

Now, $R\sin Y'Y\cos XX'$ is the declination of Y; as $Y'Y$ is on *rāśi-kūṭa-vṛtta,* it corresponds to declination of a planet at Y whose latitude is zero (*avikṣipta-graha*). Denoting it by δ_Y, (9.19) reduces to

$$|R\sin\delta| = |R\sin\delta_Y|\cos YP + R\cos XX'\cos Y'Y\sin YP. \tag{9.20}$$

Also, the declination of X is

$$|R\sin\delta_X| = R\sin XX' = R\sin\epsilon\ \sin ZX, \tag{9.21a}$$

and the declination of Y is

$$R\sin\delta_Y = R\sin\epsilon\sin ZY = R\sin\epsilon\ \cos ZX. \tag{9.21b}$$

From (9.21a) and (9.21b), we get

$$R^2\sin^2\delta_X + R^2\sin^2\delta_Y = R^2\sin^2\epsilon. \tag{9.22}$$

Subtracting both sides from the square of *trijyā*, we get

$$
\begin{aligned}
R^2 - R^2(\sin^2\delta_X + R^2\sin^2\delta_Y) &= R^2 - R^2\sin^2\epsilon. && (9.23a)\\
\text{or} \qquad R^2\cos^2\delta_X - R^2\sin^2\delta_Y &= R^2\cos^2\epsilon. && (9.23b)
\end{aligned}
$$

But,

$$
\begin{aligned}
R^2\sin^2\delta_Y &= R^2\sin^2 Y'Y\cos^2 XX'\\
&= R^2\sin^2 Y'Y\cos^2\delta_X. && (9.23c)
\end{aligned}
$$

Using (9.23c) in (9.23b), we get

$$
\begin{aligned}
R^2\cos^2\epsilon &= R^2\cos^2\delta_X - R^2\sin^2 Y'Y\cos^2\delta_X\\
&= R^2\cos^2(Y'Y)\cos^2\delta_X\\
&= R^2\cos^2(Y'Y)\cos^2(XX'). && (9.24a)
\end{aligned}
$$

Hence,

$$
R\cos YY'\cos XX' = R\cos\epsilon. \qquad (9.24b)
$$

Substituting (9.24b) in (9.19), we have

$$
\begin{aligned}
|R\sin\delta| &= |R\sin\delta_Y|\cos YP + R\cos\epsilon\sin YP\\
&= \textit{krāntijyā}\text{ of } Y \times \textit{vikṣepa-koṭi} +\\
&\qquad \textit{paramakrānti-koṭi} \times \textit{vikṣepa}, && (9.25a)
\end{aligned}
$$

where $\sin YP$ is *vikṣepa* (*jyā*), and $\cos YP$ is the *vikṣepa-koṭi* of a planet P with latitude.

In Figure 9.13*a*, all the arcs are measured westwards. Also, X, Y and P are south of the celestial equator. Let λ, β and δ be the longitude, latitude and the declination of P respectively. In terms of these, we have (since λ is also the longitude of Y)

$$
\begin{aligned}
R\sin\delta &= R(\sin(\delta_Y)\cos\beta + \cos\epsilon\sin\beta)\\
&= R(\sin\epsilon\sin\lambda\cos\beta + \cos\epsilon\sin\beta). && (9.25b)
\end{aligned}
$$

This result is exact and is same as the expression for the declination of a planet with latitude in modern spherical astronomy, as we shall explain below.

9.12 *Apakrama-koṭi*

Apakrama-koṭi refers to the distance between the planet and the *dakṣiṇottara-vṛtta* (north-south circle). In Figure 9.13a, the *dakṣiṇottara-nata-vṛtta* is perpendicular to the *rāśi-kūṭa-vṛtta* and the *dakṣiṇottara-vṛtta*. The maximum divergence between the latter two circles occurs on the former and is equal to DV'. Further

$$VV' = V'D + DX + XV = 180°,$$

as the *dakṣiṇottara-nata-vṛtta* is bisected by the *dakṣiṇottara-vṛtta*. Also $DX = 90°$, as X is the pole of the *rāśi-kūṭa-vṛtta*. Hence, the distance between D and *dakṣiṇottara-vṛtta* is $R\sin DV'$, where

$$\begin{aligned} R\sin DV' &= R\cos VX \\ &= \textit{koṭi of iṣṭāpakrama-koṭi}, \end{aligned} \tag{9.26}$$

as *iṣṭāpakrama-koṭi* or *iṣṭakrānti-koṭi* $= R\sin VX$, as was noted earlier.

Now the problem is to determine the distance of the planet P from the *dakṣiṇottara-vṛtta*. Let the *rāśi-kūṭa-vṛtta* passing through P intersect the *dakṣiṇottara-vṛtta* at B and C as shown in Figure 9.13a. D, which is 90° away from the intersection point of *dakṣiṇottara-vṛtta* and *rāśi-kūṭa-vṛtta*, is at a distance of $R\cos VX$ from the *dakṣiṇottara-vṛtta*. Hence the distance of P from the *dakṣiṇottara-vṛtta* is $R\sin(PC)\cos VX$. But $PC = YC - YP$, where YP is the latitude of the planet P. Hence,

$$\begin{aligned} \textit{Apakrama-koṭi} &= R\sin(YC - YP)\cos VX \\ &= R\sin YC\cos VX\cos YP \\ &\qquad - R\cos YC\cos VX\sin YP. \end{aligned} \tag{9.27}$$

Now the distance of Y from the *dakṣiṇottara-vṛtta* is $R\sin YC\cos VX$. This can also be calculated in a different way. The maximum divergence between the ecliptic and the *dakṣiṇottara-vṛtta* occurs on the *Laṅkā-kṣitija* or *ayanānta-viparīta-vṛtta* as shown in Figure 9.13a, and is equal to $R\sin(90 - \epsilon) = R\cos\epsilon$, as the two circles are inclined to each other at an angle $90 - \epsilon$, as is clear from Figure 9.13a. Hence, the distance of Y from the *dakṣiṇottara-vṛtta* is $R\cos\epsilon\sin YZ'$. (If λ is the longitude of the planet P, $YZ' = \lambda - 180°$). Therefore,

$$R\sin YC\cos VX = R\cos\epsilon\,\sin YZ'. \tag{9.28}$$

Now, we have to simplify $R\cos YC\cos VX$ in the second term of the RHS of (9.27). For this, consider

$$\begin{aligned}
R^2\cos^2 YC\cos^2 VX &= R^2\cos^2 VX\,(1-\sin^2 YC) \\
&= R^2\cos^2 VX - R^2\sin^2 YC\cos^2 VX \\
&= R^2 - R^2\sin^2 VX \\
&\qquad - R^2\cos^2\epsilon\sin^2 YZ' \qquad \text{[using (9.28)]} \\
&= R^2 - R^2\cos^2\epsilon\sin^2 ZX \\
&\qquad - R^2\cos^2\epsilon\sin^2 YZ', \qquad (9.29)
\end{aligned}$$

as $R\sin VX = R\cos\epsilon\sin ZX$ (the *iṣṭa-dorjyā-koṭi*). But, $ZX + YZ' = 90°$, as $ZZ' = 180°$ and $XY = 90°$. Hence, ZX and YZ' are *bhujā* and *koṭi* of each other, and

$$R^2\sin^2 ZX + R^2\sin^2 YZ' = R^2.$$

Using the above, (9.29) reduces to

$$\begin{aligned}
R^2\cos^2 YC\ \cos^2 VX &= R^2 - R^2\cos^2\epsilon \\
&= R^2\sin^2\epsilon. \qquad (9.30a)
\end{aligned}$$

Therefore,

$$R\cos YC\cos VX = R\sin\epsilon. \qquad (9.30b)$$

Using (9.27), (9.28) and (9.30), we obtain the distance of the planet P with latitude YP from the *dakṣiṇottara-vṛtta* to be

$$\begin{aligned}
\textit{Apakrama-koṭi} &= R\cos\epsilon\sin YZ'\cos YP - R\sin\epsilon\sin YP \\
&= \frac{1}{R}(\textit{Apakrama-koṭi}\text{ of } Y \times \textit{vikṣepa-koṭi} \\
&\qquad - \textit{Paramāpakrama} \times \textit{vikṣepa}). \qquad (9.31)
\end{aligned}$$

We now find the expression for the *kālajyā*. For this, consider the great circle in Figure 9.15 passing through the planet P and the north and south poles P_1 and P_2. Let it intersect the celestial equator at A. Then $R\sin AZ'$ is the *kāla-jyā* or *kāla-dorguṇa*. This is termed so, as AZ' is an arc on the celestial equator and hence related to the time. In fact, $AZ' = \alpha - 180°$, where α is the Right Ascension of the planet P. PB is a section of the diurnal path of the planet, which is a small circle parallel to the equator. $PA = -\delta$, where δ is the declination of P, and $PP_2 = 90° - PA$. Hence,

$$R\sin PP_2 = R\cos\delta = \textit{dyujyā}.$$

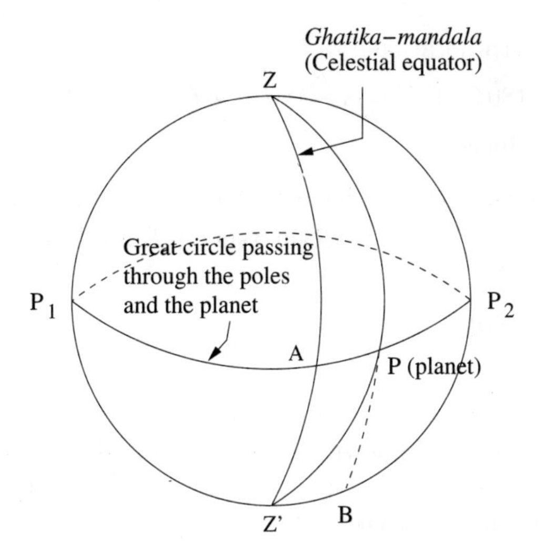

Figure 9.15: Determination of the *kālajyā*.

Now the maximum separation between the *dakṣiṇottara-vṛtta* and the great circle P_1PP_2 is AZ', as the celestial equator is perpendicular to P_1PP_2 and $AP_2 = 90°$. Hence, the distance of P from the *dakṣiṇottara-vṛtta* is given by

$$\begin{aligned} \textit{Apakrama-koṭi} &= R \sin PP_2 \sin AZ' \\ &= \frac{1}{R}(\textit{dyujyā} \times \textit{kālajyā}). \end{aligned} \qquad (9.32)$$

This is already given by (9.31). Equating the two, we get

$$R \sin PP_2 \sin AZ' = R(\cos\epsilon \sin YZ' \cos YP - \sin\epsilon \sin YP). \qquad (9.33)$$

Or,

$$\begin{aligned} R \sin AZ' &= \textit{kālajyā} \\ &= \frac{R(\cos\epsilon \sin YZ' \cos YP - \sin\epsilon \sin YP)}{\sin PP_2}. \end{aligned} \qquad (9.34)$$

It may be noted that the RHS of the above equation is

$$\frac{\textit{Apakrama-koṭi of } Y \times \textit{vikṣepa-koṭi} - \textit{Paramāpakrama} \times \textit{vikṣepa}}{R \times \textit{dyujyā}}.$$

If we use the modern notation,

$$AZ' = \alpha - 180°,\;\; YZ' = \lambda - 180°,\;\; YP = -\beta,\;\; PP_2 = 90° - \delta,$$

equation (9.32) reduces to

$$\begin{aligned} \textit{Apakrama-koṭi} &= R\cos\delta\sin\alpha \\ &= -R\sin\beta\sin\epsilon + R\cos\beta\cos\epsilon\sin\lambda. \end{aligned} \quad (9.35)$$

Supplementary Note

Since these results for the declination and right ascension of a planet with latitude are not commonly known, we sketch a simple spherical trigonometrical derivation of these results in the following. In Figure 9.16, X is the planet with longitude λ and latitude β.

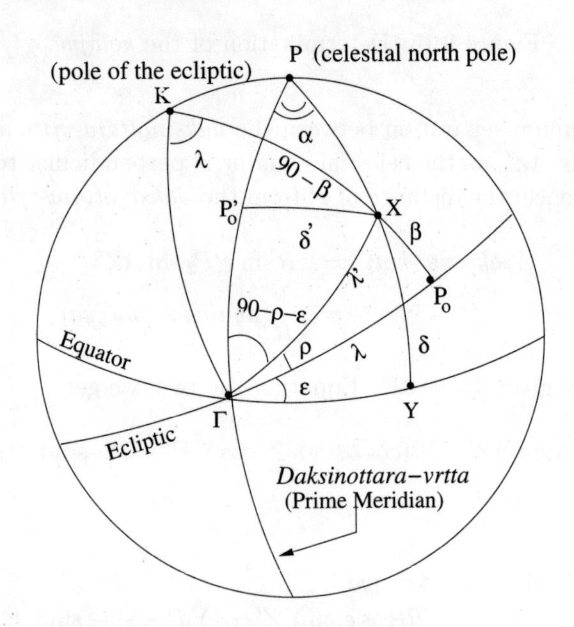

Figure 9.16: Declination and the Right Ascension of a planet X with longitude λ and latitude β.

Expression for Declination

Consider the spherical triangle KPX. Here,

$$KX = 90 - \beta,\;\; KP = \epsilon,\;\; PX = 90 - \delta \text{ and } P\hat{K}X = 90 - \lambda.$$

Applying the cosine formula, we get

$$\cos(90 - \delta) = \cos\epsilon\cos(90 - \beta) + \sin\epsilon\sin(90 - \beta)\cos(90 - \lambda). \qquad (9.36a)$$

Hence,

$$\sin\delta = \sin\epsilon\sin\lambda\cos\beta + \cos\epsilon\sin\beta. \qquad (9.36b)$$

This is the distance of X from the celestial equator which is same as (9.25*b*).

Expression for Right Ascension

In Figure 9.16, $XP_o' = \delta'$ is perpendicular to the *dakṣiṇottara-vṛtta*. Then the distance of X from the plane of *dakṣiṇottara-vṛtta* is $R\sin\delta'$. Let $\Gamma X = \lambda'$, and $X\hat{\Gamma}P_o = \rho$. Note that $K\hat{\Gamma}P_o = 90°$ and $K\hat{\Gamma}P = \epsilon$. Hence,

$$P\hat{\Gamma}X = 90° - \rho - \epsilon; \qquad X\hat{\Gamma}K = 90° - \rho.$$

Applying the sine formula to the spherical triangle $K\Gamma X$, we get

$$\frac{\sin\Gamma X}{\sin P\hat{K}\Gamma} = \frac{\sin KX}{\sin X\hat{\Gamma}K}.$$

Therefore,

$$\frac{\sin\lambda'}{\sin\lambda} = \frac{\sin(90 - \beta)}{\sin(90 - \rho)} = \frac{\cos\beta}{\cos\rho},$$

or,

$$\sin\lambda'\cos\rho = \sin\lambda\cos\beta. \qquad (9.37a)$$

In the spherical triangle $X\Gamma P_0$, XP_0 is perpendicular ΓP_0. Therefore,

$$\sin\beta = \sin\rho\sin\lambda'. \qquad (9.37b)$$

Consider the spherical triangle $XP_0'\Gamma$, where $X\hat{\Gamma}P_0' = P\hat{\Gamma}X = 90 - \rho - \epsilon$. Using the sine formula, we get

$$\sin XP_0' = \sin\lambda'\sin(90 - \rho - \epsilon). \qquad (9.38a)$$

That is,

$$\begin{aligned}\sin\delta' &= \sin\lambda'\cos(\rho + \epsilon)\\ &= \sin\lambda'(\cos\rho\cos\epsilon - \sin\rho\sin\epsilon)\\ &= \cos\epsilon\sin\lambda'\cos\rho - \sin\epsilon\sin\lambda'\sin\rho. \qquad (9.38b)\end{aligned}$$

Using (9.37a) and (9.37b) in the above, we get

$$\sin\delta' = \cos\epsilon\sin\lambda\cos\beta - \sin\epsilon\sin\beta. \qquad (9.39)$$

This is the distance of X from the *dakṣiṇottara-vṛtta*. Now, consider the spherical triangle PXP'_o. Here $X\hat{P}P'_0 = Arc(\Gamma Y) = \alpha$, which is the Right Ascension of X. Hence,

$$\frac{\sin(XP'_o)}{\sin(X\hat{P}P'_o)} = \frac{\sin(PX)}{\sin(P\hat{P_o}'X)}, \qquad (9.40a)$$

or,

$$\frac{\sin\delta'}{\sin\alpha} = \frac{\sin(90-\delta)}{\sin 90^\circ}. \qquad (9.40b)$$

Therefore,

$$\sin\delta' = \sin\alpha\cos\delta. \qquad (9.40c)$$

Using (9.40c) in (9.39), we have

$$\cos\delta\sin\alpha = \cos\epsilon\sin\lambda\cos\beta - \sin\epsilon\sin\beta, \qquad (9.41)$$

which is the same as (9.35).

Chapter 10

The Fifteen Problems

10.1 The fifteen problems

The seven great circles which are frequently employed in deriving various results in this chapter are listed in Table 10.1. These circles are indicated by solid lines in Figure 10.1. Three more circles which are referred to later in the chapter are indicated by dashed lines. In Table 10.1, the second column gives the names of the circles in Sanskrit. The third column gives their modern equivalents. In the last column we have listed the poles (visible ones with ref. to Figure 10.1) of these great circles.

No.	Circle	Description in modern terms	Pole/s
1	*Apakrama-vṛtta*	Ecliptic	K_1
2	*Dakṣiṇottara-vṛtta*	Prime meridian	W
3	*Dakṣiṇottara-nata-vṛtta*	Secondary to the prime meridian passing through the celestial body X	B, C
4	*Laṅkā-kṣitija*	Horizon for equatorial observer	Z, Z'
5	*Ghaṭikā-vṛtta*	Celestial equator	P_1, P_2
6	*Ghaṭikā-nata-vṛtta*	Secondary to the celestial equator passing through the celestial body X	Y'
7	*Rāśi-kūṭa-vṛtta*	Secondary to the ecliptic intersecting it at points which are at 90° away from the celestial body X	X

Table 10.1

In Figure 10.1, for the sake of convenience, the celestial sphere has been drawn for an equatorial observer. The position of the ecliptic is chosen such that the equinoxes coincide with the zenith and the nadir. This does not result in any loss of generality, as only (terrestrial) latitude-independent quantities are discussed in this chapter. X is a celestial body whose longitude ($ZX = \Gamma X$) is λ, declination (south) is δ and right ascension is α.

With reference to the seven great circles listed in Table 10.1, six quantities, which are primarily related to the motion of a celestial object, are defined below (Table 10.2). When any two of them are known, the other four can be determined. We know that, given six independent quantities, two of them can be chosen in 15 different ways. Hence the title of the chapter.

No.	Quantity	Description	Notation
1	*parama-krānti*	Maximum declination	$R\sin\epsilon$
2	*iṣṭa-krānti*	Desired declination	$R\sin\delta$
3	*iṣṭāpakrama-koṭi*	Distance of the celestial body from prime meridian	$R\cos\epsilon\sin\lambda$
4	*dorjyā*	Rsine longitude	$R\sin\lambda$
5	*kālajyā*	Rsine of the Right Ascension	$R\sin\alpha$
6	*natajyā*	Max. separation between the celestial equator and the Secondary to the meridian passing through the body	$R\sin z_v = \frac{R\sin\delta}{\sqrt{R^2 - R^2\cos^2\epsilon\sin^2\lambda}}$

Table 10.2

The following table, would be useful in identifying the six quantities, with reference to the seven great circles shown in Figure 10.1:

Number	Quantity	Representation in Figure 10.1		
1	*parama-krānti*	$R\sin\epsilon$	=	$R\sin X'\hat{Z}X$
2	*iṣṭa-krānti*	$R\sin\delta$	=	$R\sin XX'$
3	*iṣṭāpakrama-koṭi*	$R\cos\epsilon\sin\lambda$	=	$R\sin VX$
4	*dorjyā*	$R\sin\lambda$	=	$R\sin ZX$
5	*kālajyā*	$R\sin\alpha$	=	$R\sin ZX'$
6	*natajyā*	$R\sin z_v$	=	$R\sin ZV$

Table 10.3

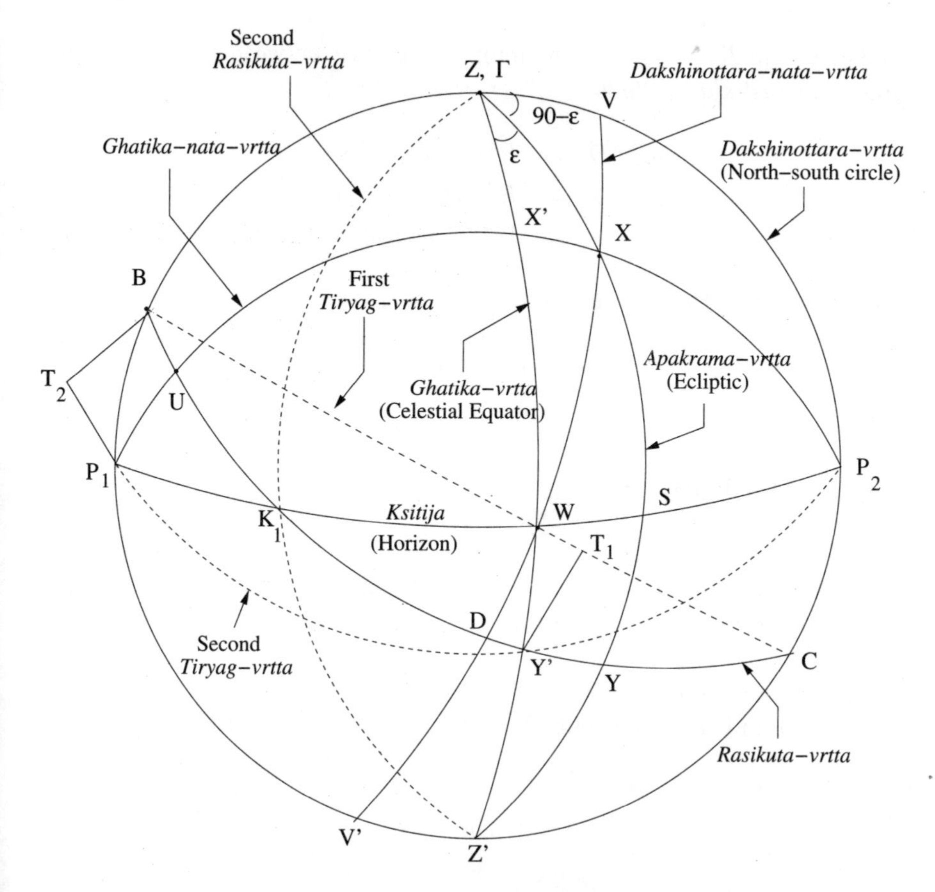

Figure 10.1: The seven great circles and their intersections.

The seven circles depicted in Figure 10.1, have already been explained in chapter 9 in connection with Figure 9.13a on page 683. Now we give some relations which would be used in later discussions. In Figure 10.1, it may be noted that $X'Y' = XY = XD = 90°$. Hence,

$$Y'Z' = 90° - ZX', \qquad YZ' = 90° - ZX, \qquad V'D = 90° - VX.$$

Since X is the pole of the *rāśi-kūṭa-vṛtta*, $XU = XD = 90°$; Hence, $UX' = 90° - XX' = 90° - \delta$; Also, $XP_2 = 90° - \delta$. The *R*sine of the maximum divergence between celestial equator and the *rāśi-kūṭa-vṛtta* is

$$R\sin UX' = R\sin(90° - \delta) = R\cos\delta = dyujy\bar{a}. \qquad (10.1)$$

Similarly, the Rsine of the maximum divergence between the north-south circle and *rāśi-kūṭa-vṛtta* is

$$\begin{aligned} R\sin V'D &= R\sin(90^\circ - VX) \\ &= R\cos VX \\ &= \textit{koṭi of iṣṭāpakrama-koṭi}. \end{aligned} \tag{10.2}$$

Also $BV = 90°$. Hence, $ZV = 90° - BZ$, so that

$$\textit{natajyā} = R\sin ZV = R\cos BZ. \tag{10.3}$$

10.2 Problem 1

The maximum declination $R\sin\epsilon$ (*parama-krānti*), and the actual declination $R\sin\delta$ (*iṣṭa-krānti*), are given.

It may be noted that the first two items listed in Table 10.2 are given and we have to find the other four. Now, from the given quantities,

$$\begin{aligned} R\cos\epsilon &= \sqrt{R^2 - (R\sin\epsilon)^2} \\ &= \textit{parama-krānti-koṭi}, \end{aligned} \tag{10.4}$$

and

$$\begin{aligned} R\cos\delta &= \sqrt{R^2 - (R\sin\delta)^2} \\ &= \textit{iṣṭa-krānti-koṭi or dyujyā}, \end{aligned} \tag{10.5}$$

are trivially found. The other four are determined as follows.

1. *Dorjyā* : The relation between δ, λ and ϵ is determined as before (Eq. (9.4)), by rule of three

$$R\sin\lambda = \frac{R.\, R\sin\delta}{R\sin\epsilon}. \tag{10.6}$$

Since the RHS is known, *iṣṭa-dorjyā* is found.

2. *Iṣṭāpakrama-koṭi* : It is defined by

$$\begin{aligned} \textit{iṣṭāpakrama-koṭi} &= R\sin VX \\ &= R\cos\epsilon\sin\lambda. \end{aligned} \tag{10.7}$$

Since both the factors in the RHS have been found, *iṣṭāpakrama-koṭi* is known. The rationale for the above expression is as follows. For the arc $ZS = 90°$, the divergence between the ecliptic and *dakṣiṇottara-vṛtta* is

$$R\sin SP_2 = R\sin(90 - \epsilon) = R\cos\epsilon.$$

Hence, for the arc $ZX = \lambda$, the divergence is given by

$$R\sin XV = R\cos\epsilon\sin\lambda.$$

3. *Nata-Jyā* : This refers to $R\sin ZV$ which is the maximum divergence between the celestial equator and *dakṣiṇottara-nata-vṛtta*, measured along the *dakṣiṇottara-vṛtta* corresponding to the arc $WV = 90°$. Hence, the divergence corresponding to the arc $WX = 90° - VX$ on the *nata-vṛtta* is given by

$$R\sin ZV\sin(90° - VX) = R\sin ZV\cos VX.$$

But this is $R\sin XX' = R\sin\delta$. Hence,

$$\begin{aligned} R.\,R\sin\delta &= R\sin ZV\ R\cos VX \\ &= R\sin ZV\sqrt{R^2 - R^2\sin^2 VX}. \end{aligned} \tag{10.8}$$

Using (10.7) in (10.8), we have

$$\begin{aligned} R\sin ZV &= \frac{R.\,R\sin\delta}{\sqrt{R^2 - (R\cos\epsilon\sin\lambda)^2}} \\ &= \frac{\textit{trijyā} \times \textit{iṣṭakrānti}}{\sqrt{\textit{trijyā}^2 - \textit{iṣṭāpakrama-koṭi}^2}}. \end{aligned} \tag{10.9}$$

Since all the terms in the RHS are known, *nata-jyā* is known.

4. *Laṅkodaya-jyā* : Consider the divergence between the *ghaṭikā-nata-vṛtta* and the *dakṣiṇottara-vṛtta*. *Laṅkodaya-jyā* or *kālajyā*, $R\sin ZX'$, is the maximum divergence corresponding to the arc $P_2X' = 90°$. The

iṣṭāpakrama-koṭi, as given by (10.7), is the divergence corresponding to the arc $P_2X = 90° - \delta$. Hence, by the rule of three, we get

$$R\sin\alpha = R\sin ZX' = \frac{R\sin(VX)R}{R\sin(P_2X)} = \frac{R\cos\epsilon\sin\lambda\, R}{R\cos\delta}. \quad (10.10)$$

$$\textit{Laṅkodaya-jyā} = \frac{\textit{iṣṭāpakrama-koṭi} \times \textit{trijyā}}{\textit{dyujyā}}.$$

Considering the divergence between *ghaṭikā-nata* and *kṣitija*, *Laṅkodaya-jyā-koṭi* is given by

$$R\cos\alpha = R\sin X'W = \frac{R\sin(XS)R}{R\sin(P_2X)} = \frac{R\cos\lambda\, R}{R\cos\delta}. \quad (10.11)$$

That is,

$$\textit{Laṅkodaya-jyā-koṭi} = \frac{\textit{dorjyā-koṭi} \times \textit{trijyā}}{\textit{dyujyā}}.$$

Similarly, considering the divergence between the *dakṣiṇottara-nata* and *kṣitija*, *nata-jyā-koṭi* is obtained. It is given by

$$R\sin VP_2 = \frac{R\sin(XS)R}{R\sin(XW)} = \frac{R\cos\lambda\, R}{\sqrt{R^2 - R^2\cos^2\epsilon\sin^2\lambda}}. \quad (10.12)$$

10.3 Problem 2

The maximum declination, $R\sin\epsilon$ (*parama-krānti*), and *iṣṭāpakrama-koṭi* = $R\cos\epsilon\sin\lambda$, are given.

Using the rule of three

$$R\sin SP_2 \;:\; R\sin ZS = R\sin XV \;:\; R\sin ZX,$$

$$\text{or} \qquad \frac{R\cos\epsilon}{R} = \frac{R\cos\epsilon\sin\lambda}{\textit{dorjyā}}.$$

Hence,

$$\textit{dorjyā} = R\sin\lambda = \frac{R\,.\,R\cos\epsilon\sin\lambda}{R\cos\epsilon}. \quad (10.13)$$

The other quantities are obtained as in problem 1.

10.4 Problem 3

The maximum declination $= R\sin\epsilon$ (*parama-krānti*), and *dorjyā* $= R\sin\lambda$, are given.

By considering the divergence between the *apakarama* and *ghaṭikā-vṛtta*-s, we find

$$\begin{aligned} \textit{iṣṭāpakrama} &= R\sin\delta \\ &= R\sin(XX') \\ &= \frac{R\sin(WS)\,R\sin(ZX)}{R\sin ZS} \\ &= \frac{R\sin\epsilon\,R\sin\lambda}{R}. \end{aligned} \tag{10.14}$$

Similarly, by considering the divergence between the *apakarama* and *dakṣiṇottara-vṛtta*-s, we find

$$\begin{aligned} \textit{iṣṭāpakrama-koṭi} &= R\sin(VX) \\ &= \frac{R\sin(SP_2)\,R\sin(ZX)}{R\sin ZS} \\ &= \frac{R\cos\epsilon\,R\sin\lambda}{R}. \end{aligned} \tag{10.15}$$

The rest (*kālajyā* and *natajyā*) are obtained as before.

10.5 Problem 4

The maximum declination $= R\sin\epsilon$ (*parama-krānti*), and *kālajyā* $= R\sin\alpha$, are given.

Now,

$$\textit{kālajyā} = R\sin ZX' = R\sin\alpha.$$

By construction, $X'Y' = 90°$. Therefore, $ZX' = WY'$. Hence,

$$\begin{aligned} R\cos\alpha = \textit{kāla-koṭi} &= R\cos ZX' \\ &= R\sin(90 + ZX') \\ &= R\sin ZY' \\ &= R\sin Y'Z'. \end{aligned} \tag{10.16}$$

The distance between *ghaṭikā* and *apakrama-vṛtta* on *rāśi-kūṭa-vṛtta* is

$$\begin{aligned} R\sin Y'Y &= R\sin\epsilon\sin Y'Z' \\ &= R\sin\epsilon\cos\alpha \qquad (10.17) \\ &= \textit{Kālakoṭi-apakrama}. \end{aligned}$$

Consider a second *rāśi-kūṭa-vṛtta* ZK_1Z', passing through the zenith and the pole of the ecliptic K_1. By construction, the angle between this second *rāśi-kūṭa-vṛtta* and the equator is $90 - \epsilon$. Therefore, the distance between Y' and the second *rāśi-kūṭa-vṛtta* will be

$$\begin{aligned} &= R\sin(90-\epsilon)\sin(ZY') \\ &= R\cos\epsilon\cos\alpha \\ &= \sqrt{R^2\cos^2\alpha - R^2\sin^2\epsilon\cos^2\alpha} \\ &= \sqrt{(\textit{kālakoṭi-jyā})^2 - (\textit{kālakoṭi-apakrama})^2}. \qquad (10.18) \end{aligned}$$

Now, K_1 being the pole of ecliptic, $K_1Y = 90°$. Therefore, $K_1Y' + Y'Y = 90°$. And

$$\begin{aligned} R^2\sin^2 K_1Y' &= R^2\cos^2 Y'Y \\ &= R^2 - R^2\sin^2 Y'Y. \qquad (10.19) \end{aligned}$$

Using (10.17) in the above equation we have

$$R^2\sin^2 K_1Y' = R^2 - R^2\sin^2\epsilon\cos^2\alpha. \qquad (10.20)$$

Consider the two *rāśi-kūṭa-vṛtta*-s passing through K_1. It can be seen that

$$\frac{\text{Distance between } Y' \text{ and second } \textit{rāśi-kūṭa-vṛtta}}{R\sin K_1Y'} =$$

$$\frac{\text{Max. divergence between the two } \textit{rāśi-kūṭa-vṛtta}\text{-s}}{R\sin K_1Y} = \frac{R\sin YZ'}{R}.$$

Hence,

$$R\sin YZ' = \frac{R.\ R\cos\epsilon\cos\alpha}{\sqrt{R^2 - R^2\sin^2\epsilon\cos^2\alpha}}. \qquad (10.21)$$

Using the relation, $YZ' = 90° - ZX = 90° - \lambda$, in the above equation, we have

$$R\cos\lambda = \frac{R.\ R\cos\epsilon\cos\alpha}{\sqrt{R^2 - R^2\sin^2\epsilon\cos^2\alpha}}. \qquad (10.22)$$

The *koṭi* of this is $R \sin ZX = R \sin \lambda$. Other quantities can be determined as before.

Note: The above relation can also be derived using $\cos \alpha = \frac{\cos \lambda}{\cos \delta}$ (10.11). Using this in RHS of (10.22), we have

$$\begin{aligned}
\frac{\cos \epsilon \cos \alpha}{\sqrt{1 - \sin^2 \epsilon \cos^2 \alpha}} &= \frac{\cos \epsilon \cos \lambda}{\cos \delta \sqrt{1 - \sin^2 \epsilon \frac{\cos^2 \lambda}{\cos^2 \delta}}} \\
&= \frac{\cos \epsilon \cos \lambda}{\sqrt{\cos^2 \delta - \sin^2 \epsilon \cos^2 \lambda}} \\
&= \frac{\cos \epsilon \cos \lambda}{\sqrt{1 - \sin^2 \delta - \sin^2 \epsilon \cos^2 \lambda}} \\
&= \frac{\cos \epsilon \cos \lambda}{\sqrt{1 - \sin^2 \epsilon \sin^2 \lambda - \sin^2 \epsilon \cos^2 \lambda}} \\
&= \frac{\cos \epsilon \cos \lambda}{\sqrt{1 - \sin^2 \epsilon}} \\
&= \cos \lambda. \qquad (10.23)
\end{aligned}$$

10.6 Problem 5

The maximum declination $= R \sin \epsilon$ (*parama-krānti*), and the *natajyā* $= R \sin ZV$, are given.

It is stated[1] that

$$\textit{nata-koṭi} = R \cos ZV = R \sin Z'C. \qquad (10.24)$$

Now, the maximum separation between the *apakrama* and *dakṣiṇottara-vṛtta*

[1]This can be derived once we note the following:

- X is the pole of the *rāśi-kūṭa-vṛtta* and hence X is at 90° from C.
- W is the pole of the *dakṣiṇottara-vṛtta* and hence that is also at 90° from C.

Therefore, C is pole of the great circle through X and W. This implies that $VC = 90°$. But, $ZV + VC + Z'C = 180°$. Hence, $ZV + Z'C = 90°$. Therefore,

$$R \cos ZV = R \sin Z'C.$$

is $R\sin SP_2 = R\cos\epsilon$. Therefore, the distance of C from the *apakrama-vṛtta*,

$$\begin{aligned} R\sin YC &= R\sin(ZC)\cos\epsilon \\ &= R\cos(ZV)\cos\epsilon. \end{aligned} \tag{10.25}$$

The angle between the second *rāśi-kūṭa-vṛtta* (ZK_1Z') and the *dakṣiṇottara-vṛtta* is ϵ. Hence, the distance of C from the second *rāśi-kūṭa-vṛtta* will be

$$\begin{aligned} &= R\sin\epsilon\ \sin(Z'C) \\ &= R\sin\epsilon\ \cos ZV \\ &= \sqrt{R^2\cos^2 ZV - R^2\cos^2 ZV\cos^2\epsilon}. \end{aligned} \tag{10.26}$$

Considering the divergence between the two *rāśi-kūṭa-vṛtta*-s, the above is the *pramāṇa-phala* or distance, which corresponds to the arc

$$K_1C = K_1Y + YC = 90° + YC,$$

the Rsine of which is the *pramāṇa* given by

$$R\sin K_1C = R\cos YC = \sqrt{R^2 - R^2\cos^2 ZV\cos^2\epsilon}.$$

The *icchā-phala* is the maximum divergence between the two *rāśi-kūṭa-vṛttas*, which is $R\sin YZ'$. This corresponds to the arc $K_1Y = 90°$, the Rsine of which is the *icchā* $= R$. Applying the rule of three in the form

$$\frac{\textit{icchā-phala}}{\textit{icchā}} = \frac{\textit{pramāṇa-phala}}{\textit{pramāṇa}},$$

we have

$$\frac{R\sin YZ'}{R} = \frac{\sqrt{R^2\cos^2 ZV - R^2\cos^2 ZV\cos^2\epsilon}}{\sqrt{R^2 - R^2\cos^2 ZV\cos^2\epsilon}}. \tag{10.27}$$

From this, $R\sin YZ'$ is found. The *koṭi* of this is $R\sin ZX$ (as $ZX + YZ' = 90°$), which is $\sin\lambda$ (*iṣṭa-dorjyā*). This is how the *iṣṭa-dorjyā* is determined in terms of *parama-krānti* and the *nata-jyā*. The rest is as in the earlier problems.

10.7 Problems six to nine

In problems 1 – 5, one of the two quantities given was *parama-krānti*, (item 1 in Table 10.2). We now move on to the next set of four problems (6 – 9) in which one of the quantities given is *iṣṭa-krānti*, the second of the six quantities listed in Table 10.2.

10.7.1 Problem 6

The actual declination $= R\sin\delta$ (*iṣṭa-krānti*), and the *iṣṭāpakrama-koṭi* $= R\cos\epsilon\sin\lambda$, are given.

Here, the *dorjyā* $= R\sin\lambda$ is simply obtained from the square-root of the sum of the squares of the given quantities. That is,

$$R\sin\lambda = \sqrt{R^2\sin^2\delta + R^2\cos^2\epsilon\sin^2\lambda}, \tag{10.28}$$

since δ and λ are related by the relation (10.6). All the other quantities are determined as earlier.

10.7.2 Problem 7

The actual declination $= R\sin\delta$ (*iṣṭa-krānti*), and the *dorjyā* $= R\sin\lambda$, are given.

It is straightforward to find all the four quantities.

10.7.3 Problem 8

The actual declination $= R\sin\delta$ (*iṣṭa-krānti*), and the *kālajyā* $= R\sin\alpha$, are given.

First the cosines $R\cos\delta$ (*dyujyā*) and $R\cos\alpha$ (*kālakoṭi-jyā*) of the given quantities are determined. For this, consider the divergence between the *ghaṭikā-nata* and *kṣitija*. Here, $R\sin X'W = R\cos\alpha$, the *pramāṇa-phala*, and $R\sin XS = R\cos\lambda$, the *icchā-phala*, are the distances of X' and X corresponding to the arcs $X'P_2$, the Rsine of which is R (*pramāṇa*) and XP_2, the Rsine of which is $R\cos\delta$ (*icchā*). Now, the *dorjyā-koṭi* $= R\cos\lambda$ is found using the principle of rule of three. Thus, we have

$$\frac{R\cos\alpha}{R} = \frac{R\cos\lambda}{R\cos\delta}. \tag{10.29}$$

From the above $\cos\lambda$ can be found. With this, the *dorjyā* and the other quantities can be determined.

10.7.4 Problem 9

The actual declination = $R\sin\delta$ (*iṣṭa-krānti*), and *natajyā* = $R\sin ZV$, are given.

When $R\sin ZV$ is the separation between the *yāmyottara-nata* and the equator, $R\sin VP_2 = R\cos ZV$ is the separation between *yāmyottara-nata* and the horizon. When

$$R\sin\delta = R\sin XX', \tag{10.30}$$

is the separation between the first two, the distance between the other two, $R\sin XS$, is given by the rule of three:

$$\frac{R\sin XS}{R\sin\delta} = \frac{R\cos ZV}{R\sin ZV}. \tag{10.31}$$

Since $ZX + XS = 90°$ and $ZX = \lambda$,

$$R\cos\lambda = \frac{R\cos ZV}{R\sin ZV} R\sin\delta. \tag{10.32}$$

In the language of the Text, the above equation may be written as,

$$\textit{dorjyā-koṭi} = \frac{\textit{natajyā-koṭi}}{\textit{natajyā}} \times \textit{iṣṭāpakrama}.$$

The *koṭi* of (10.32) is the *dorjyā* = $R\sin\lambda$. The rest are found as earlier. The result (10.32) can also be obtained using standard spherical trigonometry. Considering the triangle ZVX and applying the four-parts formula,

$$\cos ZV \cos(90° - \epsilon) = \sin ZV \cot\lambda - \sin(90° - \epsilon)\cot 90°.$$

Simplifying the above, and using the result $\sin\delta = \sin\epsilon\sin\lambda$, we have

$$R\cos\lambda = \frac{R\cos ZV}{R\sin ZV} R\sin\delta,$$

which is the same as (10.32).

10.8 Problems ten to twelve

In problems 6 – 9, one of the two given quantities was *iṣṭa-krānti*. We now move on to the next set of three problems in which one of the quantities given is *iṣṭāpakrama-koṭi*, the third of the six quantities listed in Table 10.2.

10.8.1 Problem 10

The *iṣṭāpakrama-koṭi* $= R\cos\epsilon\sin\lambda$, and *dorjyā* $= R\sin\lambda$, are given.

By finding the difference of the squares of the given quantities and taking the square root, we get the *iṣṭa-krānti*

$$R\sin\delta = \sqrt{R^2\sin^2\lambda - R^2\cos^2\epsilon\sin^2\lambda} = R\sin\lambda\sin\epsilon. \tag{10.33}$$

From $R\sin\lambda$ and $R\sin\delta$, the rest can be obtained.

10.8.2 Problem 11

The *iṣṭāpakrama-koṭi* $= R\cos\epsilon\sin\lambda$, and *kālajyā* $= R\sin\alpha$, are given.

From *kālajyā*, the *kālakoṭi*, $R\cos\alpha$ $(R\sin X'W)$ is obtained. Consider the separation of X' and X on the *ghaṭikā-nata-vṛtta* from the *dakṣiṇottara-vṛtta*. Using the rule of three, we have

$$\begin{aligned} \frac{R\sin(ZX')}{R\sin X'P_2} &= \frac{R\sin(VX)}{R\sin(XP_2)}, \\ \text{or} \qquad \frac{R\sin\alpha}{R} &= \frac{R\sin(VX)}{R\cos\delta}. \end{aligned} \tag{10.34}$$

Therefore,

$$R\cos\delta = R\,\frac{R\sin VX}{R\sin\alpha} \tag{10.35}$$

$$= \textit{trijyā} \times \frac{\textit{iṣṭāpakrama-koṭi}}{\textit{kālajyā}}. \tag{10.36}$$

From this, the *iṣṭa-krānti* $= R\sin\delta$ is obtained.

Again, consider the separation of X and X' on the *ghaṭikā-nata-vṛtta* from the horizon. Then,

$$\begin{aligned} R\sin XS &= \frac{R\sin(XP_2)\, R\sin(X'W)}{R\sin X'P_2}, \\ \text{or} \qquad R\cos\lambda &= \frac{R\cos\delta\, R\cos\alpha}{R}. \end{aligned} \tag{10.37}$$

This is the *dorjyā-koṭi*, from which the *dorjyā* $(R\sin\lambda)$ is obtained. From $R\sin\lambda$ and $R\sin\delta$, the rest are obtained.

10.8.3 Problem 12

The *iṣṭāpakrama-koṭi*, $R \sin VX = R \cos \epsilon \sin \lambda$, and *natajyā* $= R \sin ZV$, are given.

The maximum separation between *yāmyottara-nata-vṛtta* and the horizon is

$$R \sin VP_2 = R \cos ZV = \sqrt{R^2 - R^2 \sin^2 ZV}. \tag{10.38}$$

Also,

$$R \sin XW = R \cos VX = \sqrt{R^2 - R^2 \sin^2 VX}. \tag{10.39}$$

Then, $R \sin XS = R \cos \lambda$ (*dorjyā-koṭi*), which is the separation between X and the horizon, is given by

$$R \cos \lambda = \frac{R \sin(XW)\, R \sin(VP_2)}{R}. \tag{10.40}$$

From this, the *dorjyā* is obtained. Again, from the *iṣṭāpakrama-koṭi* and *dorjyā*, $R \cos \epsilon$, and hence *parama-krānti* ($R \sin \epsilon$), can be obtained. With them, the rest can be determined.

10.9 Problems thirteen and fourteen

In problems 10 – 12, one of the two given quantities was *iṣṭāpakrama-koṭi*. We now move on to the next set of two problems in which one of the quantities given is *dorjyā*, the fourth of the six quantities listed in Table 10.2.

10.9.1 Problem 13

The *dorjyā* $= R \sin \lambda$, and *kālajyā* $= R \sin \alpha$, are given.

From this,

$$R \sin XS = R \cos \lambda = \sqrt{R^2 - R^2 \sin^2 \lambda}, \tag{10.41}$$

and

$$R \sin X'W = R \cos \alpha = \sqrt{R^2 - R^2 \sin^2 \alpha}, \tag{10.42}$$

are found. Also we have

$$\frac{R\sin(XS)}{R\sin(X'W)} = \frac{R\sin(XP_2)}{R\sin(X'P_2)} = \frac{R\cos\delta}{R}. \tag{10.43}$$

Using (10.40) and (10.41) in the above equation, $R\cos\delta$ is determined. From this, $R\sin\delta$ is found and with the knowledge of $R\sin\lambda$ and $R\sin\delta$, the rest are obtained.

10.9.2 Problem 14

The *dorjyā* $= R\sin\lambda$, and *natajyā* $= R\sin ZV$, are given.

From them,

$$R\sin XS = R\cos\lambda = \sqrt{R^2 - R^2\sin^2\lambda}, \tag{10.44}$$

and

$$R\sin VP_2 = R\cos ZV = \sqrt{R^2 - R^2\sin^2 ZV}, \tag{10.45}$$

are found. Now, consider the separation of X and V on the *dakṣiṇottara-nata-vṛtta* from *kṣitija*. We have

$$\frac{R\sin XW}{R\sin XS} = \frac{R\sin VW}{R\sin VP_2} = \frac{R}{R\sin VP_2}. \tag{10.46}$$

Using the previous two equations in the above equation, $R\sin XW$ is obtained. The *koṭi* of this is $R\sin VX$ (*krānti-koṭi*). From $R\sin\lambda$ and $R\sin VX$, others are obtained.

10.10 Problem 15

This is the last problem in which the last two quantities in Table 10.2, namely the *kālajyā* $= R\sin ZX'$, and *nata-jyā* $= R\sin ZV$, are given.

Now,

$$\begin{aligned} R\sin WY' &= \textit{kālajyā}, \\ \text{and} \quad R\sin WD = R\sin VX &= \textit{krānti-koṭi}. \end{aligned} \tag{10.47}$$

Further, $R\sin X'W =$ *kāla-koṭi* and $XD = VW = 90°$. Now,

$$\begin{aligned} R\sin P_1B &= \textit{nata-jyā}, \\ \text{and} \qquad R\sin P_1U &= \textit{krānti}, \end{aligned} \tag{10.48}$$

which is the divergence between the *rāśi-kūṭa-vṛtta* and the horizon along the *ghaṭikā-nata-vṛtta*. Also, $UY' = CD = 90°$, and $P_1Z = 90°$.

Now, the maximum divergence between the *ghaṭikā* and *yāmyottara-nata-vṛtta* is the *nata-jyā* $= R\sin ZV = R\sin Z'V'$. Hence, the divergence between these two *vṛtta*-s on the *rāśi-kūṭa-vṛtta*, which is $R\sin Y'D$ corresponding to the arc $WY' = ZX'$, is given by

$$R\sin Y'D = R\sin ZV \sin ZX'. \tag{10.49}$$

Similarly, the maximum divergence between the *ghaṭikā-nata-vṛtta* and the north-south circle is the *kālajyā* $= R\sin ZX'$. Hence, $R\sin BU$ which is the distance between B on the north-south circle and the *ghaṭikā-nata-vṛtta* corresponding to the arc P_1B is given by

$$\begin{aligned} R\sin BU &= R\sin P_1B \sin ZX' \\ &= R\sin ZV \sin ZX' \\ &= R\sin Y'D. \end{aligned} \tag{10.50}$$

That is, the two *icchā-phala*-s are equal. Now $Y'C = CD - Y'D = 90° - Y'D$. Hence, the divergence between the *ghaṭikā-vṛtta* and the north-south circle along the *rāśi-kūṭa-vṛtta*, is $R\sin Y'C$ given by the expression

$$\begin{aligned} R\sin Y'C &= \sqrt{(R^2 - R^2\sin^2 Y'D)} \\ &= \sqrt{(R^2 - R^2\sin^2 ZV \sin^2 ZX')}. \end{aligned} \tag{10.51}$$

Then, the *iṣṭāpakrama-koṭi* $R\sin VX$ and *iṣṭāpakrama* $= R\sin\delta = R\sin P_1U$, which are considered as *icchā-phala*-s, are obtained from the relations (based on the rule of three) :

$$\begin{aligned} \frac{R\sin VX}{R} &= \frac{\sqrt{(R^2\sin^2 ZX' - R^2\sin^2 Y'D)}}{R\sin Y'C}, \\ \frac{R\sin P_1U}{R} &= \frac{\sqrt{(R^2\sin^2 ZV - R^2\sin^2 Y'D)}}{R\sin Y'C}. \end{aligned} \tag{10.52}$$

In the above expressions, the LHS is nothing but the ratio of *icchā-phala* to *icchā*. These can be derived in the following manner.

Consider the 'first *tiryag-vṛtta*', which is the great circle through B, W and C. As B is the pole of the *yāmyottara-nata-vṛtta* the maximum divergence between this *vṛtta* and *rāśi-kūṭa-vṛtta* is

$$R\sin WD = R\sin VX = \textit{iṣṭāpakrama-koṭi}.$$

Consider the divergence between these two *vṛtta*-s at Y' which is $R\sin Y'T_1$ ($Y'T_1$ being perpendicular to this *tiryag-vṛtta*). Therefore,

$$R\sin Y'T_1 = R\sin(WD)\sin(BY'). \tag{10.53}$$

Since, $BY' = 90^\circ + BU = 90^\circ + Y'D$, we have

$$\begin{aligned} R\sin Y'T_1 &= R\sin(WD)\cos(Y'D) \\ &= \left(\frac{1}{R}\right) R\sin(WD)\sqrt{R^2 - R^2\sin^2(Y'D)}. \end{aligned} \tag{10.54}$$

Now, the angle between the *ghaṭikā-vṛtta* and the first *tiryag-vṛtta* is

$$\begin{aligned} Y'\hat{W}T_1 &= 90^\circ - Y'\hat{W}D \\ &= 90^\circ - V'Z' \\ &= 90^\circ - ZV. \end{aligned} \tag{10.55}$$

Therefore,

$$\begin{aligned} R\sin Y'T_1 &= R\sin WY'\sin(90^\circ - ZV) \\ &= R\sin WY'\cos ZV \\ &= \sqrt{R^2\sin^2 WY' - R^2\sin^2 WY'\sin^2 ZV} \\ &= \sqrt{R^2\sin^2 WY' - R^2\sin^2 Y'D}. \end{aligned} \tag{10.56}$$

From (10.54) and (10.56), we have

$$R\sin WD\sqrt{R^2 - R^2\sin^2 Y'D} = R\sqrt{R^2\sin^2 WY' - R^2\sin^2 Y'D},$$

Or,

$$\begin{aligned} &R\sin VX\sqrt{R^2 - R^2\sin^2(ZX')\sin^2 ZV} = \\ &\qquad \sqrt{R^2\sin^2(ZX') - R^2\sin^2(ZX')\sin^2 ZV}. \end{aligned} \tag{10.57}$$

From this, the *iṣṭāpakrama-koṭi* ($R\sin VX$) is obtained in terms of the *kālajyā* ($R\sin ZX'$) and *natajyā* ($R\sin ZV$).

Consider the 'second *tiryag-vṛtta*', which is the great circle through P_1, Y' and P_2. As $Y'P_1 = UY' = 90°$, the maximum divergence between this *vṛtta* and the *rāśi-kūṭa-vṛtta* is

$$R\sin P_1U = R\sin\delta = \textit{iṣṭāpakrama}. \tag{10.58}$$

Consider the divergence between the two *vṛtta*-s at B which is $R\sin BT_2$ (BT_2 being perpendicular to this *tiryag-vṛtta*). Therefore,

$$R\sin BT_2 = R\sin(P_1U)\sin(BY'). \tag{10.59}$$

Since, $BY' = 90° + BU = 90° + Y'D$, we have

$$\begin{aligned} R\sin BT_2 &= R\sin(P_1U)\cos(Y'D) \\ &= \left(\frac{1}{R}\right) R\sin(P_1U)\sqrt{R^2 - R^2\sin^2 Y'D} \\ &= \left(\frac{1}{R}\right) R\sin\delta\sqrt{R^2 - R^2\sin^2 Y'D}. \end{aligned} \tag{10.60}$$

Now, the angle between the *yāmyottara-vṛtta* and the second *tiryag-vṛtta* is

$$\begin{aligned} B\hat{P}_1T_2 = Y'Z' &= 90° - WY' \\ &= 90° - ZX'. \end{aligned} \tag{10.61}$$

Therefore,

$$\begin{aligned} R\sin BT_2 &= R\sin P_1B\sin(90° - ZX') \\ &= R\sin ZV\cos ZX' \\ &= \sqrt{R^2\sin^2 ZV - R^2\sin^2 ZX'\sin^2 ZV} \\ &= \sqrt{R^2\sin^2 ZV - R^2\sin^2 Y'D}. \end{aligned} \tag{10.62}$$

Equating the two expressions for $R\sin BT_2$, we get

$$R\sin\delta\sqrt{R^2 - R^2\sin^2 Y'D} = R\sqrt{R^2\sin^2 ZV - R^2\sin^2 Y'D},$$

or
$$\begin{aligned} &R\sin\delta\sqrt{R^2 - R^2\sin^2(ZX')\sin^2 ZV} = \\ &\qquad R\sqrt{R^2\sin^2(ZV) - R^2\sin^2(ZX')\sin^2 ZV}. \end{aligned} \tag{10.63}$$

From this, the *iṣṭāpakrama* ($R\sin\delta$) is obtained in terms of the *kālajyā* ($R\sin ZX'$) and *natajyā* ($R\sin ZV$).

In summary, the formulas for *iṣṭāpakrama-koṭi* ($R\sin VX = R\cos\epsilon\sin\lambda$), and *iṣṭāpakrama* ($R\sin\delta$), in terms of the *kālajyā* ($R\sin\alpha$), and *natajyā* ($R\sin ZV$), are:

$$\begin{aligned} R\sin VX\sqrt{R^2 - R^2\sin^2\alpha\sin^2 ZV} &= R\sqrt{R^2\sin^2\alpha - R^2\sin^2\alpha\sin^2 ZV}, \\ R\sin\delta\sqrt{R^2 - R^2\sin^2\alpha\sin^2 ZV} &= R\sqrt{R^2\sin^2 ZV - R^2\sin^2\alpha\sin^2 ZV}. \end{aligned}$$

Then, from *kālajyā*, *natajyā* and the above relations, the other quantities can be obtained.

Chapter 11

Gnomonic Shadow

Apart from providing the rationale behind different procedures, this chapter also summarizes and synthesizes all the problems related to the diurnal motion of the Sun and shadow measurements carried out with a simple instrument called *śaṅku* (gnomon).[1] Since a major portion of the chapter deals with the measurement of shadow (*chāyā*) cast by gnomon, the choice of the title of the chapter, '*Chāyā-prakaraṇam*' (chapter on gnomic shadow) seems quite natural and appropriate.

The chapter commences with a discussion of the method of identifying the four directions using the forenoon and afternoon shadows of a gnomon. A few corrections, such as the one due to the finite size of the Sun, the effect of parallax etc., that need to be incorporated for making the measured values more accurate, are discussed in the next few sections. The theoretical background for the procedures involved in finding the latitude of a place and estimating the time from shadow are also presented.

The Text then goes on to an important topic called *Daśa-praśnāḥ* (Ten Problems), wherein among the five quantities related to the diurnal motion, the method to derive two of them given the other three is discussed. This is followed by a detailed discussion of topics related to the calculation of the orient ecliptic point, called *udaya-lagna* or simply *lagna*. Then, the effect of parallax on longitudes and latitudes is discussed. The chapter ends with an interesting discussion on the calculation of gnomic shadow of Moon when it has a latitudinal deflection.

[1]Gnomon is essentially a stick of suitable thickness and height, usually taken to be 12 units, with one of its edges sharpened to facilitate taking fine measurements of the tip of the shadow cast by a celestial body.

11.1 Fixing directions

Draw a circle with a suitable radius, on a flat surface and place the gnomon at its centre. The centre of the circle is represented by O in Figure 11.1(a). This is the base of the gnomon (*śaṅku* OA).

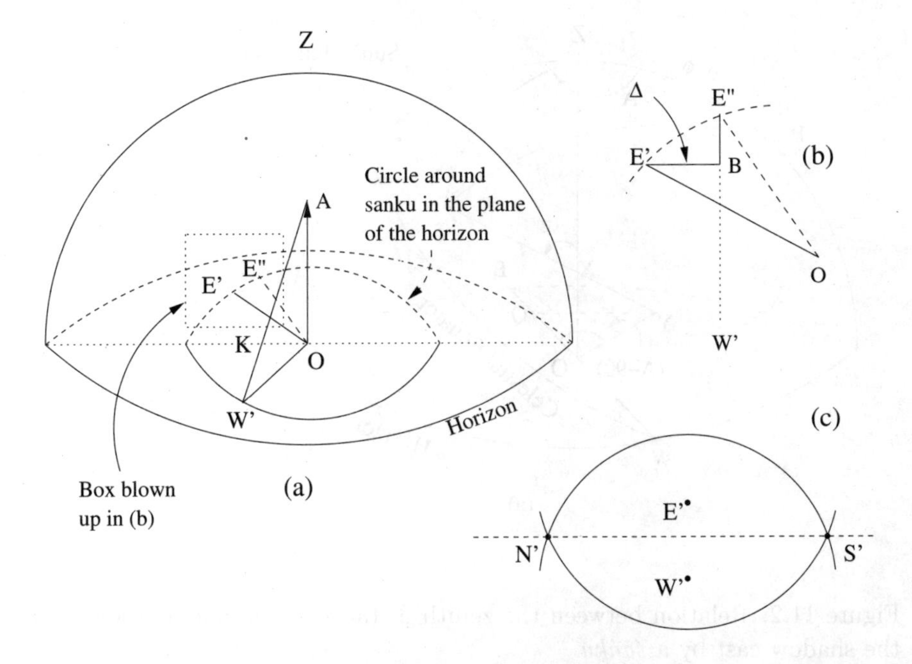

Figure 11.1: Fixing the directions through shadow measurements.

Let the tip of the shadow be at W' and E'' in the forenoon and afternoon respectively, on the circle. If the declination of the Sun were to be constant during the course of the day, then $W'E''$ would be the west-east line. However, due to the northward or southward motion of the Sun, the declination (δ) changes. Consequently, the tip of the eastern shadow point would have got shifted towards south (to the point E'', as shown Figure 11.1(a)), if the Sun has northward motion (δ increases) or north if the Sun has southward motion (δ decreases). So a correction Δ, which is equal to $E'B$ (see Figure 11.1(b)), has to be applied to E'' to get the true east-point E'. If the change in the declination be from δ_1 to δ_2, then the magnitude of the correction, Δ

is stated to be

$$\Delta = \frac{K(R\sin\delta_2 - R\sin\delta_1)}{R\cos\phi}, \tag{11.1}$$

where K is the hypotenuse of the shadow in *aṅgula*-s (the gnomon being taken to be 12 *aṅgula*-s) and ϕ is the latitude of the place. The expression for Δ given here is the same as the one found in *Siddhāntaśiromaṇi* and *Tantrasaṅgraha* and may be understood as follows.

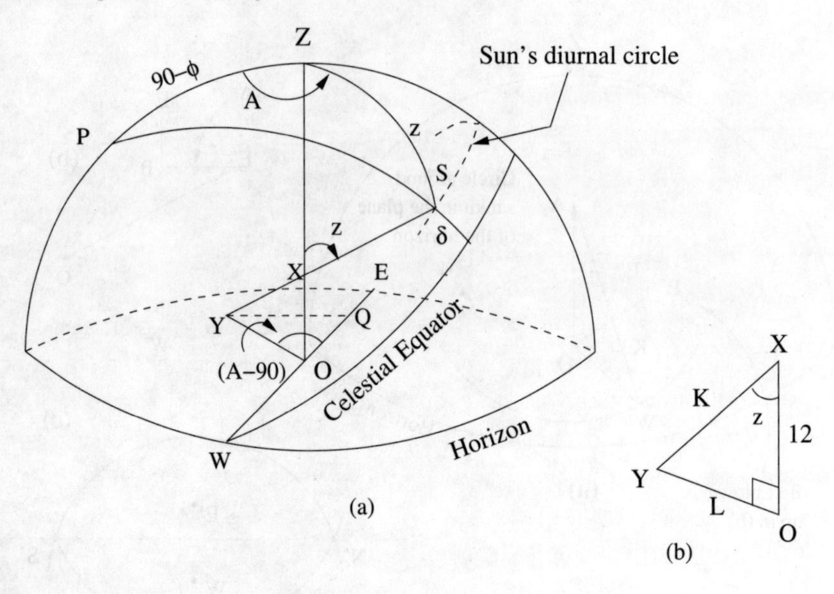

Figure 11.2: Relation between the zenith distance of the Sun and length of the shadow cast by a *śaṅku.*

Consider the situation when the Sun has declination δ, zenith distance z and azimuth A (refer Figure 11.2). OX is the gnomon, the length of whose shadow is L given by

$$L = OY = XY\sin z = K\sin z, \tag{11.2}$$

where $K = XY$ is the *chāyā-karṇa* (shadow-hypotenuse). For future purposes, we also note that

$$12 = K\cos z \quad \text{or} \quad K = \frac{12}{\cos z}, \tag{11.3}$$

as the gnomon $OX = 12$, and the shadow will be

$$L = 12\frac{\sin z}{\cos z}. \tag{11.4}$$

Arkāgrāṅgula YQ is the distance of Y, the tip of the shadow of the Sun, from the east-west line. It is given by

$$YQ = L\sin(A - 90) = L\cos A. \tag{11.5}$$

The declination of the Sun, δ, is given by the expression below, a formula similar to which will be derived later:

$$\sin\delta = \cos z\sin\phi + \sin z\cos\phi\cos A. \tag{11.6}$$

Now the shadow-lengths corresponding to W' and E'' being the same, their zenith distances are the same. When the declination of the Sun changes from δ_1 to δ_2, we have

$$\begin{aligned}\sin\delta_1 &= \cos z\sin\phi + \sin z\cos\phi\cos A_1\\ \sin\delta_2 &= \cos z\sin\phi + \sin z\cos\phi\cos A_2.\end{aligned} \tag{11.7}$$

Therefore,

$$\sin\delta_2 - \sin\delta_1 = \sin z\cos\phi\left(\cos A_2 - \cos A_1\right). \tag{11.8}$$

Rewriting the above, we get

$$\begin{aligned}\frac{K\left(\sin\delta_2 - \sin\delta_1\right)}{\cos\phi} &= K\sin z\left(\cos A_2 - \cos A_1\right)\\ &= L\left(\cos A_2 - \cos A_1\right),\end{aligned} \tag{11.9}$$

which is the difference in "*arkāgrāṅgula*" or 'amplitude' corresponding to δ_1 and δ_2. Hence,

$$\Delta = \frac{K\left(\sin\delta_2 - \sin\delta_1\right)}{\cos\phi}. \tag{11.10}$$

Then the true east point E' is the point on the circle which is at a distance Δ from the line $E''W'$. The true east-west line is $E'W'$. The north-south line is the perpendicular bisector of this, and is determined by the standard fish-figures.

The fish-figure is constructed as follows. With E' and W' as centres draw two circles of equal radii. These circles instersect at two points N' and S'. The region of intersection forms a fish figure as illustrated in Figure 11.1(c). The line passing through N' and S' is the north-south line. By construction, it is perpendicular to the east-west line through E' and W'.

11.2 Latitude and co-latitude

On the equinoctial day, the declination at sunrise and sunset are equal and opposite, and the Sun would be on the equator at noon. Let the shadow of the gnomon ($OX = 12$) be OY on that day (see Figure 11.3). Then the shadow hypotenuse is

$$K = XY = \sqrt{OX^2 + OY^2} = \sqrt{12^2 + OY^2}. \tag{11.11}$$

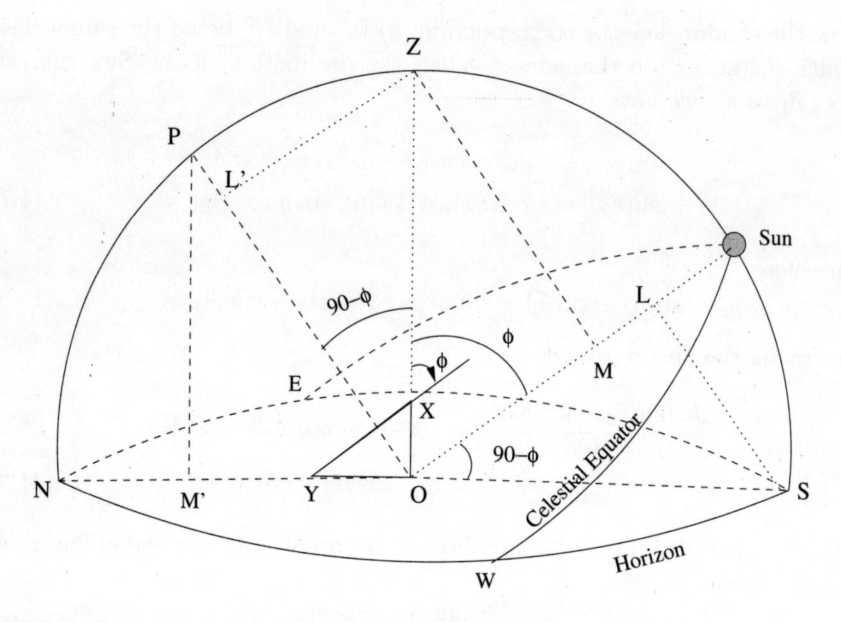

Figure 11.3: Determination of latitude through shadow measurements.

It is obvious that

$$\begin{aligned} OX &= K\cos\phi \\ OY &= K\sin\phi. \end{aligned} \tag{11.12}$$

Therefore,

$$R\sin\phi = \frac{OY \times R}{K}, \tag{11.13}$$

is the latitude (*akṣa*), and

$$R\cos\phi = \frac{OX \times R}{K}, \tag{11.14}$$

is the co-latitude (*lambana*). The equinoctial shadow is

$$OY = \frac{OX \sin\phi}{\cos\phi} = \frac{12 \sin\phi}{\cos\phi}. \tag{11.15}$$

If the radius of the celestial sphere is R, the distance between the zenith and the celestial equator is $ZM = R\sin\phi$. This is the same as the distance PM' between the pole star *Dhruva* and horizon, and is referred to as the *akṣa*. Similarly the *lambana* is the perpendicular distance SL between the *ghaṭikā-maṇḍala* (celestial equator) and the horizon, or the distance ZL' between the zenith and the *Dhruva*, both of which are $R\cos\phi$.

11.3 Time after sunrise or before sunset

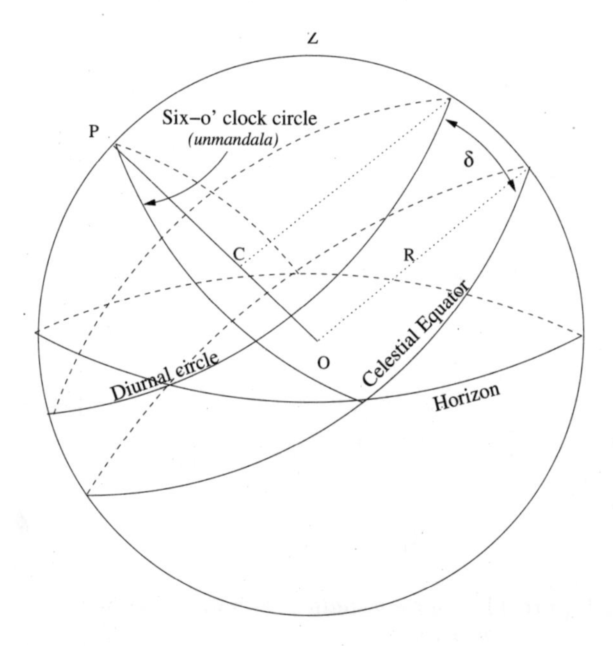

Figure 11.4: The role of *unmaṇḍala* in the determination of time.

On any day, the Sun moves on the diurnal circle (*svāhorātra-vṛtta*), all of whose parts are at a constant distance $R\sin\delta$ from the celestial equator (ignoring the change in the declination during the course of the day). This circle is parallel to the celestial equator (see Figure 11.4). Its centre C is on the polar axis of the celestial sphere and the radius is $R\cos\delta$ (*iṣṭa-dyujyā*, day radius). This circle is divided into four quadrants using the north-south circle and the six-o′ clock circle (*unmaṇḍala*) and into 21,600 divisions, being the number of *prāṇa*-s in a day (1 *prāṇa* = 4 seconds). The rate of motion of the *Pravaha* wind is constant. Hence it is possible to calculate correctly the position of a planet on the diurnal circle, given the time elapsed after it has risen or the time yet to elapse before setting.

11.4 *Unnata-jyā*

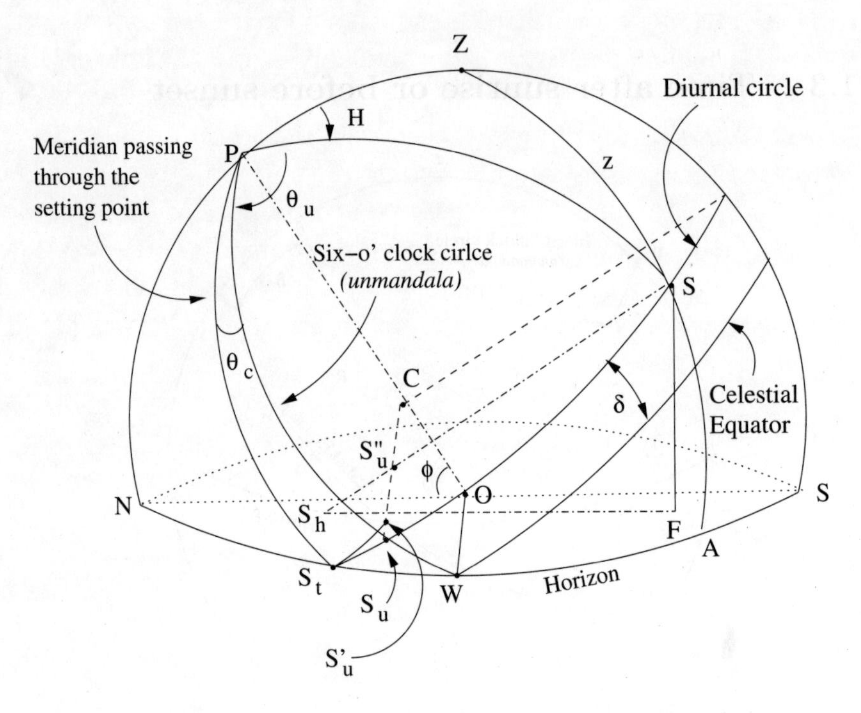

Figure 11.5: The *unnata-prāṇa* and *cara-prāṇa*.

In Figure 11.5, the diurnal circle of the Sun with C as the centre is indicated. It is divided into 21,600 equal divisions, each of which is a *prāṇa*. S is the position of the Sun at some instant. The Sun sets at S_t. Then the arc $SS_t = \theta$, on the diurnal circle corresponds to the 'time to elapse' before sunset. The *unmaṇḍala* or the six-o′ clock circle and the diurnal circle intersect at S_u. The arc $S_uS_t = \theta_c$ corresponds to the *cara-prāṇa*. $SS_u = \theta_u$ is termed the *unnata-prāṇa*. Both are measured in the *prāṇa* measure of the diurnal circle. Clearly,

$$SS_t = SS_u\ (unnata\text{-}prāṇa) + S_uS_t\ (cara\text{-}prāṇa),$$

or,

$$\theta = \theta_u + \theta_c. \tag{11.16}$$

Now drop perpendiculars $S_tS'_u$ and SS''_u from S_t and S on CS_u. CS_u is clearly parallel to OW, the east-west line. Also, let SS''_u be extended to meet the horizon at S_h. SS_h is the *Unnata-jyā*. It may be noted that

$$\begin{aligned} SS_h &= SS''_u + S_hS''_u \\ &= SS''_u + S_tS'_u, \end{aligned} \tag{11.17}$$

or,

$$Unnata\text{-}jyā\ (north) = R\cos\delta\left(\sin\theta_u + \sin\theta_c\right), \tag{11.18}$$

where $R\cos\delta$ or the *dyujyā*, is the radius of the diurnal circle. This is true when the declination of the Sun is north. When it is south, one can see that $\theta = \theta_u - \theta_c$ and

$$Unnata\text{-}jyā\ (south) = R\cos\delta\left(\sin\theta_u - \sin\theta_c\right). \tag{11.19}$$

Note: Considering the spherical triangle S_tPW, it can be shown that

$$R\sin\theta_c = \frac{R\sin\phi}{\cos\phi}\frac{\sin\delta}{\cos\delta} = R\tan\phi\tan\delta, \tag{11.20}$$

which is the well known relation for the *cara-jyā*. Also, $\theta_u = 90^o - H$, where H is the hour angle in modern parlance. Hence,

$$Unnata\text{-}jyā = R\cos\delta(\cos H + \tan\phi\tan\delta). \tag{11.21}$$

Though this relation is not stated here, we mention it as it will be useful later.

11.5 *Mahā-śaṅku* and *Mahācchāyā*: Great gnomon and great shadow

In Figure 11.5, let F be the foot of the perpendicular from S to the horizon. Then SF, 'the perpendicular from the Sun to the horizon' is the *mahā-śaṅku*.

Now, SS_h is a straight line in the plane of the diurnal circle perpendicular to the east-west line. Also the diurnal circle is inclined to the horizon at an angle $90 - \phi$, equal to the co-latitude of the place.

Clearly $SF = SS_h \cos\phi$. Therefore,

$$Mah\bar{a}\text{-}\acute{s}a\dot{n}ku = Unnata\text{-}jy\bar{a} \times \cos\phi. \tag{11.22}$$

Note: If z is the zenith distance of the planet S, the *mahā-śaṅku* $SF = R\cos z$. This can also be seen as follows. From (11.21) and (11.22),

$$Mah\bar{a}\text{-}\acute{s}a\dot{n}ku = R(\cos\delta\cos\phi\cos H + \sin\phi\sin\delta). \tag{11.23}$$

Applying the cosine formula to the side ZS (which is the zenith distance z) in the spherical triangle PZS, where $PZ = 90° - \phi$, $PS = 90 - \delta$ and $Z\hat{P}S = H$, we get

$$R\cos z = R(\cos\delta\cos\phi\cos H + \sin\phi\sin\delta). \tag{11.24}$$

Thus we see that *mahā-śaṅku* is same as the Rcosine of the zenith distance of the Sun, $R\cos z$. The *koṭi* of this, or $R\sin z$, is called *mahācchāyā*. The reason for this nomenclature could be as follows. The '*chāyā*' and '*śaṅku*' are equal to $K\sin z$ and $K\cos z$ respectively, where K is the *chāyā-karṇa* (shadow-hypotenuse). When K is replaced by the *trijyā* R, we obtain the *mahā-śaṅku*, $R\cos z$, and *mahācchāyā*, $R\sin z$.

11.6 *Dṛṅmaṇḍala* or *Dṛgvṛtta*

The *dṛṅmaṇḍala* is the vertical circle ZSA (refer Figure 11.5) passing through the zenith and the planet. Clearly, *mahā-śaṅku* and *mahācchāyā* are the sine and cosine of the arc AS on this circle. The centre of *dṛṅmaṇḍala* is O, which is the centre of the Earth-sphere.

11.7 *Dṛggolacchāyā*

Bhagola is the celestial sphere with the centre of the Earth C as the centre and *dṛggola* is the celestial sphere with the observer O as the centre (as

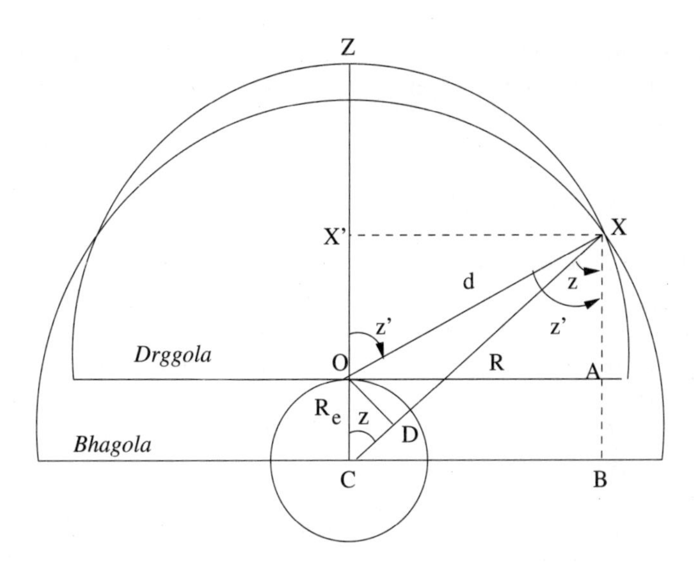

Figure 11.6: The *bhagola* and *dṛggola.*

in Figure 11.6). The *śaṅku* and *chāyā* are different for these two, when we consider an object at a finite distance. The distance between the two centres, $OC = R_e$ is the radius of the Earth. Let X be an object at a distance R from Earth's centre. Further, let d be the distance of X from the observer at O. Then,

$$\textit{Bhagola-śaṅku} = CX' = R\cos z, \tag{11.25}$$

$$\textit{Dṛggola-śaṅku} = OX' = d\cos z', \tag{11.26}$$

where z and z' are the zenith distances of X for *bhagola* and *dṛggola.* The relation between the two is given by

$$\begin{aligned}\textit{Dṛggola-śaṅku} &= OX' \\ &= CX' - OC \\ &= \textit{Bhagola-śaṅku} - \text{Earth-radius}.\end{aligned} \tag{11.27}$$

It may be noted that the linear measure of the *chāyā-śaṅku* , which is CB with reference to *bhagola* and OA with reference to *dṛggola*, is the same. In other words,

$$d\sin z' = R\sin z, \tag{11.28}$$

where d, the *dṛkkarṇa*, is the distance of the object from the observer on the surface of the Earth and is given by

$$\begin{aligned} d &= \sqrt{OA^2 + OX'^2} \\ &= \sqrt{R^2 \sin^2 z + (R \cos z - R_e)^2}. \end{aligned} \quad (11.29)$$

Moreover, the procedure for obtaining *dṛkkarṇa* is the same as that for the computation of *manda-karṇa* using *pratimaṇḍala*, with the radius of the Earth playing the role of *ucca-nīca-vṛtta-vyāsārdha* (radius of the epicycle). When R and R_e are in *yojanā*-s, d is called the *sphuṭa-yojana-karṇa.* As is clear from the figure, the *dṛkkarṇa* d is smaller than R. Hence, the zenith distance z' for O, is larger than that for C which is z, since $d \sin z' = R \sin z$. For future purposes, we note that

$$d \approx R - R_e \cos z, \quad (11.30)$$

up to first order in $\frac{R_e}{R}$. When the observer takes the distance between him and the object X as the *trijyā* R, the shadow in the *dṛggola* is

$$R \sin z' = R \sin z \frac{R}{d}. \quad (11.31)$$

This is the *dṛggolacchāyā.*

11.8 *Chāyā-lambana*

In Figure 11.6 above, drop a perpendicular OD from O on CX. Now

$$OD = d \sin(z' - z) = R_e \sin z. \quad (11.32)$$

Hence,

$$R \sin(z' - z) = R \sin z \frac{R_e}{d}. \quad (11.33)$$

Therefore,

$$z' - z = (R \sin)^{-1} \left[R \sin z \frac{R_e}{d} \right], \quad (11.34)$$

or

$$\textit{Chāyā-lambana} = \textit{cāpa} \left[\frac{\textit{bhagolacchāyā} \times \textit{bhū-vyāsārdha}}{\textit{sphuṭa-yojana-karṇa}} \right]. \quad (11.35)$$

Chāyā-lambana is the difference between the zenith distances in minutes as measured from the surface of the Earth and its centre. These are the arcs corresponding to *dṛggolacchāyā* and *bhagolacchāyā*. Therefore,

$$\textit{Dṛggolacchāyā} = \textit{Bhagolacchāyā} + \textit{Chāyā-lambana}, \qquad (11.36)$$

where it is understood that the entities refer to the corresponding arcs. The procedure is the same as the determination of *cāpa* corresponding to *manda-phala* in the *manda-saṃskāra*.

11.9 Earth's radius and *Chāyā-lambana*

It may be noted that the radius of the Earth plays the role of *antya-phala*, when *karṇa* is taken to be *trijyā*. When d is taken to be *trijyā* and the shadow $R\sin z$ is also *trijyā*, then,

$$\begin{aligned} R_e &= R\sin(z' - z) \\ &\simeq R(z' - z) \\ &= z' - z \qquad \text{(in min.)}, \end{aligned} \qquad (11.37)$$

when $z' - z$ is small. Hence, the radius of the Earth in *yojanā*-s is the *chāyā-lambana* in minutes. Also, there is not much difference between the *sphuṭa-yojana-karṇa* d (distance between the observer and the planet), and the *madhya-yojana-karṇa* R (distance between the planet and the centre of the Earth). For the Sun, it is stated that

$$\frac{R_e}{R} = \frac{1}{863}. \qquad (11.38)$$

Essentially this is the horizontal parallax. Then *chāyā-lambana* is the product of the above and the shadow, $R\sin z$, when d is approximated by R in the denominator. If *chāyā-lambana* of the *dṛṅmaṇḍala* is taken as the hypotenuse, then as we shall see later (section 11.37), its sine and cosine will be '*nati*' and '*lambana*'.

11.10 Corrected shadow of the 12-inch gnomon

Here the correction to the shadow and the gnomon (*mahā-śaṅku*) due to the finite size of the Sun is described. In Figure 11.7, PSQ represents the solar

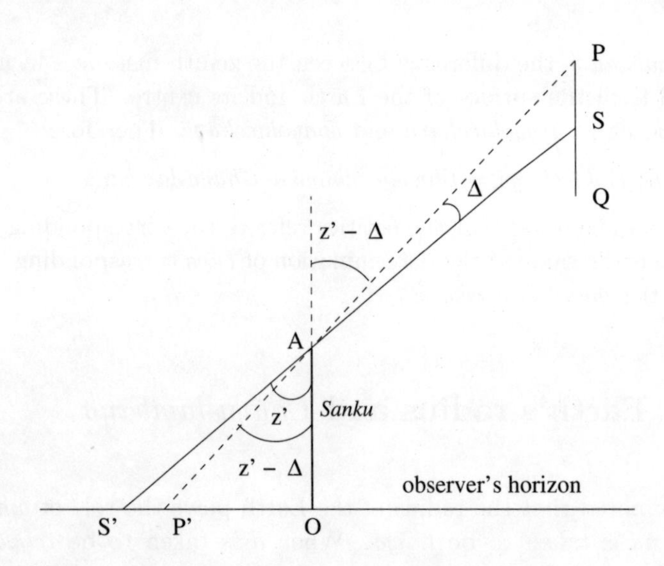

Figure 11.7: The correction to the shadow where, the source of light is an extended object.

disc, where S is the centre and P and Q are upper and lower points of the disc. $\Delta = PS$ is the angular semi-diameter of the Sun. OS' is the shadow corresponding to the centre and OP' is the shadow corresponding to the point P, which is what is observed. Then the corrected shadow (*chāyā*) and the gnomon (*mahā-śaṅku*) are given by

$$R\sin(z' - \Delta) = R\sin z' - \Delta(R\cos z'), \tag{11.39}$$

and

$$R\cos(z' - \Delta) = R\cos z' + \Delta(R\sin z'), \tag{11.40}$$

where the second terms are the differentials of the sine and cosine functions, the '*khanda-jyā*-s'. The corrected *mahā-śaṅku* and *chāyā* are stated to be pertaining to the *dṛg-viṣaya* (actually observed) i.e., related to what is 'actually' observed.

The increase in size of the *mahā-śaṅku* can also be viewed in the following manner (illustrated in Figure 11.8). If the Sun were a point object at S, then the length of the *śaṅku* is OB corresponding to the observed length of the shadow OC. Since it is actually the upper limb P which corresponds to the tip of the shadow at C, the *mahā-śaṅku* is effectively increased to OA.

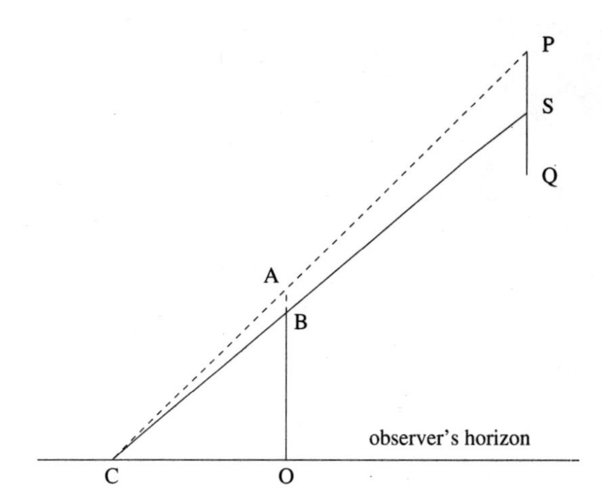

Figure 11.8: Another rationale for the increase in *mahā-śaṅku.*

In the above, the *khaṇḍa-jyā*-s are evaluated at z', the tip of the arc PS, whereas they should be evaluated at a point midway between P and S for better accuracy. However, the difference between the two would be small. So, from the *lambana* and the *khaṇḍa-jyā*-s corresponding to the radius of the Sun, the corrected gnomon and shadow corresponding to the upper point of the solar disc in the *dṛggola* are obtained. The corrected shadow of the 12-inch gnomon will be the corrected shadow (*chāyā*) as obtained above, multiplied by 12 and divided by the gnomon (*mahā-śaṅku*).

11.11 *Viparītacchāyā* : Reverse shadow

The procedure to obtain the time to elapse before sunset or the time elapsed after sunrise from the observed shadow of the 12-inch gnomon is termed *viparītacchāyā* or the reverse shadow. Obviously, the process is the reverse of obtaining the actual shadow from the time, which was indicated in the previous sections (11.3–10).

If L is the length of the shadow corresponding to the 12-inch gnomon, the *chāyā-karṇa* is

$$K = \sqrt{12^2 + L^2}, \tag{11.41}$$

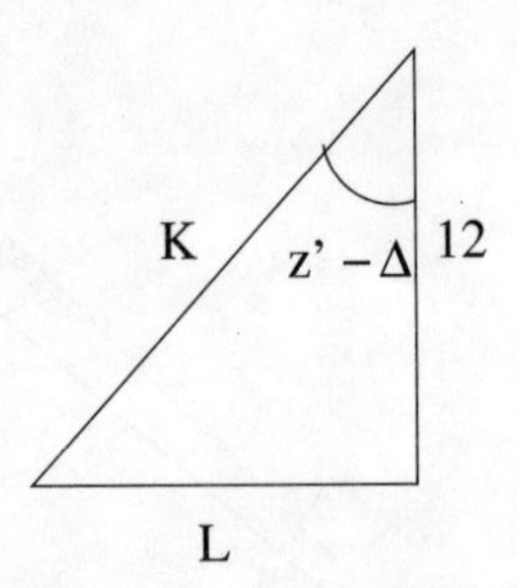

Figure 11.9: Relation between *śaṅku* and its shadow.

and

$$\begin{aligned} L &= K \sin(z' - \Delta) \\ 12 &= K \cos(z' - \Delta). \end{aligned} \tag{11.42}$$

Then the *mahācchāyā* and *mahā-śaṅku* are obtained as:

$$R \sin(z' - \Delta) = \frac{R}{K}.K \sin(z' - \Delta) = \frac{R}{K} L, \tag{11.43}$$

and

$$R \cos(z' - \Delta) = \frac{R}{K}.K \cos(z' - \Delta) = \frac{R}{K} 12, \tag{11.44}$$

respectively. These are *dṛg-viṣaya* and correspond to the upper limb of the Sun. The same quantities corresponding to the centre of the Sun are

$$\begin{aligned} R \sin z' &= R \sin(z' - \Delta) + \Delta\, R \cos(z' - \Delta) \\ R \cos z' &= R \sin(z' - \Delta) - \Delta\, R \sin(z' - \Delta). \end{aligned} \tag{11.45}$$

These correspond to the *dṛggola*. We have to obtain $R \sin z$ and $R \cos z$ corresponding to *bhagola*. These are stated to be

$$R \sin z = R \sin z' - R \sin z' \left(\frac{1}{863}\right), \tag{11.46}$$

and

$$R \cos z = R \cos z' + R_e. \qquad (\textit{in min.}) \tag{11.47}$$

Actually,

$$\begin{aligned} R \sin z &= d \sin z' \\ &\approx (R - R_e \cos z) \sin z' \\ &\approx R \sin z' - R \sin z' \, \frac{R_e}{R} \cos z, \end{aligned} \tag{11.48}$$

where terms up to first order in $\frac{R_e}{R}$ are considered. As was noted earlier, the Text takes $\frac{R_e}{R} = \frac{1}{863}$. Hence a factor of $\cos z$ (or $\cos z'$ to this order) is missing in the given correction term for $R\sin z$ in (11.46)

Again,

$$\begin{aligned} R\cos z &= d\cos z' + R_e \\ &\approx (R - R_e\cos z')\cos z' + R_e \\ &\approx (R\cos z') + R_e\sin^2 z'. \end{aligned} \tag{11.49}$$

Hence a factor of $\sin^2 z'$ is missing in the correction term for $R\cos z$ given in (11.47).

The same procedure is to be adopted for computing the latitude and the colatitude of the place also.

Now $R\cos z$ is given by the expression

$$R\cos z = R\cos\delta\cos\phi\cos H + R\sin\phi\sin\delta. \tag{11.50}$$

Thus,

$$R\sin\theta_u = R\cos H = \frac{R\cos z.R^2}{R\cos\delta R\cos\phi} - \frac{R.R\sin\phi R\sin\delta}{R\cos\phi R\cos\delta}. \tag{11.51}$$

Here θ_u is the time to elapse before the Sun reaches the *unmaṇḍala* or six-o′ clock circle. Since all the quantities in the RHS are known, θ_u can be determined. The second term in the RHS of the above equation is the *cara-jyā* ($R\sin\theta_c$). This is calculated separately and from that the arc θ_c, the *cara*, corresponding to time interval between six-o′ clock circle and sunset, is determined. The sum of θ_u and θ_c, or the difference between them[2] gives the time to elapse before sunset from the given instant, in angular measure.

11.12 Noon-time shadow

The distance between the celestial equator and the zenith on the north-south circle is the latitude ϕ. The declination δ is the distance between the planet and the celestial equator. The meridian zenith distance is z.

[2]When the Sun's declination is north the *cara* has to be added and when it is south it has to be subtracted.

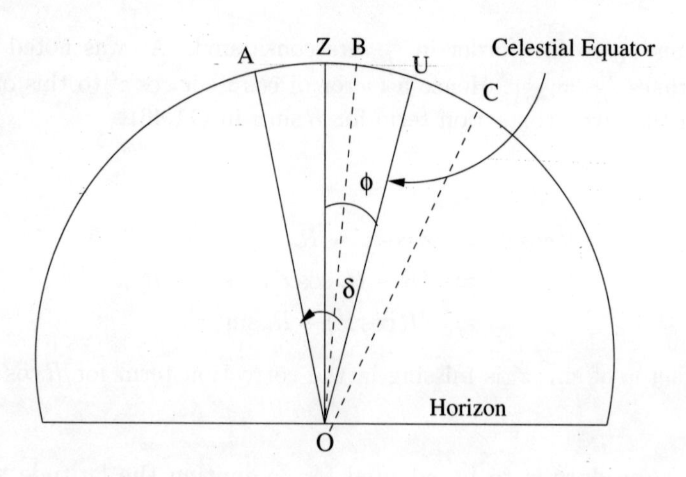

Figure 11.10: Relation between ϕ, δ and z Sun at noon.

In Figure 11.10, when the Sun is at A or B or C at noon,

$$z = \delta - \phi,\ \phi - \delta \text{ or } \phi + \delta, \tag{11.52}$$

and correspondingly,

$$\begin{aligned} \delta &= z + \phi,\ \phi - z \text{ or } z - \phi, \\ \phi &= \delta - z,\ z + \delta \text{ or } z - \delta. \end{aligned} \tag{11.53}$$

When any two of the three quantities z at noon, δ and ϕ are known, the other can be found.

11.13 *Chāyā-bhujā*, *Arkāgrā* and *Śaṅkvagrā*

Chāyā-bhujā (sine-shadow) is the distance between the planet and the prime vertical (*sama-maṇḍala*). This is represented by FR in Figure 11.11. If a is the angle between ZS and the prime vertical (i.e., azimuth $(P\hat{Z}S) - 90°$), then

$$\begin{aligned} \textit{Chāyā-bhujā} &= FR \\ &= OF \sin a \\ &= R \sin z \sin a, \end{aligned} \tag{11.54}$$

where z is the zenith distance of S. *Chāyā-koṭi* is the distance between S and the prime meridian (north-south circle) and is given by

$$Chāy\bar{a}\text{-}ko\d{t}i = R \sin z \cos a. \tag{11.55}$$

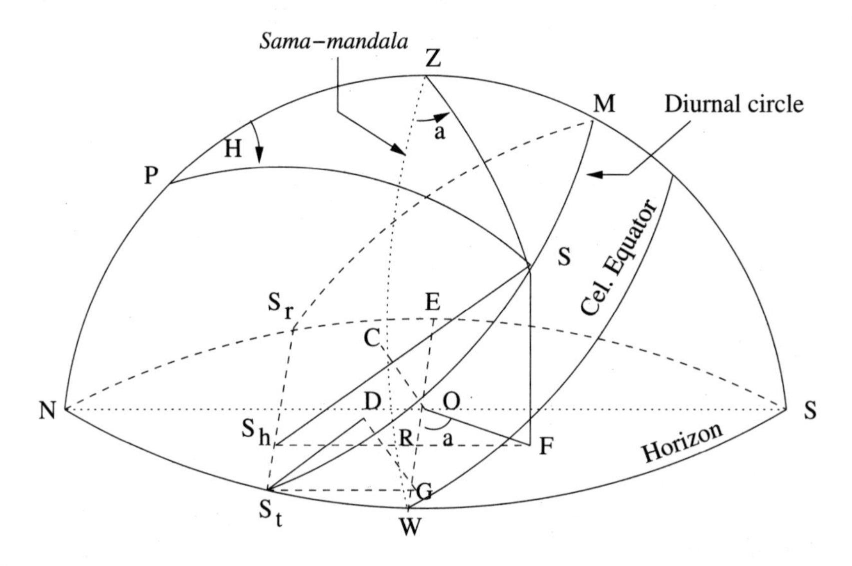

Figure 11.11: Relation between *chāyā-bhujā*, *arkāgrā* and *śaṅkvagrā*.

The distance of the rising or setting point of the Sun from the east-west line is the *arkāgrā*. In Figure 11.11, the Sun sets at S_t and S_tG is the perpendicular from S_t to EW line and

$$Ark\bar{a}gr\bar{a} = S_tG. \tag{11.56}$$

SF is the gnomon which is perpendicular to the horizon. S_rS_t is the line connecting the rising and setting points of the Sun on the diurnal circle. This is clearly parallel to the EW line.

The distance of the foot of the gnomon, F, from the line S_rS_t, is *śaṅkvagrā*. That is,

$$\acute{S}a\dot{n}kvagr\bar{a} = S_hF. \tag{11.57}$$

The foot of the gnomon has shifted from S_r to F during the diurnal motion. Hence the name *śaṅkvagrā*.

11.14 Some allied correlations

Consider Figure 11.11. Draw a perpendicular GD from G to the plane of the diurnal circle. It can be easily seen that S_tD is the sine on the diurnal circle intercepted between the horizon and the *unmaṇḍala*. This is the *kṣiti-jyā* ($R\tan\phi\sin\delta$) which is the product of *cara-jyā* and $\cos\delta$. Also,

$$GD = OC = R\sin\delta. \tag{11.58}$$

Consider the planar triangle S_tGD. This is a right angled triangle where the angle $G\hat{S_t}D$ is the inclination between the diurnal circle and the horizon which is $90° - \phi$, and the hypotenuse is the *arkāgrā* $= S_tG$. Therefore,

$$\frac{GD}{S_tG} = \frac{GD}{Ark\bar{a}gr\bar{a}} = \sin(90 - \phi) = \cos\phi. \tag{11.59}$$

Using (11.58) in the above equation, we have

$$Ark\bar{a}gr\bar{a} = \frac{R\sin\delta}{\cos\phi}. \tag{11.60}$$

Further, the triangle SS_hF is also a latitudinal triangle with the *śaṅkvagrā* S_hF as the *bhujā*, the gnomon SF as the *koṭi* and the *unnata-jyā* SS_h as the hypotenuse. The angle $S\hat{S_h}F$ is of course the co-latitude $(90° - \phi)$. Hence,

$$\frac{\acute{S}a\dot{n}kvagr\bar{a}}{\acute{S}a\dot{n}ku} = \frac{S_hF}{SF = R\cos z} = \frac{\cos(90 - \phi)}{\sin(90 - \phi)}. \tag{11.61}$$

Therefore,

$$\acute{S}a\dot{n}kvagr\bar{a} = R\cos z\,\frac{\sin\phi}{\cos\phi}. \tag{11.62}$$

Now, both the *śaṅkvagrā* S_hF and the *arkāgrā* S_tG are north-south lines. Also, the distance of F from the east-west line is the *chāyā-bhujā* given by

$$\begin{aligned} Ch\bar{a}y\bar{a}\text{-}bhuj\bar{a} &= FR \\ &= OF\sin a \\ &= R\sin z\sin a. \end{aligned} \tag{11.63}$$

Now,

$$\begin{aligned} S_hF &= S_hR + RF \\ &= S_tG + RF, \end{aligned}$$

or

$$\textit{Śaṅkvagrā} = \textit{Arkāgrā} + \textit{Chāyā-bhujā}. \qquad (11.64)$$

This translates into

$$R\cos z \frac{\sin\phi}{\cos\phi} = R\sin z \sin a + R\frac{\sin\delta}{\cos\phi},$$

or

$$\sin\delta = \cos z \sin\phi - \sin z \cos\phi \sin a. \qquad (11.65)$$

Note: The above relation is what would result when we apply the cosine formula to the side PS $(= 90^\circ - \delta)$ in the spherical triangle PZS, where $PZ = 90 - \phi$, $ZS = z$ and the spherical angle $P\hat{Z}S = A = 90^\circ + a$.

When the declination is south, it is easily seen that

$$\textit{Chāyā-bhujā} = \textit{Śaṅkvagrā} + \textit{Arkāgrā}. \qquad (11.66)$$

However, *chāyā-bhujā* $(= R\sin z \sin a)$ is also the distance between the planet on the *dṛṅmaṇḍala* and the *sama-maṇḍala* (prime vertical) as was noted earlier in (11.54). When this is considered as *bhujā*, and *chāyā* $(R\sin z)$ as the hypotenuse, the corresponding *koṭi* is *chāyā-koṭi* $= R\sin z \cos a$, which is same as in (11.55).

Now, the *chāyā-koṭi* is the sine of the hour angle (*nata-prāṇa*) on the diurnal circle (whose radius is $R\cos\delta$), or

$$R\sin H = \frac{R\sin z \cos a}{\cos\delta}, \qquad (11.67)$$

where $H = Z\hat{P}S$ is the '*nata-prāṇa*' in degrees. This can be seen as follows. Let the diurnal circle intersect the north-south circle at M. The north-south circle is inclined to PS and ZS by H and $90^\circ - a$, respectively. Then,

$$R\sin SM = R\sin H \sin(PS) = R\sin(ZS)\sin(90 - a), \qquad (11.68)$$

which leads to (11.67), as $PS = 90 - \delta$ and $ZS = z$.

Agrāṅgula is defined to be

$$\begin{aligned} \textit{Agrāṅgula} &= \textit{Chāyā-karṇa} \times \frac{\textit{Arkāgrā}}{\textit{Trijyā}} \\ &= K\frac{\sin\delta}{\cos\phi} \\ &= \frac{12\sin\delta}{\cos z\cos\phi}, \end{aligned} \tag{11.69}$$

as *dvādaśāṅgula-śaṅku* $= 12 = K\cos z$. Also, since

$$\begin{aligned} \textit{Śaṅkvargrā} &= R\cos z\frac{\sin\phi}{\cos\phi}, \\ \textit{Dvādaśāṅgula-śaṅkvagrā} &= 12\frac{\sin\phi}{\cos\phi}, \end{aligned}$$

which is the *viṣuvacchāyā* (equinoctial shadow) in *aṅgula*-s. We had

$$\textit{Śaṅkvagrā} - \textit{Arkāgrā} = \textit{Chāyā-bhujā}, \tag{11.70}$$

or,

$$\frac{R\cos z\sin\phi}{\cos\phi} - \frac{R\sin\delta}{\cos\phi} = R\sin z\sin a. \tag{11.71}$$

Multiplying by 12 and dividing by $R\cos z$, we have

$$\frac{12\sin\phi}{\cos\phi} - \frac{12\sin\delta}{\cos z\cos\phi} = \frac{12\sin z}{\cos z}\sin a. \tag{11.72}$$

Now *chāyā* in *aṅgula*-s is

$$L = K\sin z = \frac{12\sin z}{\cos z}, \tag{11.73}$$

and *chāyā-bhujā* in *aṅgula*-s is (see Figure 11.2(b) on page 716)

$$YQ = L\sin a = \frac{12\sin z}{\cos z}\sin a, \tag{11.74}$$

is the difference between *śaṅkvagrā* and *arkāgrā* in *aṅgula*-s. If the declination is south, we have to add these two. In both the cases, the direction of the *chāyā-bhujā* will be clearly opposite to that of *bhujā* of *mahācchāyā* (see Figure 11.11).

11.15 Determination of the directions

Here is described a method to find the east-west and the north-south directions from the *chāyā* (OY), *chāyā-bhujā* and *chāyā-koṭi*. In Figure 11.12, OX is the *śaṅku* whose length is taken to be 12 units (*dvādaśāṅgula*). Now,

$$\begin{aligned} Dv\bar{a}da\acute{s}\bar{a}\dot{n}gulacch\bar{a}y\bar{a} = OY &= \frac{12\ \sin z}{\cos z} \\ &= \frac{12 \times Mah\bar{a}cch\bar{a}y\bar{a}}{Mah\bar{a}\text{-}\acute{s}anku}. \end{aligned} \tag{11.75}$$

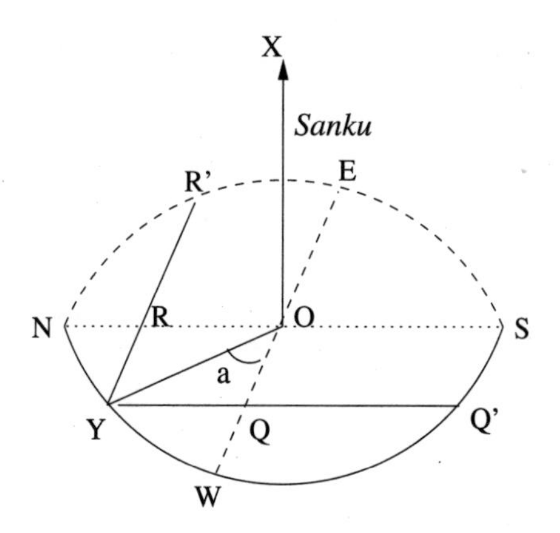

Figure 11.12: Determination of the directions from the *chāyā*, *chāyā-bhujā* and *chāyā-koṭi*.

The *bhujā* of the above, or the *chāyā-bhujā* in *aṇgula*-s is given by

$$Ch\bar{a}y\bar{a}\text{-}bhuj\bar{a} = \frac{12 \sin z}{\cos z} \sin a,$$

which is obtained from *śaṅkvagrā*, and *arkāgrā* (in *aṇgula*-s) is found from (11.72). The *chāyā-koṭi* in *aṇgula*-s is found from the above two.

Now, draw a circle with the radius equal to *chāyā* in *aṅgula*-s with the gnomon at the centre. Let the tip of the shadow be at Y at some instant. Place two rods equal to twice the *chāyā-bhujā* ($YQ' = 2\,YQ$) and twice the *chāyā-koṭi* ($YR' = 2\,YR$) at Y such that their other ends touch the circle. Then YQ' is the north-south direction and YR' is the east-west direction.

11.16 *Sama-śaṅku*: Great gnomon at the prime vertical

In Figure 11.13, *sama-maṇḍala* is the prime vertical passing through E, Z and W. Celestial equator is the great circle passing through E, W and a point U on the north-south circle, such that ZU is the latitude of the place (ϕ). When the declination of the Sun is zero, the celestial equator is the diurnal circle. When the declination is northerly and less than the latitude, corresponding to D_1 in the figure, the rising and setting will be to the north of E and W respectively and the midday will be to the south of the zenith. Then, the diurnal circle cuts the *sama-maṇḍala* at two points, once before the noon and once after. The *mahā-śaṅku* at the time corresponding to the Sun at S (and S' not shown in the figure) on D_1 in Figure 11.13 is termed the '*sama-śaṅku*'.

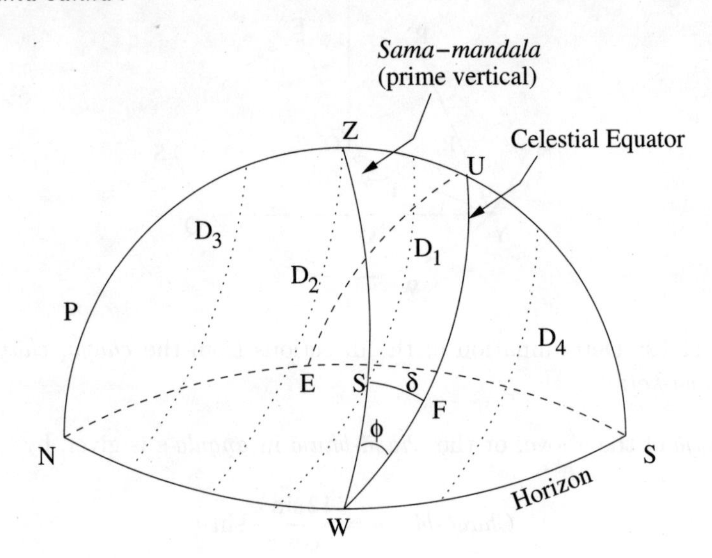

Figure 11.13: The *sama-śaṅku*.

When the declination $\delta = \phi$, the diurnal circle (D_2) touches the *sama-maṇḍala* at Z and there is no midday shadow, and the *sama-śaṅku* is equal to *trijyā* R (as zenith distance $z = 0$). The *sama-śaṅku* does not occur during the days when the declination is northerly and greater than the latitude (diurnal circle D_3 in the figure), and also when the declination is southerly (as in D_4 in the figure).

The angle between the *sama-maṇḍala* and the *ghaṭikā-maṇḍala* is equal to the latitude of the place ($Z\hat{W}U = \phi$). The *sama-śaṅku* ($R\cos ZS$), when the northerly declination δ is less than the latitude, is given by the relation

$$R\sin\delta = \frac{R\sin\phi \; R\cos z_s}{R}, \tag{11.76}$$

or,

$$R\cos z_s = \frac{R \; R\sin\delta}{R\sin\phi}. \tag{11.77}$$

(Here $z_s = ZS$ on the *sama-maṇḍala*, when the Sun is at S in Figure 11.13). This is obvious from the spherical triangle WSF, where $WS = 90° - z_s$, $SWF = \phi$, $SF = \delta$ and $SFW = 90°$. Alternately, the northerly declination δ and the longitude λ can be obtained from the *sama-śaṅku*.

11.17 *Samacchāyā*

Samacchāyā is the shadow (*chāyā*) when the Sun is on the *sama-maṇḍala*. The hypotenuse of the *samacchāyā* of the 12-inch *śaṅku* is

$$\frac{12R}{R\cos z_s} = \frac{12.R.R\sin\phi}{R.R\sin\delta} = 12\frac{R\sin\phi}{R\sin\delta}, \tag{11.78}$$

where we have used (11.77). Now,

$$\text{Equinoctial Shadow} = \frac{12R\sin\phi}{R\cos\phi}, \tag{11.79}$$

or,

$$12.R\sin\phi = \text{Equinoctial Shadow} \times R\cos\phi. \tag{11.80}$$

So, the *samacchāyā-karṇa* is also given by

$$\frac{\text{Equinoctial Shadow} \times R\cos\phi}{R\sin\delta}. \tag{11.81}$$

The *samacchāyā* occurs when δ is north and noon-shadow is less than the equinoctial shadow ($\delta < \phi$). Therefore, the difference between the equinoc-

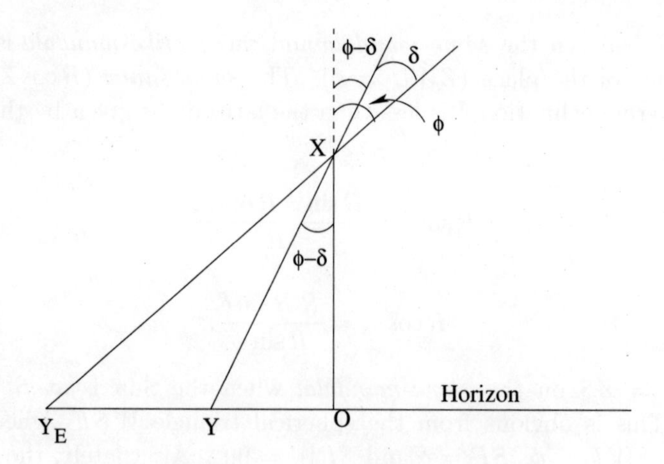

Figure 11.14: The noon shadow.

tial shadow and the noon shadow is given by

$$\begin{aligned} OY_E - OY &= \frac{12\sin\phi}{\cos\phi} - \frac{12\sin(\phi-\delta)}{\cos(\phi-\delta)} \\ &= \frac{12[\sin\phi\cos(\phi-\delta) - \cos\phi\sin(\phi-\delta)]}{\cos\phi\cos(\phi-\delta)} \\ &= \frac{12\sin\delta}{\cos\phi\cos(\phi-\delta)}. \end{aligned} \tag{11.82}$$

From the previous expression for *agrāṅgula* (11.69), we get

$$Madhy\bar{a}hna\text{-}agr\bar{a}\dot{n}gula = \frac{12\sin\delta}{\cos z\cos\phi} = \frac{12\sin\delta}{\cos\phi\cos(\phi-\delta)}, \tag{11.83}$$

as $z = \phi - \delta$ at noon. Hence,

$$Madhy\bar{a}hna\text{-}agr\bar{a}\dot{n}gula = \text{Equinoctial shadow} - \text{Noon shadow}.$$

On the day when the Sun passes through the zenith at noon ($\delta = \phi$),

$$Madhy\bar{a}hna\text{-}agr\bar{a}\dot{n}gula = \frac{12\sin\phi}{\cos\phi} = \text{Equinoctial shadow}. \tag{11.84}$$

On this day,

$$Madhyacch\bar{a}y\bar{a}\text{-}kar\d{n}a = Samacch\bar{a}y\bar{a}\text{-}kar\d{n}a = 12. \tag{11.85}$$

This is because *karṇa* $= OX = 12$, as the rays are travelling from the zenith, vertically down and there is no shadow.

When δ is very small, *madhyāhna-agrāṅgula* is very small, *madhyacchāyā-karṇa* is $\frac{12}{\cos(\phi-\delta)}$ and *samacchāyā -karṇa* which is $\frac{12\sin\phi}{\sin\delta}$ is very large. Now,

$$\frac{\dfrac{12\sin\phi}{\cos\phi}\times\dfrac{12}{\cos(\phi-\delta)}}{\dfrac{12\sin\phi}{\cos\phi}-\dfrac{12\sin(\phi-\delta)}{\cos(\phi-\delta)}}=\frac{12\sin\phi}{\sin\delta},$$

or,

$$Samacch\bar{a}y\bar{a}\text{-}kar\d{n}a = \frac{Vi\d{s}uvacch\bar{a}y\bar{a}\times Madhyacch\bar{a}y\bar{a}\text{-}kar\d{n}a}{Madhy\bar{a}hna\text{-}agr\bar{a}\dot{n}gula}. \tag{11.86}$$

11.18 The *Sama-śaṅku*-related triangles

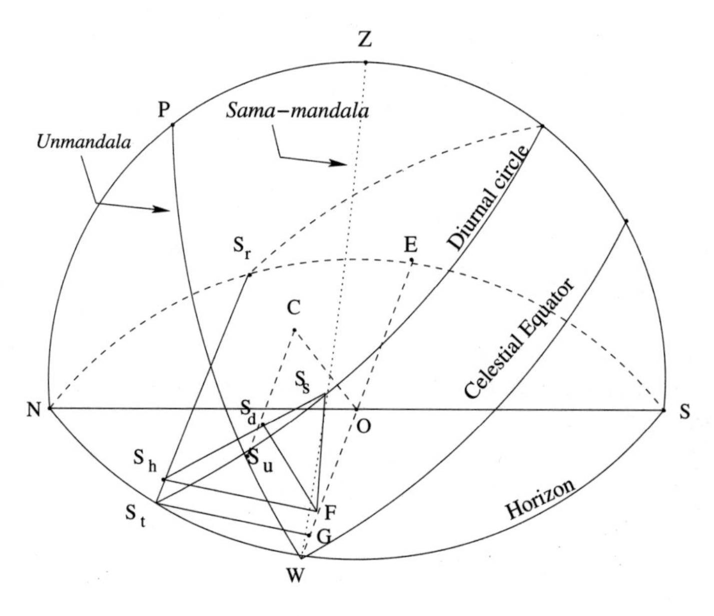

Figure 11.15: The latitudinal triangle formed by *sama-śaṅku*.

Let the Sun be on the *sama-maṇḍala* at S_s on a day when the declination is δ (see Figure 11.15). S_sF is drawn perpendicular to the horizon. F is on the east-west line, as the Sun is on the prime vertical. The *sama-śaṅku* will be

$$S_sF = \frac{R\sin\delta}{\sin\phi}.$$

From S_s draw S_sS_h perpendicular to the line S_tS_r passing through the rising and setting points and is parallel to the east-west line. S_hF is also perpendicular to the east-west line and is equal to

$$Arkāgrā = \frac{R\sin\delta}{\cos\phi}.$$

S_sF (*sama-śaṅku*), S_hF (*arkāgrā*) and S_sS_h (portion of the diurnal circle between the horizon and the *sama-maṇḍala*) form a right angled triangle with one angle being $S_h\hat{S}_sF = \phi$, the latitude. Hence, it is a latitudinal triangle.

If the Sun is at S_u on the *unmaṇḍala* (six-o'clock circle), CS_u is parallel to the east-west line, where C is the centre of the diurnal circle. Let S_sS_h, which is also in the plane of the diurnal circle, cut this line at S_d. S_dF is perpendicular to CS_u and is equal to $R\sin\delta$. S_dF is parallel to CO and perpendicular to the plane of the diurnal circle and hence perpendicular to S_sS_d and S_sS_h. In the triangle S_hS_dF, $S_h\hat{S}_dF = 90°$. Further,

$$\begin{aligned} S_dF &= R\sin\delta \\ \text{and} \quad S_hF &= arkāgrā = \frac{R\sin\delta}{\cos\phi}, \end{aligned} \tag{11.87}$$

is the hypotenuse. Here the angle at F is the latitude, and S_hS_dF is a latitudinal triangle. S_dS_sF is also a latitudinal triangle with $S_s\hat{S}_dF = 90°$. As before,

$$\begin{aligned} S_dF &= R\sin\delta \\ \text{and} \quad S_sF &= Sama\text{-}śaṅku = \frac{R\sin\delta}{\sin\phi}, \end{aligned} \tag{11.88}$$

is the hypotenuse. In this case, the angle at S_s is the latitude.

These three triangles are shown in Figure 11.16. The fourth triangle shown is the standard latitudinal triangle with the sides R, $R\sin\phi$ and $R\cos\phi$.

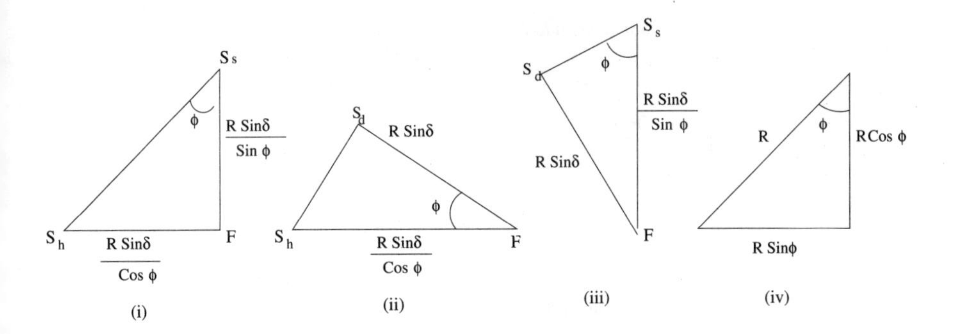

Figure 11.16: The different latitudinal triangles.

11.19 The ten problems

In Figure 11.17a, two circles with a common radius R and a common centre, O intersect at points X and X'. Let i be the angle of inclination between the two circles. It may be noted that the maximum separation between the two circles "given by $CD = R\,i$" occurs when $CX = DX = 90°$.

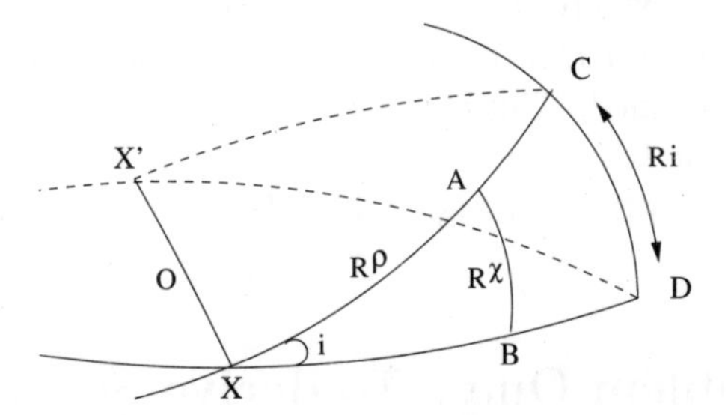

Figure 11.17a: Measure of the arc connecting two intersecting circles.

Consider a point A on one of the circles such that arc $XA = R\rho$. Draw a great circle arc $AB = R\chi$ such that it is perpendicular to the second circle XDX' at B. Then $R\sin\chi$ is the perpendicular distance between A and the second circle and is given by

$$R\sin\chi = R\sin i\ \sin\rho. \tag{11.89}$$

This can be found if the arc $R\rho$ is given; conversely, the arc $R\rho$ can be found when the perpendicular distance $R\sin\chi$ is known. This is the '*trairāśika*' that is being referred to and was discussed in detail in chapter 9. The applications of this follow.

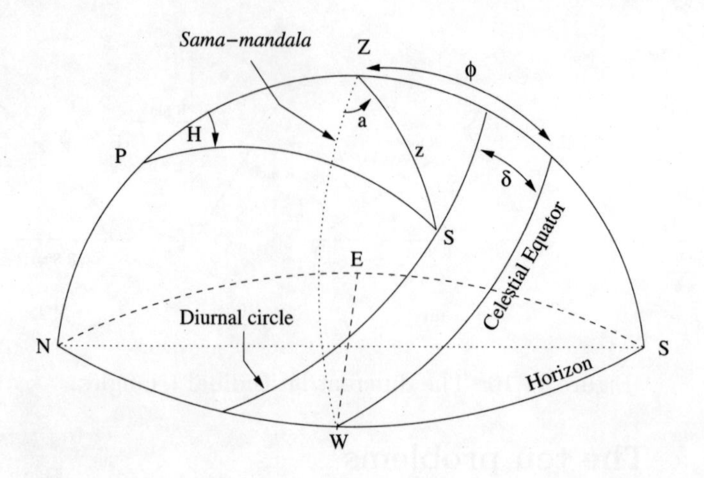

Figure 11.17*b*: Representation of the five quantities discussed in the "ten problems".

Now, there are five quantities: (i) *śaṅku* (gnomon) $R\cos z$, (ii) *nata-jyā* (Rsine hour angle) $R\sin H$, (iii) *apakrama* (declination) $R\sin\delta$, (iv) *āśāgrā* (amplitude) $R\sin a$, where $a = 90° \sim A$, A being the azimuth, and (v) *akṣajyā* (Rsine latitude) $R\sin\phi$. When three of them are known, the other two are to be determined. This can happen in ten different ways, and so the section is titled 'The ten problems'. The angles/arcs corresponding to these five quantities are depicted in Figure 11.17*b*.

11.20 Problem One : To derive *Śaṅku* and *Nata*

11.20.1 Shadow and gnomon at a desired place

We now discuss the method to derive the *śaṅku* and the *nata-jyā*, when the declination, *āśāgrā* and latitude are known.

In Figure 11.18 X is the planet. The great circle through Z and X is the *iṣṭa-digvṛtta*, cutting the horizon at A. If $WA = a$ is the arc between the west point and A, the *aśāgrā* is $R\sin a$. Let B be between N and W, at $90°$ from A. Then the great circle through Z and B is the *viparīta-digvṛtta*. Consider the great circle through B and the north celestial pole P. This is the *tiryag-vṛtta* which is perpendicular to both the *iṣṭa-digvṛtta* and the

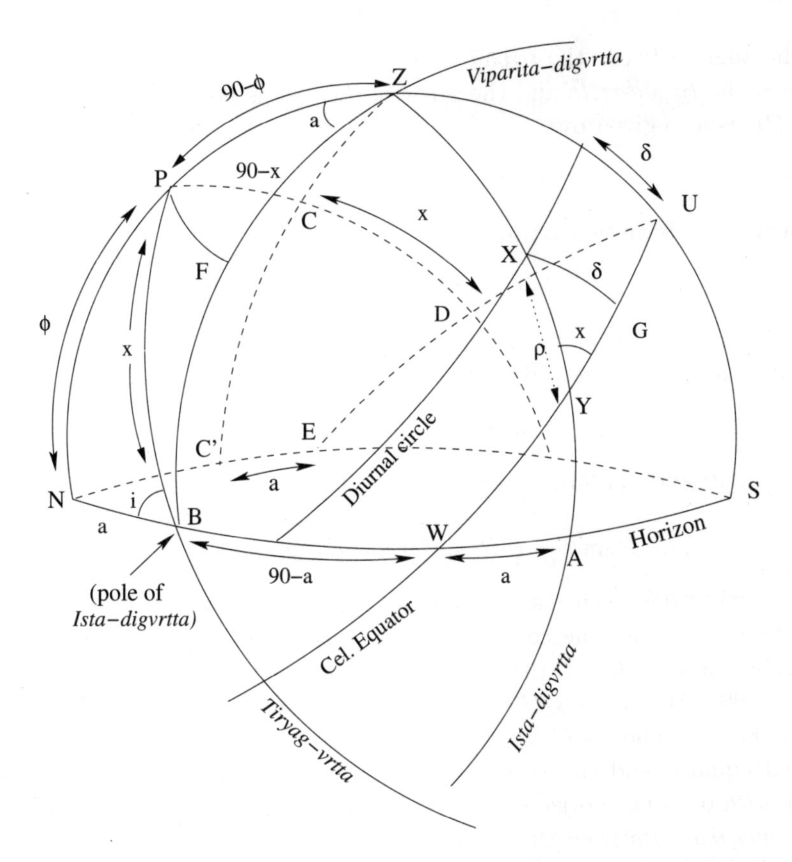

Figure 11.18: The important circles and their secondaries considered in the "ten problems".

celestial equator. This is so because this circle passes through the poles of both the *digvṛtta* and the celestial equator (B and P respectively). Let the *tiryag-vṛtta* intersect the *iṣṭa-digvṛtta* and the celestial equator at C and D respectively. Let the arc $BP = x$. Then, as B is the pole of the *iṣṭa-digvṛtta*, $BC = 90°$ or $PC = 90° - x$. As $PD = 90°$, $CD = x$. This is indeed the angle between the *digvṛtta* and the celestial equator at Y ($X\hat{Y}U$). The distance between P on the meridian and the *viparīta-digvṛtta* ZB is given by

$$R \sin PF = R \sin a \; \cos\phi, \tag{11.90}$$

as $PZ = 90° - \phi$, and $P\hat{Z}B$, the inclination of the *viparīta-digvṛtta* with the meridian is a.

Let the angle between the *tiryag-vṛtta* and the horizon be i. Then the angle between the *tiryag-vṛtta* and the *viparīta-digvṛtta* is $90° - i$. It follows that $R\sin PF$ is also given by

$$R\sin PF = R\sin x \cos i. \tag{11.91}$$

Equating the above two expressions,

$$R\sin x \cos i = R\sin a \cos\phi. \tag{11.92}$$

Now $PN = \phi$ is the perpendicular arc from P on the *tiryag-vṛtta*, on the horizon, which is inclined to it at angle i. Therefore,

$$R\sin x \sin i = R\sin\phi. \tag{11.93}$$

From (11.92) and (11.93), we get

$$R\sin x = \sqrt{R^2\sin^2 a\cos^2\phi + R^2\sin^2\phi}, \tag{11.94}$$

which is what has been stated. This is the maximum separation between the *iṣṭa-digvṛtta* and the celestial equator, as the angle betweeen them is x. Now the arc BC on the *tiryag-vṛtta* and the arc BC' on the horizon are both $90°$. Hence arc $CC' = i$, the angle between the two *vṛtta*-s. Then $CZ = 90° - i$, and as C is at $90°$ from Y, the intersection between the celestial equator and the *iṣṭa-digvṛtta*, $ZY = i$. Hence, the ascent of the *tiryag-vṛtta* from the horizon on the *digvṛtta* $= i$, is the same as the descent of the *ghaṭika-vṛtta* from the zenith on the *digvṛtta*. Let the arc $XY = \rho$. XG is the perpendicular arc from X on the *digvṛtta* on the celestial equator.

$$\begin{aligned} R\sin(XG) = R\sin\delta &= R\sin(XY)\ \sin x \\ &= R\sin\rho\sin x. \end{aligned} \tag{11.95}$$

Now the perpendicular arc from Z on the *digvṛtta* on the celestial equator $= ZU = \phi$. Therefore,

$$\begin{aligned} R\sin ZU = R\sin\phi &= R\sin(ZY)\sin x \\ &= R\sin i\sin x. \end{aligned} \tag{11.96}$$

$R\sin\rho$ and $R\sin i$ are called the *sthānīya*-s or the 'representatives' of the *apakrama* and *akṣajyā* on the *digvṛtta*. Now the zenith distance[3]

$$\begin{aligned} z = ZX &= ZY - XY \\ &= i - \rho. \end{aligned} \tag{11.97}$$

[3]If the declination is southern, $z = i + \rho$.

Therefore,

$$\begin{aligned} R\sin z = R\sin(i-\rho) &= R\sin i\cos\rho - R\cos i\sin\rho \\ &= \frac{(R\sin\phi\cos\rho - R\sin\delta\cos i).R}{R\sin x} \end{aligned} \qquad (11.98)$$

Consider the *koṭi*-s of the $R\sin\phi$ and $R\sin\delta$ on a circle of radius $R\sin x$ (which are denoted as *koṭi'*):

$$\begin{aligned} koṭi'(\phi) &= \sqrt{R^2\sin^2 x - R^2\sin^2\phi} \\ &= \sqrt{R^2\sin^2 x - R^2\sin^2 i\sin^2 x} \\ &= R\cos i\sin x. \end{aligned} \qquad (11.99)$$

Similarly,

$$\begin{aligned} koṭi'(\delta) &= \sqrt{R^2\sin^2 x - R^2\sin^2\delta} \\ &= \sqrt{R^2\sin^2 x - R^2\sin^2\rho\sin^2 x} \\ &= R\cos\rho\sin x. \end{aligned} \qquad (11.100)$$

Hence, we have

$$R\sin z = \frac{(R\sin\phi\ koṭi'(\delta) - R\sin\delta\ koṭi'(\phi))R}{R^2\sin^2 x}. \qquad (11.101)$$

This is the shadow $R\sin z$ at the desired place which is expressed in terms of the declination δ, latitude ϕ and the *āśāgrā*, as x is given in terms of ϕ and a by

$$R\sin x = \sqrt{R^2\sin^2 a\cos^2\phi + R^2\sin^2\phi}. \qquad (11.102)$$

The gnomon $R\cos z$ is given by

$$\begin{aligned} R\cos z &= R\cos(i-\rho) \\ &= R(\cos i\cos\alpha + \sin i\sin\rho) \\ &= \frac{(koṭi'(\phi)koṭi'(\delta) + R\sin\phi\ R\sin\delta)R}{R^2 sin^2 x}. \end{aligned} \qquad (11.103)$$

When the declination δ is south and $\delta > 90° - \phi$, the diurnal circle is below the horizon and there is no gnomon. When the northern declination is greater than the latitude, the midday would be to the north of the zenith and there will be gnomon in the southern direction. However, in this case, gnomon will occur only when *āśāgrā* is north i.e., A is north of W.

The different possible cases (for northerly declinations) are depicted in Figures 11.19(a)–(c).

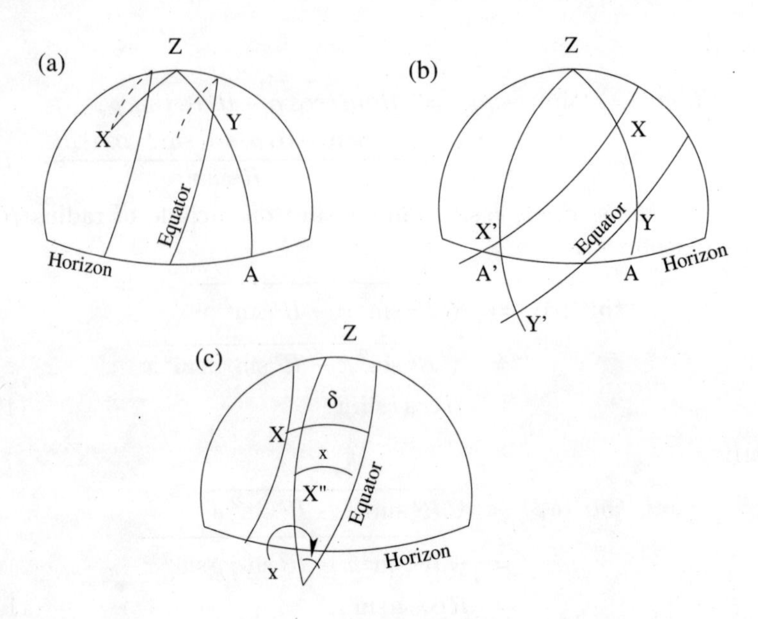

Figure 11.19: The different possible cases of northerly declinations.

(a) In this case, the sum of the representatives of the co-latitude, YA, and the declination XY, is greater than *trijyā*. Even then, there is a gnomon $R\cos(ZX)$.

(b) On a given day, consider the verticals through X and X' corresponding to southern and northern *āśāgrā*-s. Then $XA = XY + YA =$ sum of representatives of co-latitude and declination that figure in the expression for the gnomon at X, and $X'A' = X'Y' - Y'A' =$ difference of representatives of co-latitude and declination that figure in the expression for the gnomon at X'.

(c) Corresponding to some *āśāgrā* the declination δ will be greater than x. In such a case, there is no gnomon. This corresponds to a point X'' with $x < \delta$, which cannot lie on the declination circle. For points on the declination circle with $\delta > \phi$, *āśāgrā* a should be such that $R\sin x \geq R\sin\delta$.

11.20.2 *Koṇa-śaṅku* (Corner Shadow)

The term *koṇa* means corner. In this context, it refers to the corner between any two cardinal directions, such as north-east, south-west etc. Technically, *koṇa-śaṅku* or the corner shadow occurs when the *āśāgrā* $= 45°$. In this case, from (11.101) and (11.102) we have

$$R\sin x = \sqrt{\frac{1}{2}R^2\cos^2\phi + R^2\sin^2\phi} \tag{11.104}$$

$$R\sin z\sin x = \frac{R\sin\phi R\cos'\delta - R\sin\delta\, R\cos'\phi}{R\sin x}. \tag{11.105}$$

In the RHS of the above equation, $R\sin x$ is given by (11.104), and

$$\begin{aligned} R\cos'\delta &\equiv koṭi'\delta = \sqrt{R^2\sin^2 x - R^2\sin^2\delta}, \\ R\cos'\phi &\equiv koṭi'\phi = \sqrt{R^2\sin^2 x - R^2\sin^2\phi}. \end{aligned} \tag{11.106}$$

Similarly,

$$R\cos z\sin x = \frac{(R\cos'\phi\, R\cos'\delta + R\sin\phi\, R\sin\delta)}{R\sin x}. \tag{11.107}$$

Now in the case of *koṇa-śaṅku*, we have

$$\sin^2 x = \frac{1}{2}\cos^2\phi + \sin^2\phi.$$

Comparing the above result with (11.106), we have

$$(\cos'\phi)^2 = \frac{1}{2}\cos^2\phi.$$

Using this and (11.106), we get

$$\begin{aligned} \frac{1}{2}\cos^2\phi - \frac{\frac{1}{2}\cos^2\phi\sin^2\delta}{\sin^2 x} &= \frac{\frac{1}{2}\cos^2\phi\,(\sin^2 x - \sin^2\delta)}{\sin^2 x} \\ &= \frac{\frac{1}{2}\cos^2\phi\,(\cos'\delta)^2}{\sin^2 x} \\ &= \frac{(\cos'\phi)^2(\cos'\delta)^2}{\sin^2 x}. \end{aligned} \tag{11.108}$$

Therefore,

$$\frac{R^2(\cos'\phi)^2\; R^2(\cos'\delta)^2}{(R\sin x)^2} = \frac{1}{2}R^2\cos^2\phi - \frac{\frac{1}{2}R^2\cos^2\phi\; R^2\sin^2\delta}{R^2\sin^2 x}. \tag{11.109}$$

'One part' of the *koṇa-śaṅku*, viz.,

$$\frac{R\cos'\phi\; R\cos'\delta}{R\sin^2 x}, \tag{11.110a}$$

is got this way and the other part is

$$\frac{R\sin\phi\; R\sin\delta}{R\sin^2 x}. \tag{11.110b}$$

Now $R\sin\delta$ can be written as

$$\frac{R\sin\delta}{\cos\phi}\cos\phi = \frac{Ark\bar{a}gr\bar{a} \times Lambaka}{R}, \tag{11.111}$$

and the second part can be expressed in terms of *arkāgrā.*

It may be noted that in the denominator of (11.109) and (11.110) we have $R\sin x$. From (11.104), we get

$$\sin x = \sqrt{\frac{1}{2}\cos^2\phi + \sin^2\phi}. \tag{11.112}$$

Also,

$$\begin{aligned} \frac{L}{12} &= \frac{\sin\phi}{\cos\phi} \\ \text{or} \qquad \sin\phi &= \frac{L}{K}, \quad \cos\phi = \frac{12}{K}, \end{aligned} \tag{11.113}$$

where L is the equinoctial shadow and $K = \sqrt{L^2 + 12^2} =$ *karṇa*. Using (11.113) in (11.112), we have

$$\sin x = \frac{\left(\sqrt{\frac{1}{2}12^2 + L^2}\right)}{K}. \tag{11.114}$$

Thus, instead of $\sin\phi$ and $\cos\phi$, the equinoctial shadow L and the 12 inch gnomon can be used in the various expressions.

11.20.3 Derivation of *Nata-jyā*

In Figure 11.20, X is the planet whose declination is δ. Let H be the hour angle. Since $PX = 90 - \delta$, the distance between X and the north-south

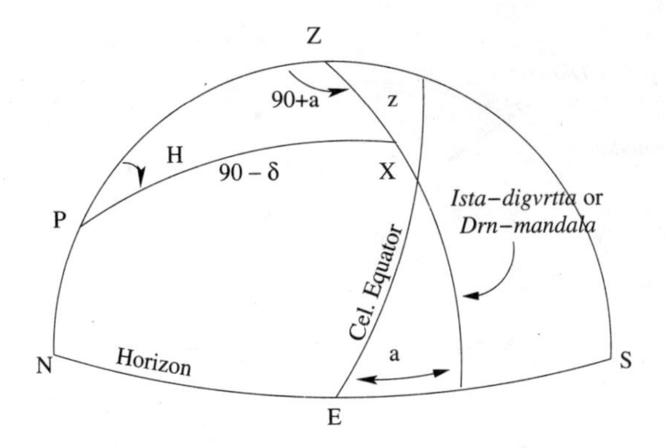

Figure 11.20: The *iṣṭa-digvṛtta* passing through a planet.

circle will be

$$\begin{aligned} &= R \sin H \sin(90 - \delta) \\ &= R \sin H \cos \delta. \end{aligned} \tag{11.115}$$

But the maximum angle between the north-south circle and *iṣṭa-digvṛtta* on which X is situated at a distance z from the zenith is $90 + a$. Therefore the distance between X and north-south circle is also

$$\begin{aligned} &= R \sin z \sin(90 + a) \\ &= R \sin z \cos a = \textit{chāyā-koti}. \end{aligned} \tag{11.116}$$

Equating the two expressions, we get

$$R \sin H \cos \delta = R \sin z \cos a = \textit{chāyā-koti}.$$

Therefore, the *nata-jyā* is given by

$$R \sin H = \frac{\textit{chāyā-koṭi}}{\cos \delta} = \frac{\textit{chāyā-koṭi} \times \textit{trijyā}}{\textit{dyujyā}}. \tag{11.117}$$

11.21 Problem two: *Śaṅku* and *Apakrama*

Here, the *śaṅku* and *krānti* (*apakrama*) are to be derived in terms of the *nata-jyā*, *āśāgrā* and *akṣa*.

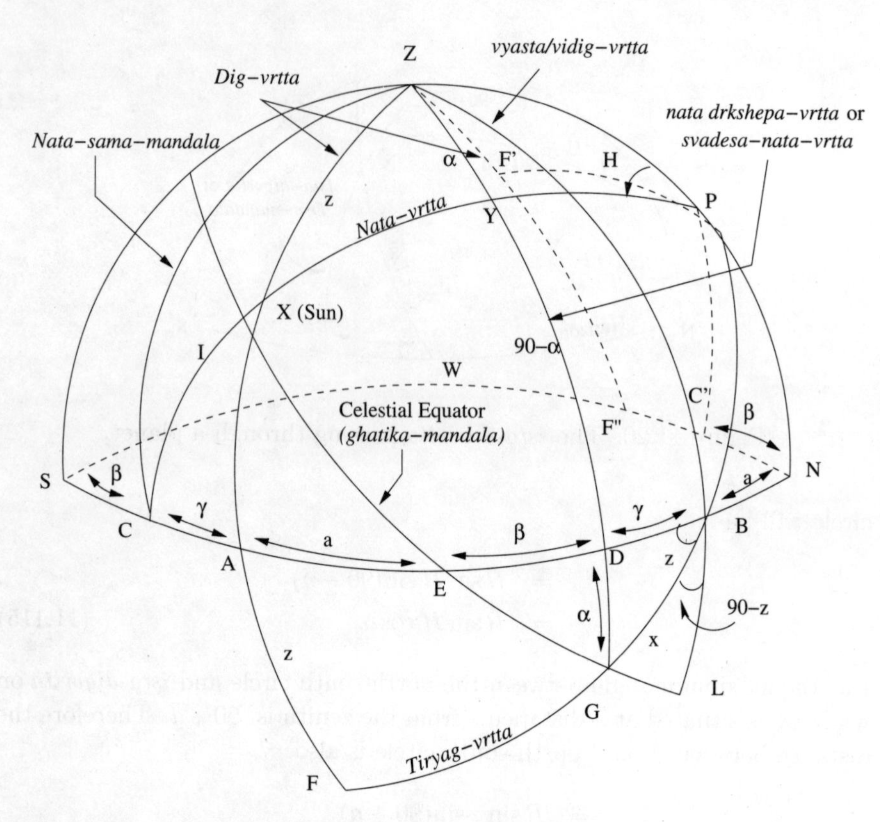

Figure 11.21: Some important great circles and their secondaries.

11.21.1 Derivation of *Śaṅku*

In Figure 11.21, *nata-vṛtta* is the great circle passing through P and X (Sun) which intersects the horizon at C. Now, draw the *nata-samamaṇḍala* which is a vertical through Z and C. D is a point on the horizon at 90° from C. *Nata-dṛkkṣepa-vṛtta* or *svadeśa-nata* is the vertical through D and *iṣṭa-digvṛtta* is the vertical through X intersecting the horizon at A. B is a point 90° from A and the vertical through B is the '*vyasta*' or *viparīta* or *vidig-vṛtta*. The point of intersection of *ghaṭikā-maṇḍala* and the *nata-dṛkkṣepa-vṛtta* is denoted by G.

Consider the great circle (*tiryag-vṛtta*) through B and G. We show that BG is perpendicular to both the *nata-vṛtta* and *digvṛtta*. The *tiryag-vṛtta* and the *iṣṭa-digvṛtta* intersect at F. Y is the point of intersection of *nata-dṛkkṣepa-vṛtta* and *nata-vṛtta*. Let $ZY = \alpha$. $R\sin ZY = R\sin\alpha$ is the *svadeśa-nata-jyā*. $YD = 90^\circ - \alpha$, $R\sin(YD) = R\cos\alpha$ is the *svadeśa-nata-koṭi*. Since B is at 90° from Z and A, it is the pole of the *iṣṭa-digvṛtta*. Therefore $BF = BX = 90^\circ$. Similarly, C is the pole of the *nata-dṛkkṣepa-vṛtta*, since $CD = CZ = 90^\circ$. Therefore G is at 90° from C. G being on the celestial equator is at 90° from P. Therefore G is the pole of *nata-vṛtta*. Hence BG passes through the poles of *nata-vṛtta* and *digvṛtta*. Thus, BG is the perpendicular to both the *nata-vṛtta* and *iṣṭa-digvṛtta*.

Now X is the pole of *tiryag-vṛtta*, as it is at 90° from B and G.[4] Therefore $XF = 90^\circ$. But $XA = 90^\circ - z$. Hence, $AF = z$, where z is the maximum separation between the horizon and the *tiryag-vṛtta* (as $BA = BF = 90^\circ$). Therefore, $z = D\hat{B}G$. The *tiryag-vṛtta* meets the *iṣṭa-digvṛtta* also at F'. Then,

$$
\begin{aligned}
180^\circ = FF' &= ZF' + ZF \\
&= ZF' + ZA + AF \\
&= ZF' + 90 + z.
\end{aligned}
$$

Therefore, $ZF' = 90 - z$ or $F'F'' = z$. This is the elevation of the *tiryag-vṛtta* from the horizon on the *iṣṭa-digvṛtta*. As this maximum separation occurs at 90°, $BF' = 90^\circ$. It is clear from the figure that the angle between the *tiryag-vṛtta* and the *vidig-vṛtta* is $90^\circ - z$.

Now C is the pole of ZD. Therefore $CY = 90^\circ$, and the angle at Y is 90°. Since the angle between ZP and YP is H and $ZP = 90^\circ - \phi$, the sine of the zenith distance of the point Y, denoted by α, is

$$
\begin{aligned}
\sin\alpha &= \sin(90-\phi)\sin H \\
&= \cos\phi\sin H. \qquad (11.118)
\end{aligned}
$$

Therefore,

$$
\cos\alpha = \sqrt{1-\cos^2\phi\sin^2 H}. \qquad (11.119)
$$

Let $CS = \beta$ be the distance between north-south circle and *nata-vṛtta* at the horizon. It is easy to see that $NC' = ED = \beta$, where C' is the point on the horizon diametrically opposite to C.

[4] The point X is at 90° from G, since G is the pole of *nata-vṛtta*.

Note:

(i) C being the pole of ZDG, $DY = 90° - \alpha$ is the angle between *nata-vṛtta* and the horizon. Therefore

$$\sin\phi = \sin PN = \sin(90 - \alpha)\sin(PC).$$

Hence,

$$\sin PC = \frac{\sin\phi}{\cos\alpha}. \qquad (11.120)$$

(ii) Now H is the angle between the north-south circle and the *nata-vṛtta.* Therefore,

$$\sin\beta = \sin(SC) = \sin H \sin PC.$$

Using (11.120) in the above equation, we get

$$\sin\beta = \frac{\sin H \sin\phi}{\cos\alpha} \qquad (11.121)$$

$$= \frac{\sin\phi\sin H}{\sqrt{1-\cos^2\phi\sin^2 H}}, \qquad (11.122)$$

using (11.119). This result would be used later.

Again, in Figure 11.21, $AE = a$ is *iṣṭāgrā.* The angle between the *nata-sama-vṛtta* and *digvṛtta* on the horizon is given by $CA = \gamma$. It may be noted that this is also equal to the angle between *nata-dṛkkṣepa-vṛtta* and *vyasta-dṛkkṣepa-vṛtta.* Since B is the pole of the the *digvṛtta*, clearly $\gamma = 90° - \beta - a$. Therefore,

$$\sin\gamma = \sin(90° - \beta - a)$$

$$= \cos(\beta + a)$$

$$= (\cos\beta\cos a - \sin\beta\sin a). \qquad (11.123)$$

When *āśāgrā* a is to the north of east, $\gamma = 90° - \beta + a$ and $\sin\gamma = \cos\beta\cos a + \sin\beta\sin a$. Thus $\sin\gamma$ is determined in terms of known quantities, since $\sin a$ is given and $\sin\beta$ is known from (11.122).

Now, let $GB = x$ and GL be the perpendicular arc from G to *vidig-vṛtta.* Then $\sin DG$, which is the same as $\sin ZY$, is given by

$$\sin\alpha = \sin z \sin x. \qquad (11.124)$$

Also

$$\sin GL = \sin x \cos z, \tag{11.125}$$

as z and $90 - z$ are the angles between *tiryag-vṛtta*and horizon, and *tiryag-vṛtta* and *vidig-vṛtta*, respectively. But the angle between ZG and ZL is γ and $ZG = 90^\circ + \alpha$. (For, $GY = 90^\circ$, G being the pole of *nata-vṛtta*). Therefore,

$$\begin{aligned}\sin GL &= \sin(90+\alpha)\sin\gamma \\ &= \sin\gamma\cos\alpha.\end{aligned} \tag{11.126}$$

Equating the two expressions for $\sin GL$, we get

$$\sin x \cos z = \sin\gamma\cos\alpha. \tag{11.127}$$

We had

$$\sin x \sin z = \sin\alpha. \tag{11.128}$$

From (11.127) and (11.128), we get

$$\sin x = \sqrt{\sin^2\alpha + \sin^2\gamma\cos^2\alpha}. \tag{11.129}$$

Using the above in (11.127) and (11.128), we have

$$\cos z = \frac{\sin\gamma\cos\alpha}{\sqrt{\sin^2\alpha + \sin^2\gamma\cos^2\alpha}}, \tag{11.130}$$

$$\text{and} \qquad \sin z = \frac{\sin\alpha}{\sin x}. \tag{11.131}$$

Now

$$\sin\beta = \frac{\sin\phi\sin H}{\cos\alpha}. \tag{11.132}$$

Therefore,

$$\begin{aligned}\cos\beta &= \sqrt{1-\sin^2\beta} \\ &= \sqrt{1-\frac{\sin^2\phi\sin^2 H}{\cos^2\alpha}} \\ &= \frac{\sqrt{\cos^2\alpha - \sin^2\phi\sin^2 H}}{\cos\alpha} \\ &= \frac{\sqrt{1-\cos^2\phi\sin^2 H - \sin^2\phi\sin^2 H}}{\cos\alpha} \\ &= \frac{\cos H}{\cos\alpha}.\end{aligned} \tag{11.133}$$

Hence from (11.123), (11.132) and (11.133), we have

$$\begin{aligned} \sin\gamma\cos\alpha &= (\cos\beta\cos a - \sin\beta\sin a)\cos\alpha \\ &= \cos H\cos a - \sin\phi\sin H\sin a. \end{aligned} \tag{11.134}$$

We have already shown that

$$\sin\alpha = \cos\phi\sin H. \tag{11.135}$$

Substituting these in (11.130), we obtain the following expression for *śaṅku* in terms of *natajyā*, *āśāgrā* and *akṣa*:

$$R\cos z = \frac{(R\cos H\cos a - R\sin\phi\sin H\sin a)R}{\sqrt{R^2\cos^2\phi\sin^2 H + (R\cos H\cos a - R\sin\phi\sin H\sin a)^2}}. \tag{11.136}$$

Similarly substituting in (11.131), we have

$$R\sin z = \frac{(R\cos\phi\sin H)R}{\sqrt{R^2\cos^2\phi\sin^2 H + (R\cos H\cos a - R\sin\phi\sin H\sin a)^2}}. \tag{11.137}$$

These are the gnomon and the shadow respectively.

11.21.2 Derivation of *Apakrama*

Now X is at the intersection of the *nata-vṛtta* and *digvṛtta* which make angles H and $90° - a$, respectively, with the north-south circle. $PX = 90° - \delta$ and $ZX = z$. Equating the two expressions for the distance between X and the north-south circle, we get

$$R\cos\delta\sin H = R\sin z\cos a. \tag{11.138}$$

Hence,

$$R\cos\delta = \frac{R\sin z\ R\cos a}{R\sin H}, \tag{11.139}$$

or

$$Dyujy\bar{a} = \frac{ch\bar{a}y\bar{a} \times \bar{a}\acute{s}\bar{a}gr\bar{a}\text{-}ko\d{t}i}{nata jy\bar{a}},$$

from which the *apakrama* can be obtained as

$$R\sin\delta = \sqrt{R^2 - R^2\cos^2\delta}. \tag{11.140}$$

11.22 Problem three: *Śaṅku* and *Āśāgrā*

Now the problem is to find $R\sin z$ and $R\sin a$ given $R\sin H$, $R\sin\delta$ and $R\sin\phi$.

11.22.1 Derivation of *Śaṅku*

Consider the product of *dyujyā* and the *koṭi* of the hour angle divided by *trijyā*, that is, $R\cos\delta\cos H$. To this, we add or subtract *kṣitijyā* (Rsine of the ascensional difference on the diurnal circle) given by

$$\frac{R\sin\phi\sin\delta}{\cos\phi},$$

depending upon whether the declination is positive or negative. Multiply by $\frac{R\cos\phi}{R}$. This is the *śaṅku*. In other words, we have

$$R\cos z = \cos\phi\left(R\cos\delta\cos H + \frac{R\sin\phi\sin\delta}{\cos\phi}\right). \tag{11.141}$$

This expression for *mahā-śaṅku* has already been proved in section 11.5, after deriving the expression for *unnata-jyā*. In the modern notation, this will be

$$\cos z = \sin\phi\sin\delta + \cos\phi\cos\delta\cos H.$$

11.22.2 Derivation of *Āśāgrā*

The shadow $R\sin z$ can be found from the *śaṅku*, $R\cos z$. In the previous section, we had derived the relation

$$R\sin H\cos\delta = R\sin z\cos a,$$

or

$$R\cos a = \frac{(R\sin H)(R\cos\delta)}{R\sin z}. \tag{11.142}$$

This is the *āśāgrā-koṭi*. From this, we find the *āśāgrā*, $R\sin a$.

11.23 Problem four: *Śaṅku* and *Akṣa*

Given *nata* ($R \sin H$), *krānti* ($R \sin \delta$) and *āśāgrā* ($R \sin a$), to derive the *śaṅku* ($R \cos z$) and *akṣa* ($R \sin \phi$):

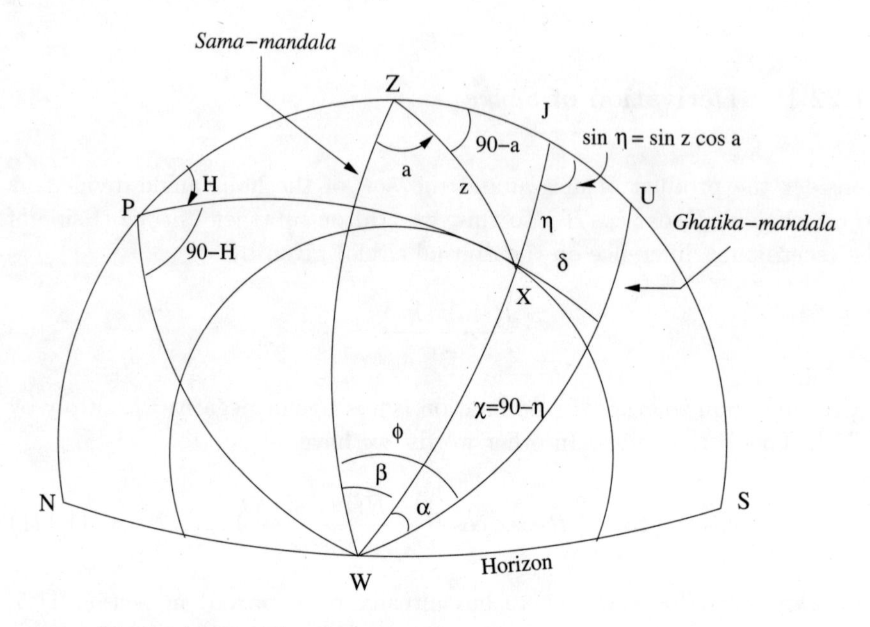

Figure 11.22: The '*koṭi* circle' passing through the planet.

11.23.1 Derivation of *Śaṅku*

We have already shown that

$$R \sin z = \frac{R \sin H.\ R \cos \delta}{R \cos a}.$$

From this, we have the *śaṅku*

$$R \cos z = \sqrt{R^2 - R^2 \sin^2 z}, \tag{11.143}$$

and the *chāyā-koṭi*

$$R \sin z \cos a = \frac{(R \sin H.\ R \cos \delta)}{R}. \tag{11.144}$$

11.23.2 Derivation of the *Akṣa*

Consider Figure 11.22. The distance of the planet X from the north-south circle is

$$R \sin \eta = \textit{chāyā-koṭi} = R \sin z \cos a. \qquad (11.145)$$

Now, draw a small circle through X parallel to the north-south circle. This will bear the same relation to the north-south circle, as the diurnal circle does to the equator. Here, $R \sin \eta = R \sin z \cos a$, is the equivalent of the *krānti* $= R \sin \delta$. The radius of this circle called '*koṭi* circle' is equivalent to *dyujyā*, $R \cos \delta$. The arc $WX = \chi =$ *koṭi* $= 90 - \eta$ and the radius of the '*koṭi* circle' is equal to

$$R \sin \chi = R \sin (90 - \eta) = \sqrt{R^2 - R^2 \sin^2 z \cos^2 a}. \qquad (11.146)$$

This is the *chāyā-koṭi-koṭi*. If we conceive of a right-angled triangle with the radius of the *koṭi* circle as the hypotenuse, and *chāyā-bhujā* $R \sin z \sin a$ as the *bhujā*, *koṭi* of this on the *koṭi* circle is the *śaṅku*, because,

$$\sqrt{R^2 - R^2 \sin^2 z \cos^2 a - R^2 \sin^2 z \sin^2 a} = R \cos z. \qquad (11.147)$$

Similarly, if the *apakrama* $R \sin \delta$ is the *bhujā*, then the *koṭi* of this (*apakrama-koṭi*) on this circle is $R \cos \delta \cos H$, since,

$$\begin{aligned} \sqrt{R^2 - R^2 \sin^2 z \cos^2 a - R^2 \sin^2 \delta} &= \sqrt{R^2 - R^2 \sin^2 H \cos^2 \delta - R^2 \sin^2 \delta} \\ &= R \cos \delta \cos H, \qquad (11.148) \end{aligned}$$

where we have used $R \sin z \cos a = R \sin H \cos \delta$.

It may be noted that $R \cos \delta \cos H$ is the distance between the planet X and the *unmaṇḍala* PW. It can also be visualized as Rsine of $90^\circ - H$ on the diurnal circle (whose radius is $R \cos \delta$).

The *akṣa*, $R \sin \phi$, is then obtained from the relation

$$\begin{aligned} R \sin \phi &= \frac{\textit{apakrama} \times \textit{śaṅku} + \textit{chāyā-bhujā} \times \textit{apakrama-koṭi}}{(\textit{chāyā-koṭi-koṭi})^2} \times (\textit{trijyā}) \\ &= \frac{R \sin \delta.\ R \cos z + R \sin z \sin a.\ R \cos \delta \cos H}{(R^2 - R^2 \sin^2 z \cos^2 a)} \times R. \qquad (11.149) \end{aligned}$$

This can be understood as follows. The latitude ϕ is the angle between the *sama-maṇḍala* and the *ghaṭikā-maṇḍala* and is the sum of two angles α and

β, where α is the angle between XW and *ghaṭikā-maṇḍala*, and β is the angle between XW and *sama-maṇḍala*, as shown in Figure 11.22.

Now,

$$R\sin\chi\sin\alpha = R\sin\delta,$$

or

$$R\sin\alpha = \frac{R\sin\delta}{\sin\chi}. \tag{11.150}$$

Hence,

$$\begin{aligned} R\cos\alpha &= \sqrt{R^2 - R^2\sin^2\alpha} \\ &= \frac{\sqrt{R^2\sin^2\chi - R^2\sin^2\delta}}{\sin\chi} \\ &= \frac{\sqrt{R^2 - R^2\sin^2 z\cos^2 a - R^2\sin^2\delta}}{\sin\chi} \\ &= \frac{R\cos\delta\cos H}{\sin\chi}, \end{aligned} \tag{11.151}$$

where we have used (11.148). The distance of the planet X from *sama-maṇḍala* is

$$R\sin z\sin a = R\sin\chi\sin\beta.$$

Therefore,

$$R\sin\beta = \frac{R\sin z\sin a}{\sin\chi}. \tag{11.152}$$

Hence,

$$\begin{aligned} R\cos\beta &= \sqrt{R^2 - R^2\sin^2\beta} \\ &= \frac{\sqrt{R^2\sin^2\chi - R^2\sin^2 z\sin^2 a}}{\sin\chi} \\ &= \frac{\sqrt{R^2 - R^2\sin^2 z\cos^2 a - R^2\sin^2 z\sin^2 a}}{\sin\chi} \\ &= \frac{R\cos z}{\sin\chi}. \end{aligned} \tag{11.153}$$

Now,

$$\begin{aligned} R\sin\phi &= R\sin(\alpha+\beta) \\ &= \frac{R\sin\alpha.\ R\cos\beta + R\cos\alpha.\ R\sin\beta}{R}. \end{aligned} \tag{11.154}$$

Using (11.150) – (11.153) in the above, we get

$$
\begin{aligned}
R\sin\phi &= \frac{R\sin\delta\ R\cos z + R\sin z\sin a.\ R\cos\delta\cos H}{R\sin^2\chi} \\
&= \frac{(R\sin\delta\ R\cos z + R\sin z\sin a.\ R\cos\delta\cos H)R}{(R^2 - R^2\sin^2 z\cos^2 a)}, \qquad (11.155)
\end{aligned}
$$

which is the desired expression for the *akṣa* as given in (11.149). This is true when declination δ and *āśāgrā* a are in opposite directions from X. But, when they are in the same direction (when X is to the north of *sama-maṇḍala*), the second term in the numerator of (11.155) is negative. However, when the planet is between the *unmaṇḍala* and the horizon ($H > 90°$, a is to the north of *sama-maṇḍala*), it is positive. Note that the *akṣa* on the *koṭi* circle is

$$
\frac{R\sin\delta.\ R\cos z + R\sin z\sin a.\ R\cos\delta\cos H}{\sqrt{R^2 - R^2\sin^2 z\cos^2 a}}.
$$

Note: The modern way of deriving the expression for $\sin\phi$ would be to start from

$$
\cos z = \sin\phi\sin\delta + \cos\phi\cos\delta\cos H, \qquad (11.156)
$$

or,

$$
\sqrt{1 - \sin^2\phi}\ \cos\delta\cos H = \cos z - \sin\phi\sin\delta.
$$

Squaring, we get

$$
(1 - \sin^2\phi)\cos^2\delta\cos^2 H = \cos^2 z + \sin^2\phi\sin^2\delta - 2\cos z\sin\phi\ \sin\delta,
$$

or,

$$
\sin^2\phi\ (\sin^2\delta + \cos^2\delta\cos^2 H) - 2\cos z\sin\delta.\ \sin\phi + \cos^2 z - \cos^2\delta\cos^2 H = 0.
$$

Solving this quadratic equation, we get the same expression as stated above.[5]

11.24 Problem five: *Nata* and *Krānti*

The first four problems involved the calculation of *śaṅku*. Now the fifth problem is to find *nata* ($R\sin H$) and *krānti* ($R\sin\delta$) from *akṣa* ($R\sin\phi$), *āśāgrā* ($R\sin a$), and *śaṅku* ($R\cos z$).

[5] In the course of simplification we need to use (11.138).

Referring to Figure 11.22 again, we may note that the angle between XW and the *ghaṭikā-maṇḍala* is $\alpha = \phi - \beta$. Using this, we have

$$\begin{aligned} R\sin\alpha &= R\sin(\phi-\beta) \\ &= \frac{R\sin\phi\, R\cos\beta - R\cos\phi\, R\sin\beta}{R}. \end{aligned} \qquad (11.157)$$

Hence,

$$\begin{aligned} R\sin\delta &= R\sin\alpha.\ \sin\chi \\ &= \frac{(R\sin\phi\sin\chi)R\cos\beta}{R} - \frac{(R\cos\phi\sin\chi)R\sin\beta}{R}. \end{aligned} \qquad (11.158)$$

Using (11.152) and (11.153), we obtain

$$\begin{aligned} \textit{Apakrama} &= \frac{(\text{Latitude on the } \textit{koṭi}\text{-circle}) \times \textit{śaṅku}}{\text{Radius of the } \textit{koṭi}\text{-circle}} \\ &\quad - \frac{(\text{Co-latitude on the } \textit{koṭi}\text{-circle}) \times \textit{chāyā-bhujā}}{\text{Radius of the } \textit{koṭi}\text{-circle}}, \end{aligned}$$

which is same as the relation

$$\textit{Apakrama} = \frac{\text{Latitude} \times \textit{Śaṅku} - \text{ Co-latitude} \times \textit{Chāyā-bhujā}}{\textit{Trijyā}}. \qquad (11.159)$$

We see that the above relation is equivalent to

$$\sin\delta = \sin\phi\cos z - \cos\phi\sin z\sin a, \qquad (11.160)$$

which is the result obtained by applying the cosine formula to the spherical triangle PZX, where $PZ = 90-\phi$, $ZX = z$, $PX = 90-\delta$ and $P\hat{Z}X = 90 + a$. When the planet X is to the north of *sama-maṇḍala*, the second term is positive and we have to add the two quantities in the numerator of (11.159). From *apakrama*, we find *dyujyā* $= R\cos\delta$. Then, we have

$$\begin{aligned} R\sin H &= \frac{(R\sin z\cos a)\times R}{R\cos\delta}, \\ \text{or} \quad \textit{Nata-jyā} &= \frac{\textit{Chāyā-koṭi} \times \textit{Trijyā}}{\textit{Dyujyā}}. \end{aligned} \qquad (11.161)$$

11.25 Problem six: *Nata* and *Āśāgrā*

To find *āśāgrā* ($R\sin a$) and *nata* ($R\sin H$) from *śaṅku* ($R\cos z$), *apakrama* ($R\sin\delta$) and *akṣa* ($R\sin\phi$):

It was shown in section 11.13 that

$$\textit{Chāyā-bhujā} = \textit{Śaṅkvagrā} - \textit{Arkāgrā}.$$

That is,

$$R\sin z \sin a = \frac{R\cos z.\ \sin\phi}{\cos\phi} - \frac{R\sin\delta}{\cos\phi}. \tag{11.162}$$

When the declination is south, the second term is positive. In either case, the RHS is known. Then, *āśāgrā* is given by

$$\begin{aligned} R\sin a &= \frac{R\sin z \sin a.\ R}{R\sin z} \\ &= \frac{\textit{Chāyā-bhujā} \times \textit{Trijyā}}{\textit{Chāyā}}. \end{aligned} \tag{11.163}$$

From this, *āśāgrā-koṭi* $= R\cos a$ is found. Then, *nata-jyā* is given by

$$\begin{aligned} R\sin H &= \frac{R\sin z \cos a}{\cos\delta} \\ &= \frac{R\sin z\ R\cos a}{R\cos\delta} \\ &= \frac{\textit{Chāyā} \times \textit{Āśāgrā-koṭi}}{\textit{Dyujyā}}. \end{aligned} \tag{11.164}$$

11.26 Problem seven: *Nata* and *Akṣa*

To find *nata* ($R\sin H$) and *akṣa* ($R\sin\phi$) from *śaṅku* ($R\cos z$), *apakrama* ($R\sin\delta$) and *āśāgrā* ($R\sin a$):

Nata is found by the method described earlier (Eq. 11.161) and is given by

$$R\sin H = \frac{(R\sin z\ R\cos a)}{(R\cos\delta)}. \tag{11.165}$$

Now, consider Figure 11.23. For definiteness, we consider the planet to be in the 'northern hemisphere', or the declination to be north, $\delta > 0$. It has been noted that Rsine of the arc (XX_u) between the planet and the *unmaṇḍala* on the diurnal circle is equal to $R\cos\delta\cos H$, which is given by

$$R\cos\delta\cos H = \sqrt{R^2\cos^2\delta - R^2\sin^2 z\cos^2 a}. \tag{11.166}$$

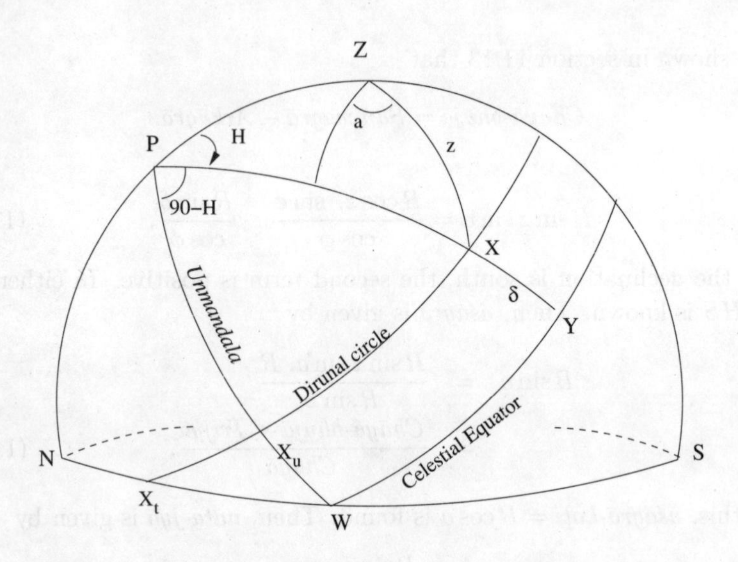

Figure 11.23: The *unmaṇḍala* and the diurnal circle.

This is nothing but the root of difference between the squares of *dyujyā* and *chāyā-koṭi*. This is equal to the difference between the *unnata-jyā*, which is the Rsine of the arc (XX_t) between the planet and its setting point on the diurnal circle and the *kṣitijyā*, which is the Rsine of the arc (X_uX_t) between the *unmaṇḍala* and the setting point on the diurnal circle. (The latter is equal to ascensional difference multiplied by *dyujyā* and divided by *trijyā* or $\frac{R \sin\phi \sin\delta}{\cos\phi}$).

Now *unnata-jyā*, *śaṅku* and *śaṅkvagrā* $\left(= R\cos z \frac{\sin\phi}{\cos\phi}\right)$ form a latitudinal triangle (see sections 11.13, 11.14). Similarly, *arkāgrā* $\left(= \frac{R\sin\delta}{\cos\phi}\right)$, *apakrama* and *kṣitijyā* $\left(= \frac{R\sin\phi\sin\delta}{\cos\phi}\right)$, with *arkāgrā* as the *karṇa* and *kṣitijyā* as the *bhujā*, form another latitudinal triangle. We can consider a third triangle whose *karṇa* and *bhujā* are the sum of the *karṇa*-s and *bhujā*-s of the aforesaid latitudinal triangles. It is clear that this is also a latitudinal triangle.

The latitudinal triangles involved in finding *akṣa* are depicted in Figure 11.24. In the third latitudinal triangle,

$$\begin{aligned} Karṇa &= Unnata\text{-}jyā + Arkāgrā \\ Bhujā &= Kṣitijyā + Śaṅkvagrā. \end{aligned}$$

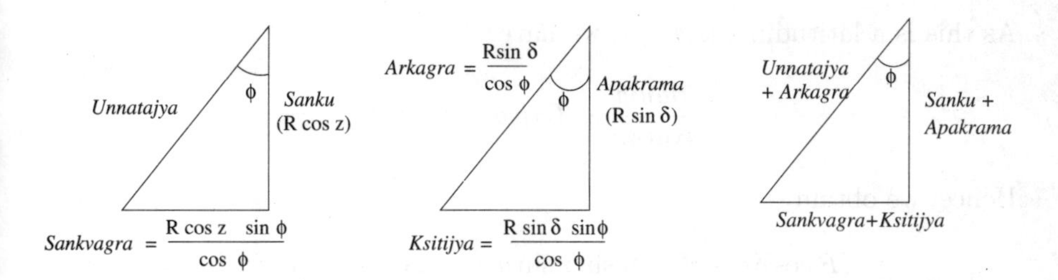

Figure 11.24: The different latitudinal triangles.

Therefore,

$$Karṇa - Bhuj\bar{a} = Unnata\text{-}jy\bar{a} - kṣitijy\bar{a} - (\acute{S}a\dot{n}kvagr\bar{a} - Ark\bar{a}gr\bar{a}).$$

Since,

$$\begin{aligned} Unnata\text{-}jy\bar{a} - Kṣitijy\bar{a} &= R\cos\delta\cos H, \\ \text{and} \quad \acute{S}a\dot{n}kvagr\bar{a} - Ark\bar{a}gr\bar{a} &= Ch\bar{a}y\bar{a}\text{-}bhuja \\ &= R\sin z \sin a, \end{aligned}$$

we have,

$$Karṇa - Bhuj\bar{a} = R\cos\delta.\cos H - R\sin z\sin a. \tag{11.167}$$

Now, in this triangle,

$$\begin{aligned} Karṇa^2 - Bhuj\bar{a}^2 &= Koṭi^2 \\ &= (\acute{S}a\dot{n}ku + Apakrama)^2 \\ &= (R\cos z + R\sin\delta)^2. \end{aligned} \tag{11.168}$$

From (11.167) and (11.168), we get

$$\begin{aligned} Karṇa + Bhuj\bar{a} &= \frac{(Karṇa^2 - Bhuj\bar{a}^2)}{Karṇa - Bhuj\bar{a}} \\ &= \frac{(R\cos z + R\sin\delta)^2}{(R\cos\delta\cos H - R\sin z\sin a)}. \end{aligned} \tag{11.169}$$

Adding and subtracting (11.167) and (11.169), we have

$$\begin{aligned} Karṇa &= \frac{1}{2}\frac{(R\cos\delta\cos H - R\sin z\sin a)^2 + (R\cos z + R\sin\delta)^2}{(R\cos\delta\cos H - R\sin z\sin a)}, \\ Bhuj\bar{a} &= \frac{1}{2}\frac{(R\cos\delta\cos H - R\sin z\sin a)^2 - (R\cos z + R\sin\delta)^2}{(R\cos\delta\cos H - R\sin z\sin a)}. \end{aligned}$$

As this is a latitudinal triangle, we have

$$\frac{Bhuj\bar{a}}{Kar\d{n}a} \times Trijy\bar{a} = R\sin\phi.$$

Hence, we obtain

$$R\sin\phi = \frac{(R\cos\delta\cos H - R\sin z\sin a)^2 - (R\cos z + R\sin\delta)^2}{(R\cos\delta\cos H - R\sin z\sin a)^2 + (R\cos z + \sin\delta)^2}, \qquad (11.170)$$

where the RHS is a function of z, a and δ, when we recall that

$$R\cos\delta\ \cos H = \sqrt{R^2\cos^2\delta - R^2\sin^2 z\cos^2 a}.$$

When the declination is south, we should consider

$$\begin{aligned} Kar\d{n}a + Bhuj\bar{a} &= Unnata\text{-}jy\bar{a} + K\d{s}itijy\bar{a} + \acute{S}a\dot{n}kvagr\bar{a} + Ark\bar{a}gr\bar{a} \\ &= R\cos\delta\cos H + R\sin z|\sin a|. \end{aligned} \qquad (11.171)$$

Also, *Chāyā-bhujā* = *śaṅkvagrā* + *arkāgrā* and *Unnata-jyā* + *kṣitijyā* = $R\cos\delta\cos H$, in this case. Here,

$$\begin{aligned} Kar\d{n}a - Bhuj\bar{a} &= \frac{(Kar\d{n}a^2 - Bhuj\bar{a}^2)}{(Kar\d{n}a + Bhuj\bar{a})} \\ &= \frac{(R\cos z + |R\sin\delta|)^2}{R\cos\delta\cos H + R\sin z|\sin a|}. \end{aligned} \qquad (11.172)$$

Karṇa and *bhujā* are now found and finally we have the same expression as (11.170) for $R\sin\phi$, with $-\sin z\sin a$ replaced by $\sin z|\sin a|$ and $\sin\delta$ replaced by $|\sin\delta|$.

11.27 Problem eight: *Apakrama* and *Āśāgrā*

To find *apakrama* ($R\sin\delta$) and *āśāgrā* ($R\sin a$) from *śaṅku*, *akṣa* and *nata*:

Refer to Figure 11.21 on page 750. The Rsine of the angle between the *nata-vṛtta* and the horizon, which is the *koṭi* of the *svadeśa-nata-vṛtta*, is given by

$$R\sin(90 - \alpha) = R\cos\alpha = Pram\bar{a}\d{n}a. \qquad (11.173)$$

The divergence between the *svadeśa-nata-vṛtta* and the horizon on the *nata-vṛtta* is

$$R\sin(CY) = R\sin 90^\circ = R = \textit{Pramāṇa-phala}. \tag{11.174}$$

The distance between planet at X on the *nata-vṛtta* and the horizon is

$$R\sin AX = R\sin(90 - z) = R\cos z = \textit{Śaṅku} = \textit{Icchā} \tag{11.175}$$

Distance between planet at X and C on the horizon, along the *nata-vṛtta*, is

$$R\sin CX = \textit{Icchā-phala}. \tag{11.176}$$

Using the rule of three,

$$R\sin CX = R\frac{R\cos z}{R\cos\alpha} = \frac{R\cos z}{\cos\alpha}. \tag{11.177}$$

With *pramāṇa* and *pramāṇa-phala* being the same, we now take the *icchā* to be the distance between the north pole P and N which is *Dhruva-nati* $=$ $R\sin\phi$. Then *icchā-phala*, which is the distance between the P and C along *nata-vṛtta*, is given by

$$R\sin PC = \frac{R.\ R\sin\phi}{R\cos\alpha} = \frac{R\sin\phi}{\cos\alpha}. \tag{11.178}$$

The arc corresponding to *dyujyā*

$$\begin{aligned} PX &= PC - CX \\ &= (R\sin)^{-1}(R\sin PC) - (R\sin)^{-1}(R\sin CX). \end{aligned} \tag{11.179}$$

When the planet X is to the north of the intersection between the *svadeśa-nata-vṛtta* and the *nata*, $CX > 90^0$. Then,

$$(R\sin)^{-1}(R\sin CX) = 180^\circ - CX,$$

as the *cāpa* is always less than 90° (when derived from the *jyā*). Similarly $PC > 90^\circ$ when X is above the horizon. Then,

$$(R\sin)^{-1}(R\sin PC) = 180^\circ - PC.$$

Using the above in (11.159), we have

$$\begin{aligned} PX = 90^\circ - \delta &= (180^\circ - CX) - (180^\circ - PC) \\ &= \text{Difference of the } \textit{cāpa}\text{-s}. \end{aligned} \tag{11.180}$$

Hence,

$$R\sin(90^\circ - \delta) = R\sin\left[(R\sin)^{-1}(R\sin CX) - (R\sin)^{-1}(R\sin PC)\right]. \tag{11.181}$$

Using (11.177) and (11.178), the above reduces to

$$R\cos\delta = R\sin\left[(R\sin)^{-1}\left(\frac{R\cos z}{\cos\alpha}\right) - (R\sin)^{-1}\left(\frac{R\sin\phi}{\cos\alpha}\right)\right]. \tag{11.182}$$

In the above expression, the RHS is known, since $\cos\alpha$ is known (refer to (11.119) in section 11.21). From $R\cos\delta$, the *apakrama* $R\sin\delta$ is determined. Now, *āśāgrā-koṭi* is determined from the relation

$$R\cos a = \frac{R\sin H\cos\delta}{\sin z}, \tag{11.183}$$

as usual. From this, *āśāgrā* is calculated.

When X is to the south of the intersection between *svadeśa-nata-vṛtta* and *nata-vṛtta*, $CX < 90^\circ$ (as in Figure 11.21). Then $(R\sin)^{-1}(R\sin CX) = CX$, whereas $(R\sin)^{-1})(R\sin PC)$ continues to be $180^\circ - PC$. Then the distance between X and south pole, Q (not shown in the figure), will be

$$XQ = CX + 180^\circ - PC.$$

It may be noted that along the *nata-vṛtta*, $PQ = 180^\circ = PC + CQ$. Therefore, $CQ = 180^\circ - PC$. Hence,

$$XQ = CX + CQ = CX + 180^\circ - PC. \tag{11.184}$$

Now, the arc corresponding to *dyujyā* is given by

$$\begin{aligned} 90^\circ + \delta &= CX + (180^\circ - PC) \\ &= \text{sum of the } c\bar{a}pa\text{-s}. \end{aligned} \tag{11.185}$$

Therefore,

$$R\sin(90^\circ + \delta) = R\cos\delta = R\sin(CX + 180^\circ - PC),$$

where

$$\begin{aligned} R\sin CX &= \frac{R\cos z}{\cos\alpha}, \\ \text{and} \quad R\sin(180^\circ - PC) &= \frac{R\sin\phi}{\cos\alpha}. \end{aligned} \tag{11.186}$$

From this, the *apakrama*, $R\sin\delta$, is determined and *āśāgrā* follows from (11.183).

11.28 Problem nine: *Krānti* and *Akṣa*

To determine $R\sin\delta$ and $R\sin\phi$ from $R\cos z$, $R\sin a$ and $R\sin H$:

The *dyujyā* $R\cos\delta$ is determined using the relation

$$R\cos\delta = \frac{R\cos a \sin z}{\sin H}, \tag{11.187}$$

and from that the *apakrama*, $R\sin\delta$, is found. Then *akṣa* is determined from the method outlined in problem four (section 11.23) or problem six (section 11.26).

11.29 Problem ten: *Āśāgrā* and *Akṣa*

To determine $R\sin a$ and $R\sin\phi$ from $R\sin z$, $R\cos\delta$ and $R\sin H$:

Given the *śaṅku*, *apakrama* and *nata-jyā*, *āśāgrā-koṭi* is obtained using the relation

$$\begin{aligned} R\cos a &= \frac{R\cos\delta\sin H}{\sin z} \\ &= \frac{Dyujyā \times Nata\text{-}jyā}{Chāyā}. \end{aligned} \tag{11.188}$$

From this, *āśāgrā*, $R\sin a$, is determined. Then *akṣa*, $R\sin\phi$, is derived as in section 11.26. Thus the solutions to all the ten problems have been discussed.

11.30 *Iṣṭadik-chāyā*: Another method

The term *iṣṭādik-chāyā* essentially refers to the Rsine zenith distance of the planet (having non-zero declination), denoted by $R\sin z$. To determine this, first the *chāyā* of a corresponding point on the *iṣṭa-diṅmaṇḍala*, with a given *āśāgrā* and which is located on the equator, is obtained. As noted in section 11.1, the 12-inch gnomic shadow is

$$\frac{12\ \sin z}{\cos z}. \tag{11.189}$$

And, the *viṣuvacchāyā*, equinoctial shadow, is

$$\frac{12\ \sin\phi}{\cos\phi}. \tag{11.190}$$

When $\delta = 0$, *arkāgrā* $= 0$. We denote z by z_0 in this case. Hence, from (11.70), we obtain

$$\begin{aligned} \textit{Śaṅkvagrā} &= \textit{Chāyā-bhujā} \\ \frac{R\cos z_0\ \sin\phi}{\cos\phi} &= R\sin z_0 \sin a, \end{aligned} \tag{11.191}$$

or

$$\frac{\sin z_0}{\cos z_0}\sin a = \frac{\sin\phi}{\cos\phi}. \tag{11.192}$$

Therefore

$$\begin{aligned} \textit{Dvādaśāṅgula-chāyā-bhujā} &= \frac{12\sin z_0}{\cos z_0}\sin a \\ &= \frac{12\sin\phi}{\cos\phi} \\ &= \textit{Viṣuvacchāyā}. \end{aligned} \tag{11.193}$$

Hence,

$$\begin{aligned} \textit{Dvādaśāṅgula-chāyā-koṭi} &= \frac{12\sin z_0}{\cos z_0}\cos a \\ &= \frac{12\sin\phi}{\cos\phi}\,\frac{\cos a}{\sin a}. \end{aligned} \tag{11.194}$$

Therefore *Dvādaśāṅgulacchāyā*[6] (l) is given by

$$\begin{aligned} l &= \frac{12\sin z_0}{\cos z_0} \\ &= \sqrt{\left(\frac{12\sin z_0}{\cos z_0}\cos a\right)^2 + \left(\frac{12\sin z_0}{\cos z_0}\sin a\right)^2} \\ &= \sqrt{\left(\frac{12\sin\phi}{\cos\phi}\,\frac{\cos a}{\sin a}\right)^2 + \left(\frac{12\sin\phi}{\cos\phi}\right)^2}. \end{aligned} \tag{11.195}$$

[6] Though the term literally means shadow corresponding to 12 *aṅgula*-s, in the present context it refers to the shadow of a *śaṅku* whose height is 12 inches, taking an *aṅgula* to be equivalent to an inch.

From this, the *karṇa*, $K = \sqrt{l^2 + 12^2}$, can be obtained. But, the *karṇa* is also given by

$$K = \frac{12}{\cos z_0}. \tag{11.196}$$

Therefore, the shadow in the *trijyā-vṛtta*, which is the ratio of the *dvādaśāṅgula-chāyā* and *karṇa* multiplied by *trijyā*,

$$R \sin z_0 = \frac{R\dfrac{12 \sin z_0}{\cos z_0}}{\dfrac{12}{\cos z_0}}, \tag{11.197}$$

can be determined.

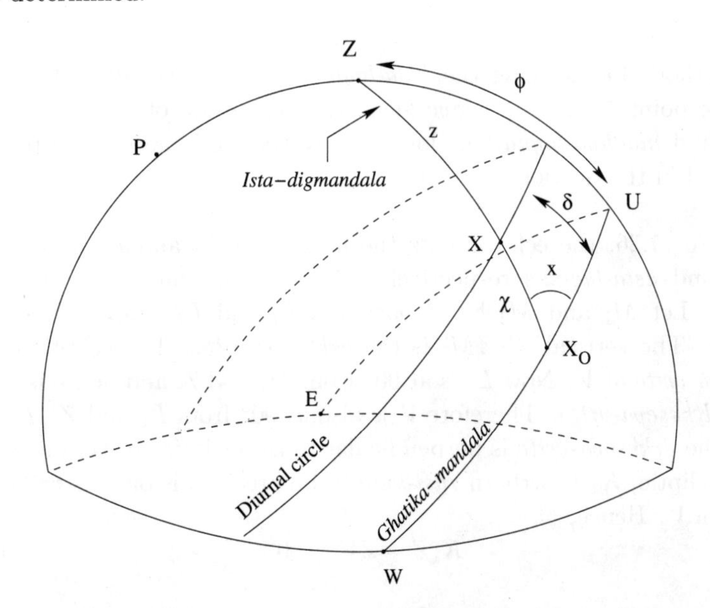

Figure 11.25: *Iṣṭa-diṅmaṇḍala* passing through the planet.

In Figure 11.25, let the planet with declination δ be at X on its diurnal circle. *Iṣṭa-diṅmaṇḍala* is a vertical passing through the planet. Let it intersect the *ghaṭikā-maṇḍala* at X_0. The angle between these two circles is denoted by x. The zenith distance of this point (z_0), has already been obtained. This is called the representative of the latitude on the *dṛṅmaṇḍala*. Similarly $\chi = XX_0$ is called the representative of the declination. If x is the angle between the *dṛṅmaṇḍala* and the equator, it is clear that

$$\begin{aligned} R\sin\phi &= R\sin x \, \sin z_0 \\ R\sin\delta &= R\sin x \, \sin\chi. \end{aligned} \tag{11.198}$$

Therefore,

$$R\sin\chi = \frac{R\sin z_0 \, R\sin\delta}{R\sin\phi}, \tag{11.199}$$

from which the *cāpa* χ can be calculated. Then the desired zenith distance $z = z_0 - \chi$.[7] The shadow (*iṣṭadik-chāyā*) is $R\sin z$.

11.31 *Kāla-lagna*, *Udaya-lagna* and *Madhya-lagna*

The methods for deriving the *kāla-lagna* (time elapsed after the rising of the first point of Aries), *udaya-lagna* (the longitude of the orient ecliptic point) and *madhya-lagna* (the longitude of the meridian ecliptic point) are explained in this section.

In Figure 11.26, the ecliptic cuts the horizon at L_1 and L_2 which are the *udaya* and *asta-lagna*-s respectively. *Lagna-sama-maṇḍala* is the vertical L_1ZL_2. Let M_1 and M_2 be at 90° from L_1 and L_2, respectively, on the horizon. The vertical M_1ZM_2 is the *dṛkkṣepa-vṛtta*. The ecliptic cuts the *dṛkkṣepa-vṛtta* at V. Now L_1 is at 90° from M_1 and Z, and hence is the pole of the *dṛkkṣepa-vṛtta*. Therefore V is also at 90° from L_1 and $Z\hat{V}L_1 = 90°$. Thus, the *dṛkkṣepa-vṛtta* is perpendicular to the ecliptic. Naturally, the pole of the ecliptic, K_1 (northern *rāśi-kūṭa* to be precise), is on this *vṛtta* and at 90° from V. Hence,

$$K_1Z + ZV = 90°. \tag{11.200}$$

But,

$$K_1Z + K_1M_2 = 90°. \tag{11.201}$$

Therefore, $K_1M_2 = ZV$. In other words,

$$\text{Altitude of the } \textit{rāśi-kūṭa}\ (K_1) = z_v, \tag{11.202}$$

where z_v is the zenith distance of V (*vitribha-lagna* or *dṛkkṣepa-lagna*). Further it may be noted that when the vernal equinox is at E, K_1 is on the

[7] When the declination is south, it is clear that $z > z_0$ and $z = z_0 + |\chi|$.

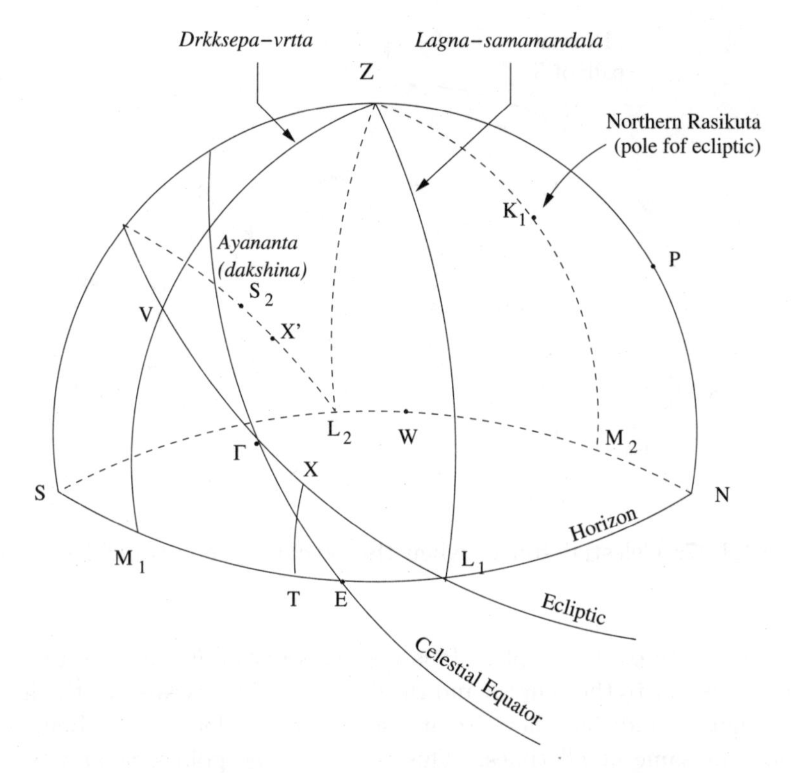

Figure 11.26: The *lagna-sama-maṇḍala* and *dṛkkṣepa-vṛtta*.

north-south circle. At the given time, Γ and K_1 are as indicated in the figure.[8] Hence, at any given instant, $Z\hat{P}K_1 = H$ is the time after the rise of Γ and is the *kāla-lagna*. Note that the maximum divergence between *lagna-sama-maṇḍala* and the ecliptic (or the angle between them) is $ZV = z_v$, the *dṛkkṣepa*. S_2 at 90° from the equinox Γ, is the southern solstice which is at the maximum distance from the equator.

In Figure 11.27, we consider the situation at the equator ($\phi = 0$) when the vernal equinox Γ is at the zenith. The northern and southern solstices S_1 and S_2 and the northern and southern *rāśi-kūṭa*-s K_1 and K_2 are indicated in the figure. Here,

$$ES_1 = NK_1 = WS_2 = SK_2 = \epsilon, \tag{11.203}$$

[8]It may be recalled that their position in the celestial sphere keeps continuously changing due to diurnal motion.

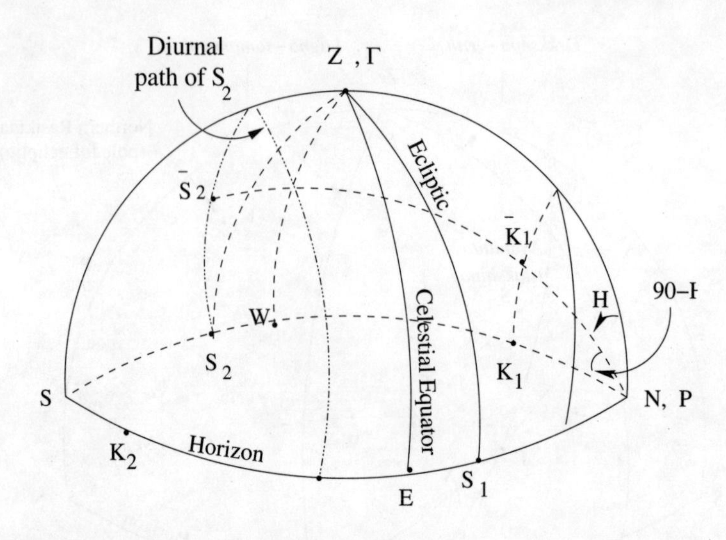

Figure 11.27: Celestial sphere when the vernal equinox coincides with the zenith.

is the obliquity of the ecliptic. It is easy to see that K_1 and S_2 are on the same secondary to the ecliptic and the equator (which is same as the horizon in the figure), and that they rise and set together. Hence, their hour angles remain the same at all times. This is true of the points S_1 and K_2 also. The diurnal circle of S_2 has radius $R\sin\epsilon$, and that of K_1 has radius $R\cos\epsilon$. Consider the situation when the northern *rāśi-kūṭa* and the southern solstice are at $\bar{K}_1$ and $\bar{S}_2$ respectively. Now, the *kāla-lagna* is given by

$$Z\hat{P}\bar{K}_1 = Z\hat{P}\bar{S}_2 = H, \tag{11.204}$$

where P is the north celestial pole coinciding with the north point of the horizon. $\bar{K}_1\hat{P}W = 90^\circ - H$, is the angle between the secondary to equator through $\bar{K}_1$ (northern *rāśi-kūṭa*) and the horizon. The gnomon of the northern *rāśi-kūṭa* is the perpendicular distance between $\bar{K}_1$ and the horizon and is equal to $R\sin\epsilon\sin(90 - H) = R\sin\epsilon\cos H$, as $N\bar{K}_1 = \epsilon$.

Now we consider a place with latitude ϕ. It has been shown that the altitude $(90 - z)$ of northern *rāśi-kūṭa* is equal to the zenith distance of *dṛkkṣepa* (11.202). Hence,

$$\textit{dṛkkṣepa} \;=\; \textit{Śaṅku} \text{ of the } \textit{rāśi-kūṭa} = R\sin z_v, \tag{11.205}$$

where $z_v = ZV = K_1M_2$ in Figure 11.26 on page 771.

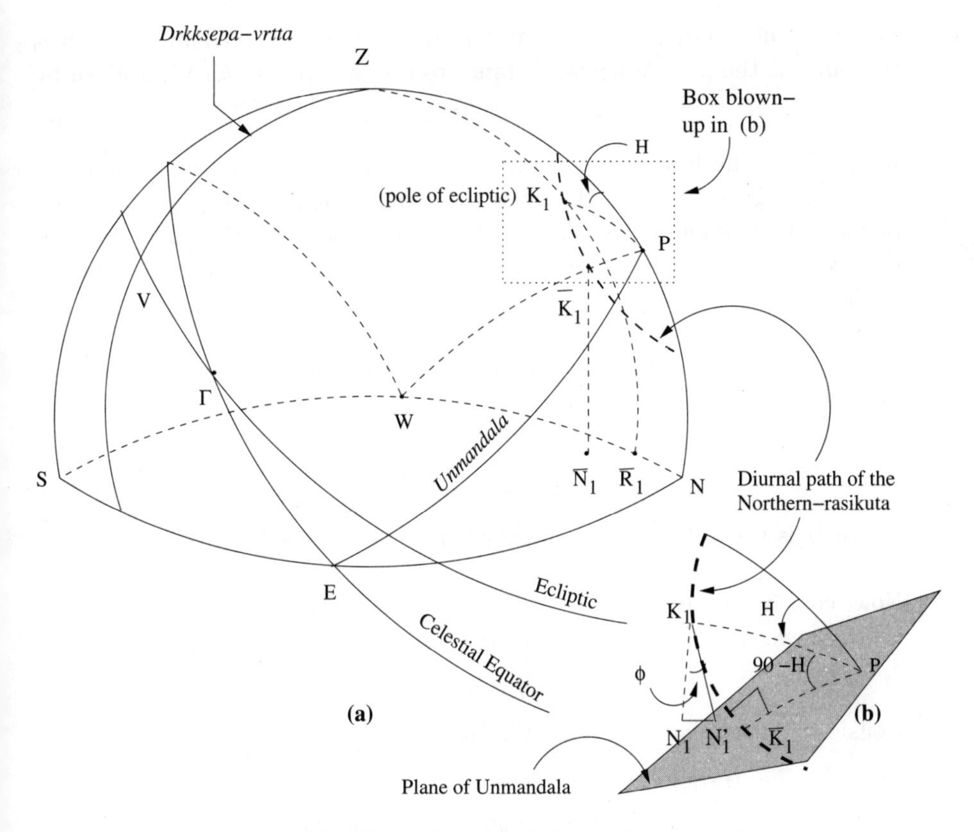

Figure 11.28: The dirunal path of the nothern *rāśi-kūṭa*.

In Figure 11.28, consider the diurnal path of the northern *rāśi-kūṭa*, K_1. Let it intersect the *unmaṇḍala* at $\bar{K}_1$. Since,

$$W\bar{K}_1 = 90 - \epsilon,\ WP = 90^\circ,\ \text{and}\ PN = \phi, \tag{11.206}$$

the *śaṅku* corresponding to this point is

$$\begin{aligned} \bar{K}_1\bar{N}_1 &= R\sin\,(\bar{K}_1\bar{R}_1) \\ &= R\sin\phi\sin\,(90-\epsilon) \\ &= R\sin\phi\cos\epsilon. \end{aligned} \tag{11.207}$$

This is the portion of the *śaṅku* of the northern *rāśi-kūṭa*, between the *unmaṇḍala* and the *kṣitija*.

When the northern *rāśi-kūṭa* is at K_1, its gnomon at the equator, which is the same as the perpendicular distance to the *unmaṇḍala* K_1N_1', is given by

$$K_1N_1' = R\sin\epsilon\cos H, \tag{11.208}$$

where N_1' is the foot of the perpendicular from K_1 on the *unmaṇḍala*, as shown in Figure 11.28(b). This has to be multiplied by $\cos\phi$ to obtain the portion of the gnomon, K_1N_1, above the *unmaṇḍala*, as the angle between the *unmaṇḍala* and the horizon at the desired place is equal to the latitude of the place (ϕ).

Hence, the gnomon of the northern *rāśi-kūṭa* at the desired place, which is the *dṛkkṣepa* $R\sin ZV$, is given by

$$\begin{aligned} R\sin ZV &= K_1N_1 + \bar{K}_1\bar{N}_1 \\ &= R(\cos\phi\sin\epsilon\cos H + \sin\phi\cos\epsilon), \end{aligned} \tag{11.209}$$

where H is the *kāla-lagna* or the hour angle of the northern *rāśi-kūṭa* K_1.

Now, consider Figure 11.26. As $VL_1 = L_1M_1 = 90°$, the maximum divergence between the horizon and the ecliptic is

$$VM_1 = 90° - ZV. \tag{11.210}$$

Considering the planet at X on the ecliptic, we have

$$\begin{aligned} R\sin VM_1 &= R\cos ZV = \textit{Pramāṇa}, \\ R &= \textit{Trijyā} = \textit{Pramāṇa-phala}, \\ R\sin XT &= \text{Gnomon of the planet} = \textit{Icchā}. \end{aligned}$$

$R\sin XL_1$ is the distance between the horizon and the planet on the ecliptic, and is also the *icchā-phala*.

Therefore,

$$\frac{R\sin XL_1}{R\sin XT} = \frac{R}{R\sin VM_1},$$

or,

$$R\sin XL_1 = \frac{R.\, R\sin XT}{R\sin VM} = \frac{R.\, R\sin XT}{R\cos ZV}. \tag{11.211}$$

We have to find XL_1 from this. *Udaya-lagna* is the longitude of L_1. Therefore,

$$\begin{aligned} \textit{Udaya-lagna} &= \Gamma L_1 \\ &= \Gamma X + XL_1 \\ &= \lambda_p + XL_1, \end{aligned} \tag{11.212}$$

where $\lambda_p = \Gamma X$ is the longitude of the planet. Thus the *udaya-lagna* has been determined in terms of the gnomon of the planet and the *dṛkkṣepa*, which involves the *kāla-lagna* H.

It has already been shown (recall the discussion in sections 11.4 and 11.5) that the gnomon corresponding to the planet is given by

$$\begin{aligned} R\sin XT &= \cos\phi \times \textit{Unnata-jyā} \\ &= R\cos\phi\cos\delta\left(\cos H_p + \frac{\sin\phi\sin\delta}{\cos\phi\cos\delta}\right), \end{aligned} \qquad (11.213)$$

where $H_p = Z\hat{P}X$ is the 'hour angle' of the planet. The second term in the bracket corresponds to the ascensional difference.[9]

When the planet is in the western hemisphere at X', as shown in Figure 11.26, $\Gamma X' = \Gamma L_2 + X'L_2$, where $\Gamma X'$ and ΓL_2 are measured eastwards. Therefore,

$$\begin{aligned} \textit{Asta-lagna} &= \Gamma L_2 \\ &= \Gamma X' - X'L_2 \\ &= \lambda_p - X'L_2, \end{aligned} \qquad (11.214)$$

where $X'L_2$ is measured as described earlier. The same considerations apply when the planet is below the horizon.

Consider the motion of the planet (Sun) as shown in Figure 11.29. Here it transits the north-south circle below the horizon at X_m, rises at X_r and reaches the *unmaṇḍala* at X_u. The angle $X_m\hat{P}X_r$ (or arc X_mX_r on the diurnal circle) corresponds to half the duration of the night, and the angle $X\hat{P}X_r$ corresponds to the portion of the night yet to pass. The difference between them, $X_m\hat{P}X$, is the hour angle H_p and $X\hat{P}X_u = 90 - H_p$. Find $R\cos H_p$ (cosine of the hour angle). The *cara* corresponds to $X_r\hat{P}X_u$ and is given by

$$R\sin(X_r\hat{P}X_u) = \frac{R\sin\phi\sin\delta}{\cos\phi\cos\delta}. \qquad (11.215)$$

Then the *śaṅku* of the planet X is given by

$$R\sin(XT) = R\cos\phi\cos\delta\left(\cos H_p - \frac{R\sin\phi\sin\delta}{\cos\phi\cos\delta}\right). \qquad (11.216)$$

[9] H_p is found from the time after sunrise. $H = H_p + R.A.$ of X, where R.A. (Right Ascension) is obtained readily from the longitude. Hence the *udaya-lagna* would be related to known quantities.

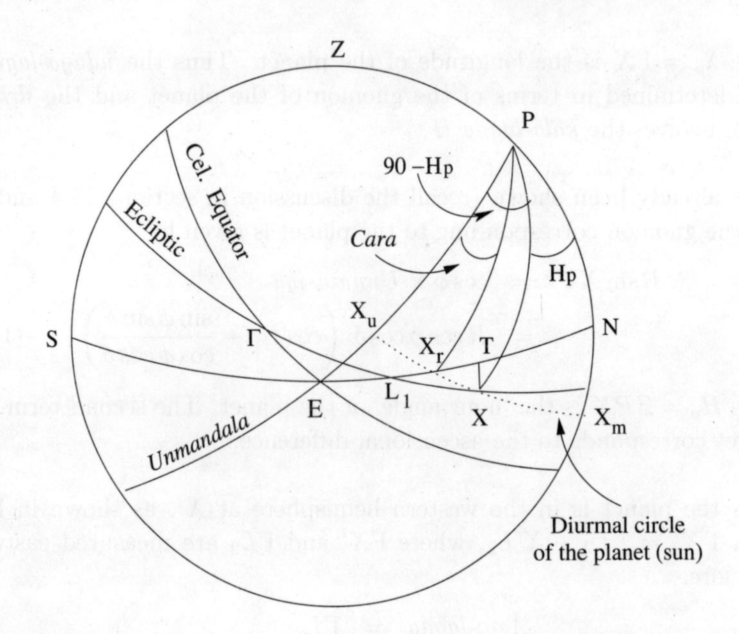

Figure 11.29: The transit of the Sun across the north-south circle below the horizon.

When the declination is south, the *cara* has to be added. As shown earlier (11.211), if L_1 is the *udaya-lagna*,

$$R\sin XL_1 = \frac{R.\,R\sin(XT)}{R\cos ZV}. \tag{11.217}$$

In the RHS of the above equation, while the numerator is known from (11.216), the denominator has to be calculated from $R\sin ZV$ (*dṛkkṣepa*), which in turn is given by (11.209). Then the *udaya-lagna* is given by

$$\begin{aligned}\Gamma L_1 &= \Gamma X - XL_1 \\ &= \lambda_p - XL_1,\end{aligned} \tag{11.218}$$

where λ_p is the longitude of the planet. The *asta-lagna* is also determined in a similar manner and is given by

$$\Gamma L_2 = \lambda_p + X'L_2. \tag{11.219}$$

In this case, X' is in the western hemisphere below the horizon. The *dṛkkṣepa-lagna* is exactly midway between the *udaya* and *asta-lagna*-s and is at the intersection of the *dṛkkṣepa-vṛtta* with the ecliptic.

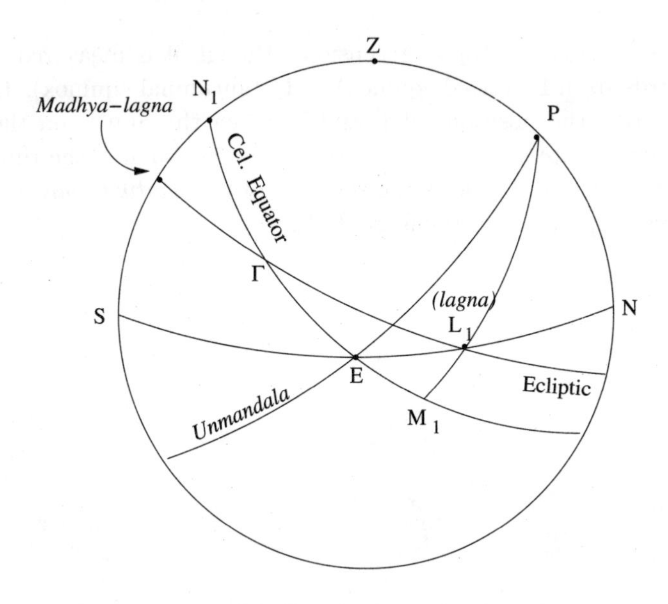

Figure 11.30: *Madhya-lagna* and *udaya-lagna.*

Madhya-lagna is the longitude of the meridian ecliptic point, at any instant. *Madhya-kāla* is defined to be the difference in the time of rising of *madhya-lagna* and the vernal equinox, Γ.

$$\begin{aligned} \textit{Madhya-kāla} &= \text{Time of rising of } \textit{madhya-lagna} - \text{Time of rising of } \Gamma \\ &= -N_1\Gamma. \end{aligned} \qquad (11.220)$$

The presence of the negative sign indicates that *madhya-lagna* has risen before the equinox. Now we find the relation between *madhya-lagna* and *kāla-lagna* The latter, which is the time after the rise of Γ, is given by

$$\begin{aligned} \Gamma E &= \Gamma M_1 - EM_1 \\ &= N_1 M_1 - EM_1 - N_1\Gamma \\ &= N_1 E - N_1\Gamma \\ &= 90^\circ - N_1\Gamma \\ &= 90^\circ + \textit{Madhya-kāla}. \end{aligned} \qquad (11.221)$$

11.32 *Kāla-lagna* corresponding to sunrise

Here the aim is to determine the *kāla-lagna* at sunrise, which is the same as the time interval between sunrise and the rise of the vernal equinox Γ. *Bhujā-*

prāṇa is essentially the Right Ascension - though it is measured eastwards or westwards from Γ (vernal equinox) or Γ' (autumnal equinox). In Figures 11.31(a) – (d), the positions of Γ and Γ' when the Sun is on the horizon (that is sunrise) are shown. In all cases, the *kāla-lagna* is the time elapsed after the rise of Γ at E and is the segment of the *ghaṭikā-maṇḍala* between E and Γ corresponding to the angle $\Gamma\hat{P}E$.

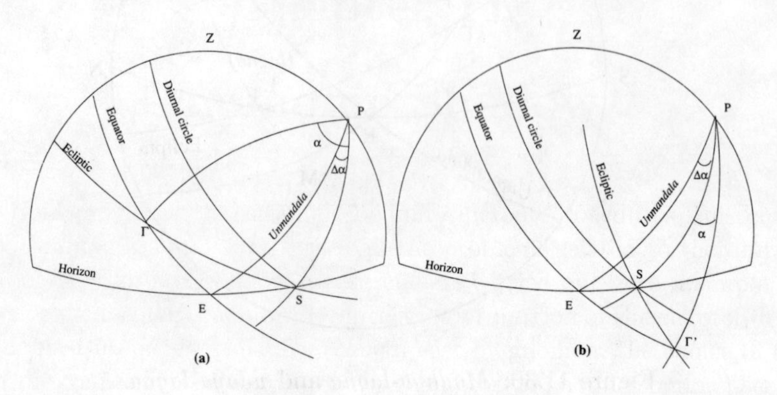

Figure 11.31*a:* *Kāla-lagna* when the *sāyana* longitude of the Sun is $< 180°$.

(a) When the Sun is in the first-quadrant, that is, $0 \leq \lambda_s \leq 90°$ [Figure 11.31(a)],

$$\begin{aligned} \textit{Kāla-lagna} &= \Gamma\hat{P}E \\ &= \Gamma\hat{P}S - E\hat{P}S \\ &= \alpha - \Delta\alpha, \end{aligned} \tag{11.222}$$

where α is the *bhujā-prāṇa* (Right Ascension) and $\Delta\alpha$ is the *cara-prāṇa* given by

$$R\sin\Delta\alpha = \frac{R\sin\phi\sin\delta}{\cos\phi\cos\delta}. \tag{11.223}$$

(b) When the Sun is in the second quadrant, that is, $90° \leq \lambda_s \leq 180°$ [Figure 11.31(b)],

$$\begin{aligned} \Gamma'\hat{P}E &= \Gamma'\hat{P}S + E\hat{P}S \\ &= \alpha + \Delta\alpha. \end{aligned} \tag{11.224}$$

But $\Gamma'\hat{P}E + \Gamma\hat{P}E = 180°$, as Γ and Γ' are $180°$ apart and E is between Γ

and Γ'. Therefore,

$$\begin{aligned} Kāla\text{-}lagna &= \Gamma\hat{P}E \\ &= 180^\circ - \Gamma'\hat{P}E \\ &= 180^\circ - (\alpha + \Delta\alpha). \end{aligned} \quad (11.225)$$

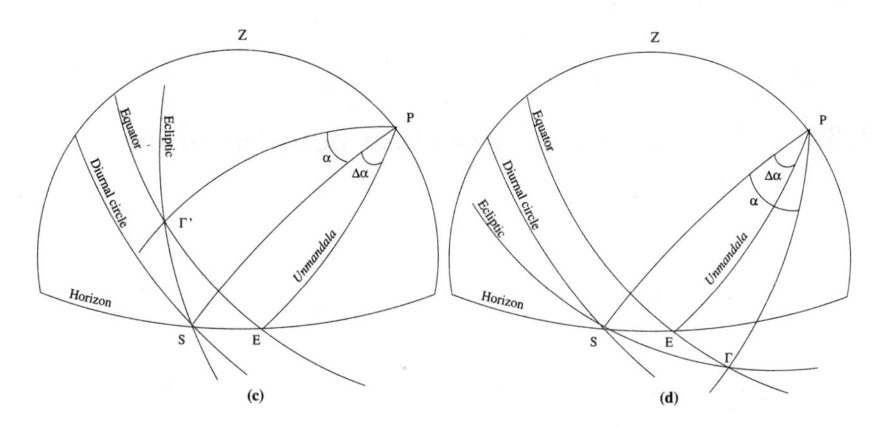

Figure 11.31*b*: *Kāla-lagna* when the *sāyana* longitude of the Sun is > 180°.

(c) When the Sun is in the third quadrant, that is, $180^\circ \leq \lambda_s \leq 270^\circ$ [Figure 11.31(c)],

$$\begin{aligned} \Gamma'\hat{P}E &= \Gamma'\hat{P}S + EPS \\ &= \alpha + \Delta\alpha. \end{aligned} \quad (11.226)$$

Now $\Gamma\hat{P}E = \Gamma'\hat{P}E + 180^\circ$, as Γ' is between E and Γ. Therefore,

$$\begin{aligned} Kāla\text{-}lagna &= \Gamma\hat{P}E \\ &= 180^\circ + \alpha + \Delta\alpha. \end{aligned} \quad (11.227)$$

(d) When the Sun is in the fourth quadrant, that is, $270^\circ \leq \lambda_s \leq 360^\circ$ [Figure 11.31(d)], Γ is below the horizon at sunrise and

$$\begin{aligned} \Gamma\hat{P}E &= \Gamma\hat{P}S - E\hat{P}S \\ &= \alpha - \Delta\alpha. \end{aligned} \quad (11.228)$$

As Γ is below the horizon,

$$\begin{aligned} Kāla\text{-}lagna &= 360^\circ - \Gamma\hat{P}E \\ &= 360^\circ - (\alpha - \Delta\alpha). \end{aligned} \quad (11.229)$$

This is the way to determine *kāla-lagna* corresponding to sunrise, when the Sun is in various quadrants. In this manner, *kāla-lagna* corresponding to the beginning of each *rāśi* can be calculated. The time taken by a particular *rāśi* to rise above the horizon is the difference between the *kāla-lagna*-s corresponding to the beginning and end of that *rāśi*. This can be calculated for each *rāśi*.

11.33 *Madhya-lagna*: Meridian ecliptic point

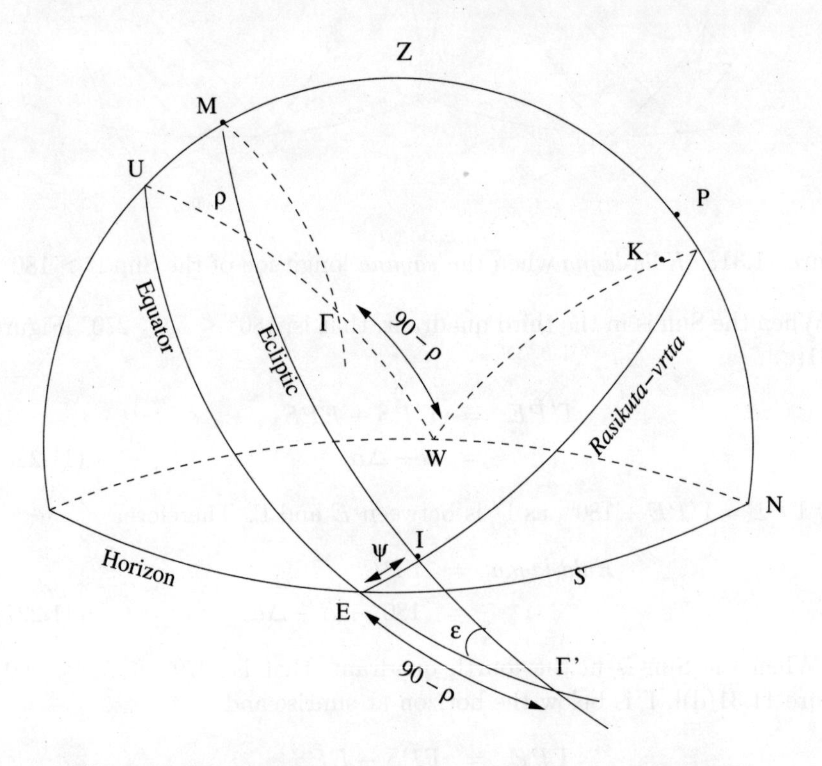

Figure 11.32: Determination of the meridian ecliptic point.

Kāla-lagna at any desired instant is the *kāla-lagna* at sunrise (discussed in detail in the preceeding section) plus the time elapsed after sunrise. When 90° is subtracted from the *kāla-lagna* in degrees, the resultant point U represents the point of contact of the celestial equator and the north-south circle.

In Figure 11.32,

$$\begin{aligned}\Gamma U &= \textit{Madhya-kāla} \\ &= \textit{Kāla-lagna} - 90^\circ \\ &= \rho. \end{aligned} \tag{11.230}$$

In other words, *madhya-kāla* is the time elapsed after the meridian transit of Γ. Clearly,

$$\begin{aligned}\Gamma W &= \Gamma' E \\ &= 90^o - \Gamma U, \end{aligned} \tag{11.231}$$

where ΓU has been obtained in (11.230). The ecliptic cuts the meridian at M. The longitude of this point, represented by ΓM, is the *madhya-lagna*. Consider the *rāśi-kūṭa-vṛtta* $WKIE$ in Figure 11.32 passing through east and west points and intersecting the ecliptic at I. K is the northern *rāśi-kūṭa*, the pole of the ecliptic. M is at 90° from both E and K. Hence, it is the pole of this *rāśi-kūṭa-vṛtta*. In that case, the arc EI is perpendicular to the ecliptic at I, and $EI = \psi$ is given by

$$\begin{aligned}R \sin \psi &= R \sin \epsilon \sin{(90 - \rho)} \\ &= R \sin \epsilon \cos \rho. \end{aligned} \tag{11.232}$$

Normally, we draw perpendiculars from points on the ecliptic and calculate the *bhujā-prāṇā*-s along the equator. Here, we do the reverse. $R\sin\psi$ and $R\cos\psi$ are the equivalents of *apakrama-jyā* and *dyujyā*. Corresponding to $\Gamma' E$ on the equator ($= 90^o - \rho$), we find the *bhujā-prāṇā* $\Gamma' I$ along the ecliptic (using the formula for the *bhujā-prāṇā*) as follows:

$$R\cos(\Gamma' I) = R\cos(\Gamma' E)\frac{R}{R\cos\psi}. \tag{11.233}$$

From this we find $\Gamma' I$. Then *madhya-lagna*

$$\Gamma M = 90^\circ - \Gamma' I. \tag{11.234}$$

This is the method of deriving the *madhya-lagna* or the meridian ecliptic point.

11.34 *Dṛkkṣepa* and *Koṭi*

The aim is to determine the *dṛkkṣepa* from the *udaya-lagna* and *madhya-lagna*. Refer to Figure 11.33. L_1, L_2 and M are the *udaya-lagna*, *asta-lagna* and *madhya-lagna* respectively. EL_1 is related to the azimuth of L_1. The *udaya-lagna-jyā*, $R\sin(EL_1)$ is found in the same way as *arkāgrā*. Just as in the case of the *arkāgrā*, *udaya-lagna-jyā* is given by

$$R\sin EL_1 = \frac{R\sin(\delta_{L_1})}{\cos\phi}, \tag{11.235}$$

where δ_{L_1} is the declination of L_1 (determined from $R\sin\delta_{L_1} = R\sin\lambda_{L_1}\sin\epsilon$). The *madhya-jyā* is the Rsine of the zenith distance of the *madhya-lagna*.

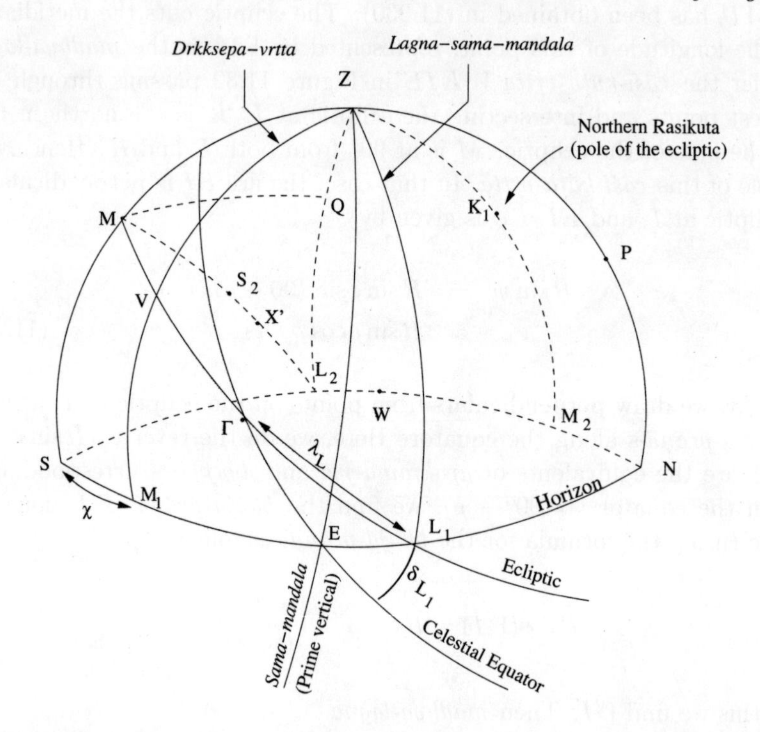

Figure 11.33: Determination of *dṛkkṣepa* from the *udaya-lagna* and *madhya-lagna*.

As the *madhya-lagna* M is on the meridian, the *madhya-jyā* $R\sin ZM$ is found in the same manner as the *madhyāhnacchāyā* (noon shadow) of the

Sun, which is the meridian zenith distance of the Sun:

$$\begin{aligned} \textit{Madhya-jyā} &= R\sin ZM \\ &= R\sin(\phi \pm |\delta_M|), \end{aligned} \tag{11.236}$$

δ_M being the declination of the *madhya-lagna*. This can be found as λ_M has been determined.

Now the maximum divergence between the *dṛkkṣepa-vṛtta* and the north-south circle is

$$SM_1 = EL_1 = \chi. \tag{11.237}$$

This is also equal to the angle between prime vertical EZ and *lagna-sama-maṇḍala* or *dṛkkṣepa-sama-maṇḍala*, L_1Z. Now,

$$\begin{aligned} \textit{Udaya-jyā} &= R\sin EL_1 \\ &= R\sin SM_1 \\ &= R\sin\chi. \end{aligned} \tag{11.238}$$

Using the rule of three, we have

$$\frac{R\sin MV}{R\sin SM_1} = \frac{R\sin ZM}{R\sin ZS = R}. \tag{11.239}$$

Therefore,

$$\begin{aligned} R\sin MV &= R\sin(ZM)\ \sin\chi \\ &= \frac{\textit{Udaya-jyā} \times \textit{Madhya-jyā}}{R}. \end{aligned} \tag{11.240}$$

This is the interval between the *madhya-lagna* and the *dṛkkṣepa* on the ecliptic and is termed *bhujā*. Now V is the *dṛkkṣepa*, L_2 is the *asta-lagna*. Therefore $VL_2 = 90°$ and

$$ML_2 = VL_2 - VM = 90 - VM.$$

Therefore, $R\sin ML_2 = R\cos MV$. This is the Rsine of the portion of the ecliptic between the north-south circle and the horizon.

Consider the spherical triangle ZMQ, where MQ is the perpendicular arc from the *madhya-lagna* M to the *lagna-sama-maṇḍala*. Now,

$$\begin{aligned} M\hat{Z}Q &= V\hat{Z}Q - V\hat{Z}M \\ &= 90 - \chi. \end{aligned} \tag{11.241}$$

Therefore,

$$\begin{aligned} R\sin MQ &= R\sin ZM \sin(90-\chi) \\ &= R\sin ZM \cos\chi. \end{aligned} \tag{11.242}$$

But, we had $R\sin MV = R\sin ZM \sin\chi$. Therefore,

$$\begin{aligned} R\sin MQ &= \sqrt{R^2\sin^2 ZM - R^2\sin^2 ZM \sin^2\chi} \\ &= \sqrt{R^2\sin^2 ZM - R^2\sin^2 MV}. \end{aligned} \tag{11.243}$$

This is the distance between the *madhya-lagna* and *dṛkkṣepa-sama-maṇḍala* or *lagna-sama-maṇḍala.* Consider the quadrants L_2V and L_2Z. MQ and VZ are perpendicular arcs from M and V on L_2V to L_2Z. Therefore, using the rule of three,[10] we have

$$\frac{R\sin ZV}{R\sin L_2V = R} = \frac{R\sin MQ}{R\sin L_2M = R\cos MV}. \tag{11.244}$$

Therefore,

$$R\sin ZV = \frac{R\sin(MQ)\, R}{R\cos(MV)}. \tag{11.245}$$

This gives the *dṛkkṣepa* $R\sin ZV$ in terms of *madhya-jyā* and *udaya-jyā* which are in turn determined from *udaya-lagna* and *madhya-lagna.* This is the maximum divergence between the ecliptic and the *lagna-sama-maṇḍala.*

Consider the quadrants L_2V and L_2M_1 (along the ecliptic and horizon). Again, applying the rule of three, we get

$$\frac{R\sin VM_1}{R\sin L_2M_1 = R} = \frac{R\sin MS = R\cos MZ}{R\sin L_2M = R\cos MV}. \tag{11.246}$$

Therefore,

$$R\sin VM_1 = \frac{R\cos MZ \times R}{R\cos MV}. \tag{11.247}$$

This is the *dṛkkṣepa-koṭi* which is also the maximum divergence between the ecliptic and the horizon. This is also the *dṛkkṣepa-śaṅku*, as it is equal to $R\cos(ZV)$.

[10] Here the rule of three is of the form

$$\frac{\textit{icchā-phala}}{\textit{icchā}} = \frac{\textit{pramāṇa-phala}}{\textit{pramāṇa}},$$

11.35 Parallax in latitude and longitude

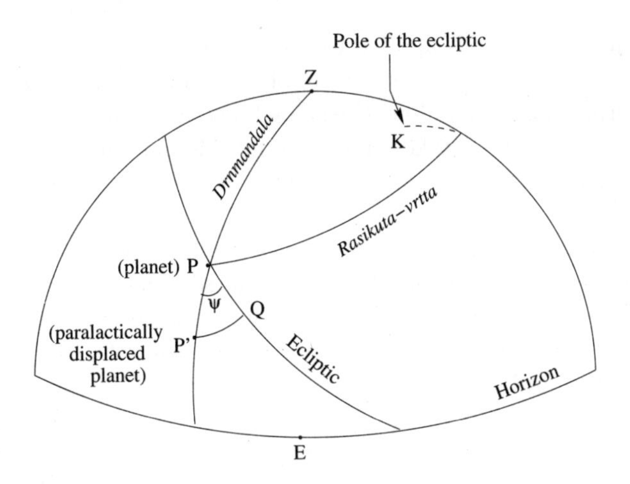

Figure 11.34: Deflection of the planet along the vertical due to parallax.

In Figure 11.34, P is the planet and P' represents the position of the planet displaced due to parallax along *dṛṅmaṇḍala*. The displacements due to parallax are given by

$$\begin{aligned} QP' &= \textit{Nati} \text{ (parallax in latitude)}, \\ PQ &= \textit{Lambana} \text{ (parallax in longitude)}, \\ PP' &= \textit{Chāyā-lambana}. \end{aligned}$$

If ψ is the angle between the *dṛṅmaṇḍala* and the ecliptic, it can be easily seen that

$$Nati = P'Q = PP' \sin\psi, \tag{11.248}$$

$$Lambana = PQ = PP' \cos\psi. \tag{11.249}$$

In obtaining the above relations, the triangle $PP'Q$ has been considered to be small and hence planar.[11] It is seen that *nati* is the *bhujā* of the *chāyā-lambana* and the *lambana* is the *koṭi* of the *chāyā-lambana*.

[11]This is true in reality since the shift due to parallax is of the order of a few minutes at the most, though this has been exaggerated in Figure 11.34 for the purposes of clarity.

11.36 Second correction for the Moon

Here a second correction is applied to the Moon to obtain the *dvitīya-sphuṭa* of the Moon with respect to the centre of the Earth. This is essentially the 'Evection term', calculated along the same lines as in *Tantrasaṅgraha*, except for a modification which takes into account Moon's latitude. The *chāyā-lambana* is then calculated taking the above correction also into account.

The procedure for the second correction is similar to the calculation of the *manda-sphuṭa* with the centre of the *bhagola* serving as the *ucca*, which is taken to be in the direction of the Sun. The distance between this and the centre of the Earth, which is the radius of the epicycle, is a continuously varying quantity and is given by

$$\frac{R}{2}\cos(\lambda_S - \lambda_U), \tag{11.250}$$

in *yojanā*-s, where λ_S and λ_U are the longitudes of the Sun and the apogee of Moon (*candrocca*). Here, the mean distance between the Moon and the centre of the *bhagola* is $10R = 34380$ *yojanā*-s. The actual distance between the same points is $10K$ where K is the *manda-karṇa* in minutes.

For the present we ignore Moon's latitude. In Figure 11.35, C is the centre of the Earth, separated from the centre of the *bhagola* (C_Z) by a distance

$$r = \frac{R}{2}\cos(\lambda_S - \lambda_U) \qquad \text{(in } yojan\bar{a}\text{-s)}. \tag{11.251}$$

A is the *Meṣādi*, and $A\hat{C}C_Z = \lambda_S$ (Sun's longitude). The *manda-sphuṭa* of Moon is at M_1. Hence $A\hat{C}_Z M_1 = \lambda_M$ (Moon's *manda-sphuṭa*). $C_Z M_1 = 10K$, where K is the *manda-karṇa* in minutes. It is clear that $C\hat{C}_Z N = \lambda_M - \lambda_S$.

CM_1, the *dvitīya-sphuṭa-karṇa* in *yojanā*-s, is the distance between the *manda-sphuṭa* and the centre of the Earth. The *bhujā-phala* and *koṭi-phala* are given by

$$\begin{aligned} CN &= r\sin(\lambda_M - \lambda_S) \\ &= \frac{R}{2}\cos(\lambda_S - \lambda_U)\sin(\lambda_M - \lambda_S), \end{aligned} \tag{11.252}$$

$$\begin{aligned} \text{and} \quad C_Z N &= r\cos(\lambda_M - \lambda_S) \\ &= \frac{R}{2}\cos(\lambda_S - \lambda_U)\cos(\lambda_M - \lambda_S). \end{aligned} \tag{11.253}$$

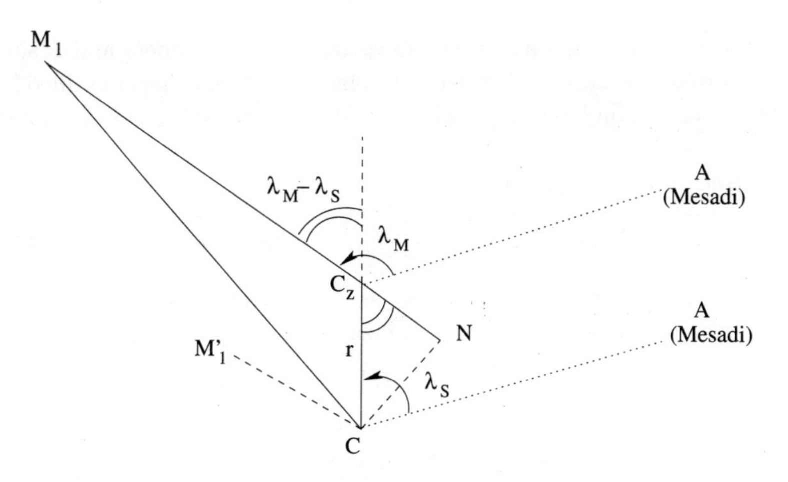

Figure 11.35: The second correction for the Moon.

Then, *dvitīya-sphuṭa-karṇa* is given by

$$\begin{aligned} CM_1 &= \sqrt{(M_1N)^2 + CN^2} \\ &= \sqrt{(M_1C_Z + C_ZN)^2 + CN^2} \\ &= \sqrt{(\textit{manda-karṇa} + \textit{koṭi-phala})^2 + \textit{bhujā-phala}^2} \\ &= \left[\left(10K + \frac{R}{2}\cos(\lambda_S - \lambda_U)\cos(\lambda_M - \lambda_S)\right)^2 \right. \\ &\quad \left. + \left(\frac{R}{2}\cos(\lambda_S - \lambda_U)\sin(\lambda_M - \lambda_S)\right)^2\right]^{\frac{1}{2}}. \end{aligned} \tag{11.254}$$

When the Moon has a latitude β, both the *manda-karṇa*

$$C_ZM_1 = 10K, \tag{11.255}$$

and the *koṭi-phala*

$$C_ZN = \frac{R}{2}\cos(\lambda_S - \lambda_U)\cos(\lambda_M - \lambda_S), \tag{11.256}$$

have to be reduced to the ecliptic. This is achieved by replacing the *manda-karṇa* K by

$$K\cos\beta = \sqrt{K^2 - K^2\sin^2\beta} = \frac{K}{R}\sqrt{R^2 - R^2\sin^2\beta}, \tag{11.257}$$

where the *vikṣepa* is $K\sin\beta$ in the measure of *pratimaṇḍala* and $R\sin\beta$ in the measure of the *manda-karṇa-vṛtta*. The *koṭi-phala* is also modified in the same manner (by multiplying it with $\cos\beta$). The *bhujā-phala* is not affected.

The *dvitīya-sphuṭa-karṇa* with *vikṣepa* is given by

$$\sqrt{\cos^2\beta\,(manda\text{-}karṇa + koṭi\text{-}phala)^2 + bhujā\text{-}phala^2}. \tag{11.258}$$

Now, the true longitude of the Moon, *dvitīya-sphuṭa*, is $\lambda_{M'} = A\hat{C}M_1$. By drawing CM_1' parallel to $C_Z M_1$, it is clear that

$$\begin{aligned}\lambda_M - \lambda_{M'} &= A\hat{C}_Z M_1 - A\hat{C}M_1 \\ &= A\hat{C}M_1' - A\hat{C}M_1 \\ &= M_1\hat{C}M_1' \\ &= C\hat{M}_1 C_Z.\end{aligned} \tag{11.259}$$

Therefore,

$$\begin{aligned}R\sin(\lambda_M - \lambda_{M'}) &= R\sin(C\hat{M}_1 C_Z) \\ &= \frac{R\times CN}{CM_1} \\ &= \frac{R\times bhujā\text{-}phala}{dvitīya\text{-}sphuṭa\text{-}karṇa}.\end{aligned} \tag{11.260}$$

Hence,

$$\begin{aligned}\lambda_M - \lambda_{M'} &= manda\text{-}sphuṭa - dvitīya\text{-}sphuṭa \\ &= (R\sin)^{-1}\left[\frac{trijyā\times bhujā\text{-}phala}{dvitīya\text{-}sphuṭa\text{-}karṇa}\right].\end{aligned} \tag{11.261}$$

Thus the *dvitīya-sphuṭa* is obtained. The sign of the RHS is determined by $(\lambda_S - \lambda_U)$ and $(\lambda_M - \lambda_S)$. When $(\lambda_S - \lambda_U)$ is in first or fourth quadrant, $\cos(\lambda_S - \lambda_U)$ is positive. Then the RHS is positive if $(\lambda_M - \lambda_S)$ is in first or second quadrant (the bright fortnight) and negative if $(\lambda_M - \lambda_S)$ is in third or fourth quadrant (the dark fortnight). When $(\lambda_S - \lambda_U)$ is in second or third quadrant, it is the other way round.

The distance of the planet from the centre of the Earth is actually the *dvitīya-sphuṭa-karṇa*, instead of $10K$. Hence, the mean motion of *dvitīya-sphuṭa* is

$$\frac{10K \times \text{mean motion of Moon}}{dvitīya\text{-}karṇa}.$$

Thus, the true Moon on the circle with its centre at the centre of the Earth has been calculated.

11.37 *Chāyā-lambana* : Parallax of the gnomon

Next, the *chāyā*, $R\sin z$ of the true Moon, is calculated.

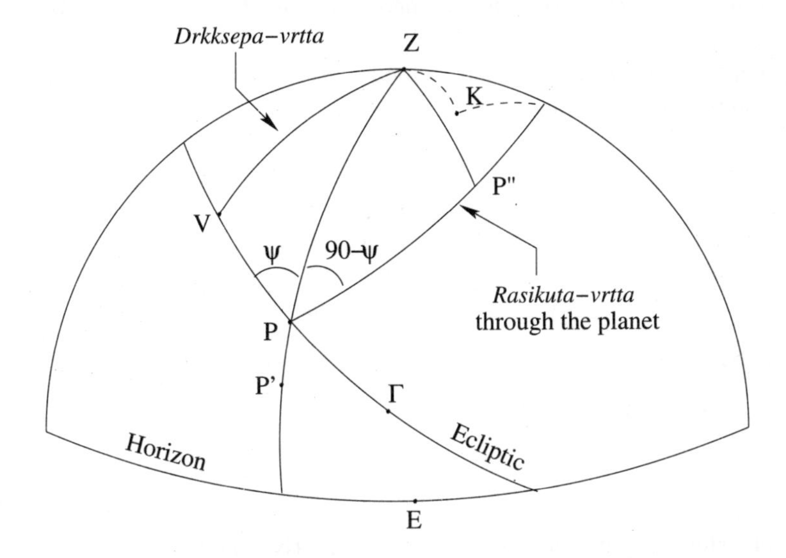

Figure 11.36: Determination of *chāyā-lambana*.

In Figure 11.36, V is the *dṛkkṣepa-lagna*, P is the planet and K is the pole of the ecliptic. P'' is the foot of the perpendicular arc from Z to the *rāśi-kūṭa-vṛtta* through the planet. The maximum divergence between the *dṛkkṣepa-vṛtta* and this *rāśi-kūṭa-vṛtta* is $R\sin(\lambda_P - \lambda_V)$, where λ_P and λ_V are the longitudes of the planet and the *dṛkkṣepa-lagna*. This corresponds to the arc $KV = 90°$. Hence, the divergence $R\sin(ZP'')$ called *dṛggati-jyā* or *dṛggati* corresponding to the arc $KZ = 90° - ZV$, is given by

$$\begin{aligned} \textit{Dṛggati-jyā} &= \frac{R\sin(\lambda_P - \lambda_V) \times R\sin(90° - ZV)}{R} \\ &= R\sin(\lambda_P - \lambda_V)\cos ZV. \end{aligned} \tag{11.262}$$

If z is the zenith distance of the planet P along the *dṛṅmaṇḍala* passing through it, and ψ is the angle between the *dṛṅmaṇḍala* and the ecliptic, then

$$\textit{Dṛkkṣepa} = R\sin(ZV) = R\sin\psi\sin z, \tag{11.263}$$

$$\textit{Dṛggati-jyā} = R\sin(ZP'') = R\cos\psi\sin z. \tag{11.264}$$

Hence

$$R\sin z = Ch\bar{a}y\bar{a} \;=\; \sqrt{(D\d{r}kk\d{s}epa)^2 + (D\d{r}ggati\text{-}jy\bar{a})^2}. \tag{11.265}$$

Thus the *chāyā* ($R\sin z$) is determined in terms of λ_P, λ_V and *dṛkkṣepa*.

Now *chāyā-lambana* PP' is determined in terms of the *chāyā* ($R\sin z$) and other quantities. We have

$$\begin{aligned} Nati &= PP' \times \sin\psi \\ &= \frac{PP' \times R\sin z \sin\psi}{R\sin z} \\ &= \frac{Ch\bar{a}y\bar{a}\text{-}lambana \times D\d{r}kk\d{s}epa}{Ch\bar{a}y\bar{a}}. \end{aligned} \tag{11.266}$$

Similarly,

$$\begin{aligned} Lambana &= PP' \times \sin\psi \\ &= \frac{Ch\bar{a}y\bar{a}\text{-}lambana \times D\d{r}ggati}{Ch\bar{a}y\bar{a}}. \end{aligned} \tag{11.267}$$

In fact, the *nati* and *lambana* can be directly calculated by multiplying *dṛkkṣepa* and *dṛggati*, respectively, by the ratio of the radius of the Earth and *dṛkkarṇa* (the actual distance between the observer and the planet). This can be understood as follows. By definition, *chāyā-lambana* is the difference in the zenith distances measured by an observer on the surface of Earth and as measured from the center of the Earth (see Figure 11.37). That is,

$$Ch\bar{a}y\bar{a}\text{-}lambana = z' - z = p. \tag{11.268}$$

From the planar triangle OCP, we have

$$\frac{R\sin p}{R_e} = \frac{R\sin z}{d}. \tag{11.269}$$

Therefore, *chāyā-lambana* in minutes is given by

$$\begin{aligned} Ch\bar{a}y\bar{a}\text{-}lambana &= R \times p \\ &\approx R\sin p \\ &= \frac{R_e \times R\sin z}{d} \\ &= \frac{\text{Radius of the Earth} \times Ch\bar{a}y\bar{a}}{D\d{r}kkar\d{n}a}. \end{aligned} \tag{11.270}$$

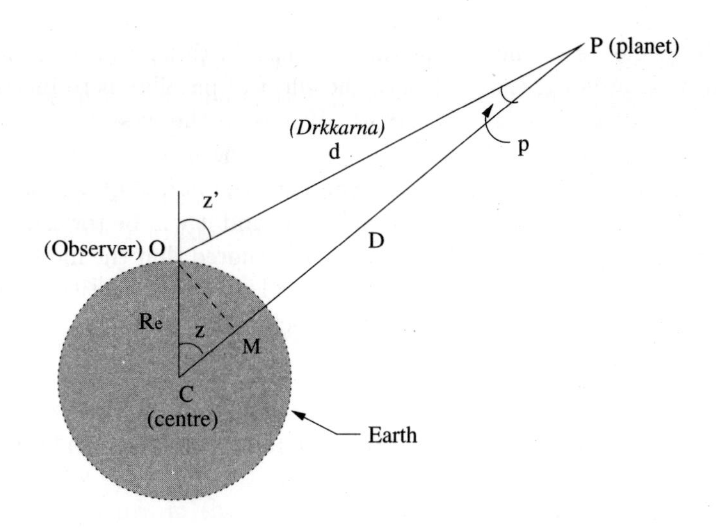

Figure 11.37: Change in the zenith distance due to the effect of parallax.

Using the above relation in (11.266) and (11.267), we have

$$Nati = \frac{\text{Radius of the Earth} \times \textit{Dṛkkṣepa}}{\textit{Dṛkkarṇa}}, \tag{11.271}$$

and

$$Lambana = \frac{\text{Radius of the Earth} \times \textit{Dṛggati}}{\textit{Dṛkkarṇa}}. \tag{11.272}$$

The procedure for calculating the *dṛkkarṇa* in terms of *dvitīya-karṇa* is described in the next section.

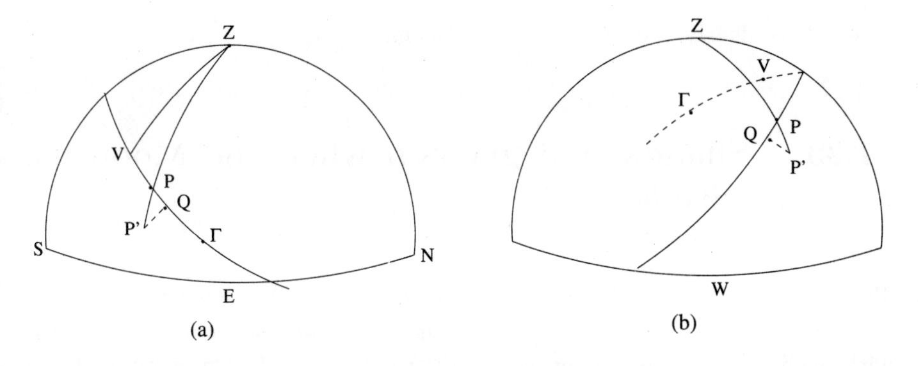

Figure 11.38: The increase and the decrease in the longitude due to parallax.

When the planet is to the east of the *dṛkkṣepa* V, the parallax in longitude PQ is also towards the east. That is, the effect of parallax is to increase the longitude as shown in Figure 11.38(a). If it is to the west of the *dṛkkṣepa* as in Figure 11.38(b), the parallax PQ is also towards the west and hence the apparent longitude will decrease. Similarly the *nati* $P'Q$ will be towards south, if the planet is in southern hemisphere, and it will be towards north if it is in northern hemisphere. (Here it should be noted that the increase or decrease in the latitude of the planet will depend upon the relative orientation of the vertical through the planet and the ecliptic).

11.38 *Dṛkkarṇa* when the Moon has no latitude

When the Moon has no latitude, we had seen in (11.265) that the *chāyā* ($R \sin z$) was given by

$$R \sin z = \sqrt{(Dṛkkṣepa)^2 + (Dṛggati\text{-}jyā)^2}. \qquad (11.273)$$

Chāyā-śaṅku ($R \cos z$) is the *koṭi* of this. Clearly $OM = R_e \sin z$ and $CM = R_e \cos z$, in Figure 11.37, are the *chāyā* and *śaṅku* in *yojanā*-s. Then the *dṛkkarṇa*, $OP = d$, is given by

$$\begin{aligned} d = OP &= \sqrt{(MP)^2 + (OM)^2} \\ &= \sqrt{(CP - CM)^2 + (OM)^2} \\ &= \sqrt{(D - R_e \cos z)^2 + (R_e \sin z)^2}, \end{aligned} \qquad (11.274)$$

where D is clearly the *dvitīya-sphuṭa-karṇa*.

11.39 Shadow and gnomon when the Moon has latitude

The procedure for calculating the *śaṅku* and *chāyā* of a planet with latitude is similar to the procedure for calculating the *śaṅku* and *chāyā* of an object with declination at any given time. Figure 11.39 is drawn keeping this in mind, where P is the planet with latitude β. RPP_t is the *vikṣepa-koṭi-vṛtta*,

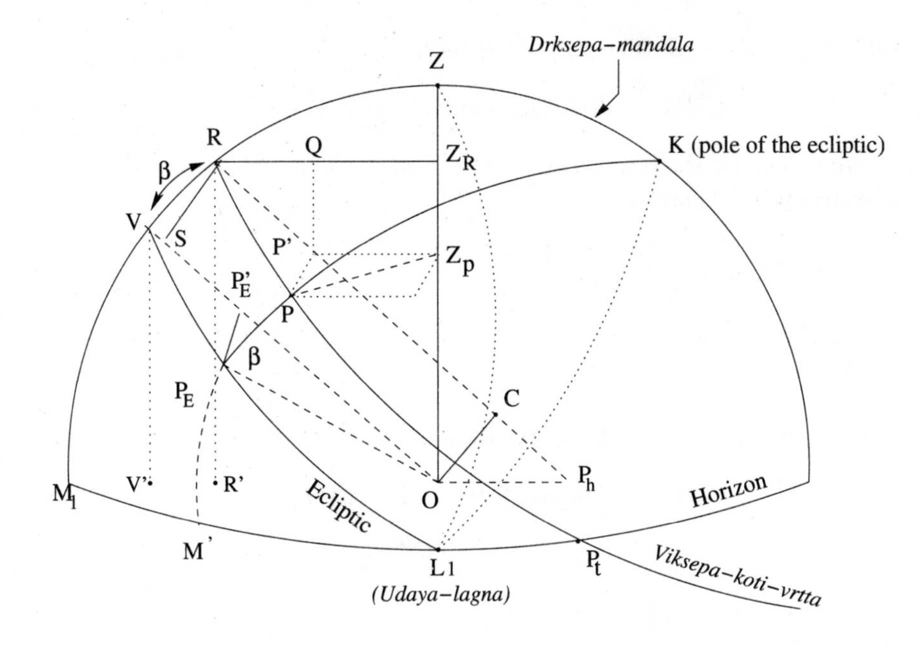

Figure 11.39: The *vikṣepa-koṭi-vṛtta* passing through the planet.

which is a small circle passing through P and parallel to the ecliptic. O is the centre of the celestial sphere and C is the centre of the *vikṣepa-koṭi-vṛtta*, with $OC = R\sin\beta$ and the radius of the *vikṣepa-koṭi-vṛtta* is $RC = R\cos\beta$. If P_E is the projection of the planet on the ecliptic, then $P_E L_1 = \lambda_{L_1} - \lambda_P$, which is the difference between the longitudes of the *lagna* and the planet. The *śaṅku* of P_E (planet with no latitude) is $R\sin P_E M'$, where M' is the foot of the vertical through P_E on the horizon. We have the *trairāśika* (rule of three)

$$\frac{R\sin VM_1 = R\cos ZV}{\textit{trijyā} = R} = \frac{R\sin(P_E M')}{R\sin(P_E L_1)}, \tag{11.275}$$

or,

$$R\sin(P_E M') = \frac{R\cos ZV \times R\sin(\lambda_{L_1} - \lambda_P)}{R}. \tag{11.276}$$

The interstice between the zenith and the *vikṣepa-koṭi-vṛtta* along *dṛkkṣepa-vṛtta* is the *nati*. This is the equivalent of the Rsine of the meridian zenith distance. Thus,

$$\begin{aligned} Nati &= R\sin(RZ) \\ &= R\sin(ZV - \beta), \end{aligned} \tag{11.277}$$

since $RZ = ZV - \beta$. The *koṭi* of this is the *parama-śaṅku* and is given by

$$R\cos(RZ) = R\sin(RM_1). \tag{11.278}$$

Parama-śaṅku is the equivalent of noon shadow. Then the *śaṅku*, $R\cos(PZ)$, is stated to be equal to

$$R\cos(RZ) - \frac{R\cos\beta \times R\cos(ZV)}{R} \times \frac{R - R\cos(VP_E)}{R}, \tag{11.279}$$

where $R - R\cos(VP_E)$ is the *śara* corresponding to the arc VP_E in the ecliptic.

This can be derived as follows. We draw $P_E P_E'$ perpendicular to OV. Then,

$$OP_E' = OP_E\cos(VP_E) = R\cos(VP_E). \tag{11.280}$$

Then,

$$\acute{S}ara = VP_E' = OV - OP_E' = R - R\cos(VP_E). \tag{11.281}$$

Similarly, draw PP' perpendicular to RC. Then RP' is parallel to VP_E' (and is in the plane of the *dṛkkṣepa-maṇḍala*), and it is the *śara* reduced to the *vikṣepa-koṭi-vṛtta* as given by

$$RP' = (\cos\beta)(R - R\cos(VP_E)). \tag{11.282}$$

Draw $P'Z_p$ and RZ_R perpendicular to OZ. $P'Q$ is perpendicular to RZ_R. It is easy to see that $P_E P_E'$ and PP' are parallel to the plane of the horizon (in fact parallel to OL_1), and OZ_P is the *śaṅku* corresponding to P. Therefore,

$$\acute{S}a\dot{n}ku = OZ_P = OZ_R - Z_P Z_R. \tag{11.283}$$

Now $OZ_R = R\cos(RZ)$. As the inclination of the ecliptic with the 'prime vertical' is the angle corresponding to the arc ZV, we have

$$\begin{aligned} Z_P Z_R &= P'Q \\ &= RP'\cos(ZV) \\ &= R\cos\beta \times \frac{R\cos(ZV)}{R} \times \frac{R - R\cos(VP_E)}{R}. \end{aligned} \tag{11.284}$$

Hence,

$$\acute{S}a\dot{n}ku = R\cos RZ - R\cos\beta \times \frac{R\cos(ZV)}{R} \times \frac{R - R\cos(VP_E)}{R}. \tag{11.285}$$

As $RZ = ZV - \beta$,

$$\cos(RZ) = \cos(ZV)\cos\beta + \sin(ZV)\sin\beta. \tag{11.286}$$

Hence,

$$\begin{aligned} \acute{S}a\dot{n}ku &= R\cos(PZ) \\ &= R[\sin(ZV)\sin\beta + \cos(ZV)\cos\beta \times \cos(VP_E)]. \end{aligned} \tag{11.287}$$

This is similar to the standard relation

$$R\cos z = R\,[\sin\phi\sin\delta + \cos\phi\cos\delta\cos H], \tag{11.288}$$

when it is realized that ZV is the equivalent of *akṣa* ϕ, latitude β is the equivalent of the declination δ, and $ZK\hat{P} = VP_E$ is the equivalent of the hour angle. The *parama-śaṅku* $R\cos(RZ)$ is the equivalent of the noon-gnomon.

When $R\cos(RZ)$ is used as the multiplicand in the second term (instead of $R\cos(ZV)$) it is stated that the divisor should not be *trijyā*, but *vikṣepa-koṭi* corrected by the difference between the horizon and the *unmaṇḍala*. This correction is the equivalent of the Rsine of the ascensional difference on the diurnal circle, and is given by

$$\frac{R\sin(ZV)\sin\beta \times \cos\beta}{\cos(ZV)\cos\beta}.$$

So, the divisor should be

$$R\cos\beta + \frac{R\sin(ZV)\sin\beta}{\cos(ZV)}. \tag{11.289}$$

This can be understood from the relation

$$\begin{aligned} \frac{R\cos(RZ)}{R\cos\beta + \dfrac{R\sin(ZV)\sin\beta}{\cos(ZV)}} &= \frac{R\cos(ZV) \times \cos(RZ)}{R(\cos(ZV)\cos\beta + \sin(ZV)\sin\beta)} \\ &= \frac{R\cos(ZV)}{R}, \end{aligned} \tag{11.290}$$

as $RZ = ZV - \beta$. Geometrically this can be seen as follows. Let V', R' be the feet of the perpendiculars from V and R on the plane of the horizon. Let RC meet the plane of the horizon at P_h. Then, OCP_h is a right angled triangle in the plane of the *dṛkkṣepa-vṛtta*. $C\hat{O}P_h$ is the angle between the great circle[12] in the case of the celestial equator. through K and L_1 and the horizon. This angle is the same as ZV which is the equivalent of *akṣa*. Then,

$$\begin{aligned} CP_h &= \frac{\sin(ZV) \times OC}{\cos(ZV)} \\ &= \frac{R \sin\beta \sin(ZV)}{\cos(ZV)}. \end{aligned} \tag{11.291}$$

As $RC = R\cos\beta$, we get

$$RP_h = R\cos\beta + \frac{R\sin(ZV)\sin\beta}{\cos(ZV)}. \tag{11.292}$$

Also, $RR' = R\cos(RZ)$ and $VV' = R\cos(ZV)$. Further,

$$\frac{RR'}{RP_h} = \frac{VV'}{OV}. \tag{11.293}$$

Therefore,

$$\frac{R\cos(RZ)}{R\cos\beta + \dfrac{R\sin(ZV)\sin\beta}{\cos(ZV)}} = \frac{R\cos(ZV)}{R}. \tag{11.294}$$

We now consider the *chāyā*. It may be noted that

$$\begin{aligned} PP' &= \cos\beta\, P_E P'_E \\ &= \cos\beta\, R\sin(VP_E). \end{aligned} \tag{11.295}$$

This is the *bhujā*. Further,

$$P'Z_p = QZ_R = RZ_R - RQ. \tag{11.296}$$

But,

$$\begin{aligned} RZ_R &= R\sin(RZ), \\ \text{and } RQ &= RP'\sin(ZV) \\ &= \frac{R\sin(ZV)}{R} \times R\cos\beta \times \frac{(R - R\cos(VP_E))}{R}, \end{aligned} \tag{11.297}$$

[12]This great circle perpendicular to the ecliptic may be thought of as the equivalent of *unmaṇḍala*

from (11.282). Hence,

$$P'Z_p = R\sin(RZ) - \frac{R\sin(ZV)}{R} \times R\cos\beta \times \frac{(R - R\cos(VP_E))}{R}. \quad (11.298)$$

This is the distance between the planet and the vertical circle ZL, in the diagram and is termed *bāhu*. Then the shadow, (*chāyā*) is PZ_P and is given by

$$\begin{aligned} \mathit{Ch\bar{a}y\bar{a}} &= R\sin(PZ) \\ &= PZ_P \\ &= \sqrt{(P'Z_P)^2 + (PP')^2} \\ &= \sqrt{\mathit{b\bar{a}hu}^2 + \mathit{bhuj\bar{a}}^2}. \end{aligned} \quad (11.299)$$

As

$$\mathit{\acute{S}a\dot{n}ku}^2 + \mathit{Ch\bar{a}y\bar{a}}^2 = \mathit{Trijy\bar{a}}^2, \quad (11.300)$$

it is sufficient to calculate any one of them.

Chapter 12

Eclipse

12.1 Eclipsed portion at required time

The *dṛkkarṇa* d in *yojanā*-s is calculated in terms of the gnomon ($R\cos z$), and the shadow ($R\sin z$), as

$$d = \sqrt{(D - R_e \cos z)^2 + (R_e \sin z)^2}, \tag{12.1}$$

where D is the *dvitīya-sphuṭa-yojana-karṇa* and R_e is the radius of the Earth (Refer to Figure 11.37 and equation (11.274)). The *lambana*-s of the Sun and Moon should be applied, to obtain their true longitudes (for the observer). When the true longitudes are the same, it is the mid-eclipse. Now, we had

$$\begin{aligned} \textit{Lambana} &= \frac{R_e}{d} \times \textit{Dṛggati} \\ &\approx \frac{R_e}{D} \times \textit{Dṛggati}, \end{aligned} \tag{12.2}$$

where, we approximate d by D, the true distance from the centre of the Earth in the denominator (essentially ignoring the higher order terms in $\frac{R_e}{D}$).

Let $\bar{D}$ be the mean distance from the centre of the Earth. Now the rate of angular motion is inversely proportional to the distance (as the linear velocity is assumed to be constant). Hence

$$\frac{D}{\bar{D}} = \frac{\text{Mean motion}}{\text{True motion}}. \tag{12.3}$$

Therefore

$$\textit{Lambana}\ (\text{in min}) = \frac{R_e}{\bar{D}} \times \frac{\text{True motion}}{\text{Mean motion}} \times \textit{Dṛggati}. \tag{12.4}$$

Now

$$\frac{\bar{D} \times \text{Mean motion (in minutes)}}{R_e}$$

is stated to be 51,770.[1] *Tantrasaṅgraha* gives the number of revolutions of the Moon in a *Mahāyuga* with 1,57,79,17,500 *yuga-sāvana-dina*-s as 57,753,320. $\bar{D}$ for the Moon is 10 times *trijyā* or 34,380 *yojanā*-s. The circumference of the Earth is 3300 *yojanā*-s from which $R_e = 1050.42$ *yojanā*-s taking $\pi = \frac{355}{113}$ as stated in Śaṅkara Vāriyar's *Laghu-vivṛti*. Then,

$$\begin{aligned} \frac{\bar{D} \times \text{Mean motion}}{R_e} &= \frac{34380 \times 57753320 \times 360 \times 60}{1050.42 \times 1577917500} \\ &= 51751.06591. \end{aligned} \tag{12.5}$$

In the text, this is taken to be 51,770. Hence,

$$Lambana \text{ (in min.)} = \frac{Dṛggati}{51770} \times \text{(True daily motion)}. \tag{12.6}$$

The assumption made in many Indian texts that the horizontal parallax is equal to $\frac{1}{15}$ of daily motion is not being made here. Therefore, the difference in *lambana* of the Moon and the Sun is given by

$$\frac{Dṛggati}{51770} \times \text{(Difference in daily motion)}. \tag{12.7}$$

Here, the value of the difference in daily motions in minutes of arc, corresponds to 60 *nāḍikā*-s. Therefore, the difference in *lambana* of the Moon and Sun in *nāḍikā*-s is given by

$$\frac{Dṛggati}{51770} \times 60. \tag{12.8}$$

This has to be applied to the *parvānta* or the middle of the eclipse to obtain the true mid-eclipse. The *lambana* is again calculated at this value of the *parvānta* and applied to the original *parvānta* to obtain the true *parvānta* corresponding to the second iteration. This iterative process is carried on till the successive values of the *parvānta* are the same (to the desired accuracy). As the Text notes:

> 'Only by knowing the correct *lambana* can the *samalipta-kāla* be ascertained and only by knowing the *samalipta-kāla* can the *lambana* be ascertained.'

[1] This would be the same for all celestial bodies as the linear velocity is constant.

At the true middle of the eclipse, the longitudes of the Sun and the Moon are the same. That is, $\lambda_M = \lambda_S$. However, the difference in *nati*-s of the Sun and the Moon and the *vikṣepa* (latitudinal deflection) of the Moon have to be taken into account.

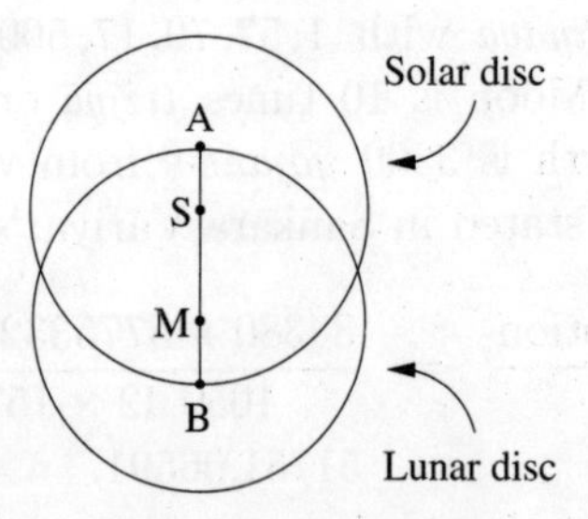

Figure 12.1: Mid-eclipse.

Let S and M be the centres of the solar and lunar discs (see Figure 12.1). The distance between them at the mid-eclipse is given by

$$SM = Nati + Vik\d{s}epa = \beta', \tag{12.9}$$

where *nati* stands actually for the difference in *nati*-s of Moon and Sun. The eclipsed portion is given by

$$\begin{aligned} AB &= SB + SA \\ &= SB + MA - SM \\ &= \frac{1}{2}(\text{Sum of orbs of Moon and Sun}) - \beta'. \end{aligned} \tag{12.10}$$

It can be seen from Figure 12.2(a) that when

$$SM = SC + CM = \frac{1}{2}(\text{Sum of orbs of Sun and Moon}), \tag{12.11}$$

the eclipse commences. Similarly, it is obvious that there is no eclipse when $SM > \frac{1}{2}$(Sum of orbs of Sun and Moon) (see Figure 12.2(b)).

The distance between the centres of the solar and lunar discs (*bimbāntara*) is given by

$$SM = \sqrt{(\lambda_M - \lambda_S)^2 + \beta'^2}. \tag{12.12}$$

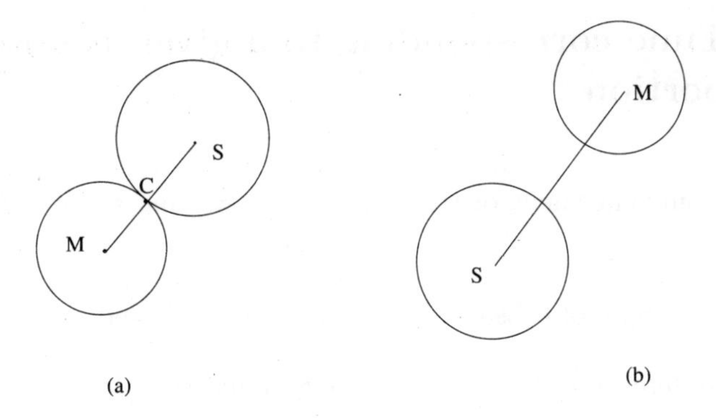

Figure 12.2: (a) Orbs at the commencement of eclipse; (b) Orbs when there is no eclipse.

The eclipsed portion at that time (see Figure 12.3) is given by

$$\begin{aligned} AB &= AM - BM \\ &= AM - (SM - SB) \\ &= AM + SB - SM \\ &= \frac{1}{2}(\text{Sum of orbs of Moon and Sun}) - \textit{Bimbāntara}. \end{aligned} \tag{12.13}$$

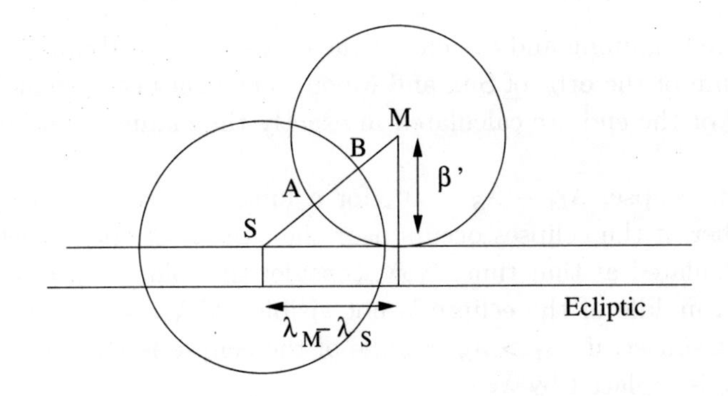

Figure 12.3: The portion of the Moon eclipsed.

12.2 Time corresponding to a given eclipsed portion

After the commencement of the eclipse, at any time t, the *bimbāntara* is given by

$$SM = \frac{1}{2}(\text{Sum of orbs of Moon and Sun}) - \text{Eclipsed portion.} \quad (12.14)$$

Then, the difference in longitudes of the Sun and the Moon is given by

$$\lambda_M - \lambda_S = \sqrt{Bimb\bar{a}ntara^2 - \beta'^2}. \quad (12.15)$$

The time interval, Δt corresponding to this *sphuṭāntara* is readily carlculated from the fact that difference in daily motions corresponds to 60 *nāḍikā*-s. Then, the desired time t is given by

$$t = t_m \pm \Delta t,$$

where t_m corresponds to the *parvānta* or the 'middle of the eclipse'. Now in calculating Δt, β' is involved. But, this is unknown for it is the value of *nati* + *vikṣepa* at the desired instant $t_m \pm \Delta t$. Hence, an iterative process is used to find Δt. First find β' at t_m. From this Δt is calculated, as explained above. This is the first approximation to Δt. From this β' is calculated at $t_m \pm \Delta t$, and Δt is determined from this. That would be the second approximation to Δt. The iterative process is carried out till Δt determined does not vary significantly[2] in successive iterations. Then, as mentioned earlier, $t_m \pm \Delta t$ is the desired time corresponding to a given eclipsed portion.

Both at the beginning and the end of the eclipse, the *bimbāntara* is equal to half the sum of the orbs of Sun and Moon. The times corresponding to the beginning or the end are calculated in exactly the manner described above.

For a solar eclipse, $\lambda_M - \lambda_S = 0°$, for a lunar eclipse, $\lambda_M - \lambda_S = 180°$. When either of the eclipses occurs near the sunrise or the sunset, λ_M and λ_S are calculated at that time. Now, consider the solar eclipse. If $\lambda_M > \lambda_S$ at sunrise, middle of the eclipse is not visible. If $\lambda_M < \lambda_S$, the middle is visible. At sunset, if $\lambda_M > \lambda_S$, middle of the eclipse is visible. For a lunar eclipse, λ_S is replaced by $\lambda_S + 180°$.

[2]The accuracy is set to desired value, which could be as gross a one-hundredth or as fine as one-billionth part.

12.3 Computation of *Bimbāntara*

Now, the angular radius of the Sun or Moon in minutes is given by

$$\text{Angular radius} = \frac{\text{Linear radius} \times R}{\text{Distance from Earth}}. \tag{12.16}$$

Here the denominator is actually *dṛkkarṇa*. So, a 'reverse rule of three' is being used in calculating the angular radius.

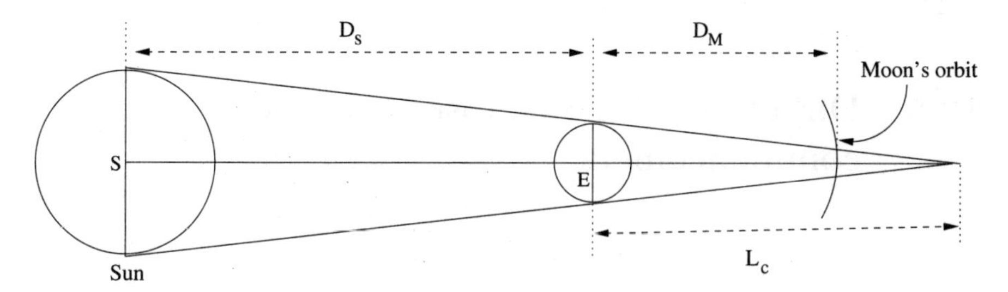

Figure 12.4: Lunar eclipse.

Figure 12.4 depicts the lunar eclipse. Let the diameters of the Sun, the Earth and that of the shadow (*chāyā*) at the Moon's orbit be d_S, d_E and d_c respectively. Further, let D_S and D_M be the true distances of the Sun and the Moon from the Earth, and L_c be the length of Earth's shadow. Then from Figure 12.4, it is clear that

$$\frac{L_c}{d_E} = \frac{L_c + D_S}{d_S}, \tag{12.17}$$

or

$$\frac{L_c}{d_E} = \frac{D_S}{d_S - d_E}. \tag{12.18}$$

From this, L_c is determined. Further,

$$\frac{L_c - D_M}{L_c} = \frac{d_c}{d_E}. \tag{12.19}$$

Therefore,

$$d_c = \frac{L_c - D_M}{L_c} \times d_E. \tag{12.20}$$

12.4 Orb measure of the planets

The shadow diameter at the Moon's orbit in minutes is given by

$$\frac{d_c}{D_M} \times R. \tag{12.21}$$

In the above expression d_c is in *yojanā*-s. Having obtained the orb-measures of the eclipsed and eclipsing planets, the extent of the eclipse at the mid-point and at any desired time can be calculated.

12.5 Direction of the eclipses and their commencement

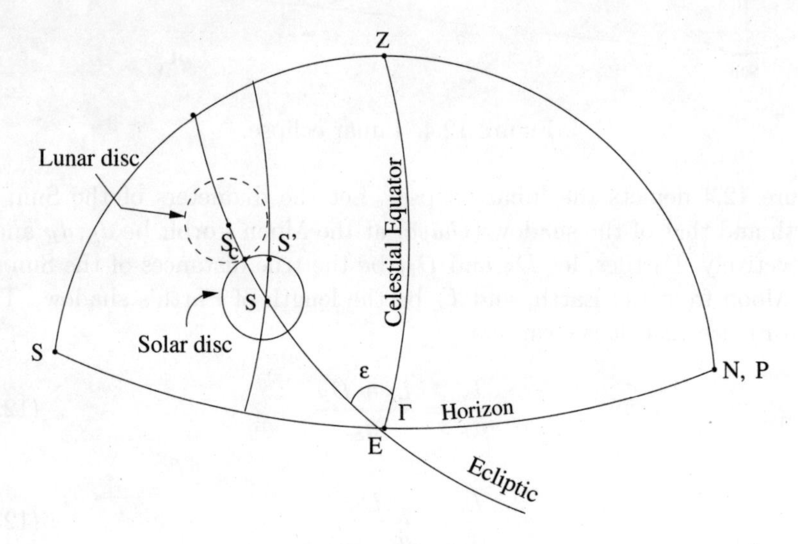

Figure 12.5: *Āyana-valana*.

Consider the small circle which is parallel to the prime vertical passing through the centre of the solar disc. This is the *chāyā-koṭi-vṛtta* or the east-west small circle all of whose points are at a distance of

$$Ch\bar{a}y\bar{a}\text{-}bhuj\bar{a} = R \sin z \sin A,$$

from the prime vertical. At the beginning of the solar eclipse, the solar and lunar discs touch each other at a point. The separation of this point from this small circle (*chāyā-koṭi-vṛtta*) is the *valana.* It consists of three components. (i) *āyana-valana*, which is due to the inclination of the ecliptic with the diurnal circle (which coincides with the small circle at the equator) (ii) *ākṣa-valana*, which is due to the inclination of the diurnal circle with the small circle, and (iii) *valana* due to the *vikṣepa* of the Moon.

12.6 *Āyana-valana*

Consider a place on the equator. Without loss of generality, we take the vernal equinox, Γ at the east point (see Figure 12.5), the winter solstice on the meridian, and the Sun situated between them on the ecliptic. The "Moon without latitude" is also situated on the ecliptic and touches the solar disc at S_c on the ecliptic. S' is the point where the solar disc intersects the ecliptic. The eclipse starts at S_c and $S'S_c$ is the *āyana-valana* and it is southwards in the figure. The angle between the ecliptic and the celestial equator (prime vertical) is ϵ. Let $S\Gamma = \lambda$, be the longitude of the Sun. Let r_s be the angular radius of the solar disc.

The distances of the centre of the Sun S and the point S_c from the celestial equator are given by

$$R \sin \delta_s = R \sin \lambda \sin \epsilon. \tag{12.22}$$

$$R \sin \delta_{s_c} = R \sin(\lambda + r_s) \sin \epsilon. \tag{12.23}$$

Therefore $S'S_c$, which is the difference between the two, is given by

$$\begin{aligned} R \sin \delta_{s_c} - R \sin \delta_s &= R[\sin(\lambda + r_s) - \sin \lambda] \sin \epsilon. \\ &= r_s \, R \cos \bar{\lambda} \, \sin \epsilon, \end{aligned} \tag{12.24}$$

where $\bar{\lambda}$ corresponds to a point midway between S and S_c, and is given by

$$\begin{aligned} \bar{\lambda} &= \lambda + \frac{r_s}{2} \\ &= \lambda + \frac{d_s}{4}, \end{aligned} \tag{12.25}$$

where d_s is the angular diameter of the solar disc. (The Text states that the *bhujā-khaṇḍa*, $R \sin(\lambda + r_s) - R \sin \lambda$, is to be obtained from the *koṭi-jyā* $R \cos \bar{\lambda}$ at the *cāpa-khaṇḍa-madhya*).

12.7 *Ākṣa-valana*

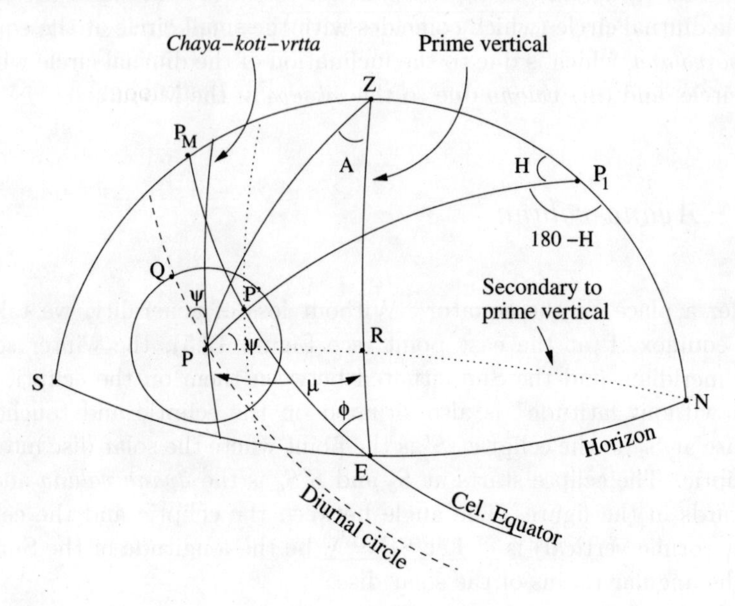

Figure 12.6: *Akṣa-valana.*

Now we consider a place with latitude, ϕ. Here the diurnal circle is inclined to the *chāyā-koṭi-vṛtta* (the small circle parallel to the prime vertical) at an angle ψ as shown in Figure 12.6. Then *ākṣa-valana*, $r_s \sin\psi$, is the distance along the north-south direction from the point on the diurnal circle to the *chāyā-koṭi-vṛtta.* Here r_s represents the radius of the Sun's disc (PQ, as shown in the figure). The expression for $R\sin\psi$ given in the Text is

$$R\sin\psi = R\sin\phi\sin H. \tag{12.26}$$

It is further mentioned that the *akṣāṃśa* of the *natotkrama-jyā* is given by

$$R(1-\cos H)\sin\phi.$$

This is the distance between P_M on the celestial equator and the vertical small circle passing through P', a point on the celestial equator corresponding to hour angle H. This may be understood as follows. The distance between P_M and the prime vertical is $R\sin\phi$. Similarly, the distance between P' and the prime vertical is $R\sin\phi\cos H$, as $P'E = 90° - H$. Therefore, the distance between P_M and the small vertical circle is

$$R\sin\phi(1-\cos H).$$

The correct expression for $R\sin\psi$ may be obtained as follows. It may be noted that ψ is also the angle between the secondaries to celestial equator and prime vertical (whose pole is N) from P. In Figure 12.6, let $PR = \mu$ (that is, $PN = 90° + \mu$). In the spherical triangle PP_1N,

$$\frac{\sin(PN)}{\sin(180 - H)} = \frac{\sin\phi}{\sin\psi}. \tag{12.27}$$

Therefore,

$$\sin\psi = \frac{\sin\phi\ \sin H}{\cos\mu}. \tag{12.28}$$

The denominator in the RHS of the above equation could be determined using the *chāyā-bhujā* given by[3]

$$\sin\mu = \sin z \sin A. \tag{12.29}$$

In any case, $R\sin\psi$ is the *valana* in the *trijyā* circle. Therefore, *ākṣa-valana*, which is the *valana* corresponding to radius of the solar disc, is given by

$$\frac{R\sin\psi}{R}\, r_s. \tag{12.30}$$

In Figure 12.7(b), A_y is the *āyana-valana* and A_k is the *ākṣa-valana*. If V be the total *valana*, when A_y and A_k are in the same direction, it is given by[4]

$$V = A_y + A_k. \tag{12.31}$$

If they are in opposite direction, then

$$V = A_y \sim A_k. \tag{12.32}$$

So far, we have considered the case when there is no *vikṣepa*. Considering *vikṣepa*, it may be noted that *vikṣepa*+*nati*= β', as shown in the Figure 12.7(a). This corresponds to the *bimbāntara*. Hence, *vikṣepa-valana* at the circumference of the disc is

$$\beta' \times \frac{\text{Radius of solar disc}}{\textit{Bimbāntara}}. \tag{12.33}$$

The direction of the *valana*-s at the time of *mokṣa* (release) will be opposite to those at *sparśa* (contact).

[3]For details the reader may refer to section 11.14.

[4]However, A_y should be multiplied by $\cos\phi$ to obtain *valana* in the north-south direction.

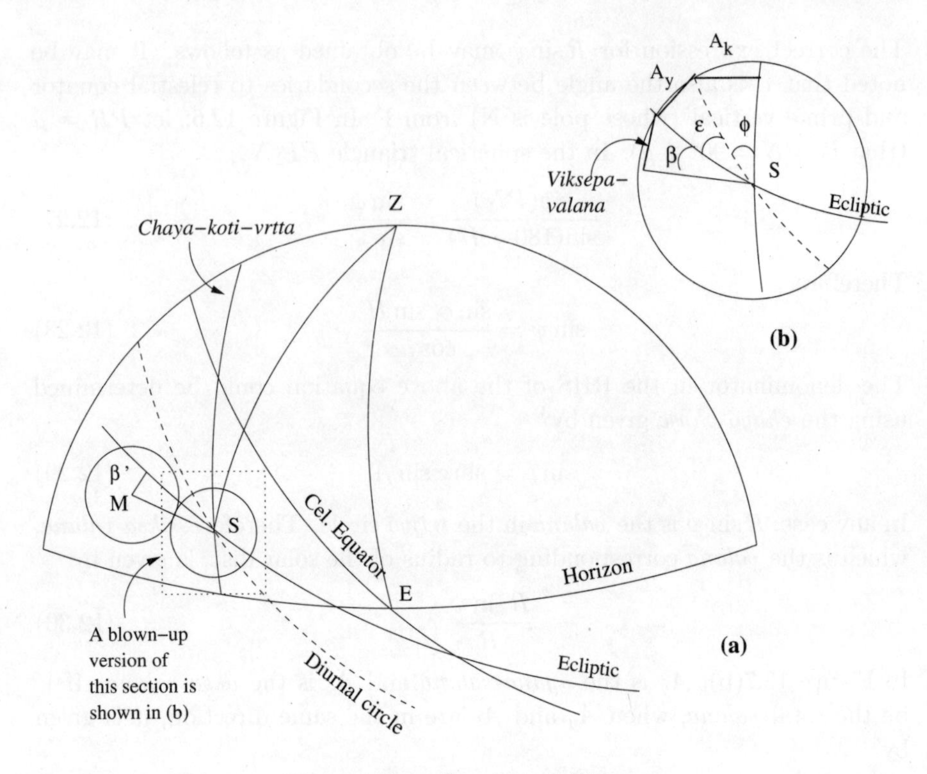

Figure 12.7: Combined *Valana*.

12.8 Graphical chart of the eclipse

Valana is calculated for the times of commencement and culmination of the eclipse, as well as for any other desired instant. Then, the eclipsed orb (solar disc in the solar eclipse) is drawn and the local east-west line (*chāyā-koṭi-vṛtta*) is drawn through its centre (as in Figure 12.8). Choose a point at a distance of *valana* from the point on the eclipsed orb which is on the local east-west line. The *valana* line passes through the chosen point and the centre of the eclipsed orb. Draw the orb of the eclipsing planet with its centre on the *valana* line at a distance of *bimbāntara* from the centre of the eclipsed orb. Then, the eclipsed and bright portion of the eclipsed orb can be easily found as indicated in Figure 12.8. Here it is not mandatory that the *valana* corresponding to the actual radius of the eclipsed orb should be calculated. It can be calculated for any suitable radius, and the *valana* line can be drawn suitably.

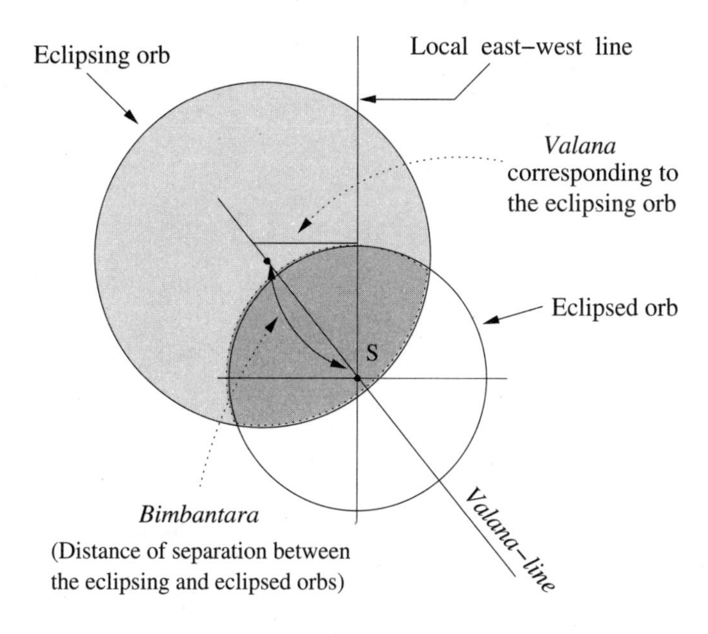

Figure 12.8: Graphical representation of eclipse.

12.9 Lunar eclipse

In the lunar eclipse, the Moon's orb is being eclipsed and the Earth's shadow is the eclipser. The diameter of the Earth's shadow at the path of the Moon is called *tamo-biṃba* (orb of darkness). As the Earth's shadow and the Moon's orb are at the same distance from the Earth, the *nati* and *lambana* are the same for the eclipser and the eclipsed. Hence, they cancel out and do not figure in the calculation. All the other rules are the same for the solar and lunar eclipses.

Thus the procedures for the computation of eclipses have been stated. It is noted that there is a correction called *paridhi-sphuṭa* for both the Sun and the Moon. Nothing more is stated about its magnitude or nature, except that it would affect the longitudes of the Sun and the Moon and thereby the time of the eclipse.

Chapter 13

Vyatīpāta

13.1 Occurence of *Vyatīpāta*

Vyatīpāta is said to occur when the (magnitudes of) declinations of the Sun and Moon are equal, and when one of them is increasing and the other is decreasing. This can happen when one of these bodies is in an odd quadrant, and the other is in an even quadrant.

13.2 Derivation of declination of the Moon

A method of computing the declination of the Moon (which has a latitude) has already been described. Here, a new method to compute the same is described. The declination of the Sun is determined with the knowledge of the intersection point (Γ in Figure 13.1) and the maximum divergence $R\sin\epsilon$ of the ecliptic and the celestial equator. Similarly, the declination of the Moon can be determined if we know (i) the point where the celestial equator and the *vikṣepa-vṛtta* (lunar orbit) intersect, (ii) the maximum divergence between them, and (iii) the position of the Moon on the *vikṣepa-vṛtta*.

13.3 *Vikṣepa*

The *vikṣepa-vṛtta* will intersect the ecliptic at *Rāhu* (ascending node of the Moon) and *Ketu* (descending node) and diverge northwards and southwards respectively, from those points. A method to determine the intersection point of the celestial equator and the *vikṣepa-vṛtta*, and their maximum divergence, is described first in qualitative terms. For this, four distinct cases are discussed.

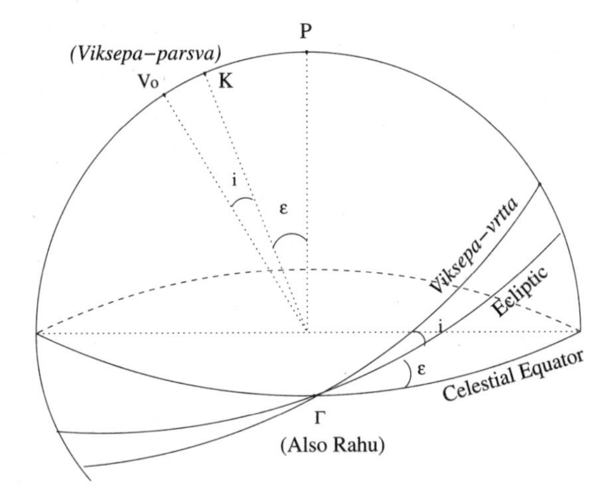

Figure 13.1: Moon's orbit when the node *Rāhu* coincides with the vernal equinox Γ.

Case 1: *Rāhu* at the vernal equinox:

Here, the maximum declination (ϵ) on the ecliptic and maximum *vikṣepa* (i) on the *vikṣepa-vṛtta* are both on the north-south circle as shown in Figure 13.1. The maximum possible declination of the Moon on that day will be equal to the sum of these two ($\epsilon + i$). Then, the declination of the Moon can be determined with the knowledge of its position on the *vikṣepa-vṛtta*, as the inclination of *vikṣepa-vṛtta* with the equator is ($\epsilon + i$). The *vikṣepa-pārśva*[1] is the northern pole (V_0) of the *vikṣepa-vṛtta*. When the *Rāhu* is at the vernal equinox, the distance between this and the north celestial pole is equal to ($\epsilon + i$).

The *vikṣepa-pārśva* is the (north) pole of the *vikṣepa-vṛtta*, just as the north celestial pole is the pole of the celestial equator or the *rāśi-kūṭa* is the pole of the ecliptic. Whatever be the position of *Rāhu*, the distance between the celestial pole and the *vikṣepa-pārśva* is equal to the maximum divergence between the equator and the *vikṣepa-vṛtta*.

[1]Though generally the term *pārśva* refers to a side, in the present context it is used to refer to the pole.

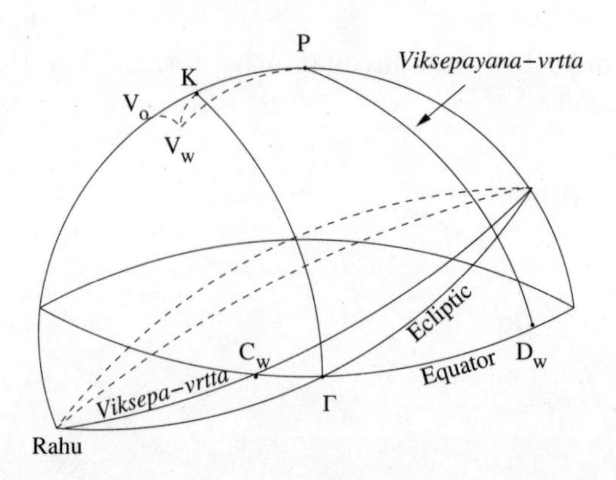

Figure 13.2: Moon's orbit when the node *Rāhu* coincides with the winter solstice.

Case 2: *Rāhu* at the winter (southern) solstice:

In this case, the *vikṣepa-vṛtta* would be deflected towards the north from the vernal equinox by the measure of maximum *vikṣepa* as shown in Figure 13.2. The *vikṣepa-pārśva*[2] would be deflected towards west from V_0 and would be at V_W, with the arc length $KV_W = i$. The distance between the (celestial) pole P and V_W is the *vikṣepāyanānta* (I). The great circle passing through P and V_W is called *vikṣepāyana-vṛtta.* Its intersection point (D_W) with the celestial equator would be deflected west from the north-south circle by the angle $K\hat{P}V_W$. The point of intersection of the *vikṣepa* and *vikṣepāyana-vṛtta*-s corresponds to maximum declination of the Moon in this set-up. The *vikṣepa-viṣuvat* is the point of intersection of the *vikṣepa-vṛtta* and the celestial equator and is denoted by C_W. C_W is at 90° from D_W. $C_W\Gamma = K\hat{P}V_W$ is called *vikṣepa-calana.* C_W is situated west of the vernal equinox when *Rāhu* is at winter solstice.

Case 3: *Rāhu* at the autumnal equinox:[3]

As depicted in Figure 13.3, the *vikṣepa-vṛtta* would intersect the north-south circle at a point north of the winter solstice by i, which is taken to be $4\frac{1}{2}^\circ$. The *vikṣepa-pārśva*, now at V', would also be deflected towards north from K, and the distance between V' and P would be $\epsilon - i = 19\frac{1}{2}^\circ$. It is easy to

[2]It may be noted that this point V_W lies on the other side of the celestial sphere.

[3]Autumnal equinox was approximately at the middle of the *Kanyā-rāśi* at the time of composition of *Yuktibhāṣā.*

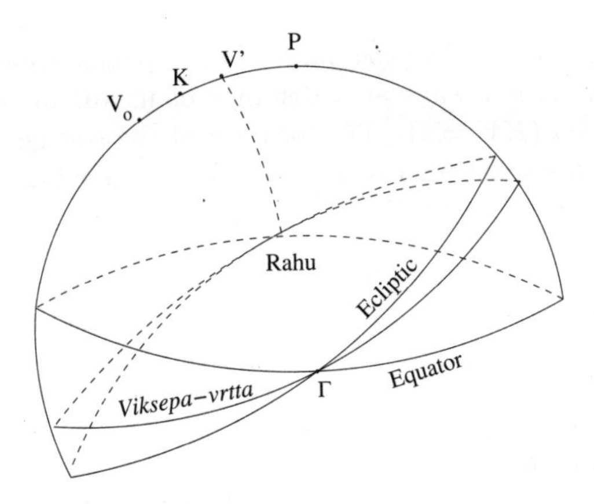

Figure 13.3: Moon's orbit when the node *Rāhu* coincides with the autumnal equinox.

see that *vikṣepa-viṣuvat* would coincide with the equinox now and there will be no *vikṣepa-calana.*

Case 4: *Rāhu* at the summer (northern) solstice:

This situation is depicted in Figure 13.4. Here, the *vikṣepa-pārśva* V_E is deflected towards the east from V_0, with $KV_E = i$. The *vikṣepāyana-vṛtta* touches the equator at D_E, which is deflected east from the north-south circle. The *vikṣepa-viṣuvat* is at C_E and is east of the vernal equinox Γ.

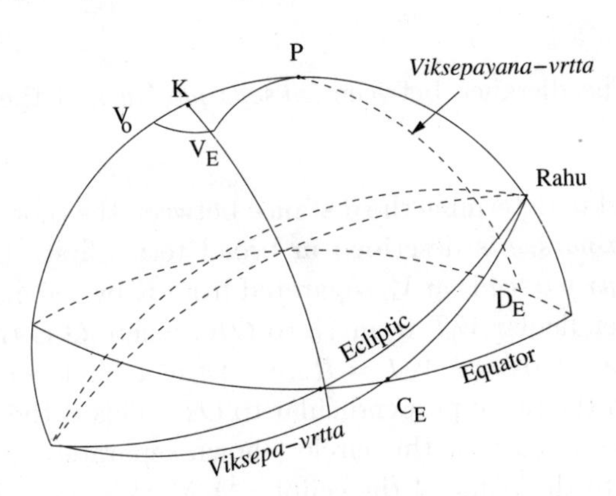

Figure 13.4: Moon's orbit when the node *Rāhu* coincides with the summer solstice.

Thus the location of the *vikṣepa-pārśva*, V, depends upon the position of *Rāhu.* However, it is always at a distance of maximum *vikṣepa* from the northern *rāśi-kūṭa* ($KV = i$). The location of the southern *vikṣepa-pārśva* with respect to the southern *rāśi-kūṭa* can be discussed along similar lines.

13.4 *Vikṣepa-calana*

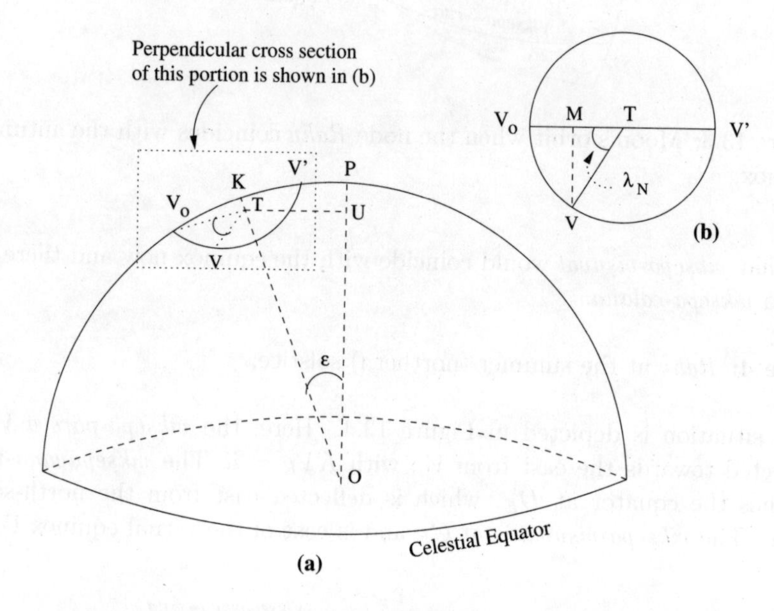

Figure 13.5: The distance between *vikṣepa-pārśva* and the north celestial pole.

Here the method to determine the distance between the (north) celestial pole and the *vikṣepa-pārśva* is described in broad terms first. Consider Figure 13.5. The *vikṣepa-pārśva* is at V_0 separated from K by the maximum *vikṣepa* i. Drop a perpendicular V_0T from V_0 to OK, where O is the centre of the sphere. As the arc $V_0K = i$, $V_0T = R\sin i$. Draw a circle with radius $R\sin i$ centered at T in the plane perpendicular to OT. This is the *vikṣepa-pārśva-vṛtta.* It may be noted that this circle (shown separately in Figure 13.5(b) will be parallel to the plane of the ecliptic.[4]) Mark a point V on this circle

[4]In the figure VM is along the east-west line and is perpendicular to the plane of the figure.

such that the angle corresponding to the arc V_0V is the longitude of *Rāhu*, λ_N. Drop a perpendicular TU from T to the *akṣa-daṇḍa* OP. Conceive a circle with U as the centre and TU as the radius in the plane perpendicular to OP. The relationship between this circle and the *vikṣepa-pārśva-vṛtta* is the same as that of the *kakṣyā-vṛtta* and the *ucca-nīca-vṛtta.* Now,

$$\begin{aligned} OT &= R\cos i, \\ \text{and } TU &= OT\sin\epsilon = R\cos i\sin\epsilon, \end{aligned}$$

is the radius of the *kakṣyā-vṛtta.* Draw VM perpendicular to V_0T. Then, $VM = R\sin i\sin\lambda_N$ and $MT = R\sin i\cos\lambda_N$ play the role of *bhujā-phala* and *koṭi-phala* respectively in the determination of VU, which is the *karṇa.* It must be noted that VM is along the east-west direction and perpendicular to the plane of the figure. It is the distance between V and the north-south circle. When the *Rāhu* is between *Makarādi* and *Karkyādi* (or equivalently λ_N is between 270° and 90°), the *koṭi-phala* has to be added to the representative of *trijyā* which is TU. Similarly, when it is between *Karkyādi* and *Makarādi* (λ_N is between 90° and 270°), the *koṭi-phala* is to be subtracted. (Actually the *koṭi-phala* has to be projected along TU before this is done; this becomes clear in the next section). When *Rāhu* is at the vernal equinox, *vikṣepa-pārśva* is at V_0 and VP would be maximum. Similarly, when *Rāhu* is at the autumnal equinox, *vikṣepa-pārśva* is at V' and VP is minimum.

The *vikṣepa-pārśva* is in the eastern part of the sphere (or to the east of the north-south circle), when *Rāhu* moves from the vernal equinox to the autumnal equinox (or λ_N is between 0° and 180°). Then the *vikṣepa-viṣuvat* is situated east of the equinox, and *vikṣepa-calana* is to be subtracted (from the longitude of the Moon) while calculating Moon's declination. Similarly, the *vikṣepa-viṣuvat* is situated west of the equinox, when the *Rāhu* moves from the autumnal equinox to the vernal equinox (or λ_N is between 180° and 360°), and *vikṣepa-calana* is to be added (to the longitude of the Moon) while calculating Moon's declination.

13.5 *Karṇānayana*

In Figure 13.6, the points V_0, V, T (centre of the *vikṣepa-pārśva-vṛtta*), M and U have the same significance as in Figure 13.5. MV is perpendicular to the plane of the figure. Draw MU' from M, perpendicular to the *akṣa-daṇḍa*,

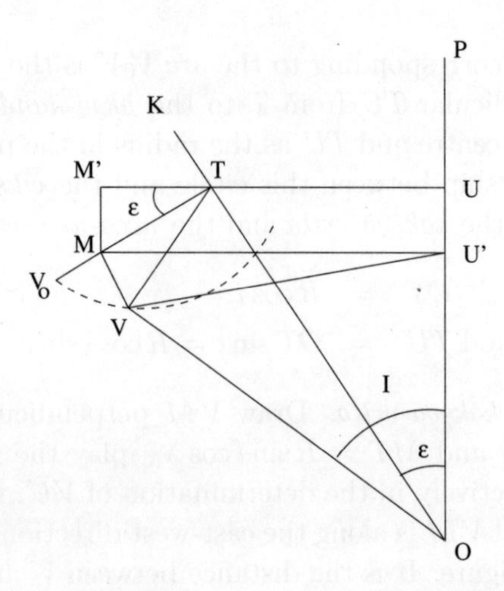

Figure 13.6: The inclination of the Moon's orbit with the equator.

OP. VM is perpendicular to the plane of the figure and hence to OP, and MU' is also perpendicular to OP. Hence $VU'M$ is a triangle, right angled at M, and in a plane perpendicular to OP. Therefore, VU' is perpendicular to OP and is the desired distance, $R\sin I$, between V and *akṣa-daṇḍa*. Let MM' be perpendicular to UM' which is the extension of UT. The angle between TM' and TM is ϵ. It is clear that $MU' = M'U$. Therefore,

$$\begin{aligned} M'U &= M'T + TU \\ &= MT\cos\epsilon + R\cos i \sin\epsilon \\ &= R\sin i \cos\lambda_N \cos\epsilon + R\cos i \sin\epsilon, \end{aligned} \tag{13.1}$$

where MT is the *koṭi-phala* discussed in the previous section. It may be seen that $MV = R\sin i \sin\lambda_N$, is the *bhujā-phala*. Then,

$$\begin{aligned} VU' &= \sqrt{(MV)^2 + (MU')^2} \\ &= \sqrt{(R\sin i \sin\lambda_N)^2 + (R\sin i\cos\lambda_N\cos\epsilon + R\cos i\sin\epsilon)^2}. \end{aligned} \tag{13.2}$$

Clearly $VU' = R\sin I$, where I is the angle corresponding to the arc VP. Hence,

$$R\sin I = \sqrt{(R\sin i \sin\lambda_N)^2 + (R\sin i\cos\lambda_N\cos\epsilon + R\cos i\sin\epsilon)^2}. \tag{13.3}$$

This is the maximum declination, or the maximum divergence between the equator and the *vikṣepa-vṛtta* (*vikṣepāyanānta*).

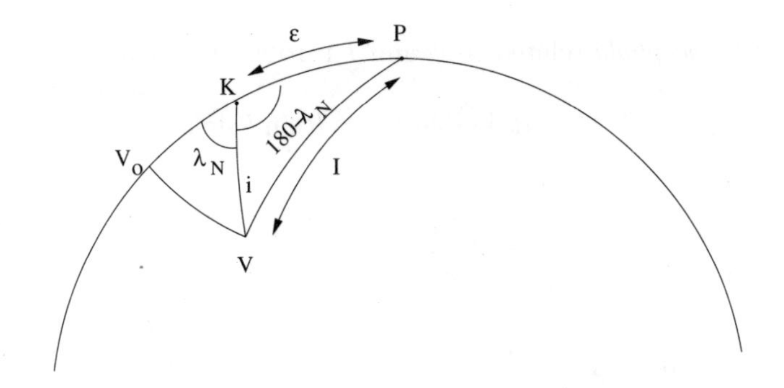

Figure 13.7: Spherical trigonometric derivation of the inclination.

Note: Equation (13.3) can be derived using spherical trigonometrical results as follows: In Figure 13.7, consider the spherical triangle VKP, with $KV = i$, $KP = \epsilon$, $VP = I$ and the spherical angle at K being $180° - \lambda_N$ (as $V_0\hat{K}V = \lambda_N$). Then, applying the cosine formula to the side VP,

$$\begin{aligned}\cos I &= \cos i \cos\epsilon + \sin i \sin\epsilon \cos(180 - \lambda_N) \\ &= \cos i \cos\epsilon - \sin i \sin\epsilon \cos\lambda_N. \qquad (13.4)\end{aligned}$$

From this, it can be easily shown that

$$\sin I = \sqrt{(\sin i \sin\lambda_N)^2 + (\sin i \cos\lambda_N \cos\epsilon + \cos i \sin\epsilon)^2}, \qquad (13.5)$$

which is the same as (13.3).

13.6 Determination of *Vikṣepa-calana*

In Figure 13.8, the *vikṣepāyana-vṛtta* cuts the equator at D. The *vikṣepa-viṣuvat* is at C which is at 90° from D. Hence the *vikṣepa-calana* is $\Gamma C = DN$, which is the arc corresponding to $K\hat{P}V = \psi$. Now ψ is the inclination of *vikṣepāyana-vṛtta* with the north-south circle. The distance between V and the north-south circle is VM[5] and is given by

$$VM = R \sin i \sin\lambda_N. \qquad (13.6)$$

[5]The point M is the foot of perpendicular from V to the plane of the north-south circle, which is the same as the plane of the paper.

This is the *bhujā-phala* related to ψ and I through the relation

$$VM = R\sin i \sin \lambda_N = R\sin I \sin \psi. \tag{13.7}$$

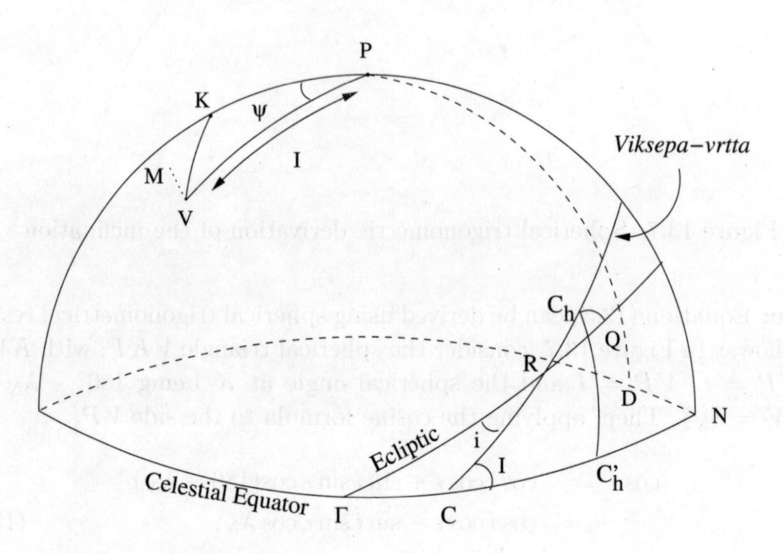

Figure 13.8: The change in the deflection or *Vikṣepa-calana*.

Hence, *vikṣepa-calana* is the arc corresponding to

$$R\sin \Gamma C = R\sin \psi = \frac{R\sin i \sin \lambda_N}{\sin I}. \tag{13.8}$$

Vikṣepa-calana is to be applied to the *sāyana-candra* to obtain the distance between the *vikṣepa-viṣuvat* (C) and the Moon on the *vikṣepa-vṛtta* (C_h), that is CC_h. Then the declination of the Moon, $R\sin\delta_M$, is given by

$$R\sin\delta_M = R\sin C'_h C_h = R\sin I \sin(CC_h), \tag{13.9}$$

as I is the inclination and CC_h is the arc.

Here, it is not specified how the *vikṣepa-calana* is actually applied to the *sāyana-candra* to obtain the arc CC_h along the lunar orbit (*vikṣepa-vṛtta*). In *Tantrasaṅgraha* (VI, 3–6), the declination of the Moon is stated to be $R\sin(\lambda_M - \Gamma C)\sin I$, where λ_M is the *sāyana* longitude of the Moon (ΓQ in Figure 13.8) and ΓC is the *vikṣepa-calana*. Perhaps, this is what is implied here also. This could be understood as follows, when the inclination of

Moon's orbit is taken to be very small.

$$\begin{aligned} CC_h &= C_hR + RC, \qquad \text{(where R is } \textit{Rāhu}\text{)} \\ &= C_hR + \Gamma R + CR - \Gamma R \\ &\approx RQ + \Gamma R - (\Gamma R - CR) \\ &= \Gamma Q - (\Gamma R - CR) \\ &\approx \Gamma Q - \Gamma C \\ &= \lambda_M - \Gamma C. \end{aligned} \tag{13.10}$$

Here we have taken $RC_h \approx RQ$ and $\Gamma R - CR \approx \Gamma C$. These are fairly good approximations, to the first order in the inclination i, as $\cos i$ is taken to be 1 in both the cases. Hence,

$$R\sin\delta_M = R\sin I \sin(\lambda_M - \Gamma C). \tag{13.11}$$

13.7 Time of *Vyatīpāta*

13.8 Derivation of *Vyatīpāta*

As already stated, *vyatīpāta* occurs when the declinations of the Moon (calculated as above, taking into account *vikṣepa-calana*) and the Sun are equal, and when one of them is in the odd quadrant, and the other in the even quadrant. First, the instant of *vyatīpāta* is estimated in an approximate manner from the longitudes of the Sun and Moon on any day. This approximate instant is the zeroth approximation and is denoted by t_0. The declination of the Sun is given by

$$R\sin\delta_s = R\sin\epsilon\sin\lambda_S. \tag{13.12}$$

The declination of the Moon is calculated using the procedure described in the previous sections, and that is equated to the declination of the Sun as follows:

$$R\sin\delta_M = R\sin I \sin(\lambda_M - \Gamma C) = R\sin\epsilon\sin\lambda_s. \tag{13.13}$$

From this, the longitude of the Moon, λ_M is calculated from the arc corresponding to the expression below (and adding the *vikṣepa-calana*):

$$R\sin(\lambda_M - \Gamma C) = \frac{R\sin\epsilon\sin\lambda_s}{\sin I}. \tag{13.14}$$

λ_M, calculated in this manner from the Sun's longitude (and other quantities), would not coincide with λ_M calculated directly, as the instant of *vyatīpāta* is yet to be found. If λ_M(from Sun) $> \lambda_M$ (direct), and the Moon is in the odd quadrant, the declination of the Moon is less than that of the Sun and the *vyatīpāta* is yet to occur. Similarly, the *vyatīpāta* has already occurred if λ_M(from Sun) $< \lambda_M$ (direct), with the Moon in the odd quadrant. In the even quadrant, it is the other way round.

In any case, $\Delta\theta_1 = \lambda_M$(Sun) - λ_M (direct), is found. This is the angle to be covered. As the Sun and Moon are moving in opposite directions for *vyatīpāta*, the above is divided by the sum of the daily motions of the Sun and the Moon to obtain the instant at which *vyatīpāta* will occur as a first approximation. That is, the approximation for the instant of *vyatīpāta* is $t_1 = t_0 + \Delta t_1$, where

$$\Delta t_1 = \frac{\Delta\theta_1}{\dot{\lambda}_M + \dot{\lambda}_S} = \frac{\lambda_M(\text{Sun}) - \lambda_M(\text{direct})}{\dot{\lambda}_M + \dot{\lambda}_S}, \tag{13.15}$$

where $\dot{\lambda}_M$ and $\dot{\lambda}_S$ are the daily motions of the Moon and Sun respectively at t_0. The above result which is in units of days has to be multiplied by 60 to obtain it in *nāḍikā*-s.

The longitudes of the Sun or Moon are calculated at t_1 by multiplying Δt_1 by $\dot{\lambda}_S$ or $\dot{\lambda}_M$ and adding the results to λ_S or λ_M at t_0. In the case of Moon's node, Δt_1 should be multiplied by $\dot{\lambda}_N$ (daily motion of the node) and subtracted from λ_N at t_0, as the motion of Moon's node is retrograde. Again, the longitude of the Moon is calculated from that of the Sun by equating its declination with that of the Sun, and $\Delta\theta_2 = \lambda_M(\text{Sun}) - \lambda_M$ (direct) is found. Then, the next approximation for the instant of *vyatīpāta* is

$$t_1 + \Delta t_2 = t_0 + \Delta t_1 + \Delta t_2, \tag{13.16}$$

where,

$$\Delta t_2 = \frac{\Delta\theta_2}{\dot{\lambda}_M + \dot{\lambda}_S}. \tag{13.17}$$

This iteration procedure is carried on till the longitude of the Moon as calculated from that of the Sun (by equating the declination) and that obtained directly, are equal (to the desired accuracy). Thus,

$$t = t_0 + \Delta t_1 + \Delta t_2 \dots \ , \tag{13.18}$$

is the instant of *vyatīpāta* where the declinations of the Sun and Moon are equal. At any stage, Δt is positive or negative, when λ_M(Sun) is greater or less than λ_M (direct), when the Moon is in an odd quadrant (that is, when its declination keeps increasing with time). It is the other way round, when Moon is in an even quadrant. Thus, it is clear that *vyatīpāta* can occur only when the declination circle of some part of Moon's orb is identical with that of a part of Sun's orb at the same instant.[6]

[6]Towards the end of the chapter, it is stated that the duration of *vyatīpāta* is 4 *nāḍikā*-s. What this means is not clear and perhaps this cannot be true. The procedure for calculating the half-duration of *vyatīpāta* is described in *Tantrasaṅgraha*.

Chapter 14
Mauḍhya and Visibility Correction to Planets

Here, the *lagna* corresponding to the rising and setting of a planet having a latitudinal deflection (*vikṣepa*), is calculated. The visibility correction (*darśana-saṃskāra*) is the correction that should be applied to the longitude of the planet to obtain the *lagna* corresponding to the rising and setting of the planets.

14.1 Computation of visibility correction

Consider the situation in Figure 14.1 when the point L on the ecliptic, having the same longitude as the planet P, is on the horizon, or L is the *lagna*. The planet P has a (northern) latitude β and PP' is the *vikṣepa-koṭi-vṛtta* parallel to the ecliptic, with C as the centre. K_1PL and $K_1P'L'$ are the arcs of the *rāśi-kūṭa-vṛtta*-s passing through P and P' (point in the *vikṣepa-koṭi-vṛtta* on the horizon) respectively. Here K_1 is the northern *rāśi-kūṭa*. V is the *dṛkkṣepa* whose zenith distance is $ZV = z_v$, also referred to as *dṛkkṣepa*. ZK_1MM' is a vertical circle and it is clear that

$$ZK_1 = 90° - z_v, \quad K_1M = z_v \text{ and } \quad MM' = P'\hat{L}L' = 90° - z_v.$$

As L is at 90° from both Z and K_1, it is the pole of the vertical Z_1KMM'. Hence, $LM = LK_1 = 90°$ and $K_1\hat{L}P' = z_v$.

Now, drop a perpendicular PF from P to the plane of the horizon. Clearly, PF is the *śaṅku* of P, whose zenith distance is $ZP = z$. Also, the arc $LP = \beta$ (latitude of the planet) is inclined at an angle, $P\hat{L}M = K_1\hat{L}M = z_v$, with the arc $L\hat{P}'M$. Therefore,

$$PF = R\cos z = R\sin z_v \sin\beta. \tag{14.1}$$

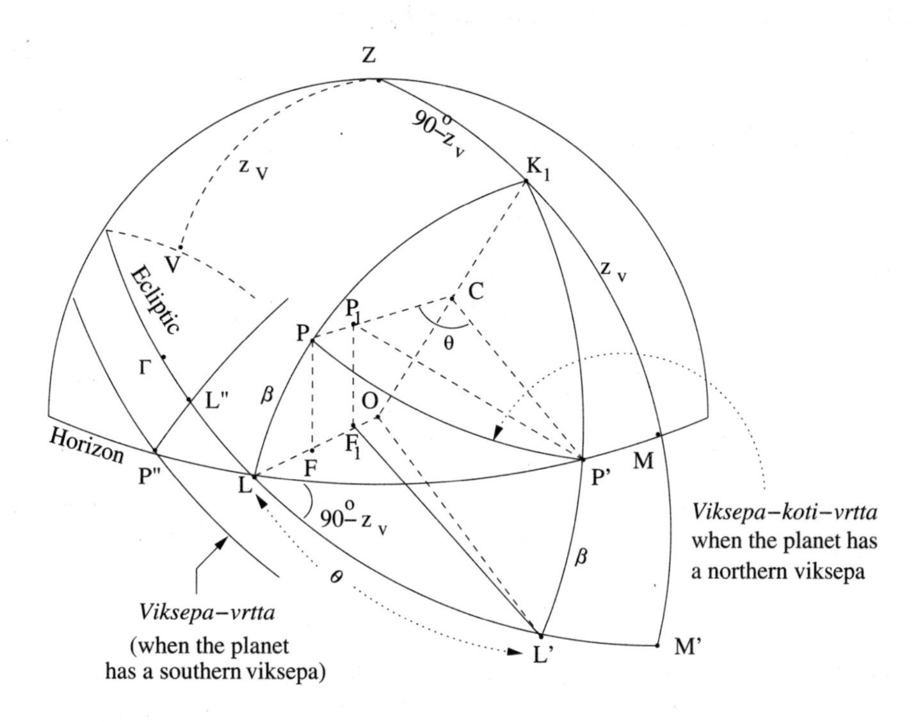

Figure 14.1: Visibility correction when the planet has latitude and the northern K_1 is above the horizon.

Draw $P'P_1$ perpendicular to CP and P_1F_1 perpendicular to OL. It is clear that $P_1F_1 = PF = R\cos z$. Now, P_1F_1P' is a right angled triangle with $P_1\hat{P'}F_1 = 90° - z_v$, and this is the angle between the *vikṣepa-koṭi-vṛtta* and the horizon, which is the same as the angle between the ecliptic and the horizon. Hence,

$$\begin{aligned} P'P_1 &= \frac{P_1F_1}{\sin(90° - z_v)} \\ &= \frac{R\cos z}{\cos z_v} \\ &= \frac{R\sin z_v \sin\beta}{\cos z_v}. \end{aligned} \tag{14.2}$$

This is the distance between the planet and the horizon on the *vikṣepa-koṭi-vṛtta*.

14.2 Rising and setting of planets

Consider the angle θ between the *rāśi-kūṭa-vṛtta*-s K_1PL and $K_1P'L'$ in Figure 14.1. It corresponds to the arc LL' on the ecliptic and

$$\theta = P\hat{C}P' = L\hat{O}L'. \tag{14.3}$$

Now, the planes of the *vikṣepa-koṭi-vṛtta* and ecliptic are parallel and, just as $P'P_1$ is perpendicular to CP, $L'F_1$ is perpendicular to OL. Hence,

$$L'F_1 = R\sin\theta. \tag{14.4}$$

As the radius of the *vikṣepa-koṭi-vṛtta* is $R\cos\beta$,

$$P'P_1 = CP'\sin\theta = R\cos\beta\sin\theta. \tag{14.5}$$

Hence,

$$R\sin\theta = \frac{P'P_1}{\cos\beta} = \frac{R\sin z_v \sin\beta}{\cos z_v \cos\beta}. \tag{14.6}$$

From this, the arc $LL' = \theta$ is obtained. This formula is the same as the one for *cara*, with z_v replacing the latitude of the place, and β replacing the declination δ.

Now consider the situation when the planet is at P' on the horizon, i.e., it is rising. Then, the *lagna* ΓL is given by

$$\begin{aligned} \Gamma L &= \Gamma L' - LL' \\ &= \text{Longitude of the planet} - \theta, \end{aligned} \tag{14.7}$$

where the arc $LL' = \theta$ is calculated as above. So, the visibility correction (θ), is subtracted in this case. When the planet has a southern latitude, which is also shown in Figure 14.1, the visibility correction θ, which is the angle between the *rāśi-kūṭa-vṛtta*-s passing through the planet at P'' and the *lagna* at L, is calculated in the same manner. In this case, the *lagna* is given by

$$\begin{aligned} \Gamma L &= \Gamma L'' + LL'' \\ &= \text{Longitude of the planet} + \theta, \end{aligned} \tag{14.8}$$

and hence, the correction has to be added.

At the setting, there is reversal of addition and subtraction. In fact, the same figure can be used, with the only difference being that ΓL and $\Gamma L'$ are westwards now. As the longitude is always measured eastwards, it is clear that when the *vikṣepa* is north, the *lagna* will be greater than the longitude of the planet (visibility correction is added). Similarly, the visibility correction is subtracted when the latitude is south.

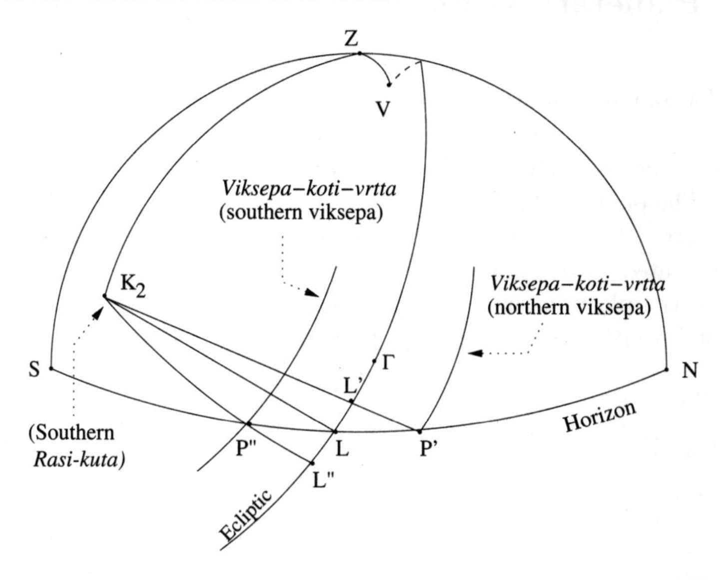

Figure 14.2: Visibility correction when the planet has latitude and the southern K_2 is above the horizon.

It may be noted that the *dṛkkṣepa* V is south, when the northern *rāśi-kūṭa* K_1 is above the horizon (Figure 14.1). In Figure 14.2, the situation when the southern *rāśi-kūṭa* K_2 is above the horizon is displayed. Here, the *dṛkkṣepa* is north. P' and P'' correspond to raising points of a planet with northern and southern latitudes respectively, and L is the *lagna*. Then the 'visibility correction' is LL' when the planet has a northern latitude and it has to be added to the longitude of the planet ($\Gamma L'$) to obtain the *lagna* (ΓL). Similarly, the visibility correction is LL'', when the planet has a southern latitude and it has to be subtracted from the longitude of the planet($\Gamma L''$) to obtain the *lagna* (ΓL). Hence, this is the reverse of the situation when the northern *rāśi-kūṭa* is above the horizon.

In both the cases, the *darśana-saṃskāra* (visibility correction) should be added to the planet's longitude when the directions of the *vikṣepa* and the *dṛkkṣepa* are the same, and subtracted from it when the directions of these two are opposite, to obtain the *lagna.* At the setting we have the reverse situation.

14.3 Planetary visibility

Having determined the *lagna* at the rising and setting of a planet, the corresponding *kāla-lagna* is determined (as described in Chapter 11). The difference in *kāla-lagna*-s of the planet and the Sun (in terms of minutes of time) is found. The planet is visible only when this difference is more than a specified measure.[1] The method for obtaining the *madhya-lagna* of a planet with *vikṣepa* is stated to be similar. The *madhya-lagna* does not depend on the latitude of the place, as it is the meridian ecliptic point, and the meridian or the north-south circle is the same for places with or without latitude.

[1]This measure is not specified in this Text, whereas in chapter 7 of *Tantrasaṅgraha*, the minimum angular separation in degrees for visibility are specified to be 12, 17, 13, 11, 9 and 15 for the Moon, Mars, Mercury, Jupiter, Venus and Saturn respectively.

Chapter 15

Elevation of the Moon's Cusps

Though the title of this short chapter is *candra-śṛṅgonnati* or 'Elevation of the Moon's Cusps', it is exclusively devoted to the computation of the distance between the centres of the lunar and solar discs (*bimbāntara*). The *bimbāntara* of course figures prominently in the computations of the Moon's phase and the elevation of its cusps, but these are not discussed in the Text as available.

15.1 The *Dvitīya-sphuṭa-karṇa* of the Sun and the Moon

In chapter 11, prior to the discussion on *chāyā-lambana*, the calculation of *dvitīya-sphuṭa-karṇa* (section 11.36) which is the actual distance of the Sun and the Moon from the centre of the Earth, after taking into account the 'second correction' (essentially, the evection term), was discussed. The second correction has to be applied to the *manda-sphuṭa* of the Moon, to obtain the true longitude. (In the case of the Sun, there is no correction to the *manda-sphuṭa*, as the *mandocca* of second correction is in the same direction as *manda-sphuṭa* of the Sun). Here the view of *Siddhāntaśekhara* (of Śrīpati) is quoted.[1]

The *dṛkkarṇa*, or the distance of the planet from the observer on the surface of the Earth, is obtained from the *dvitīya-sphuṭa-karṇa* as in chapter 11. The *nati* (parallax in latitude) and *lambana* (parallax in longitude) of the Sun

[1] It is also stated that according to *Laghumānasa* (by Muñjāla) "the *antya-phala* of the Moon is to be multiplied by Moon's *manda-karṇa* and five and divided by *trijyā*".

and Moon are found from the *dṛkkarṇa*. The longitudes are corrected for *lambana*. From the corrected longitudes and the *nati*, the distance between the centres of the solar and lunar spheres is to be computed; as outlined below.

15.2 Distance between the orbs of the Sun and the Moon

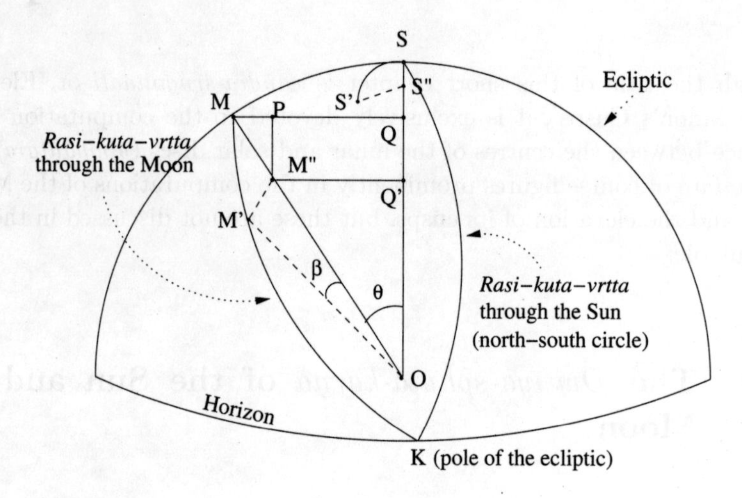

Figure 15.1: Calculation of *bimbāntara*, the distance between the orbs of the Sun and the Moon.

In Figure 15.1, the Sun S without *nati* is conceived to be at the zenith and the ecliptic is conceived as the prime vertical with its poles on the North and South points. The *rāśi-kūṭa-vṛtta* through the Sun will be the north-south circle. (In the figure, the ecliptic which is the prime vertical is in the plane of the paper). If O is the center of the sphere, OS is the vertical line.

Case 1: Consider the Sun without *nati* at S, and the Moon without *vikṣepa* at M. Draw MQ perpendicular to the vertical line. If θ is the difference in their longitudes,

$$\begin{aligned} MQ &= R\sin\theta = \textit{Bhujā-jyā} \\ SQ &= R(1-\cos\theta) = \textit{Śara}. \end{aligned} \tag{15.1}$$

In this case, it is clear that

$$\begin{aligned} Bimb\bar{a}ntara = SM &= \sqrt{MQ^2 + SQ^2} \\ &= \sqrt{(Bhuj\bar{a}\text{-}jy\bar{a})^2 + (\acute{S}ara)^2}. \end{aligned} \tag{15.2}$$

Case 2: Now consider the Moon with latitude β at M' and the Sun without *nati*.[2] From M' draw $M'M''$ perpendicular to OM. Clearly,

$$\begin{aligned} M\hat{O}M' &= \beta = \text{latitude } (c\bar{a}pa \text{ of } vik\d{s}epa), \\ M'M'' &= R\sin\beta = vik\d{s}epa, \\ MM'' &= R(1-\cos\beta) = vik\d{s}epa\text{-}\acute{s}ara, \\ OM'' &= R\cos\beta. \end{aligned} \tag{15.3}$$

Now we have to find the distance between M' and S. From M'' drop perpendiculars $M''P$ and $M''Q'$ on MQ and OS respectively. Then,

$$M''Q' = PQ = R\cos\beta\sin\theta. \tag{15.4}$$

Equation (15.4) can also be obtained as follows:

$$\begin{aligned} MP &= MM''\sin\theta \\ &= R(1-\cos\beta)\sin\theta \\ &= Bhuj\bar{a}\text{-}phala \text{ of } vik\d{s}epa\text{-}\acute{s}ara. \end{aligned} \tag{15.5}$$

Hence,

$$\begin{aligned} M''Q' = PQ &= MQ - MP \\ &= R\sin\theta - R(1-\cos\beta)\sin\theta \\ &= R\cos\beta\sin\theta. \end{aligned} \tag{15.6}$$

Now,

$$\begin{aligned} QQ' = PM'' &= MM''\cos\theta \\ &= R\cos\theta(1-\cos\beta) \\ &= ko\d{t}i\text{-}phala \text{ of } vik\d{s}epa\text{-}\acute{s}ara, \end{aligned} \tag{15.7}$$

[2]The difference in the longitudes of the Sun and Moon is assumed to be less than 90°.

which can also be computed from

$$\begin{aligned} QQ' = PM'' &= \sqrt{MM''^2 - PM^2} \\ &= \sqrt{R^2(1-\cos\beta)^2 - R^2(1-\cos\beta)^2\sin^2\theta} \\ &= R\cos\theta(1-\cos\beta). \end{aligned} \quad (15.8)$$

SQ' is the distance between the Sun and the foot of the *bhujā-jyā* (Q'), which is drawn from the foot of the *vikṣepa* (M''). It is given by

$$\begin{aligned} SQ' &= SQ + QQ' \\ &= R(1-\cos\theta) + R\cos\theta(1-\cos\beta) \\ &= R(1-\cos\theta\cos\beta). \end{aligned} \quad (15.9)$$

Then, SM'' which is the distance between the Sun and the foot of the *vikṣepa*, is given by

$$\begin{aligned} SM'' &= \sqrt{SQ'^2 + M''Q'^2} \\ &= \sqrt{R^2(1-\cos\theta\cos\beta)^2 + R^2\sin^2\theta\cos^2\beta}. \end{aligned} \quad (15.10)$$

Now the *vikṣepa* $M'M''$ is perpendicular to the plane of the ecliptic and is perpendicular to SM''. Therefore,

$$\begin{aligned} SM'^2 &= SM''^2 + M'M''^2 \\ &= R^2(1-\cos\theta\cos\beta)^2 + \\ &\qquad R^2\sin^2\theta\cos^2\beta + R^2\sin^2\beta. \end{aligned} \quad (15.11)$$

SM' which is the square root of this is the distance between the Sun, S (without *nati*) and the Moon M' (with *vikṣepa*).

Case 3: Now consider the case when the Sun has *nati* and it is at S', separated from the ecliptic by the arc $SS' = \eta_s$. Drop a perpendicular $S'S''$ from S' to the vertical, OS. $S'S''$ is perpendicular to the ecliptic and to OS. Then,

$$S'S'' = R\sin\eta_s = nati, \quad (15.12)$$

$$\begin{aligned} \text{and} \quad SS'' &= R(1-\cos\eta_s) \\ &= \textit{Arkonnati-śara}. \end{aligned} \quad (15.13)$$

Then,

$$\begin{aligned} S''Q &= SQ - SS'' \\ &= \textit{Sphuṭa-śara} - \textit{Arkonnati-śara} \\ &= R(1-\cos\theta) - R(1-\cos\eta_s), \end{aligned} \quad (15.14)$$

is the vertical distance between the base of the *nati-śara* and the base of the *bhujā-jyā*. It may be noted that

$$\begin{aligned} QQ' &= \textit{koṭi-phala} \text{ of Moon's } \textit{kṣepa-śara} \\ &= R\cos\theta(1-\cos\beta). \end{aligned} \quad (15.15)$$

Then,

$$\begin{aligned} S''Q' &= S''Q + QQ' \\ &= R(1-\cos\theta) - R(1-\cos\eta_s) + R\cos\theta(1-\cos\beta) \\ &= r_1, \end{aligned} \quad (15.16)$$

is one quantity (*rāśi*), which is the vertical distance between the Sun and the Moon.

The horizontal distance $M''Q'$ between the Sun and the Moon in the plane of the ecliptic is the second quantity given by,

$$r_2 = M''Q' = R(1-\cos\theta\cos\beta). \quad (15.17)$$

Clearly,

$$S''M'' = \sqrt{r_1^2 + r_2^2}. \quad (15.18)$$

The sum or difference of the *nati*-s of the Sun and Moon is the third quantity, r_3. This is the distance between the Sun and the Moon along the line perpendicular to the ecliptic (plane of the paper in Figure 15.1). If the *nati*-s are in the same direction with respect to the ecliptic, the difference is to be considered. If they are in the opposite directions, the sum of the *nati*-s is to be taken. In Figure 15.1, where both the *nati*-s are above the plane of the paper, we have

$$r_3 = M'M'' - S'S'' = R\sin\beta - R\sin\eta_s. \quad (15.19)$$

Then the *bimbāntara* or the distance $S'M'$ between the centres of the lunar and solar discs is given by

$$\textit{Bimbāntara} = \sqrt{r_1^2 + r_2^2 + r_3^2}. \quad (15.20)$$

This can be understood as follows:

In Figure 15.2, $S''M''$ is in the plane of the ecliptic (of the paper). $S''S'$ and $M''M'$ are perpendicular to the plane of the paper. Let

$$M''T = S''S' = \textit{Nati} \text{ of Sun.}$$

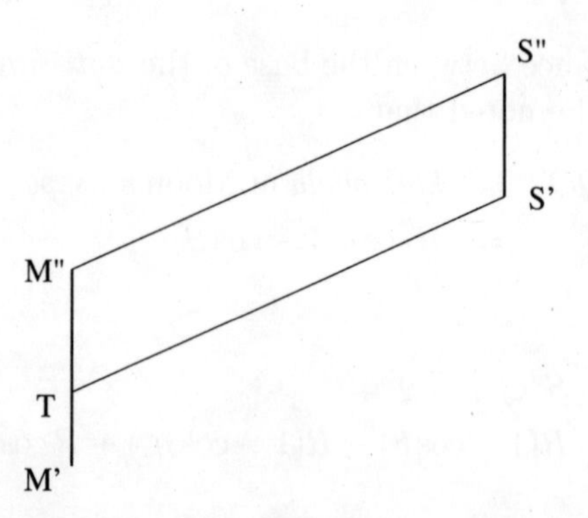

Figure 15.2: The actual distance between the apparent Sun and the Moon.

Then,

$$\begin{aligned} TM' &= M''M' - M''T \\ &= \text{Difference in } nati-s \\ &= r_3. \end{aligned} \tag{15.21}$$

$S'T$ is a line parallel and equal in length to $S''M''$ and TM' is perpendicular to it. Hence,

$$\begin{aligned} S'M' &= \sqrt{S'T^2 + TM'^2} \\ &= \sqrt{S''M''^2 + TM'^2} \\ &= \sqrt{r_1^2 + r_2^2 + r_3^2}. \end{aligned} \tag{15.22}$$

Hence, r_1, r_2 and r_3 are essentially the differences in the coordinates of the Sun and the Moon along the vertical, east-west (horizontal direction in the plane of the paper), and the north-south directions respectively. This is the rationale for the expression for the *bimbāntara*.

Case 4: Now consider the case when the difference between the longitudes of the Sun and Moon is more than 90°. In this case, the treatment is similar except that the zenith Z is conceived to be situated exactly midway between the Sun and the Moon, without *vikṣepa* or *nati*, at S and M respectively (see Figure 15.3). The line SM cuts the vertical at Q. As the arcs ZM and ZS are both equal to half the difference in longitudes, we have

$$\begin{aligned} MQ = QS &= R\sin\left(\frac{\theta}{2}\right) \\ &= \textit{Bhujā-jyā} \text{ of Moon/Sun}. \end{aligned} \tag{15.23}$$

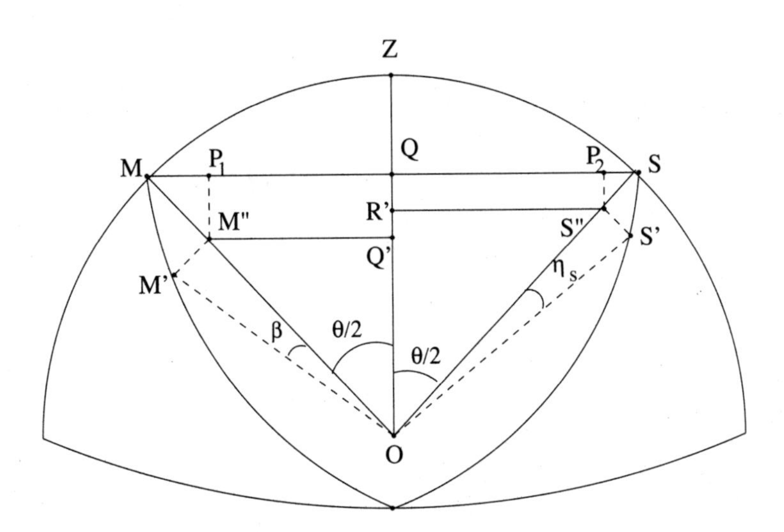

Figure 15.3: The distance between the Sun and the Moon when their difference in longitudes is 90°.

In Figure 15.3, M' and S' are the true Moon and Sun with *vikṣepa* and *nati*. M'' and S'' are at the feet of the *vikṣepa* (arc β) and *nati* (arc η_s) on the *sūtra*-s of Moon and Sun respectively. $M''Q'$ and $S''R'$ are the perpendiculars from M'' and S'' respectively on the vertical OZ. $M''P_1$ and $S''P_2$ are perpendiculars from M'' and S'' on MS. Now,

$$\begin{aligned} QQ' &= \textit{Koṭi-phala} \text{ of Moon's } \textit{śara} \\ &= R\cos\left(\frac{\theta}{2}\right)(1-\cos\beta), \qquad (15.24) \\ \text{and} \quad QR' &= \textit{Koṭi-phala} \text{ of Sun's } \textit{śara} \\ &= R\cos\left(\frac{\theta}{2}\right)(1-\cos\eta_s). \qquad (15.25) \end{aligned}$$

$R'Q'$, which is the difference between the feet of the perpendiculars from S'' and M'' on the vertical line, is given by the relation:

$$\begin{aligned} R'Q' &= QQ' - QR' \\ &= r_1. \qquad (15.26) \end{aligned}$$

This is the difference between *koṭi-phala*-s of the *śara*-s of Sun and Moon, and is the first quantity.

Now,

$$\begin{aligned} M''Q' &= MQ - MP_1 \\ &= R\sin\left(\frac{\theta}{2}\right) - R\sin\left(\frac{\theta}{2}\right)(1-\cos\beta) \\ &= R\sin\left(\frac{\theta}{2}\right)\cos\beta, \end{aligned} \tag{15.27}$$

is the Rsine of half the longitude difference from which the *dorjyā-phala* of Moon's *śara* has been subtracted.

Similarly,

$$\begin{aligned} S''R' &= SQ - SP_2 \\ &= R\sin\left(\frac{\theta}{2}\right) - R\sin\left(\frac{\theta}{2}\right)(1-\cos\eta_s) \\ &= R\sin\left(\frac{\theta}{2}\right)\cos\eta_s, \end{aligned} \tag{15.28}$$

is the Rsine of half the longitude difference from which the *dorjyā-phala* of Sun's *śara* has been subtracted. The sum of the above two is the second quantity:

$$\begin{aligned} r_2 &= M''Q' + S''R' \\ &= R\sin\left(\frac{\theta}{2}\right)\cos\beta + R\sin\left(\frac{\theta}{2}\right)\cos\eta_s. \end{aligned} \tag{15.29}$$

The third quantity (r_3), is the sum or difference in *nati*-s of the Sun and Moon:

$$\begin{aligned} r_3 &= M'M'' \sim S'S'' \qquad \text{(same direction)}, \\ r_3 &= M'M'' + S'S'' \qquad \text{(opposite directions)}. \end{aligned} \tag{15.30}$$

Then *bimbāntara*, $S'M'$, is given by the square root of the sum ofv the squares of the above three quantities r_1, r_2 and r_3:

$$S'M' = \sqrt{r_1^2 + r_2^2 + r_3^2}. \tag{15.31}$$

Here the third quantity r_3, which is the sum or difference of the *nati*-s, is the 'north-south' separation between the Sun and the Moon. The second quantity r_2, which is the sum of $M''Q'$ and $S''R'$ (*nati-phala-tyāga-viśiṣṭāntara-ardhajyānāṃ yogaṃ*), is the 'east-west' separation between them. The first quantity r_1, which is the distance between the feet of the perpendiculars from the Sun and the Moon on the vertical (*nati-śarāṇāṃ koṭi-phalāntaram ūrdhvādhobhāgīya-antarālam*), is the 'vertical' separation between them. Hence,

$$Bimb\bar{a}ntara = S'M' = \sqrt{r_1^2 + r_2^2 + r_3^2}. \tag{15.32}$$

The same procedure is used in the derivation of the separation of the orbs in the computation of eclipses.

The Text (as presently available) ends at this point without going further into the details of the calculation of *Śṛṅgonnati*, which may be found in other texts such as *Tantrasaṅgraha*.

Epilogue

Revision of Indian Planetary Model by Nīlakaṇṭha Somayājī (c. 1500 AD)*

It is now generally recognised that the Kerala School of Indian astronomy,[1] starting with Mādhava of Saṅgamagrāma in the fourteenth century, made important contributions to mathematical analysis much before this subject developed in Europe. The Kerala astronomers derived infinite series for π, sine and cosine functions and also developed fast convergent approximations to them. Here, we shall discuss how the Kerala School also made equally significant discoveries in astronomy, in particular, planetary theory.

Mādhava's disciple Parameśvara of Vaṭasseri (c. 1380-1460) is reputed to have made continuous and careful observations for a period of over fifty-five years. He is famous as the originator of the *Dṛg-gaṇita* system, which replaced the older *Parahita* system. Nīlakaṇṭha Somayājī of Tṛkkaṇṭiyur (c. 1444-1550), disciple of Parameśvara's son Dāmodara, carried out an even more fundamental revision of the traditional planetary theory. In his treatise *Tantrasaṅgraha* (c. 1500), Nīlakaṇṭha Somayājī presents a major revision of the earlier Indian planetary model for the interior planets Mercury and Venus. This led Nīlakaṇṭha Somayājī to a much better formulation of the equation of centre and the latitude of these planets than was available either in the earlier Indian works or in the Islamic or the Greco-European traditions

*The material in this Epilogue is based on the following sources, which may be consulted for details: (i) K. Ramasubramanian, M. D. Srinivas and M. S. Sriram, 'Modification of the Earlier Indian Planetary Theory by the Kerala Astronomers (c. 1500 AD) and the Implied Heliocentric Picture of Planetary Motion', Current Science **66**, 784-790, 1994; (ii) M. S. Sriram, K. Ramasubramanian and M. D. Srinivas (eds.), *500 years of Tantrasaṅgraha: A Landmark in the History of Astronomy*, Shimla 2002, p. 29-102.

[1]For the Kerala School of Astronomy, see for instance, K. V. Sarma, *A Bibliography of Kerala and Kerala-based Astronomy and Astrology*, Hoshiarpur 1972; K. V. Sarma, *A History of the Kerala School of Hindu Astronomy*, Hoshiarpur 1972.

of astronomy till the work of Kepler, which was to come more than a hundred years later.

Nīlakaṇṭha Somayājī was the first savant in the history of astronomy to clearly deduce from his computational scheme and the observed motion of the planets – and not from any speculative or cosmological arguments – that the interior planets go around the Sun and the period of their motion around Sun is also the period of their latitudinal motion. He explains in his *Āryabhaṭīya-bhāṣya* that the Earth is not circumscribed by the orbit of the interior planets, Mercury and Venus; and the mean period of motion in longitude of these planets around the Earth is the same as that of the Sun, precisely because they are being carried around the Earth by the Sun. In his works, *Golasāra* and *Siddhāntadarpaṇa*, Nīlakaṇṭha Somayājī describes the geometrical picture of planetary motion that follows from his revised model, where the five planets Mercury, Venus, Mars, Jupiter and Saturn move in eccentric orbits around the mean Sun, which in turn goes around the Earth. Most of the Kerala astronomers who succeeded Nīlakaṇṭha Somayājī, such as Jyeṣṭhadeva, Acyuta Piṣāraṭi, Putumana Somayājī, etc., seem to have adopted this revised planetary model.

1 The conventional planetary model of Indian astronomy

In the Indian astronomical tradition, at least from the time of Āryabhaṭa (499 AD), the procedure for calculating the geocentric longitudes for the five planets, Mercury, Venus, Mars, Jupiter and Saturn involves essentially the following steps.[2] First, the mean longitude (called the *madhyama-graha*) is calculated for the desired day by computing the number of mean civil days elapsed since the epoch (this number is called the *ahargaṇa*) and multiplying it by the mean daily motion of the planet. Then, two corrections namely the *manda-saṃskāra* and *śīghra-saṃskāra* are applied to the mean planet to obtain the true longitude.

[2]For a general review of Indian astronomy, see D. A. Somayājī, *A Critical Study of Ancient Hindu Astronomy*, Dharwar 1972; S. N. Sen and K. S. Shukla (eds), *A History of Indian Astronomy*, New Delhi 1985 (Rev. Edn. 2000); B. V. Subbarayappa, and K. V. Sarma (eds.), *Indian Astronomy: A Source Book*, Bombay 1985; S. Balachandra Rao, *Indian Astronomy: An Introduction*, Hyderabad 2000.

In the case of the exterior planets, Mars, Jupiter and Saturn, the *manda-saṃskāra* is equivalent to taking into account the eccentricity of the planet's orbit around the Sun. Different computational schemes for the *manda-saṃskāra* are discussed in Indian astronomical literature. However, the *manda* correction in all these schemes coincides, to first order in eccentricity, with the equation of centre as currently calculated in astronomy. The *manda*-corrected mean longitude is called *manda-sphuṭa-graha.* For the exterior planets, the *manda-sphuṭa-graha* is the same as the true heliocentric longitude.

The *śīghra-saṃskāra* is applied to this *manda-sphuṭa-graha* to obtain the true geocentric longitude known as *sphuṭa-graha.* The *śīghra* correction is equivalent to converting the heliocentric longitude into geocentric longitude. The exterior and interior planets are treated differently in applying this correction. We shall now briefly discuss the details of the *manda-saṃskāra* and the *śīghra-saṃskāra* for the exterior and the interior planets respectively.

1.1 Exterior planets

For the exterior planets, Mars, Jupiter and Saturn, the mean heliocentric sidereal period is identical with the mean geocentric sidereal period. Thus, the mean longitude calculated prior to the *manda-saṃskāra* is the same as the mean heliocentric longitude of the planet as we understand today. As the *manda-saṃskāra*, or the equation of centre, is applied to this longitude to obtain the *manda-sphuṭa-graha*, the latter will be the true heliocentric longitude of the planet.

The *manda-saṃskāra* for the exterior planets can be explained using a simple epicycle model[3] as shown in Figure 1. Here O is the centre of the concentric circle called *kakṣyā-maṇḍala.* P_0 is the mean planet on the concentric, and P is the true planet (*manda-sphuṭa*) on the epicycle. OU is the direction of *mandocca* or the aphelion. $PP_0 = OU = r$, is the radius of the epicycle and $OP_0 = R$ is the radius of the concentric. $OP = K$ is the *manda-karṇa.* The longitudes are always measured in Indian astronomy with respect to a fixed point in the zodiac known as the *nirayaṇa-meṣādi* denoted by A in the figure.

[3]Equivalently this can be explained with an eccentric model also.

$$A\hat{O}P_0 = \theta_0 \quad \text{(mean longitude of the planet)}$$
$$A\hat{O}U = \theta_u \quad \text{(longitude of } \textit{mandocca}\text{)}$$
$$A\hat{O}P = \theta_{ms} \quad (\textit{manda-sphuṭa}).$$

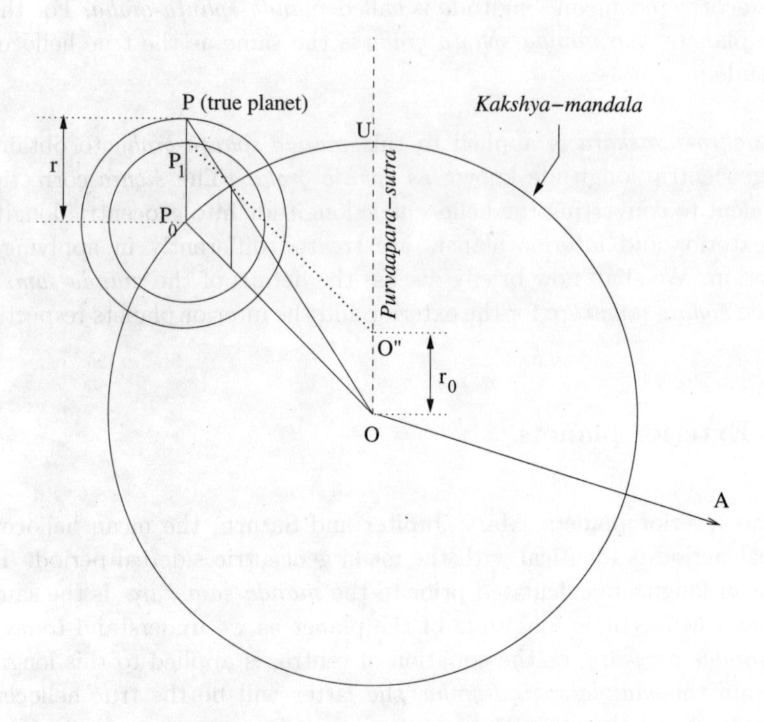

Figure 1: The *manda-saṃskāra.*

The difference between the longitudes of the mean planet and the *mandocca*, namely,

$$\mu = \theta_0 - \theta_u, \tag{1}$$

is called the *manda-kendra* (anomaly) in Indian astronomy. From the triangle OP_0P we can easily obtain the result

$$\sin(\theta_{ms} - \theta_0) = \frac{r}{K} \sin \mu. \tag{2}$$

An important feature of the Indian planetary models, which was specially emphasised in the texts of the Āryabhaṭan School, is that the radius of the epicycle r is taken to vary in the same way as the *manda-karṇa* K, so that their ratio is constant

$$\frac{r}{K} = \frac{r_0}{R},$$

where r_0 is the tabulated or the mean radius of the epicycle. Eaquation (2) therefore reduces to

$$\sin(\theta_{ms} - \theta_0) = \frac{r_0}{R} \sin \mu. \tag{3}$$

The *manda-sphuṭa* θ_{ms} can be evaluated without calculating the true radius of the epicycle r or the *manda-karṇa* K. The texts however give a process of iteration by which the *manda-karṇa* K (and hence the epicycle radius r also) can be evaluated to any given degree of accuracy.[4] The for the exterior planets can be explained with reference to Figure 2. The *nirayaṇa-meṣādi* denoted by A in the figure, E is the Earth and P the planet. The mean Sun S is referred to as the *śīghrocca* for exterior planets. We have

$$\begin{array}{lcll} A\hat{S}P & = & \theta_{ms} & (\textit{manda-sphuṭa}) \\ A\hat{E}S & = & \theta_s & (\text{longitude of } \textit{śīghrocca} \text{ (mean Sun)}) \\ A\hat{E}P & = & \theta & (\text{geocentric longitude of the planet}). \end{array}$$

The difference between the longitudes of the *śīghrocca* and the *manda-sphuṭa*, namely,

$$\sigma = \theta_s - \theta_{ms}, \tag{4}$$

is called the *śīghra-kendra* (anomaly of conjunction) in Indian astronomy. From the triangle EPS we can easily obtain the result

$$\sin(\theta - \theta_{ms}) = \frac{r \sin \sigma}{[(R + r\cos\sigma)^2 + r^2 \sin^2 \sigma]^{\frac{1}{2}}}, \tag{5}$$

which is the *śīghra* correction formula given by Indian astronomers to calculate the geocentric longitude of an exterior planet.

From the figure it is clear that the *śīghra-saṃskāra* transforms the true heliocentric longitudes into true geocentric longitudes. This will work only if $\frac{r}{R}$ is equal to the ratio of the Earth-Sun and planet-Sun distances and is indeed very nearly so in the Indian texts. But (5) is still an approximation as it is based upon the mean Sun and not the true Sun.

[4] Nīlakaṇṭha, in his *Tantrasaṅgraha*, has given an exact formula due to Mādhava by which the *manda-karṇa* can be evaluated without resorting to successive iterations.

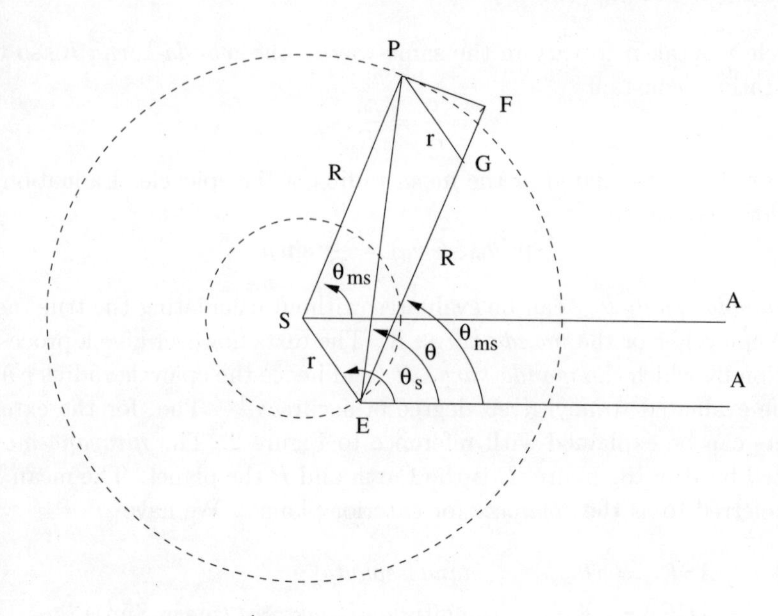

Figure 2: *Śīghra* correction for exterior planets.

1.2 Interior planets

For the interior planets Mercury and Venus, ancient Indian astronomers, at least from the time of Āryabhaṭa, took the mean Sun as the *madhyama-graha* or the mean planet. For these planets, the mean heliocentric sidereal period is the period of revolution of the planet around the Sun, while the mean geocentric sidereal period is the same as that of the Sun. The ancient astronomers prescribed the application of *manda* correction or the equation of centre characteristic of the planet, to the mean Sun, instead of the mean heliocentric planet as is done in the currently accepted model of the solar system. However, the ancient Indian astronomers also introduced the notion of the *śīghrocca* for these planets whose period is the same as the mean heliocentric sidereal period of these planets. Thus, in the case of the interior planets, it is the longitude of the *śīghrocca* which will be the same as the mean heliocentric longitude of the planet as understood in the currently accepted model for the solar system.

The *śīghra-saṃskāra* for the interior planets can be explained with reference to Figure 3. Here E is the Earth and S (*manda*-corrected mean Sun) is the *manda-sphuṭa-graha* and P corresponds to the planet. We have,

$$\begin{array}{rcll} A\hat{E}S & = & \theta_{ms} & (\textit{manda-sphuṭa}) \\ A\hat{S}P & = & \theta_s & (\text{longitude of } \textit{śīghrocca}) \\ A\hat{E}P & = & \theta & (\text{geocentric longitude of the planet}). \end{array}$$

Again, the *śīghra-kendra* is defined as the difference between the *śīghrocca* and the *manda-sphuṭa-graha* as in (4). Thus, from the triangle EPS we get the same formula

$$\sin(\theta_s - \theta_{ms}) = \frac{r \sin \sigma}{[(R + r\cos\sigma)^2 + r^2 \sin^2 \sigma]^{\frac{1}{2}}}, \tag{6}$$

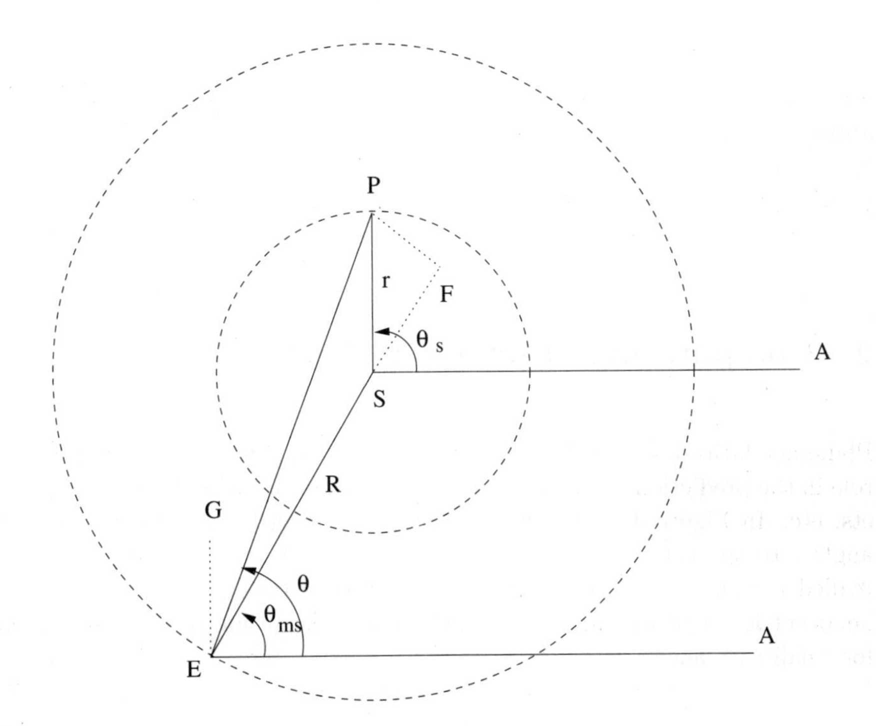

Figure 3: *Śīghra* correction for interior planets.

which is the *śīghra* correction given in the earlier Indian texts to calculate the geocentric longitude of an interior planet. For the interior planets also, the value specified for $\frac{r}{R}$ is very nearly equal to the ratio of the planet-Sun and Earth-Sun distances. In Table 1, we give Āryabhaṭa's values for both the exterior and interior planets along with the modern values based on the mean Earth-Sun and Sun-planet distances.

Table 1: Comparison of $\frac{r}{R}$ in *Āryabhaṭīya* with modern values

Planet	*Āryabhaṭīya*	Modern value [5]
Mercury	0.361 to 0.387	0.387
Venus	0.712 to 0.737	0.723
Mars	0.637 to 0.662	0.656
Jupiter	0.187 to 0.200	0.192
Saturn	0.100 to 0.113	0.105

Since the *manda* correction or equation of centre for an interior planet was applied to the longitude of the mean Sun instead of the mean heliocentric longitude of the planet, the accuracy of the computed longitudes of the interior planets according to the ancient Indian planetary models would not have been as good as that achieved for the exterior planets.

2 Computation of planetary latitudes

Planetary latitudes (called *vikṣepa* in Indian astronomy) play an important role in the prediction of planetary conjunctions, occultation of stars by planets, etc. In Figure 4, P denotes the planet moving in an orbit inclined at angle i to the ecliptic, intersecting the ecliptic at the point N, the node (called *pāta* in Indian astronomy). If β is the latitude of the planet, θ_h its heliocentric longitude, and θ_0 the heliocentric longitude of the node, then for small i we have

$$\sin\beta = \sin i \ \sin(\theta_h - \theta_0). \tag{7}$$

This is also essentially the rule for calculating the latitude, as given in Indian texts, at least from the time of Āryabhaṭa.[6] For the exterior planets, it was

[5] Ratio of the mean values of Earth-Sun and planet-Sun distances for the exterior planets and the inverse ratio for the interior planets.

[6] Equation (7) actually gives the heliocentric latitude and needs to be multiplied by the ratio of the geocentric and heliocentric distances of the planet to get the geocentric latitude. This feature was implicit in the traditional planetary models.

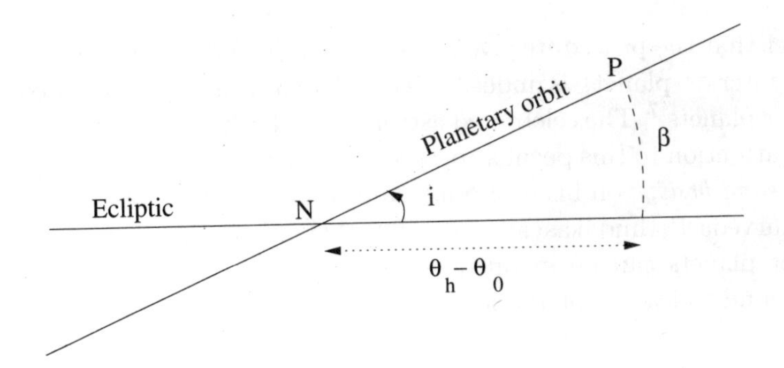

Figure 4: Heliocentric latitude of a planet.

stipulated that

$$\theta_h = \theta_{ms}, \tag{8}$$

the *manda-sphuṭa-graha*, which as we saw earlier, coincides with the heliocentric longitude of the exterior planet. The same rule applied for interior planets would not have worked, because according to the traditional Indian planetary model, the *manda*-corrected mean longitude for the interior planet has nothing to do with its true heliocentric longitude. However, all the older Indian texts on astronomy stipulated that, in the case of the interior planets, the latitude is to be calculated from (7) with

$$\theta_h = \theta_s + manda \text{ correction}, \tag{9}$$

the *manda*-corrected longitude of the *śīghrocca*. Since the longitude of the *śīghrocca* for an interior planet, as we explained above, is equal to the mean heliocentric longitude of the planet, (9) leads to the correct identification so that, even for an interior planet, θ_h in (7) becomes identical with the true heliocentric longitude.

Thus, we see that the earlier Indian astronomical texts did provide a fairly accurate theory for the planetary latitudes. But they had to live with two entirely different rules for calculating latitudes, one for the exterior planets – given by (8), where the *manda-sphuṭa-graha* appears – and an entirely different one for the interior planets given by (9), which involves the *śīghrocca* of the planet, with the *manda* correction included.

This peculiarity of the rule for calculating the latitude of an interior planet was repeatedly noticed by various Indian astronomers, at least from the time of Bhāskarācārya I (c. 629), who in his *Āryabhaṭīya-bhāṣya* drew attention to the fact that the procedure given in *Āryabhaṭīya*, for calculating the latitude of an interior planet, is indeed very different from that adopted for the exterior planets.[7] The celebrated astronomer Bhāskarācārya II (c. 1150) also draws attention to this peculiar procedure adopted for the interior planets, in his *Vāsanā-bhāṣya* on his own *Siddhāntaśiromaṇi*, and quotes the statement of Caturveda Pṛthūdakasvāmin (c. 860) that this peculiar procedure for the interior planets can be justified only on the ground that this is what has been found to lead to predictions that are in conformity with observations.[8]

3 Planetary model of Nīlakaṇṭha Somayājī

Nīlakaṇṭha Somayājī (c. 1444-1550), the renowned Kerala astronomer, appears to have been led to his important reformulation of the conventional planetary model, mainly by the fact that it seemingly employed two entirely different rules for the calculation of planetary latitudes. As he explains in his *Āryabhaṭīya-bhāṣya*,[9] the latitude arises from the deflection of the planet (from the ecliptic) and not from that of a *śīghrocca*, which is different from the planet. Therefore, he argues that what was thought of as being the *śīghrocca* of an interior planet should be identified with the mean planet itself and the *manda* correction is to be applied to this mean planet, and not to the mean Sun. This, Nīlakaṇṭha argues, would render the rule for calculation of latitudes to be the same for all planets, exterior or interior.

Nīlakaṇṭha has presented his improved planetary model for the interior planets in his treatise *Tantrasaṅgraha* which, according to Nīlakaṇṭha's pupil Śaṅkara Vāriyar, was composed in 1500 AD.[10] We shall describe here, the main features of Nīlakaṇṭha's model in so far as they differ from the conventional Indian planetary model for the interior planets.[11]

---

[7] *Āryabhaṭīya*, with the commentary of Bhāskara I and Someśvara, K. S. Shukla (ed.) New Delhi 1976, p. 32, 247.

[8] *Siddhāntaśiromaṇi* of Bhāskarācārya, with *Vāsanābhāṣya* and *Vāsanāvārttika* of *Nṛsiṃha Daivajña*, Muralīdhara Caturveda (ed.), Varanasi 1981, p. 402.

[9] *Āryabhaṭīya* with the *bhāṣya* of Nīlakaṇṭha Somayājī, *Golapāda*, S. K. Pillai (ed.), Trivandrum 1957, p. 8.

[10] *Tantrasaṅgraha* of Nīlakaṇṭha Somayājī with the commentary *Laghuvivṛtti* of Śaṅkara Vāriyar, S. K. Pillai (ed.), Trivandrum 1958, p. 2.

[11] For more details concerning Nīlakaṇṭha's model see, M. S. Sriram *et al, 500 years of*

In the first chapter of *Tantrasaṅgraha*, while presenting the mean sidereal periods of planets, Nīlakaṇṭha gives the usual values of 87.966 days and 224.702 days (which are traditionally ascribed to the *śīghrocca*-s of Mercury and Venus), but asserts that these are '*svaparyaya*-s', i.e., the mean revolution periods of the planets themselves.[12] As these are the mean heliocentric periods of these planets, the *madhyama-graha* or the mean longitude as calculated in Nīlakaṇṭha's model would be equal to the mean heliocentric longitude of the planet, for both the interior and exterior planets.

In the second chapter of *Tantrasaṅgraha*, Nīlakaṇṭha discusses the *manda* correction or the equation of centre and states[13] that this should be applied to the *madhyama-graha* as described above to obtain the *manda-sphuṭa-graha*. Thus, in Nīlakaṇṭha's model, the *manda-sphuṭa-graha* will be equal to the true heliocentric longitude for both the interior and exterior planets.

Subsequently, the *sphuṭa-graha* or the geocentric longitude is to be obtained by applying the *śīghra* correction. While Nīlakaṇṭha's formulation of the *śīghra* correction is the same as in the earlier planetary theory for the exterior planets, his formulation of the *śīghra* correction for the interior planets is different. According to Nīlakaṇṭha, the mean Sun should be taken as the *śīghrocca* for interior planets also, just as in the case of exterior planets. In Figure 5, P is the *manda*-corrected planet. E is the Earth and S the *śīghrocca* or the mean Sun. We have,

$$\begin{aligned} A\hat{E}S &= \theta_s && \text{(longitude of } \textit{śīghrocca}\text{)} \\ A\hat{S}P &= \theta_{ms} && \text{(longitude of } \textit{manda-sphuṭa}\text{)} \\ A\hat{E}P &= \theta && \text{(geocentric longitude of the planet).} \end{aligned}$$

The *śīghra-kendra* is defined in the usual way (4) as the difference between the *śīghrocca* and the *manda-sphuṭa-graha*. Then from triangle ESP, we get

---

Tantrasaṅgraha, cited earlier, p. 59-81.

[12] *Tantrasaṅgraha*, cited above, p. 8. It is surprising that, though *Tantrasaṅgraha* was published nearly fifty years ago, this crucial departure from the conventional planetary model introduced by Nīlakaṇṭha seems to have been totally overlooked in most of the studies on Indian Astronomy. For instance, Pingree in his review article on Indian Astronomy presents the mean rates of motion of Mercury and Venus given in *Tantrasaṅgraha* as the rates of motion of their *śīghrocca*-s (D. Pingree, 'History of Mathematical Astronomy in India', in *Dictionary of Scientific Biography*, Vol.XV, New York 1978, p. 622).

[13] *Tantrasaṅgraha*, cited above, p. 44-46.

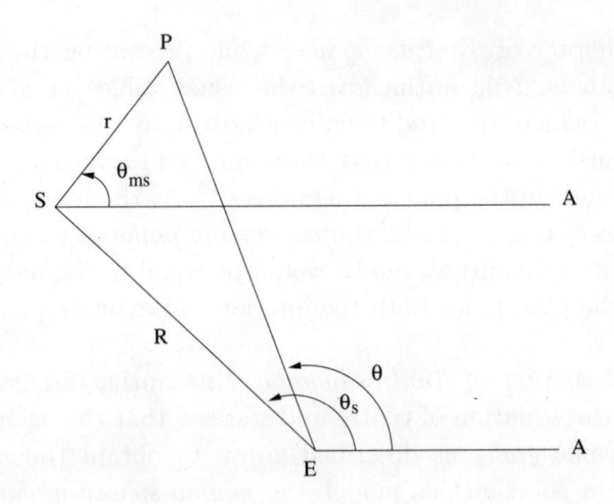

Figure 5: *Śīghra* correction for interior planets according to Nīlakaṇṭha

the relation:

$$\sin(\theta - \theta_s) = \frac{r \sin \sigma}{[(R + r \cos \sigma)^2 + r^2 \sin^2 \sigma]^{\frac{1}{2}}}, \tag{10}$$

which is the *śīghra* correction given by Nīlakaṇṭha for calculating the geocentric longitude of the planet. Comparing (10) with (6), and Figure 5 with Figure 3, we notice that they are the same except for the interchange of the *śīghrocca* and the *manda-sphuṭa-graha*. The *manda* correction or the equation of centre is now associated with P whereas it was associated with S earlier.

In the seventh chapter of *Tantrasaṅgraha*, Nīlakaṇṭha gives formula (7) for calculating the latitudes of planets,[14] and prescribes that for all planets, both exterior and interior, θ_h in (7) should be the *manda-sphuṭa-graha*. This is as it should be for, in Nīlakaṇṭha's model, the *manda-sphuṭa-graha* (the *manda*-corrected mean longitude) coincides with the true heliocentric longitude, for both the exterior and interior planets. Thus Nīlakaṇṭha, by his modification of traditional Indian planetary theory, solved the long-standing problem in Indian astronomy, of there being two different rules for calculating the planetary latitudes.

In this way, perhaps for the first time in the history of astronomy, Nīlakaṇṭha, by 1500 AD, had arrived at a consistent formulation of the equation of centre

[14] *Tantrasaṅgraha*, cited above, p. 139.

and a reasonable planetary model that is applicable also to the interior planets. As in the conventional Indian planetary model, the ancient Greek planetary model of Ptolemy and the planetary models developed in the Islamic tradition during the 8th-15th centuries also postulated that the equation of centre for an interior planet should be applied to the mean Sun, rather than to the mean heliocentric longitude of the planet as we understand today. In fact, Ptolemy seems to have compounded the confusion by clubbing together Venus along with the exterior planets and singling out Mercury as following a slightly deviant geometrical model of motion.[15] Further, while the ancient Indian astronomers successfully used the notion of the *śīghrocca* to arrive at a satisfactory theory of the latitudes of the interior planets, the Ptolemaic model is totally off the mark when it comes to the question of latitudes of these planets.[16]

Even the celebrated Copernican Revolution brought about no improvement in the planetary theory for the interior planets. As is widely known now, the Copernican model was only a reformulation of the Ptolemaic model (with some modifications borrowed from the Maragha School of Astronomy of Nasir ad-Din at-Tusi (c. 1201-74), Ibn ash-Shatir (c. 1304-75) and others) for a heliocentric frame of reference, without altering its computational scheme in any substantial way for the interior planets. As a recent study notes:

> 'Copernicus, ignorant of his own riches, took it upon himself for the most part to represent Ptolemy, not nature, to which he had nevertheless come the closest of all.' In this famous and just assessment of Copernicus, Kepler was referring to the latitude theory of Book V [of *De Revolutionibus*] , specifically to the 'librations' of the inclinations of the planes of the eccentrics, not

[15] See for example, The *The Almagest* by Ptolemy, Translated by G. J. Toomer, London 1984. For the exterior planets, the ancient Indian planetary model and the model described by Ptolemy are very similar except that, while the Indian astronomers use a variable radius epicycle, Ptolemy introduces the notion of an equant. Ptolemy adopts the same model for Venus also, and presents a slightly different model for Mercury. In both cases the equation of centre is applied to the mean Sun.

[16] As a well known historian of astronomy has remarked: "In no other part of planetary theory did the fundamental error of the Ptolemaic system cause so much difficulty as in accounting for the latitudes, and these remained the chief stumbling block up to the time of Kepler." (J. L. E. Dreyer, *A History of Astronomy from Thales to Kepler*, New York 1953, p. 200)

> in accordance with the motion of the planet, but ...the unrelated motion of the earth. This improbable connection between the inclinations of the orbital planes and the motion of the earth was the result of Copernicus's attempt to duplicate the apparent latitudes of Ptolemy's models in which the inclinations of the epicycle planes were variable. In a way this is nothing new since Copernicus was also forced to make the equation of centre of the interior planets depend upon the motion of the earth rather than the planet.[17]

Indeed, it appears that the correct rule for applying the equation of centre for an interior planet to the mean heliocentric planet (as opposed to the mean Sun), and a satisfactory theory of latitudes for the interior planets, were first formulated in the Greco-European astronomical tradition only in the early 17th century by Kepler.

4 Geometrical model of planetary motion

It is well known that the Indian astronomers were mainly interested in successful computation of the longitudes and latitudes of the Sun, Moon and the planets, and were not much worried about proposing models of the universe. The Indian astronomical texts usually present detailed computational schemes for calculating the geocentric positions of the Sun, Moon and the planets. Their exposition of planetary models, is by and large analytical and the corresponding geometrical picture of planetary motion is rarely discussed especially in the basic texts.

However, the Indian astronomers do discuss the geometrical model implied by their computations at times in the commentaries. The renowned Kerala astronomer Parameśvara of Vaṭasseri (c. 1380-1460) has discussed the geometrical model implied in the conventional planetary model of Indian astronomy. In his super-commentary *Siddhānta-dīpikā* (on Govindasvāmin's commentary) on *Mahābhāskarīya* of Bhāskarācārya-I, Parameśvara gives a detailed exposition of the geometrical picture of planetary motion as implied

[17] N. M. Swerdlow and O. Neugebauer, *Mathematical Astronomy in Copernicus' De Revolutionibus*, Part I, New York 1984, p. 483.

by the conventional model of planetary motion in Indian astronomy.[18] A shorter version of this discussion is available in his commentary *Bhaṭadīpikā* on *Āryabhaṭīya*.[19]

Following Parameśvara,[20] Nīlakaṇṭha has also discussed in detail the geometrical model of motion as implied by his revised planetary model. Nīlakaṇṭha is very much aware that the geometrical picture of planetary motion crucially depends on the computational scheme employed for calculating the planetary positions. In his *Āryabhaṭīya-bhāṣya*, Nīlakaṇṭha clearly explains that the orbits of the planets, and the various auxiliary figures such as the concentric and eccentric circles associated with the *manda* and *śīghra* processes, are to be inferred from the computational scheme for calculating the *sphuṭa-graha* (true geocentric longitude) and the *vikṣepa* (latitude of the planets).[21]

Nīlakaṇṭha's revision of the traditional computational scheme for the longitudes and latitudes of the interior planets, Mercury and Venus, was based on his clear understanding of the latitudinal motion of these planets. It is this understanding which also leads him to a correct geometrical picture of the motion of the interior planets. The best exposition of this revolutionary discovery by Nīlakaṇṭha is to be found in his *Āryabhaṭīya-bhāṣya*, which is reproduced below:

> Now he [Āryabhaṭa] explains the nature of the orbits and their locations for Mercury and Venus... In this way, for Mercury, the increase of the latitude occurs only for 22 days and then in the next 22 days the latitude comes down to zero. Thus Mercury moves on one side of the *apamaṇḍala* (the plane of the ecliptic) for 44 days and it moves on the other side during the next 44 days. Thus one complete period of the latitudinal motion is completed in 88 days only, as that is the period of revolution of the *śīghrocca* [of Mercury].

[18] *Siddhāntadīpikā* of Parameśvara on *Mahābhāskarīya-bhāṣya* of Govindasvāmin, T. S. Kuppanna Sastri (ed.), Madras 1957, p. 233-238.

[19] *Bhaṭadīpikā* of Parameśvara on *Āryabhaṭīya*, H. Kern (ed.), Laiden 1874, p. 60-1. It is surprising that this important commentary, published over 125 years ago, has not received any scholarly attention.

[20] Dāmodara the son and disciple of Parameśvara was the teacher of Nīlakaṇṭha. Nīlakaṇṭha often refers to Parameśvara as *Paramaguru*.

[21] *Āryabhaṭīya-bhāṣya* of Nīlakaṇṭha, *Kālakriyāpāda*, K. Sambasiva Sastri (ed.), Trivandrum 1931, p. 70.

> The latitudinal motion is said to be due to that of the *śīghrocca*. How is this appropriate? Isn't the latitudinal motion of a body dependent on the motion of that body only, and not because of the motion of something else? The latitudinal motion of one body cannot be obtained as being due to the motion of another body. Hence [we should conclude that] Mercury goes around its own orbit in 88 days... However this also is not appropriate because we see it going around [the Earth] in one year and not in 88 days. True, the period in which Mercury completes one full revolution around the *bhagola* (the celestial sphere) is one year only [like the Sun]...
>
> In the same way Venus also goes around its orbit in 225 days only...
>
> All this can be explained thus: The orbits of Mercury and Venus do not circumscribe the earth. The Earth is always outside their orbit. Since their orbit is always confined to one side of the [geocentric] celestial sphere, in completing one revolution they do not go around the twelve *rāśi*-s (the twelve signs).
>
> For them also really the mean Sun is the *śīghrocca*. It is only their own revolutions, which are stated to be the revolutions of the *śīghrocca* [in ancient texts such as the *Āryabhaṭīya*].
>
> It is only due to the revolution of the Sun [around the Earth] that they (i.e., the interior planets, Mercury and Venus) complete their movement around the twelve *rāśi*-s [and complete their revolution of the Earth]... Just as in the case of the exterior planets (Jupiter etc.), the *śīghrocca* (i.e., the mean Sun) attracts [and drags around] the *manda-kakṣyā-maṇḍala* (the *manda* orbits on which they move), in the same way it does for these [interior] planets also.[22]

The above passage exhibits the clinching argument employed by Nīlakaṇṭha. From the fact that the motion of the interior planets is characterised by two different periods, one for their latitudinal motion and another for their motion in longitude, Nīlakaṇṭha arrived at what may be termed a revolutionary discovery concerning the motion of the interior planets: That they go around the Sun in orbits that do not circumscribe the Earth in a period that

[22] *Āryabhaṭīya-bhāṣya* of Nīlakaṇṭha, *Golapāda*, cited above, p. 8-9.

corresponds to the period of their latitudinal motion (which is the period assigned to their *śīghrocca*-s in the traditional planetary model), and that they go around the zodiac in one year as they are dragged around the Earth by the Sun.

It was indeed well known to the ancients that the exterior planets, Mars, Jupiter and Saturn, also go around the Sun in the same mean period as they go around the Earth, as they clearly placed the geocentric orbits of these planets outside that of the Sun. Nīlakaṇṭha was the first savant in the history of astronomy to rigourously derive from his computational scheme and the observed motion of the planets, and not from any speculative or cosmological argument, that the interior planets go around the Sun in a period of their latitudinal motion. The fact that the mean period of their motion in longitude around the Earth is the same as that of the Sun is explained as being due to their being carried around the Earth by the Sun. Nīlakaṇṭha also wrote a tract called *Grahasphuṭānayane vikṣepavāsanā*, where he has set forth his latitude theory in detail. There he has given the qualitative nature of the orbits of the Sun, Moon and the five planets in a single verse, which may be cited here:

> The Moon and the planets are deflected along their *manda-kakṣyā* (*manda* orbit) from the ecliptic both to the North and the South by amounts depending on their [longitudinal] separation from their nodes. For the Moon the centre of *manda-kakṣyā* is also the centre of the ecliptic. For Mars and other planets, centre of their *manda-kakṣyā* [which is also the centre of their *manda* deferent circle], is the mean Sun that lies on the orbit of the Sun on the ecliptic.[23]

Nīlakaṇṭha presents a clear and succinct statement of the geometrical picture of the planetary motion as implied by his revised planetary model in two of his small tracts, *Siddhānta-darpaṇa* and *Golasāra*. We present the version given in *Siddhāntadarpaṇa*:

> The [eccentric] orbits on which planets move (*graha-bhramaṇa-vṛtta*) themselves move at the same rate as the apsides (*ucca-gati*)

[23] *Grahasphuṭānayane vikṣepavāsana* of Nīlakaṇṭha, in *Gaṇitayuktayaḥ*, K. V. Sarma (ed.), Hoshiarpur 1979, p. 63.

> on *manda-vṛtta* [or the *manda* epicycle drawn with its centre coinciding with the centre of the *manda* concentric]. In the case of the Sun and the Moon, the centre of the Earth is the centre of this *manda-vṛtta.*
>
> For the others [namely the planets Mercury, Venus, Mars, Jupiter and Saturn] the centre of the *manda-vṛtta* moves at the same rate as the mean Sun (*madhyārka-gati*) on the *śīghra-vṛtta* [or the *śīghra* epicycle drawn with its centre coinciding with the centre of the *śīghra* concentric]. The *śīghra-vṛtta* for these planets is not inclined with respect to the ecliptic and has the centre of the celestial sphere as its centre.
>
> In the case of Mercury and Venus, the dimension of the *śīghra-vṛtta* is taken to be that of the concentric and the dimensions [of the epicycles] mentioned are of their own orbits. The *manda-vṛtta* [and hence the *manda* epicycle of all the planets] undergoes increase and decrease in size in the same way as the *karṇa* [or the hypotenuse or the distance of the planet from the centre of the *manda* concentric].[24]

The geometrical picture described above is presented in Figures 6, 7. It is important to note that Nīlakaṇṭha has a unified model for the both the exterior and interior planets and the same is reflected in his formulation of the corresponding geometrical picture of planetary motion. Nīlakaṇṭha's description of the geometrical picture of the planetary motions involves the notions of *manda-vṛtta* and *śighra-vṛtta*, which are nothing but the *manda* and *śīghra* epicycles drawn with the centre of their concentric as the centre.

An important point to be noted is that the geometrical picture of planetary motion as discussed in *Siddhānta-darpaṇa*, deals with the orbit of each of the planets individually and does not put them together in a single geometrical model of the planetary system. Each of the exterior planets have different *śīghra-vṛtta*-s, which is in the same plane as the ecliptic, and we have to take the point where the *āditya-sūtra* (the line drawn from the centre in the direction of the mean Sun) touches each of these *śīghra-vṛtta*-s as the centre of the corresponding *manda-vṛtta.* On this *manda-vṛtta* the *mandocca* is to be located, and with that as the centre the *graha-bhramaṇa-vṛtta* or the planetary orbit is drawn with the standard radius (*trijyā* or $R \sin 90$). In the

[24] *Siddhāntadarpaṇa* of Nīlakaṇṭha, K. V. Sarma (ed.), Hoshiarpur 1976, p. 18.

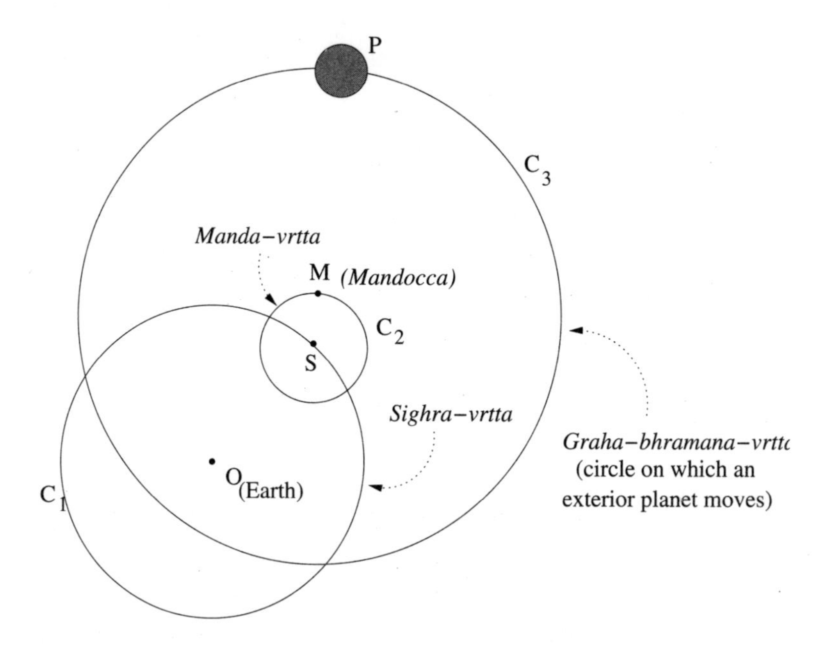

Figure 6: Nīlakaṇṭha's geometrical model for an exterior planet

case of the interior planets, *Nīlakaṇṭha* says that the *śīghra-vṛtta* has to be drawn with the standard radius (*trijyā* or $R\sin 90$) and the *graha-bhramaṇa-vṛtta* is to be drawn with the given value of the *śīghra* epicycles as the radii. In this way, we see that the two interior planets can be represented in the same diagram, as the *śīghra-vṛtta* is the same for both of them.

The integrated model involving all the planets in a single diagram adopting a single scale, that can be inferred from *Nīlakaṇṭha*'s discussions at several places, is essentially following: the five planets, Mercury, Venus, Mars, Jupiter and Saturn move in eccentric orbits (of variable radii) around the mean Sun, which goes around the Earth. The planetary orbits are tilted with respect to the orbit of the Sun or the ecliptic, and hence cause the motion in latitude. Since it is well known that the basic scale of distances are fairly accurately represented in the Indian astronomical tradition, as the ratios of the radius of the *śīghra* epicycle to the radius of the concentric *trijyā* is very nearly the mean ratio of the Earth-Sun and the Earth-planet distances (for exterior planets) or the inverse of it (for interior planets), the planetary picture will also be fairly accurate in terms of the scales of distances.

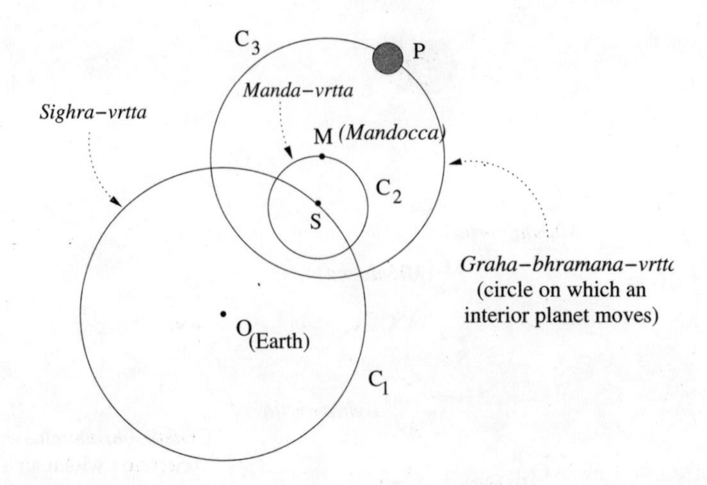

Figure 7: Nīlakaṇṭha's geometrical model for an interior planet

Nīlakaṇṭha's modification of the conventional planetary model of Indian astronomy seems to have been adopted by most of the later astronomers of the Kerala School. This is not only true of Nīlakaṇṭha's pupils and contemporaries such as Citrabhānu (c. 1530), Śaṅkara Vāriyar (c. 1500-1560) and Jyeṣṭhadeva (c. 1500-1600), but also of later astronomers such as Acyuta Piṣāraṭi (c. 1550-1621), Putumana Somayājī (c. 1660-1740) and others. Incidentally, it may be of interest to note that the well-known Oriya astronomer of 19th century, Candraśekhara Sāmanta, who was trained solely in traditional Indian astronomy, wrote a treatise *Siddhānta-darpaṇa*, in 1869, wherein he has also discussed a model of planetary motion in which the five planets, Mercury, Venus, Mars, Jupiter and Saturn, go around the Sun.[25]

[25] *Siddhāntadarpaṇa*, of Candraśekhara Sāmanta, J. C. Roy (ed.), Calcutta 1897, V.36.

ജ്യേഷ്ഠദേവകൃതമായ

ഗണിതയുക്തിഭാഷാ

അദ്ധ്യായം VIII - XV.

ഗണിതയുക്തിഭാഷാ

രണ്ടാം ഭാഗം

അദ്ധ്യായം എട്ട്

ഗ്രഹഗതിയും സ്ഫുടവും

1. ഗ്രഹങ്ങളുടെ ഗതി

[1]ഇവിടെ ഗ്രഹങ്ങളെല്ലാം ഒരു വൃത്തമാർഗ്ഗേണ ഗമിക്കും. ദിവസത്തിൽ വൃത്തത്തിന്റെ ഇത്ര അംശം ഗമിക്കുമെന്നു നിയതം താനും. അവിടെ ദിവസത്തിൽ ഇത്ര[2] യോജന ഗമിക്കുമെന്നുള്ള യോജനഗതി എല്ലാ[3] ഗ്രഹത്തിനും സമം. അവിടെ ചെറിയ വൃത്തത്തിങ്കൽ[4] ഗമിക്കുന്നവറ്റിനു കുറഞ്ഞോരു കാലം കൊണ്ടു വട്ടം കൂടും. വലിയ വൃത്തത്തിങ്കൽ[5] ഗമിക്കുന്നവറ്റിനു പെരികെ കാലം കൂടിയേ വട്ടം തികയൂ. എന്നിട്ടു ചന്ദ്രന് ഇരുപത്തെട്ടു[6] ദിവസം കൊണ്ടു പന്ത്രണ്ടു രാശിയിങ്കലും ഗമിച്ചു കൂടും മുപ്പതിറ്റാണ്ടു കൂടിയേ ശനി നടന്നുകൂടൂ[7]. വൃത്തത്തിന്റെ വലിപ്പത്തിന്നു തക്കവണ്ണം കാലത്തിന്റെ പെരുപ്പം. അതതു ഗ്രഹം തന്റെ[8] വൃത്തത്തിങ്കൽ ഒരു വട്ടം ഗമിച്ചുകൂടുന്നതിന്ന് 'ഭഗണം' എന്നു പേർ. ചതുർയുഗത്തിങ്കൽ എത്ര ആവൃത്തി ഗമിക്കും തന്റെ വൃത്തത്തിങ്കൽ അതു തന്റെ തന്റെ യുഗഭഗണമാകുന്നത്[9].

ഇവിടെ ചന്ദ്രനെ ഒരുനാൾ ഒരു നക്ഷത്രത്തോടു കൂടെ കണ്ടാൽ പിറ്റേനാൾ അതിന്റെ കിഴക്കേ നക്ഷത്രത്തോടുകൂടി കാണാം. ഇതിനെക്കൊണ്ടു

---

1. 1. C.ആരംഭം: ഹരിഃ ശ്രീഗണപതയെ നമഃ, അവിഘ്നമസ്തു.
2. B.അത്ര
3. D.എല്ലാ
4. B.വൃത്തത്തിന്മെ
5. B.വൃത്തത്തിന്മേൽ
6. B.27
7. H.കൂടു
8. G.തന്റെ
9. H.om.യുഗ

ഗതിയുണ്ടെന്നും കിഴക്കോട്ടു[10] ഗതി എന്നും കൽപ്പിക്കാം. കിഴക്കോട്ട് രാശി ക്രമമെന്നും കല്പിക്കാം. ഈ വൃത്തങ്ങൾക്ക് എല്ലാറ്റിന്നും കൂടി ഒരു പ്രദേശത്തെ ആദിയെന്നു കൽപ്പിക്കുമാറുണ്ട്. അവിടത്തിനു മേഷരാശിയുടെ ആദി എന്നുപേർ. ഈ ഗോളത്തിങ്കൽ[11] കല്പിക്കുന്ന വൃത്തങ്ങളെ എല്ലാറ്റേയും ഇരുപത്തോരായിരത്തറുനൂറു ഖണ്ഡമായി[12] വിഭജിക്കുമാറുണ്ട്. ഇതിൽ ഓരോ 'ഖണ്ഡം' ഇലിയാകുന്നത്. ഇവ വലിയ വൃത്തത്തിങ്കൽ വലുത്[13] ചെറിയ വൃത്തത്തിങ്കൽ ചെറുത്. സംഖ്യ എല്ലാറ്റിനും[14] ഒക്കും. അതതു ഗ്രഹം തന്റെ തന്റെ വൃത്തത്തിങ്കൽ ഇത്ര ഇലി ഗമിക്കും ഓരോ ദിവസം[15] എന്നു നിയതം. ഗ്രഹം ഗമിക്കുന്ന വൃത്തത്തിന്റെ കേന്ദ്രത്തിങ്കൽ ഇരുന്നു നോക്കും ദ്രഷ്ടാവ് എങ്കിൽ, നിത്യവും ഗതി ഒക്കും ഈ ഗ്രഹത്തിന് എന്നു തോന്നും. ഭൂമധ്യത്തിങ്കേന്നു[16] ഒട്ടു മേലു ഗ്രഹവൃത്തകേന്ദ്രം. ഭൂമീങ്കലു ദ്രഷ്ടാവ്. ഈ ദ്രഷ്ടാവിങ്കൽ കേന്ദ്രമായിട്ട് ഒരു വൃത്തം കൽപിപ്പൂ ഗ്രഹത്തോട് സ്പർശിക്കുമാറ്. ആ വൃത്തത്തിൽ[17] എത്ര ചെന്നിട്ടിരിക്കും[18] ഗ്രഹം അത്ര ചെന്നൂ മേഷാദിയിങ്കൽ നിന്നു ഗ്രഹം എന്നു തോന്നും ഈ[19] ദ്രഷ്ടാവിന്. ഇത് അറിയുംപ്രകാരം സ്ഫുടക്രിയയാകുന്നത്. ഇതിനെ ചൊല്ലുന്നൂ.[20] അനന്തരം വിശേഷമുള്ളതിനെ പിന്നെ ചൊല്ലുന്നുണ്ട്[21].

2. ഭഗോളം

അവിടെ[1] 'ഭഗോളമധ്യം' എന്നുണ്ടൊരു[2] പ്രദേശം. യാതൊരു പ്രദേശത്തിങ്കേന്നു സാമാന്യം നക്ഷത്രങ്ങളെല്ലാം അകലം ഒക്കുന്നൂ ആ പ്രദേശം അത്. അവിടെ ഭൂമധ്യവും ഈ ഭഗോളമധ്യവും[3] മിക്കവാറുമൊന്നേ എന്നു തോന്നും[4]. വിശേഷമുള്ളതിനെ പിന്നെ ചൊല്ലുന്നുണ്ട്.

1. 10. B.കിഴക്കോട്ടേക്കു
11. F. ഈ ഗോളത്തിൽ
12. C. ഖണ്ഡമാക്കി
13. B. വലിയ ഇലികൾ; C.G. വലുത് ഇലികൾ
14. B.C.D എല്ലായിങ്കലും
15. H. om. ഓരോ ദിവസം എന്ന്
16. F. ഇതുമുതൽ "ഇവിടെ ഉച്ചനീചവൃത്തം" എന്നു തുടങ്ങുന്ന ഭാഗം വരെ f-ൽ കാണാനില്ല.
17. D.G വൃത്തത്തിങ്കൽ
18. B യെന്നിരിക്കും
19. D.G. om. ഈ
20. B.C.G ചൊല്ലുന്നുണ്ട്
21. D. om. അനന്തരം (to) ചൊല്ലുന്നുണ്ട്

2. 1. G reads അനന്തരം ഇവിടെ
2. H. എന്നൊരു
3. H. ഈ ഭഗോളമദ്ധ്യവും
4. D. മിക്കതും ഒന്നേതാനും; G. om. തോന്നും

3. ഗ്രഹങ്ങളുടെ മധ്യഗതി – ഒന്നാം പ്രകാരം

അവിടെ ആദിത്യചന്ദ്രൻമാരുടെ സ്ഫുടത്തെ നടേ ചൊല്ലുന്നു, എളുപ്പമുണ്ടതിന് എന്നിട്ട്. ഇവിടെ ഭൂഗോളമധ്യം കേന്ദ്രമായിട്ട് ഒരു വൃത്തത്തെ കൽപിക്കൂ[1]. ഇത് ഗ്രഹം ഗമിക്കുന്ന വൃത്തത്തേക്കാൾ പെരികെ ചെറുത്. ഈ വൃത്തത്തിന്റെ നേമിയിങ്കൽ കേന്ദ്രമായിട്ടുള്ള 'ഗ്രഹഭ്രമണവൃത്തം'. ഈ ചെറിയ വൃത്തത്തിന് 'മന്ദോച്ചനീചവൃത്ത'മെന്നു പേർ. ഗ്രഹഭ്രമണവൃത്തത്തിന് 'പ്രതിമണ്ഡല'മെന്നു പേർ. പ്രതിമണ്ഡലകേന്ദ്രം ഉച്ചനീചവൃത്തത്തിന്മേൽ ഗമിക്കും. മന്ദോച്ചത്തിന്റെ ഗതി ഇതിനു ഗതിയാകുന്നത്[2]. പ്രതിമണ്ഡലത്തിന്മേൽ ഗ്രഹത്തിന്റെ ഗതി 'മധ്യമഗതി' ആകുന്നത്[3]. വൃത്തങ്ങൾ കേന്ദ്രത്തോടും നേമിയോടും ഇടയിൽ പഴുതുകൂടാതെ തൂർന്നിരിക്കുമാറും കൽപിക്കണം.

ഇവിടെ സ്ഫുടന്യായത്തിങ്കൽ ഒരു 'വൃത്തനേമി'യിങ്കൽ ഒരു വൃത്തത്തിന്റെ കേന്ദ്രം ഭ്രമിക്കുമാറ് കൽപിക്കുമ്പോൾ ഈ ഭ്രമിക്കുന്ന വൃത്തത്തിന്റെ 'പൂർവ്വാപരരേഖ' എല്ലായ്പോഴും കിഴക്കുപടിഞ്ഞാറായിട്ടുതന്നെ ഇരിക്കണം. ഇതിനു വിപരീതമായിരിക്കുന്ന ദക്ഷിണോത്തരരേഖ താൻ ഊർദ്ധ്വാധോരേഖ താൻ ഒന്ന്[4]. അത് സദാ അവ്വണ്ണം തന്നെ ഇരിക്കണം. ദിഗ്ഭേദം വരൊല്ലാ. അവ്വണ്ണം വേണം ഭ്രമണത്തെ കൽപിക്കാൻ[5]. അവ്വണ്ണമാകുമ്പോൾ ഈ വൃത്തകേന്ദ്രം എത്ര വലിയൊരു വൃത്തത്തിന്മേൽ ഭ്രമിക്കുന്നൂ, ഈ ഭ്രമിക്കുന്ന വൃത്തത്തിന്റ എല്ലാ അവയവവും അത്ര വലിയൊരു വൃത്തത്തിന്മേൽ ഭ്രമിക്കുന്നൂ എന്നു വരും. കേന്ദ്രഭ്രമണത്തിന് ഒരാവൃത്തി കഴിയുമ്പോൾ അവയവാന്തരങ്ങൾ എല്ലാറ്റിനും ഒരു ഭ്രമണം കഴിഞ്ഞുകൂടും. ഇവിടെ വൃത്തനേമിയിങ്കൽ ഇരിക്കുന്ന ഗ്രഹത്തിനു തനിക്കു ഗതി ഇല്ലാതിരിക്കുന്നതാകിലും തനിക്ക്[6] ആധാരമായിരിക്കുന്ന വൃത്തത്തിന്റ കേന്ദ്രത്തിനു[7] തക്കവണ്ണം, താനിരിക്കുന്ന നേമിപ്രദേശം യാതൊരു വൃത്തപ്രദേശത്തിങ്കൽ ഭ്രമിക്കുന്നൂ, ആ ഗ്രഹവും

---

3. 1. G. കല്പ്പിക്കൂ
2. G. om. പ്രതിമണ്ഡല....... ആകുന്നത്
3. B. ആകുന്നതും
4. G. om. ഒന്ന്
5. G. കല്പ്പിപ്പാൻ
6. B. സ്വാധാര
7. G. ഭ്രമണത്തിനു

ആ വൃത്തപ്രദേശത്തിങ്കൽ ആ മന്ദോച്ചത്തിന്റെ ഗതിയായിട്ടു ഗമിക്കുന്നു എന്ന് ഫലം കൊണ്ടു വന്നിരിക്കും. വാഹനത്തിന്മേലേ ഗമിക്കുന്നവരുടെ ഗതിപോലെ.

എന്നാൽ പ്രതിമണ്ഡലകേന്ദ്രഗതിക്ക് അധീനമായിരിക്കുന്ന ഗ്രഹഗതി ഇത്. ഇങ്ങനെ അർക്കചന്ദ്രന്മാർക്ക് ഭഗോളമധ്യം മദ്ധ്യമായിട്ട് ഒരു മന്ദനീചോച്ച വൃത്തമുണ്ട്. ഇതിന്റെ നേമിയിങ്കൽ കേന്ദ്രമായിട്ട് ഒരു ഗ്രഹഭ്രമണവൃത്തമുണ്ട്. ഈ ഗ്രഹഭ്രമണവൃത്തത്തിന്റെ കേന്ദ്രം മന്ദോച്ചഗതിക്കു തക്കവണ്ണം ഈ മന്ദോച്ചനേമിയിങ്കൽ ഗമിക്കും. പിന്നെ ഈ ഗ്രഹവൃത്തത്തിങ്കൽ മധ്യഗതി ക്കുതക്കവണ്ണം ഗ്രഹവും ഗമിക്കും. ഇങ്ങനെ ഗ്രഹത്തിന്റേയും ഗ്രഹഭ്രമണ വൃത്തത്തിന്റേയും ഗതിപ്രകാരം കല്പിക്കണം. ഈവണ്ണം വസ്തുസ്ഥിതി.

4. ഗ്രഹങ്ങളുടെ മധ്യഗതി – രണ്ടാം പ്രകാരം

പിന്നെ മറ്റൊരു പ്രകാരം കല്പിച്ചാലും ഫലസാമ്യമുണ്ട്. അവിടെ ഭഗോളമധ്യം കേന്ദ്രമായിട്ട് ഗ്രഹഭ്രമണവൃത്തത്തോളം പോന്നൊരു[1] വൃത്തത്തെ കൽപിപ്പൂ. ഇതിനു 'കക്ഷ്യാവൃത്ത'മെന്നു പേർ. ഇതിന്റെ നേമീങ്കൽ കേന്ദ്രമായിട്ട് ഒരു ഉച്ചനീചവൃത്തത്തെ കൽപിപ്പൂ. മുമ്പു ചൊല്ലി യതിനോളം ഉച്ചനീചവൃത്തത്തിന്റെ വലിപ്പം. ഇക്കക്ഷ്യാവൃത്തനേമീങ്കൽ ഇക്ക ല്പിച്ച ഉച്ചനീചവൃത്തകേന്ദ്രം[2] ഗ്രഹമധ്യത്തിന്റെ ഗതിയോളം ഗതിയായിട്ട് ഗമിക്കും. ഈ ഉച്ചനീചവൃത്തനേമീങ്കൽ മന്ദോച്ചത്തിന്റെ ഗതിയോളം ഗതി യായിട്ടു ഗ്രഹവും ഗമിക്കും. ഇവിടെ ഉച്ചനീചവൃത്തം ഗ്രഹഭ്രമണത്തിന് ആധാരമാകുന്നത്. എന്നിട്ടു മുമ്പിൽ പ്രതിമണ്ഡലവൃത്തത്തിങ്കൽ ഗ്രഹ ത്തിനു ചൊല്ലിയ ഗതി ഇപ്പോൾ ഉച്ചനീചവൃത്തകേന്ദ്രത്തിനു കൽപിപ്പൂ. മുമ്പിൽ പ്രതിമണ്ഡലകേന്ദ്രത്തിനു ചൊല്ലിയ ഗതി കക്ഷ്യാഭ്രമണവൃത്തനേ മിയിങ്കൽ കേന്ദ്രമായി കല്പിച്ച് ഉച്ചനീചവൃത്തത്തിന്റെ നേമീങ്കലേ ഗ്രഹ ത്തിന്റെ ഗതിയായിട്ടു കല്പിപ്പൂ. എന്നാലും ഫലസാമ്യം വരും. ഇവിടെ പ്രതിമണ്ഡലത്തോളം വലിയൊരു കക്ഷ്യാവൃത്തത്തിന്റെ[3] നേമിയിങ്കൽ ഉച്ച

4\. 1. C. പോന്നിട്ട്
2. G. വൃത്തത്തിന്റെ കേന്ദ്രം
3. G. adds ഈ

നീചവൃത്തത്തിന്റെ കേന്ദ്രം ഗമിക്കുമ്പോൾ. ഈ നീചോച്ചവൃത്തത്തിന്റെ എല്ലാ അവയവവും അക്കക്ഷ്യാവൃത്തത്തോളം പോന്നൊരു വൃത്തത്തിന്മേൽ ഗമിക്കും. എന്നാൽ ഉച്ചനീചവൃത്തനേമീങ്കലേ ഗ്രഹവും തനിക്ക് ആധാരമായിരിക്കുന്ന[4] വൃത്തഭ്രമണം കൊണ്ടുതന്നെ അത്ര പോന്നൊരു പ്രതിമണ്ഡലവൃത്തത്തിന്മേൽ ഭ്രമിക്കുന്നൂ എന്നു ഫലിച്ചിരിക്കും. ഇവിടെ ഉച്ചനീചവൃത്തത്തിന്റെ കേന്ദ്രഭ്രമണത്തിന് ആധാരമായിട്ടിരിക്കുന്ന കക്ഷ്യാമണ്ഡലത്തിനു യാതൊരിടത്തു കേന്ദ്രം, ഇവിടുന്ന് ഉച്ചനീചവ്യാസാർദ്ധത്തോളം അകന്നേടത്തു കേന്ദ്രമായിട്ടുള്ളൂ. ഉച്ചനീചവൃത്തത്തിന്റെ നേമിഭ്രമണത്തിന് ആധാരമായിട്ടിരിക്കുന്ന പ്രതിമണ്ഡലത്തിന്റെ കേന്ദ്രം.

ഈ സ്ഫുടപ്രകരണത്തിൽ കല്പിക്കുന്ന വൃത്തഭ്രമണത്തിങ്കൽ എല്ലാടവും ഭ്രമിക്കുന്ന വൃത്തത്തിന്റെ ദിഗ്രേഖയ്ക്ക് ദിഗ്ഭേദം വരാതെ ഇരിക്കുമാറ് ഭ്രമണം കല്പിക്കുന്നു. എന്നിട്ട് ഈ ഭ്രമണത്തിനു വൃത്തത്തിന്റെ കേന്ദ്രം എത്ര വലിയൊരു വൃത്തത്തിങ്കൽ ഭ്രമിക്കുന്നൂ, മറ്റ് എല്ലാ അവയവവും അത്ര വലിയൊരു വൃത്തത്തിന്മേൽ ഭ്രമിക്കും എന്ന് നിയതമായിരിക്കുന്നു. എന്നാൽ കക്ഷ്യാവൃത്തനേമീങ്കലേ നീചോച്ചവൃത്തകേന്ദ്രത്തിനുതാൻ ഇതിന്റെ നേമിഭ്രമണത്തിനാധാരമായിട്ടിരിക്കുന്ന പ്രതിമണ്ഡലത്തിന്മേൽ ഗ്രഹത്തിനുതാൻ മധ്യഗതി കല്പിക്കാം, രണ്ടു പ്രകാരവും ഫലസാമ്യം ഉണ്ടാകയാൽ. ഇവിടെ ഭഗോളമദ്ധ്യത്തിങ്കൽ കേന്ദ്രമായിട്ടിരിക്കുന്ന കക്ഷ്യാമണ്ഡലവും ഇതിന്റെ നേമീങ്കലേ ഉച്ചനീചവൃത്തവും, ഇവ രണ്ടേ മതി സ്ഫുടയുക്തി നിരൂപിപ്പാൻ എന്നാകിലും ഇച്ചൊല്ലിയ നാലുവൃത്തങ്ങളും കൂടി കല്പിക്കുമാം.

5. ചന്ദ്രതുംഗന്റെ സ്ഥാനം

ഈവണ്ണമാകുമ്പോൾ ചന്ദ്രന് ഇഷ്ടകാലത്തിങ്കലേക്കു ത്രൈരാശികം കൊണ്ടു വരുത്തിയ "തുംഗൻ" എന്നു പേരാകുന്ന ഉച്ചം മേഷാദിയിങ്കേന്നു തുടങ്ങീട്ടു എത്ര ചേർന്നിരിക്കുന്നു ഭഗോളമദ്ധ്യം കേന്ദ്രമായിരിക്കുന്ന ഉച്ചനീചവൃത്തത്തിങ്കൽ[1] കേന്ദ്രമായിട്ടു പ്രതിമണ്ഡലത്തെ കല്പിപ്പൂ. പ്രതിമണ്ഡലനേമീങ്കൽ ഗ്രഹത്തേയും കല്പിപ്പൂ. ത്രൈരാശികം കൊണ്ടു വന്ന മധ്യമം

4. 4. C. E. G മായിട്ട് ഇരിക്കുന്ന
5. 1. G. adds ആ പ്രദേശത്തിങ്കൽ

യാതൊരു പ്രദേശത്ത്, അവിടെ കല്പിക്കുന്നു ഗ്രഹത്തെ. പിന്നെ കക്ഷ്യാവൃത്തനേമീങ്കൽ ഉച്ചനീചവൃത്തകേന്ദ്രത്തേയും തല്ക്കാലമധ്യമം ചെന്നിരിക്കുന്നു യാതൊരിടത്ത്, അവിടെ കല്പിപ്പൂ. പിന്നെ ഉച്ചനീചവൃത്തനേമീങ്കൽ തല്ക്കാലതുംഗൻ യാതൊരിടത്ത്, അവിടെ ഗ്രഹത്തേയും കല്പിപ്പൂ. ഈവണ്ണം കല്പിക്കുമ്പോൾ കക്ഷ്യാവൃത്തനേമീങ്കലേ ഉച്ചനീചവൃത്തനേമിയും പ്രതിമണ്ഡലനേമിയും യാതൊരിടത്തു തങ്ങളിൽ ഉച്ചാസന്നമായിരിക്കുന്ന പ്രദേശത്തിങ്കൽ സ്പർശിക്കുന്നൂ അവിടെ ഗ്രഹത്തിന്റെ സ്ഥിതി. ഈ വൃത്തനേമികൾക്ക് രണ്ടു[2] പ്രദേശത്തിങ്കൽ സംപാതമുണ്ട്. അവിടെ ഉച്ചപ്രദേശത്തിങ്കലെ നേമീസംപാതത്തിങ്കൽ ഗ്രഹത്തിന്റെ സ്ഥിതി സംഭവിച്ചിരിക്കും[3].

6. ഉച്ചമധ്യമാന്തരവും സ്ഫുടമധ്യമാന്തരവും

ഇവിടെ യാതൊരിക്കൽ ത്രൈരാശികാനീതമായിരിക്കുന്ന ഉച്ചവും മധ്യവും തുല്യമായിട്ടിരിക്കുന്നു, അപ്പോൾ ഒരു സൂത്രത്തിങ്കലേ ഇരിക്കും നാലു വൃത്തങ്ങളുടേയും കേന്ദ്രങ്ങൾ. ഇവ യാതൊരിക്കൽ പൂർവ്വസൂത്രത്തിങ്കൽ സംഭവിക്കുന്നൂ, അവിടം ആദിയായി വൃത്തഭ്രമണത്തേയും ഗ്രഹഭ്രമണത്തേയും കൂടെ നിരൂപിച്ചിട്ട് ഉച്ചമദ്ധ്യമാന്തരത്തെ കല്പിക്കും പ്രകാരത്തെ ചൊല്ലുന്നു.

അവിടെ കക്ഷ്യാമണ്ഡലകേന്ദ്രവും ഉച്ചനീചകേന്ദ്രവും ഭഗോളമദ്ധ്യത്തിങ്കൽ തന്നെ കല്പിപ്പൂ. പിന്നെ ഈ ഉച്ചനീചവൃത്തത്തിന്റെ പൂർവ്വസൂത്രാഗ്രത്തിങ്കൽ പ്രതിമണ്ഡലകേന്ദ്രത്തെ. ഈ പൂർവ്വസൂത്രത്തിങ്കൽ തന്നെ കക്ഷ്യാവൃത്തനേമീങ്കൽ കേന്ദ്രമായിട്ടു മറ്റൊരു ഉച്ചനീചവൃത്തത്തേയും കല്പിപ്പൂ. ഇതിന്റെ പൂർവ്വസൂത്രാഗ്രവും പ്രതിമണ്ഡലത്തിന്റെ പൂർവ്വസൂത്രാഗ്രവും തങ്ങളിൽ സ്പർശിച്ചിരിക്കും[1]. ഉച്ചനീചവൃത്തത്തിന്റേയും പ്രതിമണ്ഡലത്തിന്റേയും നേമീസ്പർശം പൂർവ്വസൂത്രാഗ്രത്തിങ്കൽ തന്നെ ആകയാൽ ഗ്രഹവും പൂർവ്വസൂത്രാഗ്രത്തിങ്കൽ തന്നെ ഇരിക്കും. ഇന്നേരത്തു കക്ഷ്യാമണ്ഡലകേന്ദ്രത്തിങ്കേന്നും പ്രതിമണ്ഡലകേന്ദ്രത്തിങ്കേന്നും തുടങ്ങി ഗ്രഹത്തെ സ്പർശിക്കുന്ന

5. 2. H. ഒരു; B. om രണ്ടു
3. B. G. സംഭവിക്കും
6. 1. G. സ്പർശിക്കും

സൂത്രം ഒന്നേ ആകയാൽ ഗ്രഹത്തിന്റെ സ്ഫുടമദ്ധ്യമങ്ങൾക്കു ഭേദമില്ലാ. പിന്നെ മദ്ധ്യത്തിന്റെ ഉച്ചയോഗത്തിങ്കേന്നു തുടങ്ങിയിട്ട് സ്ഫുടമദ്ധ്യമങ്ങളുടെ ഭേദമുണ്ടാകുന്നു.

7. സൂര്യസ്ഫുടവും സ്ഫുടമദ്ധ്യാന്തരാളവും

ഇവിടെ ആദിത്യന്റെ സ്ഫുടത്തെ നിരൂപിക്കേണ്ടൂ. നടേ അവിടെ പ്രതിമണ്ഡലകേന്ദ്രത്തിന്റെ ഗതി അതിമന്ദമാകയാൽ ഇല്ല എന്നപോലെ കല്പിച്ചിട്ടു നിരൂപിക്കാം. എന്നാൽ ഗ്രഹത്തിന് ഒന്നിനേയെല്ലോ ഗതി കല്പിക്കേണ്ടു എന്ന് ഒരു എളുപ്പമുണ്ട്. ഇങ്ങനെ നടേത്തെ പക്ഷം. രണ്ടാം പക്ഷത്തിൽ പിന്നെ കക്ഷ്യാവൃത്തത്തിന്റെ നേമീങ്കലെ ഉച്ചനീചവൃത്തകേന്ദ്രത്തിനേ ഗതിയുള്ളു എന്നു കല്പിപ്പൂ. എന്നാലും[1] ഫലസാമ്യമുണ്ട്. രണ്ടു പ്രകാരമുള്ള ഗതി കൂടി ഒരിക്കലെ നിരൂപിക്കേണ്ടൂ എന്ന് എളുപ്പമാകുന്നത്. അനന്തരം ഉച്ചയോഗത്തിങ്കേന്നു മധ്യമം മൂന്നു രാശി ചെല്ലുമ്പോൾ കക്ഷ്യാനേമീങ്കലു[2] ഉച്ചനീചവൃത്തകേന്ദ്രം. ഈ ഉച്ചനീചവൃത്തത്തിന്റെ പൂർവ്വസൂത്രാഗ്രം പ്രതിമണ്ഡലത്തിന്റെ ദക്ഷിണോത്തരസൂത്രാഗ്രത്തെ സ്പർശിച്ചിട്ടുമിരിക്കും. അവിടത്ത് അന്നേരത്തെ ഗ്രഹം. ഇവിടെ പ്രതിമണ്ഡലനേമീങ്കൽ[3] ഗ്രഹത്തിന്നും കക്ഷ്യാനേമീങ്കൽ ഉച്ചനീചവൃത്തകേന്ദ്രത്തിന്നും[4] തുല്യമായിട്ട് ഇരിപ്പൊന്നു ഗതി. ഒരിക്കലേ ഒരു ദിക്കിൽ തന്നെ തുടങ്ങി സമങ്ങളായിരിക്കുന്ന രണ്ടു വൃത്തങ്ങളിൽ സമമായി ഗമിക്കുന്നവ രണ്ടും താൻ ഗമിക്കുന്ന വൃത്തത്തിങ്കൽ തുല്യങ്ങളായിരിക്കുന്ന[5] അംശങ്ങളെക്കൊണ്ട് ഗമിച്ചിരിക്കും. എന്നിട്ടു തന്റെ തന്റെ വൃത്തത്തിങ്കൽ നാലൊന്നു ഗമിച്ചിരിക്കുമ്പോൾ ഗ്രഹവും ഉച്ചനീചവൃത്തകേന്ദ്രവും അതതിങ്കൽ ഉത്തരസൂത്രാഗ്രത്തിങ്കൽ ഇരിക്കും. ഇവിടെ കക്ഷ്യാപ്രതിമണ്ഡലങ്ങൾ രണ്ടിന്നും കൂടിയുള്ള പൂർവ്വസൂത്രത്തിന്ന് 'ഉച്ചനീചസൂത്ര'മെന്നു പേർ, ഭഗോളമധ്യത്തിങ്കേന്നു പ്രതിമണ്ഡലനേമിയിങ്കൽ എല്ലായിലുമകന്ന പ്രദേശത്തിങ്കലും അണഞ്ഞ പ്രദേശത്തിങ്കലും

7\. 1. C. എന്നാകിലും
2. H. കക്ഷ്യാനേമീങ്കലേ ഉത്തരസൂത്രഭാഗം പ്രതിമണ്ഡലത്തിന്റെ ഉത്തരസൂത്രാഗ്രത്തെ സ്പർശിച്ചിട്ടുമിരിക്കും
3. D. നേമീങ്കലെ
4. D. വൃത്തത്തിനും
5. B.G. തുല്യങ്ങളാകുന്ന

സ്പർശിച്ചിരിക്കയാൽ. ഇവിടെ പ്രതിമണ്ഡലത്തിന്റെ പൂർവ്വസൂത്രാഗ്രത്തി ങ്കേന്ന് ഇതിന്മേൽ മൂന്നു രാശി ചെന്നതു "മധ്യമ"മാകുന്നത്. ഭഗോളമധ്യം കേന്ദ്രമായിട്ടു ഗ്രഹത്തെ സ്പർശിക്കുന്ന സൂത്രം കൊണ്ടു വൃത്തം വീശി[6] ആ വൃത്തത്തിങ്കൽ എത്രചെന്നു അതു "സ്ഫുട"മാകുന്നത്[7]. ഇവിടെ കക്ഷ്യാവൃത്തത്തിന്റെ ഉത്തരസൂത്രാഗ്രത്തിൽ ഗ്രഹമിരിക്കുമ്പോൾ സ്ഫുടം ഉച്ചത്തിങ്കേന്നു മൂന്നു രാശി ചെന്നിരിക്കും. എന്നാൽ മധ്യമം മൂന്നു രാശി ചെല്ലുമ്പോൾ കക്ഷ്യാവൃത്തത്തിന്റെ ഉത്തരസൂത്രാഗ്രത്തിങ്കേന്ന് ഉച്ചനീച വ്യാസാർദ്ധത്തോളം കിഴക്കു ഗ്രഹം. എന്നിട്ട് ഉച്ചനീചവ്യാസാർദ്ധം അന്നേ രത്തു മധ്യമസ്ഫുടാന്തരമാകുന്നത്. എന്നിട്ട് മൂന്നു രാശി തികയുന്നേടത്തീന്ന് ഉച്ചനീചവ്യാസാർദ്ധത്തോളം കുറയും സ്ഫുടം.

ഇവിടെ ഭഗോളമധ്യത്തിങ്കൽ കേന്ദ്രമായിട്ട് ഗ്രഹത്തോളമുള്ള സൂത്രം വ്യാസാർദ്ധമായിട്ടുള്ള വൃത്തത്തിന്നു 'കർണ്ണവൃത്ത' മെന്നു പേർ. ഇതിനും കക്ഷ്യാമണ്ഡലത്തിനും കേന്ദ്രം ഒരിടത്താകയാൽ ഇലികൾ രണ്ടിങ്കലും ഒന്നേ എന്നിട്ടു കക്ഷ്യാവൃത്തത്തിന്റെ ഉത്തരസൂത്രാഗ്രത്തിങ്കലേ ഉച്ചനീചവൃത്ത കേന്ദ്രം മധ്യമഗ്രഹം എന്നു കല്പിച്ചിരിക്കുന്നതിനെ കർണ്ണവൃത്തത്തിന്റെ ഉത്തരസൂത്രാഗ്രത്തിങ്കൽ കല്പിച്ചിട്ട് അവിടുന്ന് ഗ്രഹത്തോളമുള്ള അന്ത രാളം "സ്ഫുടമധ്യമാന്തരാളചാപം" എന്നിരിക്കും. ആകയാൽ ഉച്ചനീചവ്യാ സാർദ്ധത്തെ കർണ്ണവൃത്തത്തിങ്കലെ ജ്യാവ് എന്നു കല്പിച്ചു ചാപിച്ചാൽ ഉണ്ടാകും സ്ഫുടമദ്ധ്യമാന്തരാളചാപം. ഇതിനെ ഉച്ചരേഖയാകുന്ന പൂർവ്വ സൂത്രത്തിങ്കേന്നു ചെന്നു മധ്യമം. മൂന്നുരാശി അതിങ്കേന്നു കളഞ്ഞാൽ കർണ്ണ വൃത്തത്തിങ്കൽ ഗ്രഹത്തിങ്കേന്നു ഉച്ചസൂത്രത്തോടുള്ള അന്തരാളം ശേഷി ക്കും. അതിൽ ഉച്ചത്തെക്കൂട്ടിയാൽ മേഷാദിഗ്രഹസ്ഫുടം വരും. പിന്നെ മദ്ധ്യമത്തിങ്കേന്നു തന്നെ സ്ഫുടമധ്യമാന്തരാളമാകുന്ന കർണ്ണവൃത്തത്തി ങ്കലേ ചാപഭാഗത്തെ കളഞ്ഞാലും കർണ്ണവൃത്തത്തിങ്കൽ ഇത്രചെന്നു ഗ്രഹ സ്ഫുടം എന്ന് ഉണ്ടാകും. ഇങ്ങിനെ പൂർവ്വസൂത്രത്തിങ്കൽ ഉച്ചവും ഉച്ചത്തി ങ്കൽ മധ്യമവും എന്നു കല്പിക്കുമ്പോൾ പ്രതിമണ്ഡലത്തിന്റെ പൂർവ്വസൂ ത്രാഗ്രത്തിങ്കൽ ഗ്രഹവും ഉച്ചനീചവൃത്തകേന്ദ്രവും എന്നിരിക്കുമ്പോൾ കക്ഷ്യാപ്രതിമണ്ഡലങ്ങളിൽ രണ്ടിങ്കലുമൊന്നേ പൂർവ്വസൂത്രം. എന്നിട്ട് ആ ഇലി ഒന്നേ ആയിട്ടിരിക്കും രണ്ടിങ്കലും അന്നേരത്ത്. എന്നിട്ട് സ്ഫുടമധ്യ

7. 6. D. വിയി; C.E.F വീയിയ
7. G. adds എന്നിട്ട് മൂന്നു രാശിചെന്നേടത്ത്

ങ്ങൾ ഒന്നു തന്നെ അന്നേരത്ത്. പിന്നെ തുല്യഗതികളായിരിക്കുന്ന ഗ്രഹവും ഉച്ചനീചവൃത്തകേന്ദ്രവും മൂന്നു രാശി ഗമിക്കുമ്പോൾ പ്രതിമണ്ഡലോത്തര സൂത്രാഗ്രത്തിങ്കൽ ഗ്രഹം. കക്ഷ്യാവൃത്തത്തിന്റെ ഉത്തരസൂത്രാഗ്രത്തിങ്കൽ ഉച്ചനീചവൃത്തകേന്ദ്രവും. ഈവണ്ണം[8] ഉച്ചനീചവൃത്തകേന്ദ്രം ഗമിക്കുമ്പോൾ ദിഗ്ഭേദം വരാതെയിരിക്കുമാറ് കല്പിക്കുമ്പോൾ ഉച്ചനീചവൃത്തത്തിന്റെ പൂർവ്വ സൂത്രാഗ്രത്തിങ്കേന്നു ഗ്രഹം ഒരിക്കലും വേർപെടുകയില്ല. എന്നിട്ടു സ്ഫുട മധ്യമാന്തരാളം ഉച്ചനീചവ്യാസാർദ്ധമായിട്ടിരിക്കും. അപ്പോൾ പിന്നെ അവി ടുന്ന് ഒരു വൃത്തപാദം ഗമിക്കുമ്പോൾ പശ്ചിമസൂത്രാഗ്രത്തിങ്കൽ പ്രതിമ ണ്ഡലത്തിങ്കൽ ഗ്രഹവും കക്ഷ്യാവൃത്തത്തിങ്കൽ ഉച്ചനീചവൃത്തകേന്ദ്രവു മായിട്ടിരിക്കും. അപ്പോളും ഉച്ചനീചവൃത്തത്തിന്റെ പൂർവ്വസൂത്രാഗ്രത്തിങ്കല് ഗ്രഹം എന്നിരിക്കും. ഇവിടെ കക്ഷ്യയിങ്കേന്നും പ്രതിമണ്ഡലത്തിങ്കേന്നു മുള്ള പശ്ചിമസൂത്രം ഒന്നേയാകയാൽ ആ പ്രദേശത്തിങ്കൽ രണ്ടിന്റേയും ഇലി ഒന്നേ ആകയാൽ സ്ഫുടമധ്യമങ്ങൾ ഒന്നേ അന്നേരത്തും. ഇങ്ങനെ നീചവും മധ്യമവും സമം ആകുമ്പോളും സ്ഫുടമധ്യമാന്തരാളമില്ല. പിന്നെ ഇവിടന്ന് ഒരു വൃത്തപാദം ഗമിക്കുമ്പോൾ ദക്ഷിണസൂത്രാഗ്രത്തിങ്കൽ രണ്ടും. ഇവിടേയും ഉച്ചനീചവൃത്തത്തിന്റെ പൂർവ്വസൂത്രാഗ്രത്തിങ്കൽ ഗ്രഹം. എന്നിട്ടു ഗ്രഹത്തിങ്കേന്നു ഉച്ചനീചവ്യാസാർദ്ധത്തോളം പടിഞ്ഞാറു ഉച്ചനീചവൃത്ത ത്തിന്റെ കേന്ദ്രം. എന്നിട്ട് ഇവിടെ ഉച്ചനീചവ്യസാർദ്ധചാപം കൂട്ടണം. മധ്യ ത്തിങ്കൽ അതു സ്ഫുടമാകുന്നത്. പിന്നെയും മൂന്നുരാശി ഗമിച്ചിട്ട് ഉച്ചത്തി ങ്കൽ[9] ചെല്ലുമ്പോൾ സ്ഫുടമദ്ധ്യമാന്തരമില്ല. എന്നിങ്ങനെ ഉച്ചയോഗത്തി ങ്കേന്നു തുടങ്ങീട്ട് മദ്ധ്യമത്തിന്റെ പദത്തിന്നു തക്കവണ്ണം സ്ഫുടമധ്യമാന്തര ങ്ങളുടെ വൃദ്ധിഹ്രാസങ്ങൾ തികയുന്നു എന്നു വന്നു. എന്നാൽ പ്രതിമണ്ഡ ലത്തിങ്കലെ ഉച്ചമധ്യമാന്തരാളഭുജാജ്യാവിനെ ത്രൈരാശികം ചെയ്ത് ഉച്ച നീചവൃത്തത്തിങ്കലാക്കിയതു സ്ഫുടമദ്ധ്യമാന്തരജ്യാവായിട്ടിരിക്കും എന്നു വിശേഷം.

അത് എങ്ങനെയെങ്കിൽ അവിടെ കക്ഷ്യാവൃത്തകേന്ദ്രത്തിങ്കേന്നു അതിന്റെ നേമീങ്കലെ ഉച്ചനീചവൃത്തകേന്ദ്രത്തൂടെ അതിന്റെ പുറത്തെ നേമീങ്കലോളം ചെല്ലുന്ന സൂത്രം യാതൊന്ന് ഇത് മധ്യമഗ്രഹത്തിങ്കലെ ഇലി ആകുന്നത്.

---

7. 8. H. ഈവണ്ണം ഉച്ചനീചവ്യാസാർദ്ധമായിരിക്കും. അപ്പോൾ പിന്നെ അവിടുന്നു ഒരു വൃത്തപാദം ഗമി
9. H. ഉച്ചത്തിങ്കൽ

എന്നേ ഉച്ചനീചവൃത്തത്തിന്റെ നേമീങ്കലേ പൂർവ്വസൂത്രാഗ്രം സ്പർശിക്കുന്ന[10] പ്രദേശവും പ്രതിമണ്ഡലനേമിയും തങ്ങളിലുള്ള സംപാതം യാതൊരിടത്ത് അവിടത്ത് എല്ലൊ ഗ്രഹം. ആ ഗ്രഹത്തോടു മധ്യമലിപ്തയോടുള്ള അന്ത രാളം മധ്യമസ്ഫുടാന്തരമാകുന്നത്. അവിടെ കക്ഷ്യാവൃത്തകേന്ദ്രത്തിങ്കേന്ന് ഉച്ചനീചവൃത്തത്തിന്റെ എല്ലായിലും അകന്ന പ്രദേശത്തിങ്കലും അണഞ്ഞ പ്രദേശത്തിങ്കലും സ്പർശിച്ചിരിക്കും മദ്ധ്യമസൂത്രം. എന്നിട്ട് ഈ സൂത്രാഗ്ര ത്തിങ്കല് ഉച്ചനീചവൃത്തത്തിങ്കലെ ഉച്ചപ്രദേശമാകുന്നത്. എന്നിട്ട് ഉച്ചനീച വൃത്തത്തിങ്കലെ ഉച്ചപ്രദേശത്തോട് പൂർവ്വസൂത്രത്തോടുള്ള അന്തരാളചാ പഭാഗത്തിങ്കലെ ഉച്ചനീചവൃത്തഗതജ്യാവ് മദ്ധ്യമസ്ഫുടാന്തരമാകും. അവിടെ കക്ഷ്യാവൃത്തനേമീങ്കലു പൂർവ്വസൂത്രത്തോടുള്ള അന്തരാളചാപഭാഗത്തി ങ്കലെ ഉച്ചനീചവൃത്തഗതജ്യാവ് മധ്യമസ്ഫുടാന്തരമാകുന്നത്[11]. അവിടെ കക്ഷ്യാവൃത്തനേമീങ്കലെ പൂർവ്വസൂത്രത്തിങ്കല് ഉച്ചനീചവൃത്തകേന്ദ്രമെങ്കിൽ ഉച്ചനീചവൃത്തത്തിങ്കലും പൂർവ്വസൂത്രാഗ്രം ഉച്ചപ്രദേശമാകുന്നത്. പിന്നെ കക്ഷ്യാനേമിയിങ്കൽ ഈശകോണിങ്കൽ ഉച്ചനീചവൃത്തകേന്ദ്രമെങ്കിൽ ഉച്ച നീചവൃത്തത്തിങ്കൽ ഈ ഈശകോണം ഉച്ചനീചപ്രദേശമാകുന്നത്. ഉത്തര സൂത്രത്തിങ്കൽ കേന്ദ്രത്തിങ്കൽ അവിടം ഉച്ചപ്രദേശമാകുന്നത്[12]. എന്നാൽ ഉച്ചസൂത്രമായിട്ടു കല്പിച്ചിരിക്കുന്ന പൂർവ്വസൂത്രത്തിങ്കേന്നു കക്ഷ്യാനേമി യിങ്കൽ എത്ര ചെന്നേടുത്ത് ഉച്ചനീചവൃത്തകേന്ദ്രം വർത്തിക്കുന്നൂ, ഉച്ചനീ ചവൃത്തത്തിങ്കലൂ ഉച്ചപ്രദേശത്തോടുള്ള അന്തരവും തന്റെ അംശം കൊണ്ടു അത്ര ഉണ്ടായിരിക്കും. എന്നാൽ കക്ഷ്യാവൃത്തനേമീങ്കലേ ഉച്ചമധ്യാന്ത രാളപ്രദേശത്തിന്റെ ജ്യാവിനെ ത്രൈരാശികം കൊണ്ട് ഉച്ചനീചവൃത്തത്തി ങ്കലെ ജ്യാവാക്കിക്കൊണ്ടാൽ ഉച്ചമധ്യാന്തരാളജ്യാവായിട്ടിരിക്കും. അവിടെ പ്രതിമണ്ഡലത്തിങ്കലെ ഉച്ചപ്രദേശത്തോടു ഗ്രഹത്തോടുള്ള അന്തരാളജ്യാ വിനെ ത്രൈരാശികം ചെയ്താലും വരും ഈ മധ്യമസ്ഫുടാന്തരാളജ്യാവാ കുന്നത്. അവിടെ പ്രതിമണ്ഡലത്തിങ്കലെ ഉച്ചസൂത്രഗ്രഹാന്തരാളവും ഉച്ച നീചവൃത്തത്തിങ്കലെ ഉച്ചസൂത്രഗ്രഹാന്തരാളവും തുല്യം. എന്നിട്ട് ഇവിടെ പ്രതിമണ്ഡലത്തിങ്കൽ പൂർവ്വസൂത്രാഗ്രം എന്നു കല്പിച്ച് ഉച്ചപ്രദേശത്തി ങ്കന്ന് തുടങ്ങി ഗ്രഹം ഒരാവർത്തി ഭ്രമിക്കുമ്പോൾ ഉച്ചനീചവൃത്തത്തിങ്കലെ ഉച്ചസൂത്രഗ്രഹാന്തരാളവും തന്റെ അംശം കൊണ്ടു തുല്യം. എന്നാൽ ഉച്ച

7. 10. C. adds സൂത്രത്തിന്റെ
11. G. om. അവിടെ കക്ഷ്യാവൃത്തനേമിങ്കലു to സ്ഫുടാന്തരമാകുന്നത്.
12. B.H.om. പിന്നെ to പ്രദേശമാകുന്നത്.

മധ്യമാന്തരാളജ്യാവിനെ ഉച്ചനീചവ്യാസാർദ്ധം കൊണ്ടു ഗുണിച്ച് ത്രിജ്യ കൊണ്ടു ഹരിച്ചാൽ മധ്യമസ്ഫുടാന്തരാളജ്യാവായിട്ടു വരും[13]. പിന്നെ അതിനെ അന്നേരത്തെ കർണ്ണവൃത്തത്തിലെ ജ്യാവെന്നു കല്പിച്ചു ചാപിച്ച് മധ്യമത്തിൽ സംസ്കരിച്ചാൽ സ്ഫുടം വരും[14].

8. കർണ്ണാനയനം

അനന്തരം[1] കർണ്ണവൃത്തത്തിങ്കലെ ജ്യാവാകും പ്രകാരത്തെ ചൊല്ലുന്നു. അവിടെ കക്ഷ്യാവൃത്തകേന്ദ്രത്തിങ്കേന്നു പ്രതിമണ്ഡലകേന്ദ്രത്തിൽകൂടി പ്രതിമണ്ഡലനേമീങ്കൽ സ്പർശിക്കുന്ന സൂത്രം യാതൊന്ന് അത് 'ഉച്ചസൂത്ര'മാകുന്നത്. അതിനെ ഇവിടെ പൂർവ്വസൂത്രമെന്നു കല്പിച്ചത് എന്നു മുമ്പിൽ ചൊല്ലിയെല്ലോ. ആ സൂത്രശേഷമായിരിക്കുന്ന പ്രത്യക്സൂത്രം 'നീചസൂത്ര'മാകുന്നത്. എന്നിട്ട് ഈ സൂത്രത്തിന്നൊക്കെയും 'ഉച്ചനീചസൂത്ര' മെന്നു പേർ. ഇത് ഉച്ചനീചസൂത്രത്തോടു ഗ്രഹത്തോടുള്ള അന്തരാളം പ്രതിമണ്ഡലഭാഗത്തിങ്കലെ ജ്യാവ് യാതൊന്ന് അത് മധ്യമത്തിങ്കേന്ന് ഉച്ചം വാങ്ങിയ ശേഷത്തിങ്കലേ ഭുജാജ്യാവ്. ഇതിന്നു ഗ്രഹത്തിങ്കൽ അഗ്രം, ഉച്ചനീചസൂത്രത്തിങ്കൽ മൂലം, എന്നിങ്ങനെ കല്പിക്കും പ്രകാരം. ഇതിവിടെ കർണ്ണവൃത്തവ്യാസാർദ്ധം വരുത്തുന്നേടത്തേക്ക് ഭൂജാജ്യാവാകുന്നത്. ഭുജാമൂലത്തോടു കക്ഷ്യാകേന്ദ്രത്തോടുള്ള അന്തരാളം കോടി ആകുന്നത്. കക്ഷ്യാകേന്ദ്രത്തോട് ഗ്രഹത്തോടുള്ള അന്തരാളം കർണ്ണമാകുന്നത്. പിന്നെ പ്രതിമണ്ഡലത്തിന്റെ നീചഭാഗത്തിലൂ ഗ്രഹം എന്നിരിക്കുന്നൂതാകിൽ മധ്യമത്തിന്റെ കോടിയിങ്കൽ ഉച്ചനീചവ്യാസാർദ്ധം കൂട്ടിയത് കോടിജ്യാവാകുന്നത്[2].

ഇവിടെ പ്രതിമണ്ഡലോച്ചഭാഗത്തിങ്കലു ഗ്രഹം എന്നിരിക്കുന്നൂതാകിൽ ഉച്ചോനമധ്യമകോടിയും ഉച്ചനീചവ്യാസാർദ്ധവും തങ്ങളിലന്തരിച്ചതു ഭുജാമൂലത്തോടു കക്ഷ്യാകേന്ദ്രത്തോടുള്ള അന്തരാളമാകുന്നത് കോടിജ്യാവാകുന്നത്. ഇവിടെ പ്രതിമണ്ഡലകേന്ദ്രത്തോടു ഭുജാമൂലത്തോടുള്ള അന്തരാളം യാതൊന്ന് അത് ഉച്ചമധ്യമാന്തരാളം കോടിജ്യാവാകുന്നത്. പ്രതിമ

7. 13. B. ജ്യാവായിവരും; F. ജ്യാവെന്നു വരും
14. G. add എന്നു സ്ഥിതമായി

8. 1. G. പിന്നെ അതിനെ അന്നേരത്തെ കർണ്ണവൃത്തത്തിങ്കലെ ജ്യാവാ...........
2. C. om. പിന്നെ ....to.... കോടിജ്യാവാകുന്നത്.

ണ്ഡലകേന്ദ്രത്തോടു കക്ഷ്യാകേന്ദ്രത്തോടുള്ള അന്തരാളം ഉച്ചനീചവ്യാസാർദ്ധമാകുന്നത്. ഇവിടെ പ്രതിമണ്ഡലകേന്ദ്രത്തിങ്കലെ ദക്ഷിണോത്തരസൂത്രത്തിന്റെ[3] കിഴക്ക് ഗ്രഹമെന്നിരിക്കിൽ കേന്ദ്രകോടിയിൽ ഉച്ചനീചവ്യാസാർദ്ധം കൂട്ടി കർണ്ണവൃത്തകോടി ഉണ്ടാക്കേണ്ടൂ. പിന്നെ പ്രതിമണ്ഡലത്തിന്റെ ദക്ഷിണോത്തരസൂത്രത്തിനു പടിഞ്ഞാറു ഗ്രഹം എന്നിരിക്കുന്നൂതാകിൽ ഉച്ചനീചവ്യാസാർദ്ധത്തിന്റെ അന്തർഭാഗത്തിങ്കലായിട്ടിരിക്കും ഭുജാമൂലം.

ഇവിടെ കക്ഷ്യാമദ്ധ്യത്തിങ്കലേ ഉച്ചനീചവൃത്തത്തിന്റെ ദക്ഷിണോത്തരസൂത്രത്തിനു കിഴക്കെപ്പുറത്ത് ഭുജാമൂലമെന്നിരിക്കുന്നൂതാകിൽ ഭുജാമൂലത്തോട് ഉച്ചനീചനേമിയോടുള്ള അന്തരാളം കേന്ദ്രകോടി ആകുന്നത്. ഇതിനെ ഉച്ചനീചവ്യാസാർദ്ധത്തിങ്കലെ ഉച്ചനീചവൃത്തത്തിങ്കലെ[4] വ്യാസാർദ്ധത്തിങ്കന്ന് കളഞ്ഞാൽ ശേഷം കേന്ദ്രത്തോട് ഭുജാമൂലത്തോടുള്ള അന്തരാളം കർണ്ണവൃത്തകോടിയാകുന്നതു വരും. പിന്നെ കക്ഷ്യാമദ്ധ്യത്തിങ്കലെ ഉച്ചനീചവൃത്തത്തിങ്കലെ ദക്ഷിണോത്തരസൂത്രത്തിങ്കേന്ന് പടിഞ്ഞാറു ഭുജാമൂലം എന്നിരിക്കുന്നൂതാകിൽ കേന്ദ്രകോടിയിങ്കേന്ന് ഉച്ചനീചവ്യാസാർദ്ധത്തെ കളഞ്ഞ ശേഷം കർണ്ണവൃത്തകോടി ആകുന്നത്. ഉച്ചോനമദ്ധ്യത്തെ ഇവിടെ കേന്ദ്രമെന്നു ചൊല്ലുമാറുണ്ട്. ഇങ്ങനെ കർണ്ണവൃത്തത്തിങ്കലെ ഭുജാകോടികളെ ഉണ്ടാക്കി വർഗ്ഗിച്ചു കൂട്ടി മൂലിച്ചാൽ കക്ഷ്യാകേന്ദ്രത്തോടു ഗ്രഹത്തോടുള്ള അന്തരാളം കർണ്ണവൃത്തവ്യാസാർദ്ധം പ്രതിമണ്ഡലകലകളെക്കൊണ്ട് അളന്നത് ഉണ്ടാകും. ഇതിനെത്തന്നെ കർണ്ണവൃത്തകലകളേക്കൊണ്ട് അളക്കുമ്പോൾ ത്രിജ്യാതുല്യമായിട്ടിരിക്കും. അതതു വൃത്തത്തെ ഇരുപത്തോരായിരത്തറുനൂറായി വിഭജിച്ചതിൽ ഒരംശം തന്റെതന്റെ[5] കലാമാനമാകുന്നത്. അതിനെക്കൊണ്ട് തന്റെതന്റെ വ്യാസാർദ്ധം ത്രിജ്യാതുല്യമായിട്ടിരിക്കും എന്നു കർണ്ണവൃത്തകലകളെക്കൊണ്ടു ത്രിജ്യാതുല്യം എന്മാൽ ഹേതു. ഇവിടെ മന്ദോച്ചനീചവൃത്തത്തിന്നു മന്ദകർണ്ണവശാൽ വൃദ്ധിഹ്രാസമുണ്ടാകയാൽ സർവ്വദാ കർണ്ണവൃത്തകലാമിതം ഇത്. എന്നിട്ട് ഈ കർണ്ണത്തെ അവിശേഷിച്ചേ പ്രതിമണ്ഡലകലാമിതമാവൂ. ഇങ്ങനെ പ്രതിമണ്ഡലകലകളെക്കൊണ്ട് കർണ്ണവൃത്തമാനത്തെ അറിയും പ്രകാരം.

---

8. 3. D.G.സൂത്രത്തിങ്കന്ന്; F. സൂത്രത്തിനു
4. F. നീചഭാഗത്തിങ്കലും
5. B. അതിലൊരംശം സ്വസ്വകലാമാനമാകുന്നത്.

9. കർണ്ണാനയനം – പ്രകാരാന്തരം

പിന്നെ പ്രകാരാന്തരേണ[1] അറിയും പ്രകാരം. അവിടെ കക്ഷ്യാകേന്ദ്രത്തിങ്കേന്നു തുടങ്ങി കക്ഷ്യാനേമീങ്കലേ ഉച്ചനീചകേന്ദ്രത്തൂടെ ഇതിന്റെ നേമീങ്കൽ സ്പർശിക്കുന്ന സൂത്രം യാതൊന്ന് അതിന്നു 'മധ്യമസൂത്ര'മെന്നു പേർ എന്നു മുമ്പിൽ ചൊല്ലി. ഈ മധ്യമസൂത്രത്തോടു ഗ്രഹത്തോടുള്ള അന്തരാളം മദ്ധ്യമസ്ഫുടാന്തരമാകുന്നത്. ഇതിന്നു 'ഭുജാഫല'മെന്നു പേർ. ഇതിനെ ഗ്രഹത്തിങ്കൽ അഗ്രമായി മധ്യമസൂത്രത്തിങ്കൽ മൂലമായിട്ടു കൽപിക്കേണ്ടു. ഇവിടെ ഭുജാമൂലത്തോടു കക്ഷ്യാവൃത്തനേമീങ്കലെ ഉച്ചനീചവൃത്തകേന്ദ്രത്തോടുള്ള അന്തരാളം കോടിഫലമാകുന്നത്. ഇവിടെ കക്ഷ്യാനേമീങ്കേന്നു പുറത്തകപ്പെട്ടിരിക്കുന്ന പ്രതിമണ്ഡലനേമീങ്കലു ഗ്രഹം എന്നിരിക്കുന്നൂതാകിൽ ദോഃഫലമൂലം കക്ഷ്യാനേമിയുടെ പുറത്തേ അകപ്പെട്ടിരിക്കും. അപ്പോൾ കോടിഫലത്തെ കക്ഷ്യാവ്യാസാർദ്ധത്തിൽ കൂട്ടിയാൽ ദോഃഫലമൂലത്തോടു കക്ഷ്യാകേന്ദ്രത്തോടുള്ള അന്തരാളമുണ്ടാകും. പിന്നെ കക്ഷ്യാനേമിയുടെ അന്തർഭാഗത്തിങ്കലേ പ്രതിമണ്ഡലനേമീങ്കലു ഗ്രഹം എന്നിരിക്കുന്നൂതാകിൽ കക്ഷ്യാനേമിയുടെ അന്തർഭാഗത്തിങ്കലായിട്ടിരിക്കും ദോഃഫലമൂലം. അപ്പോൾ കോടിഫലത്തെ കക്ഷ്യാവ്യാസാർദ്ധത്തിങ്കേന്നു കളഞ്ഞു ശേഷം ദോഃഫലമൂലത്തോടു കക്ഷ്യാകേന്ദ്രത്തോടുള്ള അന്തരാളം കോടിയായി ദോഃഫലം ഭുജാകോടിയായും കല്പിച്ചു രണ്ടിന്റേയും വർഗ്ഗയോഗമൂലം ചെയ്താൽ കക്ഷ്യാകേന്ദ്രത്തോട് ഗ്രഹത്തോടുള്ള അന്തരാളം പ്രതിമണ്ഡലവൃത്തകലാമിതമായിട്ടു മുമ്പിൽ വരുത്തിയ കർണ്ണം തന്നെ വരും. ഇങ്ങനെ കർണ്ണവൃത്തവ്യാസാർദ്ധം രണ്ടു പ്രകാരം വരുത്താം. ഇവിടെ പ്രതിമണ്ഡലത്തിങ്കൽ ഇത്ര ചെന്നു ഗ്രഹം എന്നു മദ്ധ്യമം കൊണ്ടറിഞ്ഞതു കർണ്ണവൃത്തത്തിങ്കൽ ഇത്ര ചെന്നു ഗ്രഹം എന്നറിക വേണ്ടിയിരിക്കുന്നത്. അതിന്നു സാധനം ഈ കർണ്ണം.

10. വിപരീതകർണ്ണം

അനന്തരം കർണ്ണവൃത്തകലകളെക്കൊണ്ടു കക്ഷ്യാവ്യാസാർദ്ധം എത്രയെന്നു താൻ പ്രതിമണ്ഡലവ്യാസാർദ്ധം എത്രയെന്നു താൻ അറിയും പ്രകാരത്തെ ചൊല്ലുന്നു. കർണ്ണാനയനം വിപരീതക്രിയകൊണ്ടു വരികയാൽ

---

9. 1. D.adds കക്ഷ്യാകേന്ദ്രം ആ

'വിപരീതകർണ്ണ'മെന്നു പേരുണ്ട് ഇതിന്ന്. ഇവിടെ മന്ദസ്ഫുടത്തിങ്കൽ മദ്ധ്യ സ്ഫുടാന്തരം മന്ദകർണ്ണവൃത്തകലകളെ കൊണ്ടു അളന്നതായിട്ടുള്ളു. എന്നാൽ മദ്ധ്യസ്ഫുടാന്തരജ്യാവാകുന്ന ദോഃഫലത്തെ വർഗ്ഗിച്ചു ത്രിജ്യാ ഫലത്തിങ്കേന്നു കളഞ്ഞ് മൂലിച്ചാൽ ദോഃഫലമൂലത്തോടു കക്ഷ്യാകേന്ദ്ര ത്തോടുള്ള അന്തരാളം വരും. ഇതിങ്കേന്നു കോടിഫലം കളവൂ ദോഃഫല മൂലം കക്ഷ്യാനേമീങ്കേന്നു പുറത്തു എന്നാൽ[1], അകത്തെങ്കിൽ കൂട്ടൂ[2]. അതു കക്ഷ്യാവ്യാസാർദ്ധം കർണ്ണവൃത്തകലാമിതമായിട്ടിരിപ്പോന്ന്[3].

11. വിപരീതകർണ്ണം – പ്രകാരാന്തരം

ഇനി പ്രകാരാന്തരേണ[1] പ്രതിമണ്ഡലവ്യാസാർദ്ധത്തെ കർണ്ണവൃത്തകല കളേക്കൊണ്ട് ഇത്ര[2] എന്നറിയും പ്രകാരത്തെ ചൊല്ലുന്നു. ഇവിടെ ഉച്ച സ്ഫുടാന്തരദോർജ്യാവു കർണ്ണവൃത്തകലാമിതമായിട്ടുള്ളു. കർണ്ണവൃത്തത്തി ങ്കൽ ഇത്ര ചെന്നൂ ഗ്രഹം എന്നല്ലോ സ്ഫുടമാകുന്നത്. നീചോച്ചസൂത്ര ത്തിങ്കൽ മൂലമായി ഗ്രഹത്തിങ്കൽ അഗ്രമായിട്ടിരിപ്പൊന്ന് ഈ ഭുജാജ്യാവ്. ഭുജാമൂലത്തോടു കക്ഷ്യാകേന്ദ്രത്തോടുള്ള അന്തരാളം സ്ഫുടോച്ചാ ന്തരകോടിജ്യാവാകുന്നത്. ഇതിങ്കേന്നു[3] കക്ഷ്യാപ്രതിമണ്ഡലാന്തരങ്ങളുടെ കേന്ദ്രാന്തരാളമാകുന്ന ഉച്ചനീചവ്യാസാർദ്ധത്തെ കളവൂ. ഉച്ചനീചവൃത്തനേ മീങ്കന്നു പുറത്തു ദോർജ്യാമൂലമെങ്കിൽ, അല്ലായ്കിൽ[4] കൂട്ടൂ. ശേഷിച്ച്[5] കോടി യേയും ദോർജ്യാവിനേയും വർഗ്ഗിച്ചു കൂട്ടി മൂലിച്ചാൽ പ്രതിമണ്ഡലകേന്ദ്ര ത്തിങ്കേന്നു ഗ്രഹത്തോടുള്ള അന്തരാളം പ്രതിമണ്ഡലവ്യാസാർദ്ധം കർണ്ണ വൃത്തകലാമിതമായിട്ടുണ്ടാകും.

---

10. 1. B. ആകിൽ; E. G എങ്കിൽ
 2. F. Om. അകത്തെങ്കിൽ കൂട്ടു
 3. F. ആയിരിപ്പൊന്ന്
11. 1. B. അഥ പ്രകാരാന്തരേണ
 2. F. എത്ര
 3. C. F ഇതിൽ
 4. B. D. അല്ലെങ്കിൽ
 5. B. C.om. ശേഷിച്ച; D. ആ കോടിയേയും

12. വിപരീതകർണ്ണാനയനം – പ്രകാരാന്തരം

അനന്തരം സ്ഫുടോച്ചാന്തരദോഃകോടിഫലങ്ങളെക്കൊണ്ട് ഈ വ്യാസാർദ്ധത്തെ വരുത്തും പ്രകാരം. ഇവിടെ കർണ്ണവൃത്തത്തിങ്കലെ ഉച്ച സൂത്രവും ഗ്രഹസൂത്രവും തങ്ങളിലുള്ള അന്തരാളഭാഗത്തിങ്കലെ ജ്യാവു യാതൊന്ന് അതു സ്ഫുടോച്ചാന്തരദോർജ്യാവാകുന്നത്. ഈ സൂത്രാന്തര ജ്യാവു തന്നെ കർണ്ണവൃത്തകേന്ദ്രത്തിങ്കലേ ഉച്ചനീചവൃത്തത്തിങ്കൽ കല്പിക്കുമ്പോൾ സ്ഫുടോച്ചാന്തരദോഃഫലമായിട്ടിരിക്കും. ഈ ദോഃഫലത്തെ[1] പ്രതിമണ്ഡലകേന്ദ്രത്തിങ്കൽ അഗ്രവും ഗ്രഹസൂത്രത്തിങ്കൽ മൂലവുമായിട്ടു കല്പിക്കേണ്ടൂ. ഈ[2] ദോഃഫലമൂലത്തോടു കർണ്ണവൃത്തകേന്ദ്രത്തോടുള്ള അന്തരാളം[3] ഗ്രഹസൂത്രത്തിങ്കലേത് ഇവിടെ[4] അക്കോടിഫലമാകുന്നത്. കോടിഫലം പോയ സൂത്രശേഷം കോടിയാകുന്നത്. ദോഃഫലം ഭുജങ്ങളിലെ വർഗ്ഗയോഗമൂലം പ്രതിമണ്ഡലകേന്ദ്രത്തോട് ഗ്രഹത്തോടുള്ള അന്തരാളം പ്രതിമണ്ഡലവ്യാസാർദ്ധം കർണ്ണവൃത്തകലാമിതമായിട്ടുണ്ടാകും.

ഇങ്ങനെ ഉച്ചഭാഗത്തിങ്കൽ ഗ്രഹമിരിക്കുമ്പോൾ. നീചഭാഗത്തിങ്കൽ[5] വിശേഷമുണ്ട്. ഇവിടെ നീചസൂത്രത്തോടു ഗ്രഹസൂത്രത്തോടുള്ള അന്തരാളം കർണ്ണവൃത്തത്തിങ്കലേത് സ്ഫുടോച്ചാന്തരദോർജ്യാവാകുന്നത്. ഈ അന്തരാളം നീചോച്ചവൃത്തത്തിങ്കലേത് ദോഃഫലമാകുന്നത്. ഇവിടെ നീചസൂത്രത്തിന്റെ ശേഷം ഉച്ചസൂത്രമായിട്ടുണ്ടല്ലോ, ഗ്രഹസൂത്രവും അവ്വണ്ണം കർണ്ണവൃത്ത കേന്ദ്രത്തൂടെ മറ്റേപുറത്തു നീട്ടി കല്പിപ്പൂ. അവിടെ ഈ ഗ്രഹസൂത്രപുച്ഛത്തോട് ഉച്ചസൂത്രത്തോടുള്ള അന്തരാളം ഈ ദോഃഫലം തന്നെയായിട്ടിരിക്കും. ഇവിടേയും പ്രതിമണ്ഡലകേന്ദ്രത്തിങ്കൽ അഗ്രമായി ഗ്രഹസൂത്രശേഷത്തിങ്കൽ മൂലമായിട്ട് ദോഃഫലത്തെ കല്പിപ്പൂ. ദോഃഫലത്തോടു കർണ്ണവൃത്തത്തോടുള്ള അന്തരാളം ഗ്രഹസൂത്രശേഷത്തിങ്കലേതു കോടിഫലമാകുന്നത്. ഈ കോടിഫലം കർണ്ണവൃത്തവ്യാസാർദ്ധമാകുന്ന ഗ്രഹസൂത്രത്തിങ്കൽ കൂട്ടിയാൽ ഗ്രഹത്തോട് ഇക്കല്പിച്ച ദോഃഫലമൂലത്തോടുള്ള അന്തരാളമുണ്ടാകും. ഇതിന്റെ വർഗ്ഗത്തിൽ ദോഃഫലവർഗ്ഗം കൂട്ടി മൂലിച്ചാൽ പ്രതി

---

12. 1. B. F. ദോഃഫലകോടിഫലങ്ങളെ
2. C. H. om. ഈ
3. C. D. G. സൂത്രാന്തരാള
4. F. ഇവിടത്ത്, G. ഇവിടത്തേക്ക്
5. B. adds പിന്നെ

മണ്ഡലകേന്ദ്രത്തോട് ഗ്രഹത്തോടുള്ള അന്തരാളം പ്രതിമണ്ഡലവ്യാസാർദ്ധം കർണ്ണവൃത്തകലാമിതമായിട്ടുണ്ടാകും. അന്തരാളജ്യാക്കളെ തലപകർന്നു കല്പിച്ചതുകൊണ്ടു മാനഭേദമുണ്ടാകയില്ല. ഇങ്ങനെ കർണ്ണവൃത്തകലാമിതമായിട്ടു പ്രതിമണ്ഡലവ്യാസാർദ്ധത്തേയും കക്ഷ്യാമണ്ഡലവ്യാസാർദ്ധത്തേയും ഉണ്ടാക്കും പ്രകാരം. ഇങ്ങനെ ഉണ്ടായ ഇതിന്നു 'വിപരീതകർണ്ണ' മെന്നുപേർ. പ്രതിമണ്ഡലകലകളേക്കൊണ്ടു കർണ്ണവൃത്തവ്യാസാർദ്ധത്തെ മാനം ചെയ്തതിനെയെല്ലോ കർണ്ണമെന്നു ചൊല്ലുന്നു. അതിങ്കേന്നു വൈപരീത്യമുണ്ടാകയാൽ വിപരീതകർണ്ണമിത്. ഈ വിപരീതകർണ്ണത്തെക്കൊണ്ട് വ്യാസാർദ്ധവർഗ്ഗത്തെ ഹരിച്ചാൽ ഫലം പ്രതിമണ്ഡലകലകളെക്കൊണ്ടു കർണ്ണവൃത്തവ്യാസാർദ്ധത്തെ മാനം ചെയ്തിരിക്കുന്ന കർണ്ണമാകുന്നത് ഉണ്ടാകും. ഇവിടെ പ്രതിമണ്ഡലവ്യാസാർദ്ധം തന്റെ വൃത്തത്തിന്റെ 'അനന്തപുരാം'ശം കൊണ്ട് ത്രിജ്യാതുല്യം, കർണ്ണവൃത്തകലകളെക്കൊണ്ട് വിപരീതകർണ്ണതുല്യം കർണ്ണവൃത്തവ്യാസാർദ്ധം. പിന്നെ തന്റെ കലകളെക്കൊണ്ട് ത്രിജ്യാതുല്യം. പ്രതിമണ്ഡലകലകളെക്കൊണ്ട് എത്ര എന്ന് ത്രൈരാശികം. ഫലം പ്രതിമണ്ഡലകലകളേക്കൊണ്ടു മാനം ചെയ്തു കർണ്ണവൃത്തവ്യാസാർദ്ധമായിട്ടിരിക്കും.

13. മന്ദസ്ഫുടം

പിന്നെ മധ്യമത്തിൽ ഉച്ചം വാങ്ങിയ ശേഷത്തിന്റെ ഭുജാജ്യാവു യാതൊന്ന് അത് ഉച്ചനീചസൂത്രത്തിങ്കേന്നു ഗ്രഹത്തോടുള്ള അന്തരാളത്തിങ്കലെ പ്രതിമണ്ഡലഭാഗത്തിങ്കലെ ജ്യാവായിട്ടിരിക്കും. ഈ ജ്യാവിനെത്തന്നെ കർണ്ണവൃത്തകലകളേക്കൊണ്ടു ഇത്രയെന്നറിഞ്ഞു ചാപിച്ചാൽ ഉച്ചനീചസൂത്രത്തോടു ഗ്രഹത്തോടുള്ള അന്തരാളത്തിങ്കലേ കർണ്ണവൃത്തഭാഗമാകും. ഇച്ചാപത്തെ ഉച്ചത്തിൽ താൻ നീചത്തിൽ താൻ സംസ്കരിച്ചാൽ കർണ്ണവൃത്തത്തിങ്കൽ ഇത്ര[1] ചെന്നൂ ഗ്രഹം എന്നു വരും. അതു സ്ഫുടഗ്രഹമാകുന്നത്.

ഇവിടെ ഇങ്ങനെ ഇരിപ്പൊന്നു ത്രൈരാശികം. കർണ്ണവൃത്തവ്യാസാർദ്ധം പ്രതിമണ്ഡലകലാമിതമായിട്ടിരിക്കുന്നത് കർണ്ണതുല്യം. ഇതു പ്രമാണമാകുന്നത്. ഇതു കർണ്ണവൃത്തകലാമിതമാകുമ്പോൾ ത്രിജ്യാതുല്യം. ഇതു പ്രമാ

13. 1. F. എത്ര

ണഫലമാകുന്നത്. ഉച്ചനീചസൂത്രഗ്രഹാന്തരാളത്തിങ്കലെ പ്രതിമണ്ഡലഭാഗ[2] ജ്യാവ് ഇച്ഛാ. ഇതിനേത്തന്നെ കർണ്ണവൃത്തകലാമിതമാക്കി കർണ്ണവൃത്തഭാഗജ്യാവായി കല്പിച്ചിരിക്കുന്നത് ഇച്ഛാഫലം. ഇതിനെ അണവിന്നു തക്കവണ്ണം ഉച്ചത്തിൽ താൻ നീചത്തിൽ താൻ സംസ്കരിച്ചത് സ്ഫുടഗ്രഹമാകുന്നത്. ഈ സ്ഫുടപ്രകാരത്തിനു "പ്രതിമണ്ഡലസ്ഫുട"മെന്നു പേർ. ഇവിടെ സ്ഫുടത്തിങ്കേന്നു ഉച്ചം വാങ്ങിയ ശേഷത്തിന്നു ജ്യാവുകൊണ്ടാൽ ഇത് ഇവിടെ ചൊല്ലിയ ഇച്ഛാഫലമായിട്ടിരിക്കും. മദ്ധ്യമത്തിങ്കേന്ന് ഉച്ചം വാങ്ങിയ[3] ശേഷത്തിങ്കേന്ന് ഉണ്ടാക്കിയ ഭുജാജ്യാവ് ഇവിടെ ചൊല്ലിയതു ഇച്ഛാരാശിയാകുന്നത്.

എന്നാൽ ഇച്ഛാഫലത്തെ പ്രമാണമെന്നും ഇച്ഛാരാശിയെ പ്രമാണഫലമെന്നും കല്പിച്ച് കർണ്ണവൃത്തവ്യാസാർദ്ധമായിരിക്കുന്ന ത്രിജ്യയെ ഇച്ഛാരാശിയെന്നു കല്പിച്ചുണ്ടാക്കുന്ന ഇച്ഛാഫലം ഇവിടെ ചൊല്ലിയ കർണ്ണമായിട്ടിരിക്കും. ഇവിടെ മദ്ധ്യമത്തിങ്കേന്ന് ഉച്ചം വാങ്ങിയ ശേഷത്തിന്റെ ജ്യാവ് യാതൊന്ന് അതു പ്രതിമണ്ഡലൈകദേശത്തിങ്കലേ ജ്യാവാകയാൽ പ്രതിമണ്ഡലകലാമിതം. പിന്നെ സ്ഫുടത്തിങ്കേന്ന് ഉച്ചം വാങ്ങിയ ശേഷത്തിന്റെ ജ്യാവാകുന്നതും ഇതുതന്നെയത്രേ. ഉച്ചനീചസൂത്രത്തോടു വിപരീതമായി ഗ്രഹത്തോട് ഉച്ചനീചസൂത്രത്തോട്[4] അന്തരാളപ്രദേശമായിട്ടിരിക്കയാൽ ജ്യാക്കൾ രണ്ടുമൊന്നേ. മാനഭേദം കൊണ്ടു സംഖ്യാഭേദമേയുള്ളു. സ്ഫുടകേന്ദ്രജ്യാവിന് ചാപമാകുന്നതു കർണ്ണവൃത്തത്തിന്റെ ഏകദേശമാകയാൽ കർണ്ണവൃത്തകലാമിതം. സ്ഫുടകേന്ദ്രജ്യാവ് എന്നാൽ ഈ ജ്യാവ് കർണ്ണവൃത്തകലകളേക്കൊണ്ടു സ്ഫുടകേന്ദ്രജ്യാവ്. ഇതുതന്നെ പ്രതിമണ്ഡലകലാമിതമാകുമ്പോൾ മദ്ധ്യമകേന്ദ്രജ്യാവ്. അപ്പോൾ കർണ്ണവൃത്തകലകളേക്കൊണ്ടു വ്യാസാർദ്ധതുല്യമാകുമ്പോൾ പ്രതിമണ്ഡലകലകളേക്കൊണ്ട് എത്രയെന്നു മുമ്പിൽചൊല്ലിയ കർണ്ണം വരും.

ഇങ്ങനേയും വരുത്താം കർണ്ണം. ഇവിടെ മദ്ധ്യമകേന്ദ്രജ്യാവ് പ്രതിമണ്ഡലകലാമിതമാകുമ്പോൾ മദ്ധ്യമകേന്ദ്രഭുജാഫലം എങ്ങനെ കർണ്ണവൃത്തകലാമിതമാകുന്നു എന്ന ശങ്കയ്ക്ക്[5] ഉത്തരം-കർണ്ണം വലുതാകുമ്പോൾ മന്ദ

13. 2. B. C. F. G. om. ഭാഗ
3. F. പോയ
4. F.adds ഉള്ള
5. B. എന്നു ശങ്ക

നീചവൃത്തവും അതിന്നു തക്കവണ്ണം കൂടി വലുതാകും. ത്രിജ്യയേക്കാൾ ചെറുതാകുമ്പോൾ മന്ദോച്ചനീചവൃത്തവും അതിന്നു തക്കവണ്ണം ചെറുതാകും. എന്നിട്ട് ഈ വൃത്തത്തിലെ ജ്യാവ് എല്ലായ്പ്പോഴും കർണ്ണവൃത്തകലാമിതമായിട്ടിരിക്കുമത്രേ. എന്നിട്ട് മന്ദകർണ്ണം ഇച്ചൊല്ലിയവണ്ണം വരുത്തേണ്ടതും, മധ്യകേന്ദ്രഭുജാഫലത്തെ മധ്യമത്തിൽ സംസ്കരിപ്പാനായികൊണ്ടു കർണ്ണവൃത്തകലാമിതമാക്കുവാൻ ത്രൈരാശികം ചെയ്യേണ്ടിവരാഞ്ഞതും. ഇങ്ങനെ മന്ദകർണ്ണത്തിനു തക്കവണ്ണം മന്ദോച്ചനീചവൃത്തത്തിനു വൃദ്ധിഹ്രാസമുണ്ടാകയാൽ മന്ദകർണ്ണത്തിനും മന്ദഭുജാഫലം കൊണ്ടുള്ള സ്ഫുടത്തിന്നും വിശേഷമുണ്ട്. ശീഘ്രഫലത്തിങ്കേന്നു ശീഘ്രനീചോച്ചവൃത്തത്തിൽ തന്റെ കർണ്ണത്തിനു തക്കവണ്ണം വൃദ്ധിഹ്രാസമില്ല. ഇങ്ങനെ മന്ദസ്ഫുടത്തിങ്കലേ ക്രിയ.

14. ശീഘ്രസ്ഫുടം

അനന്തരം ശീഘ്രസ്ഫുടപ്രകാരത്തെ ചൊല്ലുന്നു. ഇവിടെ ചന്ദ്രാദിത്യന്മാരുടെ[1] മന്ദനീചോച്ചവൃത്തത്തിന്റെ കേന്ദ്രം ഭഗോളമദ്ധ്യത്തിങ്കലു എന്നിട്ട് ചന്ദ്രാദിത്യന്മാർക്കു മന്ദസ്ഫുടം ചെയ്തതു തന്നെ ഭഗോളഗതിയാകുന്നത്. ചൊവ്വാ തുടങ്ങിയുള്ളവറ്റിനും ഭഗോളമധ്യം കേന്ദ്രമായി ഗ്രഹത്തെ സ്പർശിച്ചിട്ടു ഒരു വൃത്തത്തെ കല്പിച്ചാൽ അതിങ്കൽ എത്ര ചെന്നൂ എന്നുള്ളത് ഭഗോള ഗതിയാകുന്നത്. അവറ്റിന്നു വിശേഷമാകുന്നത് - ഭഗോളമധ്യം കേന്ദ്രമായിട്ട് ഒരു ശീഘ്രനീചോച്ചവൃത്തമുണ്ട്. അതിന്റെ നേമീങ്കൽ ശീഘ്രോച്ചത്തിന്റെ ഗതിയായിട്ടിരുന്നൊന്ന്[2] മന്ദനീചോച്ചവൃത്തം. എന്നിട്ടു തല്ക്കാലത്തിങ്കൽ ശീഘ്രോച്ചനീചവൃത്തനേമീങ്കൽ യാതൊരിടത്തു ശീഘ്രോച്ചം വർത്തിക്കുന്നൂ[3], അവിടെ കേന്ദ്രമായിട്ടിരിപ്പൊന്നു മന്ദനീചോച്ചവൃത്തം. ഈ വൃത്തത്തിങ്കൽ മന്ദോച്ചത്തിന്റെ ഗതി. ആ മന്ദോച്ചം യാതൊരിടത്ത് അവിടം കേന്ദ്രമായിട്ടിരിപ്പൊന്ന് മന്ദനീചോച്ചവൃത്തത്തിങ്കൽ പ്രതിമണ്ഡലമെന്നും കല്പിച്ച് ആ പ്രതിമണ്ഡലനേമീങ്കൽ ഗ്രഹബിംബം ഗമിക്കുന്നു എന്നും കല്പിപ്പൂ. പിന്നെ പ്രതിമണ്ഡലത്തിങ്കൽ മേഷാദീങ്കേന്നു തുടങ്ങീട്ട് ഇത്ര

14. 1. B. C. F. ആദിത്യചന്ദ്രന്മാരുടെ
2. G. ഇവിടെ അവസാനിക്കുന്നു
3. D. ഉച്ചമാകുന്ന രവിമദ്ധ്യം വർദ്ധിക്കുന്നു

ചെന്നു ഇപ്പോൾ ഗ്രഹമെന്നു മധ്യമം കൊണ്ടറിയുന്നത്. പിന്നെ മന്ദനീചോച്ചവൃത്തത്തിന്റെ കേന്ദ്രം തന്നെ കേന്ദ്രമായിട്ടു ഗ്രഹബിംബത്തെ സ്പർശിപ്പോരു[4] വൃത്തത്തെ കല്പിപ്പൂ. ഇതിന്നു 'മന്ദകർണ്ണവൃത്ത'മെന്നു പേർ. ആ മന്ദകർണ്ണവൃത്തത്തിങ്കൽ പ്രതിമണ്ഡലമെന്നും കല്പിച്ചു മേഷാദീങ്കേന്നു തുടങ്ങി എത്രേടം[5] ചെന്നിരിക്കുന്നൂ ഗ്രഹം എന്നു മന്ദസ്ഫുടം കൊണ്ടറിയുന്നത്. പിന്നെ ശീഘ്രോച്ചനീചവൃത്തത്തിന്റെ കേന്ദ്രം തന്നെ കേന്ദ്രമായിട്ടു ഗ്രഹബിംബത്തെ സ്പർശിപ്പോരു വൃത്തത്തെ കല്പിപ്പൂ. ഇതിനു 'ശീഘ്രകർണ്ണവൃത്ത"മെന്നു പേർ. ഈ വൃത്തത്തിങ്കൽ മേഷാദിയിങ്കേന്നു തുടങ്ങി രാശ്യാദി എത്ര ചെന്നു എന്നതു 'ശീഘ്രസ്ഫുടം" കൊണ്ടറിയുന്നത്. ഈ ശീഘ്രസ്ഫുടത്തിങ്കൽ മന്ദകർണ്ണവൃത്തത്തെ പ്രതിമണ്ഡലമെന്നു കല്പിച്ചു മന്ദസ്ഫുടഗ്രഹത്തെ മദ്ധ്യമമെന്നും കല്പിച്ചു മന്ദസ്ഫുടത്തിങ്കലേപ്പോലെ ക്രിയചെയ്താൽ ശീഘ്രകർണ്ണവൃത്തത്തിങ്കൽ മേഷാദിയിങ്കേന്നു തുടങ്ങി രാശ്യാദി എത്ര ചെന്നൂ എന്നതുണ്ടാകും.

ഇവിടെ ശീഘ്രസ്ഫുടത്തിങ്കൽ വിശേഷമാകുന്നതു പിന്നെ. ശീഘ്രഭുജാഫലത്തെ ഉണ്ടാക്കിയാൽ പിന്നെ അതിനെ ശീഘ്രകർണ്ണകലാപ്രമിതമാക്കിയാൽ ശീഘ്രകർണ്ണവൃത്തത്തിലെ ജ്യാവാകും. അതിനെ ചാപിച്ചു സംസ്കരിച്ചാൽ ഗ്രഹം ശീഘ്രകർണ്ണവൃത്തത്തിങ്കൽ ഇത്ര ചെന്നു എന്നു വരും. ഇതിന്നായിക്കൊണ്ടു ശീഘ്രഭുജാഫലത്തെ ത്രിജ്യകൊണ്ടു ഗുണിച്ച് ശീഘ്രകർണ്ണം കൊണ്ടു ഹരിക്കണം. എന്നാൽ ശീഘ്രഭൂജാഫലം മന്ദകർണ്ണവൃത്തപ്രമിതമായിട്ടു വരുവാൻ ത്രിജ്യകൊണ്ട് ഗുണിച്ച് ശീഘ്രകർണ്ണം കൊണ്ടു ഹരിക്കണം. എന്നാൽ ശീഘ്രകർണ്ണവൃത്തകലാപ്രമിതമായിട്ടു വരും ശീഘ്രഭുജാഫലം. ഇവിടെ മന്ദഭുജാഫലം മന്ദകർണ്ണവൃത്തകലാപ്രമിതമായിട്ടു വരുവാൻ ഈ ത്രൈരാശികം ചെയ്യേണ്ട. മന്ദകേന്ദ്രജ്യാക്കളെ മന്ദോച്ചനീചവൃത്തവ്യാസാർദ്ധം കൊണ്ടു ഗുണിച്ച് ത്രിജ്യ കൊണ്ടു ഹരിച്ചാൽ തന്നെ മന്ദകർണ്ണവൃത്തകലാമിതമായിട്ടു[6] വരും. അതിന്നു ഹേതു - മന്ദകർണ്ണം വലുതാകുമ്പോൾ മന്ദോച്ചനീചവൃത്തവും കൂടി വലുതാകും. ആ കർണ്ണത്തിനു തക്കവണ്ണം കർണ്ണം ചെറുതാകുമ്പോൾ ചെറുതാവൂതും ചെയ്യും. എന്നിട്ടു മന്ദഭുജാകോടിഫലങ്ങൾ എല്ലായ്പ്പോഴും മന്ദകർണ്ണവൃത്തകലാപ്രമിതമായിട്ടിരിക്കുന്നു. ശീഘ്രോച്ചനീചവൃത്തത്തിനു പിന്നെ ശീഘ്രകർണ്ണ

14. 4. D. സ്പർശിച്ചിരിപ്പോരു
5. B. എത്രോളം
6. D. കലാപ്രമിത

ത്തിനു തക്കവണ്ണം വൃദ്ധിഹ്രാസമില്ല. എന്നിട്ടു ശീഘ്രഫലകോടിഭുജാഫലങ്ങൾ പ്രതിമണ്ഡലകലാപ്രമിതമായിട്ടേ ഇരിക്കുമത്രേ. എന്നിട്ട് ഇവറ്റെ ശീഘ്രകർണ്ണ വൃത്തകലാപ്രമിതങ്ങൾ ആക്കുവാൻ ത്രൈരാശികാന്തരം വേണം.

നടേ പഠിക്കുമ്പോൾ തന്റെ തന്റെ[7] പ്രതിമണ്ഡലകലാമാനം കൊണ്ട് ഇത്ര ഉണ്ട് മന്ദനീചോച്ചവൃത്തം, ഇത്ര ഉണ്ട് ശീഘ്രനീചോച്ചവൃത്തമെന്നു പഠി ക്കുന്നത്. എന്നിട്ടു പ്രതിമണ്ഡലകലാമിതമായിട്ടു നടേ ഉണ്ടാക്കുന്നു. എന്നിട്ട്[8] മന്ദനീചോച്ചവൃത്തത്തിന്നു വൃദ്ധിഹ്രാസമുണ്ട്. ശീഘ്രനീചോച്ചവൃത്തത്തിന്നു വൃദ്ധിഹ്രാസങ്ങൾ ഇല്ല എന്നു വിശേഷമാകുന്നത്. ഇവിടെ മന്ദസ്ഫുടഗ്ര ഹവും ശീഘ്രോച്ചവും തങ്ങളിലുള്ള അന്തരത്തിന്നുള്ള ജ്യാക്കൾ ശീഘ്ര കേന്ദ്രജ്യാക്കൾ എന്നു പേരാകുന്നവ മന്ദകർണ്ണവൃത്തത്തിലെ ജ്യാവാകയാൽ മന്ദകർണ്ണകലാപ്രമിതങ്ങൾ. ശീഘ്രവൃത്തം പിന്നെ പ്രതിമണ്ഡലകലാപ്രമി തമാകയാൽ ശീഘ്രോച്ചനീചവൃത്തത്തേയും അതിന്റെ വ്യാസാർദ്ധമാകുന്ന ശീഘ്രാന്ത്യഫലത്തേയും ത്രിജ്യയെക്കൊണ്ടു ഗുണിച്ചു മന്ദകർണ്ണം കൊണ്ടു ഹരിച്ചാൽ ഫലങ്ങൾ മന്ദകർണ്ണവൃത്തകലാപ്രമിതങ്ങളായിട്ടിരിക്കുന്ന ശീഘ്രോച്ചനീചവൃത്തവും അതിന്റെ വ്യാസാർദ്ധവും ആയിട്ടു വരും. ഇവ പ്രമാണഫലങ്ങളായിട്ടു മന്ദകർണ്ണവൃത്തം സ്വകലാപ്രമിതങ്ങളായിട്ടിരിക്കു ന്നതും അതിന്റെ വ്യാസാർദ്ധവും പ്രമാണഫലങ്ങളായിട്ടു കല്പിച്ചു ശീഘ്ര കേന്ദ്രഭുജാകോടിജ്യാക്കളെ ഇച്ഛാരാശിയായിട്ടും കല്പിച്ച് ഉണ്ടാകുന്ന ഇച്ഛാ ഫലങ്ങൾ മന്ദകർണ്ണവൃത്തകലാപ്രമിതങ്ങളായിട്ടിരിക്കുന്ന ശീഘ്രകോടിഭുജാ ഫലങ്ങളായിട്ടുവരും. പിന്നെ സ്വകലാപ്രമിതമായി[9] ത്രിജ്യാതുല്യമായിട്ടിരി ക്കുന്ന മന്ദകർണ്ണവൃത്തവ്യാസാർദ്ധത്തിങ്കൽ ഈ കോടിഫലം സംസ്ക്കരിച്ച് പിന്നെ ഇതിന്റെ വർഗ്ഗവും ഭുജാഫലവർഗ്ഗവും കൂട്ടി മൂലിച്ചാൽ ശീഘ്രോച്ച നീചവൃത്തകേന്ദ്രമായിട്ടിരിക്കുന്ന ഭഗോളമദ്ധ്യത്തിങ്കേന്നു ഗ്രഹത്തോടുള്ള ഇട മന്ദകർണ്ണവൃത്തകലാപ്രമിതമായിട്ടുണ്ടാകും. ഇതു 'ശീഘ്രകർണ്ണ'മാ കുന്നത്.

---

14. 7. B.സ്വ സ്വ for തന്റെ തന്റെ
 8. B. എന്നേടത്ത്
 9. D. കലാമിതമായി

ഇങ്ങനെ പലപ്രകാരം വരുത്താം—

(i) ഇവിടെ മന്ദകർണ്ണവൃത്തകലാപ്രമിതമായിട്ടിരിക്കുന്ന ശീഘ്രാന്ത്യ ഫലത്തെ ശീഘ്രകോടിജ്യാവിങ്കൽ മകരാദികർക്ക്യാദിക്കു തക്കവണ്ണം യോഗം താനന്തരം താൻ ചെയ്ത് ഇതിനേയും ശീഘ്രകേന്ദ്രഭുജാജ്യാവിനേയും വർഗ്ഗിച്ചു കൂട്ടി മൂലിച്ചാലുമുണ്ടാകും മുമ്പിൽ ചൊല്ലിയ ശീഘ്രകർണ്ണം. ഇവിടെ മന്ദസ്ഫുടഗ്രഹവും[10] ശീഘ്രോച്ചമാകുന്ന ആദിത്യമധ്യമവും തങ്ങളിലന്തരി ച്ചശേഷം ശീഘ്രകേന്ദ്രമാകുന്നത്. ഇതിന്റെ ഭുജാകോടിജ്യാക്കൾ മന്ദകർണ്ണ വൃത്തകലാപ്രമിതങ്ങൾ ആ വൃത്തത്തിലെ ജ്യാക്കളാകയാൽ അവറ്റെ പിന്നെ മന്ദകർണ്ണം കൊണ്ടു ഗുണിച്ചു ത്രിജ്യകൊണ്ടു ഹരിച്ചാൽ മന്ദകർണ്ണവൃത്ത ത്തിലെ ജ്യാക്കൾ പ്രതിമണ്ഡലകലാപ്രമിതങ്ങളായിട്ടു വരും.

(ii) പിന്നെ ഇവറ്റേ കേവലം പ്രതിമണ്ഡലകലാപ്രമിതമായി പഠിച്ചിരി ക്കുന്ന ശീഘ്രാന്ത്യഫലത്തെക്കൊണ്ടു ഗുണിച്ച് മന്ദകർണ്ണം കൊണ്ടു ഹരി ച്ചാൽ ശീഘ്രകോടിഭുജാഫലങ്ങൾ പ്രതിമണ്ഡലകലാപ്രമിതമായിട്ടുണ്ടാകും. പിന്നെ ഈ കോടിഫലത്തെ മന്ദകർണ്ണത്തിൽ സംസ്കരിച്ച് അതിന്റെ വർഗ്ഗവും ഈ ഭുജാഫലവർഗ്ഗവും കൂട്ടി മൂലിച്ചാൽ ശീഘ്രകർണ്ണം പ്രതിമ ണ്ഡലകലാപ്രമിതമായിട്ടുണ്ടാകും.

(iii) പിന്നെ പ്രതിമണ്ഡലകലാപ്രമിതങ്ങളായിട്ടിരിക്കുന്ന ശീഘ്രകേന്ദ്ര കോടിജ്യാവും അന്ത്യഫലവും തങ്ങളിൽ യോഗം താനന്തരം താൻ ചെയ്ത് പിന്നെ അതിന്റെ വർഗ്ഗവും ഈവണ്ണമിരിക്കുന്ന ഭുജാജ്യാവർഗ്ഗവും കൂട്ടി മൂലിച്ചാൽ ശീഘ്രകർണ്ണം പ്രതിമണ്ഡലകലാപ്രമിതമായിട്ടുണ്ടാകും.

(iv) പിന്നെ ശീഘ്രകേന്ദ്രഭുജാജ്യാവിനെ ത്രിജ്യകൊണ്ടു ഗുണിച്ച് കർണ്ണം കൊണ്ടു ഹരിച്ചാൽ ഫലം ശീഘ്രോച്ചനീചസൂത്രത്തോടു ഗ്രഹത്തോടുള്ള അന്തരാളജ്യാവു ശീഘ്രകർണ്ണവൃത്തകലാപ്രമിതമായിട്ടുണ്ടാകും. ഇതിനെ ചാപിച്ച് ശീഘ്രോച്ചത്തിൽ സംസ്കരിച്ചാൽ ഭഗോളമധ്യം കേന്ദ്രമായിട്ടിരി ക്കുന്ന ശീഘ്രകർണ്ണവൃത്തത്തിൽ ഗ്രഹം ഇത്ര ചെന്നു എന്നുണ്ടാകും. പിന്നെ ഭുജാഫലത്തെ ത്രിജ്യകൊണ്ടു ഗുണിച്ചു കർണ്ണം കൊണ്ടു ഹരിച്ചു ചാപിച്ച തിനെ മന്ദസ്ഫുടഗ്രഹത്തിൽ സംസ്കരിച്ചാൽ ഈ സ്ഫുടഗ്രഹം തന്നെ വരും. ഇവിടെ മന്ദകർണ്ണകലാപ്രമിതമായിട്ടിരിക്കുന്ന ഭുജാജ്യാവിനേത്താൻ ഭുജാഫലത്തെത്താൻ ത്രിജ്യകൊണ്ടു ഗുണിക്കിൽ മന്ദകർണ്ണകലാപ്രമിത

14. 10. D. മന്ദസ്ഫുടവും; F. മന്ദഗ്രഹസ്ഫുടവും

മായിട്ടിരിക്കുന്ന ശീഘ്രകർണ്ണം കൊണ്ടു ഹരിക്കേണ്ടൂ. പ്രതിമണ്ഡലകലാ പ്രമിതത്തെ എങ്കിൽ പ്രതിമണ്ഡലകലാപ്രമിതമായിട്ടിരിക്കുന്ന ശീഘ്രകർണ്ണം കൊണ്ടു ഹരിക്കേണ്ടൂ എന്നേ വിശേഷമുള്ളു[11]. ഇവിടെ യാതൊരു പ്രദേശ ത്തിങ്കൽ കേന്ദ്രമായിട്ടിരിക്കുന്ന വൃത്തത്തിന്റെ നേമീങ്കലേ[12] ഗ്രഹഗതി അറി ഞ്ഞത്, ആ വൃത്തത്തിന്ന് "ജ്ഞാതഭോഗഗ്രഹ"മെന്നു പേർ. ഇതിനെ പ്രതി മണ്ഡലമെന്നു[13] കല്പിക്കേണ്ടൂ. പിന്നെ യാതൊരിടം കേന്ദ്രമായി നേമിയി ങ്കൽ ഗ്രഹസ്പർശമായിട്ടിരിക്കുന്ന വൃത്തത്തിങ്കൽ ഗ്രഹം ചെന്നൂ എന്നു അറിയേണ്ടുവത് ആ വൃത്തത്തിന്നു "ജ്ഞേയഭോഗഗ്രഹ" എന്നു പേർ. ഇതിനെ കർണ്ണവൃത്തമെന്നു കല്പിക്കേണ്ടൂ. പിന്നെ ജ്ഞേയഭോഗഗ്രഹവൃ ത്തകേന്ദ്രത്തിങ്കൽ കേന്ദ്രമായിട്ട് ജ്ഞാതഭോഗഗ്രഹവൃത്തകേന്ദ്രത്തിങ്കൽ നേമിയായിട്ട് ഒരു വൃത്തം കല്പിപ്പൂ. ഇത് "ഉച്ചകേന്ദ്രവൃത്ത"മാകുന്നത്. ഇങ്ങനെ മൂന്ന് വൃത്തങ്ങളെ കല്പിച്ചു ശീഘ്രന്യായത്തിനു തക്കവണ്ണം കർണ്ണം വരുത്തി ഇച്ചൊല്ലിയവണ്ണം സ്ഫുടിച്ചാൽ ഇഷ്ടപ്രദേശത്തിങ്കൽ കേന്ദ്രമായി ഗ്രഹത്തിങ്കൽ[14] നേമിയായി കല്പിച്ചിരിക്കുന്ന വൃത്തത്തിലെ ഗ്രഹഗതി അറിയാം. ഇങ്ങനെ സാമാന്യം സ്ഫുടന്യായം.

പിന്നെ യാതൊരിടത്തു കക്ഷ്യാവൃത്തവും കക്ഷ്യാനേമീങ്കലെ ഉച്ചനീച വൃത്തവും കല്പിക്കുന്നു, അവിടെ കക്ഷ്യാവൃത്തം കല്പിക്കും പ്രകാരം. ജ്ഞേയഭോഗഗ്രഹവൃത്തത്തിന്റെ കേന്ദ്രം തന്നെ കേന്ദ്രമായിട്ടു ജ്ഞാതഭോ ഗഗ്രഹവൃത്തത്തോടു തുല്യമായിട്ട് ഒരു വൃത്തം കല്പിപ്പൂ. അതു കക്ഷ്യാ വൃത്തമാകുന്നത്. അതിന്റെ നേമീങ്കൽ ഉച്ചനീചവൃത്തം കല്പിപ്പൂ. ജ്ഞാത ജ്ഞേയകേന്ദ്രങ്ങളുടെ അന്തരാളവ്യാസാർദ്ധം മാനമായി കല്പിച്ചിട്ട്. ഇവിടെ ജ്ഞാതഭോഗഗ്രഹവൃത്തത്തിങ്കൽ യാതൊരിടത്തു ഗ്രഹം, ഈ കല്പിച്ച കക്ഷ്യാവൃത്തത്തിങ്കൽ അവിടം കേന്ദ്രമായിട്ട് ഉച്ചനീചവൃത്തകേന്ദ്രം കല്പി ക്കേണ്ടൂ. ഇങ്ങനെ അഞ്ചുവൃത്തം കല്പിച്ചിട്ടു സ്ഫുടോപപത്തി നിരൂപി ക്കേണ്ടൂ. ഇച്ചൊല്ലിയവണ്ണം[15] കുജഗുരുമന്ദന്മാരുടെ സ്ഫുടപ്രകാരം.

---

14. 11. B. ഹരിക്കണം എന്നു വിശേഷം
12. H. add. പിന്നെ യാതൊരിടം കേന്ദ്രമായി നേമിയിങ്കൽ ഗ്രഹസ്പർശമായിരിക്കുന്ന വൃത്തത്തിങ്കൽ
13. F. പ്രതിമണ്ഡലത്തിന്റെ നേമിയിങ്കലേടം എന്നു പേർ കല്പിക്കേണ്ടു
14. F. ഇഷ്ടഗ്രഹത്തിങ്കൽ
15. B. ഏവം

15. ബുധശുക്രന്മാരുടെ ശീഘ്രസ്ഫുടം

ബുധശുക്രന്മാർക്ക് വിശേഷമുണ്ട്. അവിടേയും മന്ദസ്ഫുടം ഈവണ്ണം തന്നെ. ശീഘ്രസ്ഫുടത്തിങ്കൽ ശീഘ്രോച്ചനീചവൃത്തം വലുത്. മന്ദകർണ്ണ വൃത്തം ചെറുത് ആകയാൽ. മന്ദകർണ്ണവൃത്തനേമീടെ പുറത്ത് അകപ്പെടും ശീഘ്രോച്ചനീചവൃത്തത്തിന്റെ കേന്ദ്രം. ഇങ്ങനെ യാതൊരിടത്ത് ജ്ഞാത ഭോഗഗ്രഹമായിരിക്കുന്ന വൃത്തത്തിന്റെ വ്യാസാർദ്ധത്തേക്കാട്ടിൽ ജ്ഞാത ജ്ഞേയഗ്രഹവൃത്തകേന്ദ്രാന്തരം ആകുന്ന ഉച്ചനീചവൃത്തത്തിന്റെ വ്യാസാർദ്ധം വലിയൂ എന്നിരിക്കുന്നിടത്ത് ഈ ഉച്ചനീചവൃത്തസ്ഥാനീയമാ കുന്നതിനെ കക്ഷ്യാവൃത്തമെന്നും പ്രതിമണ്ഡലസ്ഥാനീയമായിരിക്കുന്ന ജ്ഞാതഭോഗഗ്രഹവൃത്തത്തെ കക്ഷ്യാവൃത്തനേമീങ്കലേ ഉച്ചനീചവൃത്ത മെന്നും കല്പിപ്പൂ. കക്ഷ്യാവൃത്തകേന്ദ്രം തന്നെ കേന്ദ്രമായിട്ടു നേമീങ്കൽ ഗ്രഹസ്പർശം വരുമാറു ജ്ഞേയഭോഗഗ്രഹവൃത്തമെന്നു പേരാകുന്ന കർണ്ണ വൃത്തത്തേയും കല്പിച്ചിട്ട് സ്ഫുടക്രിയയെ നിരൂപിക്കേണ്ടൂ.

ഇവിടെ പിന്നെയും രണ്ടു വൃത്തത്തെ കല്പിക്കേണ്ടുകിൽ ഈ കല്പിച്ച കക്ഷ്യാനേമീങ്കലെ ജ്ഞാതഭോഗഗ്രഹവൃത്തത്തോടു തുല്യമായിട്ടു കക്ഷ്യാ കേന്ദ്രത്തിങ്കൽ[1] കേന്ദ്രമായി ഒരു വൃത്തത്തെ കല്പിപ്പൂ. ഇതിനെ ജ്ഞാത ഭോഗഗ്രഹവൃത്തത്തോടു തുല്യമായി ജ്ഞേയഭോഗഗ്രഹവൃത്തകേന്ദ്രമായി രിക്കയാൽ കക്ഷ്യാവൃത്തം എന്നു മുമ്പിലേ ന്യായം കൊണ്ടു കല്പിക്കേ ണ്ടു. എങ്കിലും ജ്ഞാതഭോഗഗ്രഹവൃത്തത്തോടു സ്പർശമില്ലായ്കയാൽ ഉച്ചനീചവൃത്തമെന്ന് കല്പിച്ചുകൊള്ളൂ. പിന്നെ ജ്ഞാതഭോഗഗ്രഹവൃത്തത്തി ങ്കൽ മന്ദസ്ഫുടഗ്രഹം എത്ര ചെന്നിരിക്കുന്നു. ഇക്കല്പിച്ച ഉച്ചനീചവൃത്ത ത്തിങ്കൽ[2], അക്കലയിങ്കൽ[3] കേന്ദ്രമായിട്ടു കല്പിച്ച് കക്ഷ്യാവൃത്തത്തോളം പോന്നൊരു പ്രതിമണ്ഡലത്തെ കല്പിപ്പൂ. ഇങ്ങനെ കല്പിക്കുമ്പോൾ കക്ഷ്യാ പ്രതിമണ്ഡലങ്ങളെ ഉച്ചനീചവൃത്തമെന്നും ഉച്ചനീചവൃത്തങ്ങളെ കക്ഷ്യാ പ്രതിമണ്ഡലങ്ങളെന്നും കല്പിച്ചതായി. ജ്ഞേയവൃത്തത്തെ[4] കർണ്ണവൃത്ത മെന്നുതന്നെയും കല്പിപ്പൂ. ആകയാൽ കല്പിതപ്രതിമണ്ഡലത്തിന്റെ കേന്ദ്രം ഗ്രഹഗതിയായിട്ടു ഗമിക്കുന്നു എന്നിരിക്കുന്നു. എങ്കിലും അതിനെ ഉച്ചഗതി

---

15. 1. D.adds കല്പിതകക്ഷ്യവൃത്തകേന്ദ്ര
2. B. വൃത്തത്തിന്റെ
3. B. ആ കലയിങ്കൽ
4. D. add ജ്ഞേയഭോഗഗ്രഹ വൃത്തത്തെ

എന്നും ഇതിന്റെ കേന്ദ്രമിരിക്കുന്നേടം ഉച്ചമെന്നും കല്പിക്കേണം. പിന്നെ ജ്ഞാതഭോഗഗ്രഹവൃത്തത്തിന്റെ കേന്ദ്രഗതി ഉച്ചഗതി എന്നിരിക്കുന്നതാകിലും ഗ്രഹഗതി എന്നു കല്പിപ്പൂ. ജ്ഞാതഭോഗഗ്രഹവൃത്തത്തെ ഉച്ചനീചവൃത്തമെന്നു കല്പിക്കയാൽ. പിന്നെ ജ്ഞാതഭോഗഗ്രഹവൃത്തത്തിന്റെ കേന്ദ്രം[5] യാതൊരിടത്തു കല്പിച്ചു കക്ഷ്യാവൃത്തനേമീങ്കൽ അവിടെ മദ്ധ്യമഗ്രഹം തുല്യകലാ എന്നു കല്പിപ്പൂ, എന്നാൽ അതിന്നു തക്കവണ്ണം കല്പിച്ചു പ്രതിമണ്ഡലത്തിങ്കൽ ഗ്രഹം ഗമിക്കും. ഇതു ജ്ഞാതഭോഗഗ്രഹവൃത്തത്തിന്റേയും കല്പിതപ്രതിമണ്ഡലത്തിന്റേയും നേമികളുടെ യോഗങ്ങളിൽ ഉച്ചപ്രദേശം, അടുത്തുള്ള യോഗത്തിങ്കല് എല്ലായ്പോഴും ഗ്രഹം ഇരിക്കയാൽ. പിന്നെ ജ്ഞാതഭോഗഗ്രഹവൃത്തത്തിങ്കലെ ഗ്രഹഗതി കക്ഷ്യാവൃത്തനേമീങ്കൽ കല്പിക്കുന്ന ഉച്ചനീചവൃത്തനേമീങ്കൽ കല്പിക്കുന്ന ഗ്രഹഗതി എന്നു കല്പിപ്പൂ. ഈവണ്ണമാകുമ്പോൾ ഗ്രഹത്തെ ഉച്ചമെന്നും ഉച്ചത്തെ ഗ്രഹമെന്നും കല്പിച്ചതായി എന്നു വന്നിരിക്കും. ആകയാൽ മന്ദസ്ഫുടത്തിങ്കൽ സംസ്കരിക്കുന്ന ശീഘ്രഭുജാഫലത്തെ ശീഘ്രോച്ചത്തിലും ശീഘ്രകർണ്ണകലാപ്രമിതമായിരിക്കുന്ന ശീഘ്രകേന്ദ്രഭുജാജ്യാവിനെ മന്ദസ്ഫുടഗ്രഹത്തിങ്കലും[6] സംസ്കരിപ്പൂ. എന്നാൽ ബുധശുക്രന്മാരുടെ ഗോളഗതി[7] ഉണ്ടാകും.

ഇവിടേയ്ക്ക് അപേക്ഷയുണ്ടാകയാൽ മുമ്പിൽ അഞ്ചു വൃത്തങ്ങളിൽ കൂടീട്ടു കല്പിച്ച ഗ്രഹോച്ചഗതികളേയും സ്ഫുടോപപത്തിയേയും കാട്ടി. എന്നിട്ട് അതിനെ സാവധാനമായി നിരൂപിച്ചാൽ ഇപ്രകാരം ബുദ്ധ്യധിരൂഢമാകും[8]. ഇവിടെ കുജാദികൾക്കും പ്രതിമണ്ഡലകലകളേക്കൊണ്ടു മാനം ചെയ്തിട്ട് ഇത്ര എന്നു പഠിച്ചിരിക്കുന്നു, മന്ദവൃത്തത്തേയും ശീഘ്രവൃത്തത്തേയും. ബുധശുക്രന്മാർക്ക് പിന്നെ ശീഘ്രവൃത്തം വലുതാകയാൽ ഇതിന്റെ കല കൊണ്ടു പ്രതിമണ്ഡലത്തെ മാനം ചെയ്തതിനെ[9] ശീഘ്രവൃത്തമായിട്ടു പഠിച്ചിരിക്കുന്നു, ഇത്തന്ത്രസംഗ്രഹത്തിങ്കൽ. ഇതിനെ ഒഴിച്ചുള്ള ഗ്രന്ഥങ്ങളിൽ ബുധശുക്രന്മാരുടെ മന്ദവൃത്തത്തേയും ശീഘ്രവൃത്തമാനം കൊണ്ട് അളന്നു പഠിച്ചിരിക്കുന്നു. ഇത്തന്ത്രസംഗ്രഹത്തിങ്കൽ പിന്നെ പ്രതിമണ്ഡല

15. 5. B. വൃത്തകേന്ദ്രം
6. B.D. മന്ദസ്ഫുടത്തിലും
7. D. ഭഗോളഗതി
8. D. ബുദ്ധ്യാരൂഢമാകും; ബുദ്ധിരൂഢമാകും
9. D. ചെയ്തിട്ട് അതിനെ

കലാമാനം കൊണ്ടു തന്നെ അളന്നു മന്ദനീചോച്ചവൃത്തങ്ങളെ പഠിച്ചിരിക്കുന്നു. എന്നിട്ട്[10] സ്വമദ്ധ്യത്തിങ്കേന്നു മന്ദോച്ചം വാങ്ങിയാൽ[11] മന്ദസ്ഫുടന്യായേന ഉണ്ടാക്കിയ മന്ദഫലത്തെ തന്റെ[12] മധ്യമത്തിൽ തന്നെ സംസ്കരിച്ച് ആ മന്ദസ്ഫുടത്തെ പിന്നെ ശീഘ്രോച്ചമെന്നു കല്പിച്ച് ആദിത്യമധ്യമത്തെ തന്റെ മധ്യമമെന്നു കല്പിച്ച് ശീഘ്രസ്ഫുടം ചെയ്യുന്നു. പിന്നെ മന്ദോച്ചനീചവൃത്തം പ്രതിമണ്ഡലത്തേക്കാൾ ചെറുത് ഇവറ്റിന്ന് എന്നിട്ട് മന്ദസ്ഫുടം കഴിവോളം സാമാന്യം ന്യായമത്രേ ബുധശുക്രന്മാർക്കും. ശീഘ്രസ്ഫുടത്തിങ്കൽ തന്നെ ഗ്രഹോച്ചങ്ങളേയും ഇവറ്റിന്റെ ഗതികളേയും ഇവറ്റിന്റെ വൃത്തങ്ങളേയും പകർന്നു കല്പിക്കേണ്ടൂ. അവിടെ മന്ദകർണ്ണത്തെ ശീഘ്രാന്ത്യഫലം കൊണ്ടു ഗുണിച്ച് ത്രിജ്യകൊണ്ടു ഹരിച്ചാൽ ശീഘ്രവൃത്തകലാമിതമായിരിക്കുന്ന മന്ദകർണ്ണവൃത്തവ്യാസാർദ്ധമുണ്ടാകും. പ്രതിമണ്ഡലത്തെ മുമ്പിൽ ശീഘ്രകർണ്ണവൃത്തമായിട്ടു പഠിച്ചിരിക്കുന്നു. മന്ദകർണ്ണവൃത്തത്തെ ശീഘ്രകർണ്ണവൃത്തമായിട്ടു കല്പിക്കേണ്ടൂ എന്നിതു ഹേതുവാകുന്നത്. ഇത്രേ വിശേഷമുള്ളൂ ബുധശുക്രന്മാർക്ക്. ഇങ്ങനെ ചൊല്ലി താരാഗ്രഹങ്ങളുടെ സ്ഫുടത്തെ വിക്ഷേപമില്ലാത്ത നേരത്തേക്ക്.

16. സവിക്ഷേപഗ്രഹത്തിൽ ശീഘ്രസംസ്കാരം

അനന്തരം വിക്ഷേപമുള്ള നേരത്തേക്കു വിശേഷമുണ്ട്. അതിനെ ചൊല്ലുന്നു[1]. ഇവിടെ[2] ഭഗോളമദ്ധ്യത്തിങ്കൽ 'അപക്രമം' എന്നുണ്ടൊരു വൃത്തം ഉള്ളൂ[3]. ഇതിന്റെ കാലദേശങ്ങൾക്കു തക്കവണ്ണമുള്ള സംസ്ഥാനഭേദത്തെ ഇവിടേയ്ക്ക് അപേക്ഷയില്ലായ്കയാൽ നേരേ മേൽകീഴായി കിഴക്കുപടിഞ്ഞാറായി ഇരിപ്പൊന്നു ഈ[4] അപക്രമവൃത്തം എന്ന് കല്പിപ്പൂ. ഇതിന്റെ നേമീങ്കൽ പന്ത്രണ്ട് അംശം കല്പിച്ച് ഇവറ്റിന്റെ[5] വൃത്താർദ്ധങ്ങളിൽ അകപ്പെടുന്ന ഈരണ്ട് അംശങ്ങളിൽ സ്പർശിക്കുമാറ് ആറുവൃത്തങ്ങളെ കല്പിപ്പൂ.

---

15. 10. F. om. സ്വ
11. D.F. വാങ്ങി
12. B. സ്വ for തന്റെ

16. 1. B. അഥ വിക്ഷേപമുള്ളപ്പോഴേക്ക്
2. B. om. ഇവിടെ
3. B. om. ഉള്ളൂ; F. ഉള്ളത്
4. B. E. om. ഈ
5. B. ഇവറ്റിൽ

അപക്രമവൃത്തത്തിന്റെ കേന്ദ്രത്തിങ്കേന്ന്[6] നേരെ തെക്കും വടക്കും ഇവ തങ്ങളിലുള്ള യോഗം. ഈ രണ്ടു യോഗത്തിന്നും[7] 'രാശികൂടങ്ങൾ' എന്നു പേർ. ആറു വൃത്തങ്ങളേക്കൊണ്ടു പന്ത്രണ്ട് അന്തരാളങ്ങൾ ഉണ്ടാകും. ഈ രണ്ടു വൃത്തങ്ങളുടെ പഴുതുകൾ പന്ത്രണ്ടു രാശികളാകുന്നത്. ഈ രാശികൾക്ക് അപക്രമവൃത്തത്തിങ്കൽ നടുവേ രാശികൂടങ്ങളിൽ രണ്ട് അഗ്രങ്ങളും. അവിടെ നടുവേ പെരികെ ഇടമുണ്ടായി ഇരുതലയും കൂർത്തിരിപ്പോ ചില രാശികൾ. പിന്നെ ഈ രാശികളേക്കൊണ്ട് ഈവണ്ണം തന്നെ അംശിച്ചു തീയതികൾ ഇലികൾ തുടങ്ങി[8] കല്പിച്ചുകൊള്ളൂ.

ഈവണ്ണമിരിക്കുന്നേടത്ത് ഈ അപക്രമവൃത്തത്തിന്റെ കേന്ദ്രംതന്നെ കേന്ദ്രമായി നേമിയും. അതിന്റെ മാർഗ്ഗത്തിങ്കൽ തന്നെ ആയിരുന്നൊന്ന് ശീഘ്രവൃത്തം. കേന്ദ്രത്തിന്നടുത്തുള്ളിടത്തിന്ന് അപക്രമമണ്ഡലത്തിന്നു 'ശീഘ്രോച്ചനീചവൃത്തം' എന്നു പേർ എന്ന് ഓർക്കിലുമാം. ഓരോ ഗ്രഹത്തിന്ന് ഓരോ പ്രകാരം ശീഘ്രവൃത്തത്തിന്റെ വലിപ്പം എന്നേ വിശേഷമുള്ളൂ. സംസ്ഥാനഭേദമില്ല; എല്ലാറ്റിനും ഒരു പ്രകാരം തന്നെ.

പിന്നെ ഈ ശീഘ്രവൃത്തനേമീങ്കൽ യാതൊരു പ്രദേശത്ത് ആദിത്യമദ്ധ്യമം, അവിടെ കേന്ദ്രമായിട്ടു മന്ദോച്ചനീചവൃത്തം. ഇത് എല്ലാറ്റിനും. ഈവണ്ണം. ഈ മന്ദോച്ചനീചവൃത്തത്തിന്റെ നേമീങ്കൽ പ്രതിലോമമായിട്ടു പാതന്റെ ഗതി. ഈ പാതൻ യാതൊരിടത്ത് മന്ദോച്ചനീചവൃത്തിന്റെ ആ പ്രദേശം അപക്രമമണ്ഡലത്തെ സ്പർശിക്കും. ഈ പാതങ്കേന്നു[9] തുടങ്ങി അപക്രമമണ്ഡലമാർഗ്ഗത്തിങ്കേന്നു വടക്കേപുറമേ ഇരിക്കും മന്ദോച്ചനീചവൃത്തത്തിന്റെ പാതിയും. പാതങ്കേന്നു വൃത്താർദ്ധം ചെല്ലുന്നേടം പിന്നെയും അപക്രമമണ്ഡലമാർഗ്ഗത്തെ സ്പർശിക്കും. പിന്നത്തെ അർദ്ധം അപക്രമമണ്ഡലമാർഗ്ഗത്തിന്റെ തെക്കേ പുറമേ. ഇവിടെ അപക്രമമണ്ഡലമാർഗ്ഗത്തിങ്കേന്നു എല്ലായിലുമകലുന്നേടം മന്ദോച്ചനീചവൃത്തത്തിന്റ[10] കലകൊണ്ട് അതതു ഗ്രഹത്തിന്റെ പരമവിക്ഷേപത്തോളം അകലും. പിന്നെ ഈ നീചോച്ചവൃത്തമാർഗ്ഗം

---

16. 6. C. B. അപക്രമകേന്ദ്രത്തിങ്കൽ
7. B.C.F.ഈ രണ്ടു യോഗത്തിന്നും രാശികൂടമെന്ന; D. ഈ രണ്ടു വൃത്തസമൂഹയോഗത്തി
8. B. മുതലായി For തുടങ്ങി എന്നു തുടങ്ങി
9. F. ഇതിങ്കേന്ന്
10. D. മന്ദോച്ചനീചവൃത്തകലകളെകൊണ്ട്; B.F.മന്ദോച്ചവൃത്തകലകൊണ്ട്

തന്നെ പ്രതിമണ്ഡലത്തിന്നു മാർഗ്ഗമാകുന്നത്. എന്നിട്ട് പ്രതിമണ്ഡലവും കൂടി നീചോച്ചവൃത്തത്തിനു തക്കവണ്ണം അപക്രമമണ്ഡലമാർഗ്ഗത്തിങ്കേന്നു തെക്കും വടക്കും ചരിഞ്ഞിരിക്കും. ഈവണ്ണം തന്നെ മന്ദകർണ്ണവൃത്തവും ചരിഞ്ഞിരിക്കും. എന്നിട്ടു മന്ദസ്ഫുടത്തിങ്കേന്നു വിക്ഷേപം കൊള്ളേണ്ടൂ.

അവിടെ മന്ദകർണ്ണവൃത്തത്തിനു മന്ദോച്ചവൃത്തത്തിന്റെ കേന്ദ്രം തന്നെ കേന്ദ്രമാകയാലും അതിന്നു തക്കവണ്ണം അപക്രമമണ്ഡലമാർഗ്ഗത്തിങ്കേന്നു[11] തെക്കും വടക്കും ചരിഞ്ഞിരിക്കയാലും മന്ദകർണ്ണവൃത്തത്തിന്റെ നേമി അപ ക്രമമണ്ഡലമാർഗത്തിങ്കേന്നു എല്ലായ്പോഴും[12] അകന്നേടം മന്ദകർണ്ണവൃത്ത കലകളേക്കൊണ്ടു പരമവിക്ഷേപത്തോളം അകന്നിരിക്കും. ആകയാൽ പാതോനമന്ദസ്ഫുടത്തിന്റെ ജ്യാവിനെ പരമവിക്ഷേപം കൊണ്ടു ഗുണിച്ചാൽ ത്രിജ്യകൊണ്ട് ഹരിക്കേണ്ടൂ, ഫലം മന്ദകർണ്ണവൃത്തത്തിങ്കലേ ഗ്രഹത്തിന്റെ ഇഷ്ടവിക്ഷേപം. ഇച്ചരിവിന്നു 'വിക്ഷേപ'മെന്നു പേരാകുന്നു.

ഈവണ്ണമിരിക്കുന്നിടത്തു യാതൊരിക്കൽ ഗ്രഹം അപക്രമമണ്ഡലമാർഗ്ഗ ത്തിങ്കേന്നു നീങ്ങിയിരിക്കുന്നൂ അപ്പോൾ ശീഘ്രോച്ചനീചവൃത്തത്തിന്നും മന്ദകർണ്ണവൃത്തത്തിന്നും ദിക്ക് ഒന്നേ അല്ലായ്കയാൽ മന്ദകർണ്ണവൃത്തം ശീഘ്രസ്ഫുടത്തിങ്കൽ പ്രതിമണ്ഡലമായി കല്പിപ്പാൻ യോഗ്യമല്ല. പിന്നെ പാതനോടു മന്ദസ്ഫുടഗ്രഹത്തിന്നു യോഗമുള്ളപ്പോൾ മന്ദകർണ്ണവൃത്തത്തെ ശീഘ്രോച്ചനീചവൃത്തത്തിന്നും നേരേ കല്പിച്ചുകൊള്ളാം. ഗ്രഹത്തിന്നു വിക്ഷേപമില്ലാത്തപ്പോൾ ഈ മന്ദകർണ്ണവൃത്തത്തെതന്നെ വിക്ഷേപിച്ചു കല്പിക്കേണ്ടാ. എന്നാൽ ഗ്രഹമിരിക്കുന്ന പ്രദേശം മന്ദകർണ്ണവൃത്തത്തിന് അപക്രമമണ്ഡലമാർഗ്ഗത്തിങ്കേന്ന് എല്ലായിലുമകന്നേടമെന്ന് കൽപിച്ച് അവി ടുന്ന് വൃത്തപാദം ചെല്ലുന്നേടത്ത് അപക്രമമണ്ഡലത്തോട് യോഗമെന്നും കല്പിച്ചാൽ ഈ വിക്ഷേപത്തെകൊണ്ടു ചെരിവ് ഉണ്ടാക്കാം. മന്ദകർണ്ണവൃ ത്തത്തിന് വിക്ഷേപമില്ലാത്തപ്പോൾ പിന്നെ മന്ദകർണ്ണവൃത്തവ്യാസാർദ്ധവർഗ്ഗ ത്തിൽ വിക്ഷേപവർഗ്ഗത്തെക്കളഞ്ഞ് മൂലിച്ച് വിക്ഷേപകോടിയെ ഉണ്ടാക്കാം. ഇതു ഗ്രഹത്തിങ്കൽ അഗ്രമായി മന്ദകർണ്ണവൃത്തകേന്ദ്രത്തിങ്കേന്ന് വിക്ഷേപ ത്തോളം അകന്നേടത്തു മൂലമായിരിപ്പൊന്ന് ഈ വിക്ഷേപകോടി. ഈ

---

16. 11. F. ക്രമമണ്ഡലത്തിങ്കന്ന്
12. H. എല്ലായ്പോഴും

വിക്ഷേപകോടിക്കു നേരെയുള്ള വ്യാസാർദ്ധമായിട്ട് ഒരു വൃത്തത്തെ കല്പിപ്പൂ. ഇത് അപക്രമമണ്ഡലമാർഗ്ഗത്തിങ്കേന്ന് എല്ലാ അവയവവുമൊപ്പമകന്ന് ഘടികാമണ്ഡലത്തിന്ന് യാതൊരു പ്രകാരം സ്വാഹോരാത്രവൃത്തം എന്നപോലെ ഇരുന്നൊന്ന് ഈ വിക്ഷേപകോടിവൃത്തം. ഇതിന്നു നേരേ നീക്കി കല്പിപ്പൂ ശീഘ്രോച്ചനീചവൃത്തം.

അപ്പോൾ ശീഘ്രോച്ചനീചവൃത്തമാർഗ്ഗത്തിന്നു നേരേ ഇരിക്കയാൽ വിക്ഷേപകോടിവൃത്തം പ്രതിമണ്ഡലമാകും ശീഘ്രസ്ഫുടത്തിങ്കലേക്ക്[13]. ഇവിടെ വിക്ഷേപത്തെ പ്രതിമണ്ഡലകലാപ്രമിതമാക്കി വർഗ്ഗിച്ചു മന്ദകർണ്ണവർഗ്ഗത്തിങ്കേന്നു കളഞ്ഞു മൂലിച്ച് വിക്ഷേപകോടിയെ ഉണ്ടാക്കാം. എന്നാൽ പ്രതിമണ്ഡലകലാപ്രമിതം വിക്ഷേപകോടി എന്നറിഞ്ഞിട്ട് ശീഘ്രഫലങ്ങളെ ഉണ്ടാക്കണം. മുമ്പിൽ ചൊല്ലിയ വിക്ഷേപകോടിയെ വ്യാസാർദ്ധമെന്നും ഇതിനെ മന്ദകർണ്ണമെന്നും കല്പിച്ച് മുമ്പിലേപ്പോലെ ശീഘ്രസ്ഫുടം ചെയ്വൂ. എന്നാൽ അപക്രമമണ്ഡലകേന്ദ്രത്തിങ്കേന്നു വിക്ഷപത്തോളം തെക്കുതാൻ വടക്കുതാൻ നീങ്ങിയ പ്രദേശം കേന്ദ്രമായി ഗ്രഹത്തെ സ്പർശിക്കുന്ന നേമിയോടു കൂടിയിരിക്കുന്ന ശീഘ്രകർണ്ണവൃത്തത്തിങ്കലെ ഗ്രഹസ്ഫുടമുണ്ടാകും. ഇതു തന്നെയത്രേ അപക്രമമണ്ഡലഭാഗത്തിങ്കലേ സ്ഫുടമാകുന്നതും. അപക്രമമണ്ഡലത്തിന്റെ ഇരുപുറവും ഉള്ള കോടിവൃത്തത്തിങ്കലും ഇലികൾ അപക്രമമണ്ഡലത്തോട് തുല്യങ്ങൾ, കോടിവൃത്തത്തിങ്കൽ കലാദികൾ ചെറുത് എന്നിട്ടു സംഖ്യാസാമ്യമുണ്ട്. പിന്നെ സ്വാഹോരാത്രവൃത്തത്തിങ്കൽ യാതൊരു പ്രകാരം പ്രമാണങ്ങൾ വലിയ ഘടികാവൃത്തത്തിങ്കലോളം സംഖ്യ ഉണ്ടാക്കിയിരിക്കുന്നു, അവ്വണ്ണം വിക്ഷേപകോടി[14] വൃത്തത്തിങ്കൽ ഇലികൾ. ഇതു മേലിൽ വ്യക്തമാകും.

പിന്നെ ശീഘ്രകർണ്ണവർഗ്ഗത്തിൽ വിക്ഷേപവർഗ്ഗം കൂട്ടി മൂലിച്ചാൽ അപക്രമമണ്ഡലകേന്ദ്രത്തിങ്കേന്ന് ഗ്രഹത്തോളമുള്ള അന്തരാളമുണ്ടാകും. ഇതിന്നു 'ഭൂതാരാഗ്രഹവിവര'മെന്നു പേർ. പിന്നെ മുമ്പിൽ ചൊല്ലിയ വിക്ഷേപത്തെ ത്രിജ്യകൊണ്ടു ഗുണിച്ച് ഭൂതാരാഗ്രഹവിവരം കൊണ്ടു ഹരിച്ച് ഉണ്ടായ വിക്ഷേപം ഭഗോളത്തിങ്കലേ വിക്ഷേപമാകുന്നത്. അപക്രമമണ്ഡലകേന്ദ്രം

16. 13. B. സ്പുടത്തിങ്കലേയും
14. D. om. കോടി

തന്നെ കേന്ദ്രമായിരിക്കുന്ന ഭൂതാരാഗ്രഹവിവരവൃത്തത്തിന്റെ നേമി അപക്രമമണ്ഡലമാർഗ്ഗത്തിങ്കേന്ന് എത്ര അകലമുണ്ട് എന്നതു ഭഗോളവിക്ഷേപമാകുന്നത്. ഈ ഭൂതാരാഗ്രഹവിവരം വേണ്ടാ സ്ഫുടിച്ചാൽ രാശികൂടത്തോട് അണവിന്നു തക്കവണ്ണം ചെറുതായിരിക്കുന്ന ഇലി, എന്നിട്ടു കോടിവൃത്തത്തിങ്കലും അപക്രമവൃത്തത്തിങ്കലും രാശ്യാദികൾ യാതൊരു പ്രകാരം സ്വാഹോരാത്രവൃത്തത്തിങ്കലും ഘടികാവൃത്തത്തിങ്കലും പ്രാണങ്ങൾ സംഖ്യകൊണ്ടു സമങ്ങളായിരിക്കുന്നൂ. അവ്വണ്ണമാകയാൽ ഭൂതാരാഗ്രഹവിവരം വേണ്ടാ ശീഘ്രഭുജാഫലത്തെ ഉണ്ടാക്കുവാൻ. ഇങ്ങനെ[15] ചൊല്ലിയതായി സ്ഫുടക്രിയ.

അനന്തരം[16] ശീഘ്രോച്ചനീചവൃത്തത്തിന്നും അപക്രമമണ്ഡലമാർഗ്ഗത്തിങ്കേന്നു വിക്ഷേപമുണ്ട്. അതു മന്ദകർണ്ണവൃത്തത്തിന്റെ മാർഗ്ഗത്തിനു തക്കവണ്ണമല്ലാ താനും വിക്ഷേപം, ഈ ശീഘ്രവൃത്തത്തിങ്കേന്ന്[17] മന്ദകർണ്ണവൃത്തം മറ്റൊരുപ്രകാരം വിക്ഷേപിച്ച് ഇരിക്കുന്നു എന്നും ഇരിപ്പൂ എങ്കിൽ ഇങ്ങനെ സ്ഫുടത്തേയും വിക്ഷേപത്തേയും അറിയേണ്ടൂ എന്നതിനെ കാട്ടുന്നൂ. ഇവിടെ ശീഘ്രോച്ചനീചവൃത്തത്തിന്നു തന്റെ പാതസ്ഥാനം എവിടത്ത്[18] എന്നും ഇതിന്നു പരമവിക്ഷേപം എത്രയെന്നും അറിഞ്ഞു[19a] പിന്നെ ഇതിങ്കലേ മന്ദകർണ്ണവൃത്തകേന്ദ്രത്തിന്നും തല്ക്കാലത്തിങ്കൽ എത്ര വിക്ഷേപം എന്നറിയൂ. ഇതിനായിക്കൊണ്ടു ശീഘ്രോച്ചത്തിങ്കേന്നു ശീഘ്രവൃത്തപാതനെ വാങ്ങി ശേഷത്തിന്റെ ഭുജയ്ക്കു ജ്യാവുകൊണ്ട് തന്റെ പരമവിക്ഷേപം കൊണ്ടും ഗുണിച്ചു ത്രിജ്യകൊണ്ടു ഹരിച്ചാൽ ഫലം ശീഘ്രോച്ചനീചവൃത്തനേമീങ്കലേ മന്ദകർണ്ണവൃത്തകേന്ദ്രപ്രദേശം അപക്രമമണ്ഡലമാർഗ്ഗപ്രദേശത്തിങ്കേന്ന് ഇത്ര വിക്ഷേപിച്ചു എന്നതുണ്ടാകും. ഇഷ്ടാപക്രമം പോലെ ഈ വിക്ഷേപത്തെ വർഗ്ഗിച്ച് ത്രിജ്യാവർഗ്ഗത്തിൽ നിന്നു കളയണം[19b]. ശേഷത്തിന്റെ മൂലം വിക്ഷേപകോടി. അനന്തരം ഈ വിക്ഷേപകോടിയെ വ്യാസാർദ്ധമായി കല്പിച്ച് അപക്രമമണ്ഡലമാർഗ്ഗത്തിലൂടെ ഒരു വൃത്തം

---

16. 15. B. ഇനി സ്ഫുടക്രിയ
16. B. അഥഃ; D. om. അനന്തരം
17. D. ശീഘ്രോച്ചനീചവൃത്തത്തിങ്കേന്ന്
18. B. എവിടെ
19a F. അറിഞ്ഞു ഇരിപ്പു എങ്കിൽ ഇങ്ങനെ സ്ഫുടത്തെ വിക്ഷേപത്തേയും അറിഞ്ഞു
19b D. വർഗ്ഗത്തിങ്കേന്നു കളഞ്ഞ് മൂലിച്ചതിൽ വിക്ഷേപകോടി പിന്നെ വ്യാസാർദ്ധമായി ഒരു വൃത്തത്തെ കല്പിപ്പു. വിക്ഷേപത്തോളം നീങ്ങിയെന്ന് പിന്നെ.

വരയ്ക്കുക. പിന്നെ അത് അപക്രമമണ്ഡലമാർഗ്ഗത്തിങ്കന്ന് ഇഷ്ടവിക്ഷേപ ത്തോളം നീങ്ങിയേടത്തു പിന്നെ ഇതിനെ വിക്ഷേപകോടിയെ ശീഘ്രാന്ത്യ ഫലം കൊണ്ടു ഗുണിച്ചു ത്രിജ്യകൊണ്ടു ഹരിച്ചാൽ പ്രതിമണ്ഡലകലാപ്ര മിതമായിട്ടിരിക്കുന്ന വിക്ഷേപകോടി[20]വ്യാസാർദ്ധമുണ്ടാകും. പിന്നെ ഈ വിക്ഷേപകോടിവൃത്തത്തെ[21] ശീഘ്രനീചോച്ചവൃത്തമെന്നു കല്പിച്ച്, പിന്നെ മുമ്പിൽ ചൊല്ലിയ മന്ദകർണ്ണവൃത്തവിക്ഷേപകോടിവൃത്തത്തെ പ്രതിമണ്ഡലം എന്നു കല്പിച്ച് ശീഘ്രഭുജാഫലത്തെ ഉണ്ടാക്കി മന്ദസ്ഫുടത്തിൽ സംസ്ക രിപ്പൂ എന്നിങ്ങനെ വേണ്ടി വരും. ശീഘ്രോച്ചനീചവൃത്തത്തിനു വിക്ഷേപം വേറെ ഒരു മാർഗ്ഗത്തിങ്കൽ ഉണ്ടായിട്ടിരിക്കുന്നൂതാകിൽ പിന്നെ വിക്ഷേപി ച്ചിരിക്കുന്ന ശീഘ്രനീചോച്ചവൃത്തത്തിങ്കേന്നു ഇത്ര വിക്ഷേപിച്ചിരിക്കുന്നു മന്ദകർണ്ണവൃത്തമെന്ന് ആ പക്ഷത്തിങ്കൽ വരും. അപ്പോൾ അപക്രമമണ്ഡ ലമാർഗ്ഗത്തിങ്കേന്നു വടക്കോട്ടു വിക്ഷപിച്ചിരിക്കുന്ന പ്രദേശത്തിങ്കൽ ശീഘ്ര വൃത്തനേമീങ്കൽ കേന്ദ്രമായിരിക്കുന്ന മന്ദകർണ്ണവൃത്തത്തിങ്കൽ ഈ ശീഘ്ര വൃത്തനേമീങ്കേന്ന് നേരേ തെക്കോട്ട് വിക്ഷേപിച്ചിരിക്കുന്ന പ്രദേശത്തിങ്കല് ഗ്രഹം എന്നുമിരിപ്പൂ. എങ്കിൽ ഈ ശീഘ്രനീചോച്ചവൃത്തത്തിന്റേയും മന്ദകർണ്ണവൃത്തത്തിന്റേയും ഇഷ്ടവിക്ഷേപങ്ങളുടെ അന്തരം അപക്രമമണ്ഡ ലമാർഗ്ഗത്തിങ്കേന്നു തൽക്കാലത്തിങ്കൽ ഗ്രഹത്തിനു വിക്ഷേപമാകുന്നത്. വിക്ഷേപങ്ങൾ രണ്ടുംകൂടി ഉത്തരംതാൻ ദക്ഷിണംതാൻ എന്നിരിക്കുന്ന താകിൽ വിക്ഷേപങ്ങളുടെ യോഗം തല്ക്കാലത്തിങ്കൽ ഗ്രഹത്തിന്നു വിക്ഷേ പമാകുന്നത്. അപക്രമമണ്ഡലമാർഗ്ഗത്തിങ്കേന്ന് ഉള്ള വിക്ഷേപം താനും അത്.

ഇങ്ങനെ[22] ജ്ഞാതഭോഗഗ്രഹത്തിനും ജ്ഞാതജ്ഞേയാന്തരാള[23]രൂപ മായിട്ടിരിക്കുന്ന[24] ഉച്ചനീചവൃത്തത്തിന്നും രണ്ടു മാർഗ്ഗത്തൂടെ വിക്ഷേപമു ണ്ട് എന്നിരിക്കുന്നൂതാകിൽ സ്ഫുടത്തിന്റേയും വിക്ഷേപത്തിന്റേയും പ്രകാ രത്തെ ചൊല്ലീതായി. സ്ഫുടപ്രകാരമിങ്ങനെയെല്ലാം സംഭവിക്കുമെന്നീ ന്യായത്തെ കാട്ടുവാനായിക്കൊണ്ട് ചൊല്ലുകയത്രേ ഇതിനെ ചെയ്തത്. ഇങ്ങനെ ഉണ്ടായിട്ടല്ല. പിന്നെ ഇവിടെ ഭഗോളമധ്യമം കേന്ദ്രമായിരിക്കുന്ന

---

16. 20. H. കോടിവൃത്തവ്യാസാർദ്ധ
21. H. കോടിവൃത്തത്തെ
22. F. ഈ
23. F. ഭുജാന്തരാള
24. B. മുണ്ടായിരിക്കുന്ന

വൃത്തത്തിങ്കൽ ഇത്രചെന്നൂ (ചൊവ്വ?) ചന്ദ്രബിംബഘനമധ്യം കേന്ദ്രമായിട്ടിരിക്കുന്ന വൃത്തത്തിങ്കൽ എത്രചെന്നു എന്നറിയേണ്ടുകിൽ ചന്ദ്രകക്ഷ്യാവൃത്തത്തെ ഉച്ചനീചവൃത്തമാക്കി കല്പിച്ച് സ്ഫുടക്രിയയെ നിരൂപിക്കുമ്പോൾ ഇപ്രകാരം സംഭവിക്കും. മറ്റേ പ്രകാരമെങ്കിലും കണക്കു ചന്ദ്രബിംബഘനമദ്ധ്യത്തിങ്കലേക്കു അറിഞ്ഞ് ഭഗോളമദ്ധ്യം കേന്ദ്രമായിട്ടിരിക്കുന്ന വൃത്തത്തിങ്കലേക്കു അറിയേണ്ടൂ എന്നിരിക്കുന്നതാകിലും, ഈവണ്ണമോർക്കണം.

17. സ്ഫുടത്തിൽ നിന്ന് മധ്യമാനയനം.

അനന്തരം[1] സ്ഫുടത്തെക്കൊണ്ടു മദ്ധ്യമത്തെ വരുത്തും പ്രകാരത്തെ ചൊല്ലുന്നു. അവിടെ ചന്ദ്രാദിത്യന്മാർക്ക് സ്ഫുടത്തിങ്കേന്ന് ഉച്ചംവാങ്ങിയ ശേഷത്തിന്റെ ഭുജാജ്യാവാകുന്നത്, ഉച്ചനീചസൂത്രത്തോടു ഗ്രഹത്തോടുള്ള അന്തരാളത്തിങ്കലേ ജ്യാവ്. ഇതിനെ കർണ്ണംകൊണ്ടു ഗുണിച്ച് ത്രിജ്യകൊണ്ടു ഹരിച്ച് പ്രതിമണ്ഡലകലാമിതമാക്കിയാൽ പ്രതിമണ്ഡലഭാഗത്തിന്റെ ജ്യാവായിട്ടു വരും. ഇതിനെ ചാപിച്ചാൽ പ്രതിമണ്ഡലൈകദേശത്തിന്റെ ജ്യാവ്. ഇതിനെ ഉച്ചത്തിൽതാൻ നീചത്തിൽതാൻ സംസ്കരിച്ചാൽ പ്രതിമണ്ഡലത്തിങ്കൽ ഇത്രചെന്നൂ ഗ്രഹം എന്നു വരും. പിന്നെ ദോഃഫലത്തെയെങ്കിലും ഈവണ്ണം പ്രതിമണ്ഡലകലാപ്രമിതമാക്കി ചാപിച്ച് മേഷതുലാദി വിപരീതമായി സ്ഫുടത്തിൽ സംസ്കരിപ്പൂ. എന്നാലും മധ്യമം വരും. ഇവിടെ യാതൊരുപ്രകാരം ത്രിജ്യാകർണ്ണങ്ങളുടെ അന്തരമിരിക്കുന്നൂ, ഉച്ചസ്ഫുടാന്തരദോർജ്യാവും ഉച്ചമധ്യാന്തരദോർജ്യാവും തങ്ങളിലുള്ള അന്തരവും അവ്വണ്ണമിരിക്കും, പ്രമാണതൽഫലങ്ങളും ഇച്ഛാതൽഫലങ്ങളും തങ്ങളിൽ.

17. 1. B. അഥ

18. സ്ഫുടത്തിൽനിന്ന് മധ്യമാനയനം- പ്രകാരാന്തരം

പിന്നെ ദോഃഫലത്തെ അവിശേഷിച്ചാലും സ്ഫുടം കൊണ്ടു മധ്യമത്തെ വരുത്താം. അതിന്റെ പ്രകാരമാകുന്നത് — സ്ഫുടത്തിങ്കേന്ന് ഉച്ചം വാങ്ങിയ ദോഃഫലത്തെ ഉണ്ടാക്കി മേഷതുലാദി വിപരീതമായി[1] സ്ഫുടത്തിൽതന്നെ സംസ്കരിച്ചാൽ മധ്യമം വരും സ്ഥൂലമായിട്ട്. പിന്നെ ഈ മധ്യമത്തിങ്കേന്ന് ഉച്ചം വാങ്ങി ദോഃഫലത്തെ ഉണ്ടാക്കി മുമ്പിലത്തെ[2] സ്ഫുടത്തിൽ തന്നെ സംസ്കരിച്ചു പിന്നെയും ഈ മധ്യമത്തിങ്കേന്ന് ഉച്ചം വാങ്ങി ദോഃഫലത്തെക്കൊണ്ടു നടേത്തേ[3] സ്ഫുടത്തിൽതന്നെ സംസ്കരിപ്പൂ. ഇങ്ങനെ അവിശേഷം വരുവോളം[4] എന്നാൽ മധ്യമം സൂക്ഷ്മമാകും. കർണ്ണമുണ്ടാക്കേണ്ടാ താനും ഒരിക്കലും മന്ദസ്ഫുടത്തിങ്കൽ. ഇവിടെ സ്ഫുടോച്ചാന്തരദോഃഫലത്തെ കർണ്ണം കൊണ്ടു ഗുണിച്ചു ത്രിജ്യകൊണ്ടു ഹരിക്കുന്നേടത്ത് ത്രിജ്യാകർണ്ണാന്തരം കൊണ്ടു സ്ഫുടദോഃഫലത്തെ ഗുണിച്ച് ത്രിജ്യകൊണ്ടു ഹരിച്ചാൽ ഫലാന്തരമുണ്ടാകും[5]. ഇതിനെ സ്ഫുടദോഃഫലത്തിൽ കൂട്ടൂ മകരാദിയിൽ, കളയൂ കർക്ക്യാദിയിങ്കൽ. എന്നാൽ മദ്ധ്യോച്ചാന്തരദോഃഫലമായിട്ടു വരും. ഇവിടെ ത്രിജ്യാകർണ്ണാന്തരമാകുന്നതു കോടിഫലം[6] മിക്കവാറും ദോഃഫലവർഗ്ഗയോഗം കൊണ്ടുള്ള വിശേഷം കുറയുമെല്ലോ. എന്നിട്ട് ഇവിടെ[7] ദോഃഫലത്തെ കോടിഫലം കൊണ്ടു ഗുണിച്ചു ത്രിജ്യകൊണ്ടു ഹരിച്ച ഫലം സ്ഫുടമദ്ധ്യമകേന്ദ്രദോഃഫലങ്ങളുടെ അന്തരമാകുന്നത്. ഇതു മിക്കവാറും സ്ഫുടദോഃഫലത്തിന്റെ വർത്തമാനഖണ്ഡജ്യാക്കളായിട്ടിരിക്കും[8]. കോടിജ്യാവിനെ അനുസരിച്ചെല്ലോ ഭുജാഖണ്ഡമിരിപ്പൂ. എന്നിട്ടു കോടിഫലത്തിന്നു തക്കവണ്ണം ഭുജാഫലഖണ്ഡമിരിപ്പൂ. ഭുജാഫലചാപത്തെ മന്ദജ്യാവെന്നു[9] കല്പിച്ചു കോടിഫലം കൊണ്ടു ഗുണിച്ചു

---

18. 1. H. adds മുമ്പിലത്തെ
2. D. മുന്നിലെ
3. C.F.om. ക്കൊണ്ടു നടേത്തെ
4. C. F. വരുമ്പോൾ
5. D. ഫലം സ്ഫുടം മധ്യമദോഃഫലങ്ങളുടെ അന്തരാളമാകുന്നത്
6. D. കോടിഫലസമം
7. C.F. അവിടെ
8. B. ആയിരിക്കും; C. യിട്ടിരിക്കും, F. ആയിരിക്കുന്ന
9. H. മന്ദജ്യാ

ത്രിജ്യകൊണ്ടു ഹരിച്ചാൽ ഭുജാഫലഖണ്ഡം വരും. ഇവിടെ ഭുജാഫലത്തെ ഖണ്ഡജ്യാവിനേക്കൊണ്ടു ഗുണിച്ച് ചാപത്തേക്കൊണ്ടു ഹരിച്ചാലും ഭുജാ ഫലത്തിന്റെ[10] ഭുജാഫലഖണ്ഡം വരും. ഇതു പിന്നെ തങ്കേന്നു കൊണ്ട ഭുജാഫലത്തെ സംസ്കരിച്ചിരിക്കുന്ന കേന്ദ്രത്തിങ്കേന്നുകൊണ്ട[11] ഭുജാഫല ത്തിലും ഈ ഭുജാഫലത്തിന്റെ ഭുജാഫലഖണ്ഡം ഏറിത്താൻ കുറഞ്ഞു താൻ ഇരിക്കും. എന്നാൽ സ്ഫുടകേന്ദ്രഭുജാഫലത്തെ വിപരീതമായി സംസ്കരിച്ച് അവിശേഷിച്ചാൽ മധ്യമകേന്ദ്രഭുജാഫലം വരും. അതിനെ സ്ഫുടത്തിൽ സംസ്കരിച്ചാൽ മധ്യമം വരും. ഇങ്ങനെ അർക്കേന്ദുക്കളുടെ സ്ഫുടത്തേക്കൊണ്ടു മധ്യമത്തെ വരുത്താം.

19. മറ്റു ഗ്രഹങ്ങളുടെ ശീഘ്രമധ്യമാനയനം.

ഈവണ്ണം മറ്റുള്ളവരുടെ[1] മന്ദസ്ഫുടം കൊണ്ടു മധ്യമത്തെ വരുത്താം. പിന്നെ ശീഘ്രസ്ഫുടകേന്ദ്രഭുജാഫലത്തെക്കൊണ്ടു[2] മന്ദസ്ഫുടത്തെ വരു ത്തുംപ്രകാരവും[3] ഈവണ്ണം തന്നെ. അവിടെ വിശേഷമുണ്ട്. അവിശേഷി ക്കേണ്ട[4]. കർണ്ണഗുണനവും ത്രിജ്യാഹരണവും വേണ്ടാ. ശീഘ്രസ്ഫു ടത്തെ കേന്ദ്രഭുജാജ്യാവിനെ വൃത്തം കൊണ്ടു ഗുണിച്ച് അശീതി കൊണ്ടു ഹരിച്ച് ശീഘ്രോച്ചനീചവൃത്തത്തിലേ ജ്യാവാക്കി ചാപിച്ച് മേഷതുലാദി വിപ രീതമായി ശീഘ്രസ്ഫുടത്തിൽ സംസ്കരിച്ചാൽ മന്ദസ്ഫുടമായിട്ടുവരും.

ഇവിടെ യാതൊരിടത്തു മധ്യമം കൊണ്ടു സ്ഫുടം വരുത്തുവാ നായിക്കൊണ്ടു ഭുജാഫലത്തെ ഉണ്ടാക്കുന്നേടത്തു കർണ്ണം കൊണ്ടു ത്രൈരാ ശികം ചെയ്യേണ്ടാത്തു, അങ്ങനെ[5] ഇരിക്കുന്ന മന്ദസ്ഫുടം കൊണ്ടു മധ്യമത്തെ വരുത്തുമ്പോൾ ഭുജാഫലത്തെ അവിശേഷിക്കേണം എന്നതിന്റെ ഉപപത്തിയെ ചൊല്ലിയെല്ലോ. ഇതു കൊണ്ടുതന്നെ വരും മന്ദസ്ഫുടത്തി

18. 10. D. ഭൂജാഫലാംശത്തിന്റെ
11. A. om. കൊണ്ടു
19. 1. B. അന്യേഷാം
2. B. ശീഘ്രം കൊണ്ട്
3. C.F. പ്രകാരം
4. C. അവശേഷിക്കുകയും വേണ്ട
5. B. ഇങ്ങനെ

ങ്കേന്നു ശീഘ്രസ്ഫുടത്തെ ഉണ്ടാക്കുന്നേടത്തു[6] കർണ്ണാപേക്ഷ ഉണ്ടാകയാൽ ശീഘ്രസ്ഫുടത്തിങ്കേന്നു മന്ദസ്ഫുടത്തെ വരുത്തുവാൻ കർണ്ണം വേണ്ടാ. ആകയാൽ ഭുജാഫലത്തെ അവിശേഷിക്കയും വേണ്ടാ, ന്യായം തുല്യമാകയാൽ[7]. ഈവണ്ണമാകുമ്പോൾ മന്ദസ്ഫുടം കൊണ്ടു ശീഘ്രസ്ഫുടത്തെ വരുത്തുന്നേടത്തു കർണ്ണംകൂടാതെ ശീഘ്രഭുജാഫലത്തെ അവിശേഷിച്ചു സംസ്കരിച്ചാൽ ശീഘ്രസ്ഫുടം വരും. ഇങ്ങനെ കർണ്ണംകൂട്ടി ത്രൈരാശികം ചെയ്തു ഭുജാഫലത്തെ ഉണ്ടാക്കുന്നേടത്തു കർണ്ണം കൂടാതെ അവിശേഷിക്കിലും ഭുജാഫലം തുല്യമായിട്ടിരിക്കും. ഇവിടെ കോടിഫലവും ത്രിജ്യയും കൂടീട്ടുള്ള ത്രൈരാശികത്തിങ്കലേ ഇച്ഛാഫലത്തെ ഭുജാഫലഖണ്ഡവും[8] ചാപഖണ്ഡവും കൂട്ടീട്ട് ഉണ്ടാക്കാം. ഇതിനെ ജ്യാപ്രകരണത്തിങ്കൽ വിസ്തരിച്ചു ചൊല്ലിയിരിക്കുന്നു. അവിടന്നു കണ്ടുകൊള്ളൂ.

പിന്നെ ഈ ന്യായം കൊണ്ടുതന്നെ മന്ദകർണ്ണവശാൽ യാതൊന്നു ശീഘ്രപരിധിക്കു വിശേഷം വരുന്നത്, അതുകൊണ്ടു യാതൊന്നു പിന്നെ ശീഘ്രഭുജാഫലത്തിങ്കൽ ഭേദം വരുന്നത്, അതിനെ മന്ദഫലഖണ്ഡമായിട്ടുണ്ടാക്കാം. ഇതിന്നായിക്കൊണ്ടു നടേ മദ്ധ്യമത്തിങ്കേന്നു ശീഘ്രോച്ചം വാങ്ങിയ ശീഘ്രഭുജാഫലത്തെ ഉണ്ടാക്കി മദ്ധ്യമത്തിൽ സംസ്കരിച്ച് അതിങ്കേന്ന് മന്ദോച്ചം വാങ്ങി മന്ദഫലത്തെ[9] വരുത്തുമ്പോൾ ആ മന്ദഫലത്തിൽ[10] ശീഘ്രഭുജാഫലഭാഗത്തിന്റെ മന്ദഫലഖണ്ഡജ്യാക്കൾ ഏറീട്ടിരിക്കും, കുറഞ്ഞിട്ടുതാൻ. പിന്നെ ഈ ഫലത്തെ[11] ഇങ്ങനെ വരുത്തുമ്പോൾ മന്ദകർണ്ണവശാൽ യാതൊന്നു ശീഘ്രഫലത്തിങ്കൽ വിശേഷമുണ്ടാകുന്നത് അത് ഇവിടെ കൂടി വന്നിരിക്കും. എന്നാൽ മന്ദഫലത്തെ മദ്ധ്യമത്തിൽ സംസ്കരിക്കുമ്പോൾ മന്ദകർണ്ണവശാൽ ശീഘ്രഭുജാഫലത്തിങ്കൽ ഉണ്ടാകുന്ന ഫലഭേദത്തെക്കൂട്ടി സംസ്കരിച്ചതായിട്ടുവരും.

ഇവിടെ മന്ദകർണ്ണവശാൽ യാതൊന്നു ശീഘ്രഭുജാഫലത്തിങ്കൽ വിശേഷമുണ്ടാകുന്നത്. അതിനെ ശീഘ്രഭുജാഫലത്തിങ്കേന്ന് വേറേ ഉണ്ടാകുമാറു

---

19. 6. D. വരുന്നേടത്തു
7. C. F. തുല്യമാക കൊണ്ട്
8. C. F. ഭുജാഫലവും
9. D. മന്ദഭുജാഫലത്തെ
10. B. F. ഏറിയിരിക്കും കുറഞ്ഞിരിക്കും താൻ
11. C. F. മന്ദഫല

നിരൂപിക്കുമ്പോൾ അതിന്നു രണ്ടു ത്രൈരാശികം ഉണ്ട്. അതിന്റെ നടേത്തേതിൽ ശീഘ്രഭുജാഫലത്തെ ത്രിജ്യ കൊണ്ടു ഗുണിച്ച് മന്ദകർണ്ണം കൊണ്ടു ഹരിപ്പൂ എന്ന്. പിന്നെ അതിനേയും ത്രിജ്യ കൊണ്ടു ഗുണിച്ച് ശീഘ്രകർണ്ണം കൊണ്ടു ഹരിപ്പൂ എന്നതു രണ്ടാമത്. പിന്നെ ഇതിനെ ശീഘ്രകേന്ദ്രത്തിനു തക്കവണ്ണം സംസ്കരിപ്പൂ എന്നിതും.

ഇങ്ങനെ ഈ രണ്ടു ത്രൈരാശികഫലവും ധനർണ്ണവ്യവസ്ഥയും മൂന്നും കൂടി എങ്ങനെ വരുന്നൂ, നടേ ശീഘ്രദോഃഫലത്തെ സംസ്കരിച്ചേടത്തിന്നു മന്ദഫലത്തെ കൊണ്ടാൽ എന്നതിനെ ചൊല്ലുന്നു. അവിടെ യാതൊന്നു ശീഘ്രദോഃഫലത്തെ ത്രിജ്യകൊണ്ടു ഗുണിച്ച് മന്ദകർണ്ണം കൊണ്ടു ഹരിച്ചാലേ ഫലം, ഇതും കേവലശീഘ്രദോഃഫലവും തങ്ങളിൽ ഉള്ള അന്തരം യാതൊന്ന് ഇത് നടേത്തേ ത്രൈരാശികത്തിന്റെ ഇച്ഛാതൽഫലാന്തരം. ഇതു നടത്തേ ഗുണ്യത്തെ തന്നെ ഗുണഹാരാന്തരത്തെക്കൊണ്ടു ഗുണിച്ച് ഹാരകം കൊണ്ടു ഹരിച്ചാലുണ്ടാകും. പിന്നെ മന്ദകോടിഫലം കൊണ്ടു ഗുണിച്ച് ത്രിജ്യകൊണ്ടു ഹരിച്ചാലുണ്ടാകും ഈ ഫലം മിക്കവാറും. ഇവിടെ നടേ ശീഘ്രദോഃഫലത്തെ സംസ്കരിച്ചിട്ട് പിന്നെ മന്ദദോഃഫലം കണ്ടാൽ അതിൽകൂടി ശീഘ്രദോഃഫലത്തിന്റെ മന്ദഖണ്ഡജ്യാക്കളുണ്ടാകും. ഇതും ശീഘ്രദോഃഫലത്തിങ്കേന്നു മന്ദകർണ്ണവശാൽ ഉള്ള വിശേഷമായിട്ടിരിക്കും. ഇങ്ങനെ ശീഘ്രദോഃഫലത്തിങ്കലേ നടേത്തേ ത്രൈരാശികത്തിന്റെ ഫലം മന്ദദോഃഫലത്തിൽ കൂട്ടി ഉണ്ടാക്കിക്കൊള്ളാം. പിന്നെ കേവലമധ്യത്തിങ്കൽ നിന്ന് ഉള്ള മന്ദദോഃഫലവും ശീഘ്രദോഃഫലവും സംസ്കരിച്ചേടത്തു കേവലമധ്യമത്തിങ്കേന്നുണ്ടാക്കിയ ശീഘ്രദോഃഫലവും മന്ദദോഃഫലവും തങ്ങളിൽ അന്തരം യാതൊന്ന് അതു മന്ദകർണ്ണവശാൽ ശീഘ്രദോഃഫലത്തിങ്കലുള്ള വിശേഷമാകുന്നത്. ഇങ്ങനെ നടേത്തെ ത്രൈരാശികഫലം.

പിന്നെ കേവലമധ്യമത്തിങ്കന്നുകൊണ്ട മന്ദഫലം സംസ്കരിച്ചു മധ്യമത്തിങ്കേന്നുണ്ടാക്കിയ ശീഘ്രദോഃഫലം യാതൊന്ന്, പിന്നെ ശീഘ്രദോഃഫലം സംസ്കരിച്ചേടത്തിന്നുകൊണ്ട് മന്ദദോഃഫലം സംസ്കരിച്ചു കേവലമധ്യമത്തിങ്കേന്നുണ്ടാക്കിയ[12] ശീഘ്രദോഃഫലവും തങ്ങളിലുള്ള അന്തരം യാതൊന്ന്

19. 12. B.C.D.F.om. കേവലമദ്ധ്യ.....to......ശീഘ്രദോഃഫലവും

അതു രണ്ടാം ത്രൈരാശികം കൊണ്ടുണ്ടാകുന്ന ഫലമാകുന്നത്. ശീഘ്ര കർണ്ണവശാലുണ്ടാകുന്ന വിശേഷം ശീഘ്രകർണ്ണഭുജാഖണ്ഡങ്ങളായിട്ടുണ്ടാകും. മന്ദകർണ്ണവശാലുണ്ടാകുന്ന ഫലം മന്ദഭുജാഖണ്ഡങ്ങളായിട്ടുണ്ടാകും.

ഇവിടെ കർണ്ണത്തെ ത്രിജ്യ എന്നും ത്രിജ്യാകർണ്ണാന്തരത്തെ കോടിഫലമെന്നും ദോഃഫലചാപത്തെ സമസ്തജ്യാവെന്നും ചാപഖണ്ഡാഗ്രത്തിങ്കലേ കോടിഫലത്തെ മധ്യമത്തിങ്കലേത് എന്നും കല്പിച്ചിട്ട് ഇച്ചൊല്ലിയപ്രകാരം ഇതു കൊണ്ടുണ്ടാകുന്ന സ്ഥൌല്യത്തെ ഉപേക്ഷിപ്പൂതും ചെയ്വൂ. പിന്നെ മന്ദകർണ്ണവശാൽ ശീഘ്രദോഃഫലത്തിലുണ്ടാകുന്ന വിശേഷത്തെ മന്ദദോഃഫലത്തിൽ കൂട്ടി ഉണ്ടാക്കുമ്പോൾ മന്ദകേന്ദ്രത്തിന് തക്കവണ്ണം സംസ്കാരം സംഭവിക്കേണ്ട[13], ശീഘ്രകേന്ദ്രത്തിന്നു തക്കവണ്ണം സംസ്കരിക്കണം[14].

ഇതിനെ മന്ദകേന്ദ്രവശാൽ സംസ്കരിച്ചാലും ഫലസാമ്യം വരും എന്നതിനെ കാട്ടുന്നു. ഇവിടെ മന്ദകർണ്ണവശാൽ ശീഘ്രദോഃഫലത്തിങ്കലേ വൃദ്ധിക്ഷയാംശം യാതൊന്ന് അത് മന്ദകർണ്ണത്തേക്കാൾ ത്രിജ്യ വലുതാകുമ്പോൾ ഏറും, ചെറുതാകുമ്പോൾ കുറയും. മന്ദകേന്ദ്രത്തിന്റെ കർക്കിമൃഗാദിക്കു തക്കവണ്ണം ഇരിക്കുമിത്. ഈ ഫലം മുമ്പിലേ ശീഘ്രഫലത്തിന്റെ മന്ദഫലം സംസ്കരിക്കുമ്പോൾ കഴിഞ്ഞിരിക്കും. ഇവിടെ മുമ്പിലെ ശീഘ്രഫലം ധനമായിട്ടിരിക്കുമ്പോൾ യാതൊരിക്കൽ മന്ദകേന്ദ്രം മേഷാദിരാശിത്രികത്തിങ്കൽ ഇരിക്കുന്നൂ, അപ്പോൾ മന്ദകർണ്ണം വലുതാകയാൽ ഇതിന്നു തക്കവണ്ണമുണ്ടാകുന്ന ശീഘ്രഫലം ചെറുതായിട്ടിരിക്കും. എന്നാൽ മന്ദകോടിക്കു തക്കവണ്ണമുണ്ടാകുന്ന ശീഘ്രഫലം കളകവേണ്ടുവത്. മന്ദഫലവും കളകവേണ്ടുവത്. എന്നാൽ രണ്ടും കൂടിക്കളയാം. അവിടെ പിന്നെ ശീഘ്രഫലം ധനം മന്ദകർക്ക്യാദിത്രികത്തിങ്കലൂ എന്നിരിക്കുമ്പോൾ മന്ദകർണ്ണവശാലുണ്ടാകുന്ന ശീഘ്രഫലാംശം ധനമായിട്ടിരിക്കും. പിന്നെ കേവലമന്ദകേന്ദ്രത്തേക്കാട്ടിൽ ശീഘ്രഭുജാഫലം കൂട്ടിയ മന്ദകേന്ദ്രം വലുതായിട്ടിരിക്കും. അതു യുഗ്മപദമാകുമ്പോൾ മേന്മേൽചെന്നോളം ഭുജാഫലം കുറഞ്ഞിരിക്കും. ഈ ഭുജാഫലം ഋണമാകുമ്പോൾ ചെറുതാകയാൽ ശീഘ്രാംശം[15] ധനമായിട്ടു വന്നുകൂടും ഫലത്തിങ്കൽ. പിന്നെ ശീഘ്രഫലം ധനം, മന്ദകേന്ദ്രം തുലാദിത്രിക

19. 13. B.F. സംഭവിക്കും
14. B.C.F. സംസ്ക്കരിക്കേണ്ട ഇത്
15. B. ശീഘ്രഫലാംശം

ത്തിങ്കലൂ എന്നിരിക്കുമ്പോൾ[16] മന്ദകർണ്ണം ത്രിജ്യയേക്കാൾ ചെറുതാകയാൽ ഇതിനേക്കൊണ്ടുണ്ടാകുന്ന ശീഘ്രഫലമധികം. മന്ദഫലം തുലാദിയാകയാൽ ധനം താനും. മന്ദകേന്ദ്രം ഓജപദമാകയാൽ ശീഘ്രഫലം സംസ്കരിച്ചു മധ്യമത്തിങ്കേന്നു ഉണ്ടാക്കിയ മന്ദഫലം വലുതായിട്ടിരിക്കും. ഈ ഫലം തുലാദിയാകയാൽ ധനം. എന്നാൽ ഇവിടേയും മന്ദകേന്ദ്രത്തിനു തക്കവണ്ണം ശീഘ്രാംശകത്തിന്റെ സംസ്കാരമുചിതം.

പിന്നെ മന്ദകേന്ദ്രം മകരാദിത്രികത്തിങ്കലൂ, ശീഘ്രഫലം ധനം എന്നുമിരിപ്പൂ അപ്പോൾ കേവലമന്ദകേന്ദ്രത്തേക്കാൾ ശീഘ്രഫലം സംസ്കരിച്ചിരിക്കുന്ന മന്ദഫലം വലുത്. ഇത് യുഗ്മപദമായിട്ട് ഗതഭാഗം ഏറുകയാൽ ഏഷ്യഭാഗമാകുന്ന ഭുജാചാപം ചെറുത്. എന്നിട്ട് ഇതിന്റെ മന്ദഫലം കേവലകേന്ദ്രമന്ദഫലത്തേക്കാൾ കുറയും. ഇതു മദ്ധ്യമത്തിൽ കൂട്ടുമ്പോൾ കുറഞ്ഞൊന്നു കൂടി എന്നിരിക്കുമ്പോൾ ഋണമായിട്ടിരിക്കുന്ന ശീഘ്രാംശത്തിന്റെ സംസ്കാരം കൂട്ടി ഇവിടെ ഫലിച്ചിരിക്കും, ഋണമാകുന്ന മന്ദകർണ്ണം ത്രിജ്യയേക്കാൾ വലുതാകയാൽ.

ഇങ്ങനെ ശീഘ്രഫലം ധനമാകുമ്പോൾ മന്ദകേന്ദ്രത്തിന്റെ നാലു പാദത്തിങ്കലും മന്ദകേന്ദ്രത്തിനു തക്കവണ്ണം ശീഘ്രാംശത്തിന്റെ സംസ്കാരമുചിതം എന്നു വന്നു. പിന്നെ ശീഘ്രഫലം ഋണമാകുമ്പോളും ഈ ന്യായത്തിനു തക്കവണ്ണം മന്ദകേന്ദ്രവശാൽ ശീഘ്രാംശത്തിന്റെ ധനർണ്ണപ്രകാരം ഊഹിച്ചുകൊള്ളൂ. എന്നാൽ മന്ദകർണ്ണവശാലുണ്ടാകുന്ന ശീഘ്രഫലാംശം ശീഘ്രകേന്ദ്രത്തിനു തക്കവണ്ണം സംസ്കരിക്കേണ്ടൂ എന്നിരിക്കുന്നതാകിലും[17] മന്ദഫലത്തിൽ കൂട്ടി ഉണ്ടാക്കി മന്ദകേന്ദ്രത്തിനു തക്കവണ്ണം സംസ്കരിച്ചാൽ അന്തരം വരിക ഇല്ല[18] ഫലത്തിങ്കൽ എന്നു വന്നു. ഈവണ്ണമാകുമ്പോൾ ശീഘ്രഫലത്തിങ്കൽ മന്ദകർണ്ണാപേക്ഷ ഇല്ല. ആകയാൽ ശീഘ്രഫലത്തെ ഉണ്ടാക്കി പഠിച്ചേക്കാം, മന്ദഫലത്തേയും, സ്ഫുടക്രിയയുടെ ലാഘവത്തിനായിക്കൊണ്ട്. ഇവിടെ മൂന്നു ഭുജാഫലത്തെ ഉണ്ടാക്കി രണ്ടു ഭുജാഫലത്തെ സംസ്കരിച്ച് മധ്യമം സ്ഫുടമാകുന്നത്[19] എന്നിങ്ങനെ ഒരു പക്ഷം.

---

19. 16. B. adds ത്രികത്തിങ്കലുയർന്നിരിക്കുന്നു ആകിൽ
17. B. om. എന്നതാകിലും
18. B. വരിക
19. F. മദ്ധ്യസ്ഫുടമാകന്നത്

യാതൊരിടത്തു പിന്നെ മന്ദകർണ്ണത്രിജ്യാന്തരാർദ്ധത്തിന്നു തക്കവണ്ണം ശീഘ്രോച്ചനീചവൃത്തത്തിന്നു[20] വൃദ്ധിഹ്രാസമുണ്ട് എന്ന പക്ഷമാകുന്നു, അവിടെ ശീഘ്രദോഃഫലത്തെ ത്രിജ്യ കൊണ്ടു ഗുണിച്ച് മന്ദകർണ്ണത്രിജ്യാ യോഗാർദ്ധം കൊണ്ടു ഹരിക്കണം. ഈ അംശത്തെ മന്ദഫലത്തിൽ കൂട്ടി ഉണ്ടാക്കുവാൻ ശീഘ്രഫലാർദ്ധം സംസ്കരിച്ചിരിക്കുന്ന മധ്യമത്തിങ്കേന്നു മന്ദഫലമുണ്ടാക്കണം. അത്രേ വിശേഷമുള്ളൂ. ശേഷം മുമ്പിൽ ചൊല്ലിയ വണ്ണം. പരഹിതത്തിങ്കൽ ബുധശുക്രൻമാരുടെ[21] സ്ഫുടത്തെ ചൊല്ലിയതെന്നഭിപ്രായം.

മാനസകാരർക്കു പിന്നെ മന്ദകർണ്ണത്രിജ്യാന്തരാർദ്ധത്തിനു തക്കവണ്ണം മന്ദനീചോച്ചവൃത്തത്തിനും വൃദ്ധിക്ഷയങ്ങളുണ്ടെന്നു പക്ഷമാകുന്നു. ആ പക്ഷത്തിങ്കൽ മന്ദഫലത്തേയും ശീഘ്രഫലത്തേയും ത്രിജ്യകൊണ്ടു ഗുണിച്ച് മന്ദകർണ്ണത്രിജ്യായോഗാർദ്ധം കൊണ്ടു ഹരിക്കേണം. ഇങ്ങനെ ഉണ്ടാക്കി സംസ്കരിപ്പൂ മന്ദഫലത്തെ. ശീഘ്രഫലത്തെ ഈ ഗുണഹാരാന്തരങ്ങളേക്കൊണ്ടു ഗുണിച്ച് ഹരിച്ച്[22] പിന്നെയും ത്രിജ്യകൊണ്ടു ഗുണിച്ച് ശീഘ്രകർണ്ണം കൊണ്ടു ഹരിച്ചു സംസ്കരിപ്പൂ. ഇങ്ങനെ ആ പക്ഷത്തിങ്കൽ സ്ഫുടക്രിയ. ഈവണ്ണമാകക്കൊണ്ടു മാനസത്തിൽ കോട്യർദ്ധം സംസ്കരിച്ചിരിക്കുന്ന മന്ദച്ഛേദം കൊണ്ടു മന്ദഫലത്തേയും ശീഘ്രഫലത്തേയും സംസ്കരിപ്പാൻ ചൊല്ലി. ഈ പക്ഷത്തിങ്കൽ മന്ദകർണ്ണം കൂടാതെ അതിന്റെ ഫലത്തെ ഉണ്ടാക്കേണ്ടുകിൽ മന്ദഫലത്തേയും ശീഘ്രഫലത്തേയും അർദ്ധിച്ചു മദ്ധ്യമത്തിൽ സംസ്കരിച്ച് പിന്നെ ഇതിങ്കേന്ന് ഉണ്ടാക്കിയ മന്ദഫലത്തെ കേവല മധ്യമത്തിൽ സംസ്കരിച്ച് ഇതിങ്കേന്നുണ്ടാക്കിയ ശീഘ്രഫലം ഈ മന്ദസ്ഫുടത്തിൽ തന്നെ സംസ്കരിപ്പൂ. എന്നാൽ സ്ഫുടം വരും. ഈ പക്ഷത്തിനു തക്കവണ്ണം നാലു സ്ഫുടമായിട്ടു ചൊല്ലുന്നു. പലവറ്റിലും മന്ദകർണ്ണം കൂടായ്കിലേ ശീഘ്രകർണ്ണഭുജാഫലത്തെ പഠിച്ചിയേക്കാവൂ. എന്നിട്ട് ഇവിടെ മന്ദകോടിഫലാർദ്ധം കൊണ്ടു ഭുജാഫലങ്ങളെ രണ്ടിനേയും ഗുണിക്കേണ്ടുകയാൽ മന്ദശീഘ്രഫലാർദ്ധങ്ങൾ രണ്ടിനും കൂടി മന്ദഫലമുണ്ടാക്കേണ്ടുകയാൽ ഭുജാഫലങ്ങൾ രണ്ടിന്റേയും അർദ്ധം സംസ്കരിച്ചേടത്തുന്ന് മന്ദ

---

19. 20. F. നിചോച്ചത്തിന്
21. C. ശുക്രന്മാർക്ക്
22. D. adds സംസ്കരിച്ച് പിന്നെ ഇതിങ്കന്ന് ഉണ്ടാക്കിയ മന്ദഫലത്തെ കേവല മദ്ധ്യമത്തിൽ സംസ്കരിച്ച് ഇതിങ്കന്ന് ഉണ്ടാക്കിയ ശീഘ്രഫലം

ഫലമുണ്ടാക്കുന്നു എന്ന് ഇവിടേക്ക് ഹേതുവാകുന്നത്. ഇങ്ങനെ ചൊല്ലിയതായി സ്ഫുടക്രിയ.

20. ബുധശുക്രന്മാരുടെ മധ്യമസ്ഫുടം

അനന്തരം ബുധശുക്രന്മാർക്ക് മന്ദനീചോച്ചവൃത്തത്തേയും പ്രതിമണ്ഡലത്തേയും ശീഘ്രോച്ചവൃത്തകലകളെക്കൊണ്ടു പഠിച്ചിരിക്കുന്നേടത്ത് ആ സ്ഫുടക്രിയയെ ചൊല്ലുന്നു. ഇവിടെ നടേ ശീഘ്രോച്ചനീചവൃത്തത്തേയും പ്രതിമണ്ഡലത്തേയും പകർന്നു കല്പിച്ചു കുജാദികളേപ്പോലെ മന്ദസ്ഫുടത്തേയും ശീഘ്രസ്ഫുടത്തേയും ചെയ്യാം. കല്പിതസ്വമധ്യമമാകുന്ന ആദിത്യമധ്യമത്തിൽ മന്ദഫലം സംസ്കരിച്ചതു മന്ദസ്ഫുടമാകുന്നത് എന്നു കല്പിപ്പൂ. ഭഗോളമധ്യത്തിങ്കേന്നു തുടങ്ങി ശീഘ്രോച്ചനീചവൃത്തമെന്നു കല്പിച്ച് പ്രതിമണ്ഡലകേന്ദ്രത്തിങ്കൽ നേമിസ്പർശം വന്നിരിക്കുന്ന വൃത്തം മന്ദകർണ്ണവൃത്തമാകുന്നത്. മന്ദനീചോച്ചവൃത്തനേമീങ്കൽ മന്ദപ്രതിമണ്ഡലകേന്ദ്രമിരിക്കുന്നു. കക്ഷ്യാവൃത്തനേമീങ്കലേ മന്ദോച്ചനീചവൃത്തത്തിങ്കലേ ഗ്രഹം എന്നപോലെ ആകയാൽ മന്ദസ്ഫുടത്തിങ്കേന്നു ഉണ്ടാക്കിയ ശീഘ്രഫലത്തെ ത്രിജ്യകൊണ്ടു ഗുണിച്ച് മന്ദകർണ്ണം കൊണ്ടു ഹരിക്കേണ്ടു മന്ദകർണ്ണവൃത്തകലാപ്രമിതമാവാൻ. ഇതിനെ കർണ്ണം കൂടാതെ വരുത്തേണ്ടുകിൽ ഇവിടെ ആദിത്യമധ്യത്തിൽ മന്ദഫലം സംസ്കരിച്ചതെല്ലോ മന്ദസ്ഫുടം. ആ മന്ദസ്ഫുടത്തിങ്കേന്ന് ഉണ്ടാക്കിയ ശീഘ്രഫലത്തെ കേവലമധ്യമത്തിൽ സംസ്കരിപ്പൂ. പിന്നെ അതിനെ മന്ദസ്ഫുടത്തിങ്കൽ സംസ്കരിക്കേണ്ടുകയാൽ ആ ശീഘ്രസ്ഫുടത്തിങ്കേന്നു കൊണ്ട മന്ദഫലത്തെ അതിങ്കൽ തന്നെ സംസ്കരിപ്പൂ. എന്നാൽ സ്ഫുടം വരും. ശീഘ്രകർണ്ണം കൊണ്ടുള്ള വിശേഷത്തെ നടേ പഠിക്കുമ്പോളെ കൂട്ടി ഉണ്ടാക്കിയെല്ലോ പഠിക്കുന്നു. ആകയാൽ ഈവണ്ണം വേണ്ടിവരും. ഇങ്ങനെ സ്ഫുടക്രിയ.

**[ഗണിതയുക്തിഭാഷയിൽ
ഗ്രഹഗതിയും സ്ഫുടവുമെന്ന
എട്ടാമധ്യായം സമാപ്തം]**

അദ്ധ്യായം ഒൻപത്

ഭൂ-വായു-ഭഗോളങ്ങൾ

1. ഭൂഗോളം

അനന്തരം[1] ഭൂ-വായു-ഭഗോളങ്ങളുടെ സംസ്ഥാനങ്ങളേയും ഗതികളേയും കാട്ടുന്നൂ. അവിടെ നക്ഷത്രഗോളത്തിന്റെ നടുവിൽ ആകാശത്തിങ്കൽ നേരെ[2] ഉരുണ്ടു തന്റെ ശക്തികൊണ്ടുതന്നെ, മറ്റൊരു ആധാരം[3] കൂടാതെ, എല്ലാ പുറവും സ്ഥാവരജംഗമാത്മകങ്ങളാകുന്ന എല്ലാ ജന്തുക്കളേയും എല്ലാ വസ്തുക്കളേയും ഭരിച്ചു നില്പൊന്ന് ഈ 'ഭൂമി'. പിന്നെ[4] ഭൂമീടെ[5] എല്ലാ പുറത്തുമുള്ള ആകാശത്തിങ്കന്നും[6] കനത്ത വസ്തുക്കൾ ഭൂമിയിങ്കൽ വീഴുമാറു സ്വഭാവമുണ്ട്. എന്നിട്ട് ആകാശത്തിങ്കേന്ന് എല്ലാടവും കീഴു ഭൂമി. ഭൂമീടെ എല്ലാ പുറത്തു നിന്നും മേലു "ആകാശം". പിന്നെ ഭൂമീടെ തെക്കേ പാതിയിങ്കൽ വെള്ളമാകുന്ന പ്രദേശം ഏറൂ. വടക്കെ പാതിയിങ്കൽ സ്ഥലമാകുന്ന പ്രദേശം ഏറൂ, വെള്ളമാകുന്ന പ്രദേശം കുറവൂ[7]. പിന്നെ 'ഭാരതഖണ്ഡത്തെ'ക്കുറിച്ചു മീത്തെപ്പുറം എന്നു തോന്നുന്നേടത്തു ജലസ്ഥലസന്ധിയിങ്കൽ 'ലങ്ക' എന്നുണ്ട് ഒരു പുരീ[8]. അവിടന്നു കിഴക്കുപടിഞ്ഞാറു ഭൂമിയെ ചുറ്റുമാറ് വൃത്താകാരേണ ഒരു രേഖ കല്പിപ്പൂ. ഇതിങ്കൽ പടിഞ്ഞാറു 'രോമകവിഷയം', കീഴേപുറത്തു 'സിദ്ധപുരം', കിഴക്കു'യവകോടി'. ഇങ്ങനെ നാലു പുരങ്ങളുള്ളവ[9].

---

1. 1. B. അഥ
2. B. om. നേരെ
3. B. ആധാരവും
4. E. F.om. പിന്നെ
5. E. adds ഈ
6. B.F ആകാശത്തീന്ന് എല്ലാടത്തീന്നും
7. B. അധികം വെള്ളമാകുന്ന വടക്കെ പാതിയിങ്കൽ
8. B. ലങ്കയെന്നൊരു പുരിയുണ്ട്
9. B. നാലു പുരങ്ങൾ

പിന്നെ ഈവണ്ണം ലങ്കയിങ്കന്നു തെക്കും വടക്കും കീഴെപ്പുറത്തും കൂടി ഭൂമിയെ ചുറ്റുമാറ് വൃത്താകാരേണ ഒരു രേഖ കല്പിപ്പൂ. ഇതിങ്കൽ വടക്കു "മഹാമേരു[10]", തെക്കു 'ബഡവാമുഖം[11]', കീഴെ 'സിദ്ധപുരം' ഈ രേഖ 'സമരേഖ' ആകുന്നത്. ഇസ്സമരേഖയിങ്കലു 'ഉജ്ജയിനീ' എന്ന നഗരം. ഇവിടെ മുമ്പിൽ ചൊല്ലിയ പൂർവ്വാപരരേഖയിങ്കലേക്കു 'നിരക്ഷദേശം' എന്നു പേരുണ്ട്.

ആ വൃത്തമാർഗ്ഗത്തിങ്കൽ എല്ലാടത്തിന്നും 'ധ്രുവൻ' എന്ന ഒരു നക്ഷത്രത്തെ ഭൂപാർശ്വത്തിങ്കൽ തെക്കും വടക്കും അനുദയാസ്തം കാണാം. ഈ പ്രദേശത്തിങ്കന്നു വടക്കോട്ടു നീങ്ങിയാൽ വടക്കെ ധ്രുവനേ കാണാവൂ. വടക്കു നീങ്ങിയോളം ഉയർന്നിരിക്കും ഈ ധ്രുവൻ. ഈ ധ്രുവോന്നതിയെ "അക്ഷം" എന്നു ചൊല്ലുന്നൂ. തെക്കേ ധ്രുവനെ കാണരുത്, ഭൂപാർശ്വത്തിങ്കൽ താണുപോകകൊണ്ട്. ധ്രുവനെ ഉയർന്നു കാണാകുന്നേടത്തു ധ്രുവനടുത്തു ചില നക്ഷത്രങ്ങളെ ഉദയാസ്തമനം കൂടാതെ ധ്രുവന്റെ കീഴേപ്പുറമെ കിഴക്കോട്ടും മേലേപ്പുറമേ പടിഞ്ഞാറോട്ടും നീങ്ങുന്നതു കാണാം. അവ്വണ്ണമേ മറ്റേ ധ്രുവനടുത്തവറ്റെ ഒരിക്കലും കാണുകയും അരുത്, ഭൂപാർശ്വത്തിങ്കേന്നു കീഴേ പരിഭ്രമിക്കയാൽ. നിരക്ഷദേശത്തിങ്കന്നു പിന്നെ നക്ഷത്രങ്ങളെല്ലാറ്റിന്റേയും ഉദയാസ്തമനങ്ങളെ ക്രമേണ കാണാം. പിന്നെ, നേരേ കിഴക്കുന്നു എത്ര തെക്കോട്ടുതാൻ വടക്കോട്ടുതാൻ[12] നീങ്ങി ഉദിക്കുന്നൂ ഒരു നക്ഷത്രം, ദ്രഷ്ടാവിന്റെ നേരേ മേലീന്ന് അത്ര തന്നെ നീങ്ങി ഉച്ചയാം. ഉദിച്ചതിന്റെ നേരേ പടിഞ്ഞാറ് അസ്തമിപ്പൂതും ചെയ്യും[13]. ഇങ്ങനെ നിരക്ഷദേശത്തിങ്കൽ ഉദയാസ്തമനം. സാക്ഷദേശത്തിങ്കലും ഈവണ്ണം ഉച്ചയാകുന്നത്. പിന്നെ ഉദിച്ചേടത്തുന്ന് ഒട്ടു തെക്കു നീങ്ങീട്ട് ആയിരിക്കും സ്വദേശം വടക്കെങ്കിൽ.

2. വായുഗോളം

ഇവിടെ നിരക്ഷദേശത്തിങ്കൽ യാതൊരിടത്തു യാതൊരു നക്ഷത്രം, അതിന്ന് അവിടെ അവിടുന്ന് നേരേ കിഴക്കു പടിഞ്ഞാറു മേലുകീഴായി ഇരി

---

1. 10. D.E.F.om മഹാ
 11. ബഡവാഗ്നി; C.E. ബന്ദമാമുഖാഗ്നി
 12. B.C. വടക്കോട്ടുതാൻ തെക്കോട്ടുതാൻ
 13. B.C.F. അസ്തമിക്കയും ചെയ്യും

പ്പോന്ന് 'ഉദയാസ്തമയമാർഗ്ഗം' എന്നു തോന്നും. ഇവിടേയും പിന്നെ നേരേ[1] കിഴക്ക് ഉദിക്കുന്ന നക്ഷത്രത്തിന്റെ ഭ്രമണമാർഗ്ഗം എല്ലായിലും[2] വലിയൊരു വൃത്തം ആയിരിക്കും. പിന്നെ ഇതിനടുത്ത് ഇരുപുറവും[3] ഉള്ള നക്ഷത്രങ്ങളുടെ മാർഗ്ഗം അതിൽ ചെറിയ വൃത്തം ആയിട്ടായിരിക്കും. പിന്നെ ക്രമേണ ചെറുതായി രണ്ടു ധ്രുവന്റേയും അടുത്ത നക്ഷത്രങ്ങളുടെ വൃത്തം എല്ലായിലും ചെറുതായിട്ടിരിക്കും[4]. ഈവണ്ണമിരിക്കുമ്പോൾ രണ്ടു തലക്കലേയും കുറ്റികൾ ഊന്നിത്തിരിയുന്ന അച്ചുതണ്ടുപോലെ രണ്ടു ധ്രുവനേയും ഊന്നു കുറ്റിയായി തിരിയുന്നൊന്ന് ഈ 'ജ്യോതിർഗ്ഗോളം' എന്നു തോന്നും. ഇവിടെ നിരക്ഷദേശത്തിങ്കൽ നേരേ കിഴക്കും പടിഞ്ഞാറും തലക്കു നേരേ മേലും കൂടി സ്പർശിക്കുന്ന വൃത്തം യാതൊന്ന്, ഇതിന്നു 'ഘടികാവൃത്തം' എന്നു പേർ. ഇതിന്ന് ഇരുപുറവും രണ്ടു ധ്രുവനോളമുള്ള നാനാവൃത്തങ്ങൾക്കു 'സ്വാഹോരാത്രവൃത്തങ്ങൾ' എന്നു പേർ.

പിന്നെ[5] ലങ്കയിൽനിന്നു നേരേ മേലും രണ്ടു ധ്രുവങ്കലും സ്പർശിക്കുമാറ് ഒരു വൃത്തം ഉണ്ട്. ഇതിന്നു "ദക്ഷിണോത്തരം" എന്നു പേർ[6]. പിന്നെ ഭൂപാർശ്വത്തിങ്കൽ കിഴക്കും പടിഞ്ഞാറും രണ്ടു ധ്രുവങ്കലും സ്പർശിക്കുമാറ് ഒരു വൃത്തം ഉണ്ട്. അത് "ലങ്കാക്ഷിതിജം". പിന്നെ ഈ ലങ്കാക്ഷിതിജത്തിങ്കലെ ദക്ഷിണോത്തരവൃത്തത്തിങ്കേന്നു കിഴക്കെ അർദ്ധത്തെ സ്പർശിക്കുമ്പോൾ നക്ഷത്രങ്ങൾക്കു "ഉദയം" പടിഞ്ഞാറേ അർദ്ധത്തെ സ്പർശിക്കുമ്പോൾ "അസ്തമനം"; ദക്ഷിണോത്തരവൃത്തത്തെ സ്പർശിക്കുമ്പോൾ "ഉച്ച" ആകുന്നു.

ഈവണ്ണം ഘടികാ-ദക്ഷിണോത്തര-ക്ഷിതിജങ്ങൾ മൂന്നും അന്യോന്യം വിപരീതദിക്കുകളായിരിക്കുന്നു. അവ തങ്ങളിൽ സ്പർശിക്കുന്നേടത്തിന്ന് "സ്വസ്തികം" എന്നു പേരുണ്ട്. ഇവ ഇവിടെ ആറുള്ളു, ക്ഷിതിജത്തിങ്കൽ നാലുദിക്കിലും മേലും കീഴും. ഈ സ്വസ്തികങ്ങളുടെ പഴുതിൽ വൃത്തങ്ങളുടെ നാലൊന്നീതുഭാഗം എല്ലാറ്റിങ്കലും അകപ്പെടും. ആകയാൽ ഈ മൂന്നു

---

2. 1. A. നേര്
2. B.C.D.E എല്ലായിലും വലുതായിരിക്കും
3. D.E. പുറത്തെ
4. B.C.E. ആയിരിക്കുമ്പോൾ
5. H.adds ഭൂപാർശ്വത്തിങ്കൽ
6. B. ഇത് ദക്ഷിണോത്തരവൃത്തമാകുന്നു, C. ദിക്ഷിണോത്തരമെന്നു പേർ

വൃത്തങ്ങളെക്കൊണ്ടു തുല്യങ്ങളായിട്ട് എട്ടു ഗോളഖണ്ഡങ്ങളായിട്ടിരിപ്പോരു[7] പകുതികളുണ്ടാം[8]. ഇതിൽ നാലു ഖണ്ഡങ്ങളും ക്ഷിതിജത്തിന്ന്[9] കീഴ്, നാലു മേലൂ.

3. ഭഗോളം

പിന്നെ ആദിത്യന്റെ കിഴക്കോട്ടുള്ള ഗതിയുടെ മാർഗ്ഗത്തിന്ന് 'അപക്രമ മണ്ഡലം' എന്നു പേർ. ഇതു രണ്ടേടത്തു ഘടികാമണ്ഡലത്തോടു[1] സ്പർശിക്കും. വൃത്തത്തിന്റെ നാലൊന്നു ചെന്നേടത്ത് അപക്രമമണ്ഡലം ഘടികാമണ്ഡലത്തിങ്കേന്നു[2] തെക്കും വടക്കും[3] ഇരുപത്തിനാലു തിയ്യതി അക ന്നിരിക്കും. ഘടികാമണ്ഡലത്തോടുകൂടി പടിഞ്ഞാറോട്ടു ഭ്രമിപ്പൂതും[4] ചെയ്യും. ഇതിന്ന് അപക്രമമണ്ഡലത്തോടുള്ള നടേത്തെ യോഗം മേഷാദിക്കടുത്ത്. പിന്നെ അവിടന്ന് വടക്കോട്ട് അകലും. വൃത്തത്തിന്റെ പാതി ചെല്ലുന്നേ ടത്തു[5] തുലാദിയിങ്കൽ അടുത്ത രണ്ടാം യോഗം. അവിടുന്ന് ഘടികാമണ്ഡ ലത്തിന്റെ തെക്കേപ്പുറമെ അകലും പിന്നേയും വൃത്തത്തിന്റെ[6] പാതി ചെല്ലു ന്നേടത്തു കൂടും. ഈ യോഗങ്ങൾക്കു ക്രമേണ 'പുർവ്വോത്തരവിഷുവത്തു കൾ' എന്നു പേർ. പിന്നെ ഈ യോഗങ്ങൾ രണ്ടിന്റേയും നടുവേ എല്ലാ യിലും അകലുന്നേടത്തിന്ന് 'അയനസന്ധി' എന്നു പേർ.

ഇവിടെ പ്രവഹഭ്രമണത്തിനു തക്കവണ്ണം യാതൊരിക്കൽ മേഷാദി ഉദി ക്കുന്നൂ അപ്പോൾ തുലാദി അസ്തമിക്കുന്നു, മകരാദി ഖമധ്യത്തിങ്കന്നു തെക്കേ ദക്ഷിണോത്തരവൃത്തത്തെ സ്പർശിക്കും, കർക്ക്യാദി നേരേ കീഴു ഘടികാമണ്ഡലത്തിങ്കേന്നു വടക്കേ ദക്ഷിണോത്തരവൃത്തത്തെ സ്പർശി ക്കും. അന്നേരത്തു ദക്ഷിണോത്തരവൃത്തത്തിങ്കൽ ഘടികാപക്രമാന്തരം ഇരു പത്തിനാലു തിയ്യതി. എല്ലായിലും അകന്ന പ്രദേശം ആകയുമുണ്ടത്[7]. ഇതു പിന്നെ ഘടികാമണ്ഡലത്തിന്നു തക്കവണ്ണം തിരിയും. അപ്പോൾ യാതൊരി

---

2. 7. B. ഭഗോളഖണ്ഡങ്ങളുണ്ടാകാം
   8. C.E. പകുതി ഉണ്ടാകും
   9. C. ക്ഷിതിജത്തിങ്കേന്ന്
3. 1. C.D.E ഘടികാമണ്ഡലത്തിങ്കന്നു
   2. B. ഘടികാവൃത്തത്തിങ്കേന്ന്
   3. D. തെക്കോട്ടും വടക്കോട്ടും
   4. B. ഭ്രമിക്കുകയും
   5. D. ചെന്നേടത്ത്
   6. F. വൃത്തത്തിൽ
   7. B. പ്രദേശം ആകുന്നു

ക്കൽ മേഷാദി ഉച്ചയാകുന്നു, അപ്പോൾ തുലാദി കീഴു[8], മകരാദി പടിഞ്ഞാറേ സ്വസ്തികത്തിങ്കന്നു ഇരുപത്തിനാലു തിയ്യതി തെക്കു നീങ്ങി ക്ഷിതിജത്തെ സ്പർശിക്കും; കർക്ക്യാദി പൂർവ്വസ്വസ്തികത്തിങ്കേന്ന് അത്ര വടക്കു നീങ്ങി ക്ഷിതിജത്തെ സ്പർശിക്കും. ഇങ്ങനെ മേൽകീഴായി ഇരിപ്പൊന്നു അന്നേരത്തെ[9] അപക്രമമണ്ഡലത്തിന്റെ സംസ്ഥാനം. പിന്നെ മേഷാദി പടിഞ്ഞാറേ സ്വസ്തികത്തിങ്കലാകുമ്പോൾ തുലാദി കിഴക്കേ സ്വസ്തികത്തിങ്കൽ ഇരിക്കും, കർക്ക്യാദി ഖമദ്ധ്യത്തിങ്കന്നു ഇരുപത്തിനാലു തിയ്യതി വടക്ക് ഉച്ചയാകും, മകരാദി കീഴു തെക്കുനീങ്ങി ദക്ഷിണോത്തരവൃത്തത്തെ സ്പർശിക്കും. തുലാദി ഉച്ചയാകുമ്പോൾ, മേഷാദി നേരെ കീഴ്, മകരാദി പൂർവ്വസ്വസ്തികത്തിങ്കന്നു തെക്കു ക്ഷിതിജത്തെ സ്പർശിക്കും. കർക്ക്യാദി പടിഞ്ഞാറെ സ്വസ്തികത്തിങ്കേന്ന് വടക്കു ക്ഷിതിജത്തെ സ്പർശിക്കും. ഈ നേരത്തും മേൽകീഴായി ഇരിക്കും അപക്രമമണ്ഡലം. ഇങ്ങനെ ഈ ഘടികാമണ്ഡലത്തിന്റെ തിരിച്ചിലിനു തക്കവണ്ണം സംസ്ഥാനഭേദമുണ്ട് അപക്രമമണ്ഡലത്തിന്ന്. ഈ ഘടികാപക്രമങ്ങൾ തങ്ങളിൽ ഒരു പ്രകാരം കെട്ടുപെട്ടു തന്നെ ഇരിക്കും അത്രെ എന്നു ഹേതുവാകുന്നത്.

പിന്നെ ഘടികാമണ്ഡലം യാതൊരുപ്രകാരം പ്രവഹവായുഗോളത്തിന്റെ മധ്യവൃത്തമാകുന്നു, അവ്വണ്ണം അപക്രമമണ്ഡലം ഭഗോളത്തിന്റെ മധ്യവൃത്തമായിട്ടിരിക്കും. യാതൊരു പ്രകാരം ഘടികാപാർശ്വത്തിങ്കൽ ധ്രുവന്മാർ അവ്വണ്ണം അപക്രമപാർശ്വത്തിങ്കൽ രണ്ടു രാശികൂടങ്ങൾ ഉള്ളൂ. അവിടെ രാശികളുടെ തെക്കെ തല ഒക്ക ഒരിടത്തു കൂടി ഇരിക്കും, വടക്കെ തല ഒക്ക ഒരിടത്തു കൂടും. ഈ യോഗങ്ങൾ 'രാശികൂട'ങ്ങളാകുന്നത്.

ഇവിടെ പൂർവ്വവിഷുവത്ത് ഖമധ്യത്തിങ്കൽ ആകുമ്പോളേ രാശികൂടസംസ്ഥാനം എന്ന് ചൊല്ലുന്നത്. അന്നേരത്തു നേരേ മേൽകീഴായിരിക്കും അപക്രമമണ്ഡലം. കിഴക്കേ സ്വസ്തികത്തിങ്കേന്നു വടക്കും പടിഞ്ഞാറേ സ്വസ്തികത്തിങ്കേന്നു തെക്കും ക്ഷിതിജത്തിങ്കൽ സ്പർശിക്കും അപക്രമമണ്ഡലത്തിന്റെ "അയനാന്ത"ങ്ങൾ. അയനാന്തവും പൂർവ്വാപരസ്വസ്തികങ്ങളും തങ്ങളിൽ ഇരുപത്തിനാലു തിയ്യതി അകലമുണ്ട്[10]

---

2. 8. B.D.E.F. തുലാദി നേരെ കീഴും
9. B.C.D.E. അന്നേരത്ത്
10. F. തിയതി അന്തരമുണ്ട്

പിന്നെ വടക്കേ ധ്രുവങ്കന്നു ഇരുപത്തിനാലു തിയ്യതി പടിഞ്ഞാറും തെക്കേ ധ്രുവങ്കേന്ന്[11] അത്ര കിഴക്കേയും ക്ഷിതിജത്തിങ്കൽ അന്നേരത്തു രാശികൂട ങ്ങൾ. രണ്ടു രാശികൂടങ്ങളിലും ഖമദ്ധ്യത്തിങ്കലും[12] സ്പർശിച്ചിട്ട് ഒരു വൃത്തത്തെ കല്പിപ്പൂ. ഇത് ഒരു 'രാശികൂടവൃത്ത'മാകുന്നത്. പിന്നെ മേഷാ ദിയിങ്കന്നു അപക്രമമണ്ഡലത്തിങ്കൽ കിഴക്കു[13] തന്റെ പന്ത്രണ്ടാലൊന്നു ചെന്നേടത്തും രണ്ടു രാശികൂടങ്ങളിലും സ്പർശിച്ചിട്ട് ഒരു രാശികൂടവൃത്ത ത്തെ കല്പിപ്പൂ. കീഴു തുലാദിയിങ്കന്നും അത്ര പടിഞ്ഞാറെ സ്പർശിക്കും. ഇട മുപ്പതു തിയ്യതി അകലമുണ്ട്. ഇതു രണ്ടാം രാശികൂടവൃത്തമാകുന്നത്. ഇവിടെ ഖമദ്ധ്യത്തിങ്കന്നു കിഴക്ക് ഈ രാശികൂടവൃത്തങ്ങൾ രണ്ടിന്റേയും പഴുതു നീളം മേഷമാകുന്ന രാശി. കീഴേപ്പുറത്ത് ഈ വൃത്തങ്ങളുടെ[14] പഴുതു നീളം തുലാമാകുന്ന രാശി.

പിന്നെ രണ്ടാം രാശികൂടവൃത്തത്തിങ്കന്ന് ഇത്ര അംശം കിഴക്കേയും രണ്ടു രാശികൂടങ്ങളിലും കീഴുന്ന് അത്ര പടിഞ്ഞാറേയും കൂടി ഒരു രാശികൂട വൃത്തം കല്പിപ്പൂ. ഈ രണ്ടാം രാശികൂടവൃത്തത്തിന്റേയും മൂന്നാമതിന്റേയും[15] ഇടനീളം ഇടവം രാശി. കീഴേപ്പുറത്തേതു വൃശ്ചികം[16]. പിന്നെ മൂന്നാമതി ന്റേയും ക്ഷിതിജത്തിന്റേയും പഴുതുനീളം മിഥുനമാകുന്ന രാശി. കീഴേപ്പു റത്തു ഇവറ്റിന്റെ പഴുതുനീളം ധനു. ഇങ്ങനെ ആറു രാശികൾ.

പിന്നെ ഖമദ്ധ്യത്തിങ്കന്നു പടിഞ്ഞാറേ അപക്രമമണ്ഡലത്തിങ്കൽ ഈവണ്ണം അന്തരം തുല്യമാകുമാറ് രണ്ടു വൃത്തം കല്പിപ്പൂ. എന്നാൽ മറ്റേ രാശികൾ ആറും കാണാം, പ്രഥമരാശികൂടവൃത്തവും[17] ക്ഷിതിജവും കൂടി നിരൂപിക്കു മ്പോൾ. പിന്നെ രാശികളുടെ നടുവിൽ ഈവണ്ണം വൃത്തങ്ങളെ കല്പിച്ച് രാശ്യവയവങ്ങളാകുന്ന തിയ്യതിയും ഇലിയും ഓർക്കേണം. ഇവിടെ ക്ഷിതി ജത്തിന്നും ദക്ഷിണോത്തരവൃത്തത്തിന്നും പ്രവഹവശാലുള്ള തിരിച്ചിൽ ഇല്ല. എന്നിട്ട് ഈ ക്ഷിതിജതുല്യമായിട്ട് മറ്റൊരു രാശികൂടവൃത്തം കല്പിച്ചു

---

3. 11. B.E.F ഇരുപത്തിനാലു തിയതി കിഴക്കേയും
12. E.F. ങ്കലും കൂടി
13. E. reads കിഴക്കേയും ക്ഷിതിജത്തിങ്കൽ അന്നേരത്ത് രാശിക്കൂടങ്ങൾ രണ്ടിലും ഖമദ്ധ്യ ത്തിങ്കലും കൂടി സ്പർശിച്ചിട്ട് ഒരു വൃത്തത്തെ കല്പിപ്പൂ ഇത് (തന്റെ)
14. B. ഈ രാശികൂടവൃത്തങ്ങളുടെ
15. B. മൂന്നാം രാശികൂടവൃത്തത്തിന്റേയും
16. B. കീഴ് വൃശ്ചികം
17. B.C.DE.om. പ്രഥമരാശിക്കുട to നിരൂപിക്കുമ്പോൾ

കൊള്ളേണം തിരിയുമ്പോളേക്ക്. ഇങ്ങനെ പന്ത്രണ്ടു രാശികളെക്കൊണ്ടു നിറഞ്ഞിരിക്കും ഈ ജ്യോതിർഗ്ഗോളമൊക്കെ. ഈ[18] ജ്യോതിർഗ്ഗോളത്തിന്ന് അപക്രമമണ്ഡലം മദ്ധ്യമായി, രാശികൂടങ്ങൾ പാർശ്വങ്ങളായി നിരൂപിക്കുമ്പോൾ 'ഭഗോളം' എന്നു പേർ. ഘടികാമണ്ഡലം മദ്ധ്യമായി ധ്രുവന്മാർ പാർശ്വങ്ങളായി[19] നിരൂപിക്കുമ്പോൾ 'വായുഗോളം' എന്നുപേർ

ഇങ്ങനെ മേഷാദിയിങ്കലെ ഘടികാപക്രമയോഗം ഖമദ്ധ്യത്തിലാകുമ്പോൾ മിഥുനാന്തമാകുന്ന അയനസന്ധിയും ദക്ഷിണരാശികൂടവും ഉദിക്കും. ചാപാന്തവും ഉത്തരരാശികൂടവും അസ്തമിക്കും. പിന്നെ പ്രവഹഭ്രമണവശാൽ ഉദിച്ചവ ഉച്ചയാകുമ്പോൾ ദക്ഷിണോത്തരത്തെ[20] സ്പർശിക്കുമ്പോൾ അസ്തമിച്ചവ കീഴെപ്പുറത്തു ദക്ഷിണോത്തരത്തെ സ്പർശിക്കും. പിന്നെ മിഥുനാന്തവും ദക്ഷിണരാശികൂടവും അസ്തമിക്കുമ്പോൾ ചാപാന്തവും ഉത്തരരാശികൂടവും ഉദിക്കും. ഇങ്ങനെ മിഥുനാന്തത്തോടു തുല്യമായിട്ടു ദക്ഷിണരാശികൂടവും ചാപാന്തത്തോടു തുല്യമായിട്ട് ഉത്തരരാശികൂടവും ഭ്രമിക്കും. ഇവിടെ ഘടികാമണ്ഡലത്തിങ്കന്നു ഇരുപത്തിനാലു തിയ്യതി ഇരുപുറവും അകന്നേടത്ത് രണ്ട് അയനാന്തസ്വാഹോരാത്രങ്ങളുള്ളൂ. പിന്നെ രണ്ടു ധ്രുവങ്കന്നും ഇരുപത്തിനാലു തിയ്യതി അകന്നേടത്ത് രണ്ടു രാശികൂടസ്വാഹോരാത്രങ്ങളുള്ളൂ. ഈ സ്വാഹോരാത്രമാർഗ്ഗത്തൂടെ നിത്യമായിട്ട് ഭ്രമണമിവറ്റിന്ന്.

4. അയനചലനം

ഇവിടെ[1] അയനചലനമില്ലാത്ത നാൾ ഇവ്വണ്ണം കന്യാമീനാന്തങ്ങൾ ഗോളസന്ധുക്കൾ, മിഥുനചാപാന്തങ്ങൾ അയനസന്ധുക്കൾ ആയിട്ടിരിക്കും[2]. പിന്നെ അയനചലനം കൂട്ടേണ്ടുന്നാൾ ഈ സന്ധുക്കളിൽ നിന്നു നടേത്തെ രാശിയിങ്കൽ അയനചലനത്തോളം തിയ്യതി അകന്നേടത്ത് ഇസ്സന്ധുക്കൾ നാലും വർത്തിക്കും. അയനചലനം കളയേണ്ടുന്നാൾ ഇച്ചൊല്ലിയ സന്ധിയിങ്കന്നു പിന്നത്തെ രാശിയിങ്കൽ അയനചലനതിയ്യതിയോളം അകന്നേടത്ത് ഇസ്സ

3. 18. H. adds. ഈ
19. B.C.D. om. ധ്രുവന്മാർ to നിരൂപിക്കുമ്പോൾ
20. B.C.D.F.om. ദക്ഷിണോത്തരത്തെ സ്പർശിക്കുമ്പോൾ

4. 1. D. അവിടെ
2. B.C.D.E.F.om. ആയിട്ടിരിക്കും

ന്ധുക്കൾ നാലും വർത്തിക്കും. സന്ധുക്കളാകുന്നതു പിന്നെ ഘടികാപക്രമ ങ്ങൾ ഒരുമിച്ചേടവും എല്ലായിലും അകന്നേടവും. അകലം ഇരുപത്തിനാലു തിയ്യതി തന്നെ[3]. ചലിക്കുമ്പോളും ഘടികാപക്രമയോഗപ്രദേശമേ നീങ്ങൂ.

5. അയനചലനപ്രകാരം

ഇതിന്റെ ചലനപ്രകാരം പിന്നെ. അപക്രമമണ്ഡലത്തിന്റെ യാതൊരു അവ യവം ഘടികാമണ്ഡലത്തിന്റെ യാതൊരു[1] അവയവത്തോടു സ്പർശിക്കുന്നൂ അയനചലനമില്ലാത്ത നാൾ, അവിടുന്നു പിന്നെ അയനചലനം കൂട്ടേണ്ടു ന്നാൾ ഈ രണ്ടു വൃത്തങ്ങളുടേയും ആ[2] അവയവങ്ങളിൽ നിന്നു ഘടികാ പക്രമങ്ങൾ രണ്ടു വൃത്തങ്ങളിലും അന്നേ[3] അയനചലനതിയ്യതിയോളം പിമ്പിൽ നീങ്ങിയ അവയവങ്ങൾ തങ്ങളിൽ സ്പർശിക്കും, അയനചലനം കളയേണ്ടുന്നാൾ രണ്ടു വൃത്താന്തത്തിനും മുമ്പിലെ അവയവം തങ്ങളിൽ സ്പർശിക്കും. ഘടികാമണ്ഡലം താൻ നീങ്ങുകയില്ല. തങ്കലെ യോഗ[4] പ്രദേശമേ നീങ്ങൂ. അപക്രമവൃത്തം തനിക്കും ചലനമുണ്ട്[5]. അതു ഹേതുവായിട്ട് രാശികൂടങ്ങൾക്കും ചലനമുണ്ട്. അവ തൻെറ സ്വാഹോ രാത്രങ്ങളിൽ നിന്ന് അകലുകയില്ല. രാശികൂടസ്വാഹോരാത്രങ്ങളിൽ തന്നെ മുന്നോക്കിയും പിന്നോക്കിയും നീങ്ങുമത്രേ. ധ്രുവദ്വയത്തിങ്കന്ന് രാശികൂടങ്ങളും ഘടികാമണ്ഡലത്തിങ്കന്ന് അപക്രമായനാന്തങ്ങളും ഇരുപത്തിനാല് തീയതി അകലും എന്നു നിയതം. ഈ നാല് അന്തരാളങ്ങളും ഒരു അയനാന്തരാശികൂടവൃത്തത്തിന്മേൽ തന്നെ കാണാം. പിന്നെ ഒരു കർക്കടകശലകേടെ ഒരിടം ഊന്നി മററെ തലകൊണ്ടു തിരിച്ചു വൃത്തമുണ്ടാക്കുമ്പോൾ[6] ഊന്നിയ തലക്കൽ വൃത്തത്തിൻെറ നടുവ്, ആ[7] നടുവിന്നു 'നാഭി' എന്നും 'കേന്ദ്ര' എന്നും പേർ, വക്കിന്ന് 'നേമി' എന്നും

---

4. 3. B.C.D.E.F.om. അകലം ഇരുപത്തിനാലു തീയതി തന്നെ
5. 1. B.C.D.E.om. അവയവം (.....to.....) യാതൊരു
   2. B.C.D.E.om. ആ
   3. B.E.om.അന്നേ അയന (.....to.....) രണ്ടു വൃത്തത്തിന്
   4. C.F. തൽക്കാലയോഗ
   5. B. തന്നെ കുറേ ചലനമുണ്ട്
   6. D.E.F. മററെ തിരിക്കുമ്പോൾ
   7. F അവിടെ

പേർ[8] ഇവിടെ[9] ഗോളവിഷയത്തിൻ്റെ വൃത്തത്തിങ്കലെ വൃത്തകല്പനത്തിങ്കൽ വൃത്തങ്ങളെല്ലാററിനും ഭഗോളമദ്ധ്യത്തിങ്കൽ താൻ ഭൂഗോളമദ്ധ്യത്തിത്താൻ ഒരിടത്തു തന്നെ 'നാഭി' എന്നും വൃത്തത്തിൻെറ വലുപ്പം എല്ലാം ഒക്കും എന്നും കല്പിക്കുമാറ് സാമാന്യന്യായം, സ്വാഹോരാത്രങ്ങളേയും സ്ഫുടന്യായത്തിങ്കൽ കല്പിക്കുന്ന വൃത്തങ്ങളേയും ഒഴിച്ച്. എന്നാൽ ഇവിടെ അവ്വണ്ണം തുല്യനാഭികളായിരിക്കുന്ന ഘടികാപക്രമങ്ങൾക്കു രണ്ടേടത്തു തങ്ങളിൽ യോഗമുണ്ട്. പിന്നെ നാഭിമധ്യത്തൂടെ ഈ യോഗങ്ങൾ രണ്ടിങ്കലും സ്പർശിക്കുന്ന വ്യാസസൂത്രം രണ്ടിന്നും ഒന്നേ. പരമാന്തരാളങ്ങളിൽ സ്പർശിക്കുന്ന വ്യാസസൂത്രങ്ങൾ രണ്ടു വൃത്തത്തിന്നും രണ്ട്. പരമാന്തരാളമെന്ന് അകലമേറിയേടം. ഈ യോഗവ്യാസത്തിൻെറ ദിക്കുകൊണ്ട് വിപരീതദിക്കായിട്ടിരിപ്പോ ചിലവ പരമാന്തരാളവ്യാസസൂത്രങ്ങൾ. അതു ഹേതുവായിട്ടേ പരമാന്തരാളങ്ങളിൽ രണ്ടിനോടും സ്പർശിക്കുന്ന അയനാന്തരാശികൂടവൃത്തം ഈ ഘടികാപക്രമങ്ങൾ രണ്ടിനോടും വിപരീതമായിട്ടിരിക്കും. ഈ വിപരീത വൃത്തം രണ്ടിൻ്റേയും പാർശ്വത്തിങ്കൽ സ്പർശിക്കും എന്നു നിയതം. പാർശ്വസ്പൃഷ്ടമെങ്കിൽ വിപരീതം എന്നു നിയതം. എന്നാൽ അയനാന്തരാശികൂടവൃത്തം ഘടികാപക്രമങ്ങൾ രണ്ടിന്നും വിപരീതം. എന്നിട്ടു രണ്ടിൻ്റേയും പാർശ്വങ്ങളായിരിക്കുന്ന ധ്രുവരാശികൂടദ്വയങ്ങളിൽ സ്പർശിക്കും. എന്നാൽ തുല്യാന്തരങ്ങളായി ഒരു വൃത്തത്തിങ്കത്തന്നെ വർത്തിപ്പോ ചിലവ രണ്ടു വൃത്തങ്ങളുടേയും പാർശ്വാന്തരാളങ്ങളും പരമാന്തരാളങ്ങളും എന്നു സ്ഥിതം.

എന്നാൽ അയനചലനവശാൽ അയനാന്തം നീങ്ങുമ്പോൾ അയനാന്തത്തെ സ്പർശിക്കുന്ന വൃത്തം രാശികൂടത്തേയും സ്പർശിക്കും എന്നുള്ള നിയമം കൊണ്ട് അപക്രമായനാന്തം നീങ്ങിയ ദിക്കിൽ അയനാന്തരാശികൂടവും കൂടി നീങ്ങിയതായിട്ടിരിക്കും. അപക്രമമണ്ഡലത്തിൻെറ അയനാന്തപ്രദേശം ഘടികാമണ്ഡലത്തിങ്കന്നു ഇരുപത്തിനാലു തിയ്യതി അകന്നിരിക്കും എല്ലാനാളും എന്നു നിയതമാകയാൽ ഘടികാപാർശ്വങ്ങളിലെ ധ്രുവദ്വയത്തിങ്കേന്ന് അപക്രമ

---

5. 8. B.C. om. വക്കിന്നു .....to.... വൃത്തകല്പനത്തിങ്കൽ
9. F.om. ഇവിടെ .....to..... മദ്ധ്യത്തിങ്കത്താൻ

മണ്ഡലപാർശ്വങ്ങളിലെ രാശികൂടങ്ങളും അത്രതന്നെ അകന്നിരിക്കും എല്ലാ നാളും എന്നു നിയതം. എന്നിട്ട് രാശികൂടസ്വാഹോരാത്രങ്ങൾ എല്ലാ നാളും ഒന്നുതന്നെ. എന്നിട്ടു സ്വാഹോരാത്രത്തിങ്കൽ തന്നെ കിഴക്കോട്ടും പടിഞ്ഞാറോട്ടും നീങ്ങുമത്രെ അയനചലനവശാൽ രാശികൂടദ്വയങ്ങൾ എന്നു ഗ്രഹിക്കേണ്ടുവത്. പിന്നെ മേഷാദിയിങ്കന്നു എത്ര ചെന്നൂ ഗ്രഹം എന്നു ഗ്രഹസ്ഫുടം കൊണ്ടു വരുന്നത്.

അതിനെ പിന്നെ ഘടികാപക്രമയോഗത്തിങ്കന്ന് തുടങ്ങീട്ട് എത്ര ചെന്നൂ എന്നറിവാൻ അയനചനലം സംസ്കരിക്കേണം. പിന്നെ അതിനു 'ഗോളാദി' എന്നു പേർ. ഇങ്ങനെ അയനചലനപ്രകാരം.

6. അക്ഷവശാൽ സംസ്ഥാനഭേദങ്ങൾ

ഇങ്ങനെ നിരക്ഷദേശത്തിങ്കന്നു ജ്യോതിർഗ്ഗോളത്തെ കാണുമ്പോൾ വായുഗോളവശാൽ നേരേ പടിഞ്ഞാറു നോക്കി തിരിയുന്നൊന്ന് ഇത് എന്നു തോന്നും. അതിന്നു തക്കവണ്ണം ഈ വായുഗോളമധ്യവൃത്തം തുടങ്ങിയുള്ള ഘടികാവൃത്തദ്യുവൃത്തങ്ങൾ എന്നിവയും നേരേ മേൽക്കീഴായി തോന്നും എന്നതിനെ ചൊല്ലീതായി. പിന്നെ ആ വായുഗോളത്തീങ്കന്നു ഭഗോളത്തിന്നു ചെരിവുണ്ടെന്നും, കുറഞ്ഞൊരു ഗതിയുണ്ടെന്നും ചൊല്ലീതായി. അനന്തരം സാക്ഷദേശത്തിങ്കന്നു നോക്കുമ്പോൾ ആ വായുഗോളത്തിന്നും കൂടി ചെരിവു തോന്നും. അതിന്നു തക്കവണ്ണം ഭഗോളത്തിന്നും എന്നിതിനെ ചൊല്ലുന്നു.

7. ഭൂഗോളം

അവിടെ നേരേ ഉരുണ്ടതിന്നു 'ഗോളം' എന്നു പേർ. ഭൂമി ഗോളാകാരേണ ഉള്ളൂ. ഇങ്ങനെ ഇരിക്കുന്ന ഭൂമീടെ എല്ലാ പ്രദേശത്തിങ്കലും ലോകരുടെ സ്ഥിതിയുമുണ്ട്. അവിടെ താന്താനിരിക്കുന്ന പ്രദേശം ഭൂമീടെ മീത്തെപ്പുറം. അവിടെ ഭൂപ്രദേശം. നേരെ വിലങ്ങത്തിൽ തന്റെതന്റെ നിലവു നേരേ മേൽകീഴായിട്ട്[1] ഇങ്ങനെ എല്ലാർക്കും തോന്നും പ്രകാരം. ഇവിടെ

7. 1. B.E. നിലാ മേൽകീഴായിട്ട്

ആകാശമദ്ധ്യത്തിങ്കൽ ഉരുണ്ടിരിക്കുന്ന ഭൂമിക്ക് രണ്ടു പകുതി കല്പിപ്പൂ, മീത്തെപ്പാതിയും കീഴെപ്പാതിയുമെന്ന്. അവിടെ മീത്തെപ്പാതിക്കു നടുവാകുന്നതു താനിരിക്കുന്ന[2] പ്രദേശം എന്നിരിക്കും. അവ്വണ്ണമാകുമ്പോൾ യാതൊന്നു പാർശ്വമാകുന്നത് അവിടുന്നു കീഴെപ്പാതി ഭൂമിയെക്കൊണ്ടു മറഞ്ഞിരിക്കും ആകാശം. എന്നാൽ അവ്വണ്ണമിരിക്കുന്ന ഭൂപാർശ്വത്തിങ്കൽ അകപ്പെടുമ്പോൾ ജ്യോതിസ്സുകളുടെ ഉദയാസ്തമയങ്ങൾ. ഈ പ്രദേശത്തിന്നു മീത്തേടം ആകാശം കാണാം. അതിൻെറ നടുവു ഖമദ്ധ്യമാകുന്നത്. ഇത് ദ്രഷ്ടാവിൻെറ നേരെ തലക്കുമീത്തേടം. ഇവിടെ[3] യാതൊന്ന് ഘടികാമണ്ഡലമെന്ന് ചൊല്ലപ്പെടുന്നത് നിരക്ഷദേശത്തിങ്കൽ കിഴക്കു പടിഞ്ഞാറായി മേൽക്കീഴായി ഇരിപ്പൊന്ന് അത്. അതിൻെറ കേന്ദ്രം ഭൂമീടെ ഒത്ത നടുവിലായിട്ടിരിക്കും. രണ്ടു പാർശ്വത്തിങ്കലും ധ്രുവനും. ഇങ്ങനെ ഇരിക്കുന്നേടത്ത് ഈ ഭുമീടെ നടുവേകൂടി ധ്രുവങ്കൽ രണ്ടു ഗ്രഹങ്ങളും സ്പർശിക്കുമാറ് തെക്കുവടക്ക് ഒരു ദണ്ഡു കല്പിപ്പൂ. അതിന്ന് 'അക്ഷദണ്ഡം' എന്നു പേർ, അച്ചുതണ്ടുപോലെ ഇരിപ്പൊന്ന് അത്. അതിന്മേൽ കെട്ടുപെട്ട് അതു തിരിയുമ്പോൾ അതിന്നു തക്കവണ്ണം കൂടി തിരിയുന്നൊന്ന് ഈ 'ജ്യോതിർഗ്ഗോളം' എന്നു കല്പിപ്പൂ. എന്നാലുണ്ടു 'ഭൂപ്രദേശം'. ഭൂപ്രദേശഭേദത്തിന്നു തക്കവണ്ണം വായുഗോളത്തിൻെറ ചെരിവിന്നും ഭേദം എന്നറിയുന്നേടത്തേക്ക് ഒരു എളുപ്പം.

ഇവിടെ[4] നിരക്ഷദേശത്തിങ്കൽ നേരെ കിഴക്കുപടിഞ്ഞാറായി ഖമദ്ധ്യത്തെ സ്പർശിച്ചിരുന്നൊന്നു 'ഘടികാമണ്ഡലം'. അവിടെ ഭൂമീടെ സമപാർശ്വത്തിങ്കൽ ഇരിക്കുന്ന ധ്രുവങ്കൽ സ്പർശിച്ചിരുന്നൊന്നു 'നിരക്ഷക്ഷിതിജം' എന്നോ മുമ്പിൽ ചൊല്ലപ്പെട്ടുവല്ലോ. ഇവിടെ[5] ഭൂമീടെ വടക്കെ പാർശ്വത്തിങ്കലേ മേരുവിങ്കന്നു നോക്കുമ്പോൾ ധ്രുവനെ ഖമദ്ധ്യത്തിങ്കലായിട്ടു കാണാം. അപ്പോൾ നിരക്ഷക്ഷിതിജം മേൽക്കീഴായിരിക്കും. ഘടികാമണ്ഡലം ക്ഷിതിജം പോലെയുമിരിക്കും. അവിടേയ്ക്ക് എല്ലാർക്കും തന്നെ താനിരിക്കുന്ന പ്രദേശം

7. 2. B. താനിരിക്കുന്നിടം
3. B.C.D.E.F.om. ഇവിടെ യാതൊന്നു.....to.....ധ്രുവനും
4. B.C.D.E.F.om. ഇവിടെ to ചൊല്ലപ്പെട്ടുവെല്ലൊ
5. D. അവിടെ

സമതിര്യഗ്ഗതമായിരിപ്പൊന്ന് എന്നു തോന്നും. അതിങ്കൽ തൻെറ സ്ഥിതി മേൽക്കീഴായിരിപ്പൊന്ന് എന്നല്ലോ തോന്നുന്നൂ എന്നിതു ഖമദ്ധ്യത്തിന്നും ഭൂപാർശ്വത്തിന്നും പ്രതിദേശം ഭേദമുണ്ടാവാൻ ഹേതുവാകുന്നത്. ഇവ്വണ്ണമിരിക്കുമ്പോൾ നിരക്ഷദേശത്തിങ്കന്നു വടക്കുവടക്കു നീങ്ങുന്നതിന്നു തക്കവണ്ണം ഭൂപാർശ്വത്തിങ്കന്നു ധ്രുവനെ ഉയർന്നു ഉയർന്നു കാണാം. മേരുവിങ്കന്നു തെക്കുതെക്കു നീങ്ങുന്നതിന്നു തക്കവണ്ണം ഖമദ്ധ്യത്തിങ്കന്നു താണുകാണാം. നിരക്ഷദേശത്തോളം ഇങ്ങനെ നാനാ പ്രദേശത്തിങ്കലിരിക്കുന്നവർക്ക് ഖമദ്ധ്യവും ഭൂപാർശ്വവും വെവ്വേറെ. ഇവിടെ ലങ്കയിങ്കന്നു നേരേ വടക്ക് സമരേഖയിങ്കൽ ഒരേടം സ്വദേശം എന്നു കല്പിപ്പൂ. എന്നാൽ ഘടികാദക്ഷിണോത്തരമണ്ഡലയോഗത്തിന്ന് വടക്കു ദക്ഷിണോത്തരവൃത്തത്തിങ്കൽ യാതൊരിടം ഖമദ്ധ്യമാകുന്നത് അവിടേയും മുമ്പിൽ ചൊല്ലിയ പൂർവ്വാപരസ്വസ്തികങ്ങളിലും സ്പർശിച്ചിട്ട് ഒരു വൃത്തത്തെ കല്പിപ്പൂ. ഇതിനു 'സമമണ്ഡലം' എന്നു പേർ. പിന്നെ ദക്ഷിണോത്തരവൃത്തത്തിങ്കൽ എത്ര അകലമുണ്ട് ഘടികാസമമണ്ഡലാന്തരാളം ഉത്തരധ്രുവത്തിങ്കന്നു ദക്ഷിണോത്തരവൃത്തത്തിങ്കൽ അത്ര കീഴേയും ദക്ഷിണധ്രുവങ്കന്ന് അത്ര മീതേയും പൂർവ്വാപരസ്വസ്തികങ്ങളിലും കൂടി ഒരു വൃത്തത്തെ കല്പിപ്പൂ. അതിന്നു 'സ്വദേശക്ഷിതിജം' എന്നു പേർ, ഇവിടെ[6] മുമ്പിൽ ചൊല്ലിയ നിരക്ഷക്ഷിതിജം യാതൊന്ന് അതു പൂർവ്വാപരസ്വസ്തികത്തിങ്കന്നു വടക്കേടം ഈ സ്വദേശക്ഷിതിജത്തിങ്കന്ന് മീത്തെ ഇരിക്കും, തെക്കേടം കീഴേയും. ഇങ്ങനെ സ്വദേശക്ഷിതിജം വേറേ കല്പിക്കുമ്പോൾ നിരക്ഷക്ഷിതിജത്തിന്ന് 'ഉന്മണ്ഡലം' എന്നുപേർ. പിന്നെ ഇവിടെ യാതൊരു പ്രകാരം ദക്ഷിണോത്തരഘടികാനിരക്ഷക്ഷിതിജങ്ങളെക്കൊണ്ടു തുല്യാന്തരാളങ്ങളായിട്ട് ആറു സ്വസ്തികങ്ങളും സമങ്ങളായിട്ട് എട്ടു ഗോളഖണ്ഡങ്ങളും ഉണ്ടാകുന്നൂ, അവ്വണ്ണം ദക്ഷിണോത്തര-സമമണ്ഡല സ്വദേശക്ഷിതിജങ്ങളെക്കൊണ്ടും ഗോളവിഭാഗം കല്പിക്കാം.

ഇങ്ങനെ എല്ലാടവും അന്യോന്യസമതിര്യഗ്ഗതങ്ങളാകുന്ന മൂന്നു വൃത്തങ്ങളേക്കൊണ്ടു സമാന്തരങ്ങളായിരിക്കുന്ന ആറു സ്വസ്തികങ്ങളും,

---

7. 6. B.C.D.E.F.om. ഇവിടെ ....to.... ഇങ്ങനെ

സമങ്ങളായിട്ടു എട്ടു ഗോളഖണ്ഡങ്ങളും ഉണ്ടാകുമാറ് ഗോളവിഭാഗത്തെ കല്പിപ്പൂ നടേ.

പിന്നെ നാലാമത് ഒരു വൃത്തത്തെ കല്പിപ്പൂ. അത് ഇമ്മൂന്നിൽ രണ്ടു വൃത്തങ്ങളെക്കൊണ്ട് ഉണ്ടാകുന്ന രണ്ടു സ്വസ്തികകങ്ങളിലും സ്പർശിക്കുമാറുള്ളൂ. പിന്നെ ഈ വൃത്തത്തെക്കൊണ്ട് എട്ടു ഗോളഖണ്ഡങ്ങളിൽ നാലു ഗോളഖണ്ഡങ്ങളും പെളിയുമാറ് ഇരിക്കും., ഇതിനു 'വലിതവൃത്തം' എന്നു പേർ. ഈ വലിതവൃത്തത്തിങ്കന്നു മറേറ രണ്ടു വൃത്തങ്ങളോടുള്ള അകലമറിയുന്നത് ഒരു വൃത്താന്തരാള ത്രൈരാശികമാകുന്നത് എന്നു മേലിൽ വിവരിച്ചു[7] ചൊല്ലുന്നുണ്ട്. ഇങ്ങനെ ഇവിടെ ചൊല്ലീതായതു[8] വായുഗോളസ്വരൂപം.

പിന്നെ വായുഗോളത്തിങ്കന്നു ഭഗോളത്തിന്നു സംസ്ഥാനഭേദമുണ്ട് എന്നും, പിന്നെ ഭൂമി ഉരുണ്ടിരിക്കയാൽ ഭൂമിയിങ്കൽ ഓരോ പ്രദേശത്തിങ്കലിരിക്കുന്നവർക്ക് അക്ഷദണ്ഡാഗ്രത്തിങ്കലെ ധ്രുവൻെറ ഉന്നതി ഓരോ പ്രകാരം തോന്നും. എന്നിട്ട് ആ അക്ഷദണ്ഡത്തിൻെറ തിരിച്ചൽക്കു തക്കവണ്ണം[9] തിരിയുന്ന വായുഗോളവും നാനാപ്രകാരം ചെരിഞ്ഞുതിരിയുന്നു എന്നു തോന്നും എന്നും ചൊല്ലീതായി. പിന്നെ ഇവിടെ വായുഗോളത്തിൻെറ സ്വരൂപവും വായുഗോളസംസ്ഥാനത്തിങ്കന്നു ഭഗോളസംസ്ഥാനത്തിൻെറ[10] ഭേദവും ഭൂമി ഉരുണ്ടിരിക്കയും ഇവ മൂന്നുമത്രെ ഗ്രഹസ്ഫുടം കഴിഞ്ഞശേഷം ഗ്രഹവിഷയമായിരിക്കുന്ന ഗണിതങ്ങൾക്കു മിക്കവാറും ഹേതുവാകുന്നത്. എന്നിട്ട് അവററിൻെറ സ്വരൂപം ഇവിടെ നടേ ചൊല്ലി.

8. ഗോളബന്ധം

പിന്നെ ഇവിടെ കല്പിച്ച മണ്ഡലങ്ങളും ഭ്രമണപ്രകാരവും ബുദ്ധ്യാരൂഢം ആകായ്കിൽ ചില വളയങ്ങളെക്കൊണ്ടു കെട്ടി, അക്ഷദണ്ഡിൻെറ നടുവേ ഉരുണ്ടൊരു വസ്തു ഭൂമി എന്നും കല്പിച്ച് ഗോളം തിരിയുമാറ്

7. 7. E. വിസ്തരിച്ച്
8. B. മേൽ വിവരിക്കും ഇതി വായുഗോളസ്വരൂപം
9. B. തിരിച്ചിലിനു തക്കവണ്ണം
10. B. E.F. സംസ്ഥാന ഭേദവും

കണ്ടുകൊള്ളൂ. ഇവിടെ സമമണ്ഡലവും ദക്ഷിണോത്തരവും ക്ഷിതിജോന്മണ്ഡലങ്ങളും തിരിയേണ്ടാ, എന്നിട്ടിവറ്റെ വലിയോ ചില വൃത്തങ്ങളെക്കൊണ്ട് പുറമേ കെട്ടൂ. മറേറവ തിരിയുമാറ് ഇരിപ്പൂ. എന്നിട്ട് ആ വൃത്തങ്ങളെ ചെറുതായവറ്റെക്കൊണ്ട്[1] അകമേ കെട്ടൂ. സൂത്രങ്ങളെക്കൊണ്ട് ജ്യാക്കളേയും ബന്ധിപ്പൂ. ഇങ്ങനെ ഗോളസംസ്ഥാന ഭ്രമണങ്ങളെ അവധരിച്ചുകൊള്ളൂ.

9. മഹാവൃത്തങ്ങൾ

അനന്തരം വലിപ്പമൊത്ത് ഒരു പ്രദേശത്തുതന്നെ കേന്ദ്രവുമായിരിക്കുന്ന വൃത്തങ്ങളിൽ വച്ച് വലിതവൃത്തത്തിങ്കേന്നു മറേറ വൃത്തങ്ങൾ രണ്ടിന്റേയും അകലം ഇവിടെ എത്ര എന്നറിയും പ്രകാരത്തെ ചൊല്ലുന്നു. അവിടെ അപക്രമജ്യാവും അതിന്റെ കോടിയും വരുത്തും പ്രകാരത്തെക്കൊണ്ടതിനെ നടേ കാട്ടുന്നൂ. ഇതിനായിക്കൊണ്ട് നിരക്ഷദേശത്തിങ്കൽ പൂർവ്വവിഷുവത്തു ഖമദ്ധ്യത്തിങ്കലാമ്മാറു കല്പിച്ചു നിരൂപിക്കുംപ്രകാരം. വിഷുവത് പ്രദേശത്തിങ്കേന്നു ഘടികാമണ്ഡലത്തിന്നു വിപരീതമായിരിക്കുന്ന വിഷുവദ്വിപരീതവൃത്തം ദക്ഷിണോത്തരത്തോട് ഒരുമിച്ചിരിക്കും. അയനാന്തവിപരീതവൃത്തം നിരക്ഷക്ഷിതിജത്തോടൊരുമിച്ചിരിക്കും. ഇങ്ങനെ ഗോളവിഭാഗം വന്നിരിക്കുന്നേടത്തു മേലും കീഴുമുള്ള സ്വസ്തികങ്ങളിലും കിഴക്കേ സ്വസ്തികത്തിന്നു ഇരുപത്തിനാലു തീയ്യതി വടക്കേ ക്ഷിതിജത്തിങ്കലും പടിഞ്ഞാറേ സ്വസ്തികത്തിങ്കേന്നു ഇരുപത്തിനാലു തീയതി തെക്കേയും ക്ഷിതിജത്തിങ്കലും സ്പർശിക്കുമാറ് ആദിത്യന്റെ കിഴക്കോട്ടുള്ള ഗതിക്കു മാർഗ്ഗമാകുന്ന അപക്രമമണ്ഡലത്തെ കല്പിപ്പൂ. പിന്നെ വിഷുവത്പ്രദേശം പദാദിയായി അവിടെ ശരമാകുമാറ് ഖമധ്യത്തിങ്കേന്നു കിഴക്ക് അപക്രമമണ്ഡലത്തിന്റെ ഇഷ്ടപ്രദേശത്തിങ്കലഗ്രമാകുമാറ് ഒരു ഇഷ്ടജ്യാവിനെ കല്പിപ്പൂ. അത് അപക്രമമണ്ഡലത്തിലെ ഇഷ്ടചാപഭാഗത്തിൻെറ ജ്യാവിനെ വരുത്തിയാലുണ്ടാകും. പിന്നെ ഇഷ്ടജ്യാഗ്രത്തിങ്കേന്ന് ഘടികാമണ്ഡലം നേരേ തെക്കുവടക്ക് എത്ര

8. 1. B. adds തിരിയുമാറ്

അകലമുണ്ട് എന്ന് ഒന്ന്, ഈ ഇഷ്ടജ്യാഗ്രത്തിങ്കേന്നു തന്നെ ദക്ഷിണോത്തരമണ്ഡലം നേരേ കിഴക്കുപടിഞ്ഞാറ് എത്ര അകലമുണ്ട് എന്നു രണ്ടാമത് ഇവ അറിയും പ്രകാരം.

ഇവിടെ അപക്രമമണ്ഡലവും ഘടികാമണ്ഡലവും തങ്ങളിലുള്ള പരമാന്തരാളം അയനാന്തവിപരീതവൃത്തമാകുന്ന ക്ഷിതിജത്തിങ്കൽ[1] കാണാം. ഇത് ഇരുപത്തിനാലു തീയതിയുടെ ജ്യാവായിരിക്കുന്ന പരമാപക്രമം. പിന്നെ അപക്രമമണ്ഡലവും ദക്ഷിണോത്തരവും തങ്ങളിലുള്ള പരമാന്തരാളവും അയനാന്തവിപരീതവൃത്തത്തിങ്കൽ തന്നെ കാണാം[2]. ഇതിൽ പരമാപക്രമത്തിന്നു കോടിയായിട്ട് അപക്രമ മണ്ഡലത്തിന്നു ധ്രുവനോടുള്ള അന്തരാളമായിട്ടിരിക്കും. ഇതിന്ന് 'പരമസ്വാഹോരാത്ര'മെന്നു പേർ.

പിന്നെ കേന്ദ്രത്തിങ്കേന്ന് അയനാന്തവിപരീതവൃത്തനേമിയോളമുള്ള അപക്രമമണ്ഡലവ്യാസാർദ്ധം കർണ്ണമായി പ്രമാണമായി കല്പിച്ച്, ഈ പരമാന്തരാളങ്ങൾ രണ്ടിനേയും ഈ കർണ്ണത്തിൻെറ ഭുജാകോടികളായി പ്രമാണഫലങ്ങളായി കല്പിച്ച്, പിന്നെ അപക്രമമണ്ഡലത്തിൻെറ ഇഷ്ടപ്രദേശത്തിങ്കലഗ്രമായിരിക്കുന്ന ഇഷ്ടദോർജ്യാവിനെ ഇച്ഛ എന്നും കല്പിച്ച്, ത്രൈരാശികം ചെയ്താൽ ഈ ഇഷ്ടദോർജ്യാഗ്രത്തിങ്കേന്നു ഘടികാമണ്ഡലത്തോളവും ദക്ഷിണോത്തരമണ്ഡലത്തോളവും ഉള്ള അന്തരാളങ്ങൾ അപക്രമമണ്ഡലത്തിങ്കലെ ഇഷ്ടദോർജ്യാവിൻെറ ഭുജാകോടികളായി ഇച്ഛാഫലങ്ങളായിട്ട് ഉളവാകും. 'ഇഷ്ടാപക്രമവും' 'ഇഷ്ടാപക്രമകോടി'യും എന്നിവററിന്നു പേർ. ഇതത്രെ എല്ലാടവും കേന്ദ്രമൊന്നിച്ചു വലിപ്പമൊത്തിരിക്കുന്ന വൃത്തങ്ങളുടെ അന്തരാള ത്രൈരാശികത്തിങ്കലേ ന്യായമാകുന്നത്.

9\. 1. E. adds ഇതിൽ പരമാപക്രമത്തിന്ന്
2\. B. adds വിപരീതവൃത്തമാകുന്ന ക്ഷിതിജത്തിങ്കൽ തന്നെ കാണാം

10. വിഷുവദ്വിപരീതവൃത്തവും നതവൃത്തവും

പിന്നെ ഇതിനെ തന്നെ ചുരുക്കി അറിയും പ്രകാരത്തെ ചൊല്ലുന്നു. അവിടെ ഘടികാമണ്ഡലവും വിഷുവദ്വിപരീതവും അയനാന്തവിപരീതവും, ഇവ മുമ്മൂന്നും അന്യോന്യതിര്യഗ്ഗതങ്ങളായിട്ടുണ്ടല്ലോ. പിന്നെ ഘടികാമണ്ഡലത്തിങ്കേന്ന് ഒട്ടുചെരിഞ്ഞിട്ട് ഒരു അപക്രമവൃത്തവും ഇടൂ. പിന്നെ ഈ നാലും കൂടാതെ പിന്നെയും മൂന്നു വൃത്തങ്ങളെ കല്പിപ്പൂ. അവിടെ നടേ രണ്ടു ധ്രുവങ്കലും അപക്രമമണ്ഡലത്തിൻെറ ഇഷ്ടപ്രദേശത്തിങ്കലും സ്പർശിച്ചിട്ട് ഒരു വൃത്തത്തെ കല്പിപ്പൂ. ഇതിന്ന് 'ഘടികാനത'മെന്നുപേർ. ഇതിങ്കൽ നിന്നു വിഷുവദ്വിപരീതത്തിന്നും അയനാന്തവിപരീതത്തിന്നും ഉള്ള പരമാന്തരാളം ഘടികാമണ്ഡലത്തിങ്കൽ കാണാം.

പിന്നെ ഘടികാവൃത്തവും അയനാന്തവിപരീതവും തങ്ങളിൽ കൂട്ടുന്നേടത്തും അപക്രമമണ്ഡലത്തിൻെറ ഇഷ്ടപ്രദേശത്തിങ്കലും സ്പർശിച്ചിട്ട് ഒരു വൃത്തത്തെ കല്പിപ്പൂ. ഇതിന്നു 'വിഷുവദ്വിപരീതനത'മെന്നു താൻ അതിന്നു ദക്ഷിണോത്തരത്തോടൈക്യമുണ്ടാകയാൽ 'ദക്ഷിണോത്തര നത'മെന്നു താൻ പേർ. ഇതിങ്കേന്ന് അയനാന്തവിപരീതത്തിന്നും ഘടികാവൃത്തത്തിന്നുമുള്ള പരമാന്തരാളം വിഷുവദ്വിപരീതത്തിങ്കൽ[1] കാണാം.

പിന്നെ ഈ കല്പിച്ച അപക്രമമണ്ഡലസംസ്ഥാനത്തിങ്കൽ തെക്കേ ധ്രുവങ്കേന്നു ഇരുപത്തിനാലു തീയതി കിഴക്കേയും, വടക്കേ ധ്രുവങ്കേന്ന് അത്ര[2] പടിഞ്ഞാറേയും അയനാന്തവിപരീതമാകുന്ന ക്ഷിതിജത്തിങ്കൽ സ്പർശിച്ചിട്ടിരിക്കും[3]. അന്നേരത്തു രാശികൂടങ്ങൾ രണ്ടിങ്കലും ഖമധ്യത്തിങ്കേന്നു പടിഞ്ഞാറ് അപക്രമമണ്ഡലേഷ്ടപ്രദേശത്തിങ്കേന്നു വൃത്തത്തിൻെറ നാലൊന്നു ചെന്നേടത്ത് അപക്രമമണ്ഡലത്തിലും സ്പർശിച്ചിട്ട് ഒരു വൃത്തത്തെ കല്പിപ്പൂ. അതിന്നു 'രാശികൂടവൃത്ത'മെന്നു പേർ. ഈ രാശികൂടവൃത്തവും ഘടികാവൃത്തവും തങ്ങളിലുള്ള

---

10. 1. C. വിപരീതവൃത്തത്തിങ്കൽ
2. C. 24 തീയതി for അത്ര
3. C.F. സ്പർശിക്കും

യോഗത്തിങ്കേന്നു രണ്ടിങ്കേന്നും വൃത്തത്തിൻെറ[4] നാലൊന്നു ചെന്നേടത്ത് ഈ രണ്ടിന്റേയും പരമാന്തരാളമാകുന്നു. അതു ഘടികാനതവൃത്തത്തിങ്കൽ സംഭവിക്കും.

പിന്നെ ഘടികാനതവൃത്തം ധ്രുവദ്വയത്തിങ്കൽ സ്പർശിക്കയാൽ ഘടികാവിപരീതമായിട്ടിരിപ്പൊന്ന്. പിന്നെ ക്രാന്തീഷ്ടജ്യാഗ്രം യാതൊരിടത്ത് അവിടം രാശികൂടവൃത്തപാർശ്വമാകുന്നത്. അവിടെയും സ്പർശിക്കയാൽ[5] രാശികൂടവൃത്തവിപരീതമാകയുമുണ്ട് ഘടികാനതവൃത്തം. ഇങ്ങനെ ഘടികാരാശികൂടങ്ങൾ രണ്ടിന്നും വിപരീതമാകയാൽ ഇവ തങ്ങളിലെ പരമാന്തരാളം ഈ ഘടികാനതവൃത്തത്തിങ്കൽ സംഭവിക്കേണ്ടു. അത് ഇഷ്ടദ്യുജ്യാതുല്യം. ഇവ്വണ്ണം രാശികൂടവൃത്തവും വിഷുവദ്വിപരീതമാകുന്ന ദക്ഷിണോത്തരവൃത്തവും തങ്ങളിലുള്ള പരമാന്തരാളം ഇവ രണ്ടിന്നും കൂടി വിപരീതവൃത്തമാകുന്ന യാമ്യോത്തരനതത്തിങ്കൽ സംഭവിക്കേണ്ടൂ.

പൂർവ്വാപരസ്വസ്തികത്തിങ്കലും ഇഷ്ടജ്യാഗ്രത്തിങ്കലും കൂടി സ്പർശിക്കയാൽ രണ്ടിന്നും കൂടി വിപരീതമാകുന്നു യാമ്യോത്തരനതം. രണ്ടു വൃത്തങ്ങൾ തങ്ങളിൽ സ്പർശിക്കുന്ന രണ്ടു സംപാതത്തിങ്കേന്നും വൃത്തത്തിൽ നാലൊന്നു ചെന്നേടത്തു സ്പർശിക്കുന്ന മൂന്നാം വൃത്തം വിപരീതവൃത്തമാകുന്നത്. ഇതിങ്കൽ മുമ്പിലത്തേവ രണ്ടിന്റേയും പരമാന്തരാളം തങ്ങളിലുള്ളതു സംഭവിച്ചു [എന്നു] ന്യായം.

ഇവിടെ വിഷുവദ്വിപരീതമാകുന്ന ദക്ഷിണോത്തരവും അയനാന്തവിപരീതമാകുന്ന ക്ഷിതിജവും ഘടികാമണ്ഡലവും അന്യോന്യം വിപരീതങ്ങളാകുന്നവ. ഇങ്ങനെ ഈ മൂന്നു വൃത്തങ്ങളേക്കൊണ്ട് പദവ്യവസ്ഥയും ഗോളവിഭാഗവും വന്നിരിക്കുന്നേടത്ത് ഈ പദത്തിൻെറ നടുവിലുള്ള നതവൃത്തങ്ങൾ രണ്ടും അപക്രമവൃത്തവും രാശികൂടവൃത്തവും. ഇവററെക്കൊണ്ട് വൃത്താന്തരാളത്തെ പരിച്ഛേദിക്കുന്നു. അവിടെ ഘടികാപക്രമാന്തരാളം ക്ഷിതിജത്തിങ്കൽ പരമാപക്രമതുല്യം. പിന്നെ വിഷുവത്തിങ്കേന്നു തുടങ്ങി അപക്രമമണ്ഡലേഷ്ടപ്രദേശത്തിങ്കൽ അഗ്രമാകുന്നത് ദോർജ്യാവ്. അയനാന്തവിപരീതത്തിങ്കേന്നു തുടങ്ങി

---

10. 4. B. വൃത്തത്തിങ്കൽ
5. B. രാശിക്കൂടവൃത്തങ്ങൾ

ഇഷ്ടാപക്രമത്തിങ്കലഗ്രമാകുന്നത് ദോർജ്യാകോടി. നതാപക്രമ സംപാതത്തിങ്കേന്നു ഘടികാമണ്ഡലത്തോളമുള്ള നതമണ്ഡലത്തിങ്കലേ ജ്യാവ് ഇഷ്ടാപക്രമം. പിന്നെ ധ്രുവങ്കേന്നു തുടങ്ങി അപക്രമേഷ്ട പ്രദേശത്തോളമുള്ള നതമണ്ഡലത്തിന്മേലേ ജ്യാവ് ഇഷ്ടദ്യുജ്യാവാകുന്നത്.

11. വിക്ഷിപ്തഗ്രഹത്തിന്റെ അപക്രമം

ഇഷ്ടാപക്രമവും ഈ വൃത്തത്തിങ്കൽ തന്നെ ഉള്ളൂ. പിന്നെ ദക്ഷിണോത്തരവൃത്തത്തിങ്കേന്നു ദോർജ്യാഗ്രത്തിങ്കലഗ്രമായിട്ടിരിക്കുന്ന ദക്ഷിണോത്തരനതവൃത്തത്തിങ്കലേ ജ്യാവ് ഇഷ്ടാപക്രമകോടി. ഈ വൃത്തത്തിങ്കൽ തന്നെ പൂർവ്വാപരസ്വസ്തികത്തിങ്കേന്നു ദോർജ്യാഗ്രത്തിങ്കലഗ്രമാകുന്നത് അപക്രമകോടീടെ കോടി. പിന്നെ വിഷുവത്തിങ്കൽ നിന്നും ഘടികാനതസംപാതത്തിങ്കൽ അഗ്രമാകുന്നത് ലങ്കോദയജ്യാവ് കാലജ്യാവു തന്നെ. ഈ ജ്യാഗ്രത്തിങ്കലഗ്രമായി പൂർവ്വാപരസ്വസ്തികത്തിങ്കേന്നു തുടങ്ങിയതു ലങ്കോദയജ്യാവ്. കോടിഖമദ്ധ്യത്തിങ്കേന്നു തുടങ്ങി ഘടികാമണ്ഡലത്തിങ്കൽ രാശികൂടഘടികാസംപാതത്തിൽ അഗ്രമായിരിക്കുന്നതു കാലകോടിജ്യാവ്. ഈ ജ്യാഗ്രത്തിങ്കലഗ്രമായി രാശികൂടവൃത്തത്തിങ്കലപക്രമമണ്ഡലസംപാതത്തിങ്കേന്നു തുടങ്ങി കാലകോട്യപക്രമം ഘടികാമണ്ഡലത്തിങ്കൽ കർണ്ണം കല്പിച്ചിട്ടുള്ള പരമാപക്രമം കൊണ്ട് ഇതു വരുത്തേണ്ടൂ.

ഈ രാശികൂടവൃത്തക്രാന്തിസംപാതത്തിങ്കേന്നു ഒരു ഗ്രഹം വിക്ഷേപിച്ചു എങ്കിൽ[1] ഈ രാശികൂടവൃത്തത്തിന്മേൽ വിക്ഷേപിക്കയാൽ കാലകോട്യപക്രമചാപശേഷമായിട്ടിരിക്കും ആ വിക്ഷേപചാപം. ഇച്ചാപയോഗം താനന്തരം താൻ ഘടികാ-രാശികൂടവൃത്ത സംപാതത്തിങ്കേന്നു വിക്ഷേപിച്ച ഗ്രഹത്തോടുള്ള അന്തരാളമാകുന്നത്. പിന്നെ ഘടികാ-രാശികൂടവൃത്തങ്ങളുടെ പരമാന്തരാളം ഘടികാ-നതത്തിങ്കലാകുന്നു. അത് ഇഷ്ടദ്യുജ്യാതുല്യം താനും.

---

11. 1. F. ചോകിൽ

ഈ രാശികൂടവൃത്തത്തിനു നതാപക്രമസംപാതത്തിങ്കൽ പാർശ്വമാകുന്നു. സമപാർശ്വത്തിങ്കേന്നു തൻെറ സർവ്വാവയത്തിനും വൃത്തപാദം അകലമുണ്ടായിരിപ്പൂതും. എന്നിട്ട് ഘടികാനതത്തിങ്കലും ദക്ഷിണോത്തരത്തിങ്കലും രാശികൂടാപക്രമാന്തരാളം വൃത്തപാദമെന്നു വന്നു. ഈ വൃത്തപാദങ്ങളെ ഘടികാമണ്ഡലം കൊണ്ടും യാമ്യോത്തരംകൊണ്ടും രണ്ടു പകുത്തിരിക്കും. അതിൽ ഘടികാമണ്ഡലത്തിൻെറ വടക്കേക്കൂറ് ഇഷ്ടാപക്രമമായിരിക്കും, എന്നാൽ തെക്കേക്കൂറ് ഇഷ്ടദ്യുജ്യാവെന്നു വരും. ഇതുതന്നെ ഘടികാരാശികൂടങ്ങളുടെ പരമാന്തരാളമാകുന്നതും.

ഘടികാരാശികൂടങ്ങളുടെ പാർശ്വസ്പൃഷ്ടം ഘടികാനതം. ഘടികാനത പാർശ്വസ്പൃഷ്ടങ്ങൾ ഘടികാരാശികൂടങ്ങളാകയാൽ ഘടികാരാശികൂട വൃത്തസംപാതത്തിങ്കേന്നു തുടങ്ങി നതവൃത്തസംപാതത്തിങ്ക ലഗ്രമായിരിക്കുന്ന രാശികൂടവൃത്തത്തിങ്കലേ ത്രിജ്യാകർണ്ണത്തിന് ഇച്ചൊല്ലിയ പരമാന്തരാളമാകുന്ന[2] ഇഷ്ടദ്യുജ്യാവു കോടിയാകുന്നത്. അപ്പോൾ ഘടികാസംപാതത്തിങ്കേന്നു തുടങ്ങി രാശികൂടവൃത്തത്തിന്മേലേ വിക്ഷിപ്തഗ്രഹത്തിങ്കലഗ്രമായി കർണ്ണരൂപമായിരിക്കുന്ന ജ്യാവിന്ന് എന്തു കോടിയാകുന്നതെന്നു വിക്ഷിപ്തഗ്രഹത്തിങ്കേന്നു ഘടികാവൃത്തത്തോടുള്ള അന്തരാളമുണ്ടാകും. അത് വിക്ഷിപ്തഗ്രഹക്രാന്തിയാകുന്നത്.

ഇങ്ങനെ കാലകോടിക്രാന്തിചാപവും വിക്ഷേപചാപവും തങ്ങളിൽ യോഗം താനന്തരം താൻ ചെയ്ത് ജ്യാവു കൊണ്ട് ത്രൈരാശികം ചെയ്ത് വിക്ഷിപ്തഗ്രഹക്രാന്തി വരുത്തുംപ്രകാരം. ഈ ഇച്ഛാഫലത്തേയും പ്രമാണഫലത്തേയും കൂടി ത്രിഭുജകളെന്നു ചൊല്ലുകിലുമാം. ഇങ്ങനെ ചാപയോഗം ചെയ്യുന്നേടത്തു ജ്യായോഗം ചെയ്കിലുമാം. അന്യോന്യകോടി ഗുണനവും ത്രിജ്യാഹരണവും ചെയ്താൽ ഫലയോഗം താൻ അന്തരം താൻ ചെയ്ത് ഇഷ്ടദ്യുജ്യാഗുണനവും ത്രിജ്യാഹരണവും ചെയ്താൽ വിക്ഷിപ്തഗ്രഹക്രാന്തി വരും. ഇവിടെ കാലകോടിക്രാന്തിക്കു വിക്ഷേപ കോടിയും. ഇഷ്ടദ്യുജ്യയും ഗുണകാരങ്ങളാകുന്നത്. അവിടെ കാലകോടിക്രാന്തിയെ നടേ ഇഷ്ടദ്യുജ്യകൊണ്ടു ഗുണിച്ച് ത്രിജ്യകൊണ്ടു ഹരിപ്പൂ. ഫലം രാശികൂടക്രാന്തിവൃത്തസംപാതത്തിങ്കേന്നു ഘടികാവൃത്താന്തര

---

11. 2. B.C.F. പരമാപക്രമാന്തരാളമാകുന്ന

മുണ്ടാകും. അവിടെ അപക്രമമണ്ഡലത്തിൻെറ യാതൊരു പ്രദേശത്തിങ്കേന്നു ഗ്രഹം വിക്ഷേപിച്ചു ആ ഗ്രഹത്തെ വിക്ഷേപിയാതെ കല്പിക്കുമ്പോളേ അപക്രമമണ്ഡലജ്യാവായിരിക്കുമത്. പിന്നെ ദ്യുജ്യകൊണ്ടു വിക്ഷേപത്തെ ഗുണിക്കേണ്ടിയിരുന്നേടത്ത് ആ വിക്ഷേപത്തിൻെറ ഗുണകാരമാകുന്ന കാലകോടിക്രാന്തികോടിയെ ഇഷ്ടദ്യുജ്യകൊണ്ടു ഗുണിച്ചു ത്രിജ്യകൊണ്ടു ഹരിക്കാം, ഫലം രണ്ടു പ്രകാരമായാലും തുല്യം എന്നിട്ട്. അപ്പോൾ കാലകോടിക്രാന്തിയേയും അതിൻെറ കോടിയേയും ഇഷ്ടദ്യുജ്യകൊണ്ടു ഗുണിച്ച് തിജ്യകൊണ്ട് ഹരിച്ചതായിട്ടിരിക്കും. അപ്പോൾ ഫലങ്ങൾ ഇഷ്ടദ്യുജ്യാവ്യാസാർദ്ധമാകുന്ന വൃത്തത്തിങ്കലേ ഭുജാകോടികളായിട്ടിരിക്കും. എന്നാൽ ഈ ഇഷ്ടദ്യുജ്യാവ്യാസാർദ്ധത്തിങ്കലേ കാലകോടിക്രാന്തിയേയും അതിൻെറ കോടിയേയും വിക്ഷേപകോടികൊണ്ടും വിക്ഷേപം കൊണ്ടും യഥാക്രമം ഗുണിപ്പൂ. എന്നാകിലുമാം, അപ്പോൾ ഇവിടെ കാലകോടിക്രാന്തിയെ ഇഷ്ടദ്യുജ്യാവൃത്തത്തിങ്കലാക്കിയാൽ അതു വിക്ഷിപ്തഗ്രഹക്രാന്തിയാകുന്നത് എന്നു ചൊല്ലിയല്ലോ.

എന്നാൽ ആ വിക്ഷിപ്തഗ്രഹകാന്തിവർഗ്ഗത്തെ ഇഷ്ടദ്യുജ്യാവർഗ്ഗത്തിങ്കേന്നു കളഞ്ഞത് അന്തദ്യുജ്യാവർഗ്ഗതുല്യം, അതിനെ മൂലിച്ചതു ദ്യുജ്യാവൃത്തത്തിങ്കലേ കാലകോടിക്രാന്തികോടിയാകുന്നത്. അതു പരമാപക്രമകോടിയായിട്ടിരിക്കും. ഇവിടെ ഇഷ്ടദോർജ്യാക്രാന്തീടെ വർഗ്ഗത്തെ ത്രിജ്യാവർഗ്ഗത്തിങ്കേന്നു കളഞ്ഞത് ഇഷ്ടദ്യുജ്യയുടെ വർഗ്ഗമാകുന്നത്. പിന്നെ ദോർജ്യാഗ്രത്തിങ്കേന്ന് അവിക്ഷിപ്തഗ്രഹത്തെ കല്പിപ്പൂ. അപ്പോൾ അതിൻെറ ക്രാന്തി കോടി ക്രാന്തിയായിട്ടിരിക്കും. ഈ ക്രാന്തിയുടെ വർഗ്ഗവും കൂടി കളഞ്ഞാൽ കോടിക്രാന്തിവർഗ്ഗവും ഭുജാക്രാന്തിവർഗ്ഗവും കൂടി കളഞ്ഞതായിട്ടിരിക്കും. കോടിക്രാന്തിവർഗ്ഗവും ഭുജാക്രാന്തിവർഗ്ഗവും കൂട്ടിയാൽ പരമക്രാന്തിവർഗ്ഗമായിട്ടിരിക്കും. അതു കളഞ്ഞ ത്രിജ്യാവർഗ്ഗം പരമക്രാന്തികോടിവർഗ്ഗമായിട്ടിരിക്കും. അതിന്റെ മൂലം പരമക്രാന്തികോടി. എന്നാൽ പരമക്രാന്തികോടികൊണ്ടു വിക്ഷേപത്തെ ഗുണിപ്പൂ. വിക്ഷേപകോടികൊണ്ട് അവിക്ഷിപ്തഗ്രഹക്രാന്തിജ്യാവിനേയും ഗുണിപ്പൂ. തങ്ങളിൽ യോഗംതാനന്തരം താൻ ചെയ്തു ത്രിജ്യകൊണ്ടു ഹരിച്ചാൽ ഫലം വിക്ഷിപ്തഗ്രഹക്രാന്തി എന്നു വന്നു. ഇങ്ങനെ വിക്ഷിപ്തഗ്രഹക്രാന്തി വരുത്തും പ്രകാരം[3].

---

11. 3. B. ഇതി വിക്ഷിപ്തഗ്രഹക്രാന്ത്യാനയനം

12. അപക്രമകോടി

അനന്തരം വിക്ഷിപ്തഗ്രഹത്തിൻെറ അപക്രമകോടിയായിട്ട് കിഴക്കു പടിഞ്ഞാറ് വിഷുവദ്വിപരീതമാകുന്ന ദക്ഷിണോത്തരവൃത്തത്തോളമുള്ള അന്തരാളമുണ്ടാക്കുംപ്രകാരത്തെ ചൊല്ലുന്നു. അവിടെ പൂർവ്വാപര സ്വസ്തികങ്ങളിൽ ദക്ഷിണോത്തരവൃത്തപാർശ്വങ്ങളാകുന്നതു ക്രാന്തീഷ്ടജ്യാഗ്രം രാശികൂടവൃത്തത്തിൻെറ പാർശ്വമാകുന്നത്. ഈ രണ്ടു വൃത്തങ്ങളുടേയും പാർശ്വങ്ങളെ സ്പർശിക്കുന്ന ദക്ഷിണോത്തര നതവൃത്തത്തിൻെറ[1] യാതൊരു പ്രദേശം ക്രാന്തീഷ്ടജ്യാഗ്രത്തിങ്കൽ സ്പർശിച്ചത് അവിടുന്ന് വൃത്തത്തിൻെറ നാലൊന്നു ചെന്നേടം രാശികൂടവൃത്തത്തെ സ്പർശിക്കും, തൻെറ പാർശ്വത്തിങ്കേന്ന് തൻെറ എല്ലാ അവയവവും വൃത്തപാദാന്തരിതം എന്നിട്ട്[2]. ഈ വൃത്തപാദത്തെ ദക്ഷിണോത്തരവൃത്തം കൊണ്ട് രണ്ടു പകുക്കാം. അതിൽ ഇഷ്ടക്രാന്തിദോർജ്യാഗ്രത്തിങ്കേന്നു ദക്ഷിണോത്തരവൃത്താന്തരാളം ഇഷ്ടക്രാന്തികോടിയാകുന്നത്. ഇക്കോടിശേഷം[3] ദക്ഷിണോത്തരവൃത്തത്തിങ്കേന്നു തുടങ്ങി പദശേഷം രാശികൂടവൃത്തത്തോളമുള്ളത് ഇഷ്ടാപക്രമകോടീടേ കോടി. എല്ലാ വൃത്തത്തിങ്കലും പദത്തെക്കൊണ്ടു വിഭജിച്ചാൽ തങ്ങളിൽ ഭുജാകോടികളായിരിക്കും. എന്നിട്ട്, ഇഷ്ടാപക്രമകോടി രാശികൂടദക്ഷിണോത്തരങ്ങളുടെ പരമാന്തരാളമെന്നു വന്നു. ദക്ഷിണോത്തരരാശികൂടസംപാതത്തിങ്കേന്നു തുടങ്ങി രാശികൂടവൃത്തത്തിന്മേലേ ദക്ഷിണോത്തരനതവൃത്തത്തോളമുള്ള ത്രിജ്യാകർണ്ണം പ്രമാണം, ഈ പരമാന്തരാളജ്യാ പ്രമാണഫലം, ദക്ഷിണോത്തരസമ്പാതത്തിങ്കേന്നു രാശികൂടത്തിന്മേലേ വിക്ഷിപ്തഗ്രഹത്തോളമുള്ള ഭാഗം ഇച്ഛയായി കല്പിച്ചാൽ ഗ്രഹത്തിങ്കേന്നു ദക്ഷിണോത്തരവൃത്തത്തോടുള്ള അന്തരാളം ഇച്ഛാഫലമായിട്ടുണ്ടാകും.

ഇവിടെ ഇച്ഛാരാശിയെ ഉണ്ടാക്കും പ്രകാരം പിന്നെ. ഇവിടെ

---

12. 1. D. adds.രാശികൂടയാമ്യോത്തരങ്ങളുടെ പരമാന്തരാളം എന്നിരിക്കും. ദക്ഷിണോത്തരരാശിക്കൂടങ്ങളുടെ സംപാതത്തിങ്കൽ ദക്ഷിണോത്തരനതപാർശ്വവും ഇങ്ങനെ ആകുന്നു. പിന്നെ ദക്ഷിണോത്തരനതവൃത്തത്തിൻെറ യാതൊരു പ്രദേശം.
2. D. പാദാന്തരിതം എന്നു നിയതം എന്നിട്ട്
3. B. പാദശേഷം

ദക്ഷിണസ്വസ്തികത്തിലഗ്രമായിരിക്കുന്ന യാമ്യോത്തരവൃത്തവ്യാസാർദ്ധമാകുന്ന[4] വൃത്തത്തിന് അപക്രമവൃത്താന്തരാളം ക്ഷിതിജത്തിങ്കൽ അന്ത്യദ്യുജ്യാതുല്യം പ്രമാണഫലം. ഖമദ്ധ്യത്തിങ്കേന്നു തുടങ്ങി രാശികൂടവൃത്തസംപാതത്തിങ്കലഗ്രമായിരിക്കുന്ന യമ്യോത്തര വൃത്തജ്യാവിന്ന് എത്ര അപക്രമവൃത്തത്തോടുള്ള അന്തരാളമെന്ന് യാമ്യോത്തരാപക്രമവൃത്തങ്ങളുടെ അന്തരാളം രാശികുടവൃത്തത്തിങ്കൽ ഇച്ഛരാഫലമായിട്ടുണ്ടാകും. പിന്നെ ഈ ജ്യാവിനോടു വിക്ഷേപജ്യാവിൻെറ യോഗംതാനന്തരം താൻ ചെയ്വു. എന്നാൽ യാമ്യോത്തര വൃത്തസംപാതത്തിങ്കേന്നു വിക്ഷിപ്തഗ്രഹത്തിങ്കലഗ്രമായിരിക്കുന്ന രാശികൂടവൃത്തജ്യാവുണ്ടാകും. പിന്നെ ഇതിനെ രാശികൂടയാമ്യോത്തരങ്ങളുടെ പരമാന്തരാളം കൊണ്ടു ഗുണിച്ച് ത്രിജ്യകൊണ്ടു ഹരിപ്പൂ. ഫലം വിക്ഷിപ്തഗ്രഹത്തിങ്കേന്നു യാമ്യോത്തരവൃത്താന്തരാളമായിട്ടുണ്ടാകും. ഇവിടെ ഇച്ഛരാരാശിയെ ഉണ്ടാക്കുവാനായിക്കൊണ്ടു വിക്ഷേപജ്യാ യോഗാന്തരങ്ങൾ ചെയ്യുന്നേടത്ത് പരസ്പരകോടിഗുണനവും ത്രിജ്യാഹരണവും വേണം. പിന്നെ പരമാന്തരാളഗുണനവും വേണം. അവിടെ നടേ പരമാന്തരാളഗുണനം ചെയ്വൂ. പിന്നെ വിക്ഷേപകോടി കൊണ്ടു ഗുണിപ്പൂ എന്ന ക്രമം കൊള്ളുകിലുമാം, ഫലഭേദമില്ലായ്കയാൽ. അവിടെ യാമ്യോത്തരാപക്രമവൃത്താന്തരാളത്തിങ്കലെ രാശികൂടവൃത്തഭാഗജ്യാവിനെ രാശികൂടയാമ്യോത്തരവൃത്തങ്ങളുടെ പരമാന്തരാളജ്യാവിനേക്കൊണ്ടു ഗുണിച്ച് ത്രിജ്യകൊണ്ടു ഹരിച്ച ഫലം രാശികൂടാപക്രമവൃത്ത സംപാതത്തിങ്കേന്നു യാമ്യോത്തരവൃത്താന്തരാളമുണ്ടാകും. അതു അവിക്ഷിപ്തഗ്രഹജ്യാകർണ്ണമായിരിക്കുന്ന ക്രാന്തീടെ കോടികളായിട്ടു വരും[5]. പിന്നെ യാമ്യോത്തരാപക്രമവൃത്താന്തരാളജ്യാവിൻെറ കോടിയേയും യാമ്യോത്തരരാശികൂടങ്ങളുടെ പരമാന്തരാളം കൊണ്ടു ഗുണിച്ച് ത്രിജ്യകൊണ്ടു ഹരിച്ചാൽ, ഫലം വിക്ഷിപ്തഗ്രഹാപക്രമകോടിവർഗ്ഗത്തെ ഈ പരമാന്തരാളവർഗ്ഗത്തിങ്കേന്നു കളഞ്ഞു മൂലിച്ചതായിട്ടിരിക്കും. ത്രിജ്യാവൃത്തത്തിലെ ഭുജാകോടികൾ രണ്ടിനേയും ഒരു ഗുണകാരം കൊണ്ടു

12. 4. B.reads. യാമ്യോത്തരവൃത്തങ്ങളുടെ എത്ര അപക്രമത്തോടുള്ള അന്തരാളം യാമ്യോത്തരാപക്രമങ്ങളുടെ അന്തരാളം....

5. B. കോടികളായിവരും, D. കോടിയായിട്ടുവരും

തന്നെ ഗുണിച്ച്[6] ത്രിജ്യകൊണ്ടു ഹരിക്കുമ്പോൾ ഗുണകാരവ്യാസാർദ്ധ വൃത്തത്തിലെ ഭുജാകോടികളായിട്ടു വരും എന്നിട്ട്.

ഇവിടെ പിന്നെ പരമാന്തരാളവൃത്തത്തിലെ കോടി പരമാപക്രമമായിട്ടുമിരിക്കും[7]. ഇവിടെ ഇഷ്ടാപക്രമവർഗ്ഗത്തെ ഇഷ്ടദോർജ്യാവർഗ്ഗത്തിങ്കേന്നു കളഞ്ഞ ശേഷം ഇഷ്ടാപക്രമകോടിവർഗ്ഗം. ഇതിനെ ത്രിജ്യാവർഗ്ഗത്തിങ്കേന്നു കളഞ്ഞ് ശേഷം യാമ്യോത്തരരാശികൂടവൃത്ത പരമാന്തരാളവർഗ്ഗം. ഇതിങ്കേന്നു പിന്നെ അവിക്ഷിപ്തഗ്രഹക്രാന്തികോടി വർഗ്ഗത്തേയും കളവൂ. ശേഷം ഇവിടെ വരേണ്ടതു കോടിവർഗ്ഗം. അത് പരമാപക്രമവർഗ്ഗമായിട്ടിരിക്കും.

ഇവിടെ ഭുജാപക്രമകോടിവർഗ്ഗവും കോട്യപക്രമകോടിവർഗ്ഗവും കൂട്ടിയാൽ അന്ത്യാപക്രമകോടിവർഗ്ഗമായിട്ടിരിക്കും. അതു ത്രിജ്യാവർഗ്ഗത്തിങ്കേന്നു കളഞ്ഞ ശേഷം അന്ത്യാപക്രമവർഗ്ഗം. അതിൻെറ മൂലം അന്ത്യാപക്രമം. എന്നാൽ അന്ത്യാപക്രമം കൊണ്ടു വിക്ഷേപത്തേയും വിക്ഷിപ്തഗ്രഹക്രാന്തികോടികൊണ്ടു വിക്ഷേപകോടിയേയും ഗുണിച്ച് തങ്ങളിൽ യോഗം താനന്തരം താൻ ചെയ്തുകൊണ്ട് ത്രിജ്യകൊണ്ടു ഹരിച്ച ഫലം വിക്ഷിപ്തഗ്രഹത്തിങ്കേന്നു യാമ്യോത്തരവൃത്താന്തരാളമായിട്ടിരിക്കും.

ഇതിനെത്തന്നെ ത്രിജ്യകൊണ്ടു ഹരിയാതെ വിക്ഷിപ്തഗ്രഹക്രാന്തി വർഗ്ഗത്തെ ത്രിജ്യാവർഗ്ഗത്തിങ്കേന്നു കളഞ്ഞു മൂലിച്ചുണ്ടാകുന്ന വിക്ഷിപ്ത ഗ്രഹദ്യുജ്യാവു യാതൊന്ന് ഇതിനെക്കൊണ്ടു ഹരിക്കിൽ വിക്ഷിപ്ത ഗ്രഹത്തിൻെറ കാലദോർഗ്ഗുണമായിട്ടിരിക്കും. ഈ കാലദോർഗ്ഗുണമാകുന്നതു പിന്നെ മുമ്പിൽ ചൊല്ലിയത്.

കൂടാതെ വിക്ഷിപ്തഗ്രഹത്തിങ്കലും രണ്ടു ധ്രുവങ്കലും സ്പർശിച്ചിട്ട് ഒരു വൃത്തത്തെ കല്പിപ്പൂ. അതു യാതൊരിടത്തു ഘടികാവൃത്തത്തിന്മേൽ സ്പർശിക്കുന്നൂ, അവിടുന്ന തുടങ്ങി വിഷുവത്തോളമുള്ള ഘടികാമണ്ഡലത്തിന്മേലേ ജ്യാവ് ഈ കാലദോർഗ്ഗുണമാകുന്നത്. ഇതിൻെറ ചാപം പ്രാണങ്ങളായിട്ടുള്ളൂ.

ഇത്ര പ്രാണകാലം കൊണ്ട് വിഷുവത്തോടുള്ള വിക്ഷിപ്ത

---

12. 6. C.F. ഹരിച്ചാൽ
 7. C. ക്രമമായിട്ടുവരും, F. പരമമായിട്ടുവരും

ഗ്രഹത്തോടുള്ള അന്തരാളപ്രദേശം ഭ്രമിക്കും എന്നിട്ട്. കാലമാകുന്ന ഇതിൻെറ ജ്യാവ് കാലജ്യാവ്. ഈ ഘടികാവൃത്തത്തിലെ പ്രാണസംഖ്യകൾ. പന്ത്രണ്ടു രാശികളുടെ ഇലികൊണ്ടു തുല്യസംഖ്യകൾ. അത് 'അനന്തപുരം' എന്നതിനോളം കാലംകൊണ്ട് ഗോളം ഒന്ന് തിരിഞ്ഞു കൂടും. എന്നിട്ട് കാലപ്രാണങ്ങളുടെ സംഖ്യാസാമ്യം. ഈവണ്ണമാകുമ്പോൾ ഘടികാവൃത്തത്തിങ്കലേപ്പോലെ എല്ലാ സ്വാഹോരാത്രവൃത്തത്തിങ്കലും[8] തൻെറ തൻെറ അനന്തപുരാംശം ഒരു പ്രാണകാലംകൊണ്ടു ഭ്രമിക്കും. എന്നാൽ, എല്ലാ സ്വാഹോരാത്രങ്ങളേയും ചക്രകലാതുല്യസംഖ്യങ്ങളായിട്ടു വിഭജിക്കേണ്ടൂ കാലമറിയുമ്പോൾ. എന്നാൽ വിക്ഷിപ്തഗ്രഹത്തിങ്കേന്നു ദക്ഷിണോത്തരവൃത്താന്തരാളം ഈ വരുത്തിയതു തന്നെ ആയിട്ടിരിക്കും.

വിക്ഷിപ്തഗ്രഹസ്വാഹോരാത്രവൃത്തത്തിൻെറ "അനന്തപുരാംശം" കൊണ്ട് അളക്കുമ്പോൾ എത്ര സംഖ്യ അത് എന്നിട്ട് ആ സ്വാഹോരാത്ര വൃത്തജ്യാവായിട്ട് ഇരിക്കുന്നതാകിലുമാം കാലദോർഗ്ഗുണം. അതിൻെറ ചാപം വിക്ഷിപ്തസ്വാഹോരാത്രവൃത്തത്തിങ്കലേയും യാമ്യോത്തരരാശികൂട വൃത്താന്തരാളം പ്രമാണങ്ങളായിരിക്കുന്ന കാലദോസ്സ് ആകുന്നതെന്നാ കിലുമാം. എന്നാൽ വിക്ഷിപ്തഗ്രഹത്തിങ്കേന്നു ഘടികാവൃത്താന്ത രാളത്തേയും വിഷുവദ്വിപരീതവൃത്താന്തരാളത്തേയും അറിയും പ്രകാരം ഈവണ്ണം ചൊല്ലിയതായി[9]. ഇപ്രകാരം ഉണ്ട് ചൊല്ലീട്ട് **സിദ്ധാന്ത ദർപ്പണത്തിൽ** ആചാര്യൻഃ/

അന്ത്യദ്യുജ്യേഷ്ടഭക്രാന്ത്യോഃ ക്ഷേപകോടിഘ്നയോർ യുതിഃ/
വിയുതിർ വാ ഗ്രഹക്രാന്തിസ് ത്രിജ്യാപ്താ കാലദോർഗുണഃ/
അന്ത്യക്രാന്തീഷ്ടതത്കോട്യാ സ്വദ്യുജ്യാപ്താപി പൂർവ്വവത്/

(സിദ്ധാന്തദർപ്പണം, 28 -29)

ഇങ്ങനെ വിക്ഷിപ്തഗ്രഹക്രാന്തിയും കാലജ്യാവും വരുത്തുന്നതിനെ ചൊല്ലിയതുകൊണ്ടു വൃത്താന്തരാളത്രൈരാശികങ്ങളെ മുഴുവനെ വിസ്തരിച്ചു കാട്ടീതായി.

[ഗണിതയുക്തിഭാഷയിൽ ഭൂ-വായു-ഭഗോളമെന്ന ഒമ്പതാമധ്യായം സമാപ്തം]

12. 8. B. വൃത്തങ്ങളിലും
9. B. പ്രകാരം ചൊല്ലി E. adds ഇപ്രകാരം

അദ്ധ്യായം പത്ത്

പഞ്ചദശപ്രശ്നം

പിന്നെയും ഈ ന്യായാതിദേശത്തെത്തന്നെ വിസ്തരിച്ചു കാട്ടുവാനായിക്കൊണ്ട് ഈ കല്പിച്ച ഏഴു[1] വൃത്തങ്ങളുടേയും അന്തരാളങ്ങൾ തന്നെ വിഷയമായിട്ട് പഞ്ചദശപ്രശ്നോത്തരങ്ങളെ കാട്ടുന്നുണ്ട്.

അവിടെ അന്ത്യക്രാന്തി, ഇഷ്ടക്രാന്തി, ഇഷ്ടക്രാന്തികോടി, ദോർജ്യാ, കാലജ്യാ, നതജ്യാ ഇങ്ങനെ ആറു സാധനങ്ങൾ. അവററിൽ രണ്ടറിഞ്ഞാൽ മററേവ നാലിനേയും അറിയും പ്രകാരത്തെ ചൊല്ലുന്നു. അതു പതിനഞ്ചു പ്രകാരം സംഭവിക്കും. ഒന്നറിഞ്ഞാൽ അതിൻെറ കോടി മിക്കവാറും ത്രിജ്യാവർഗ്ഗത്തിൽ തൻെറ വർഗ്ഗം കളഞ്ഞു മൂലിച്ചിട്ട് അറിയേണ്ടൂ.

ഇവിടെ ഘടികാപക്രമവിഷുവദ്വിപരീതനതവൃത്തങ്ങൾ ഘടികാനത വൃത്തത്തോടു രാശികൂടവൃത്തത്തേക്കു വൃത്തപാദാന്തരിതങ്ങൾ. ഈ വൃത്തപാദങ്ങൾ വിഷുവദ്വിപരീതവൃത്തം കൊണ്ട് രണ്ടു ഖണ്ഡിക്കപ്പെട്ടിരിക്കും വിഷുവദ്വിപരീതഘടികാനതവൃത്തങ്ങൾ[2] വിഷുവദ്വിപരീതനതരാശികൂടവൃത്തങ്ങളുടെ ഇടയിൽ വൃത്ത പാദാന്തരിതങ്ങൾ[3]. ഈ ഖണ്ഡങ്ങൾ ഒക്കെ തങ്ങളിൽ ഭുജാകോടികളായി ട്ടിരിക്കും[4] പിന്നെ വൃത്തപാദം കൊണ്ടു രണ്ടു ഖണ്ഡിച്ചാൽ ആ ഖണ്ഡങ്ങൾ തങ്ങളിൽ ഭുജാകോടികൾ എന്നോ നിയതമെല്ലോ എന്നിട്ട്.

---

1. 1. B. ഏഴു വൃത്തങ്ങളെത്തന്നെ വിഷയീകരിച്ച് പഞ്ചദശപ്രശ്നം
   2. D. വിഷുവദ്വിപരീതാപക്രമഘടികാ
   3. B.D.E.F. add വൃത്തപാദങ്ങൾ ഘടികാവൃത്തം കൊണ്ട് രണ്ട് ഖണ്ഡിക്കപ്പെട്ടിരിക്കും
   4. B.F. കോടികളായിരിക്കും

2. ഒന്നാം പ്രശ്നം: അന്ത്യക്രാന്തിയും ഇഷ്ടക്രാന്തിയും

പിന്നെ ഇവിടെ അന്ത്യക്രാന്തിയും ഇഷ്ടക്രാന്തിയുമറിഞ്ഞിട്ട് മററേവ നാലും അറിയും[1] പ്രകാരം[2] ഇവിടെ ചൊല്ലുന്നത്. അന്ത്യക്രാന്തിക്കു ത്രിജ്യാവു കർണ്ണമെന്ന് ഇഷ്ടക്രാന്തിക്ക് ഏത് കർണ്ണം എന്ന് ദോർജ്യാവുണ്ടാം. പിന്നെ ഘടികാപക്രമാന്തരാളം അന്ത്യാപക്രമമാകുമ്പോൾ യാമ്യോത്തരാപ ക്രമാന്തരാളം, അന്ത്യദ്യുജ്യാ ഇഷ്ടാപക്രമമാകുമ്പോൾ എന്ത് എന്ന് ദോർജ്യാഗ്രത്തിങ്കേന്നു യാമ്യോത്തരാന്തരാളമുണ്ടാകും. പിന്നെ ഇവ മൂന്നിനും ത്രിജ്യാവർഗ്ഗാന്തരമൂലങ്ങൾ കൊണ്ടു കോടികൾ ഉണ്ടാകും. പിന്നെ പൂർവ്വാപരസ്വസ്തികത്തിങ്കേന്നു തുടങ്ങി യാമ്യോത്തരനതവൃത്തത്തിന്മേലേ ദോർജ്യാഗ്രത്തോളം ചെല്ലുമ്പോൾ ഘടികായാമ്യോത്തരനതവൃത്താന്തരാള മാകുന്നത് ഇഷ്ടാപക്രമം. അപ്പോൾ ദക്ഷിണോത്തരവൃത്തത്തിങ്കൽ ഇവററിൻെറ പരമാന്തരാളമെത്രയെന്ന് യാമ്യോത്തരനതജ്യാവ് ഉണ്ടാകും. പിന്നെ ഉത്തരധ്രുവങ്കേന്നു ദോർജ്യാഗ്രത്തിങ്കൽ നതയാമ്യോത്തര വൃത്താന്തരാളമാകുന്നത് ഇഷ്ടാപക്രമകോടി. അപ്പോൾ ഘടികാ വൃത്തത്തിങ്കൽ പരമാന്തരാളം എത്രയെന്നു ലങ്കോദയജ്യാവുണ്ടാകും. ഇവ്വണ്ണമാകെക്കൊണ്ട് ഇഷ്ടാപക്രമകോടികളാകുന്ന പ്രമാണഫലങ്ങൾക്ക് ഇതരേതരകോടികൾ പ്രമാണങ്ങളായിട്ടു വന്നു. ഈ പ്രമാണങ്ങൾക്കുതന്നെ ദോർജ്യാകോടി പ്രമാണഫലം ആകുമ്പോൾ ത്രിജ്യാവ് ഇച്ഛയുമാകുമ്പോൾ നതക്ഷിതിജാന്തരാളങ്ങൾ നതകോടിയും ലങ്കോദയജ്യാകോടിയുമായി ട്ടുണ്ടാകും. ഇങ്ങനെ നടേത്തെ[3] പ്രശ്നോത്തരം.

3. രണ്ടാം പ്രശ്നം: അന്ത്യക്രാന്തിയും ഇഷ്ടക്രാന്തികോടിയും

രണ്ടാമത് അന്ത്യക്രാന്തിയും ഇഷ്ടക്രാന്തികോടിയും കൂടീട്ട് പരമാപക്രമകോടിക്ക് ത്രിജ്യാ കർണ്ണം, ഇഷ്ടാപക്രമകോടിക്ക് എന്ത് കർണ്ണം എന്ന് ദോർജ്യാവുണ്ടാക്കൂ. പിന്നെ നടേത്തേപ്പോലെ ഊഹിച്ചുകൊള്ളൂ.

---

2. 1. B.C.F. നാലും അറിയേണ്ടും
   2. D. adds ഇഷ്ടക്രാന്തിക്ക് എന്തു കർണ്ണം
   3. B. ഒന്നാമത്തെ

4. മൂന്നാം പ്രശ്നം: അന്ത്യക്രാന്തിയും ദോർജ്യാവും

മൂന്നാമത്, അന്ത്യക്രാന്തിയും ദോർജ്യാവും കൂടീട്ട്. ഇവിടെ അയനാന്തവിപരീതത്തിങ്കൽ ക്രാന്തിവൃത്തസംപാതത്തിങ്കേന്നു ഘടികാവൃത്തത്തോളവും വിഷുവദ്വിപരീതത്തോളവും ഉള്ള പഴുതുകൾ അന്ത്യക്രാന്തിയും അന്ത്യദ്യുജ്യയും. ഇവ പ്രമാണഫലങ്ങളായി ദോർജ്യാവ്. ഇച്ഛയായിട്ട് ഇഷ്ടാപക്രമകോടികളുണ്ടാകും. ശേഷം മുമ്പിലേപ്പോലെ.

5. നാലാം പ്രശ്നം: അന്ത്യാപക്രമവും കാലജ്യാവും

പിന്നെ അന്ത്യാപക്രമവും കാലജ്യാവും കൂടീട്ടു നാലാമത്. അവിടെ വിഷുവത്തിങ്കേന്നു നതമണ്ഡലാന്തമുള്ള ഘടികാമണ്ഡലഭാഗം കാലഭുജയാകുന്നത്. വിഷുവത്തിങ്കേന്നു രാശികൂടവൃത്തത്തോളമുള്ള ഘടികാമണ്ഡലഭാഗം 'കാലകോടി'. ഇതിന്ന് എന്തപക്രമവൃത്താന്തരാളം എന്നു രാശികൂടവൃത്തത്തിങ്കൽ ഘടികാപക്രമാന്തരാളമുണ്ടാകും. ഇതു 'കാലകോട്യപക്രമം'. പിന്നെ വിഷുവത്താകുന്ന ഖമദ്ധ്യസ്പൃഷ്ടമായിട്ട്[1] ഒരു രാശികൂടവൃത്തം കല്പിപ്പൂ. ഇതും നടേത്തേ രാശികൂടവൃത്തവും തങ്ങളിൽ ഉള്ള യോഗം ക്ഷിതിജത്തിങ്കലെ രാശികൂടവൃത്തത്തിങ്കൽ. അത് ഉത്തരധ്രുവങ്കേന്ന് അന്ത്യാപക്രമത്തോളം പടിഞ്ഞാറ്, ദക്ഷിണധ്രുവങ്കേന്ന് അത്ര കിഴക്കും. പിന്നെ കാലകോട്യപക്രമവർഗ്ഗത്തെ കാലകോടി വർഗ്ഗത്തിങ്കേന്നു കളഞ്ഞു മൂലിച്ചാൽ, ഘടികാരാശികൂടസംപാതത്തിങ്കേന്നു രണ്ടാം രാശികൂട വൃത്തത്തോളമുള്ള അന്തരാളമുണ്ടാകും. പിന്നെ കാലകോട്യപ ക്രമവർഗ്ഗത്തെ ത്രിജ്യാവർഗ്ഗത്തിങ്കേന്നു കളഞ്ഞു മൂലിച്ചാൽ, ക്ഷിതിജ സംപാതത്തിങ്കേന്നു ഘടികാസംപാതത്തോടുള്ള[2] രാശികൂടവൃത്ത ഭാഗജ്യാവുണ്ടാകും. ഈ ജ്യാവ് കർണ്ണമായിട്ട് പ്രമാണമാകുമ്പോൾ മുമ്പിൽ

---

5. 1. F. സ്ഫുടമായിട്ട്
2. E. സംപാതത്തോളമുള്ള

ചൊല്ലിയ മൂലം രാശികൂടവൃത്തങ്ങൾ തങ്ങളിലുള്ള അന്തരാളം ഭുജയായി പ്രമാണഫലമായിരിക്കും. അപ്പോൾ ത്രിജ്യാവ് ഇച്ഛ. രാശികൂടങ്ങളുടെ പരമാന്തരാളം രാശികൂടവൃത്താന്തരാളം അപക്രമവൃത്തത്തിങ്കൽ ഖമദ്ധ്യത്തിങ്കേന്നു രാശികൂടവൃത്താന്തരാളജ്യാവ് ഇച്ഛാഫലമായിട്ടുണ്ടാകും. ഇതിൻെറ കോടി ഖമദ്ധ്യത്തിങ്കേന്നു നതവൃത്തത്തോളമുള്ള അപക്രമമണ്ഡലത്തിങ്കലേ ദോർജ്യാവ്. ശേഷം മുമ്പിലേപ്പോലെ.[3]

6. അഞ്ചാം പ്രശ്നം: നതജ്യാവും അന്ത്യക്രാന്തിയും

പിന്നെ[1] നതജ്യയും[2] അന്ത്യക്രാന്തിയുമറിഞ്ഞിട്ട് അഞ്ചാമത്. പിന്നെ ഖമദ്ധ്യത്തിങ്കന്നു[3] നതവൃത്തത്തോളമുള്ള യാമ്യോത്തരവൃത്തഭാഗം നതമാകുന്നത്. ഖമദ്ധ്യത്തിങ്കേന്നു രാശികൂടവൃത്തത്തോളമുള്ള യാമ്യോത്തരഭാഗം നതകോടിയാകുന്നത്. പിന്നെ ദക്ഷിണോത്തര വൃത്തത്തിങ്കേന്നു അപക്രമവൃത്താന്തരാളം ക്ഷിതിജത്തിങ്കലേത് അന്ത്യദ്യുജ്യാവ്, അപ്പോൾ നതകോട്യഗ്രത്തിങ്കേന്ന് എത്ര എന്ന് രാശികൂടവൃത്തത്തിങ്കൽ യാമ്യോത്തരാപക്രമവൃത്താന്തരാളമുണ്ടാകും. ഇതിൻെറ വർഗ്ഗത്തെ നതകോടിവർഗ്ഗത്തിങ്കേന്നും ത്രിജ്യാവർഗ്ഗത്തിങ്കേന്നും കളഞ്ഞു മൂലിച്ചാൽ യാമ്യോത്തരപ്രഥമരാശികൂടസംപാതത്തിങ്കേന്നു ദ്വിതീയരാശികൂടവൃത്താന്തരാളം പ്രമാണഫലമായിട്ടും യാമ്യോത്തര സംപാതത്തിങ്കേന്നു ക്ഷിതിജത്തോളമുള്ള രാശികൂടവൃത്തത്തിങ്കലേ ജ്യാവു കർണ്ണമായിട്ടും പ്രമാണമായിട്ടും[4] ഉണ്ടാകും[5]. ത്രിജ്യാ വിച്ഛയാകുമ്പോൾ രണ്ടു രാശികൂടങ്ങളുടേയും പരമാന്തരാളം ഇച്ഛാഫലം, നടേത്തേ പരമാന്തരാളം തന്നെ. ഇതിൻെറ കോടി ദോർജ്യാവ്. ശേഷം നടേത്തേപ്പോലെ[6]. ഇങ്ങനെ അന്ത്യക്രാന്തിയോടുകൂടിയുള്ള പ്രശ്നങ്ങളഞ്ചും.

---

5. 3. B. ശേഷം പൂർവ്വവൽ
6. 1. F.om. പിന്നെ
   2. F. adds നതവൃത്തത്തോളമുള്ള ജ്യായും.....
   3. D. adds വിഷുവത്തിങ്കേന്ന്
   4. E. കർണ്ണമായി പ്രമാണമായിട്ട്
   5. C.D.F. വരും
   6. B. ശേഷം പൂർവ്വവൽ

7. ആറ്, ഏഴ്, എട്ട്, ഒൻപത് പ്രശ്നങ്ങൾ

പിന്നെ അന്ത്യക്രാന്തി കൂടാതെ ഇഷ്ടക്രാന്തിയും, ഇഷ്ടക്രാന്തികോടിയും കൂടീട്ടാറാമത്. ഇവററിൻെറ വർഗ്ഗയോഗമൂലം ദോർജ്യാവ് കർണ്ണമായിട്ടുണ്ടാകും.

പിന്നെ ഇഷ്ടാപക്രമദോർജ്യാക്കളെ[1], അറിഞ്ഞിട്ടു നടേത്തേപ്പോലെ ഏഴാമത്.

ഇഷ്ടാപക്രമകാലജ്യാക്കളെക്കൊണ്ട് എട്ടാമത്. ഈ രണ്ടിൻേറയും വർഗ്ഗത്തെ ത്രിജ്യാവർഗ്ഗത്തിൽ കളഞ്ഞു മൂലിപ്പൂ. എന്നാൽ ഇഷ്ടദ്യുജ്യാവും നതക്ഷിതിജങ്ങളുടെ പരമാന്തരാളമാകുന്ന കാലകോടിജ്യാവും ഉണ്ടാകും. ത്രിജ്യാവു പ്രമാണവും, കാലകോടി പ്രമാണഫലവും, ഇഷ്ടദ്യുജ്യാവ് ഇച്ഛയും, ഇവിടെ ഉണ്ടാകുന്ന ഇച്ഛാഫലം ദോർജ്യാകോടി. ശേഷം നടേത്തേ പോലെ[2].

പിന്നെ ഇഷ്ടാപക്രമവും നതജ്യാവുമറിഞ്ഞിട്ട് ഒൻപതാമത്. യാമ്യോത്തരനതവൃത്തവും ഘടികാവൃത്തവും തങ്ങളിലന്തരാളം നതജ്യാവാകുമ്പോൾ നതക്ഷിതിജാന്തരാളം നതകോടിജ്യാവ്, ഇഷ്ടാപക്രമം പ്രഥമാന്തരാളമാകുമ്പോൾ ദ്വിതീയാന്തരാളമെന്തെന്ന് ദോർജ്യാകോടി. ദോജ്യാഗ്രത്തിങ്കേന്നു ക്ഷിതിജാന്തരാളം നടേത്തേതു തന്നെ. ഇങ്ങനെ ഇഷ്ടാപക്രമത്തോടുകൂടീട്ടു നാലു[3] പ്രശ്നം.

8. പത്തും പതിനൊന്നും പ്രശ്നങ്ങൾ

ഇനി ഇതു കൂടാതെ പിന്നെ ഇഷ്ടക്രാന്തികോടിയും ദോർജ്യാവും കൂടീട്ടു പത്താമത്. ഇവററിൻെറ വർഗ്ഗാന്തരമൂലം ഇഷ്ടക്രാന്തി. ശേഷം നടേത്തേപ്പോലെ[1].

പിന്നെ കാലജ്യാവും ഇഷ്ടാപക്രമകോടിയും അറിഞ്ഞിട്ടു

---

7. 1. H. ഇഷ്ടപക്രമകാലജ്യാക്കളെ
2. B. ശേഷം, പൂർവ്വവൽ
3. B. കൂടി നാല്

8. 1. B. ശേഷം പൂർവവൽ

പതിനൊന്നാമത്. കാലജ്യായ്ക്ക് ത്രിജ്യാവു കർണ്ണം, ഇഷ്ടക്രാന്തികോടിക്ക് എന്ത് എന്ന് ദ്യുജ്യാവുണ്ടാകും. പിന്നെ ദ്യുജ്യയെ കാലകോടികൊണ്ടു ഗുണിച്ച് ത്രിജ്യകൊണ്ടു ഹരിച്ചാൽ ദോർജ്യാകോടി ഉണ്ടാകും. നതയാമ്യോത്തരാന്തരാളം കൊണ്ടു നടേത്തെ ത്രൈരാശികം. നതക്ഷിതിജാന്തരാളം കൊണ്ടു രണ്ടാം ത്രൈരാശികം.

9. പന്ത്രണ്ടാം പ്രശ്നം: ഇഷ്ടക്രാന്തികോടിയും നതജ്യായും

പിന്നെ ഇഷ്ടക്രാന്തികോടിയേയും നതജ്യായേയുമറിഞ്ഞിട്ട്[1] പന്ത്രണ്ടാമത്. ഇവററിൻെറ വർഗ്ഗത്തെ ത്രിജ്യാവർഗ്ഗത്തിങ്കേന്നു കളഞ്ഞു മൂലിച്ചാൽ പൂർവ്വസ്വസ്തികത്തിങ്കേന്നു ദോർജ്യാഗ്രത്തോളമുള്ള യാമ്യോത്തരനതഭാഗജ്യാവും യാമ്യോത്തരനതക്ഷിതിജങ്ങളുടെ പരമാന്തരാളവുമുണ്ടാകും. തങ്ങളിൽ ഗുണിച്ച് ത്രിജ്യകൊണ്ടു ഹരിച്ചാൽ ദോർജ്യാകോടി ഉണ്ടാകും.

10. പതിമൂന്നാം പ്രശ്നം: ദോർജ്യാവും കാലകോടിജ്യാവും

പിന്നെ ദോർജ്യാവും, കാലജ്യാവും കൂടീട്ടു പതിമൂന്നാമത്. ഇവ രണ്ടിനേയും വർഗ്ഗിച്ച് ത്രിജ്യാവർഗ്ഗത്തിങ്കേന്നു കളഞ്ഞു മൂലിച്ചാൽ കോടികളുണ്ടാകും. പിന്നെ കാലകോടിക്കു ത്രിജ്യാവു കർണ്ണം ദോർജ്യാകോടിക്കേതു കർണ്ണമെന്നു ദ്യുജ്യാവുണ്ടാകും.

11. പതിനാലാം പ്രശ്നം: ദോർജ്യയും നതജ്യയും

ദോർജ്യയും നതജ്യയുമറിഞ്ഞിട്ട് പതിനാലാമത്. നതകോടിക്കു ത്രിജ്യാവു കർണ്ണം, ദോർജ്യാകോടിക്കേതു[1] കർണ്ണമെന്നു ദോർജ്യാഗ്രത്തിങ്കേന്നു

9. 1. B. നതജ്യായുമറിഞ്ഞിട്ട്
11. 1. B. എന്ത്

പൂർവ്വസ്വസ്തികത്തോളമുള്ള നതവൃത്തജ്യാവുണ്ടാകും. ഇതിൻെറ കോടി ഇഷ്ടക്രാന്തികോടി.

12. പതിനഞ്ചാം പ്രശ്നം: കാലജ്യാവും നതജ്യാവും

പിന്നെ കാലജ്യാവും നതജ്യാവുമറിഞ്ഞിട്ട് മറേറവ അറിയുംപ്രകാരം പതിനഞ്ചാമത്. ഇവിടെ പൂർവ്വാപരസ്വസ്തികത്തിങ്കേന്നു രാശികൂട വൃത്തസംപാതത്തോടിട ഘടികാവൃത്തത്തിങ്കലേതു കാലജ്യാവ്. പിന്നെ പൂർവ്വാപരസ്വസ്തികത്തോട് രാശികൂടവൃത്തത്തോടിട യാമ്യോത്തരനത വൃത്തത്തിങ്കലെ ഭാഗം ക്രാന്തികോടിയായിട്ടുമിരിക്കും. അവിടെ ഖമദ്ധ്യത്തിങ്കേന്നു കാലകോടിയുടെ ശേഷം കാലഭുജാക്ഷിതിജാരന്തമിത്. പിന്നെ ഘടികാനതവൃത്തത്തോടു രാശികൂടവൃത്തത്തോടിട പദമാകയുമുണ്ട് യാമ്യോത്തരനതവൃത്തത്തിങ്കൽ. പിന്നെ ഇതിങ്കേന്നു തന്നെ യാമ്യോത്തരസംപാതത്തിങ്കേന്നു ക്ഷിതിജാന്തവും പദം. എന്നിട്ട് ഈവണ്ണമിതെല്ലാമാകുന്നു.

പിന്നെ ഈവണ്ണം തന്നെ യാമ്യോത്തരസ്വസ്തികത്തോടു രാശികൂടവൃത്തത്തോട് അന്തരാളം യാമ്യോത്തരനതവൃത്തത്തിങ്കലേതു നതജ്യാവായിട്ടിരിക്കും. പിന്നെ ഘടികാനതത്തിങ്കലെ രാശികൂട വൃത്തക്ഷിതിജാന്തരാളം അപക്രമം. ഇവിടെയും ഇതരനതസംപാത ത്തിങ്കേന്നു രാശികൂടവൃത്താന്തരം പദം, ദക്ഷിണസ്വസ്തികാഘടികാ ന്തരാളവും പദം. എന്നാൽ ഈവണ്ണമിരിക്കും ഇവിടെ.

ഈവണ്ണം ത്രൈരാശികം. പശ്ചിമസ്വസ്തികത്തിങ്കേന്നു ഖമദ്ധ്യത്തോള മുള്ള ഘടികാവ്യാസാർദ്ധമാകുന്ന കർണ്ണത്തിനു യാമ്യോത്തരനതവൃത്ത ത്തിന്റെ പരമാന്തരമാകുന്നതു നതജ്യാവ്. പശ്ചിമസ്വസ്തികത്തിങ്കേന്നു രാശി കൂടവൃത്താന്തമുള്ള ഭാഗത്തിന്റെ ജ്യാവ് കാലജ്യാവ്. അതു കർണ്ണമാകു മ്പോൾ യാമ്യോത്തരനതവൃത്താന്തരമെന്ത് എന്നു രാശികൂടവൃത്തത്തിങ്കലേ ഘടികാനതാന്തരാളമുണ്ടാകും. അവ്വണ്ണമേ യാമ്യോത്തരവൃത്തത്തിൽ യാമ്യസ്വസ്തികത്തിങ്കന്നു രാശികൂടവൃത്താന്തമുള്ള നതജ്യാവ് ഇച്ഛ. ഖമ ദ്ധ്യത്തിങ്കേന്നു നതവൃത്തപരമാന്തരാളമാകുന്ന കാലജ്യാവ് പ്രമാണഫലം.

യാമ്യോത്തരവൃത്തത്തിങ്കേന്നു രാശികൂടവൃത്തത്തിന്മേലേ നതാന്തരം ഇച്ഛാഫലം. ഇതു നാടേത്തേ ഇച്ഛാഫലത്തോടു തുല്യമായിട്ടിരിക്കും. ഈ അന്തരവർഗ്ഗം ത്രിജ്യാവർഗ്ഗത്തിങ്കേന്നു കളഞ്ഞു മൂലിച്ചാൽ ഘടികായാമ്യോത്തരവൃത്താന്തരാളം രാശികൂടവൃത്തത്തിങ്കലേ ഭാഗം ഉണ്ടാകും. ഈ ഭാഗജ്യാവു കർണ്ണമായി പ്രമാണമായിരിക്കുമ്പോൾ പ്രമാണഫലങ്ങളായിട്ടു രണ്ടു വൃത്താന്തരാഗ്രങ്ങളുണ്ടാകും. പിന്നെ ഇവറ്റിന്റെ ഇച്ഛാഫലമായിരിക്കുന്ന പരമാന്തരാളങ്ങൾ നതരാശികൂടവൃത്തസംപാതത്തിങ്കേന്നു സ്വസ്തികാവധി ഉള്ള നതജ്യാക്കൾ ഇവിടെ യാമ്യോത്തരനതത്തിങ്കലേത്. ഇഷ്ടാപക്രമകോടി. ഘടികാനതത്തിങ്കലേത് ഇഷ്ടാപക്രമം.

ഇവറ്റിന്റെ പ്രമാണഫലമുണ്ടാക്കും പ്രകാരം പിന്നെ. നതഘടികാന്തരം രാശികൂടവൃത്തഭാഗത്തിങ്കലേ ജ്യാവർഗ്ഗത്തെ കാലജ്യാവർഗ്ഗത്തിങ്കേന്നു കളഞ്ഞു മൂലിച്ചു കല്പിക്കാനിരിക്കുന്നതിൽ നടേത്തേ തിര്യഗ്വൃത്താവധി ഉള്ളത് ഒന്ന്. പിന്നെ നതജ്യാവർഗ്ഗത്തിങ്കേന്നു കളഞ്ഞു മൂലിച്ചതു രണ്ടാമത്. രണ്ടാം തിര്യഗ്വൃത്താവധി ഉണ്ടാകുമത്. തിര്യഗ്വൃത്തങ്ങളെ കല്പിക്കും പ്രകാരം പിന്നെ. യാമ്യോത്തരനതപാർശ്വമായിരിക്കുന്ന രാശികൂടയാമ്യോത്തരവൃത്തസംപാതത്തിങ്കലും പൂർവ്വാപരസ്വസ്തികങ്ങളിലും സ്പർശിക്കുമാറ് നടേത്തെ തിര്യഗ്വൃത്തം രണ്ടാമത്. പിന്നെ ഘടികാനതപാർശ്വമായിരിക്കുന്ന ഘടികാരാശികൂടസംപാതത്തിങ്കലും ദക്ഷിണോത്തരസ്വസ്തികങ്ങളിലും സ്പർശിച്ചിട്ടിരിക്കും. ഇവറ്റോടു രാശികൂടവൃത്തത്തോടുള്ള പരമാന്തരാളങ്ങൾ നതവൃത്തങ്ങൾ രണ്ടിങ്കലുമായിട്ടിരിക്കും. ഇവ ഇഷ്ടാപക്രമതത്കോടികളാകുന്നവ.

ഇങ്ങനെ പതിനഞ്ചു പ്രശ്നോത്തരങ്ങൾ ചൊല്ലീതായി. ഇങ്ങനെ വൃത്താന്തരാളത്രൈരാശികാതിദേശപ്രകാരം.

**[ഗണിതയുക്തിഭാഷയിൽ
പഞ്ചദശപ്രശ്നമെന്ന
പത്താമദ്ധ്യായം സമാപ്തം]**

അദ്ധ്യായം പതിനൊന്ന്

ഛായാപ്രകരണം

1. ദിക്ജ്ഞാനം

അനന്തരം ദിക്കറിയും പ്രകാരം. അവിടെ നടേ[1] ഒരു നിലം നിരത്തിച്ചമപ്പൂ. അതു നടുവിൽ വെള്ളം വീണാൽ വട്ടത്തിൽ പരന്ന് എല്ലാപ്പുറവും ഒക്കൊക്കെ ഒഴുകുമാറ് ഇരിക്കണം. അതു സമനിലത്തിനു ലക്ഷണമാകുന്നത്. പിന്നെ ഈ നിലത്ത് ഒരു വൃത്തം വരപ്പൂ. രണ്ടഗ്രത്തിങ്കലും കുറഞ്ഞോരു വളവു ള്ളോരു ശലാകേടെ ഒരഗ്രത്തെ മദ്ധ്യത്തിങ്കലൂന്നി മറ്റേ അഗ്രത്തെ ചുറ്റും ഭ്രമിപ്പിപ്പൂ. അതിന്റെ അഗ്രം ഊന്നിയ പ്രദേശത്തിന്നു 'കേന്ദ്രം' എന്നും 'നാഭി'എന്നും പേരുണ്ട്. മറ്റേ അഗ്രഭ്രമണം കൊണ്ടുണ്ടായ രേഖയ്ക്ക് 'നേമി' എന്നു പേർ. ഇതിന്റെ കേന്ദ്രത്തിങ്കൽ സമമായി ഉരുണ്ടിരിപ്പോരു 'ശങ്കു'വിനെ നിർത്തൂ. പിന്നെ ഒരിഷ്ടദിവസത്തിങ്കൽ പ്രാത:കാലത്തി ങ്കൽ ഈ ശങ്കുവിന്റെ ഛായാഗ്രം[2] വൃത്തനേമിയിങ്കൽ യാതൊരിടത്തു സ്പർശിച്ചു വൃത്തത്തിങ്കൽ അകത്തു പൂവും[3] അപരാഹ്നത്തിങ്കൽ യാതൊ രിടത്തെ സ്പർശിച്ചിട്ട് പുറത്തു പുറപ്പെടുന്നതും, ഈ രണ്ടു പ്രദേശത്തി ങ്കലും വൃത്തത്തിങ്കൽ ഓരോ ബിന്ദുക്കളെ ഉണ്ടാക്കൂ. ഇവ തങ്ങളിൽ മിക്കവാറും കിഴക്കു പടിഞ്ഞാറായിരിക്കും. എന്നിട്ടിവറ്റിന്നു 'പൂർവ്വാപരബി ന്ദുക്കൾ' എന്നു പേർ. ഇവ തന്നെ നേരെ പൂർവ്വാപരബിന്ദുക്കളായിട്ടിരിക്കും, തെക്കുവടക്കു ഗതിയില്ലാത്ത നക്ഷത്രങ്ങളുടെ ഛായാബിന്ദുക്കളെങ്കിൽ. ആദി ത്യന്[4] പിന്നെ അയനാന്തവശാൽ തെക്കുവടക്കു ഗതിയുണ്ടാകയാൽ പടി ഞ്ഞാറേ ഛായാഗ്രബിന്ദുകാലത്തിങ്കേന്നു കിഴക്കു ബിന്ദു ഉണ്ടാകുന്ന കാല

---

1. 1. F. om. നടേ
2. B. ശങ്കുഛായാഗ്രം
3. B. പൂകുന്നു; C.D.പുവുന്നു
4. B. ആദിത്യന് ദക്ഷിണോത്തരഗതി ഉള്ളതുകൊണ്ട് പടിഞ്ഞാറേ ഛായാഗ്രഹബിന്ദുകാ ലത്തിങ്കന്ന് കിഴക്കേ ഛായാഗ്രബിന്ദു ഉണ്ടാകുന്നു.

ങ്ങളുടെ അന്തരത്തിങ്കലേ അപക്രമാന്തരത്തിനു തക്കവണ്ണം അദിത്യൻ വടക്കു നീങ്ങി എങ്കിൽ ഛായാഗ്രം തെക്കു നീങ്ങിയിരിക്കും[5]. കിഴക്കേത് എന്നാൽ അതു വടക്കോട്ടു നീക്കേണ്ടൂ നേരെ പൂർവ്വാപരങ്ങളാവാൻ[6]. ആദിത്യൻ തെക്കുനീങ്ങുകിൽ അതിന്നു തക്കവണ്ണം അതീന്ന് പൂർവ്വബിന്ദുവിനെ തെക്കോട്ടു നീക്കൂ[7]. അന്നീക്കമാകുന്നതു രണ്ടു കാലത്തിങ്കലേയും അപക്രമാന്തരത്തിന്നു[8] തക്കവണ്ണമുള്ള 'അർക്കാഗ്രാംഗുലം'. അപക്രമാന്തരത്തെ അന്നേരത്തെ ഛായാകർണ്ണാംഗുലം കൊണ്ടു ഗുണിപ്പൂ, സ്വദേശലംബകം കൊണ്ടു ഹരിപ്പൂ. ഫലം ആ ഛായാവൃത്തത്തിങ്കലേ അർക്കാഗ്രാംഗുലം. പിന്നെ അയനത്തിനു തക്കവണ്ണം പൂർവ്വബിന്ദുവിന്റെ ഈ അംഗുലങ്ങളെ അളന്ന് നിക്കൂ. ഈ നീക്കിയ ഇടത്തും പടിഞ്ഞാറേ ബിന്ദുവിങ്കലും കൂടി ഒരു സൂത്രമുണ്ടാക്കിയാൽ അതു സമപൂർവ്വാപരം, നേരെ കിഴക്കുപടിഞ്ഞാറ് പ്രത്യക്ബിന്ദുവിനെ എങ്കിൽ അയനവിപരീതമായിട്ടു നീക്കേണ്ടു. പിന്നെ ഈ സൂത്രത്തിങ്കേന്നു മത്സ്യത്തേ ഉണ്ടാക്കി തെക്കു വടക്കു സൂത്രത്തേയുമുണ്ടാക്കൂ. നക്ഷത്രങ്ങളുടെ ഉദയാസ്തമയങ്ങളും നേരേ കിഴക്കുപടിഞ്ഞാറായിട്ടിരിക്കും. അതിനേക്കൊണ്ടറിയാം ദിക്ക്.

2. അക്ഷവും ലംബവും

യാതൊരു ദിവസം ഉദയാസ്തമയങ്ങളിൽ ഭിന്നദിക്കുകളായിരിക്കുന്ന ക്രാന്തികൾ സമങ്ങളായിരിക്കുന്നൂ, അന്ന് മദ്ധ്യാഹ്നത്തിങ്കൽ വിഷുവത്തിങ്കലൂ ആദിത്യൻ എന്നിട്ട്. അന്നേരത്തെ ദ്വാദശാംഗുലശങ്കുവിന്റെ ഛായ വിഷുവച്ഛായയാകുന്നത്. ഈ ഛായ ഭുജയായി, ദ്വാദശാംഗുലശങ്കു കോടിയായി[1] രണ്ടിന്റേയും വർഗ്ഗയോഗമൂലം ചെയ്ത്. കർണ്ണത്തെ വരുത്തൂ. അക്കർണ്ണം പ്രമാണം. ഈ ശങ്കുഛായകൾ പ്രമാണഫലങ്ങൾ, ത്രിജ്യാവ് ഇച്ഛ. ഇവിടെ[2] ഇച്ഛാഫലങ്ങൾ അക്ഷാവലംബങ്ങളാകുന്നത്. ഇവറ്റിന്നു വിപരീതച്ഛായയിങ്കൽ ചൊല്ലുന്ന സംസ്കാരങ്ങൾ ചെയ്യണം. എന്നാൽ സൂക്ഷ്മങ്ങളാകും.

---

1. 5. B. നീങ്ങണം
   6. B. പൂർവ്വാപരമാവാൻ
   7. B. നീക്കുന്നു
   8. C. അന്തരാളത്തിന്നു
2. 1. F. കോടിയായിരിക്കുന്നു
   2. B.F. ഇവിടുത്തെ

ഇവിടെ ഖമദ്ധ്യത്തിങ്കേന്നു ഘടികാമണ്ഡലാന്തരം അക്ഷമാകുന്നത്. ദക്ഷിണോത്തരവൃത്തത്തിങ്കലേത്, ധ്രുവക്ഷിതിജങ്ങളുടെ അന്തരാളമാകിലുമാം. ഘടികാവൃത്തത്തിങ്കേന്ന് ക്ഷിതിജാന്തരാളം ദക്ഷിണോത്തരവൃത്തത്തിങ്കലേതു 'ലംബകം', ഖമദ്ധ്യധ്രുവാന്തരാളമാകിലുമാം.

3. ഉദയാസ്തമനകാലങ്ങൾ

അനന്തരം ഛായ. അവിടെ അപക്രമണ്ഡലത്തിങ്കലേ കിഴക്കു നോക്കി ഗമിക്കുന്ന ആദിത്യന്ന് അപക്രമമണ്ഡലത്തിന്റെ ചരിവിനു തക്കവണ്ണം തെക്കും വടക്കും നീക്കമുണ്ടായിരിക്കും. ഇങ്ങനെ ഇരിക്കുന്ന ആദിത്യൻ ഇഷ്ടകാലത്തിങ്കൽ അപക്രമമണ്ഡലത്തിന്റെ യാതൊരിടത്ത് അവിടെ സ്പർശിച്ചിട്ടു ഘടികാമണ്ഡലത്തിങ്കേന്നു എല്ലാ അവയവവും ഇഷ്ടാപക്രമത്തോളം നീങ്ങി രണ്ടു ധ്രുവങ്കലും ഭഗോളമദ്ധ്യത്തിങ്കലും സ്പർശിച്ചിരിക്കുന്ന അക്ഷദണ്ഡിങ്കൽ കേന്ദ്രമായിരിപ്പോരു വൃത്തത്തെ കല്പിപ്പൂ. ഇത് 'ഇഷ്ടകാലസ്വാഹോരാത്രം'. ഇതിന് ഇഷ്ടദ്യുജ്യാവ് വ്യാസാർദ്ധമാകുന്നത്. അവിടെ ഉന്മണ്ഡലം കൊണ്ടും ദക്ഷിണോത്തരവൃത്തം കൊണ്ടും ഇതിനു പദവിഭാഗം കല്പിക്കേണ്ടൂ. പ്രവഹവശാൽ ഈ ഇഷ്ടസ്വാഹോരാത്രത്തിന്മേലേ ഉള്ള ഗതികൊണ്ട് ഉദയാസ്തമനങ്ങളുണ്ടാകുന്നു. ഇവിടെ വായുവിന്റെ വേഗം നിയതമാകയാൽ, സ്വാഹോരാത്രം ഇത്രകാലം കൊണ്ട് ഇത്ര നീങ്ങുമെന്നുള്ളതും നിയതമാകയാൽ, ഉദിച്ചിട്ട് ഇത്ര ചെല്ലുമ്പോൾ എന്നുതാൻ, അസ്തമിക്കുന്നതിന് ഇത്ര മുമ്പേ എന്നുതാൻ ഈ ഇഷ്ടകാലമുണ്ടാകുമ്പോൾ അന്നേരത്തു സ്വാഹോരാത്രത്തിങ്കൽ ക്ഷിതിജത്തിങ്കേന്ന് ഇത്ര ഉയർന്നേടത്തു ഗ്രഹമെന്നു നിയതം.

4. സ്വാഹോരാത്രവൃത്തം

ഇങ്ങനെ ഇരുപത്തോരായിരത്തറുനൂറു പ്രാണകാലം കൊണ്ടു പ്രവഹവായുവിന്ന് ഒരു ഭ്രമണം വട്ടം കൂടും. ആകയാൽ സ്വാഹോരാത്രവൃത്തത്തിന്നും ഇക്കാലം കൊണ്ട് ഭ്രമണം തികയും എന്നിട്ട്. അതാത് സ്വാഹോ

രാത്രവൃത്തത്തേയും ചക്രകലാസമസംഖ്യമായിട്ട് വിഭജിപ്പൂ. എന്നാൽ ഓരോ പ്രാണകാലം കൊണ്ട് ഓരോ അവയവം ഭ്രമിക്കും. എന്നിട്ട് ഒരു പ്രാണ കാലം കൊണ്ടു ഗമിക്കുന്ന സ്വാഹോരാത്രാവയവത്തേയും ലക്ഷണയാ 'പ്രാണനെ'ന്നു ചൊല്ലുന്നു. എന്നാൽ ആദിത്യനുദിച്ചിട്ട് എത്ര പ്രാണങ്ങൾ ഗതങ്ങളായി അസ്തമിപ്പാൻ, ഇനി എത്രപ്രാണങ്ങളുണ്ട് ഇവറ്റിന്നു "ഗതഗന്തവ്യ" പ്രാണങ്ങളെന്നു പേർ.

എന്നിട്ട്[1] ഈ ഗതഗന്തവ്യപ്രാണങ്ങളോടു തുല്യം ക്ഷിതിജവും ആദിത്യനുമുള്ള അന്തരത്തിങ്കലേ സ്വാഹോരാത്രഭാഗത്തിങ്കലേ "അനന്തപുരാംശം" ഇതു ചാപമാകയാൽ ഇതിന്നു ജ്യാവുണ്ടാക്കൂ നേരറിവാൻ. അവിടെ യാതൊരു പ്രകാരം തെക്കുവടക്ക് അർദ്ധജ്യാക്കളെ കല്പിക്കുമ്പോൾ വൃത്തകേന്ദ്രത്തിന്നു നടുവേയുള്ള പൂർവ്വാപരസൂത്രം അവധിയാകുന്നു, കിഴക്കു പടിഞ്ഞാറു കല്പിക്കുമ്പോൾ, വൃത്തകേന്ദ്രമധ്യേയുള്ള ദക്ഷിണോത്തരസൂത്രവും അവധിയാകുന്നു ജ്യാവുകൊള്ളുമ്പോൾ. അവ്വണ്ണം സ്വാഹോരാത്രത്തിങ്കൽ[2] മേൽകീഴായിട്ടുള്ള ജ്യാവുണ്ടാകുമ്പോൾ സ്വാഹോരാത്രവൃത്തത്തിന്റെ കേന്ദ്രത്തിങ്കൽ കൂടിയുള്ള തിർയ്യക്സൂത്രം അവധിയാകണം. അതാകുന്നത് ഉന്മണ്ഡലവും സ്വാഹോരാത്രവൃത്തവും തങ്ങളിലുള്ള സംപാതം രണ്ടിങ്കലും അക്ഷദണ്ഡിങ്കലും സ്പർശിച്ചിട്ട് ഒരു സമസ്തജ്യാവുണ്ടാവൂ ഉന്മണ്ഡലത്തിന്ന്. അത് അവധിയായിട്ട് ജ്യാവുണ്ടാക്കേണ്ടൂ, ആദിത്യനുദിക്കുന്ന ക്ഷിതിജത്തിങ്കൽ. എന്നിട്ട് ക്ഷിതിജത്തിങ്കേന്നു തുടങ്ങീട്ടു ഗതഗന്തവ്യപ്രാണങ്ങളുണ്ടായി. എന്നിട്ടു ക്ഷിതിജോന്മണ്ഡലാന്തരാളത്തിങ്കലേ സ്വാഹോരാത്രവൃത്തഭാഗം ചരപ്രാണങ്ങളാകുന്നത്. ഇതിനെ കളയേണം. ഗതഗന്തവ്യപ്രാണങ്ങളിൽ നിന്ന് ഉദഗ്ഗോളത്തിങ്കൽ. അവിടെ പൂർവ്വാപരസ്വസ്തികത്തിങ്കേന്നു വടക്കേപ്പുറം ക്ഷിതിജം, ഉന്മണ്ഡലത്തിങ്കേന്ന് കീഴ്, എന്നിട്ട് ദക്ഷിണഗോളത്തിങ്കൽ ഗതഗന്തവ്യപ്രാണങ്ങളിൽ ചരപ്രാണങ്ങളേ കൂട്ടു, അവിടെ ക്ഷിതിജം മീത്തേ ആകയാൽ. എന്നാൽ ഉന്മണ്ഡലത്തിങ്കേന്ന് ആദിത്യനോളമുള്ള സ്വാഹോരാത്രഭാഗത്തിങ്കലേ ഉന്നതപ്രാണങ്ങളുണ്ടാകും. പിന്നെ ഇതിന്ന് ജ്യാവുണ്ടാക്കൂ. പിന്നെ ഈ ജ്യാവിങ്കൽ ചരജ്യാവിനെ വിപരീതമായിട്ട് സംസ്കരിപ്പൂ, ഉത്തരഗോളത്തിൽ കൂട്ടുകയും

4. 1. B. om. എന്നിട്ട്
2. B. വൃത്തകേന്ദ്രത്തിങ്കലെ മേൽ

ദക്ഷിണഗോളത്തിൽ കളയുകയും. എന്നാൽ ക്ഷിതിജത്തിങ്കേന്നു തുടങ്ങിയുള്ള ഉന്നതജ്യാവുണ്ടാകും. ഇത് ഈ സ്വാഹോരാത്രത്തിന്റെ രണ്ടു പദത്തിങ്കലും കൂട്ടിയുള്ളൊന്ന് ആകയാൽ കേവലം അർദ്ധജ്യാവല്ല. ആകയാൽ ഇവറ്റിന്റെ യോഗവിയോഗങ്ങൾക്ക് ഇതരേതരകോടിഗുണനം വേണ്ടാ, താനേ ജ്യാവിന്റെ ശേഷമായിരിക്കയാൽ. കേവലം യോഗവിയോഗം മാത്രമേ വേണ്ടൂ. എന്നാൽ ക്ഷിതിജത്തോട് ആദിത്യനോടുള്ള അന്തരത്തിങ്കലേ സ്വാഹോരാത്രഭാഗജ്യാവുണ്ടാകും. പിന്നെ ചെറിയ ഇലികളാകയാൽ ദ്യുജ്യാവിനെക്കൊണ്ടു ഗുണിച്ചു ത്രിജ്യകൊണ്ടു ഹരിക്കേണം. എന്നാൽ ത്രിജ്യാവൃത്തകലകളേക്കൊണ്ട് ഇച്ചൊല്ലിയ ഉന്നതജ്യാവിത്രയെന്നു വരും.

5. മഹാശങ്കുവും മഹാച്ഛായയും

പിന്നെ സ്വാഹോരാത്രവൃത്തം ഘടികാവൃത്തത്തെപ്പോലെ അക്ഷവശാൽ തെക്കോട്ടു ചരിഞ്ഞിട്ടിരിക്കയാൽ[1] കർണ്ണം പോലെ ഇരിക്കുന്ന ഈ ഉന്നതജ്യാവിനെ ലംബകം കൊണ്ടു ഗുണിച്ചു ത്രിജ്യകൊണ്ടു ഹരിപ്പൂ. ഫലം ആദിത്യങ്കേന്നു ക്ഷിതിജത്തോടുള്ള അന്തരാളമുണ്ടാകും. അത് 'മഹാശങ്കു'വാകുന്നത്. ഇതിന്റെ കോടി ഖമദ്ധ്യഗ്രഹാന്തരാളം. അതു 'മഹാച്ഛായ'യാകുന്നത്.

6. ദൃങ്മണ്ഡലം

പിന്നെ ഖമദ്ധ്യത്തിങ്കലും ഗ്രഹത്തിങ്കലും സ്പർശിച്ചിട്ട് ഒരു വൃത്തത്തെ കല്പിപ്പൂ. അതിന്നു 'ദൃങ്മണ്ഡലം' എന്നു പേർ. ഈ വൃത്തത്തിങ്കലേ ഭുജാകോടിജ്യാക്കൾ ഗ്രഹത്തിങ്കലഗ്രങ്ങളായിരിക്കുന്ന മഹാശങ്കുച്ഛായകളാകുന്നതവ[1]. ഇവിടെ ഘനഭൂമധ്യപാർശ്വത്തിങ്കൽ ക്ഷിതിജം, ക്ഷിതിജത്തിങ്കൽ ശങ്കുമൂലം. ആകയാൽ ഘനഭൂമധ്യം കേന്ദ്രമായിരിപ്പൊന്ന് ഈ ദൃങ്മണ്ഡലം.

---

5. 1. B. ചരിഞ്ഞിരിക്കയാൽ തെക്കോട്ട്
6. 1. B.ആകുന്നത്; F. ഛായകളാകുന്നവ

7. ദൃഗ്ഗോളച്ഛായ

ഭൂപൃഷ്ഠത്തിങ്കൽ വർത്തിച്ചിരിക്കുന്ന ലോകർ പിന്നെ തങ്ങടെ സമപാർശ്വത്തിങ്കേന്ന് ഇത്ര ഉയർന്നിരിക്കുന്നു ഗ്രഹം തലയ്ക്ക് മീത്തേലിന്ന് ഇത്ര താണുമിരിക്കുന്നു എന്നതിനെ കാണുന്നത്. എന്നാൽ ഭൂപൃഷ്ഠത്തിങ്കലിരിക്കുന്ന ദ്രഷ്ടാവിന്റെ ദൃങ്മധ്യം കേന്ദ്രമായി ഖമധ്യത്തിങ്കലും ഗ്രഹത്തിങ്കലും നേമിസ്പർശത്തോടുകൂടി ഇരുന്നൊരു ദൃങ്മണ്ഡലത്തിങ്കലുള്ള ഛായാശങ്കുക്കളെ ദ്രഷ്ടാക്കൾ കാണുന്നത്. ഇവിടെ ഘനഭൂമധ്യപാർശ്വത്തിങ്കലേ ക്ഷിതിജത്തിങ്കേന്ന് എല്ലാടവും ഭൂവ്യാസാർദ്ധത്തോളമുയർന്നിട്ട് ഭൂപൃഷ്ഠത്തിന്റെ സമപാർശ്വത്തിങ്കൽ ഒരു ക്ഷിതിജത്തെ കല്പിപ്പൂ. അതിങ്കേന്നു ഉയർന്നതു ഭൂപൃഷ്ഠവർത്തികൾക്കു ശങ്കുവാകുന്നത്. ഇതിന്നു 'ദൃഗ്ഗോള-ശങ്കു'വെന്നു പേർ. മുമ്പിൽ ചൊല്ലിയതു 'ഭഗോളശങ്കു'. അതിങ്കേന്നു ഭൂവ്യാസാർദ്ധലിപ്ത പോയതു ദൃഗ്ഗോളശങ്കുവാകുന്നത്. ആകയാൽ ശങ്കുമൂലത്തീന്നു വ്യാസാർദ്ധത്തോളമന്തരമുണ്ട്. ക്ഷിതിജാന്തരം കൊണ്ടു ഛായയ്ക്കും പിന്നെ മൂലമാകുന്നത് ഈ ഊർദ്ധ്വസൂത്രം. അതു ഭൂമധ്യത്തിങ്കേന്നുള്ളതും ഭൂപൃഷ്ഠത്തിങ്കേന്നുള്ളതും ഒന്നേ. എന്നിട്ട് ഛായയ്ക്ക് മൂലമൊരിടത്തേ ആയിട്ടിരിക്കും. എന്നിട്ടു ഛായയ്ക്ക് ഭേദമില്ല. എല്ലാടത്തും ഛായാശങ്കുക്കളുടെ അഗ്രങ്ങൾ ബിംബഘനമധ്യത്തിങ്കലാകുന്നു.

പിന്നെ ഈ ഭൂപൃഷ്ഠക്ഷിതിജം മൂലമായിട്ടിരിക്കുന്ന ശങ്കുവും ഛായയും വർഗ്ഗിച്ചു കൂട്ടി മൂലിച്ചാൽ ഭൂപൃഷ്ഠം കേന്ദ്രമായിട്ട് ഒരു കർണ്ണമുണ്ടാകും. അതിന്നു 'ദൃക്കർണ്ണ'മെന്നു പേർ. പ്രതിമണ്ഡലന്യായേന ഉണ്ടായൊരു കർണ്ണമിത്. ഇവിടെ ഭൂമണ്ഡലം കേന്ദ്രമായിട്ടുള്ളത് പ്രതിമണ്ഡലം. ഭൂപൃഷ്ഠം കേന്ദ്രമായിട്ടുള്ളത് കർണ്ണവൃത്തം. ഈ വൃത്തങ്ങളുടെ കേന്ദ്രാന്തരമാകുന്ന ഭൂവ്യാസാർദ്ധം ഇവിടെ ഉച്ചനീചവ്യാസാർദ്ധമാകുന്നത്. നീചസ്ഥാനം ഖമദ്ധ്യമാകയാൽ കർണ്ണവൃത്തകലകൾ സ്വതേ ചെറുത്. ആകയാൽ ഈ ഭൂഗോളമൂലകലകളേക്കൊണ്ടുണ്ടായ കർണ്ണവൃത്തത്തിലേ ഛായ ത്രിജ്യാവൃത്തത്തിങ്കലാകുമ്പോൾ സംഖ്യയേറും. അതിന്നു തക്കവണ്ണം ഖമധ്യത്തിങ്കേന്നുള്ള താഴ്ച ഏറെത്തോന്നും. എന്നാൽ ഭഗോളച്ഛായയെ ത്രിജ്യകൊണ്ടു ഗുണിച്ചു ദൃക്കർണ്ണം കൊണ്ടു ഹരിപ്പൂ. ഫലം ദൃഗ്ഗോളച്ഛായയായിട്ടിരിക്കും. ഇങ്ങനെ പ്രതിമണ്ഡലസ്ഫുടന്യായേന വൃത്താന്തരത്തിലെ ഛായയെ ഉണ്ടാക്കും പ്രകാരം.

8. ഛായാലംബനം

പിന്നെ ഭഗോളച്ഛായയെ ഭൂവ്യാസാർദ്ധയോജനകൊണ്ടു ഗുണിപ്പൂ. സ്ഫുട യോജനകർണ്ണം കൊണ്ടു ഹരിക്കാം. ദൃക്കർണ്ണയോജനകൊണ്ടു വേണ്ടു താനും. ആ ഫലം ഭുജാഫലസ്ഥാനീയമായിരിപ്പൊന്ന്. അതു 'ഛായാലം-ബന'മാകുന്നത്. ഇലികൾതാനും ഈ ഫലം, ഇതിനേ കൂട്ടൂ. ഭഗോള ച്ഛായയിങ്കൽ[1] ദൃഗ്ഗോളച്ഛായയാകും. ഇങ്ങനെ ഉച്ചനീചസ്ഫുടന്യായേന ഛായാ ലംബനലിപ്തയെ വരുത്തുംപ്രകാരം.

9. ഭൂവ്യാസാർദ്ധം

പിന്നെ അവിടത്തെ പ്രതിമണ്ഡലലിപ്താമാനം കൊണ്ട് ഉച്ചനീചവ്യാസാർ ദ്ധത്തെ മാനം ചെയ്ത് അന്ത്യഫലമാകുന്നത്. ഇവിടെ സ്ഫുടയോജനകർണ്ണം പ്രതിമണ്ഡലവ്യാസാർദ്ധമാകയാൽ അമ്മാനം കൊണ്ട് ഈ ഉച്ചനീചവ്യാ സാർദ്ധം ഭൂവ്യാസാർദ്ധയോജനതുല്യം. അക്കർണ്ണം ത്രിജ്യാവാകുമ്പോൾ എന്തെന്ന് അത്.

ആ ഗ്രഹത്തിന്റെ ഭൂവ്യാസാർദ്ധലിപ്ത ഉണ്ടാക്കും പ്രകാരം പിന്നെ. ത്രിജ്യാ ഇച്ഛയാകുമ്പോൾ ഭൂവ്യാസാർദ്ധയോജനലിപ്ത ലംബനമാകുന്ന ത്, ഇഷ്ടച്ഛായക്ക് എത്ര ലംബനലിപ്ത എന്നിങ്ങിനെ[2] ത്രിജ്യാവു ഛായയാ കുമ്പോൾ ഗുണകാരവും ഹാരകവുമാകയാൽ അതിനെ ഉപേക്ഷിക്കാം. എന്നാൽ[1] ഇഷ്ടച്ഛായയെ ഭൂവ്യാർസാദ്ധയോജന കൊണ്ടു ഗുണിച്ചു സ്ഫുട യോജനകർണ്ണം കൊണ്ടു ഹരിപ്പൂ. ഫലം ഛായാലംബനലിപ്ത. ഇവിടെ സ്ഫുടയോജനകർണ്ണവും മദ്ധ്യയോജനകർണ്ണവും തങ്ങളിൽ പെരികെ അന്ത രമില്ല. എന്നാകീട്ടു ഭൂവ്യാസാർദ്ധയോജനം കൊണ്ടു മദ്ധ്യമയോജനകർണ്ണത്തെ ഹരിപ്പൂ, ഫലം ആദിത്യന്ന് എണ്ണൂറ്ററുപത്തിമൂന്ന്. ഇതിനെക്കൊണ്ട് ഇഷ്ട ഛായയെ ഹരിപ്പൂ. ഫലം ഛായാലംബനലിപ്ത. ഇവിടെ ദൃങ്മണ്ഡലത്തി

---

8. 1. ഭഗോളച്ഛായയിങ്കൽ
9. 1. D. adds എന്നാൽ
   2. H. എന്നിതിനെ

ങ്കലെ ഛായാലംബനം കർണ്ണമായിട്ടിരിക്കുവോളം ഭുജാകോടികളായിട്ടിരിക്കും ഇതിനുമേലിൽ ചൊല്ലുവാനിരിക്കുന്ന നതിലംബനങ്ങൾ. ഇതിന്റെ പ്രകാരം മേലിൽ ചൊല്ലുന്നുണ്ട്. ഇങ്ങനെ ഒരു സംസ്കാരം ഛായയ്ക്ക്.

10. ദ്വാദശാംഗുലശങ്കുവിന്റെ സംസ്കരിച്ച ഛായ

ഈ ഛായാശങ്കുക്കൾ ആദിത്യന്റെ ബിംബഘനമധ്യത്തിങ്കലഗ്രങ്ങളായിട്ടിരിപ്പോ ചിലവ. പിന്നെ ആദിത്യമണ്ഡലത്തിന്റെ എല്ലാടവും രശ്മികളുണ്ടാകയാൽ മീത്തേ നേമിയിങ്കലേ രശ്മികൾ ശങ്കുവിനേക്കൊണ്ടു മറഞ്ഞിട്ടു എവിടെ എത്രേടം നിലത്തു തട്ടുന്നു. അത്രേടം[1] ആ ശങ്കുവിന്റെ ഛായ ഉണ്ട്. ബിംബഘനമധ്യത്തിലെ രശ്മികളേക്കൊണ്ട് അല്ലാ, ദ്വാദശാംഗുലശങ്കുവിന്റെ ഛായ ഒടുങ്ങുന്നു എന്നിട്ട് ബിംബത്തിന്റെ മീത്തേ നേമിയിങ്കലോളം നിളേണം ശങ്കു. അവിടുന്നു ഖമധ്യാന്തരാളം ഛായയാകുന്നത്. ഇവിടെ ബിംബഘനമദ്ധ്യത്തോട് ഊർദ്ധ്വനേമിയോടുള്ള അന്തരാളം ബിംബവ്യാസാർദ്ധം. ഇതു ദൃങ്മണ്ഡലത്തിങ്കൽ സമസ്തജ്യാവായിട്ടിരിക്കും. എന്നിട്ട് ഈ ബിംബവ്യാസാർദ്ധത്തെക്കൊണ്ടു ശങ്കുവിനേയും, ഛായയേയും ഗുണിച്ചു ത്രിജ്യകൊണ്ടു ഹരിച്ച ഫലങ്ങൾ ബിംബവ്യാസാർദ്ധത്തിങ്കലേ ഖണ്ഡജ്യാക്കൾ. അവിടെ ഛായയിങ്കേന്നുണ്ടാക്കിയ ഫലത്തെ ശങ്കുവിൽ കൂട്ടൂ. ശങ്കുവിങ്കേന്നുണ്ടാക്കിയതിനെ ഛായയിങ്കേന്നു കളവൂ. എന്നാൽ ആദിത്യന്റെ ഊർദ്ധ്വനേമിയിങ്കലഗ്രങ്ങളായിട്ടു ശങ്കച്ഛായകളുണ്ടാകും. അവ ദൃഗ്വിഷയത്തിങ്കലേക്കു സാധനങ്ങളാകുന്നവ. ഇവിടെ സമസ്തജ്യാമധ്യത്തിലഗ്രങ്ങളായിട്ടിരിക്കുന്ന ഭുജാകോടിജ്യാക്കളേകൊണ്ടു ഖണ്ഡജ്യാക്കളെ വരുത്തേണ്ടൂ, എങ്കിലും സമസ്തജ്യാഗ്രത്തിങ്കലേവറ്റോടു പെരികേ അന്തരമില്ല. എന്നിട്ട് അവറ്റേക്കൊണ്ടുണ്ടാക്കാൻ ചൊല്ലി.

ഇങ്ങനെ ലംബനത്തേയും ബിംബവ്യാസാർദ്ധഖണ്ഡജ്യാക്കളേയും സംസ്കരിച്ചാൽ ദൃഗ്ഗോളത്തിങ്കൽ ബിംബത്തിന്റെ ഊർദ്ധ്വനേമിയിങ്കലഗ്രങ്ങളായിട്ടിരിപ്പൊന്ന് ചില ശങ്കുച്ഛായകളുണ്ടാകും. പിന്നെ ഈ ഛായയെ പന്ത്രണ്ടിൽ ഗുണിച്ച് ഈ ശങ്കുകൊണ്ടു ഹരിപ്പൂ. ആ ഫലം ദ്വാദശാംഗുലശങ്കുവിന്റെ ഛായ.

---

10. 1. B. അത്രത്തോളം

11. വിപരീതച്ഛായ

അനന്തരം വിപരീതച്ഛായ. അതാകുന്നതു ദ്വാദശാംഗുലശങ്കുവിന്ന് ഇത്ര ഛായയെന്ന് അറിഞ്ഞാൽ അപ്പോൾ ഗതഗന്തവ്യപ്രാണങ്ങൾ എത്ര എന്ന റിയും പ്രകാരം. അവിടെ ദ്വാദശാംഗുലശങ്കുവിന്റെ ഛായയും ദ്വാദശാംഗുല ശങ്കുവിനേയും വർഗ്ഗിച്ചു കൂട്ടി മൂലിച്ചാൽ ഛായാകർണ്ണം അംഗുലമായിട്ടു ണ്ടാകും. പിന്നെ ഈ ഛായയേയും ശങ്കുവിനേയും ത്രിജ്യകൊണ്ടു ഗുണിച്ച് ഈ കർണ്ണാംഗുലം കൊണ്ട് ഹരിപ്പൂ. ഈ ഫലങ്ങൾ മഹാശങ്കുച്ഛായകൾ. അവ ദൃഗ്വിഷയച്ഛായയെക്കൊണ്ടുണ്ടാകയാൽ ബിംബത്തിന്റെ മീത്തേ നേമീങ്കലഗ്രങ്ങളായിട്ടുള്ളൂ. എന്നിട്ടു ബിംബവ്യാസാർദ്ധത്തെക്കൊണ്ടു ശങ്കു ച്ഛായയെ[1] വെവ്വേറെ ഗുണിച്ചു ത്രിജ്യകൊണ്ടു ഹരിച്ച ഫലങ്ങളെ ക്രമേണ ഛായയിൽ കൂട്ടുകയും ശങ്കുവിങ്കേന്നു കളയുകയും ചെയ്വൂ. എന്നാൽ ബിംബഘനമദ്ധ്യത്തിലഗ്രങ്ങളായിട്ടു വരും. പിന്നെ ഛായയിങ്കേന്നു 'ഗതിജ' (863) നെക്കൊണ്ട് ഹരിച്ചഫലത്തെ ഛായയിങ്കേന്നു കളവൂ. ശങ്കുവിൽ ഭൂവ്യാ സാർദ്ധലിപ്ത കൂട്ടൂ[2]. ഇത്രോടമുള്ള ക്രിയയെ അക്ഷാവലംബകങ്ങളിലും ചെയ്യേണം.

പിന്നെ ഈ ശങ്കുവിനെ ത്രിജ്യാവർഗ്ഗം കൊണ്ടു ഗുണിച്ച് ദ്യുജ്യാലംബക ഘാതം കൊണ്ടു ഹരിപ്പൂ. ഫലം ബിംബഘനമദ്ധ്യക്ഷിതിജാന്തരാളം. സ്വാ ഹോരാത്രവൃത്തത്തിങ്കലേ ജ്യാവ്, തന്റെ അനന്തപുരാംശം കൊണ്ടുള്ളത്. പിന്നെ ചരജ്യാവിനെ ഇതിങ്കൽ മേഷതുലാദിക്കു തക്കവണ്ണം ഋണം ധനം ചെയ്തു ചാപിച്ച് ചരപ്രാണങ്ങളെ ക്രമേണ ധനർണ്ണമായിട്ടു സാംസ്ക രിപ്പൂ. ഫലം ഗതഗന്തവ്യപ്രാണങ്ങൾ. ഇങ്ങനെ ഇഷ്ടകാലത്തിങ്കൽ ദ്വാദ ശാംഗുലശങ്കുവിന്നു കർണ്ണച്ഛായയിങ്കേന്നു ക്രമച്ഛായ, ഛായാവൈപരീത്യ ക്രിയയെക്കൊണ്ടു ഗതഗന്തവ്യപ്രാണങ്ങളുണ്ടാക്കും പ്രകാരം.

12. മധ്യാഹ്നച്ഛായ

അനന്തരം മധ്യാഹ്നച്ഛായ ഉണ്ടാക്കും പ്രകാരം. അവിടെ ഗ്രഹത്തിന്ന്

---

11. 1. B. C. D. ശങ്കുച്ഛായകളെ
2. B reads ഫലങ്ങളെ ക്രമേണ ച്ഛായയിൽ കൂട്ടുകയും ശങ്കുവിൽ കളയുകയും ചെയ്താൽ ബിംബഘനമദ്ധ്യത്തിങ്കലഗ്രങ്ങളായിട്ടു വരും.

ദക്ഷിണോത്തരവൃത്തസംപാതം വരുമ്പോൾ ഖമധ്യത്തോടു ഗ്രഹത്തോടുള്ള അന്തരാളം ദക്ഷിണോത്തരവൃത്തത്തിങ്കലേക്ക് മധ്യാഹ്നച്ഛായയാകുന്നത്. അവിടെ ഖമധ്യത്തോട് ഘടികാമണ്ഡലത്തോടുള്ള അന്തരാളം അക്ഷം. ഘടികാമണ്ഡലത്തോട് ആദിത്യനോടുള്ള അന്തരാളം. അപക്രമം. ഖമധ്യത്തിങ്കേന്ന് എല്ലായ്പ്പോഴും തെക്കു ഘടികാമണ്ഡലം. ഘടികാമണ്ഡലത്തിങ്കേന്നു ഗോളത്തിന്നു തക്കവണ്ണം തെക്കും, വടക്കും നീങ്ങുമാദിത്യൻ. എന്നാൽ ഗോളവശാൽ അക്ഷാപക്രമങ്ങളുടെ യോഗം താനന്തരം താൻ ചെയ്തത് മധ്യാഹ്നച്ഛായയാകുന്നത്. എന്നാൽ മദ്ധ്യാഹ്നച്ഛായാക്ഷങ്ങളുടെ യോഗം താനന്തരം താൻ ചെയ്തത് അപക്രമം. മധ്യാഹ്നച്ഛായാപക്രമങ്ങളുടെ യോഗം താനന്തരം താൻ അക്ഷമാകുന്നത്. ഇങ്ങനെ മൂന്നിൽ രണ്ടറിഞ്ഞാൽ[1] മറ്റേതു സിദ്ധിക്കും.

13. ഛായാഭുജ, അർക്കാഗ്രാ, ശംക്വഗ്രം എന്നിവ

അനന്തരം ഛായാഭുജ. അവിടെ ദൃങ്മണ്ഡലത്തിങ്കലേ ഛായാഗ്രത്തിങ്കേന്നു സമമണ്ഡലാന്തരാളം ഛായാഭുജയാകുന്നത്. പിന്നെ ഛായാഗ്രത്തിങ്കേന്നു ദക്ഷിണോത്തരമണ്ഡലാന്തരാളമാകുന്നതു 'ഛായാകോടി'[1]

ഇവിടെ ക്ഷിതിജേഷ്ടസ്വാഹോരാത്രസംപാതത്തിങ്കേന്നു പൂർവ്വാപരസ്വസ്തികാന്തരം ഇക്ഷിതിജത്തിങ്കലേത് 'അർക്കാഗ്ര'യാകുന്ന[2]ത്. അവിടെ ആദിത്യനുദിക്കുന്നു. പിന്നെ പ്രവഹവശാൽ ദക്ഷിണോത്തരത്തെ സ്പർശിക്കുമ്പോളുദിച്ചേടത്തിന്നു തെക്കു നിങ്ങിയിരിക്കും. അന്നീക്കത്തിന്നു 'ശംക്വഗ്ര' മെന്നു പേർ. അവിടെ ഉദയാസ്തമയങ്ങളിൽ കൂടീട്ട് ഒരു സൂത്രം കല്പിപ്പൂ. ഈ സൂത്രത്തിങ്കേന്നു ശങ്കുമൂലം എത്ര നീങ്ങി അതു ശംക്വഗ്രമാകുന്നത്. ആ ശങ്കുവിന്റെ അഗ്രവും അത്രതന്നെ നീങ്ങി ഇരിക്കും. എന്നിട്ട് ശംക്വഗ്രമെന്നു പേരുണ്ടായി.

---

12. 1. H. രണ്ടുമറിഞ്ഞാൽ
13. 1. B. adds ആകുന്നത്
2. H. അർക്കാഗ്രമാകുന്നത്

14. മറ്റു ബന്ധപ്പെട്ട വിഷയങ്ങൾ

അവിടെ അർക്കാഗ്ര ക്ഷിതിജത്തിങ്കലേ ജ്യാവ്, ഇഷ്ടാപക്രമം ഉന്മണ്ഡലത്തിങ്കലേ ജ്യാവ്, പൂർവ്വാപരസ്വസ്തികത്തോട് സ്വാഹോരാത്രത്തോടുള്ള അന്തരാളമായിട്ടിരിക്കുമിവ. പിന്നെ ക്ഷിതിജോന്മണ്ഡലാന്തരാളത്തിങ്കലേ സ്വാഹോരാത്രഭാഗജ്യാവിന്ന് 'ക്ഷിതിജ്യാ'വെന്നുപേർ[1]. ഇതു ഭുജ, അപക്രമം കോടി, ആർക്കാഗ്ര കർണ്ണം, ഇങ്ങനെ ഇരിപ്പോരു ത്ര്യശ്രം. അക്ഷവശാൽ ഉണ്ടായൊന്നിത്. ക്ഷിതിജവും ഉന്മണ്ഡലവും അക്ഷവശാൽ രണ്ടാകയാൽ ഉണ്ടായൊരു ക്ഷേത്രമിത്. ആകയാൽ ഇഷ്ടാപക്രമത്തെ ത്രിജ്യകൊണ്ടു ഗുണിച്ചു ലംബകം കൊണ്ടു ഹരിച്ച ഫലം അർക്കാഗ്രയാം.

പിന്നെ സ്വാഹോരാത്രോന്നതജ്യാവും ശങ്കുവും ശംക്വഗ്രവും, ഇങ്ങനെ ഒരു ത്ര്യശ്രം. ഇത് അക്ഷവശാൽ ഉന്നതജ്യാവിന്നു ചരിവുണ്ടാകയാൽ ഉണ്ടായൊരു ത്ര്യശ്രം. ഇവിടെ സ്വാഹോരാത്രോന്നതജ്യാവു കർണ്ണം, ശങ്കു കോടി, ഉന്നതജ്യാമൂലത്തോടു ശങ്കുമൂലത്തോടുള്ള അന്തരാളം ഭുജ. ഈ ഭുജ ശംക്വഗ്രമാകുന്നത്. ഇതു നേരേ തെക്കുവടക്കായി ഇരുന്നൊന്ന്. നിരക്ഷദേശത്തിങ്കൽ സ്വാഹോരാത്രവൃത്തം നേരേ മേൽകീഴാകയാൽ അവിടെ ഉന്നതജ്യാവും നേരെ മേൽകീഴായിരുന്നൊന്ന്. അവിടുന്നു പിന്നെ അക്ഷവശാലുള്ള ചരിവു നേരേ തെക്കു നോക്കിയാകയാൽ, ശങ്കുമൂലവും ഉന്നതജ്യാമൂലവും തങ്ങളിലന്തരാളം, നേരേ തെക്കുവടക്ക്, അർക്കാഗ്രയും നേരേ തെക്കു വടക്ക്. അപ്പോൾ രണ്ടിനും ദിക്കൊന്നാകയാൽ തങ്ങളിൽ യോഗാന്തരത്തെച്ചെയ്കേ വേണ്ടൂ, ഗോളത്തിന്നു തക്കവണ്ണം. ഇതരേതരകോടിഗുണനം വേണ്ടാ. ഇങ്ങനെ യോഗാന്തരം ചെയ്തിരിക്കുന്നത് ഛായാഭുജയാകുന്നത്. അതു പൂർവ്വാപരസൂത്രവും ശങ്കുമൂലവും തങ്ങളിലുള്ള അന്തരാളം ക്ഷിതിജത്തിങ്കലേത്. ദൃങ്മണ്ഡലത്തിങ്കലേ ഗ്രഹത്തിങ്കേന്നു സമമണ്ഡലാന്തരാളമാകിലുമാം. ഇത് ഭുജയായി ഛായാ കർണ്ണമായിട്ടിരിക്കുമ്പോളേ കോടി ഛായാകോടിയാകുന്നത്. ഗ്രഹത്തിങ്കേന്നു ദക്ഷിണോത്തരവൃത്താന്തരാളമിത്. ഇതു പിന്നെ സ്വാഹോരാത്രവൃത്തത്തിങ്കലേ ജ്യാവാകയുമുണ്ട്. ഇതിനെ തന്നിലെ അനന്തപുരാംശം കൊണ്ടു മാനം ചെയ്തു ചാപിച്ചാൽ നതപ്രാണങ്ങളുണ്ടാകും. ഇവ തന്നെ ദ്വാദശാംഗുലശങ്കുവിലേക്ക്

14. 1. B. പേരുണ്ട്

ആകയാൽ അവറ്റേക്കൊണ്ടു ദിക്കറിയാം. അതിന്ന് അർക്കാഗ്രയെ ഛായാകർണ്ണം കൊണ്ടു ഗുണിച്ച് ത്രിജ്യകൊണ്ടു ഹരിപ്പൂ. ഫലത്തിന്നു[2] 'അഗ്രാംഗുല'മെന്നു പേർ. ശംക്വഗ്രമാകുന്നത് എല്ലായ്പ്പോഴും വിഷുവച്ഛായ തന്നെയത്രേ ദ്വാദശാംഗുലശങ്കുവിന്. എന്നാൽ വിഷുവച്ഛായയും അഗ്രാം ഗുലവും തങ്ങളിൽ യോഗം താനന്തരം താൻ ചെയ്താൽ ദ്വാദശാംഗുലശ ങ്കുവിന്റെ ഛായാഭുജയുണ്ടാകും. മഹാച്ഛായ ഭുജാദിക്കിന്നു വിപരീതം[3] ഇതിന്നു ദിക്കാകുന്നത്, ആദിത്യനുള്ള ദിക്കിനു വിപരിതദിക്കിലുമെല്ലോ ഛായാഗ്രം എന്നിട്ട്.

15. ദിക്ജ്ഞാനം പരീക്ഷണത്തിലൂടെ

ഇവിടെ ഇഷ്ടകാലത്തിങ്കലേക്ക്[1] ദ്വാദശാംഗുലശങ്കുവിന്റെ ഛായ-ഭുജാ-കോ ടികൾ മൂന്നിനേയും വരുത്തിയ ഛായാതുല്യവ്യാസാർദ്ധമായിട്ട് ഒരു വൃത്തം വീയി ആ വൃത്തമദ്ധ്യത്തിങ്കൽ ശങ്കു വച്ചാൽ ആ ശങ്കുവിന്റെ ഛായാഗ്രം വൃത്തനേമിയിങ്കൽ യാതൊരിടത്തു സ്പർശിക്കുന്നൂ, അവിടെ ഒരു ബിന്ദുവു ണ്ടാക്കി[2] ആ ബിന്ദുവിങ്കലഗ്രങ്ങളായിട്ട്[3] രണ്ടു ശലാകകളെ വെയ്പൂ. അവിടെ ഛായാഭുജയിൽ ഇരട്ടി നീളമായുള്ളൊന്നിനെ തെക്കുവടക്കും, ഛായാകോടി യിൽ ഇരട്ടി നീളമുള്ളൊരു[4] ശലാകയെ കിഴക്കുപടിഞ്ഞാറായിട്ടും വയ്പൂ. മറ്റേ അഗ്രങ്ങളും പരിധിയിങ്കൽ സ്പർശിക്കുമാറ്.

ഇങ്ങനെ പ്രായികമായിട്ട് ദിക്കിനെ അറിഞ്ഞിരിക്കുമ്പോൾ ഈ ശലാക കളിൽ കോടിശലാക നേരേ പൂർവ്വാപരം, ഭുജാശലാക ദക്ഷിണോത്തരം ഇങ്ങനെയുമുണ്ടൊരു പ്രകാരം ദിഗ്വിഭാഗത്തെ അറിയുവാൻ.

---

14. 2. B; C.adds ആ
    3. B. വിപരീതമായിട്ട്
15. 1. H. ത്തിലേക്കു
    2. B. ബിന്ദുമിട്ട്
    3. B. അഗ്രമായിട്ട്
    4. B. ഇരട്ടി നീളമുള്ളൊരു

16. സമശങ്കു

അനന്തരം സമശങ്കുവിനെ ചൊല്ലുന്നു. ഇവിടെ പൂർവ്വാപരസ്വസ്തികത്തിങ്കലും ഖമദ്ധ്യത്തിങ്കലും[1] സ്പർശിച്ചിരുന്നോരു[2] സമവൃത്തം. പിന്നെ പൂർവ്വാപരസ്വസ്തികത്തിങ്കലും ഖമദ്ധ്യത്തിങ്കേന്നു അക്ഷത്തോളം തെക്കു നീങ്ങിയേടത്തു ദക്ഷിണോത്തരവൃത്തത്തിങ്കലും സ്പർശിച്ചിരുന്നോരു[3] ഘടികാവൃത്തം. ക്ഷിതിജത്തിങ്കേന്ന് ഉത്തരധ്രുവന്റെ ഉന്നതിക്കു തക്കവണ്ണം ഖമദ്ധ്യത്തിങ്കേന്നു ഘടികാമണ്ഡലത്തിന്റെ താഴ്ച[4]. ഇവിടെ യാതൊരിന്നാൾ ഘടികാമണ്ഡലം സ്വാഹോരാത്രമാകുന്നു ഗ്രഹത്തിന്ന്[5], അന്നു പൂർവ്വാപരസ്വസ്തികങ്ങളിൽ ഉദയാസ്തമയങ്ങൾ. ഖമദ്ധ്യത്തിങ്കേന്ന് അക്ഷത്തോളം തെക്കു നീങ്ങിയേടത്ത് ഉച്ചയാകുന്നു. പിന്നെ സ്വാഹോരാത്രങ്ങളൊക്കെ തെക്കുനോക്കി ചരിഞ്ഞിരിക്കും. ആകയാൽ ഉദിച്ചേടത്തുന്നു തെക്കുനീങ്ങി ഉച്ചയാകും. എന്നാൽ ഉത്തരാപക്രമം അക്ഷത്തേക്കാൾ കുറയുന്നാൾ പൂർവ്വാപരസ്വസ്തികങ്ങളിൽ നിന്ന് വടക്കേ[6] ഉദയാസ്തമയങ്ങൾ. ഖമദ്ധ്യത്തിന്നു[7] തെക്കേപ്പുറത്തു ഉച്ച. ദക്ഷിണോത്തര[8]സംപാതമാകയാൽ ഉദയത്തിന്റേയും മദ്ധ്യാഹ്നത്തിന്റേയും നടുവിലൊരിക്കൽ സമമണ്ഡലത്തെ സ്പർശിക്കും ഗ്രഹം. അവ്വണ്ണം ഉച്ചതിരിഞ്ഞാൽ അസ്തമയത്തിനിടയിലുമൊരിക്കൽ സമമണ്ഡലത്തെ സ്പർശിക്കും. അന്നേരത്തെ ശങ്കു 'സമശങ്കു'വാകുന്നത്. അന്നേരത്തു നേരേ കിഴക്കുപടിഞ്ഞാറായിരിക്കും ഛായ.

പിന്നെ ഉത്തരാപക്രമം അക്ഷത്തോടു സമമായിരിക്കുന്നാൾ ഖമദ്ധ്യത്തിങ്കൽ സമമണ്ഡലസംപാതം. സ്വാഹോരാത്രത്തിന്ന് അക്ഷത്തേക്കാൾ ഉദർക്കക്രാന്തി ഏറുന്നാൾ സ്വാഹോരാത്രവൃത്തത്തിൽ സമമണ്ഡലസ്പർശമില്ല. ആകയാൽ അന്ന് സമശങ്കുവില്ല. ദക്ഷിണക്രാന്തിയിലും സ്വാഹോരാത്രത്തിന്ന് സമമണ്ഡലസ്പർശമില്ലായ്കയാൽ അന്നും സമശങ്കുവില്ല[9].

---

16. 1. B. ഖമദ്ധ്യത്തിങ്കേന്ന്
2. E. സ്പർശിച്ചിരുന്നൊന്ന്
3. D. F സ്പർശിച്ചിരുന്നൊന്ന്
4. B. തല
5. H. om. ഗ്രഹത്തിന്ന്
6. E. വടക്കേത്
7. B. ഖമദ്ധ്യത്തിങ്കേന്ന്
8. B. add. വൃത്ത
9. D. അന്ന് സമശങ്കുവുമില്ല

ഇവിടെ ഉത്തരാപക്രമം അക്ഷത്തോടു തുല്യമാകുമ്പോൾ ഖമദ്ധ്യത്തിങ്കൽ സമമണ്ഡലസ്പർശമാകയാൽ സമശങ്കു ത്രിജ്യാതുല്യം. അപ്പോൾ അക്ഷത്തേക്കാൾ കുറഞ്ഞ ഇഷ്ടോത്തരക്രാന്തിക്ക് എത്ര സമശങ്കുവെന്ന് സമശങ്കുവുണ്ടാകും. ഇതിന്റെ വിപരീതക്രിയകൊണ്ട് സമശങ്കുവിങ്കേന്നു ഉത്തരാപക്രമമുണ്ടാകും. അതിങ്കേന്നു ഗ്രഹഭുജാജ്യാവും ഉണ്ടാകും. ഇങ്ങനെ ഒരു പ്രകാരം സമശങ്ക്വാനയനം.

17. സമച്ഛായാ

അനന്തരം സമശങ്കുവിലെ ദ്വാദശാംഗുലശങ്കുവിന്റെ കർണ്ണത്തെ വരുത്തും പ്രകാരം. ഇവിടെ അക്ഷത്തിൽ കുറഞ്ഞ ഉത്തരാപക്രമവും ത്രിജ്യയും തങ്ങളിൽ ഗുണിച്ചതിങ്കേന്ന് അക്ഷജ്യാവിനേക്കൊണ്ടു ഹരിച്ചതു സമശങ്കുവെന്നോ ചൊല്ലിയെല്ലോ. പിന്നെ ഈ സമശങ്കുവിന്നു ത്രിജ്യാ കർണ്ണം, ദ്വാദശാംഗുലശങ്കുവിന്ന് എന്തു കർണ്ണമെന്ന് സമച്ഛായാകർണ്ണമുണ്ടാകും. അപ്പോൾ ത്രിജ്യയെ പന്ത്രണ്ടിൽ ഗുണിച്ചു സമശങ്കുവിനേക്കൊണ്ടു ഹരിച്ചതല്ലോ അംഗുലാത്മകമാകുന്ന സമച്ഛായാകർണ്ണം. ഇവിടെ മഹാശങ്കു ഹാരകമാകയാൽ, അതു ത്രിജ്യാപക്രമഘാതം കൊണ്ടുണ്ടാകയാൽ ത്രിജ്യാപകക്രമഘാതം ഹാരകം, ത്രിജ്യയും പന്ത്രണ്ടും തങ്ങളിൽ ഗുണിച്ചതു[1] ഹാര്യം. അപ്പോൾ ഹാരകത്തിങ്കലും ഹാര്യത്തിങ്കലും കൂടി ത്രിജ്യയുണ്ടാകയാൽ ത്രിജ്യയുപേക്ഷിക്കാം. അക്ഷം പിന്നെ ഹാരകത്തിന്നും ഹാരകമാകയാൽ ഹാര്യത്തിന്നു ഗുണകാരമായിട്ടിരിക്കും. എന്നാൽ അക്ഷത്തെ പന്ത്രണ്ടിൽ ഗുണിച്ച് അക്ഷത്തിൽ കുറഞ്ഞ ഉത്തരാപക്രമം കൊണ്ടു ഹരിച്ചാൽ സമച്ഛായാകർണ്ണമുണ്ടാകും[2]. ഇവിടെ അക്ഷത്തെ പന്ത്രണ്ടിൽ ഗുണിച്ചതിനോടു തുല്യം ലംബകത്തെ വിഷുവച്ഛായകൊണ്ടു ഗുണിച്ചാൽ, ഇച്ഛാപ്രമാണഫലങ്ങളും പ്രമാണേച്ഛാഫലങ്ങളും തങ്ങളിലുള്ള ഘാതം തുല്യമാകയാൽ. എന്നാൽ ഇതിനെ അപക്രമംകൊണ്ടു ഹരിക്കിലുമാം, സമച്ഛായാകർണ്ണമുണ്ടാകയാൽ.

17. 1. ആ
2. B. കർണ്ണമാകും

പിന്നെ വിഷുവത്തിങ്കലെ ഗ്രഹത്തിന്റെ മധ്യാഹ്നത്തിങ്കലേ ദ്വാദശാംഗുല ശങ്കുവിന്റെ ഛായ വിഷുവച്ഛായയാകുന്നത്. ഉദർക്കക്രാന്തിയിങ്കൽ മധ്യാഹ്നച്ഛായ വിഷുവച്ഛായയേക്കാൾ കുറയുമ്പോളേ സമച്ഛായയുണ്ടാവൂതും. ഈ മധ്യാഹ്നച്ഛായയും, വിഷുവച്ഛായയും തങ്ങളിലുള്ള അന്തരം മധ്യാഹ്നാഗ്രാംഗുലമാകുന്നത്. മദ്ധ്യാഹ്നാഗ്രാംഗുലം വിഷുവച്ഛായാതുല്യമാകും. ഖമധ്യത്തിങ്കൽ ഉച്ചയാകുന്ന നാൾ. അന്നു മദ്ധ്യാഹ്നഛായാകർണ്ണം തന്നെ സമച്ഛായാകർണ്ണമാകുന്നത്. അഗ്രാംഗുലം പെരികെ കുറയുന്ന നാൾ മധ്യാഹ്നച്ഛായാകർണ്ണത്തേക്കാൾ സമച്ഛായാകർണ്ണം പെരികെ വലുത്. അഗ്രാംഗുലമേറുന്നതിന്നു തക്കവണ്ണം മദ്ധ്യാഹ്നഛായാകർണ്ണത്തോടു സമച്ഛായാകർണ്ണത്തിന് അന്തരം കുറഞ്ഞു കുറഞ്ഞിരിക്കും. എന്നിട്ടിവിടെ വ്യസ്തത്രൈരാശികം വേണ്ടുകയാൽ വിഷുവച്ഛായയും മദ്ധ്യാഹ്നകർണ്ണവും തങ്ങളിൽ ഗുണിച്ച് മദ്ധ്യാഹ്നാ ഗ്രാംഗുലം കൊണ്ടു ഹരിച്ച ഫലം സമച്ഛായാകർണ്ണം.

18. സമശങ്കുഗതക്ഷേത്രങ്ങൾ

അനന്തരം[1] സാക്ഷദേശത്ത് അക്ഷവശാൽ ഉണ്ടായ ചില ക്ഷേത്രവിശേഷങ്ങളെ കാട്ടുന്നു. സ്വാഹോരാത്രത്തീന്നു പൂർവ്വാപരസ്വസ്തികത്തിങ്കേന്നു വടക്കു ക്ഷിതിജസംപാതം. സമമണ്ഡലത്തിങ്കേന്നു തെക്കു നീങ്ങി ദക്ഷിണോത്തരസംപാതം എന്നിരിക്കും നാൾ[2] ക്ഷിതിജത്തോട് സമമണ്ഡലത്തോടുള്ള അന്തരാളത്തിങ്കലേ സ്വാഹോരാത്രവൃത്തഭാഗം കർണ്ണം, സമശങ്കു കോടി, അർക്കാഗ്ര ഭുജാ. നിരക്ഷദേശത്തിൽ സ്വാഹോരാത്രത്തിനു നമനമില്ലായ്കയാൽ ഈ ക്ഷേത്രമില്ല[3]. ഇവിടെ പൂർവ്വാപരസ്വസ്തിക സ്വാഹോരാത്രങ്ങളുടെ സംപാതങ്ങളിലുള്ള[4] അന്തരാളം[5] ക്ഷിതിജഭവം. അർക്കാഗ്ര ഉന്മണ്ഡലഭവം. അപക്രമം ക്ഷിതിജോന്മണ്ഡലാന്തരാളഭവം. സ്വാഹോരാ

---

18. 1. B. C. om. അനന്തരം
2. C. F. എന്നിരിക്കുമ്പോൾ
3. B. C.D. E. F om. നിരക്ഷദേശത്തിൽ.....(to)..... ക്ഷേത്രമില്ല
4. E. സ്വസ്തികത്തോടു സ്വാഹോരാത്ര സംപാതത്തോടുള്ള
5. B. C.E.F.read അന്തരാളം ക്ഷിതിജത്തിങ്കടെ ഉന്മണ്ഡലത്തീന്ന് മീത്തേ സ്വാഹോരാത്ര ഭാഗം കൊണ്ട് ഉന്മണ്ഡത്തിന്മേലേ അപക്രമഭാഗം ഭുജാ, സമശങ്കു കർണ്ണം (D reads: അന്തരാളം ക്ഷിതിജത്തിങ്കലേത് അർക്കാഗ്ര. ഉന്മണ്ഡലത്തിങ്കേലേത് അപക്രമം, ക്ഷിതിജോന്മണ്ഡലാന്തരാളത്തിങ്കലെ സ്വാഹോരാത്രഭാഗം ക്ഷിതിജ്യ. (ഈ ത്ര്യശ്രം)

ത്രഭാഗം ക്ഷിതിജ്യാ. ഈ ത്ര്യശ്രം അക്ഷവശാലുണ്ടാകുന്നു, ഭുജാകോടി കർണ്ണമായി[6] വിദ്യമാനമായിട്ട്. അനന്തരം ഉന്മണ്ഡലത്തിന്നു മിത്തേ സ്വാഹോരാത്രഭാഗം കോടി, ഉന്മണ്ഡലത്തിന്മേലെ അപക്രമഭാഗം ഭുജ, സമശങ്കു കർണ്ണം. ഇങ്ങനെ ഒരു ത്ര്യശ്രം. ഈ മൂന്നു ത്ര്യശ്രങ്ങളും അക്ഷാവ ലംബകത്രിജ്യകളെപ്പോലെ ഇരിപ്പോ ചിലവ. എന്നാൽ ഇന്നാലിലൊന്നു സാധനമായിട്ടു ത്രൈരാശികം കൊണ്ടു മറ്റേവ ഉണ്ടാക്കാം.

19. ദശപ്രശ്നങ്ങൾ

ദശവിധപ്രശ്നങ്ങൾ. പിന്നെയും തുല്യപരിമാണങ്ങളായി ഒരു പ്രദേശ ത്തിങ്കൽ[1] തന്നെ കേന്ദ്രമായിരിക്കുന്ന[2] രണ്ടു വൃത്തങ്ങളുടെ നേമിയോഗത്തി ങ്കൽ നിന്നിത്ര ചെന്നേടത്തു തങ്ങളിലുള്ള അകലമെത്രയെന്നും ഇത്ര അക ലമുള്ളേടത്തുന്ന് ഇത്ര അകലത്തു തമ്മിൽ ഉള്ള നേമിയോഗമെന്നു അറിവാനായിക്കൊണ്ടുള്ള ത്രൈരാശികം യാതൊന്ന് അതിന്റെ അതിദേശ പ്രകാരത്തെത്തന്നെ വിസ്തിരിച്ചു കാട്ടുവാനായിക്കൊണ്ടു ദശപ്രശ്നങ്ങളെ ചൊല്ലുന്നു.

അവിടെ ശങ്കു, നതജ്യാവ്, അപക്രമം, ഇഷ്ടാംശാഗ്രം, അക്ഷജ്യാവ് എന്നിവ അഞ്ചു വസ്തുക്കളിൽ മൂന്നിനെ ചൊല്ലിയാൽ അവ സാധനങ്ങ ളായി മറ്റേവ രണ്ടിനേയും അറിവാനുപായം ഇവിടെ ചൊല്ലുന്നത്. അവ പത്തുപ്രകാരം സംഭവിക്കും. എന്നിട്ടു ദശപ്രശ്നം.

20. ഒന്നാം പ്രശ്നം – ശങ്കുവും നതവും

20.i. സാമാന്യന്യായങ്ങൾ

അവിടെ നടേ ക്രാന്തിദിഗഗ്രാക്ഷങ്ങളെക്കൊണ്ടു ശങ്കുനതങ്ങളെ അറിയും പ്രകാരത്തെ ചൊല്ലുന്നു. അവിടെ നടേ ഖമധ്യത്തിങ്കലും ഗ്രഹത്തിങ്കലും

---

18. 6. D. കർണ്ണങ്ങൾ താനും D.om. വിദ്യമാനമായിട്ട്, അനന്തരം
19. 1. F. ത്തിങ്കന്ന്
2. D. F. കേന്ദ്രവുമായിരിക്കുന്ന

സ്പർശിച്ചിട്ട് ഒരു വൃത്തത്തെ കല്പിപൂ. അതിന്ന് 'ഇഷ്ടദിഗ്വൃത്ത'മെന്നു പേർ. 'ദൃങ്മണ്ഡല', മെന്നും തന്നെ. ഈ ഇഷ്ടദിഗ്വൃത്തവും ക്ഷിതിജവും ഉള്ള സംപാതത്തിങ്കേന്ന് പൂർവ്വാപരസ്വസ്തികത്തോടുള്ള അന്തരാളം ക്ഷിതിജത്തിങ്കലേ ജ്യാവ്, ഇഷ്ടാശാഗ്രയത്[1]. പിന്നെ[2] ഖമദ്ധ്യത്തേയും ദക്ഷിണോത്തരസ്വസ്തികത്തിങ്കേന്ന് ഇഷ്ടാശാഗ്രയോളം അന്തരിച്ചേടത്തു ക്ഷിതിജത്തിങ്കലും സ്പർശിച്ചിട്ടു ഒരു വൃത്തത്തെ കല്പിപ്പൂ. ഇതിന്നു 'വിപരീതദിഗ്വൃത്ത'മെന്നു പേർ[3]. പിന്നെ ക്ഷിതിജവും വിപരീതദിഗ്വൃത്തവും തങ്ങളിലുള്ള സംപാതത്തിങ്കലും രണ്ടു ധ്രുവത്തിങ്കലും[4] സ്പർശിച്ചിട്ടു ഒരു വൃത്തത്തെ കല്പിപൂ. ഇതിന്ന് 'തിർയ്യഗ്വൃത്ത'മെന്നുപേർ. ഇഷ്ടദിഗ്വൃത്തവും ഘടികാവൃത്തവും ഇവ രണ്ടിന്നും കൂടി തിർയ്യഗ്ഗതം ഇത് എന്നിട്ട്. ഈ വൃത്തത്തിങ്കൽ ഇഷ്ടഘടികാവൃത്തങ്ങൾ തങ്ങളിലുള്ള പരാമാന്തരാളം പിന്നെ ദക്ഷിണോത്തരവൃത്തത്തിങ്കേന്നു ക്ഷിതിജത്തിങ്കലേ വിപരീതദിഗ്വൃത്തത്തിന്റെ അന്തരം യാതൊന്ന് അത് അവറ്റിന്റെ പരമാന്തരാളമാകുന്നത്. ഖമദ്ധ്യത്തിങ്കൽ യോഗമാകയാൽ ക്ഷിതിജത്തിങ്കൽ പരമാന്തരാളം. ഇത് ഇഷ്ടാശാഗ്രാതുല്യം. ഇതു പ്രമാണഫലമായിട്ടു ദക്ഷിണോത്തരവൃത്തത്തിങ്കൽ ഖമധ്യധ്രുവാന്തരാളചാപഭാഗത്തിന്റെ ജ്യാവ് ലംബകം. ഇത് ഇച്ഛയായി ധ്രുവങ്കേന്നു വിപരീതദിഗ്വൃത്താന്തരാളജ്യാവിച്ഛാഫലമായിട്ട് ഉണ്ടാകും. ഇതു കോടിയായി, അക്ഷജ്യാവു ഭുജയായി, വർഗ്ഗയോഗമൂലം കൊണ്ടു കർണ്ണത്തെ ഉണ്ടാക്കൂ. ഇതു തിർയ്യഗ്വൃത്തത്തിങ്കലേ ക്ഷിതിജാന്തരാളജ്യാവ്, ധ്രുവങ്കലഗ്രമായിട്ടിരിക്കും. ഇഷ്ടദിങ്മണ്ഡലവും ഘടികാമണ്ഡലവും തങ്ങളിലുള്ള പരമാന്തരാളമായിട്ടിരിക്കും ഇത് തന്നെ. വിപരീതദിഗ്വൃത്തവും ക്ഷിതിജവുമുള്ള സംപാതത്തിങ്കലു ദിഗ്വൃത്തപാർശ്വങ്ങൾ[5], ധ്രുവങ്കലു ഘടികാപാർശ്വങ്ങൾ. ഇഷ്ടദിഗ്ഘടികകളുടെ നാലു പാർശ്വത്തിങ്കലും സ്പർശിച്ചിരിപ്പൊന്ന്. ഈ[6] തിർയ്യഗ്വൃത്തത്തിങ്കലേ പാർശ്വാന്തരാളത്തോടു തുല്യം ഇവറ്റിന്റെ പരമാന്തരാളമാകയാൽ ഈ ഉണ്ടായ കർണ്ണം തന്നെ ഇഷ്ടദിഗ്ഘടികാവൃത്തങ്ങ

20.1. B. ഇഷ്ടാശാഗ്രയോഗം; D.ഇഷ്ടാശാഗ്രയാകുന്നത്
2. F. reads പിന്നെ ഖമധ്യത്തിങ്കലും രണ്ടു ധ്രുവങ്കലും
3. B. അതു വിപരീതവൃത്തം
4. H. വൃത്തത്തിങ്കലും
5. F. adds. ഇവറ്റിന്റെ പാർശ്വാന്തരത്തോളം തുല്യം ധ്രുവങ്കലു
6. B.C.D.om. ഈ

ളുടെ പരമാന്തരാളമാകുന്നത്. അവിടെ സർവ്വദിക്സാധാരണമായിട്ടു നിരൂപിക്കുമ്പോൾ ക്ഷേത്രസംസ്ഥാനം ദുർഗ്രഹം. എന്നിട്ട് ഒരു ദിഗ്‌വിശേഷത്തെ ആശ്രയിച്ചു നിരൂപിക്കേണം.

അവിടെ ദക്ഷിണഗോളത്തിങ്കൽ നിരൃതികോണിലെ ശങ്കു ഇഷ്ടമാകുമ്പോളേക്കു ചൊല്ലുന്നു. അവിടെ നിരൃതികോണിലും ഈശകോണിലും സ്പർശിച്ചിരിക്കും ദിഗ്‌വൃത്തം, വായുകോണിലും അഗ്നികോണിലും ക്ഷിതിജത്തെ സ്പർശിക്കും ദിഗ്‌വൃത്തം. വായുകോണിങ്കേന്ന് ഉത്തരധ്രുവത്തോടുള്ള അന്തരാളം തിര്യഗ്‌വൃത്തത്തിങ്കലെ ജ്യാവ് ഈ ഉണ്ടാക്കിയ ഹാരകമാകുന്നത്. ഈ ഹാരകം പ്രമാണമായി, ധ്രുവന്റെ ഉയർച്ചയാകുന്ന[7] അക്ഷം പ്രമാണഫലമായി, ഈശകോണിലെ ദിഗ്‌വൃത്തത്തിങ്കലെ തിര്യഗ്‌വൃത്തവും ക്ഷിതിജവുമുള്ള പരമാന്തരാളം ഇച്ഛാഫലമായിട്ടുണ്ടാകും.

ഇവിടെ ഈശകോണിൽ ദിഗ്‌വൃത്തത്തിങ്കൽ ക്ഷിതിജത്തിങ്കേന്ന് എത്ര ഉന്നതം തിര്യഗ്‌വൃത്തസംപാതം ഖമദ്ധ്യത്തിങ്കേന്നു നിരൃതികോണിൽ ദിഗ്‌വൃത്തത്തിങ്കൽ അത്ര താണേടത്തു ഘടികാദിഗ്‌വൃത്തങ്ങളുടെ യോഗം. പിന്നെ രണ്ടു വൃത്തത്തിനും സാധാരണമായിരിക്കുന്ന തിര്യഗ്‌വൃത്തം രണ്ടിങ്കലും യാതൊരു പ്രദേശത്തിങ്കൽ സ്പർശിക്കുന്നൂ, അവിടുന്നു രണ്ടിന്മേലേയും വൃത്തപാദം ചെന്നേടത്ത്[8] തങ്ങളിലുള്ള യോഗം എന്നു നിയതമെല്ലോ. എന്നിട്ട് ഇവിടെ[9] ഈശകോണിലെ ദിഗ്‌വൃത്തസ്പർശത്തിങ്കേന്നു തിര്യഗ്‌വൃത്തത്തിന്മേലേ ഘടികാവൃത്തത്തോടുള്ള അന്തരാളം ഇവിടെ ഉണ്ടാക്കിയ ഹാരകത്തോടു തുല്യം. ഇതിനെ ഭുജയായി പ്രമാണമായി കല്പിപ്പൂ. പിന്നെ ഈശകോണിങ്കൽ ദിഗ്‌വൃത്ത[10]സംപാതത്തിങ്കേന്നു നിരൃതികോണിലെ ഘടികായോഗ[11]ത്തോടുള്ള അന്തരാളചാപം ദിഗ്‌വൃത്തത്തിങ്കലേത് വൃത്തപാദം യാതൊന്ന് അതിന്റെ ജ്യാവ് വ്യാസാർദ്ധം. ഇതു കർണ്ണമായി പ്രമാണമായി കല്പിപ്പൂ. പിന്നെ ഖമധ്യത്തിങ്കേന്നു ദക്ഷിണോത്തരവൃത്തത്തിങ്കലെ ഘടികാന്തരാളം അക്ഷം. ഇത് ഇച്ഛയാകുന്നത്. പിന്നെ ഖമധ്യത്തിങ്കേന്നു ഘടികാന്തരാളം ദൃങ്മണ്ഡലത്തിങ്കലേത് ഇച്ഛാഫലം. ഇച്ഛാ

20. 7. B. ഉയർച്ചയായിരിക്കുന്ന
8. B. ചെല്ലുന്നേടത്തു
9. C. D അവിടെ
10. B. തിര്യഗ്‌വൃത്ത
11. H. ഘടിവൃത്ത

ഭുജയായി, ഇച്ഛാഫലം കർണ്ണമായിട്ടും ഇരിപ്പൊന്ന് ഇവിടെ ദൃങ്മണ്ഡലത്തിങ്കൽ യാതൊരിടത്തു തിര്യഗ്വൃത്തസംപാതം[12], ഇവിടന്നു വൃത്തപാദം ചെന്നേടത്തു ദിങ്മണ്ഡലത്തിങ്കൽ ഘടികാവൃത്തസംപാതം. എന്നിട്ട് ക്ഷിതിജത്തിങ്കേന്നു തിര്യഗ്വൃത്തത്തിന്റെ ഉയർച്ചയും ഖമദ്ധ്യത്തിങ്കേന്നു ഘടികാമണ്ഡലത്തിന്റെ താഴ്ചയും ദിഗ്വൃത്തത്തിങ്കൽ തുല്യമായിട്ടിരിക്കുന്നു. ഇങ്ങനെ രണ്ടു പ്രകാരം നിരൂപിക്കാം. ഇച്ഛാഫലാനയനക്രിയാഭേദമില്ല. ഇത് ഇഷ്ടദിക്ശങ്കു വരുത്തുന്നേടത്തേക്ക് അക്ഷസ്ഥാനീയമാകുന്നത്. ഇതിന്റെ കോടി ഘടികാസംപാതത്തിങ്കേന്നു ക്ഷിതിജാന്തരാളം[13]. ദിങ്മണ്ഡലത്തിങ്കലേ ഇത് ലംബസ്ഥാനീയമാകുന്നത്. പിന്നെ ഇഷ്ടസ്വാഹോരാത്രവും ഘടികാവൃത്തവും തങ്ങളിലുള്ള അന്തരാളം ദക്ഷിണോത്തരവൃത്തത്തിങ്കലെ ഇഷ്ടാപക്രമമാകുന്നത്. ഇതിനെ ഭുജയെന്നും ഇച്ഛയെന്നും കല്പിച്ച് ഈ ഘടികാസ്വാഹോരാത്രാന്തരം തന്നെ ദിങ്മണ്ഡലത്തേതിനു കർണ്ണമാക്കി ഇച്ഛാഫലമാക്കി വരുത്തൂ. ഇതു അപക്രമസ്ഥാനീയമാകുന്നത്. ഇവിടെ ദക്ഷിണോത്തരവൃത്തത്തിങ്കൽ അക്ഷാപക്രമങ്ങൾ കേവലങ്ങൾ ദിങ്മണ്ഡലത്തിങ്കലൂ അക്ഷാപക്രമസ്ഥാനീയങ്ങളാകയാൽ തുല്യാന്തരത്വമുണ്ട്[14].

20.ii. ഇഷ്ടദേശത്ത് ശങ്കുച്ഛായ

എന്നിട്ട് അക്ഷസ്ഥാനീയമുണ്ടാക്കുന്ന പ്രമാണഫലങ്ങൾ തന്നെ അപക്രമസ്ഥാനീയത്തെ ഉണ്ടാക്കുവാനും സാധനമാകുന്നത്. ഇവിടെ ഖമധ്യത്തിങ്കേന്നു ദിങ്മണ്ഡലത്തിന്മേലേ ഘടികാവൃത്തത്തോളമുള്ള ഇട അക്ഷസ്ഥാനീയമാകുന്നത്. ദിങ്മണ്ഡലത്തിങ്കൽ തന്നെ ഘടികാവൃത്തത്തിങ്കേന്നു സ്വാഹോരാത്രത്തിന്റെ ഇട അപക്രമസ്ഥാനീയമാകുന്നത്. ഇവ തങ്ങളിൽ കൂട്ടുകതാൻ അന്തരിക്കതാൻ ചെയ്താൽ ഖമധ്യത്തിങ്കേന്ന് ദിങ്മണ്ഡലത്തിങ്കലെ സ്വാഹോരാത്രത്തിന്റെ ഇട ഉണ്ടാകും. അത് ഇഷ്ടദിക്ഛായയാകുന്നത്. പിന്നെ ക്ഷിതിജത്തോടു ഘടികാവൃത്തത്തോടുള്ള അന്തരാളം ദിങ്മണ്ഡലത്തിങ്കലേതു ലംബസ്ഥാനീയം. പിന്നെ ഘടികാമണ്ഡലത്തോട് സ്വാഹോരാത്രത്തോടിട ദിങ്മണ്ഡലത്തിങ്കലേത് അപക്രമസ്ഥാനീയമാകുന്നത്. ഇവറ്റിന്റെ ശേഷത്തിന് തക്കവണ്ണമുള്ള യോഗം താനന്തരം[15] താനിഷ്ടദിക്ച്ഛങ്കുവാകു

20.12. B. adds എന്നിട്ട്
13. സംപാതക്ഷിതിജാന്തരാളം
14. തുല്യാന്തരാളമുണ്ട്
15. C. ചെയ്താൽ ഇഷ്ടം; F. ചെയ്ത്

ന്നത്. യാതൊരു പ്രകാരം ദക്ഷിണോത്തരവൃത്തങ്ങളിലെ[16] അക്ഷാപക്രമ ങ്ങളുടെ താൻ ലംബകാപക്രമങ്ങളുടെ താൻ യോഗാന്തരങ്ങളെക്കൊണ്ടു മദ്ധ്യാഹ്നച്ഛായാശങ്കുക്കളുണ്ടാകുന്നു. അവ്വണ്ണമിഷ്ടദിഗ്വൃത്തത്തിങ്കലവറ്റേ ക്കൊണ്ട് ഇഷ്ടദിക്ച്ഛായാശങ്കുക്കൾ വരും[17]. ഇവിടെ ഇഷ്ടചാപങ്ങളുടെ യോഗാന്തരങ്ങൾ ചെയ്തു ജ്യാവുണ്ടാക്കാം. ജ്യാക്കൾ തങ്ങളിൽ യോഗാ ന്തരങ്ങൾ ചെയ്കിലുമാം.

എങ്കിൽ അക്ഷാപക്രമസ്ഥാനീയങ്ങളെ വർഗ്ഗിച്ചു ത്രിജ്യാവർഗ്ഗത്തിങ്കേന്നു കളഞ്ഞു മൂലിച്ചു കോടികളെ ഉണ്ടാക്കി പിന്നെ അക്ഷാപക്രമസ്ഥാനീയങ്ങളെ ഇതരേതരകോടികളെക്കൊണ്ട് ഗുണിച്ചു യോഗം താനന്തരം താൻ ചെയ്തു ത്രിജ്യകൊണ്ടു ഹരിച്ചഫലം ഇഷ്ടദിക്ഛായയാകുന്നത്. പിന്നെ ലംബകാപക്രമസ്ഥാനീയങ്ങളെ ഇതരേതരകോടികളേക്കൊണ്ടു ഗുണിച്ചു യോഗം താനന്തരം താൻ ചെയ്ത് ത്രിജ്യകൊണ്ടു ഹരിച്ചഫലം ഇഷ്ടദിക്ഛായയാകുന്നത്[18].

പിന്നെ കേവലങ്ങളായിരിക്കുന്ന അക്ഷാപക്രമങ്ങളുടെ യോഗം താനന്തരം താൻ ചെയ്തു മധ്യാഹ്നച്ഛായയെ ഉണ്ടാക്കി അതിനെ ത്രിജ്യകൊണ്ടു ഗുണിച്ച് ഇഷ്ടദിഗ്വൃത്തഘടികാവൃത്തങ്ങളുടെ പരമാന്തരാളങ്ങളാ യിരിക്കുന്ന മുമ്പിലുണ്ടാക്കിയ ഹാരകം കൊണ്ട് ഹരിച്ച് ഇഷ്ടദിക്ഛായയെ ഉണ്ടാക്കുകിലുമാം. ഇവിടെ അക്ഷാപക്രമങ്ങളുടെ യോഗാന്തരങ്ങൾ ചെയ്യുന്നതിനു മുമ്പെയും പിമ്പെയും ആ[19] ത്രിജ്യാഗുണനവും ഹാരകഹരണവും ചെയ്ക[20], ഫലഭേദമില്ലായ്കയാൽ. ഇങ്ങനെ അക്ഷാവലംബകാപക്രമങ്ങളേക്കൊണ്ടു ഇഷ്ടശങ്കുച്ഛായകളെ ഉണ്ടാക്കാം.

പിന്നെയും ഒരു പ്രകാരം ലംബാപക്രമങ്ങളുടെ യോഗാന്തരങ്ങളെക്കൊണ്ട് ഇഷ്ടശങ്കു വരുത്താം. അവിടെ അക്ഷാപക്രമങ്ങൾ രണ്ടിനേയും ത്രിജ്യകൊണ്ടു ഗുണിച്ച് ഹാരകത്തെക്കൊണ്ടു ഹരിക്കേണ്ടുകയാൽ ഹാരകത്തെ പ്രമാണമെന്നും, കേവലാക്ഷാപക്രമങ്ങളെ പ്രമാണഫലമെന്നും, ത്രിജ്യയെ ഇച്ഛയെന്നും, അക്ഷാപക്രമസ്ഥാനീയങ്ങളെ

20.16. B.D.F. വൃത്തത്തിങ്കലെ
17. B. വരുത്താം
18 B.D ഇഷ്ടദിക്ശങ്കുവാകുന്നത്.
19. B. om. ആ
20. B ത്രിജ്യാഫലഗുണനവും ചെയ്ക

ഇച്ഛാഫലമെന്നും കല്പിക്കാം. ഇവിടെ ത്രിജ്യാവ്യാസാർദ്ധവൃത്തത്തിങ്കൽ യാതൊരു പ്രകാരമിരിക്കും അക്ഷാപക്രമസ്ഥാനീയങ്ങൾ അപ്രകാരമിരിക്കും ഹാരകം, വ്യാസാർദ്ധമായിരിക്കുന്ന വൃത്തത്തിങ്കൽ കേവലാക്ഷാപക്രമങ്ങൾ എന്നാൽ കേവലാക്ഷാപക്രമങ്ങളെ വർഗ്ഗിച്ചു ഹാരകവർഗ്ഗത്തിങ്കേന്നു കളഞ്ഞു മൂലിച്ചാൽ ഹാരകവ്യാസാർദ്ധമാകുന്ന വൃത്തത്തിങ്കലെ അക്ഷാപക്രമകോടികളുണ്ടാകും. പിന്നെ ദ്യുജ്യാലംബകങ്ങളെ[21] വർഗ്ഗിച്ചു ഹാരകവർഗ്ഗത്തിങ്കേന്നു കളഞ്ഞു മൂലിച്ചാൽ ഹാരകവ്യാസാർദ്ധമാകുന്ന വൃത്തത്തിങ്കലെ അക്ഷാപക്രമകോടികളുണ്ടാകും. പിന്നെ ദ്യുജ്യാവലംബകങ്ങളെ ഹാരകത്തെക്കൊണ്ടു ഗുണിച്ച് ത്രിജ്യകൊണ്ട് ഹരിച്ചാലും ഇക്കോടികൾ തന്നെ വരും. പിന്നെ ഈ അക്ഷ കോടിയെക്കൊണ്ട് അപക്രമകോടിയേയും അപക്രമത്തെ അക്ഷം കൊണ്ടും ഗുണിച്ച് ഹാരകത്തെക്കൊണ്ടു ഹരിച്ച ഫലങ്ങളുടെ യോഗംതാനന്തരംതാൻ ചെയ്താൽ ഹാരകവ്യാസാർദ്ധമാകുന്ന വൃത്തത്തിങ്കലേ ഇഷ്ടദിക്ച്ഛങ്കുവുണ്ടാകും. ഇതിനെ ത്രിജ്യകൊണ്ടു ഗുണിച്ച് ഹാരകം കൊണ്ടു ഹരിച്ചാൽ ഇഷ്ടദിക്ച്ഛങ്കുവുണ്ടാകും. ഇവിടെ യാമ്യഗോളത്തിങ്കൽ ലംബകക്രാന്തികളുടെ യോഗം കൊണ്ടു ദക്ഷിണദിക്ച്ഛങ്കുക്കൾ ഉണ്ടാകും.

ഇവിടെ അന്തരിക്കുന്നേടത്ത് അക്ഷകോടിയേക്കാൾ അപക്രമം വലുതാകിൽ സ്വാഹോരാത്രവൃത്തത്തിന് ഇഷ്ടദിഗ്വൃത്തത്തോടുള്ള യോഗം ക്ഷിതിജത്തിന്റെ കീഴേപ്പുറത്ത് ആകയാൽ അന്ന് ഇഷ്ടദിക്ച്ഛങ്കുവില്ല. ഉത്തരാപക്രമം അക്ഷത്തേക്കാൾ ഏറുകിൽ ഖമധ്യത്തിങ്കേന്നു വടക്കേപ്പുറത്ത് ഉച്ചയാകയാൽ അന്നു ദക്ഷിണദിക്ച്ഛങ്കുവില്ല. ഇവിടെ ഉത്തരാശാഗ്ര ആകുമ്പോൾ ആ ശങ്കു ഉണ്ടാകും. ഇവിടെ ലംബകാപക്രമങ്ങൾ സ്ഥാനീയങ്ങളുടെ[22] ചാപയോഗം ത്രിരാശിയേക്കാൾ ഏറുകയാൽ ഇതിന്റെ കോടിജ്യാവ് ഉത്തരദിക്ച്ഛങ്കുവായിട്ടുണ്ടാകും. ജ്യാക്കളുടെ യോഗം കൊണ്ട് വൃത്തപാദത്തിലേറുന്നേടത്തു കോടിജ്യാവു വരും.

20.21. B.D B. F. om. വർഗ്ഗിച്ചു ......to...... ഭ്യുജ്യാവലംബങ്ങളെ
22. B.C.D. ലംബകാപക്രമീയസ്ഥാനങ്ങളെ

ഇങ്ങനെ ഉത്തരഗോളത്തിങ്കൽ അക്ഷത്തേക്കാൾ അപക്രമമേറുമ്പോൾ ഉത്തരാശാഗ്രാശങ്കു വരും. പിന്നെ ഉത്തരാപക്രമം അക്ഷത്തേക്കാൾ കുറയുമ്പോൾ ചില ആശാഗ്രാനിയമത്തിങ്കൽ ഉത്തരാശാഗ്രാശങ്കുവും ദക്ഷിണാശാഗ്രാശങ്കുവും കൂടി ഒരു ദിവസത്തിലേ ഉണ്ടാം. അവിടെ ലംബകാപക്രമയോഗം കൊണ്ടും, അന്തരം കൊണ്ടും തുല്യമായിരിക്കുന്ന ദക്ഷിണാശാഗ്രയിങ്കലും, ഉത്തരാശാഗ്രയിങ്കലും, ശങ്കുക്കൾ ഉണ്ടാകുന്നു.

പിന്നെ ഇഷ്ടാപക്രമം ഹാരത്തേക്കാൾ ഏറുമ്പോൾ അപക്രമസ്ഥാനീയം ത്രിജ്യയേക്കാൾ വലുതാകും. ഇങ്ങനത്തോരു ജ്യാവില്ലായ്കയാൽ ആ ഇഷ്ടാശാഗ്രയ്ക്കു ശങ്കു സംഭവിക്കയില്ല[23]. ഇങ്ങനെ ഇഷ്ടദിക്ഛങ്കു വരുത്തും പ്രകാരം.

20. iii കോണച്ഛായ

അനന്തരം ഇതിനോടു സൂര്യസിദ്ധാന്തത്തിങ്കൽ ചൊല്ലിയ കോണശങ്കുവിന്റെ ന്യായസാമ്യത്തെ ചൊല്ലുന്നു. ഇവിടെ കോണാഭിമുഖം ഇഷ്ടദിങ്മണ്ഡലമാകയാൽ ഒന്നര രാശീടെ ജ്യാവ് ആശാഗ്രയാകുന്നത്. പൂർവ്വാപരസ്വസ്തികകളുടേയും ദക്ഷിണോത്തരസ്വസ്തികകളുടേയും അന്തരാളത്തിന്റെ നടുവിൽ ദിങ്മണ്ഡലത്തിന്നു[24] ക്ഷിതിജസംപാതം. എന്നിട്ട് മൂന്ന് രാശീടെ സമസ്തജ്യാവിന്റെ അർദ്ധമായിട്ടിരിക്കുമിത്. അർദ്ധജ്യാശരങ്ങളുടെ വർഗ്ഗയോഗമൂലം സമസ്തജ്യാവാകുന്നത്. ചക്രപാദത്തിങ്കൽ ജ്യാബാണങ്ങൾ ത്രിജ്യാതുല്യങ്ങൾ. എന്നാൽ അവറ്റിന്റെ വർഗ്ഗയോഗം ത്രിജ്യാവർഗ്ഗത്തിങ്കൽ[25] ഇരട്ടി. എന്നാൽ അതിൽ നാലൊന്നല്ലോ ഒന്നരരാശീടെ വർഗ്ഗ[26]മെന്നിട്ട്[27]. എന്നാൽ ആ ഇഷ്ടാശാഗ്രാവർഗ്ഗമാകുന്ന ത്രിജ്യാവർഗ്ഗാർദ്ധം ലംബകവൃത്തത്തിൽ ആകുമ്പോൾ ലംബവർഗ്ഗാർദ്ധമായിട്ടിരിക്കും. എന്നാൽ ലംബവർഗ്ഗാർദ്ധത്തിങ്കൽ അക്ഷ

20.23. B.F ഇഷ്ടാശാഗ്രാശങ്കു സംഭവിക്കില്ല
24. F. ത്തിന്റെ
25. D. വർഗ്ഗത്തീന്ന്
26. H. പാതീടെ വർഗ്ഗം
27. B,C,D,E. Adds അത് ത്രിജ്യാവർഗ്ഗാർദ്ധം ഇരട്ടിയുടെ വർഗ്ഗത്തിൽ നാലൊന്ന്, F. നാലൊന്നല്ലോ പാതിയുടെ വർഗ്ഗം എന്ന്.

ജ്യാവർഗ്ഗം കൂട്ടി മൂലിച്ചാൽ അത് ഇവിടെയും ഹാരകമാകുന്നത്. പിന്നെ ക്രാന്ത്യക്ഷഘാതവും ഇവറ്റിന്റെ ഹാരകവൃത്തത്തിലെ കോടിഘാതവും തങ്ങളിൽ യോഗം താനന്തരംതാൻ ചെയ്തു ഹാരകം കൊണ്ടു ഹരിച്ചാൽ ഈ ഹാരകവൃത്തത്തിലെ കോണശങ്കുവുണ്ടാകും. ഇവിടെ ഇവറ്റിന്റെ വർഗ്ഗഘാതങ്ങളുടെ യോഗാന്തരങ്ങളെ ഹാരകവർഗ്ഗം കൊണ്ടു ഹരിക്കിൽ ശങ്കുവർഗ്ഗമുണ്ടാകും. പിന്നെ ഇതിനെ മൂലിച്ചു ത്രിജ്യകൊണ്ട് ഗുണിച്ചു ഹാരകം കൊണ്ടു ഹരിച്ചാൽ ത്രിജ്യാവൃത്തത്തിലെ ശങ്കുവായിട്ട് വരും.

ഇവിടെ ക്രാന്ത്യക്ഷകോടികളുടെ വർഗ്ഗങ്ങൾ തങ്ങളിൽ ഗുണിച്ചത് ഒരു ഹാരകമാകുന്നത്. ഇവിടെ കോടിവർഗ്ഗങ്ങളാകുന്നത് ഹാരകവർഗ്ഗത്തിങ്കേന്നു വെവ്വേറെ അക്ഷാപക്രമവർഗ്ഗങ്ങളെ കളഞ്ഞശേഷങ്ങൾ. അവിടെ അക്ഷകോടിവർഗ്ഗത്തെ ഗുണ്യമെന്നും ക്രാന്തികോടിവർഗ്ഗത്തെ ഗുണകാരമെന്നും കല്പിപ്പൂ. അപ്പോൾ ക്രാന്തിവർഗ്ഗം ഗുണഹാരാന്തരമാകുന്നത്. എന്നാൽ ക്രാന്തിവർഗ്ഗം കൊണ്ടു ലംബവർഗ്ഗത്തെ ഗുണിച്ചു ഹാരകവർഗ്ഗം കൊണ്ടു ഹരിച്ചഫലത്തെ ലംബവർഗ്ഗാർദ്ധത്തിങ്കേന്നു കളഞ്ഞാൽ അക്ഷാപക്രമങ്ങളുടെ കോടിവർഗ്ഗഘാതത്തെ ഹാരകവർഗ്ഗം കൊണ്ടു ഹരിച്ച ഫലമുണ്ടാകും. ഇങ്ങനെ കേവലലംബവർഗ്ഗാർദ്ധം ഗുണ്യമായിരിക്കുന്നതിനെ തന്നെ ഗുണ്യമെന്നു കല്പിക്കുമ്പോൾ ഈവണ്ണം ക്രിയ. ഇവിടെ പിന്നെ–

"ഇഷ്ടോനയുക്തേന ഗുണേന നിഘ്നോ-
ƒഭീഷ്ടഘ്നഗുണ്യാന്വിതവർജ്ജിതോ വാ"

(ലീലാവതി. 16)

എന്നതിനു തക്കവണ്ണം ഗുണ്യമാകുന്ന ലംബവർഗ്ഗാർദ്ധത്തിങ്കൽ ഇഷ്ടമാകുന്ന അക്ഷവർഗ്ഗത്തെ കൂട്ടി ഹാരകവർഗ്ഗതുല്യം ഗുണ്യമെന്നു കല്പിച്ചാൽ ഗുണകാരാന്തരമാകുന്ന അപക്രമവർഗ്ഗത്തെത്തന്നെ ഗുണ്യമാകുന്ന ലംബ[28]വർഗ്ഗാർദ്ധത്തിങ്കേന്ന് കളയേണ്ടുവത് എന്നു വരും. അവിടെ[29] കളയേണ്ടുന്ന ഈ അപക്രമവർഗ്ഗത്തിന് ഒരു സംസ്കാരമുണ്ട്

20.28. B.E.F. അർധജ്യാവർഗ്ഗത്തെ ഇഷ്ടത്തിങ്കേന്ന് കളയവേണ്ടുവത് തന്നെ വരും,
29. B.C.F ഇവിടെ

എന്നു വിശേഷമാകുന്നത്. ഇവിടെ കേവലം ശൂന്യമായിരിക്കുന്ന ലംബവർഗ്ഗാർദ്ധത്തിൽ അക്ഷജ്യാവർഗ്ഗത്തെ ഇഷ്ടമായി കല്പിച്ചു കൂട്ടുകയാൽ ആ അക്ഷവർഗ്ഗത്തെ ഗുണഹാരാന്തരമാകുന്ന അപക്രമവർഗ്ഗം കൊണ്ടു ഗുണിച്ച് ഹാരകവർഗ്ഗം കൊണ്ടു ഹരിച്ചഫലം സംസ്കാരമാകുന്നത്. ഇതിനെ അപക്രമവർഗ്ഗത്തിങ്കേന്നു കളകവേണ്ടുവത്. ഇഷ്ടത്തെ ശൂന്യത്തിൽ കൂട്ടുകയല്ലോ ചെയ്തത്. എന്നിട്ട് അപക്രമവർഗ്ഗത്തിങ്കേന്നു കളയേണ്ടുവത്. കേവലാപക്രമവർഗ്ഗം കളഞ്ഞിരിക്കുന്ന ലംബകവർഗ്ഗാർദ്ധത്തിലെങ്കിൽ കൂട്ടുക വേണ്ടുവത്. പിന്നെ ശേഷത്തെ മൂലിച്ചതു ശങ്കുവിന്റെ ഒരു ഖണ്ഡം. മറ്റേ ഖണ്ഡമാകുന്നത് അക്ഷാപക്രമങ്ങൾ തങ്ങളിൽ ഗുണിച്ച് ഈ വർഗ്ഗമായിരിക്കുന്ന ഹാരകത്തിന്റെ മൂലംകൊണ്ടു ഹരിച്ച ഫലം. ഇതിനെ വർഗ്ഗിച്ചാൽ മുമ്പിൽ ചൊല്ലിയ അപക്രമവർഗ്ഗസംസ്കാരമായിട്ടിരിക്കും. പിന്നെ ഈ ശങ്കുഖണ്ഡങ്ങളെ ത്രിജ്യകൊണ്ടു ഗുണിച്ച് ഹാരകം കൊണ്ടു ഹരിക്കേണം.

20. iv. ശങ്കുവിന്റെ ഖണ്ഡദ്വയരൂപേണാനയനം

ഇവിടെ ത്രിജ്യാവർഗ്ഗത്തിങ്കേന്ന് അർക്കാഗ്രാവർഗ്ഗത്തെ കളഞ്ഞശേഷത്തെ ലംബവർഗ്ഗം കൊണ്ടു ഗുണിച്ച് ത്രിജ്യാവർഗ്ഗം കൊണ്ടു ഹരിച്ചാൽ ഫലം ലംബവർഗ്ഗാർദ്ധത്തീന്ന് അപക്രമവർഗ്ഗം കളഞ്ഞതായിട്ടിരിക്കും. ഇതിനെ പിന്നെ ത്രിജ്യാവർഗ്ഗം കൊണ്ടു ഗുണിച്ച് ഹാരകവർഗ്ഗം കൊണ്ടു ഹരിച്ചഫലം ത്രിജ്യാവൃത്തത്തിലാക്കേണ്ടുകയാൽ ത്രിജ്യാവർഗ്ഗം കൊണ്ടു ഗുണനവും ഹരണവും വേണ്ടാ. ത്രിജ്യാവർഗ്ഗാർദ്ധത്തിങ്കേന്ന് അർക്കാഗ്രാവർഗ്ഗത്തെ കളഞ്ഞ ശേഷത്തെ ലംബവർഗ്ഗം കൊണ്ടു ഗുണിച്ചു ഹാരകവർഗ്ഗം കൊണ്ടു ഹരിക്കേ വേണ്ടൂ[30]. എന്നാൽ ത്രിജ്യാവൃത്തത്തിലെ ഫലമുണ്ടാകും. ഈവണ്ണം തന്നെ അക്ഷം കൊണ്ടു അപക്രമത്തെ ഗുണിക്കേണ്ടുന്നേടത്ത് അർക്കാഗ്രേ ഗുണിക്കിൽ ഈ ഗുണിച്ചതിനെ ലംബകം കൊണ്ടു ഗുണിച്ച ഹാരകം കൊണ്ടു ഹരിക്കുമ്പോൾ ത്രിജ്യാവൃത്തത്തിലെ ശങ്കുഖണ്ഡമായിട്ടിരിക്കും, യാതൊരു പ്രകാരം ത്രിജ്യാലംബകങ്ങൾ തങ്ങളിലേ സംബന്ധം അപ്രകാരമിരിക്കും അർക്കാഗ്രാപക്രമങ്ങൾ തങ്ങളിൽ എന്നിട്ട്. പിന്നെ യാമ്യോത്തരഗോളത്തിന്നു

20.30. B. വേണ്ടുകയാൽ, ത്രിജ്യാവർഗ്ഗത്തിലെ

തക്കവണ്ണം ഈ ശങ്കുഖണ്ഡങ്ങളുടെ യോഗാന്തരങ്ങളെക്കൊണ്ടു യാമ്യോത്തരദിക്കുകളിലെ കോണശങ്കുക്കളുണ്ടാകും.

ഇവിടെ അക്ഷാവലംബകങ്ങളുടെ സ്ഥാനത്തു വിഷുവച്ഛായയും ദ്വാദശാംഗുലശങ്കുവും കൊള്ളാം. അവിടെ വിഷുവച്ഛായാവർഗ്ഗത്തെ പന്ത്രണ്ടിന്റെ വർഗ്ഗത്തിന്റെ പാതി എഴുപത്തിരണ്ടിൽ കൂട്ടിയതു ഹാരകവർഗ്ഗമാകുന്നത് എന്നേ വിശേഷമുള്ളൂ. ഇങ്ങനെ നടേത്തേ പ്രശ്നത്തിൽ ഇഷ്ടശങ്കുവിനെ വരുത്തും പ്രകാരം. അതു കോണശങ്കുവെങ്കിൽ ഇങ്ങനെത്തൊരു ക്രിയാലാഘവമുണ്ട് എന്നതിനേയും ചൊല്ലീതായി.

20.v. നതജ്യാനയനം

പിന്നെ നതജ്യാവിനെ ചൊല്ലേണ്ടൂ. അവിടെ ഇഷ്ടദിങ് മണ്ഡലത്തിങ്കേന്നു ദക്ഷിണോത്തരവൃത്തത്തിന്റെ[31] പരമാന്തരാളമാകുന്നത് ഇഷ്ടാശാഗ്ര കോടി, ഛായാഗ്രത്തിങ്കൽ[32] എന്ത് എന്നിട്ടുണ്ടാകും ഛായാകോടി. ഈ ഛായാകോടി തന്നെ നതജ്യാവാകുന്നത്. ത്രിജ്യകൊണ്ടു ഗുണിച്ച് ദ്യുജ്യകൊണ്ടു ഹരിക്കേണം, സ്വാഹോരാത്രവൃത്തത്തിങ്കൽ സ്വവൃത്തകലാപ്രമിതമാവാൻ, എന്നേ വിശേഷമുള്ളൂ. എന്നാൽ ആശാഗ്രാകോടിയും ഛായയും തങ്ങളിൽ ഗുണിച്ചതും, ഛായാകോടിത്രിജ്യകൾ തങ്ങളിലും നതദ്യുജ്യകൾ തങ്ങളിലും ഗുണിച്ചാൽ, സംഖ്യകൊണ്ടു തുല്യങ്ങളായിരിക്കും. എന്നാൽ ഇവറ്റിൽ വച്ച് ഒരുഘാതത്തിങ്കേന്നു മറ്റേവ ദ്വന്ദ്വങ്ങളിൽ രണ്ടിൽ ഒന്നുകൊണ്ടു ഹരിച്ചാൽ അതിന്റെ പ്രതിയോഗിയാകുന്നതു വരും. ഈ ന്യായം ഈ പ്രശ്നങ്ങളിൽ എല്ലാടവും ഓർത്തുകൊള്ളുക. ഇങ്ങനെ ഒരു പ്രശ്നം.

20.31. H. മണ്ഡലത്തിന്റെ
32. H. ഛായയിങ്കൽ, B.C.om.ഛായാഗ്രത്തിങ്കൽ

21. രണ്ടാം പ്രശ്നം: ശങ്കുവും അപക്രമവും

21. i. സാമാന്യസ്വരൂപം

അനന്തരം രണ്ടാം പ്രശ്നം[1]. ഇവിടെ നതജ്യാശാഗ്രാക്ഷങ്ങൾ സാധനങ്ങളായിട്ടു ശങ്കുവും ക്രാന്തിയും ഉണ്ടാകുന്നത്, ഇവിടുത്തെ ക്ഷേത്രകല്പനം പിന്നെ. രണ്ടു ധ്രുവങ്കലും ഗ്രഹത്തിങ്കലും സ്പർശിച്ചിരിപ്പോരു വൃത്തത്തെ കല്പിപ്പൂ. ഇതിന്നു 'നതവൃത്ത'മെന്നു പേർ. ഈ നതവൃത്തവും ദക്ഷിണോത്തരവൃത്തവും തങ്ങളിലെ പരമാന്തരാളം ഘടികാമണ്ഡലത്തിങ്കലും, പിന്നെ നതവൃത്തവും[2] ക്ഷിതിജവുമുള്ള സംപാതത്തിങ്കലും ഖമദ്ധ്യത്തിങ്കലും സ്പർശിച്ചിട്ട് ഒരു വൃത്തത്തെ കല്പിപ്പൂ. ഇതിന്നു 'നതസമമണ്ഡല'മെന്നു പേർ. നതസമമണ്ഡലവും ക്ഷിതിജവുമുള്ള സംപാതത്തിങ്കേന്നു[3] ക്ഷിതിജത്തിന്മേലേയുള്ള വൃത്തപാദം ചെന്നേടത്തു ക്ഷിതിജത്തിങ്കലും ഖമധ്യത്തിങ്കലും, സ്പർശിച്ചിട്ട്[4] ഒരു വൃത്തത്തെ കല്പിപ്പൂ. ഇതിന്നു 'നതദൃക്ക്ഷേപവൃത്ത[5]'മെന്നു പേർ. ഈ വൃത്തത്തിങ്കലു നതവൃത്തവും നതസമവൃത്തവും തങ്ങളിലേ പരമാന്തരാളവും. നതവൃത്തവും ക്ഷിതിജവുമുള്ള പരമാന്തരാളവും[6] ഈ വൃത്തത്തിങ്കൽ തന്നെ. 'സ്വദേശനത'മെന്നും 'സ്വദേശനതകോടി'യെന്നും ഈ പരമാന്തരാളങ്ങൾക്കു പേർ. പിന്നെ ഇഷ്ടദിഗ്വൃത്തത്തേയും വ്യസ്തദിഗ്വൃത്തത്തേയും മുമ്പിൽ ചൊല്ലിയ[7] പ്രകാരം കല്പിപ്പൂ. പിന്നെ നതവൃത്തം ഖമധ്യത്തിങ്കേന്നു എത്ര നതമായിരിക്കുന്നൂ നതദൃക്ക്ഷേപമണ്ഡലത്തിങ്കലേ ഈ വൃത്തത്തിങ്കൽ തന്നെ അത്ര ഉന്നതം ക്ഷിതിജത്തിങ്കേന്നു നതമണ്ഡലപാർശ്വമെന്നിരിക്കും. ഈ നതവൃത്തപാർശ്വത്തിങ്കൽ നതദൃക്ക്ഷേപഘടികാസംപാതം.

---

21. 1. B. അഥ ദ്വിതീയപ്രശ്നം
2. C. Adds നതഹാരകവൃത്തവും
3. B. നതസമമണ്ഡലക്ഷിതിജസംപാതത്തിങ്കേന്ന്
4. F. സ്പർശിച്ചിട്ടുള്ള
5. B. നതദൃക്ക്ഷേപം എന്നുപേർ
6. B. നതവൃത്തക്ഷിതിജപരമാന്തരാളവും

പിന്നെ ക്ഷിതിജവും വ്യസ്തദിഗ്വൃത്തവും ഉള്ള സംപാതത്തിങ്കലും[8], നതദൃക്ക്ഷേപമണ്ഡലത്തിന്മേലേ നതപാർശ്വത്തിങ്കലും സ്പർശിച്ചിട്ട് ഒരു വൃത്തത്തെ കല്പിപ്പൂ. അതു നതേഷ്ടദിങ്മണ്ഡലങ്ങൾ രണ്ടിന്നും കൂടി സാധാരണമായിരിപ്പോരു തിർയ്യഗ്വൃത്തം. ഇവിടെ തിർയ്യഗ്വൃത്തം ഇഷ്ടദിഗ്വൃത്തത്തിങ്കൽ [9]ക്ഷിതിജത്തിങ്കൽ നിന്ന് എത്ര ഉന്നതം അത്ര താണിരിക്കും ഖമധ്യത്തിങ്കേന്നു ദിഗ്വൃത്തവും നതവൃത്തവും തങ്ങളിലുള്ള സംപാതം. ഈ ദിഗ്വൃത്തത്തിങ്കലേ ഖമധ്യനതവൃത്താന്തരം ഛായയാകുന്നത്. ഇതിന്റെ കോടി ശങ്കുവാകുന്നത്.

ഇവിടെ ഉത്തരധ്രുവങ്കേന്നു ദക്ഷിണോത്തരമണ്ഡലത്തിങ്കലേ ഘടികാവൃത്തത്തോളം ചെല്ലുമ്പോൾ നതവൃത്തദക്ഷിണോത്തരാന്തരാളം. നതജ്യാവ്. ധ്രുവങ്കേന്നു ഖമദ്ധ്യത്തോടുള്ള ഇട ലംബകം. ഇതീന്നു നതവൃത്താന്തരമെത്രയെന്നു സ്വദേശനതമുണ്ടാകും. പിന്നെ ഇതിനെ നതവൃത്തവും സ്വദേശനതവൃത്തവും തങ്ങളിലുള്ള സംപാതത്തിങ്കൽ അഗ്രമായിട്ടു കല്പിപ്പൂ. ഈ സ്വദേശനതജ്യാവിന്റെ കോടി[10] ഈ സംപാതത്തിങ്കേന്നു ക്ഷിതിജാന്തരാളം. അവിടെ പൂർവ്വസ്വസ്തികത്തിങ്കേന്നു തെക്കു നീങ്ങി ക്ഷിതിജത്തെ സ്പർശിക്കുമാറ് ഇഷ്ടാശാഗ്ര കല്പിപ്പൂ. ആ ദിക്കിൽ ശങ്കുവും അവ്വണ്ണമാകുമ്പോൾ ഉത്തരസ്വസ്തികത്തിങ്കേന്ന് പടിഞ്ഞാറു നീങ്ങിയും ദക്ഷിണസ്വസ്തികത്തിങ്കേന്നു കിഴക്കു നീങ്ങിയും ക്ഷിതിജസ്പർശം നതവൃത്തത്തിന്. പിന്നെ പൂർവ്വസ്വസ്തികത്തിങ്കേന്ന് ഇത്രതന്നെ വടക്കും പശ്ചിമസ്വസ്തികത്തിങ്കേന്നു അത്ര തെക്കു നീങ്ങിയേടത്തും ക്ഷിതിജത്തെ സ്പർശിച്ചിരിപ്പൊന്ന് സ്വദേശനത വൃത്തമാകുന്നത്.

പിന്നെ ദക്ഷിണസ്വസ്തികത്തിങ്കേന്ന് ഇഷ്ടാശാഗ്രയോളം പടിഞ്ഞാറു നീങ്ങിയേടത്തു വിദിങ്മണ്ഡലത്തിന്നു ക്ഷിതിജസംപാതം. അവിടുന്നു തിർയ്യഗ്വൃത്തം ഉയർന്നു തുടങ്ങുന്നു. സ്വദേശനതവൃത്തത്തോളം ചെല്ലുമ്പോൾ ക്ഷിതിജസംപാതത്തിങ്കേന്നു സ്വദേശനതജ്യാവോളം ഉയർന്നി

21. 7. B. മുൻചൊല്ലിയ, F, വിസ്തരിച്ച് ചൊല്ലിയ
8. B. ക്ഷിതവ്യസ്തദിഗ്സംപാതത്തിങ്കലും
9. F. Adds ഈ
10 B. നതജ്യാകോടി

രിക്കുന്ന നതപാർശ്വത്തിങ്കൽ [11]സ്പർശിക്കും.[12]നതവൃത്തപാർശ്വത്തോടു ക്ഷിതിജത്തോടുള്ള[13] ഇട ഇവിടേക്കു ഹാരകമാകുന്നത്. പിന്നെ ഈ തിർയ്യഗ്വൃത്തം ദിഗ്വൃത്തത്തോളം ചെല്ലുമ്പോൾ[14] ക്ഷിതിജ സംപാതത്തിങ്കേന്നു വൃത്തപാദം ചെല്ലും. ആകയാൽ ഈ ദിഗ്വൃത്തത്തിങ്കൽ ക്ഷിതിജവും തിർയ്യഗ്വൃത്തവും തങ്ങളിലുള്ള പരമാന്തരാളം. അതു ഛായാതുല്യം. പിന്നെ ഇതിന്റെ കോടി ദിഗ്വൃത്തവും തിർയ്യഗ്വൃത്തവുമുള്ള[15] സംപാതത്തിങ്കേന്നു ഖമധ്യാന്തരാളം ദിഗ്വൃത്തത്തിങ്കലേതു ശങ്കുതുല്യം. ഇതു തിർയ്യഗ്വൃത്തവും വിദിഗ്വൃത്തവുമുള്ള പരമാന്തരാളമാകുന്നത്[16]. പിന്നെ നതക്ഷിതിജാന്തരാളം സ്വദേശനതകോടിയോളമാകുമ്പോൾ അതിന്നു കർണ്ണം ത്രിജ്യാവ്, ധ്രുവോന്നതിക്ക് എന്തു കർണ്ണമെന്ന്[17] ഉത്തരധ്രുവം[18] ക്ഷിതിജവുമുള്ള അന്തരാളം നതവൃത്തത്തിങ്കലേത് ഉണ്ടാകും. പിന്നെ നതവൃത്തവും ദക്ഷിണോത്തരവൃത്തവുമുള്ള പരമാന്തരാളം നതജ്യാവ്. അപ്പോൾ നതവൃത്തത്തിങ്കലേത്[19] ധ്രുവക്ഷിതിജാന്തരാളജ്യാവിന്ന് ഇത്ര ദക്ഷിണോത്തരവൃത്താന്തരാളം, നതവൃത്തത്തിങ്കലേത് ധ്രുവക്ഷിതിജാന്തരാളജ്യാവിന്ന് എത്ര എന്നു ദക്ഷിണോത്തരവൃത്താന്തരാളം എന്ന് നതവൃത്തദക്ഷിണോത്തരവൃത്താന്തരാളം ക്ഷിതിജത്തിങ്കലേത് ഉണ്ടാകും. ഈ അന്തരാളത്തോടു തുല്യമായിട്ട് പശ്ചിമസ്വസ്തികത്തിങ്കേന്നു തെക്കു നീങ്ങീട്ട് നതദൃക്ക്ഷേപത്തിനു ക്ഷിതിജസംപാതം. ഇതിനെ ആശാഗ്രാകോടിയിങ്കേന്നു കളവൂ. എന്നാൽ സ്വദേശനതവൃത്തവും വിദിഗ്വൃത്തവും തങ്ങളിലുള്ള അന്തരാളം[20] ക്ഷിതിജത്തിങ്കലേത് ഉണ്ടാകും.

21. ii. ശങ്ക്വാനയനം

ഇവിടെ ജ്യായോഗവിയോഗം ചെയ്യേണ്ടുമ്പോൾ പരസ്പരകോടിഗുണനം ചെയ്തു തങ്ങളിൽ പരസ്പരം കൂട്ടുകതാനന്തരിക്കതാൻ ചെയ്ത്,

---

21.11 F. സ്പർശിച്ചിരിക്കും
12 C.D.F. read : തിർയ്യഗ്വൃത്തനതവൃത്ത
13 C. D. അന്തരാളം അവിടേയ്ക്ക്
14 D. ചൊല്ലുമ്പോൾ
15 B. തമ്മിലെ, C.F. തങ്ങളിലുള്ള
16 B. തീയ്യഗ്വൃത്തദിഗ്വൃത്തപരമാന്തരാളമാകുന്നത്
17 B.C.D.E, കർണ്ണം എന്ത് എന്ന്
18 B.C ഉത്തരധ്രുവനും
19 B.om. ത്
20 D. പരമാന്തരാളം

ത്രിജ്യകൊണ്ടു ഹരിക്കേണം. എന്നാൽ നതജ്യാവിനെ അക്ഷം കൊണ്ടു ഗുണിച്ച് സ്വദേശനതകോടി കൊണ്ടുഹരിച്ച ഫലത്തേയും, ഇതിന്റെ വർഗ്ഗത്തെ ത്രിജ്യാവർഗ്ഗത്തിങ്കേന്നു കളഞ്ഞു മൂലിച്ചതിനേയും, ക്രമേണ ആശാഗ്രേകൊണ്ടും ആശാഗ്രാകോടിയേക്കൊണ്ടും ഗുണിച്ച് തങ്ങളിലന്തരിപ്പൂ ദക്ഷിണദിക്കിലൂ ശങ്കുവെങ്കിൽ. ഉത്തരദിക്കിലൂ ശങ്കുവെങ്കിൽ തങ്ങളിൽ യോഗം ചെയ്വൂ. ഇതിനെ ത്രിജ്യകൊണ്ടു ഹരിച്ചാൽ വിദിഗ്‌വൃത്തത്തോട് സ്വദേശനതവൃത്തത്തോടുള്ള അന്തരാളം ക്ഷിതിജത്തിങ്കലേത് ഉണ്ടാകും. ഇവിടെ ഉച്ചയ്ക്കു മുമ്പിൽ ഉത്തരാശാഗ്രയാകുമ്പോൾ ഉത്തരസ്വസ്തികത്തിങ്കന്ന് പടിഞ്ഞാറ് ആശാഗ്രയോളം ചെന്നേടത്ത് വിദിഗ്‌വൃത്തക്ഷിതിജസംപാതം. ഇവിടുന്നു തിർയ്യഗ്‌വൃത്തത്തിന്റെ ഉയർച്ച തുടങ്ങുന്നു. അവിടുന്നു പശ്ചിമസ്വസ്തികത്തിന്റെ അന്തരാളം ആശാഗ്രാകോടി. പിന്നെ പശ്ചിമസ്വസ്തികത്തിങ്കേന്നു തെക്കു സ്വദേശനതവൃത്തക്ഷിതിജസംപാതം. എന്നിട്ട്. ആ അന്തരാളം ആശാഗ്രാകോടിയിങ്കൽ കൂട്ടൂ. എന്നാൽ വിദിഗ്‌വൃത്തത്തിങ്കേന്നു സ്വദേശനതവൃത്താന്തരാളമുണ്ടാം. ക്ഷിതിജത്തിങ്കൽ ഇതു ദിഗ്‌വൃത്തവും സ്വദേശനതവൃത്തവും തങ്ങളിലുള്ള പരമാന്തരാളം. പിന്നെ സ്വദേശനതവൃത്തത്തിങ്കൽ ഖമദ്ധ്യത്തിങ്കേന്നു നതവൃത്തപാർശ്വത്തോളം ചെല്ലുമ്പോൾ സ്വദേശനത കോടിയോളമന്തരാളമുണ്ട്. അതിന്ന് ഏതു വിദിഗ്‌വൃത്താന്തരമെന്ന് നതപാർശ്വത്തിങ്കേന്നു വിദിഗ്വൃത്താന്തരത്തെ ഉണ്ടാക്കൂ. പിന്നെ ഇതിന്റെ വർഗ്ഗവും നതപാർശ്വോന്നതിയാകുന്ന സ്വദേശനതജ്യാവിന്റെ വർഗ്ഗവും തങ്ങളിൽ കൂട്ടി മൂലിപ്പൂ. എന്നാൽ നതവൃത്തപാർശ്വത്തോടു ക്ഷിതിജത്തോടുള്ള അന്തരാളം തിർയ്യഗ്‌വൃത്തത്തിങ്കലേത് ഉണ്ടാകും. ഇതു പ്രമാണമാകുന്നത്. ക്ഷിതിജത്തിങ്കേന്നു നതപാർശ്വോന്നതിയും നതപാർശ്വത്തിങ്കേന്നു വിദിഗ്‌വൃത്താന്തരവും പ്രമാണഫലങ്ങളാകുന്നത്. ഈ പ്രമാണത്തിന് ഇവ ഛായാശങ്കുക്കളാകുന്നത്. ത്രിജ്യ ഇച്ഛയാകുന്നത്. ഇഷ്ടദിക്ച്ഛായാശങ്കുക്കൾ ഇച്ഛാഫലങ്ങളാകുന്നത്.

21. iii. ക്രാന്തിജ്യാ

പിന്നെ ഛായയും ആശാഗ്രാകോടിയും തങ്ങളിൽ ഗുണിച്ച് നതജ്യാവിനേക്കൊണ്ടു ഹരിച്ചാൽ ഇഷ്ടദ്യുജ്യാവുണ്ടാകും. ഇതിന്റെ

വർഗ്ഗവും ത്രിജ്യാവർഗ്ഗവും തങ്ങളിലന്തരിച്ചു മൂലിച്ചത് ഇഷ്ടാപക്രമം. ഇങ്ങനെ രണ്ടാം പ്രശ്നോത്തരം.

22. മൂന്നാം പ്രശ്നം: ശങ്കുവും ആശാഗ്രയും

അനന്തരം മൂന്നാം പ്രശ്നത്തിങ്കൽ നതാപക്രമാക്ഷങ്ങളേക്കൊണ്ട് ശംക്വാശാഗ്രകളെ വരുത്തുന്നു. ഇവിടെ നതജ്യാത്രിജ്യകളുടെ വർഗ്ഗാന്തരമൂലം ഗ്രഹത്തോടുള്ള അന്തരാളം സ്വാഹോരാത്ര വൃത്തത്തിങ്കലേത് ഉണ്ടാകും. ഇതു ദ്യുജ്യാവൃത്തവ്യാസാർദ്ധത്തെ ത്രിജ്യയായിട്ടു കല്പിക്കുമ്പോളുള്ളതായിട്ടിരിക്കും. പിന്നെ ഇന്നത കോടിയെ ദ്യുജ്യകൊണ്ടു ഗുണിച്ച് ത്രിജ്യകൊണ്ടു ഹരിപ്പൂ. ഫലം ത്രിജ്യാവൃത്തകലകളേക്കൊണ്ടുണ്ടാക്കുന്ന ദ്യുവൃത്തജ്യാവ്. ഇതിങ്കേന്നു ക്ഷിതിജ്യാവിനെ കളവൂ ദക്ഷിണഗോളത്തിങ്കൽ, ഉത്തരഗോളത്തിങ്കൽ കൂട്ടൂ. പിന്നെ ഇതിനെ ലംബകം കൊണ്ടു ഗുണിച്ച് ത്രിജ്യകൊണ്ടു ഹരിപ്പൂ. ഫലം ശങ്കു.

ഇതിന്റെ കോടി ഛായ. നതജ്യയും ദ്യുജ്യയും തങ്ങളിൽ ഗുണിച്ച് ഛായയെക്കൊണ്ടു ഹരിച്ചഫലം ആശാഗ്രാ കോടി.

23. നാലാം പ്രശ്നം : ശങ്കുവും അക്ഷവും

അനന്തരം നതക്രാന്ത്യാശാഗ്രകളേക്കൊണ്ട് ശങ്ക്വക്ഷങ്ങളെ വരുത്തും പ്രകാരം.

23.i. ശങ്ക്വാനയനം

നതജ്യാദ്യുജ്യാക്കൾ തങ്ങളിൽ ഗുണിച്ച് വെവ്വേറെ ആശാഗ്രാകോടികൊണ്ടും ത്രിജ്യകൊണ്ടും ഹരിച്ചാൽ ഫലങ്ങൾ ഛായയും ഛായാകോടിയുമായിട്ട് ഉണ്ടാകും. പിന്നെ ഛായാത്രിജ്യകളുടെ വർഗ്ഗാന്തരമൂലം കൊണ്ട് ശങ്കു ഉണ്ടാകും.

23. ii. അക്ഷാനയനം

അക്ഷം വരുത്തുന്നേടത്തെ ക്ഷേത്രകല്പനം പിന്നെ. ഇവിടെ ദക്ഷിണോത്തരവൃത്തത്തിങ്കേന്നു ഛായാകോടിയോളം എല്ലാ അവയവവും അകന്നു നേരേ തെക്കുവടക്കായിട്ട് ഒരു വൃത്തത്തെ കല്പിപ്പൂ. ഘടികാമണ്ഡലത്തിന്ന് സ്വാഹോരാത്രമെന്ന പോലെ ഇരിപ്പൊന്ന്. ഇതു ദക്ഷിണോത്തരവൃത്തത്തിന്നു ഗ്രഹസ്ഫുടമായിട്ടുമിരിപ്പോന്ന്. പിന്നെ ഗ്രഹത്തിങ്കലും പൂർവ്വാപരസ്വസ്തികത്തിങ്കലും സ്പർശിച്ചിട്ടൊരു വൃത്തത്തെ കല്പിപ്പൂ. ഇതിങ്കൽ ഗ്രഹത്തിങ്കേന്നു ദക്ഷിണോത്തര വൃത്തത്തോടുള്ള അന്തരാളം ഛായാകോടി. ഗ്രഹത്തിങ്കന്ന് പൂർവ്വാപരസ്വസ്തികാന്തരാളം ഛായാകോടീടെ കോടി. ഇത് ഈ[1] കല്പിച്ച കോടിവൃത്തത്തിനു വ്യാസാർദ്ധമാകുന്നത്. ഈ കോടിവ്യാസാർദ്ധത്തിൽ ഒരു ജ്യാവ് ഛായാഭുജം. അത് ഗ്രഹത്തോടു സമമണ്ഡലത്തോടുള്ള അന്തരാളം. ഇതിന്റെ കോടി ശങ്കു. പിന്നെ ഘടികാമണ്ഡലത്തോട് ഗ്രഹത്തോടുള്ള അന്തരാളം അപക്രമം. ഇതിന്റെ കോടിയാകുന്നത് ഛായാകോടിയും ദ്യുജ്യയും തങ്ങളിലുള്ള വർഗ്ഗാന്തരമൂലം. ഇതു ഗ്രഹത്തോടു ഉന്മണ്ഡലത്തോടുള്ള അന്തരാളം ഈ കോടി വൃത്തത്തിങ്കലേതായിട്ടിരിക്കും. പിന്നെ ഛായാഭുജയെ അപക്രമ കോടികൊണ്ടു ഗുണിച്ച് തങ്ങളിൽ യോഗം താനന്തരംതാൻ യുക്തിക്കു തക്കവണ്ണം ചെയ്ത് ഛായാകോടിത്രിജ്യാവർഗ്ഗാന്തരമൂല[2]മാകുന്ന ഈ കോടിവൃത്തവ്യാസാർദ്ധം കൊണ്ടു ഹരിപ്പൂ. ഫലം ഈ[3] കോടി വൃത്തത്തിങ്കലേ അക്ഷം. പിന്നെ ഈ അക്ഷത്തെ ത്രിജ്യകൊണ്ടു ഗുണിച്ച് കോടിവൃത്തവ്യാസാർദ്ധം കൊണ്ടു ഹരിപ്പൂ. ഫലം സ്വദേശാക്ഷം. ഇവിടെ ക്രാന്ത്യാശാഗ്രകൾ ഭിന്നദിക്കുകൾ എങ്കിൽ ഈ സംവർഗ്ഗങ്ങളുടെ യോഗം തുല്യം എങ്കിൽ അന്തരം വേണ്ടുവത്. ഉന്മണ്ഡലക്ഷിതിജാന്തരാളത്തിങ്കലൂ ഗ്രഹം എങ്കിലും യോഗം. ഇവിടെ ഛായയും ഛായാകോടിയും തങ്ങളിലുള്ള വർഗ്ഗാന്തരമൂലം ഛായാബാഹു. ഇങ്ങനെ ശങ്കുവിനോടുകൂടിയുള്ള പ്രശ്നോത്തരങ്ങൾ[4] നാലിനേയും ചൊല്ലിയതായി.

---

23 1. B.om. ഈ
2. B.C.D.E.F om. മൂല
3. C. om. ഈ
4. C. പ്രശ്നോത്തരങ്ങളെ

24. അഞ്ചാം പ്രശ്നം : നതവും ക്രാന്തിയും

അനന്തരം നതക്രാന്തികളുടെ ആനയനപ്രകാരം. ഇവിടെ പൂർവ്വാപരസ്വസ്തികങ്ങളിൽ ഒന്നിങ്കൽ നിന്നു ഗ്രഹത്തോടുള്ള അന്തരാളചാപത്തിന്റെ ജ്യാവ് വ്യാസാർദ്ധമായിട്ട് ഒരു വൃത്തത്തെ കല്പിപ്പൂ. അതിങ്കലേ ജ്യാവ് ഛായാഭുജയാകുന്നത്. ഈ കോടിവൃത്തത്തിങ്കലേ അക്ഷവും ത്രിജ്യാവൃത്തത്തിങ്കലേ അപക്രമവും തങ്ങളിൽ യോഗം താനന്തരം താൻ ചെയത് ഈ ഛായാഭുജയാകുന്നത് എന്നിതു നടേ ചൊല്ലി[1]. എന്നാൽ കോടിവൃത്തത്തിങ്കലേ അക്ഷവും ഛായാഭുജയും തങ്ങളിൽ യോഗം താനന്തരം താൻ ചെയ്താൽ ത്രിജ്യാവൃത്തത്തിങ്കലേ അപക്രമമുണ്ടാകും. ഇവിടെ ലംബാക്ഷങ്ങളെ കോടിവൃത്തത്തിങ്കലാക്കേണ്ടുകയാൽ കോടിവ്യാസാർദ്ധഗുണനവും ത്രിജ്യാഹരണവും വേണം ലംബാക്ഷങ്ങൾക്ക്. പിന്നെ ഇങ്ങനെയിരിക്കുന്ന ഈ ലംബാക്ഷങ്ങളെക്കൊണ്ടു ക്രമേണ ഛായാഭുജയേയും ശങ്കുവിനേയും ഗുണിച്ച് യോഗം താനന്തരം താൻ ചെയ്ത് കോടിവൃത്തവ്യാസാർദ്ധം കൊണ്ടു ഹരിപ്പൂ. ഫലം ഇഷ്ടാപക്രമം. ഇവിടെ കോടിവൃത്തഗുണനവും ഹരണവും വേണ്ട. കേവലം ലംബാക്ഷങ്ങളെക്കൊണ്ടു ക്രമേണ ഛായാഭുജയേയും ശങ്കുവിനേയും ഗുണിച്ച് യോഗം താനന്തരം താൻ ചെയ്ത്. ത്രിജ്യകൊണ്ടു ഹരിച്ചഫലം ഇഷ്ടാപക്രമം. ഇതിന്റെ കോടി ഇഷ്ടദ്യുജ്യാവ്. ഇതിനേക്കൊണ്ടു ഹരിപ്പൂ. ഛായാകോടിത്രിജ്യകൾ തങ്ങളിൽ ഗുണിച്ചതിന്റെ ഫലം നതജ്യാവ്.

25. ആറാം പ്രശ്നം : നതവും ആശാഗ്രയും

അനന്തരം നതാശാഗ്രകളെ വരുത്തുന്നു. അവിടെ നടേ ഛായാബാഹുവിനെ വരുത്തുന്നത്. അത് അർക്കാഗ്രയും ശംക്വഗ്രവും തങ്ങളിലേ യോഗം താനന്തരം താനാകുന്നത്. അവിടെ

---

24 1. F. ചൊല്ലീതായി

പൂർവ്വാപരസ്വസ്തികവും ആദിത്യന്റെ ഉദയാസ്തമയപ്രദേശവും തങ്ങളിലെ അന്തരാളം ക്ഷിതിജത്തിങ്കലേത് അർക്കാഗ്രയാകുന്നത്. ഉദിച്ച പ്രദേശത്തിങ്കേന്നു സ്വാഹോരാത്രത്തിന്റെ ചരിവിന്നു തക്കവണ്ണം എത്ര തെക്കു നീങ്ങി ഗ്രഹം ഇഷ്ടകാലത്തിങ്കൽ എന്നതു[1] ശങ്ക്വഗ്രമാകുന്നത്. ഇതു തെക്കോട്ടു നീങ്ങു[2]. എന്നിട്ട് ഇത് നിത്യദക്ഷിണം. പിന്നെ ഉദഗ്ഗോളത്തിങ്കൽ പൂർവ്വാപരസ്വസ്തികത്തിങ്കേന്നു വടക്കു നീങ്ങി ഉദിക്കും. എന്നിട്ട് അന്ന് അർക്കാഗ്ര ഉത്തരം, ദക്ഷിണഗോളത്തിങ്കൽ അർക്കാഗ്ര ദക്ഷിണം. എന്നിട്ടു തുല്യദിക്കെങ്കിൽ[3] അർക്കാഗ്രാശംക്വഗ്രങ്ങളുടെ യോഗം, ഭിന്നദിക്കിലന്തരം. എന്നാൽ സമമണ്ഡലത്തോടു ഗ്രഹത്തോടുള്ള അന്തരാളമുണ്ടാകും. അതു ഛായാഭുജയാകുന്നത്. ഇവിടെ അർക്കാഗ്രാപക്രമങ്ങൾ ത്രിജ്യാലംബകങ്ങളെപ്പോലെയും ശങ്കുശങ്ക്വഗ്രങ്ങൾ ലംബകാക്ഷങ്ങളെപ്പോലെയും ഇരിപ്പോചിലവ. എന്നിട്ട് അപക്രമത്തെ ത്രിജ്യകൊണ്ടും ശങ്കുവിനെ അക്ഷം കൊണ്ടും ഗുണിച്ച് ഗോളത്തിനു തക്കവണ്ണം യോഗം താനന്തരം താൻ ചെയ്ത് ലംബകം കൊണ്ടു ഹരിച്ചഫലം ഛായാഭുജ. ഈ ഛായാഭുജയെ ത്രിജ്യകൊണ്ടു ഗുണിച്ച് ഛായകൊണ്ടു ഹരിച്ചഫലം ആശാഗ്രയാകുന്നത്. ഛായാശാഗ്രാകോടിഘാതത്തിങ്കേന്നു ദ്യുജ്യകൊണ്ടു ഹരിച്ച ഫലം നതജ്യാവ്.

26. ഏഴാം പ്രശ്നം : അക്ഷവും നതിയും

അനന്തരം നതാക്ഷങ്ങളെ വരുത്തുന്നു. അവിടെ നതജ്യാവു മുമ്പിലേപ്പോലെ[1]. പിന്നെ ഛായാകോടിദ്യുജ്യകളുടെ വർഗ്ഗാന്തരമൂലം ഉന്മണ്ഡലത്തോടു ഗ്രഹത്തോടുള്ള അന്തരാളത്തിങ്കലേ സ്വാഹോരാത്രത്തിങ്കലേ ജ്യാവ്. ക്ഷിതിജത്തിങ്കേന്നു തുടങ്ങിയുള്ള ഈ ജ്യാവിന്ന് 'ഉന്നതജ്യാ'വെന്നു പേർ. പിന്നെ ക്ഷിതിജോന്മണ്ഡലാന്തരാളത്തിങ്കലെ സ്വാഹോരാത്രവൃത്തഭാഗജ്യാവിന്നു 'ക്ഷിതിജജ്യാ'വെന്നു

---

25 1. B. ചരിവിന്നുമാകുന്നത്
2. B. C. F. നീങ്ങുന്നു
3. F. ദിക്കാകിൽ

26 1. B. മുൻപോലെ

പേർ. എന്നാൽ ദക്ഷിണഗോളത്തിങ്കൽ ക്ഷിതിജത്തിങ്കേന്ന് ഉന്മണ്ഡലം കീഴേയാകയാൽ ക്ഷിതിജജ്യാവിനോടു കൂടിയിരിക്കുന്ന ഉന്നതജ്യാവ് ഛായാകോടിദ്യുജ്യാവർഗ്ഗാന്തരമൂലം. ഉത്തരഗോളത്തിങ്കൽ പിന്നെ ക്ഷിതിജജ്യാവു പോയി ഇരിക്കുന്ന ഉന്നതജ്യാവ് ഇത്. ഉന്നതജ്യാവു ശങ്കുശംക്വഗ്രങ്ങൾക്കു കർണ്ണമായിരിപ്പോന്ന്. 'ക്ഷിതിജജ്യാവ് ഇതിന്നു സദൃശമായിരിപ്പോരു ത്ര്യശ്രത്തിങ്ക[2]ലെ ഭുജയാകുന്നത്[3]. എന്നാൽ ദക്ഷിണ ഗോളത്തിങ്കൽ രണ്ടു ക്ഷേത്രങ്ങളുടെ ഭുജാകർണ്ണയോഗമിത്. പിന്നെ ദക്ഷിണഗോളത്തിങ്കൽ ഛായാഭുജയാകുന്നത് അർക്കാഗ്രാശംക്വഗ്രങ്ങളുടെ യോഗം. ഈ ഛായാഭുജയെ ക്ഷിതിജ്യാവു കൂടിയിരിക്കുന്ന ഉന്നതജ്യാവിൽ കൂട്ടൂ. എന്നാൽ രണ്ടു ക്ഷേത്രങ്ങളുടെ ഭുജാകർണ്ണയോഗമിത്, ഉത്തരഗോളത്തിങ്കൽ ഭുജാകർണ്ണാന്തരം. ഇത് ഛായാകോടി, ദ്യുജ്യാവർഗ്ഗാന്തരമൂലവും ഛായാഭുജയും തങ്ങളിൽ കൂട്ടിയതായിട്ടിരിക്കും.

ഇവിടെ ശങ്കു, ശങ്ക്വഗ്രം, ഉന്നതജ്യാവ് എന്നിങ്ങനെ ഒരു ത്ര്യശ്രം. അപക്രമം ക്ഷിതിജ്യാവ്, അർക്കാഗ്ര എന്നിത് ഒരു ത്ര്യശ്രം. ഈ ക്ഷേത്രങ്ങൾ രണ്ടിന്റേയും ഭുജാദ്വയത്തിന്റേയും കർണ്ണദ്വയത്തിന്റേയും യോഗം ദക്ഷിണഗോളത്തിങ്കലുണ്ടാകുന്നത്, ഉത്തരഗോളത്തിങ്കൽ കർണ്ണദ്വയ ത്തിങ്കേന്നു ഭുജാദ്വയത്തെ കളഞ്ഞുണ്ടാകുന്നത്. ഈ രണ്ടു ത്ര്യശ്രങ്ങളും തുല്യസ്വഭാവങ്ങളാകയാൽ യോഗാന്തരങ്ങൾ ചെയ്താലും ഒരു ക്ഷേത്രത്തിങ്കലേ ഭുജാകർണ്ണയോഗംതാനന്തരംതാനെന്നപോലെ ഇരിക്കുമത്രേ. സ്വഭാവം കൊണ്ടു ശങ്ക്വപക്രമയോഗം ഈ ക്ഷേത്രത്തിനു കോടിയായിട്ടിരിക്കും. എന്നാൽ ഈ ശങ്ക്വപക്രമയോഗത്തിന്റെ വർഗ്ഗത്തെ ദക്ഷിണഗോളത്തിങ്കൽ ഭുജാകർണ്ണയോഗം കൊണ്ടു ഹരിപ്പൂ. ഫലം അന്തരം. ഉത്തരഗോളത്തിങ്കൽ ഭുജാകർണ്ണാന്തരം കൊണ്ടു ഹരിപ്പൂ. ഫലം യോഗം. ഇങ്ങനെ ഭുജാകർണ്ണങ്ങളുടെ യോഗവുമന്തരവുമുണ്ടായാൽ തങ്ങളിൽ കൂട്ടി അർദ്ധിച്ചതു കർണ്ണം, അന്തരിച്ച് അർദ്ധിച്ചതു ഭുജ. പിന്നെ ഭുജയെ ത്രിജ്യകൊണ്ടു ഗുണിച്ച് കർണ്ണം കൊണ്ട് ഹരിച്ച ഫലം അക്ഷം, ലംബാക്ഷത്രിജ്യകളോടു തുല്യസ്വഭാവങ്ങൾ നടേത്തെ ത്ര്യശ്രങ്ങൾ രണ്ടും എന്നിട്ടത്.

26.2. F. ത്രിശ്രത്തിങ്കലെ
3. B. C. F. ഭുജയായിരിപ്പൊന്ന്

27. എട്ടാം പ്രശ്നം : അപക്രമവും ആശാഗ്രയും

അനന്തരം അപക്രമാശാഗ്രങ്ങളെ വരുത്തുന്നു. അവിടെ നതവൃത്തവും ക്ഷിതിജവുമുള്ള പരമാന്തരാളം സ്വദേശനതകോടി. ഇതു പ്രമാണം. സ്വദേശനതവൃത്തവും ക്ഷിതിജവുമുള്ള അന്തരാളം നതവൃത്തത്തിന്മേലേത് വൃത്തപാദം. ഇതിനു ജ്യാവ്യാസാർദ്ധം പ്രമാണഫലം, ശങ്കു ഇച്ഛാ, നതവൃത്തത്തിൽ ഗ്രഹക്ഷിതിജ്യകളുടെ അന്തരാളം ഇച്ഛാഫലം. ഈ പ്രമാണഫലങ്ങൾക്കുതന്നെ ധ്രുവോന്നതി ഇച്ഛയാകുമ്പോൾ ധ്രുവക്ഷിതിജാന്തരാളം നതവൃത്തത്തിങ്കലേതു ഉണ്ടാകും.

പിന്നെ സ്വദേശനതവൃത്തവും നതവൃത്തവും തങ്ങളിലേ സംപാതത്തിങ്കേന്ന് വടക്കു ഗ്രഹമെങ്കിൽ ഉണ്ടാക്കിയ ഇച്ഛാഫലങ്ങളുടെ ചാപങ്ങൾ തങ്ങളിൽ അന്തരിപ്പൂ. എന്നാൽ നതവൃത്തത്തിങ്കലേ ഉത്തരധ്രുവഗ്രഹാന്തരാളമുണ്ടാകും. ഇച്ചൊല്ലിയ വൃത്തസംപാതത്തിങ്കേന്നു തെക്കു ഗ്രഹമെന്നിരിക്കിൽ ഇച്ചൊല്ലിയ ഇച്ഛാഫലജ്യാക്കളുടെ ചാപങ്ങളുടെ യോഗത്തെ ചെയ്യൂ. അതു ദക്ഷിണധ്രുവനും ഗ്രഹവുമുള്ള അന്തരാളചാപം നതവൃത്തത്തിങ്കലേത് ഉണ്ടാകും. ഇതിന്റെ ജ്യാവ് ദ്യുജ്യാവ്. ഇതിന്റെ കോടി അപക്രമം. ആശാഗ്രം മുമ്പിൽ ചൊല്ലിയപോലെ.

28. ഒൻപതാം പ്രശ്നം : ക്രാന്തിയും അക്ഷവും

അനന്തരം ക്രാന്ത്യക്ഷങ്ങൾ. ക്രാന്തിയെ ദ്യുജ്യയെ മുമ്പിലുണ്ടാക്കീട്ട് ഉണ്ടാക്കിക്കൊള്ളൂ അക്ഷത്തെ നടേത്തേതിലൊരു പ്രകാരം.

29. പത്താം പ്രശ്നം : ആശാഗ്രയും അക്ഷവും

അനന്തരം ദിഗഗ്രാക്ഷങ്ങളെ വരുത്തുന്നു. അവിടെ ദ്യുജ്യാനതജ്യാക്കളുടെതാൻ ഛായാകോടിത്രിജ്യകളുടെ താൻ

ഘാതത്തിങ്കേന്നു ഛായകൊണ്ടു ഹരിച്ചഫലം ആശാഗ്രാകോടി. അക്ഷം നടേത്തേപ്പോലെ വരുത്തൂ. ഇങ്ങനെ പത്തു പ്രശ്നങ്ങളുടേയും ഉത്തരം ചൊല്ലീതായി.

30. ഇഷ്ടദിക്ഛായ : പ്രകാരാന്തരം

അനന്തരം ഇഷ്ടദിക്ഛായയിൽ തന്നെ പ്രകാരത്തെ ചൊല്ലുന്നു. അവിടെ ഘടികാമണ്ഡലത്തിങ്കൽ ഇഷ്ടദിങ്മണ്ഡലസംപാതത്തിങ്കൽ ഗ്രഹമെന്നിരിക്കുമ്പോൾ ദ്വാദശാംഗുലശങ്കുവിന്റെ ഛായ ഉണ്ടാകുന്നു. നടേ അവിടെ വിഷുവത്തിങ്കലു ഗ്രഹം എങ്കിൽ ദ്വാദശാംഗുലശങ്കുവിന്റെ ഛായാഭുജ വിഷുവച്ഛായാതുല്യം. ത്രിജ്യാവൃത്തത്തിങ്കലേ ആശാഗ്രാ ദ്വാദശാംഗുലശങ്കുച്ഛായാ വ്യാസാർദ്ധവൃത്തത്തിങ്കൽ ഛായാഭുജ യായിട്ടിരിക്കും. അതു വിഷുവച്ഛായാതുല്യമാകുമ്പോൾ എന്തു കോടി എന്ന് ആശാഗ്രാകോടിയും വിഷുവച്ഛായയും തങ്ങളിൽ ഗുണിച്ച് ആശാഗ്രാകൊണ്ടു ഹരിപ്പൂ. ഫലം ഛായാകോടി. ഇതിനേയും വിഷുവച്ഛായയേയും വർഗ്ഗിച്ചു കൂട്ടി മൂലിപ്പൂ. അത് ഘടികാമണ്ഡലത്തിങ്കലു ഗ്രഹമെന്നിരിക്കുമ്പോഴേ ദ്വാദശാംഗുലശങ്കുഛായ. പിന്നെ ഈ ഛായയെ ത്രിജ്യാവൃത്തത്തിങ്കലാക്കിയാൽ ഇഷ്ടദിങ്മണ്ഡലത്തിങ്കലേ ഘടികാന്തരാളമുണ്ടാകും. ഇത് അക്ഷസ്ഥാനീയമാകുന്നത്. ഇവിടെ ദക്ഷിണോത്തരവൃത്തത്തിങ്കലേ ഖമദ്ധ്യഘടികാന്തരാളം അക്ഷം. അതിങ്കൽ തന്നെ ഘടികാസ്വാഹോരാത്രാന്തരാളം അപക്രമം. എന്നിട്ട് അക്ഷവും അക്ഷസ്ഥാനീയവും പ്രമാണവും പ്രമാണഫലവുമായിട്ടിരിക്കും. അപക്രമമാകുന്ന ഇച്ഛയ്ക്ക് ഇച്ഛാഫലം ഘടികാസ്വാഹോരാത്രവൃത്താന്തരാളം ഇഷ്ടദിഗ്വൃത്തത്തിങ്കലേത് ഉണ്ടാകും. ഇത് അപക്രമസ്ഥാനീയം. പിന്നെ മധ്യാഹ്നച്ഛായയെ ഉണ്ടാക്കുന്നപോലെ അക്ഷാപക്രമസ്ഥാനീയങ്ങളുടെ ചാപയോഗം താനന്തരത്താൻ ചെയ്തു ജ്യാവുണ്ടാക്കിയാൽ അത് ഇഷ്ടദിക്ച്ഛായയായിട്ടിരിക്കും.

31. കാലലഗ്നവും ഉദയലഗ്നവും

അനന്തരം കാലലഗ്നത്തേയും ഉദയലഗ്നത്തേയും വരുത്തുംപ്രകാരം. ഇവിടെ പ്രവഹവശാൽ പടിഞ്ഞാറുനോക്കി ഭ്രമിക്കുന്ന രാശിചക്രത്തിന്റെ മദ്ധ്യവൃത്തമാകുന്ന അപക്രമവൃത്തം ഇഷ്ടകാലത്തിങ്കൽ പൂർവ്വാപര സ്വസ്തികങ്ങളിൽ ഒന്നിങ്കേന്നു വടക്കെയും മറ്റതിങ്കേന്നു തെക്കെയും ക്ഷിതിജത്തിങ്കൽ സ്പർശിച്ചിരിപ്പൊന്ന്. അപക്രമമണ്ഡലത്തിങ്കലെ ക്ഷിതിജസംപാതപ്രദേശത്തിനു 'ലഗ്ന'മെന്നുപേർ. ഈ ലഗ്നങ്ങളിലും ഖമധ്യത്തിങ്കലും സ്പർശിച്ചിട്ട് ഒരു വൃത്തത്തെ കല്പിപ്പൂ. ഇതിന്നു 'ലഗ്നസമമണ്ഡല'മെന്നു പേർ. പിന്നെ ഈ ലഗ്നസമമണ്ഡലം പൂർവ്വാപരസ്വസ്തികങ്ങളിൽ നിന്ന് എത്ര നീങ്ങിയിരിക്കുന്നു ദക്ഷിണോത്തരസ്വസ്തികങ്ങളിൽ നിന്ന് അത്ര നീങ്ങിയേടത്ത് ക്ഷിതിജത്തിങ്കലും ഖമധ്യത്തിങ്കലും സ്പർശിച്ചിടത്ത് ഒരു വൃത്തത്തെ കല്പിപ്പൂ. ഇതിന്നു 'ദൃക്ക്ഷേപവൃത്ത'മെന്നു പേർ. ഇതും ലഗ്നസമമണ്ഡലവും നേരേ വിപരീതദിക്കായിരിക്കും. ഈ രണ്ടു വൃത്തങ്ങളും ക്ഷിതിജവും കൂട്ടീട്ടു ഗോളത്തിങ്കലെ പദവിഭാഗം. ഇവിടെ നടുവേ അപക്രമവൃത്തം. ഇവിടെ ലഗ്നസമമണ്ഡലവും അപക്രമവൃത്തവും തങ്ങളിലേ പരമാന്തരാളം, അപക്രമവൃത്തപാർശ്വത്തിങ്കേന്നു വൃത്തപാദാന്തരിതം രാശികൂടം, ദൃക്ക്ഷേപവൃത്തത്തിങ്കൽ ഖമദ്ധ്യത്തിങ്കേന്നു എത്ര താണു അപക്രമവൃത്തസംപാതം ഈ ദൃക്ക്ഷേപവൃത്തത്തിങ്കൽ അത്ര തന്നെ ഉയർന്നിരിക്കും അപക്രമവൃത്തപാർശ്വമാകുന്ന രാശികൂടം, ക്ഷിതിജത്തിങ്കേന്നു വൃത്തപാദാന്തരിതം ഖമധ്യം എന്നിട്ട്.

പിന്നെ ഘടികാമണ്ഡലത്തിങ്കേന്ന് അപക്രമമണ്ഡലത്തിന്റെ എല്ലായിലുമകന്ന പ്രദേശം യാതൊരിടം അവിടത്തിന്ന് 'അയനാന്ത'മെന്നു പേർ. ഈ ഘടികാപക്രമവൃത്തങ്ങളുടെ പരമാന്തരാളപ്രദേശത്തെ സ്പർശിക്കുന്ന വൃത്തം യാതൊന്ന് ഈ വൃത്തത്തിങ്കൽ തന്നെ ഘടികാപക്രമങ്ങളുടെ പാർശ്വങ്ങൾ നാലും സ്പർശിക്കും. ആകയാൽ പാർശ്വാന്തരാളങ്ങൾ രണ്ടും ഈ അയനാന്തസ്പർശമുള്ള വൃത്തത്തിങ്കൽ തന്നെ അകപ്പെടും.

എന്നാൽ നിരക്ഷദേശത്തിങ്കൽ പൂർവ്വവിഷുവത്ത് ഖമദ്ധ്യത്തിങ്കലെന്നു കല്പിക്കുമ്പോൾ ഉത്തരായനാന്തം പൂർവ്വസ്വസ്തികത്തിങ്കേന്നു പരമാപക്രമാന്തരാളം വടക്കു നിരക്ഷക്ഷിതിജത്തിങ്കലു, അപരസ്വസ്തികത്തിങ്കേന്ന് അത്ര തെക്കു ദക്ഷിണായനാന്തം. ദക്ഷിണസ്വസ്തികത്തിങ്കേന്നു കിഴക്കും ഉത്തരസ്വസ്തികത്തിങ്കേന്ന് പടിഞ്ഞാറും ക്ഷിതിജത്തിങ്കൽ രാശികൂടങ്ങൾ രണ്ടും. അവിടുന്നു പ്രവഹവശാൽ ഉത്തരായനാന്തം ക്ഷിതിജത്തിങ്കേന്ന് ഉയരുമ്പോൾ ദക്ഷിണരാശികൂടവും കൂടി ഉയരും. അവ്വണ്ണമേ ദക്ഷിണോത്തരവൃത്തപ്രാപ്തിയും പടിഞ്ഞാറു ക്ഷിതിജപ്രാപ്തിയും രണ്ടിന്നും ഒക്കും.[1] ഈവണ്ണം ദക്ഷിണായനാന്തത്തിനും ഉത്തരരാശികൂടത്തിനും തുല്യകാലത്തിങ്കൽ ഉദയാസ്തമയങ്ങൾ. എന്നിട്ട് അയനാന്തോന്നതിക്കു തക്കവണ്ണം രാശികൂടോന്നതി. എന്നാൽ അയനാന്തോന്നതജ്യാവുതന്നെ രാശികൂടോന്നതജ്യാവാകുന്നത്.

പിന്നെ രാശികൂടത്തിനു ശങ്കു വരുത്തേണം. അതു ക്ഷിതിജത്തിങ്കേന്നുള്ള രാശികൂടോന്നതിയാകുന്നത്. അവിടെ രാശികൂടം ധ്രുവങ്കേന്ന് അന്ത്യാപക്രമത്തോളം അകന്നിരിക്കയാൽ അന്ത്യാപക്രമം രാശികൂടസ്വാഹോരാത്രമാകുന്നത്. എന്നിട്ട് അയനാന്തോന്നതജ്യാവിനെ പരമാപക്രമംകൊണ്ടു ഗുണിച്ച് ത്രിജ്യകൊണ്ടു ഹരിപ്പൂ. ഇതുതന്നെ നിരക്ഷദേശത്ത് രാശികൂടശങ്കു. സാക്ഷദേശത്തിങ്കൽ പിന്നെ ഇതിന്നു ചരിവുണ്ടാകയാൽ ലംബകം കൊണ്ടു ഗുണിച്ച് ത്രിജ്യകൊണ്ട് ഹരിച്ച് പിന്നെ ഈ ഫലത്തിൽ, പിന്നെ ക്ഷിതിജോന്മണ്ഡലാന്തരാളത്തിങ്കലെ ശങ്കുഭാഗത്തെ കൂട്ടേണം. ഉത്തരരാശികൂടശങ്കുഭാഗത്തിങ്കൽ, ദക്ഷിണരാശികൂടശങ്കുവിങ്കേന്നു കളയേണം. ഇതു രാശികൂടശങ്കുവാകുന്നത്.

ഇവിടെ രാശികൂടത്തിങ്കലും ഖമദ്ധ്യത്തിങ്കലും സ്പർശിച്ചിട്ട് ഒരു വൃത്തത്തെ കല്പിപ്പൂ. അതെല്ലോ 'ദൃക്ക്ഷേപവൃത്ത'മാകുന്നത്. ഇതിങ്കൽ രാശികൂടശങ്കുവോളം താണിരിക്കും ഖമദ്ധ്യത്തിങ്കേന്നു അപക്രമസംപാതം. അതു ദൃക്ക്ഷേപമാകുന്നത്. എന്നിട്ടു രാശികൂടശങ്കുതന്നെ ദൃക്ക്ഷേപമാകുന്നത് എന്നു വന്നു. ഇവിടെ പൂർവ്വാപരസ്വസ്തികത്തിങ്കേന്ന്

31 1. C. F. ഒക്കൊക്കെ

ഉന്മണ്ഡലത്തിങ്കലെ ധ്രുവങ്കലോളം ചെല്ലുമ്പോൾ ക്ഷിതിജോന്മണ്ഡലാന്തരാളശങ്കുവാകുന്നത് അക്ഷം, രാശികൂടസ്വാഹോരാത്രത്തോളം ചെല്ലുമ്പോളേ അന്ത്യദ്യുജ്യാവിന്ന് ഏത് എന്ന് ഇവിടുത്തെ ക്ഷിതിജോന്മണ്ഡലാന്തരാളത്തിങ്കലേ ശങ്കുഖണ്ഡമുണ്ടാകും.

പിന്നെ ത്രിരാശ്യൂനകാലലഗ്നഭുജാജ്യാവ് ഇവിടെ ഉന്മണ്ഡലത്തിങ്കേന്നുള്ള രാശികൂടോന്നതജ്യാവാകുന്നത്. ഇതിനെ പിന്നെ തന്റെ സ്വാഹോരാത്രത്തിങ്കലാക്കി അക്ഷവശാലുള്ള ചരിവിനെ ലംബത്തിനു തക്കവണ്ണം കളഞ്ഞാൽ രാശികൂടശങ്കുവുണ്ടാകും. ഗോളാദിയായിരിക്കുന്ന കാലലഗ്നം ത്രിരാശ്യൂനമാകുമ്പോൾ അയനാദിയായിട്ടു വരും. എന്നാൽ കാലലഗ്നകോടിക്കു ജ്യാവു കൊള്ളുകേ വേണ്ടൂ. ഈവണ്ണമുണ്ടാക്കിയിരിക്കുന്ന ദൃക്ക്ഷേപജ്യാവിന്റെ കോടി ക്ഷിതിജാപക്രമമണ്ഡലങ്ങളുടെ പരമാന്തരാളമാകുന്നത്. ഇതിനെ പ്രമാണമെന്നും, ത്രിജ്യേ പ്രമാണഫലമെന്നും കല്പിപ്പൂ. പിന്നെ അപക്രമമണ്ഡലത്തിങ്കൽ ഗ്രഹമിരിക്കുന്ന പ്രദേശത്തിങ്കേന്നു ക്ഷിതിജമെത്രയകലമുണ്ട് എന്നത് ഗ്രഹത്തിന്റെ തല്ക്കാലശങ്കുവാകുന്നത്. അത് ഇച്ഛാരാശിയാകുന്നത്. ഗ്രഹത്തോട് ക്ഷിതിജത്തോടുള്ള അന്തരാളത്തിങ്കലേ അപക്രമവൃത്തഭാഗം ഇച്ഛാഫലം. ഇതിനെ ചാപിച്ചു ഗ്രഹത്തിങ്കൽ കൂട്ടുകതാൻ കളയുകതാൻ ചെയ്താൽ ക്രാന്തിവൃത്തത്തിങ്കലേ വിഷുവത്തിങ്കേന്നു ക്ഷിതിജസംപാതത്തിങ്കേന്ന്[2] അത്രത്തോളമുള്ള ഭാഗമുണ്ടാകും. അതു പ്രത്യക്കപാലത്തിങ്കലെ അസ്തലഗ്നം, പ്രാക്കപാലത്തിങ്കലെങ്കിൽ ഉദയലഗ്നം.

ഇവിടെ ശങ്കുവിനെ വരുത്തും പ്രകാരം പിന്നെ. തല്ക്കാലസ്വാഹോരാത്രവൃത്തത്തിങ്കൽ ദക്ഷിണോത്തരവൃത്തവും ഗ്രഹവുമുള്ള അന്തരാളം നതമാകുന്നത്. ഇവിടെ എല്ലാ സ്വാഹോരാത്രവൃത്തവും ഒരു അഹോരാത്രകാലം കൊണ്ട് അനുഭ്രമിച്ചുകൂടും. അഹോരാത്രത്തിങ്കൽ പ്രാണങ്ങൾ ചക്രകലാതുല്യസംഖ്യകൾ. എന്നിട്ട് എല്ലാ സ്വാഹോരാത്രവൃത്തങ്ങളേയും കലാവയവങ്ങളായിട്ടുള്ള പ്രാണങ്ങളായിട്ടു കല്പിക്കുമ്പോൾ

31.2. B. D. ക്ഷിതിജസംപാതത്തോളമുള്ള

ചക്രകലാതുല്യമായിട്ടു വിഭജിക്കുന്നു. എന്നിട്ട് നതപ്രാണങ്ങളാകുന്നതു സ്വാഹോരാത്രഭാഗമത്രേ. ആകയാൽ നതഭാഗത്തിന്റെ ഉൽക്രമജ്യാവിനെ കളഞ്ഞ ജ്യാവ് ഉന്മണ്ഡലഗ്രഹങ്ങളുടെ അന്തരാളഭാഗത്തിങ്കലേ[3] സ്വാഹോരാത്രവൃത്തഭാഗജ്യാവ്. ഇതിൽ ചരജ്യാവിനെ സംസ്കരിച്ചാൽ അത് ക്ഷിതിജത്തിങ്കേന്നുള്ള ഉന്നതജ്യാവ്. ഇതിനെ ത്രിജ്യാവൃത്തത്തിങ്കലാക്കുവാനായിക്കൊണ്ട് ദ്യുജ്യയെക്കൊണ്ട് അക്ഷവശാലുള്ള ചരിവു കളവാനായിക്കൊണ്ടു ലംബത്തെക്കൊണ്ടു ഗുണിച്ച് ത്രിജ്യാവർഗ്ഗത്തെക്കൊണ്ടു ഹരിപ്പൂ. ഫലം സ്വാഹോരാത്രവൃത്തത്തിങ്കൽ ഗ്രഹമിരിക്കുന്ന പ്രദേശത്തോടു ക്ഷിതിജത്തോടുള്ള അന്തരാളം ഇഷ്ടദിങ്മണ്ഡലത്തിലേത് ഉണ്ടാകും. ഇത് ശങ്കുവാകുന്നത്.

സ്വാഹോരാത്രവൃത്തത്തിങ്കൽ ആദിത്യൻ നില്ക്കുന്ന പ്രദേശം അപക്രമവൃത്തത്തോടു സ്പർശിച്ചിരിക്കും എന്നിട്ട് അപക്രമേഷ്ടപ്രദേശവും ക്ഷിതിജവുമുള്ള അന്തരാളമാകുന്നത് ഈ ശങ്കുതന്നെ. എന്നിട്ട് ഈ ശങ്കു ഇച്ഛയാകുന്നു. ഗ്രഹക്ഷിതിജാന്തരാളത്തിങ്കലേ അപക്രമവൃത്ത ഭാഗജ്യാവിച്ഛയാകുന്നേടത്ത് രാത്രിയിങ്കലും ഇങ്ങനെ വരുത്തിയ ശങ്കു അപക്രമേഷ്ടപ്രദേശവും ക്ഷിതിജവുമുള്ള അന്തരാളമായിട്ടിരിക്കും എന്നിട്ട് ശങ്കു രാത്രിയിലും ഇച്ഛാരാശിയായിട്ടിരിക്കും. അവിടെ രാത്രിപ്രമാണാർദ്ധവും രാത്രിയിങ്കലെ ഗതൈഷ്യഭാഗങ്ങളാലൊന്നും തങ്ങളിലന്തരിച്ചത് നതപ്രാണനാകുന്നത് അധോഭാഗത്തിങ്കലെ ദക്ഷിണോത്തരവൃത്തവും ഗ്രഹവും തങ്ങളിലുള്ള അന്തരാളത്തിങ്കലേ സ്വാഹോരാത്രവൃത്തഭാഗം എന്നിട്ട്. ഇതിന്ന് ഉൽക്രമജ്യാവുണ്ടാക്കി ത്രിജ്യാവിങ്കേന്നു കളഞ്ഞാൽ ഗ്രഹോന്മണ്ഡലാന്തരാളത്തിങ്കലേ സ്വാഹോരാത്രഭാഗജ്യാവ് ഉണ്ടാകും. അവിടെ ക്ഷിതിജഗ്രഹാന്തരാളമാവാനായിക്കൊണ്ട് ചരത്തെ ഉത്തര ഗോളത്തിങ്കൽ കളവൂ ദക്ഷിണഗോളത്തിങ്കൽ കൂട്ടൂ. പിന്നെ മുമ്പിലേപ്പോലെ ശങ്കുവിനെ വരുത്തൂ. ആ ശങ്കുവിനെ ത്രിജ്യകൊണ്ടു ഗുണിച്ചു ദൃക്ക്ഷേപകോടികൊണ്ടു ഹരിപ്പൂ. ഫലം ക്രാന്തിവൃത്തത്തിങ്കലേ ക്ഷിതിജഗ്രഹാന്തരാളജ്യാവ്. ഇതിന്റെ ചാപത്തെ പ്രാക്കപാലത്തിങ്കൽ ഗ്രഹത്തിൽ കൂട്ടൂ, അധോമുഖശങ്കുവെങ്കിൽ ഗ്രഹത്തിങ്കേന്നു കളവൂ. അത്

31.3. B. അന്തരാളത്തിങ്കലേ

'ഉദയലഗ്ന'മാകുന്നത്. പ്രത്യക്കപാലത്തിങ്കൽ ഇതിനെ ഗ്രഹത്തിങ്കൽ വിപരീതമായിട്ടു സംസ്കരിച്ചാൽ 'അസ്തലഗ്ന'മുണ്ടാകും. ഉദയാസ്തമയങ്ങളുടെ മധ്യലഗ്നം ദൃക്ക്ഷേപലഗ്നമാകുന്നത്. അതു ദൃക്ക്ഷേപവൃത്താപക്രമമണ്ഡലസംപാതത്തിങ്കലായിരിക്കും.

32. മധ്യലഗ്നം

പിന്നെ മദ്ധ്യലഗ്നമാകുന്നത് ദക്ഷിണോത്തരവൃത്താ[1]പക്രമമണ്ഡലസംപാതം. ഇത് മുമ്പിലേ പഞ്ചദശപ്രശ്നന്യായം കൊണ്ടുണ്ടാകും. പിന്നെ മധ്യകാലമാകുന്നത്[2] ദക്ഷിണോത്തര[3]വും ഘടികാമണ്ഡലവും തങ്ങളിലുള്ള സംപാതം. ഇതു മധ്യലഗ്നന്യായം കൊണ്ടു വരും.

കാലലഗ്നമാകുന്നതു മധ്യകാലത്തിൽ[4] മൂന്നു രാശി കൂടിയത്. അതു പൂർവ്വസ്വസ്തികവും ഘടികാമണ്ഡലവും തങ്ങളിലുള്ള സംപാതപ്രദേശം. ഇതിനെ ഉണ്ടാക്കും പ്രകാരം പിന്നെ. സായനാർക്കൻ നടേത്തേ പദത്തിങ്കലൂ എങ്കിൽ ഇതിന്റെ ഭുജാപ്രാണങ്ങളെ മുമ്പിൽ ചൊല്ലിയപോലെ ഉണ്ടാക്കൂ. ഇവിടെ ആദിത്യൻ നില്ക്കുന്നിടത്ത് അപക്രമമണ്ഡലത്തിങ്കലും ധ്രുവദ്വയത്തിങ്കലും[5] സ്പർശിച്ചിട്ട് ഒരു തിർയ്യഗ്വൃത്തത്തെ കല്പിപ്പൂ. ഇതു ഘടികാവൃത്തത്തിന്റെ യാതൊരു പ്രദേശത്തിങ്കൽ സ്പർശിക്കുന്നൂ അവിടത്തെ ഘടികാമണ്ഡലത്തിങ്കലേ വിഷുവത്തോടുള്ള അന്തരാളം ഭുജാപ്രാണങ്ങളാകുന്നത്. ഇവിടെ ആദിത്യൻ ക്ഷിതിജത്തിങ്കലൂ എന്നു കല്പിക്കുമ്പോൾ ഘടികാതിർയ്യഗ്വൃത്തങ്ങളുടെ സംപാതം പൂർവ്വസ്വസ്തികത്തിങ്കേന്ന് ഒട്ടു കീഴ്, അന്തരാളം ഇഷ്ടചരത്തോടു തുല്യം. എന്നിട്ട് ഭുജാപ്രാണങ്ങളിൽ നിന്ന് ഇഷ്ടചരത്തെ കളഞ്ഞാൽ പൂർവ്വസ്വസ്തികത്തിങ്കേന്നു വിഷുവത്തോടുള്ള അന്തരാളം ഘടികാമണ്ഡലത്തിലേത് ഉണ്ടാകും. ഇതു സായനാർക്കൻ പ്രഥമപദമാകുമ്പോളേ കാലലഗ്നമാകുന്നത്.

32.1. D. om. വൃത്ത
2. H. മദ്ധ്യദക്ഷിണോ
3. D. ദക്ഷിണോത്തരവൃത്തവും
4. D. മദ്ധ്യലഗ്നത്തിൽ
5. F. ധ്രുവത്തിങ്കലും

പിന്നെ ദ്വിതീയപദത്തിങ്കലെ ആദിത്യനുദിക്കുമ്പോഴേയ്ക്ക്[6] ആദിത്യന്റെ ഭുജാപ്രാണങ്ങളെ ഉണ്ടാക്കൂ. മുമ്പിലേപ്പോലെ തിര്യഗ്വൃത്തത്തേയും കല്പിപ്പൂ. അവിടേക്കു മുമ്പിൽ ചൊല്ലിയപോലെ ഈ തിര്യഗ്വൃത്തത്തോട് ഉത്തരവിഷുവത്തോടുള്ള അന്തരാളം ഭുജാപ്രാണങ്ങളാകുന്നത്. ഇവിടെ ഭുജാപ്രാണങ്ങൾ ക്ഷിതിജത്തിങ്കേന്നു കീഴും, തിര്യഗ്വൃത്തസംപാതം പൂർവ്വസ്വസ്തികത്തിങ്കേന്നു കീഴും ആകയാൽ ചരം കൂട്ടിയ ഭുജാപ്രാണങ്ങളെ ആറു രാശിയിങ്കേന്നു കളവൂ. ശേഷം പൂർവ്വസ്വസ്തികത്തോടു പൂർവ്വവിഷുവത്തോടുള്ള അന്തരാളം ഘടികാമണ്ഡലത്തിങ്കലേത് ഉണ്ടാകും. അത് ആദിത്യോദയത്തിങ്കലെ കാലലഗ്നമാകുന്നത്.

മൂന്നാം പദത്തിങ്കലെ ആദിത്യോദയം പിന്നെ പൂർവ്വസ്വസ്തികത്തിങ്കേന്നു തെക്കെ. അവിടെ ഉന്മണ്ഡലത്തിങ്കേന്നു ആ ക്ഷിതിജം മീതേ ആകയാൽ, അവിടെ കല്പിച്ച തിര്യഗ്വൃത്തം പൂർവ്വസ്വസ്തികത്തിനു മീതെ. ആകയാൽ അവിടെ സ്വസ്തികത്തോളം ചെല്ലുവാൻ ഭുജാപ്രാണങ്ങളിൽ ചരപ്രാണങ്ങളെ കൂട്ടേണം. അത് ഉത്തരവിഷുവദാദിയായുള്ളത്. ആകയാൽ ഇതിൽ ആറുരാശിയും കൂട്ടേണം. ഇതു കാലലഗ്നമാകുന്നത്.

പിന്നെ നാലാം പദത്തിങ്കലും രണ്ടാം പദത്തിങ്കലേപ്പോലെ ഭുജ ഏഷ്യമാകയാൽ ഭുജാപ്രാണങ്ങൾ ക്ഷിതിജത്തിങ്കേന്ന് അധോഭാഗത്തിങ്കല്. തിര്യഗ്വൃത്തം പൂർവ്വസ്വസ്തികത്തിങ്കേന്നു മീതേ. ആകയാൽ ക്ഷിതിജാവധിയാവാൻ ഭുജാപ്രാണങ്ങളിൽ നിന്നു ചരപ്രാണങ്ങളെ കളയേണം. ഇതിനെ പന്ത്രണ്ടു രാശിയിങ്കേന്നു കളയേണം, ഏഷ്യമാകയാൽ. ഇതു കാലലഗ്നമാകുന്നത് ആദിത്യോദയത്തിങ്കലേക്ക്. ഈവണ്ണം പന്ത്രണ്ടു രാശ്യന്തത്തിങ്കലെ കാലലഗ്നത്തേയുമുണ്ടാക്കി മീത്തേതിങ്കേന്നു കീഴേതു കീഴേതു കളഞ്ഞു, അന്തരങ്ങൾ ക്രമേണയുള്ള രാശിപ്രമാണങ്ങളാകുന്നത്. ഇവിടെ അപക്രമമണ്ഡലത്തിങ്കൽ പൂർവ്വവിഷുവത്തിങ്കേന്നു തുടങ്ങി സമമായിട്ട് പന്ത്രണ്ടായി വിഭജിയ്ക്കൂ. ഇത് പന്ത്രണ്ടു രാശികളാകുന്നത്. അവിടെ പ്രവഹവശാൽ ഒരു രാശീടെ ആദി, ഇത് ക്ഷിതിജത്തിങ്കൽ സ്പർശിക്കുമ്പോൾ ആ രാശി തുടങ്ങുന്നു, ഒടുക്കം ക്ഷിതിജത്തെ

---

32. 6. B. ആദിത്യൻ നിൽക്കുമ്പോഴേയ്ക്ക്

സ്പർശിക്കുമ്പോൾ ആ രാശി കഴിയുന്നു. ഈ അന്തരകാലത്തിങ്കൽ ഉള്ള പ്രാണങ്ങൾ 'രാശിപ്രാണ'ങ്ങളാകുന്നത്. ഇങ്ങനെ പ്രസംഗാൽ രാശിയേയും രാശിപ്രമാണത്തേയും ചൊല്ലീതായി.

33. മധ്യലഗ്നാനയനം

ഇങ്ങനെ ആദിത്യോദയത്തിങ്കലെ കാലലഗ്നത്തെയുണ്ടാക്കി, അതിൽ പിന്നെ കഴിഞ്ഞകാലത്തേയും പ്രാണനായി കൂട്ടൂ. അത് ഇഷ്ടകാലത്തിങ്കലെ കാലലഗ്നം. ഇതിങ്കേന്നു മൂന്നുരാശി കളഞ്ഞാൽ ഘടികാദക്ഷിണോത്തര സംപാതപ്രദേശം വരും. ഇതു മധ്യകാലമാകുന്നത്.

പിന്നെ ഇതിന്റെ കോടിയാകുന്നതു വിഷുവത്തോടു പൂർവ്വാപരസ്വസ്തികത്തോടുള്ള അന്തരാളത്തിങ്കലേ ഘടികാമണ്ഡലഭാഗം. പിന്നെ ഈ കോടിക്ക് അപക്രമജ്യാവിനെ ഉണ്ടാക്കൂ. അതു പൂർവ്വാപരസ്വസ്തികങ്ങളിൽ സ്പർശിച്ചിരിക്കുന്ന രാശികൂടവൃത്തത്തിങ്കലെ ഘടികാപക്രമാന്തരാളം. പിന്നെ ഇതിന്നു കോടിജ്യാവിനേയും ദ്യുജ്യാവിനേയുമുണ്ടാക്കി ഭുജാപ്രാണങ്ങളെ ഉണ്ടാക്കൂ. അത് ഇച്ചൊല്ലിയ രാശികൂടാപക്രമസംപാതത്തിങ്കേന്നു വിഷുവത്തോട് ഇട അപക്രമവൃത്തഭാഗം. ഇതിന്റെ കോടിയാകുന്നതു പിന്നെ വിഷുവത്തോടു ദക്ഷിണോത്തരവൃത്തത്തിങ്കലേ അന്തരാളത്തോട് ഉള്ള അന്തരാളത്തിങ്കലേ അപക്രമവൃത്തഭാഗം, ഇതു മധ്യഭുജയാകുന്നത്. പിന്നെ ശേഷം പദത്തിങ്കേന്നു തക്കവണ്ണം കാലലഗ്നത്തിങ്കൽ ചൊല്ലിയപോലെ. ഇവിടെ ഘടികാമണ്ഡലത്തേപ്പോലെ അപക്രമമണ്ഡലത്തേയും അപക്രമ മണ്ഡലത്തേപ്പോലെ ഘടികാമണ്ഡലത്തേയും കല്പിക്കുന്നു എന്നേ വിശേഷമുള്ളൂ. ഇങ്ങനെ മധ്യലഗ്നാനയനപ്രകാരം.

34. ദൃക്ക്ഷേപജ്യാകോട്യാനയനം

അനന്തരം ഉദയലഗ്നവും മധ്യലഗ്നവും കൂടി ദൃക്ക്ഷേപജ്യാവിനെ വരുത്തും പ്രകാരത്തെ ചൊല്ലുന്നു. അവിടെ അപക്രമമണ്ഡലത്തേയും ദൃക്ക്ഷേപമണ്ഡലത്തേയും[1] മുമ്പിൽ ചൊല്ലിയവണ്ണം കല്പിപ്പൂ. പിന്നെ ദക്ഷിണോത്തരവൃത്തത്തിന്നു[2] കിഴക്കേപ്പുറത്ത് ക്ഷിതിജത്തോടു[3] സ്പർശിക്കുന്ന അപക്രമമണ്ഡലപ്രദേശത്തിന് 'ഉദയലഗ്ന'മെന്നു പേർ. പടിഞ്ഞാറെ പുറത്തു സ്പർശിക്കുന്ന പ്രദേശത്തിനു 'അസ്തലഗ്ന'മെന്നു പേർ. ദക്ഷിണോത്തരവൃത്തത്തെ[4] സ്പർശിക്കുന്ന പ്രദേശത്തിനു 'മദ്ധ്യലഗ്ന'മെന്നു പേർ. ഇവറ്റെ അറിയുംപ്രകാരം മുമ്പിൽ ചൊല്ലിയവണ്ണം.

പിന്നെ പൂർവ്വാപരസ്വസ്തികങ്ങളിൽ നിന്ന് എത്ര അകലത്ത് ക്ഷിതിജസ്പർശം അപക്രമവൃത്തത്തിന്[5] അയനാന്ത[6]രാളം ഉദയജ്യാവാകുന്നത്. ഉദയലഗ്നത്തെ ആദിത്യനെന്നു കല്പിച്ച് അർക്കാഗ്രേ ഉണ്ടാക്കും പോലെ ഉദയജ്യാവുണ്ടാക്കേണ്ടൂ. പിന്നെ ഖമധ്യത്തിങ്കേന്ന് എത്ര അകലത്ത് ദക്ഷിണോത്തരവൃത്തത്തെ സ്പർശിക്കുന്നൂ അപക്രമവൃത്തം, ആ അന്തരാളം മധ്യജ്യാവാകുന്നത്. മധ്യലഗ്നത്തെ ആദിത്യനെന്നു കല്പിച്ചു അർക്കാഗ്രേ[7] ഉണ്ടാക്കുംപോലെ മധ്യാഹ്നച്ഛായയെ ഉണ്ടാക്കേണ്ടൂ.

പിന്നെ സമമണ്ഡലവും ദൃക്ക്ഷേപസമമണ്ഡലവും തങ്ങളിൽ ഖമധ്യത്തിങ്കൽ യോഗം, ക്ഷിതിജത്തിങ്കൽ പരമാന്തരാളം. ഈ പരമാന്തരാളം ഉദയജ്യാവാകുന്നത്. പിന്നെ ദൃക്ക്ഷേപസമമണ്ഡലത്തിന്നു വിപരീതമായിരിപ്പൊന്ന് 'ദൃക്ക്ഷേപവൃത്തം'. ആകയാൽ ദൃക്ക്ഷേപവൃത്തവും ദക്ഷിണോത്തരവൃത്തവും തങ്ങളിലുള്ള പരമാന്തരാളവും ക്ഷിതിജത്തിങ്കൽ ഉദയജ്യാവിനോടു തുല്യമായിരിപ്പോന്ന്. ഈവണ്ണമിരിക്കുന്നിടത്തു മദ്ധ്യമജ്യാവു ദക്ഷിണാഗ്രയെങ്കിൽ, ദക്ഷിണസ്വസ്തികത്തിങ്കൽ അഗ്രമായിട്ടിരിക്കുന്ന ദക്ഷിണോത്തര

34.1 B. adds ദൃക്ക്ഷേപസമമണ്ഡലത്തേയും
2. B. F. വൃത്തത്തിങ്കേന്നു
3. B. ക്ഷിതിജത്തെ
4. D. ദക്ഷിണോത്തരപ്രദേശവൃത്തത്തെ
5. D. അപക്രമമണ്ഡലത്തിന്
6. F. om. അയനാ
7. A. മദ്ധ്യാഹ്നച്ഛായ വരുത്തുന്നപോലലെ ഇതിനെ ഉണ്ടാക്കേണ്ടൂ

വൃത്തത്തെ കർണ്ണമായി പ്രമാണമായി കല്പിപ്പൂ[8]. മദ്ധ്യജ്യാവ് ഉത്തരാഗ്രയെങ്കിൽ, ഉത്തരസ്വസ്തികത്തിലഗ്രമായിരിക്കുന്ന വ്യാസാർദ്ധത്തെ ഈവണ്ണം കല്പിപ്പൂ. പിന്നെ ഈ സ്വസ്തികത്തിങ്കേന്നു ദൃക്ക്ഷേപവൃത്താന്തരാളം ക്ഷിതിജത്തിങ്കലേത് ഉദയജ്യാ തുല്യമായിരിക്കുന്ന[9] പ്രമാണഫലം. മദ്ധ്യമജ്യാവ് ഇച്ഛ. മദ്ധ്യജാഗ്രത്തിങ്കേന്നു ദൃക്ക്ഷേപവൃത്താന്തരാളം അപക്രമവൃത്തഭാഗം ഇച്ഛാഫലം. ഇതിന്ന് 'ഭുജ' എന്നു പേർ. ഇതു മധ്യലഗ്നദൃക്ക്ഷേപ ലഗ്നാന്തരാളത്തിങ്കലെ അപക്രമമണ്ഡലഭാഗജ്യാവ്. ഇതിന്റെ വർഗ്ഗത്തെ ത്രിജ്യാവർഗ്ഗത്തിങ്കേന്നു കളഞ്ഞു മൂലിച്ചത്, ഇതിന്റെ കോടി. ഇതു ദക്ഷിണോത്തരവൃത്തവും ക്ഷിതിജവുമുള്ള അന്തരാളത്തിങ്കലെ അപക്രമമണ്ഡലഭാഗജ്യാവ്. പിന്നെ ഈ ഉണ്ടാക്കിയ ഭുജേടെ വർഗ്ഗത്തെ മധ്യജ്യാവർഗ്ഗത്തിങ്കേന്നു കളഞ്ഞു മൂലിച്ചത് മധ്യലഗ്നത്തിങ്കേന്നു ദൃക്ക്ഷേപസമമണ്ഡലാന്തരാളം. ഇതിന് പ്രമാണഫലമെന്നും പേർ. മുമ്പിൽ ചൊല്ലിയ കോടിയെ പ്രമാണമെന്നും കല്പിപ്പൂ. പിന്നെ ദൃക്ക്ഷേപലഗ്നത്തിങ്കലഗ്രമായിരിക്കുന്ന അപക്രമമണ്ഡലവ്യാസാർദ്ധത്തെ ഇച്ഛയെന്നും കല്പിച്ച്, ത്രൈരാശികം കൊണ്ടു വരുന്ന ഇച്ഛാഫലം ദൃക്ക്ഷേപജ്യാവ്. ഇത് അപക്രമമണ്ഡലവും ദൃക്ക്ഷേപമണ്ഡലവും തങ്ങളിലുള്ള പരമാന്തരാളമാകയാൽ, ഇച്ഛാഫലമായിട്ടു വരുന്നു.

പിന്നെ ഇവിടെ ചൊല്ലിയ പ്രമാണത്തെത്തന്നെ പ്രമാണമെന്നു കല്പിച്ച്, മധ്യലഗ്നക്ഷിതിജാന്തരാളം ദക്ഷിണോത്തരവൃത്തത്തിങ്കലേ ജ്യാവ്. മധ്യമജ്യാകോടിയെ പ്രമാണഫലമെന്ന് കല്പിച്ച്, ത്രിജ്യയെ ഇച്ഛയായും കല്പിച്ച്, ഉണ്ടാകുന്ന ഇച്ഛാഫലം അപക്രമക്ഷിതിജങ്ങളുടെ പരമാന്തരാളം. ദൃക്ക്ഷേപലഗ്നക്ഷിതിജാന്തരാളം ദൃക്ക്ഷേപവൃത്തത്തിങ്കലേ ജ്യാവ്. ഇതിന്നു 'ദൃക്ക്ഷേപശങ്കു'വെന്നു പേർ. 'പരശങ്കു'വെന്നും 'ദൃക്ക്ഷേപകോടി'യെന്നും കൂടി പേർ. ഇങ്ങനെ ദൃക്ക്ഷേപജ്യാകോടികളെ വരുത്തും പ്രകാരം.

35.8. B, C,F ദക്ഷിണോത്തരവൃത്തവ്യാസാർദ്ധത്തിങ്കലഗ്രമായിരിക്കും
9. F ഇരിക്കുന്നത്

35. നതിലംബനലിപ്താനയനം

അനന്തരം നതിലംബനലിപ്തകളെ വരുത്തും പ്രകാരത്തെ ചൊല്ലുന്നു, ചന്ദ്രച്ഛായാഗ്രഹണാദ്യുപയോഗത്തിന്നായിക്കൊണ്ട്. ഇവിടെ ഭഗോളമധ്യം കേന്ദ്രമായിട്ടുള്ള ദൃങ്മണ്ഡലത്തിങ്കലേ ഛായയേക്കാൾ എത്ര ഏറും ദൃങ്മദ്ധ്യം കേന്ദ്രമായിരിക്കുന്ന ദൃങ്മണ്ഡലത്തിങ്കലേ ഛായ എന്നതു ലംബനമാകുന്നത്. ഇതിനെ ഛായാപ്രകരണത്തിങ്കൽ ചൊല്ലീതായി. പിന്നെ ഈ ലംബനം കർണ്ണമായിട്ടിരിപ്പോ ചിലവ, ഇവിടെ ചൊല്ലുവാനിരിക്കുന്ന നതിലംബനങ്ങൾ. ഇതിനായിക്കൊണ്ടു ദൃക്ക്ഷേപാപക്രമദൃങ്മണ്ഡലങ്ങൾ മൂന്നിനേയും മുമ്പിൽ ചൊല്ലിയപോലെ കല്പിപ്പൂ. പിന്നെ രണ്ടു രാശികൂടങ്ങളിലും ഗ്രഹത്തിങ്കലും സ്പർശിച്ചിരിപ്പോരു രാശികൂട വൃത്തത്തേയും കല്പിപ്പൂ. ഈവണ്ണമാകുമ്പോൾ ഗ്രഹസ്പൃഷ്ടരാശികൂടം, ദൃങ്മണ്ഡലം, അപക്രമമണ്ഡലം എന്നിവ മൂന്നിന്റേയും സംപാതത്തിങ്കലൂ ഗ്രഹം.

പിന്നെ ഈ മൂന്നു വൃത്തങ്ങളേയും ലംബിയാതെയും ഗ്രഹത്തെ ലംബിച്ചിട്ടും കല്പിപ്പൂ. ദൃങ്മണ്ഡലമാർഗ്ഗത്തൂടെ കീഴ്പോട്ടു താണിരിക്കുമാറു ഗ്രഹം ലംബിക്കുന്നു. ഇവിടെ ലംബിതഗ്രഹവും വൃത്തങ്ങളുടെ സംപാതവും തങ്ങളിലുള്ള അന്തരാളം ദൃങ്മണ്ഡലത്തിങ്കലേത് ഛായാലംബനമാകുന്നത്. പിന്നെ ഈ ലംബിതഗ്രഹത്തിങ്കേന്ന് അപക്രമവൃത്തത്തിന്റെ അന്തരാളം നതിയാകുന്നത്. ഈ ലംബിതഗ്രഹത്തിങ്കേന്നു തന്നെ ഗ്രഹസ്പൃഷ്ടരാശികൂടവൃത്താന്തരാളം ഇവിടേയ്ക്കു ലംബമാകുന്നത്. ഈ നതിലംബനങ്ങൾ ഭുജാകോടികൾ. ഛായാലംബനം കർണ്ണമായിരിക്കും.

36. ഛായാലംബനം

ഇവിടെ ഛായാലംബനത്തെ വരുത്തുംപ്രകാരം. മുമ്പിൽ ചൊല്ലിയവണ്ണം കലാത്മകമായിട്ടിരിക്കുന്ന ദൃക്കർണ്ണത്തെ വരുത്തീട്ട് ഉണ്ടാക്കുകിലുമാം.

ദൃക്കർണ്ണം യോജനാത്മകമായിട്ട് ഉണ്ടാക്കീട്ടു വരുത്തുകിലുമാം. അതിന്റെ പ്രകാരത്തെ ചൊല്ലുന്നു[1].

ഇവിടെ ചന്ദ്രാർക്കന്മാരുടെ മന്ദകർണ്ണമാകുന്ന ഭഗോളമധ്യത്തോടു ഗ്രഹത്തോടുള്ള അന്തരാളം, ഇവറ്റിന്റെ തന്നെ ദ്വിതീയസ്ഫുടകർണ്ണമാകുന്നത്, ഘനഭൂമദ്ധ്യത്തിങ്കേന്നു ഗ്രഹത്തോടുള്ള അന്തരാളം. ഇവിടെ ചന്ദ്രോച്ചവും ആദിത്യനും തങ്ങളിലുള്ള അകലത്തിനു തക്കവണ്ണം ഭഗോളമധ്യവും ഘനഭൂമധ്യവും തങ്ങളിലകലും. എന്നിട്ട് ആയന്തരാളത്തെ ഉച്ചനീചവൃത്തവ്യാസാർദ്ധമായിട്ടു കല്പിപ്പു. പിന്നെ ആദിത്യബിംബഘനമധ്യത്തോട് ഭൂച്ഛായാമധ്യത്തോട് നടുവേയുള്ള സൂത്രം യാതൊന്ന് അതിന്മേൽ ഘനഭൂമധ്യവും, ഭഗോളമധ്യവുമകലുന്നു. എന്നിട്ട് ആ[2] സൂത്രം ഉച്ചനീചസൂത്രമാകുന്നത്. ഈ ഉച്ചനീചസൂത്രത്തിങ്കൽ ആദിത്യനു സദാ സ്ഥിതിയാകയാൽ ഭഗോളമധ്യം കേന്ദ്രമായിരിക്കുന്ന വൃത്തത്തിങ്കലും[3], ഘനഭൂമധ്യം കേന്ദ്രമായിരിക്കുന്ന വൃത്തത്തിങ്കലും സ്ഫുടകല[4]മെന്നത്രേ കർണ്ണഭേദമുള്ളൂ ആദിത്യന്. ചന്ദ്രനു പിന്നെ ഈ ഉച്ചനീചസൂത്രത്തിങ്കേന്നു നീക്കമുണ്ട്. അത് ആദിത്യങ്കേന്നുള്ള നീക്കമായിട്ടിരിക്കും. ആകയാൽ പ്രതിപദാദിയായി ഇഷ്ടകലാവധി[5] ഉള്ള തിഥികൾ ഉച്ചോനഗ്രഹമായിരിക്കുന്ന കേന്ദ്രമാകുന്നത്. ആകയാൽ ഭഗോളമധ്യത്തോടു ഘനഭൂമധ്യത്തോടുള്ള അന്തരാളമായിരിക്കുന്ന ഉച്ചനീചവ്യാസാർദ്ധത്തെക്കൊണ്ടും ഇഷ്ടതിഥികളുടെ[6] ഭുജാകോടി ജ്യാക്കളേക്കൊണ്ടും കൂടി ഭുജാകോടിഫലങ്ങളെ ഉണ്ടാക്കി, ഇവറ്റേയും മന്ദകർണ്ണത്തേയും കൂടി ദ്വിതീയസ്ഫുടകർണ്ണത്തെ യോജനാത്മകമായിട്ടുതാൻ കലാത്മകമായിട്ടുതാൻ ഉണ്ടാക്കൂ. പിന്നെ ഈ കർണ്ണത്തേക്കൊണ്ടു ഭുജാഫലത്തെ സംസ്കരിച്ച് ആ ഭുജാഫലത്തെ ചന്ദ്രനിലും സംസ്കരിപ്പൂ. എന്നാൽ ഘനഭൂമധ്യം കേന്ദ്രമായിരിക്കുന്ന വൃത്തത്തിങ്കലെ ചന്ദ്രസ്ഫുടമുണ്ടാകും. ഇങ്ങിനെ ശീഘ്രസ്ഫുടന്യായേന ദ്വിതീയസ്ഫുടം. ഇവിടെ ഉച്ചനീചവ്യാസാർദ്ധം നാനാരൂപമായിരിപ്പോന്ന്,

---

36.1 B. തത്പ്രകാരം
2. F. om. ആ
3. C. മധ്യകേന്ദ്രവത്തമായിരിപ്പൊന്ന്
4. F. കാല
5. F. കാലാവധി
6. B. തിഥിയുടെ

അതിന്റെ നിയമം. ഇവിടെ ഭൂച്ഛായാർക്കന്മാരിൽ കൂടിയുള്ള ഉച്ചനീചസൂത്രം യാതൊന്ന് അതിന്നു വിപരീതമായിട്ടു ഭഗോളമധ്യത്തിൽ കൂടി ഒരു സൂത്രത്തെ കല്പിപ്പൂ. ഈ സൂത്രത്തിങ്കേന്ന് ആദിത്യനുള്ളപുറത്തു ചന്ദ്രോച്ചമെങ്കിൽ അപ്പുറത്തു നീങ്ങും ഭൂമധ്യത്തിങ്കേന്നു ഭഗോളമധ്യം. അപ്പോൾ ആദിത്യങ്കൽ ഉച്ചസ്ഥാനം.

പിന്നെ ഈ കല്പിച്ച തിർയ്യക്സൂത്രത്തിങ്കേന്നു ഭൂച്ഛായയുള്ളപുറത്തു ചന്ദ്രോച്ചമെന്നിരിക്കിൽ ഭൂച്ഛായയെ നോക്കി നീങ്ങും ഘനഭൂമധ്യത്തിങ്കേന്നു ഭഗോളമധ്യം. അപ്പോൾ ഭൂച്ഛായയിങ്കൽ ഉച്ചസ്ഥാനമാകയാൽ ഇന്ദൂച്ചോനാർക്കകോടിക്കു തക്കവണ്ണം ഉച്ചനീചവ്യാസാർദ്ധത്തിന്റെ വൃദ്ധിഹ്രാസങ്ങൾ. ഇവിടേയും പിന്നെ ഇന്ദൂച്ചോനാർക്കകോടിക്കും അർക്കോനചന്ദ്രന്റെ കോടിക്കും രണ്ടിനും കൂടി മൃഗകർക്ക്യാദികൾ ഒന്നേ എങ്കിൽ അർക്കോനചന്ദ്രന്റെ കോടിഫലം മന്ദകർണ്ണത്തിങ്കൽ ധനം, അല്ലെങ്കിൽ ഋണം എന്നു നിയതം എന്നു വരും. വിക്ഷേപമുള്ളപ്പോൾ വിക്ഷേപത്തിന്റെ കോടിയിൽ സംസ്കരിക്കേണ്ടു ഈ കോടിഫലം. യാതൊരു പ്രകാരം മന്ദസ്ഫുടത്തിങ്കേന്ന് ഉണ്ടാക്കിയ വിക്ഷേപത്തെ വർഗ്ഗിച്ച്, ഇതിനെ പ്രതിമണ്ഡലകലാപ്രമിതമെങ്കിൽ മന്ദകർണ്ണവർഗ്ഗത്തിങ്കേന്ന്, മന്ദകർണ്ണവൃത്ത കലാപ്രമിതമെങ്കിൽ മന്ദകർണ്ണവൃത്തവ്യാസാർദ്ധമായിരിക്കുന്ന ത്രിജ്യയുടെ വർഗ്ഗത്തിങ്കേന്നു കളഞ്ഞു മൂലിച്ച് വിക്ഷേപകോടിയിങ്കൽ ഇതിനു സദൃശമായിരിക്കുന്ന മാനം കൊണ്ടുള്ള കോടിഫലത്തെ സംസ്കരിക്കുന്നൂ. അപ്രകാരമിവിടേയും ചന്ദ്രന്റെ പ്രഥമസ്ഫുടത്തിങ്കേന്ന് ഉണ്ടാക്കിയ വിക്ഷേപത്തെ വർഗ്ഗിച്ച് ഇതിനെ പ്രഥമകർണ്ണവർഗ്ഗത്തിങ്കേന്നു താൻ ത്രിജ്യാവർഗ്ഗത്തിങ്കേന്നു താൻ കളഞ്ഞു മൂലിച്ച് വിക്ഷേപകോടിയിങ്കൽ ദ്വിതീയസ്ഫുടകോടിഫലത്തെ സംസ്കരിപ്പൂ. ഇവിടെ ദ്വിതീയ സ്ഫുടത്തിങ്കലേ അന്ത്യഫലമാകുന്നത് ഇന്ദൂച്ചോനാർക്കകോടിജ്യാവിന്റെ അർദ്ധം. ഇതു യോജനാത്മകമായിട്ടിരിപ്പോന്ന്. ആകയാൽ ഇതിനേക്കൊണ്ട് അർക്കോനചന്ദ്രന്റെ ഭുജാകോടിജ്യാക്കളെ ഗുണിച്ച് ത്രിജ്യകൊണ്ടു ഹരിച്ചുണ്ടാകുന്ന ദ്വിതീയസ്ഫുടത്തിന്റെ ഭുജാകോടിഫലങ്ങളും യോജനാത്മകങ്ങളായിട്ടിരിപ്പോ ചിലവ[7]. ആകയാൽ വിക്ഷേപകോടിയേയും യോജനാത്മകമായി അതിങ്കൽ കോടിഫലം സംസ്കരിക്കേണം. പിന്നെ

36.7. B. F. ഇരിപ്പേന്ന്

ഇതിന്റെ വർഗ്ഗത്തിങ്കൽ ഭുജാഫലവർഗ്ഗത്തേയും കൂട്ടി മൂലിപ്പൂ. എന്നാൽ ചന്ദ്രബിംബഘനഭൂമധ്യത്തോട് ഇടയിലെ യോജനകൾ ഉണ്ടാകും. പിന്നെ ഭുജാഫലത്തെ ത്രിജ്യകൊണ്ടു ഗുണിച്ച് ഈ കർണ്ണം കൊണ്ടു ഹരിച്ചതിനെ ചന്ദ്രസ്ഫുടത്തിങ്കൽ സംസ്കരിപ്പൂ. സംസ്കാരപ്രകാരം പിന്നെ. ഇന്ദൂച്ചോനാർക്കകോടി മകരാദിയെങ്കിൽ പൂർവ്വപക്ഷത്തിങ്കൽ ചന്ദ്രങ്കേന്നു ഭുജാഫലം കളവൂ. അപരപക്ഷത്തിൽ കൂട്ടൂ. പിന്നെ കർക്ക്യാദിയിൽ പൂർവ്വപക്ഷത്തിങ്കൽ കൂട്ടൂ, അപരപക്ഷത്തിൽ കളവൂ. പിന്നെ മധ്യഗതിയെ പത്തിൽ ഗുണിച്ച് ത്രിജ്യകൊണ്ടു ഗുണിച്ച് ഈ ദ്വിതീയസ്ഫുടകർണ്ണം കൊണ്ടു ഹരിപ്പൂ. ഫലം ദ്വിതീയസ്ഫുടഗതി. ഇങ്ങനെ ദ്വിതീയസ്ഫുടപ്രകാരം. ഇതിനേക്കൊണ്ട് ഘനഭൂമധ്യത്തിങ്കൽ കേന്ദ്രമായി ചന്ദ്രബിംബഘനമധ്യത്തിങ്കൽ നേമിയായിരിക്കുന്ന വൃത്തത്തിങ്കലെ ചന്ദ്രസ്ഫുടമുണ്ടാകും. പിന്നെ ഇതിങ്കേന്നു ഭൂപൃഷ്ഠത്തിങ്കലിരിക്കുന്ന ദ്രഷ്ടാവിങ്കൽ കേന്ദ്രമായിരിക്കുന്ന വൃത്തത്തിങ്കലെ സ്ഫുടമുണ്ടാകും. നതലംബനസംസ്കാരം കൊണ്ട് അതിന്റെ പ്രകാരത്തെ ചൊല്ലുന്നു. ഇവിടെ ഛായയിങ്കൽ ചൊല്ലിയ ലംബനന്യായത്തിങ്കന്ന് കുറഞ്ഞൊന്നേ വിശേഷമുള്ളൂ. ഛായാമാർഗ്ഗത്തൂടെ ലംബിക്കുന്ന ഗ്രഹം അപക്രമമണ്ഡലാനുസാരേണ എത്ര നീങ്ങി എന്നും. ഗ്രഹസ്പൃഷ്ടരാശി കൂടവൃത്താനുസാരേണ എത്ര നീങ്ങിയെന്നും. ഇങ്ങനെ ഛായാലംബനത്തെക്കൊണ്ടു രണ്ടു പകുത്തിട്ടു നിരൂപിക്കുന്നു. അവിടെ നടേത്തേതിനു 'ലംബ'മെന്നു പേർ. അതു സ്ഫുടാന്തരമായിട്ടിരിക്കും. പിന്നത്തേതിന്നു 'നതി'യെന്നു പേർ. അതു വിക്ഷേപമായിട്ടിരിക്കും.

ഇവിടെ പ്രവഹഭ്രമണവശാൽ യാതൊരിക്കൽ അപക്രമമണ്ഡലത്തിന്റെ ഒരു പ്രദേശം ഖമദ്ധ്യത്തെ സ്പർശിക്കുന്നൂ അന്നേരത്തു വിക്ഷേപമില്ലാതെയിരിക്കുന്ന ഗ്രഹം അപക്രമമണ്ഡലത്തിങ്കൽ തന്നെ ഇരിക്കും. അന്നേരത്ത് അപക്രമമണ്ഡലം തന്നെ ദൃങ്മണ്ഡലമാകുന്നത്. എന്നിട്ട് ഛായാലംബനമാകുന്നത് അപക്രമമണ്ഡലമാർഗ്ഗത്തൂടെ ക്ഷിതിജം നോക്കി അടഞ്ഞു എന്നു തോന്നുന്നത്. പിന്നെ ദൃക്ക്ഷേപമണ്ഡലത്തിൽ ഗ്രഹമിരിക്കുന്നു എന്നിട്ട് അന്നേരത്തെ ഛായാലംബനമൊക്കെ സ്ഫുടാന്തരമായിട്ടിരിക്കും. യാതൊരിക്കൽ പിന്നെ ദൃക്ക്ഷേപമണ്ഡലത്തിൽ ഗ്രഹമിരിക്കുന്നു. അന്നേരത്ത് ഗ്രഹസ്പൃഷ്ടരാശികൂടവും

ദൃങ്മണ്ഡലവുമൊന്നേയാകയാൽ ദൃങ്മണ്ഡലമാർഗ്ഗേണ ലംബിക്കുന്ന ഛായാലംബനം അപക്രമമണ്ഡലവിപരീതമായിട്ടിരിക്കും. എന്നിട്ടു ഛായാലംബനമൊക്കെ വിക്ഷേപമായിട്ടിരിക്കും, സ്ഫുടാന്തരമൊട്ടുമില്ല[8]. യാതൊരിക്കൽ പിന്നെ ഗ്രഹസ്പൃഷ്ടരാശികൂടവും അപക്രമമണ്ഡലവും ദൃങ്മണ്ഡലവും മൂന്നും മൂന്നായിട്ടിരിക്കുന്നൂ, അന്നേരത്തു മൂന്നിന്റേയും സംപാതത്തിങ്കേന്നു ദൃങ്മണ്ഡലമാർഗ്ഗേണ [9]ലംബിക്കുന്ന ഗ്രഹം ആ ഗ്രഹസ്പൃഷ്ട[10]രാശികൂടവൃത്തമായിരിക്കുന്നതിങ്കേന്നും അപക്രമവൃത്തത്തിങ്കേന്നും[11] അകലും. അവിടെ രാശികൂടവൃത്തത്തിങ്കേന്നുള്ള അകലം സ്ഫുടാന്തരം[12], അപക്രമമണ്ഡലത്തിങ്കേന്നുള്ള അകലം വിക്ഷേപം. നടേ വിക്ഷേപമുണ്ടാക്കിയിട്ടിരിക്കിൽ വിക്ഷേപാന്തരമായിട്ടിരിക്കുമിത്.

അവിടെ രാശികൂടാപക്രമങ്ങളേക്കൊണ്ടു പദവിഭാഗം. ഇതിങ്കൽ വലിയ വൃത്തമായിട്ടു ദൃങ്മണ്ഡലത്തേയും കല്പിച്ച്, ദൃങ്മണ്ഡലത്തിലേ ഖമദ്ധ്യഗ്രഹാന്തരാളമായിരിക്കുന്ന ഛായയെ വൃത്തത്രയസംപാതത്തിങ്കൽ മൂലവും ഖമധ്യത്തിങ്കലഗ്രവുമായി കല്പിച്ച് ആ ഛായാഗ്രത്തിങ്കൽ നിന്നും എത്ര അകലമുണ്ട് അപക്രമമണ്ഡലവും ഗ്രഹസ്പൃഷ്ടരാശി[13]കൂടവൃത്തവുമെന്നറിവൂ.

അവിടെ ഖമധ്യത്തിങ്കേന്ന് അപക്രമവൃത്താന്തരാളമാകുന്നതു ദൃക്ക്ഷേപജ്യാവ്. പിന്നെ ദൃക്ക്ഷേപവൃത്തവും ഗ്രഹസ്പൃഷ്ടരാശികൂടവൃത്തവും തങ്ങളിലുള്ള യോഗം രാശികൂടങ്ങളിൽ പരമാന്തരാളം. അപക്രമമണ്ഡലത്തിങ്കൽ ഈ പരമാന്തരാളമാകുന്നതു ദൃക്ക്ഷേപലഗ്നഗ്രഹാന്തരാളം. ഇത് ഇവിടെ പ്രമാണഫലമാകുന്നത്. രാശികൂടവൃത്തത്തോടു ദൃക്ക്ഷേപലഗ്നത്തോടുള്ള അന്തരാളം ദൃക്ക്ഷേപഭാഗത്തിങ്കലേ ത്രിജ്യാവു പ്രമാണമാകുന്നത്. ഈ ദൃക്ക്ഷേപവൃത്തത്തിങ്കൽ തന്നെ ഖമധ്യരാശികൂടാന്തരാളം ദൃക്ക്ഷേപകോടിയാകുന്നത്. അത് ഇച്ഛയാകുന്നത്. ഖമധ്യത്തിങ്കേന്ന്

36.8. സ്ഫുടാന്തരാളമൊട്ടുമില്ല
9. F. adds ഗമിക്കുന്ന
10. H. സ്പൃഷ്ട
11. C. F. അപക്രമമണ്ഡലവൃത്ത
12. B. C. സ്പുടാന്തരാളം
13. H. ഗ്രഹസ്ഫുട

ഗ്രഹസ്പൃഷ്ടരാശികൂട[14]വൃത്താന്തരാളം ഇച്ഛാഫലം. ഇതിന്നു 'ദൃഗ്ഗതിജ്യാ'വെന്നു പേർ. ഈ ദൃക്ക്ഷേപദൃഗ്ഗതികൾ ഛായയ്ക്കു ഭുജാകോടികളായിരിപ്പോ ചിലവ, ഛായാകർണ്ണമായിട്ടിരിക്കും. ഈവണ്ണം വൃത്തത്രയസംപാതത്തിങ്കേന്നു മറ്റേപ്പുറത്ത് ദൃങ്മണ്ഡലഭാഗത്തിങ്കലേ ഛായാലംബനാംശമായിരിക്കുന്ന കർണ്ണത്തിന്ന് ഭുജാകോടികളായിട്ടിരിക്കും വിക്ഷേപസ്ഫുടാന്തരങ്ങൾ. ഇവിടെ ഛായാ പ്രമാണം, ദൃക്ക്ഷേപദൃഗ്ഗതികൾ പ്രമാണഫലങ്ങൾ, ഛായാലംബനമിച്ഛാ, നതിലംബനങ്ങൾ ഇച്ഛാഫലങ്ങൾ എന്നാകിലുമാം.

പിന്നെ ദൃക്ക്ഷേപദൃഗ്ഗതികളേക്കൊണ്ടുതന്നെ നതിലംബനങ്ങളെ വരുത്തുകിലുമാം. അവിടെ ഛായാ ത്രിജ്യാതുല്യയാകുമ്പോൾ ഭൂവ്യാസാർദ്ധത്തോളം ഛായാലംബനം, ഇഷ്ടച്ഛായയാൽ[15] എത്ര. എന്നപോലെ. ദൃക്ക്ഷേപദൃഗ്ഗതികൾ ത്രിജ്യാതുല്യങ്ങളാകുമ്പോൾ ഭൂവ്യാസാർദ്ധലിപ്താതുല്യങ്ങൾ നതിലംബങ്ങൾ, ഈ ഇഷ്ടദൃക്ക്ഷേപദൃഗ്ഗതികൾക്ക് എത്ര നതിലംബനങ്ങൾ എന്നാകിലുമാം.

എന്നാൽ ദൃക്ക്ഷേപദൃഗ്ഗതികളെ ഭൂവ്യാസാർദ്ധയോജനത്തെക്കൊണ്ടു ഗുണിച്ച് ദൃക്കർണ്ണയോജനം കൊണ്ടു ഹരിപ്പൂ. അവിടെ ത്രിജ്യകൊണ്ടു ഗുണിക്കയും ഹരിക്കയും ഉപേക്ഷിച്ചുകളവൂ[16], ഫലഭേദമില്ലായ്കയാൽ. ഇവിടെ ദൃക്ക്ഷേപലഗ്നത്തിങ്കേന്നു കിഴക്കു ഗ്രഹമെങ്കിൽ കിഴക്കോട്ടു താഴുകയാൽ ഭൂമധ്യസ്ഫുടത്തേക്കാൾ ഭൂപൃഷ്ഠസ്ഫുടമേറും ദൃക്ക്ഷേപലഗ്നത്തിങ്കേന്ന്[17], പടിഞ്ഞാറു ഗ്രഹമെങ്കിൽ കുറയും. പിന്നെ വിക്ഷേപം ദക്ഷിണമെങ്കിൽ തെക്കോട്ടു താഴുകയാൽ ഇവിടെ നതി ദക്ഷിണം, മറിച്ച് എങ്കിൽ ഉത്തരം. എന്നെല്ലാം യുക്തിസിദ്ധം. ഇങ്ങനെ നതിലംബനങ്ങളുടെ പ്രകാരം.

---

36. 14. D. om. ഗ്രഹസ്പൃഷ്ടരാശി
15. D,F, ഇഷ്ടഛായയ്ക്ക് എത്ര
16. B. ഉപേക്ഷിക്കാം
17. B. adds മാറി

37. ദൃക്കർണ്ണാനയനപ്രകാരം

അനന്തരം ചന്ദ്രന്നു വിക്ഷേപമുള്ളപ്പോൾ ദൃക്കർണ്ണം വരുത്തുവാനായിക്കൊണ്ടു ഛായാശങ്കുക്കളെ വരുത്തുന്നേടത്തു വിശേഷത്തെ ചൊല്ലുന്നു. ഇവിടെ ദൃക്ക്ഷേപദൃഗ്ഗതിജ്യാക്കളെ വർഗ്ഗിച്ചു കൂട്ടി മൂലിച്ചത്, ഛായാവിക്ഷേപമില്ലാത്തപ്പോൾ ഛായയുടെ കോടിശങ്കുവാകുന്നത് എല്ലായ്പ്പോഴും എന്ന് നിയതം. ഇങ്ങനെ ഛായാശങ്കുക്കളെ ഉണ്ടാക്കി ഭൂവ്യാസാർദ്ധയോജനം കൊണ്ടു ഗുണിച്ച് ത്രിജ്യകൊണ്ടു ഹരിച്ച ഫലങ്ങൾ ദൃക്കർണ്ണം വരുത്തുന്നേടത്തേക്കു ഭുജാകോടിഫലങ്ങൾ, യോജനാത്മകങ്ങൾ താനും. പിന്നെ കോടിഫലത്തെ ദ്വിതീയസ്ഫുടയോജനകർണ്ണത്തിങ്കേന്നു കളഞ്ഞ ശേഷത്തിന്റെ വർഗ്ഗത്തേയും ഭുജാഫലവർഗ്ഗത്തേയും കൂട്ടി മൂലിപ്പൂ. അതു ദൃക്കർണ്ണയോജനമാകുന്നത്.

38. വിക്ഷിപ്തചന്ദ്രന്റെ ഛായാശങ്കുക്കൾ

പിന്നെ വിക്ഷേപമുള്ളപ്പോൾ ചന്ദ്രന്റെ ഛായാശങ്കുക്കളെ വരുത്തും പ്രകാരം. ഇവിടെ അപക്രമമണ്ഡലത്തിങ്കേന്നു വിക്ഷിപ്തഗ്രഹത്തിന്റെ ഇഷ്ടവിക്ഷേപത്തോളം അകന്നിട്ട് ഒരു വൃത്തത്തെ കല്പിപ്പൂ. അപക്രമമണ്ഡലകേന്ദ്രത്തിങ്കേന്ന് ഇഷ്ടവിക്ഷേപത്തോളം അകന്നിരിക്കും കേന്ദ്രവും. യാതൊരുപ്രകാരം ഘടികാമണ്ഡലത്തിന്ന് സ്വാഹോരാത്രം, അവ്വണ്ണമിരിക്കുമിത്. ഇതിന് 'വിക്ഷേപകോടിവൃത്ത'മെന്നുപേർ. ഇതിങ്കൽ ഗ്രഹസ്പൃഷ്ടരാശികൂടവൃത്തം സ്പർശിക്കുന്നേടത്തിരിക്കും ഗ്രഹം. ഇവിടെ നടെ ഉദയാസ്തലഗ്നങ്ങളാകുന്ന അപക്രമക്ഷിതിജസംപാതങ്ങളിലും ഖമധ്യത്തിങ്കലും സ്പർശിച്ചിരിക്കുന്ന ലഗ്നസമമണ്ഡലം, ദൃക്ക്ഷേപമണ്ഡലം, ക്ഷിതിജം എന്നിവ മൂന്നു വൃത്തങ്ങളെക്കൊണ്ടു ഗോളവിഭാഗം കല്പിച്ച് അവിടെ വലിയ വൃത്തമായിട്ട് അപക്രമമണ്ഡലത്തേയും[1] കല്പിപ്പൂ. ആ അപക്രമമണ്ഡലത്തിന്ന് ലഗ്നസമമണ്ഡലത്തോടുള്ള പരമാന്തരാളം ദൃക്ക്ഷേപജ്യാവാകുന്നത്. ക്ഷിതി

38.1. H. മണ്ഡലത്തേയും

ജാപക്രമങ്ങളുടെ പരമാന്തരാളം ദൃക്ക്ഷേപകോടി. ഇതു പ്രമാണഫലം. ത്രിജ്യാവു പ്രമാണം. ക്ഷിതിജാപക്രമയോഗത്തിങ്കേന്നു ഗ്രഹത്തോടുള്ള അന്തരാളം അപക്രമമണ്ഡലത്തിങ്കലേത് ഇച്ഛ. ഗ്രഹത്തിങ്കേന്നു ക്ഷിതിജാന്തരാളം ഇച്ഛാഫലം. ഇതു വിക്ഷിപ്തഗ്രഹത്തിന്റെ ശങ്കുവാകുന്നത്. ദൃക്ക്ഷേപദൃഗ്ഗതിജ്യാവർഗ്ഗങ്ങളെ കൂട്ടി മൂലിച്ചത് ഛായയാകുന്നത്.

പിന്നെ വിക്ഷേപകോടിവൃത്തത്തിങ്കലെ ഗ്രഹത്തിന്റെ ശങ്കുച്ഛായകൾക്കുള്ള വിശേഷം. ഇവിടെ ലഗ്നസമമണ്ഡലവും അപക്രമമണ്ഡലവുമുള്ള പരമാന്തരാളമൊന്നായിട്ടിരിക്കുന്ന ഖമധ്യദൃക്ക്ഷേപലഗ്നാന്തരാളദൃക്ക്ഷേപവൃത്തഭാഗം ദൃക്ക്ഷേപമാകുന്നത്. പിന്നെ ദൃക്ക്ഷേപലഗ്നത്തിങ്കേന്നു വിക്ഷേപകോടിവൃത്താന്തരാളം ദൃക്ക്ഷേപവൃത്തത്തിങ്കലേത് വിക്ഷേപം. വിക്ഷേപദൃക്ഷേപങ്ങളുടെ യോഗംതാനന്തരം താൻ ചെയ്ത ഖമധ്യത്തിങ്കേന്നു വിക്ഷേപകോടിവൃത്തത്തിന്റെ അന്തരാളം ദൃക്ക്ഷേപവൃത്തത്തിങ്കലേത്. ഇതിന്നു 'നതി'യെന്നു പേർ. ഇതിന്റെ കോടി വിക്ഷേപകോടി. വൃത്തക്ഷിതിജങ്ങളുടെ പരമാന്തരാളം ദൃക്ക്ഷേപവൃത്തഭാഗം. ഇതിന്നു 'പരശങ്കു'വെന്നു പേർ. ഇവിടെ യാതൊരു പ്രകാരം അക്ഷാപക്രമങ്ങളുടെ യോഗാന്തരങ്ങളേക്കൊണ്ടും ലംബകാപക്രമങ്ങളുടെ യോഗാന്തരങ്ങളെക്കൊണ്ടും അന്തരാളങ്ങളെക്കൊണ്ടും ദക്ഷിണോത്തരവൃത്തത്തിങ്കലേ മദ്ധ്യാഹ്നച്ഛായാശങ്കുക്കൾ ഉണ്ടാക്കുന്നു, അവണ്ണം ദൃക്ക്ഷേപവൃത്തത്തിങ്കൽ ദൃക്ക്ഷേപവിക്ഷേപയോഗാന്തരങ്ങളേക്കൊണ്ടും ദൃക്ക്ഷേപകോടി വിക്ഷേപങ്ങളുടെ യോഗാന്തരങ്ങളേക്കൊണ്ടും നതിയും പരശങ്കുവും വരും. ഇവിടെ ലഗ്നഗ്രഹാന്തരജ്യാവ് ഇച്ഛയാക്കി കല്പിച്ചിരുന്നതിനെ പ്രമാണമാകുന്ന ത്രിജ്യയിങ്കേന്നു കളവൂ. ശേഷത്തെ ഇച്ഛയാക്കി കല്പിക്കാം. അപ്പോൾ പ്രമാണേച്ഛകളുടെ ഫലാന്തരം ഇച്ഛാഫലമായിട്ടുണ്ടാകും.

ഇവിടെ ഗ്രഹസ്പൃഷ്ടരാശികൂടവൃത്തവും ദൃക്ക്ഷേപവൃത്തവും [2]തങ്ങളിലുള്ള അന്തരാളത്തിങ്കലേ അപക്രമമണ്ഡലഭാഗത്തിന്റെ[3] ശരത്തേ

---

38. 2. B,H. തങ്ങളിലന്തരാളത്തിങ്കലെ
3. B. അപക്രമവൃത്തഭാഗത്തിന്റെ

നടേ വരുത്തുന്നത്. പിന്നെ ഈ ശരത്തെ വിക്ഷേപകോടികൊണ്ടു ഗുണിച്ച് ത്രിജ്യകൊണ്ടു ഹരിച്ചാൽ ഫലം ഗ്രഹസ്പൃഷ്ടരാശികൂടവൃത്തവും ദൃക്ക്ഷേപവൃത്തവും തങ്ങളിലുള്ള അന്തരാളത്തിങ്കലെ വിക്ഷേപകോടി വൃത്തത്തിങ്കലേ ശരമായിട്ടു വരും. പിന്നെ വിക്ഷേപകോടിവൃത്തത്തിങ്കലേ ശരത്തെ ദൃക്ക്ഷേപകോടികൊണ്ടു ഗുണിച്ചു ത്രിജ്യകൊണ്ടു ഹരിച്ച ഫലത്തെ പരശങ്കുവിങ്കേന്നു കളവൂ. ശേഷം വിക്ഷേപകോടിവൃത്തത്തിങ്കൽ നില്ക്കുന്ന ഗ്രഹത്തിന്റെ ഇഷ്ടശങ്കു. ഇവിടെ പരശങ്കുവിനേക്കൊണ്ടു ഗുണിക്കിൽ ത്രിജ്യകൊണ്ടു ഹരിക്ക യോഗ്യമല്ല. ക്ഷിതിജോന്മണ്ഡലാന്തരാള സംസ്കൃതമായിട്ടിരിക്കുന്ന വിക്ഷേപകോടിയെക്കൊണ്ടു ഹരിക്ക യോഗ്യമാകുന്നത്. യാതൊരു പ്രകാരം സ്വാഹോരാത്രവൃത്തത്തിങ്കലേ ഉന്നതജ്യാവിനേത്താൻ നതജ്യാവിന്റെ ശരത്തേത്താൻ ലംബകത്തെക്കൊണ്ടു ഗുണിച്ചു ത്രിജ്യകൊണ്ടു ഹരിച്ച ഫലം ഇഷ്ടശങ്കുതാൻ മദ്ധ്യാഹ്നേഷ്ടശംക്വന്തരം താൻ ആയിട്ടു വരുന്നു. ഇവിടെ മധ്യാഹ്നശങ്കുവിനേക്കൊണ്ടു ഗുണിക്കിൽ ത്രിജ്യകൊണ്ടല്ല ഹരിക്കേണ്ടു. ക്ഷിതിജത്തിന്നു മീത്തേ സ്വാഹോരാത്രഭാഗമായി സ്വാഹോരാത്ര വ്യാസാർദ്ധത്തിങ്കൽ ക്ഷിതിജ്യാ സംസ്കരിച്ചിരിക്കുന്നതിനെക്കൊണ്ടു ഹരിക്കേണ്ടു, അവ്വണ്ണമിവിടെ ലംബകസ്ഥാനീയമായിരിക്കുന്നത് ദൃക്ക്ഷേപകോടി, മധ്യാഹ്നശങ്കുസ്ഥാനീയമാകുന്നതു പരശങ്കു.

ഇവിടെ ഘടികാമണ്ഡലത്തിന്റെ ചരിവുപോലെ സ്വാഹോരാത്രങ്ങളുടെ ചരിവ്, അപക്രമവൃത്തത്തിന്റെ[5] ചരിവുപോലെ വിക്ഷേപകോടിവൃത്തത്തിന്റെ ചെരിവ്[6]. എന്നിട്ടു തുല്യസ്വഭാവങ്ങളാകയാൽ ന്യായസാമ്യമുണ്ട്. ഇങ്ങനെ ശങ്കു വരും.

അനന്തരം ഛായാ. അവിടെ വിക്ഷേപദൃക്ക്ഷേപയോഗം താനന്തരംതാനാകുന്നതു യാതൊന്ന് അതു വിക്ഷേപകോടി വൃത്തത്തിങ്കന്നു ലഗ്നസമമണ്ഡലത്തോടുള്ള അന്തരാളം, ദൃക്ക്ഷേപവൃത്ത ത്തിങ്കലേത്. ഇതിന്ന് 'നതി' യെന്നു പേരാകുന്നു.

---

38. 4. H. ഹരിക്കയല്ല, യോഗ്യമാകുന്നത്
5. C. അപക്രമണ്ഡലത്തിന്റെ
6. B. ന്റെയും ചരിവ്

പിന്നെ ഗ്രഹത്തോടു ദൃക്ക്ഷേപവൃത്തത്തോടുള്ള[7] അന്തരാളം വിക്ഷേപകോടി വൃത്തത്തിങ്കലേത് യാതൊന്ന് ഇതിന്റെ ജ്യാബാണങ്ങളെ ഉണ്ടാക്കൂ. ദൃക്ക്ഷേപലഗ്നചന്ദ്രാന്തരജ്യാബാണങ്ങളെ വിക്ഷേപകോടിയെക്കൊണ്ട് ഗുണിച്ച് ത്രിജ്യകൊണ്ടു ഹരിച്ച ഫലങ്ങളവ[8]. ഇതിൽ ബാണത്തെ മുമ്പിലുണ്ടാക്കി പിന്നെ ഈ ബാണത്തെ ദൃക്ക്ഷേപജ്യാവിനെക്കൊണ്ടു ഗുണിച്ച് ത്രിജ്യകൊണ്ടു ഹരിപ്പൂ. ഫലം ബാണമൂലത്തിങ്കേന്നു ബാണാഗ്രത്തിന്ന് എത്ര ചരിവുണ്ട് എന്നതായിട്ടു വരും. ഇതിനെ ദിഗ്ഭേദസാമ്യത്തിനു തക്കവണ്ണം മുമ്പിൽ ചൊല്ലിയ നതിയിങ്കേന്നു കൂട്ടുകതാൻ കളകതാൻ ചെയ്യൂ. അതു ബാണമൂലത്തിങ്കേന്നു ഖമധ്യത്തോടുള്ള അന്തരാളമുണ്ടാകും. ഇവിടെ ചൊല്ലിയ ഭുജാമൂലത്തിങ്കേന്നാകിലുമാം. ഈ ന്യായം കൊണ്ടു തന്നെ ഭുജാഗ്രത്തിങ്കലേ ഗ്രഹത്തിങ്കേന്നു ലഗ്നസമമണ്ഡലത്തോടുള്ള അന്തരാളം ആകുന്നത് ഇതുതന്നെ എന്നുവരും. ഇതിന്നു 'ബാഹു' എന്നുപേർ. പിന്നെ ഇതിനെയും മുമ്പിൽചൊല്ലിയ ഭുജയേയും വർഗ്ഗിച്ചു കൂട്ടി മൂലിപ്പൂ. എന്നാൽ ഛായ ഉണ്ടാകും. ഇങ്ങനെ ശങ്കുച്ഛായകളെ വരുത്തുംപ്രകാരം. ഇതിലൊന്നിനെ നടേ ഇവ്വണ്ണമുണ്ടാക്കി ത്രിജ്യാവർഗ്ഗാന്തരമൂലം കൊണ്ട് മറ്റേതിനേയുമുണ്ടാക്കാം.

[ഗണിതയുക്തിഭാഷയിൽ ഛായാപ്രകരണം എന്ന പതിനൊന്നാമദ്ധ്യായം സമാപ്തം]

38. 7. F. adds. വിക്ഷേപത്തോടുള്ള
8. F. ങ്ങളിവ

അദ്ധ്യായം പന്ത്രണ്ട്

ഗ്രഹണം

1. ഗ്രാഹ്യബിംബവും ഗ്രഹണകാലവും

ഇങ്ങനെ ചന്ദ്രന്റെ ശങ്കുച്ഛായകളെ ഉണ്ടാക്കി, ഇവറ്റേക്കൊണ്ട് ദൃക്കർണ്ണയോജനമുണ്ടാക്കി, ദൃക്കർണ്ണയോജനകൊണ്ട് ലംബനലിപ്തയും ഉണ്ടാക്കി, ആദിത്യന്റേയും ചന്ദ്രന്റേയും സ്ഫുടത്തിൽ തന്റെ തന്റെ ലംബനലിപ്തകളെ സംസ്കരിച്ചാൽ യാതൊരിക്കൽ സ്ഫുടസാമ്യം വരുന്നൂ അപ്പോൾ ഗ്രഹണമധ്യകാലം. പിന്നെ ദൃഗ്ഗതിയിങ്കേന്നു തന്നെ ലംബനകാലം വരുത്തുകയുമാം. അവിടെ ദൃഗ്ഗതി ത്രിജ്യാതുല്യയാകുമ്പോൾ നാലുനാഴിക ലംബനം, ഇഷ്ടദൃഗ്ഗതിക്ക് എത്ര നാഴിക ലംബനമെന്ന് ത്രൈരാശികം. ഇവിടെ ദൃക്ക്ഷേപദൃഗ്ഗതി ത്രിജ്യാതുല്യങ്ങളാകുമ്പോൾ ഭൂവ്യാസാർദ്ധതുല്യങ്ങൾ ഗതിലംബനയോജനങ്ങൾ എന്നു നിയതം. പിന്നെ മധ്യയോജനകർണ്ണത്തിനു ത്രിജ്യാതുല്യങ്ങൾ കലകളെന്നും നിയതം. എന്നാൽ ഇങ്ങനെ ഉണ്ടാകുന്ന ലംബനലിപ്തകളെ സ്ഫുടഗതികൊണ്ടു ഗുണിച്ച് മധ്യഗതികൊണ്ട് ഹരിപ്പൂ. എന്നാൽ ലംബനം ഭഗോളകലകളാം. എന്നാൽ മധ്യയോജനകർണ്ണവും മദ്ധ്യഗതിയും തങ്ങളിൽ ഗുണിച്ച് ഭൂവ്യാസാർദ്ധയോജനകൊണ്ടു ഹരിപ്പൂ. ഫലം 'അസൌ സകാമഃ' (51770) എന്ന്. പിന്നെ ദൃക്ക്ഷേപദൃഗ്ഗതികളെ സ്ഫുടഗതികൊണ്ടു ഗുണിച്ച് 'അസൌ സകാമഃ' എന്നതുകൊണ്ടു ഹരിപ്പൂ. ഫലങ്ങൾ ദൃക്ക്ഷേപദൃഗ്ഗതികൾ എന്നുമുണ്ടാക്കാം. ഇങ്ങനെ ലംബനകാലത്തെ ഉണ്ടാക്കി പർവ്വാന്തത്തിൽ സംസ്കരിപ്പൂ. പിന്നെ അന്നേരത്തെ ദൃക്ക്ഷേപലഗ്നത്തേയും ഗ്രഹത്തേയുമുണ്ടാക്കീട്ട് ലംബനകാലത്തെ ഉണ്ടാക്കി പർവ്വാന്തത്തിങ്കൽ സംസ്കരിപ്പൂ. ഇങ്ങനെ അവിടെ അവിശേഷിപ്പൂ. ഇവിടെ എത്ര ലംബനലിപ്തകൾ എന്നറിഞ്ഞേ സമലിപ്തകാലം എപ്പോളെന്നറിയാവൂ. സമലിപ്തകാലമറിഞ്ഞേ ലംബനലിപ്തയറിയാവൂ, എന്നിട്ട് അവിശേഷിക്കേണ്ടുന്നു.

ഇങ്ങനെ ഉണ്ടാക്കിയ കാലത്തിങ്കൽ സ്ഫുടാന്തരമില്ലായ്കയാൽ ചന്ദ്രാർക്കന്മാർകളിൽ കിഴക്കുപടിഞ്ഞാറന്തരമില്ല, നതിവിക്ഷേപങ്ങൾക്കു തക്കവണ്ണം തെക്കുവടക്കന്തരമേയുള്ളു. അത് എത്ര എന്നറിഞ്ഞിട്ട്, അതിനെ ബിംബാർദ്ധങ്ങളുടെ യോഗത്തിങ്കേന്നു കളവൂ. ശേഷം ഗ്രഹിച്ചിരിക്കുന്ന പ്രദേശം.

പിന്നെ ബിംബഘനമധ്യാന്തരം ബിംബയോഗാർദ്ധത്തോളം ഉണ്ടാകുമ്പോൾ ബിംബനേമികൾ തങ്ങളിൽ സ്പർശിച്ചിരിക്കും. അപ്പോൾ ഗ്രഹണത്തിന്റെ ആരംഭാവസാനങ്ങൾ. ഇതിൽ ബിംബാന്തരമേറുമ്പോൾ ഗ്രഹണമില്ല, നേമിസ്പർശം വരായ്കയാൽ.

2. ഇഷ്ടഗ്രഹണകാലം

പിന്നെ ഇഷ്ടകാലത്തിങ്കൽ ലംബനം സംസ്കരിച്ചിരിക്കുന്ന ചന്ദ്രാർക്കന്മാരുടെ സ്ഫുടാന്തരത്തേയും സ്ഫുടവിക്ഷേപത്തേയും വർഗ്ഗിച്ചു കൂട്ടി മൂലിച്ചത് തല്ക്കാലത്തിങ്കലേ ബിംബഘനമധ്യാന്തരാളം. ഇതിനെ ബിംബലിപ്തകളുടെ യോഗാർദ്ധത്തിങ്കേന്നു കളവൂ. ശേഷിച്ചത് അന്നേരത്തെ 'ഗ്രഹണപ്രദേശം'. ഇങ്ങനെ ഇന്ന നേരത്ത് ഇത്ര ഗ്രഹണമെന്നറിയും പ്രകാരം.

പിന്നെ ഗ്രഹിച്ചിരിക്കുന്ന ഭാഗം ഇത്രയാകുമ്പോൾ കാലമേത് എന്നറിവാൻ ഗ്രഹിച്ച ഭാഗത്തെ ബിംബയോഗാർദ്ധത്തിങ്കേന്നു കളഞ്ഞത് ബിംബഘനമധ്യാന്തരാളമാകുന്നത്. ഇതിന്നു 'ബിംബാന്തര'മെന്നു പേർ. ഇതിനേക്കൊണ്ടു കാലത്തെ വരുത്തൂ. ബിംബാന്തരവർഗ്ഗത്തിങ്കേന്നു സ്ഫുടവിക്ഷേപവർഗ്ഗത്തെ കളഞ്ഞു മൂലിച്ചതു സ്ഫുടാന്തരമാകുന്നത്. പിന്നെ ദിനഗത്യന്തരത്തിന്ന് അറുപതു നാഴിക, സ്ഫുടാന്തരത്തിന് എത്ര നാഴിക എന്ന് കാലത്തെ വരുത്തി, പർവ്വാന്തകാലത്തിങ്കൽ സംസ്കരിപ്പൂ. പിന്നെ അക്കാലത്തേക്കു സ്ഫുടവിക്ഷേപത്തെ വരുത്തി വർഗ്ഗിച്ച് ഇഷ്ടബിംബയോഗാർദ്ധവർഗ്ഗത്തിങ്കേന്നു കളഞ്ഞു മൂലിച്ചത് 'സ്ഫുടാന്തര'മാകുന്നത്. പിന്നെ ഇങ്ങനെ അവിശേഷിച്ചു വന്നത് ഇഷ്ടഗ്രഹണത്തിന്റെ കാലമാകുന്നത്. പിന്നെ ഗ്രഹണമധ്യകാലത്തിങ്കേന്നു

മുമ്പിലും പിമ്പിലും ബിംബയോഗാർദ്ധത്തോളം ബിംബാന്തരമാകുമ്പോളേക്ക് ഈവണ്ണം കാലത്തെ വരുത്തൂ. നതിലംബനവിക്ഷേപങ്ങളെ അവിശേഷിച്ച് അവ സ്പർശമോക്ഷകാലങ്ങളാകുന്നവ.

ഇവിടെ ഗ്രഹണത്തിങ്കൽ സ്ഫുടസാമ്യം വരുന്ന കാലം നടേ അറിയേണ്ടുവത്. അവിടെ ആദിത്യങ്കൽ നിന്ന് ചന്ദ്രൻ ആറു രാശി അകന്നേടത്തു പൌർണ്ണമാസ്യാന്തം. ആ ചന്ദ്രനെ ഭൂച്ഛായ മറയ്ക്കുന്നതു 'ചന്ദ്രഗ്രഹണ'മാകുന്നത്.

അമാവാസ്യാന്ത്യത്തിങ്കൽ ചന്ദ്രൻ സൂര്യനെ മറയ്ക്കുന്നത് സൂര്യഗ്രഹണമാകുന്നത്. അവിടെ ആദിത്യാസ്തമയത്തിങ്കൽ അടുത്ത് ഗ്രഹണങ്ങളാലൊന്നെങ്കിൽ അവിടേക്കു ചന്ദ്രാർക്കന്മാരേ വരുത്തൂ. ഉദയത്തിനടുത്തെങ്കിലവിടേക്കു[1] വരുത്തൂ. അവിടെ ചന്ദ്രൻ ഏറുകിൽ മേൽമേൽ അകലമേറിയേറി വരും. ചന്ദ്രസ്ഫുടം കുറകിൽ മേൽമേൽ അണവു വരും. പിന്നെ ഗത്യന്തരത്തേക്കൊണ്ടു യോഗകാലത്തെ വരുത്തൂ[2].

3. ബിംബാന്തരാനയനം

അനന്തരം ബിംബാന്തരങ്ങളെ[1] വരുത്തും പ്രകാരം. അവിടെ അർക്കചന്ദ്രതമസ്സുകളുടെ ബിംബങ്ങൾ ഭൂമിയോടണയുന്നേരത്തു വലുത് എന്നു തോന്നും. അകലുന്നേരത്തു ചെറുതെന്നു തോന്നും. ഖഭൂമ്യന്തരകർണ്ണത്തിന്റെ വലിപ്പത്തിനു തക്കവണ്ണം ബിംബത്തിന്റെ അകലം. ഭൂമിയിങ്കേന്ന് അകലുമ്പോൾ ചെറുപ്പം[2]. കർണ്ണത്തിന്റെ വലുപ്പത്തിനു തക്കവണ്ണം ബിംബത്തിന്റെ ചെറുപ്പം. ആകയാൽ കർണ്ണത്തേക്കൊണ്ടു ബിംബത്തെ വരുത്തുന്നേടത്ത് വിപരീതത്രൈരാശികം വേണ്ടുവത്. അവിടെ ബിംബകലകൾക്കു പ്രതിക്ഷണം ഭേദമാകുന്നു. ബിംബയോജന എല്ലായ്പ്പോഴും ഒന്നുതന്നെ. അവിടെ സ്ഫുടയോജനകർണ്ണത്തിങ്കൽ ത്രിജ്യാതുല്യങ്ങൾ കലകൾ, ബിംബയോജനത്തിങ്കൽ എത്രയെന്നു

2. 1. B. ഉദയത്തേക്കു
   2. F reads അനന്തരം അന്തരത്തെക്കൊണ്ടു യോഗഫലത്തെ
3. 1. B. ബിംബാന്തരകല
   2. H. വലുപ്പം

ത്രൈരാശികമാകുന്നത്. അവിടെ അർക്കചന്ദ്രബിംബങ്ങളുടെ യോജനവ്യാസങ്ങളെ[3] ത്രിജ്യകൊണ്ടു ഗുണിച്ചു യോജനാത്മകമായിരിക്കുന്ന ഖഭൂമ്യന്തരകർണ്ണം[4] കൊണ്ടു ഹരിപ്പൂ. ഫലം കലാത്മകമായിരിക്കുന്ന ബിംബവ്യാസം[5]. ഇവിടെ ദൃക്കർണ്ണം കൊണ്ടു ഹരിക്കേണ്ടു വ്യസ്തത്രൈരാശികമാകയാൽ. [വ്യസ്തത്രൈരാശികഫലമിച്ഛാഭക്തം പ്രമാണഫലഘാത] ബ്രഹ്മസ്ഫുടസിദ്ധാന്തം, ഗണിത, 11) എന്നാണല്ലോ വിധി.

4. ബിംബമാനം

അവിടെ തേജോരൂപിയായി[1] ഉരുണ്ടു പെരികെ വലിയൊന്നായിട്ടിരിപ്പോന്ന് ആദിത്യബിംബം[2]. ഇതിനേക്കാൾ ചെറുതായിട്ടിരിക്കുന്നൊന്നു[3] ഭൂബിംബം. ഇതിന്ന് അർക്കാഭിമുഖമായിരിക്കുന്ന[4] പാതി പ്രകാശമായിരിക്കും. മറ്റേപ്പാതി തമസ്സായിട്ടിരിക്കും. ഇത് 'ഭൂച്ഛായ'യാകുന്നത്[5]. ഇതു ചുവടു വലുതായി അഗ്രം കൂർത്തിരിപ്പൊന്ന്[6]. അവിടെ ആദിത്യബിംബം വലുതാകയാൽ ഭൂമീടെ പുറമേ പോകുന്ന രശ്മികൾ ആദിത്യബിംബനേമീങ്കലേവ, രശ്മികളൊക്കെ തങ്ങളിൽ കൂടും. അവിടെ ഭൂച്ഛായയുടെ അഗ്രം. ഇതിന്ന് ആദിയിങ്കൽ ഭൂവ്യാസാർദ്ധത്തോടു തുല്യം വ്യാസം. പിന്നെ ക്രമത്താലെ ഉരുണ്ടു കൂർത്തിരിപ്പോന്ന്. ആദിത്യന്റെ നേമിയിങ്കലെ രശ്മികൾ ഭൂപാർശ്വത്തിങ്കൽ സ്പർശിച്ചു ഭൂമീടെ മറുപുറത്തു തങ്ങളിൽ കൂടും. അവിടെ ആദിത്യങ്കേന്നു ഭൂമി ഖഭൂമ്യന്തരകർണ്ണയോജനത്തോളമകലത്ത്. ഈ അകലത്തിനു ഭൂവ്യാസത്തോളമണഞ്ഞു ബിംബനേമീങ്കേന്നു പുറപ്പെട്ടു രശ്മികൾ. എന്നിട്ട് അർക്കഭൂവ്യാസാന്തരത്തോളം സംകുചിതമാവാൻ യോജനാകർണ്ണത്തോളം

---

3. 3. H വ്യാസാർദ്ധങ്ങളെ
   4. F. അന്തരാളകർണ്ണം
   5. B. കലാത്മകബിംബവ്യാസം
4. 1.. B.C.E. രൂപമായി
   2. B. വലിയൊരു വസ്തു സൂര്യബിംബം
   3. B. ചെറിയൊന്ന്
   4. B. മുഖമായ
   5. B. ഇതു ഭൂച്ഛായ
   6. B. കൂർത്തിരിക്കും

അകലം, ഭൂവ്യാസത്തോളം സംകുചികതമാവാൻ എത്ര അകലമെന്നു ഭൂച്ഛായയുടെ നീളമുണ്ടാകും. പിന്നെ ഭൂച്ഛായയ്ക്ക് അഗ്രത്തിങ്കേന്നു മൂലത്തിങ്കൽ ഭൂവ്യാസത്തോളം യോജനവ്യാസം, അഗ്രത്തിങ്കേന്നു ചന്ദ്രമാർഗ്ഗത്തിങ്കലോളം ചെന്നേടത്ത് എത്ര ഭൂച്ഛായായോജനവ്യാസമെന്ന് ചന്ദ്രകർണ്ണം ഊനമായിരിക്കുന്ന[7] ഭൂച്ഛായാദൈർഘ്യത്തെ ഭൂവ്യാസം കൊണ്ടു ഗുണിച്ചു ഭൂച്ഛായാദൈർഘ്യം കൊണ്ടു ഹരിപ്പൂ. ഫലം ചന്ദ്രമാർഗ്ഗത്തിങ്കലേ ഭൂച്ഛായായോജനവ്യാസം. ഇതിന്നു ചന്ദ്രനെപ്പോലെ ലിപ്താവ്യാസത്തെ വരുത്തൂ. ഇങ്ങനെ ഗ്രാഹ്യഗ്രാഹകബിംബങ്ങളെ വരുത്തും പ്രകാരം. ഈ ബിംബങ്ങളേക്കൊണ്ടു മുമ്പിൽ ചൊല്ലിയ സ്പർശമധ്യേഷ്ടഗ്രഹണങ്ങളെ അറിയേണ്ടൂ.

5. ഗ്രഹണാരംഭവും സംസ്ഥാനവും

അനന്തരം ഏതുപുറത്തു ഗ്രഹണം തുടങ്ങുന്നൂ, എങ്ങനെ ഇഷ്ടകാലത്തിങ്കൽ സംസ്ഥാനം എന്നതിനേയും അറിയും പ്രകാരം. അവിടെ സൂര്യഗ്രഹണം തുടങ്ങുന്നേരത്തു ചന്ദ്രൻ പടിഞ്ഞാറേ പുറത്തിന്നു കിഴക്കോട്ടു നീങ്ങിട്ട് ആദിത്യബിംബത്തിന്റെ പടിഞ്ഞാറേപ്പുറത്തു നേമിയിങ്കൽ ഒരിടം മറയും. അത് എവിടം എന്നു നിരൂപിക്കുന്നത്. അവിടെ ചന്ദ്രവിക്ഷേപമില്ല[1] എന്നിരിക്കുമ്പോൾ ചന്ദ്രബിംബഘനമധ്യത്തിങ്കലും ആദിത്യബിംബഘനമധ്യത്തിങ്കലും കൂടി സ്പർശിച്ചിരുന്നൊന്ന് അപക്രമമണ്ഡലം. അവിടെ ആദിത്യബിംബഘനമധ്യത്തിങ്കേന്നു തന്റെ പടിഞ്ഞാറു പാർശ്വത്തിങ്കൽ യാതൊരിടത്ത് അപക്രമമണ്ഡലം പുറപ്പെടുന്നു, അവിടം വിക്ഷേപമില്ലാത്ത ചന്ദ്രന്റെ ബിംബം കൊണ്ടു നടേ മറയുന്നത്. അവിടെ ആദിത്യന്റെ സ്വാഹോരാത്രവും തല്ക്കാലസ്വാഹോരാത്രവൃത്തവും ബിംബഘനമധ്യത്തിങ്കൽ സ്പർശിച്ചിരിപ്പൊന്ന്. അതു നിരക്ഷദേശത്തിങ്കൽ നേരെകിഴക്കു പടിഞ്ഞാറായിട്ടിരിക്കുന്നൊന്ന്. ആകയാൽ അവിടെ നേരേ പടിഞ്ഞാറു പുറത്തു സ്വാഹോരാത്രവൃത്തത്തിന്റെ പുറപ്പാട്.

---

4. 7. H.കർണ്ണമായിരിക്കുന്ന
5. 1. B. C. D ചന്ദ്രന് വിക്ഷേപം ഇല്ല

6. അയനവലനം

പിന്നെ[1] സ്വാഹോരാത്രവൃത്തത്തിങ്കേന്ന് അപക്രമവൃത്തത്തിനു ചരിവുണ്ടാകയാൽ, നേരേ പടിഞ്ഞാറേ പുറത്തിന് ഒട്ടു തെക്കുതാൻ വടക്കുതാൻ നീങ്ങിയേടത്ത് അപക്രമവൃത്തത്തിന്റെ പുറപ്പാട്. ആകയാൽ ആദിത്യബിംബത്തിന്റെ[2] നേരേ പടിഞ്ഞാറേപ്പുറത്തൂന്ന് അത്ര നീങ്ങിയേടത്തു അന്നേരത്ത് ഗ്രഹണസ്പർശം. ഈ നീക്കത്തിന് 'അയനവലന'മെന്നു പേർ.

പിന്നെ ഇതെത്രയെന്നറിവാൻ. അവിടെ അപക്രമവൃത്തത്തിങ്കലേ ദക്ഷിണായനാന്തം ദക്ഷിണോത്തരവൃത്തത്തിങ്കൽ സ്പർശിച്ചു മധ്യലഗ്നമായിട്ടിരിക്കുമാറ് പൂർവ്വവിഷുവത്ത, ഉദയലഗ്നമാകുമാറായിട്ട് ദക്ഷിണായനാന്തത്തിങ്കേന്ന് ഒരു രാശി ചെന്നേടത്തു പൂർവ്വകപാലത്തിങ്കൽ ആദിത്യൻ, ഇങ്ങനെ കല്പിച്ചിട്ടു നിരൂപിക്കുന്നു. അവിടെ അപക്രമവൃത്തവും സ്വാഹോരാത്രവൃത്തവും തങ്ങളിലുള്ള സംപാതം ആദിത്യബിംബഘനമധ്യത്തിങ്കൽ. അവിടുന്നു നേരേ പടിഞ്ഞാറോട്ടു സ്വാഹോരാത്രവൃത്തത്തിന്റെ പുറപ്പാട്. അവിടന്ന് ഒട്ടു തെക്കു നീങ്ങീട്ട് അപക്രമവൃത്തത്തിന്റെ പുറപ്പാട്. ഈ അന്തരം എത്ര എന്നറിയേണ്ടുവത്. അവിടെ ഘനഭൂമധ്യത്തിങ്കേന്നു ദക്ഷിണോത്തരവൃത്തവും അപക്രമവൃത്തവും തങ്ങളിലുള്ള സംപാതത്തിങ്കൽ അഗ്രമായിരിക്കുന്ന വ്യാസാർദ്ധത്തിങ്കൽ ശരമായിരിപ്പോ ചിലവ അപക്രമവൃത്തത്തിങ്കലേ കോടിജ്യാക്കളാകയാൽ ആ സൂത്രത്തിങ്കൽ കോടിജ്യാക്കളുടെ മൂലങ്ങൾ.

പിന്നെ ദക്ഷിണോത്തരാപക്രമവൃത്തങ്ങളുടെ സംപാതത്തിങ്കേന്നു തുടങ്ങീട്ട് ആദിത്യബിംബഘനമധ്യത്തിലഗ്രമായിട്ട് ഇരിക്കുന്ന കോടിചാപത്തിന് ഒരു ജ്യാവിനെ കല്പിപ്പൂ. പിന്നെ ആദിത്യബിംബത്തിന്റെ പടിഞ്ഞാറേ പാർശ്വത്തിങ്കൽ അപക്രമവൃത്തം പുറപ്പെടുന്നേടത്ത് അഗ്രമായിട്ട് ഒരു കോടിജ്യാവിനെ കല്പിപ്പൂ. എന്നാൽ ഇവ രണ്ടിന്റേയും മൂലം മധ്യലഗ്നത്തിങ്കലഗ്രമായിരിക്കുന്ന വ്യാസസൂത്രത്തിങ്കൽ സ്പർശിക്കും.

---

6. 1. ഈ
2. സൂര്യബിംബത്തിന്

അവിടെ ബിംബഘനമധ്യത്തിങ്കലഗ്രമായിരിക്കുന്ന [3]കീഴേനേമിയെ സ്പർശിക്കും. വ്യാസസൂത്രത്തിങ്കലേ കോടിമൂലാന്തരം നേമീങ്കലഗ്രമായിരിക്കുന്ന മീത്തേ സ്പർശിക്കും. ഈ കോടിജ്യാമൂലാന്തരം കോടിഖണ്ഡമെന്നു നിയതം. പിന്നെ ഘനഭൂമധ്യത്തിങ്കേന്നു ഖമധ്യത്തിങ്കലഗ്രമായിരിപ്പോരു ഊർദ്ധ്വസൂത്രത്തെ കല്പിപ്പൂ. ഇവിടെ അയനാന്തം. ദക്ഷിണോത്തരവൃത്തത്തെ സ്പർശിച്ചിരിക്കുമ്പോളേ അയനാന്തരവും ഊർദ്ധ്വസൂത്രവും തങ്ങളിലുള്ള പരമാന്തരാളം പരമാപക്രമമായിട്ടിരിക്കും.

പിന്നെ ബിംബഘനമധ്യാഗ്രമായിരിക്കുന്ന കോടിജ്യാമൂലം അയനാന്തസൂത്രത്തിങ്കൽ യാതൊരിടത്തു സ്പർശിക്കുന്നൂ, അവിടന്ന് ഊർദ്ധ്വസൂത്രത്തോടുള്ള അന്തരാളം ഇഷ്ടാപക്രമതുല്യം. പിന്നെ ബിംബനേമിയിങ്കലഗ്രമായിരിക്കുന്ന കോടിജ്യാവിന്റെ മൂലത്തിങ്കേന്ന് ഊർദ്ധ്വസൂത്രാന്തരം ഇഷ്ടാപക്രമത്തേക്കാൾ ഏറീട്ടിരിക്കും. ഈ ഏറിയഭാഗം ഭുജാഖണ്ഡത്തിന്റെ അപക്രമമായിട്ടിരിക്കും. ഈ ഭുജാഖണ്ഡാപക്രമത്തോടു തുല്യമായിരിക്കും അയനവലനം. ബിംബപ്രത്യഗ്ഭാഗത്തിങ്കലേ നേമിയിങ്കലേ സ്വാഹോരാത്ര വൃത്താപക്രമവൃത്തസംപാതങ്ങളുടെ പുറപ്പാടിന്റെ അന്തരാളം യാതൊന്ന്, അതായിട്ടിരിക്കും ഈ ഭുജാഖണ്ഡാപക്രമം. ഇവിടെ ചാപഖണ്ഡ മധ്യത്തിങ്കലഗ്രമായിട്ടിരിക്കുന്ന കോടിജ്യാവിനെക്കൊണ്ടു ഭുജാഖണ്ഡത്തെ വരുത്തേണ്ടൂ. ഘനമധ്യത്തിങ്കലഗ്രമായിട്ടിരിക്കുന്നതു സ്ഫുടകോടി. ഘനമധ്യത്തോടു നേമിയോടുള്ള അന്തരാളം ചാപഖണ്ഡമാകയാൽ ബിംബചതുരംശം പോയ സ്ഫുടകോടിചാപജ്യാവിനെക്കൊണ്ടു ബിംബാർദ്ധം സമസ്തജ്യാവായി ഇച്ഛാരാശിയായിരിക്കുന്നതിനെ ഗുണിച്ച്, വ്യാസാർദ്ധം കൊണ്ടു ഹരിച്ച ഫലം ഭുജാജ്യാഖണ്ഡമായിട്ടുണ്ടാകും[4]. പിന്നെ ഇതിനെ പരമാപക്രമം കൊണ്ടു ഗുണിച്ച്, വ്യാസാർദ്ധം കൊണ്ട് ഹരിച്ചഫലം, 'ആയനവലന'മാകുന്നത്. അവിടെ കോടിജ്യാവിനെ നടേ പരമാപക്രമം

---

6. 3. C.D.reads. കീഴുസ്പർശിക്കുന്ന വ്യാസസൂത്രത്തിങ്കലെ കോടിമുലാന്തരം ഭുജാഖണ്ഡ നേമീങ്കലഗ്രമായിരിക്കുന്നത്, മിത്തേ സ്പർശിക്കും. ഈ കോടിജ്യാമൂലവ്യാസസൂത്രാന്തര ത്തിങ്കലേത് ഭുജാജ്യാമൂലഖണ്ഡമായിരിക്കും. ഭുജാമൂലാന്തരം കോടിഖണ്ഡമെന്നു നിയതം.
4. C. ഭുജാഖണ്ഡമായിട്ടുണ്ടാകും

കൊണ്ടുഗുണിച്ച് വ്യാസാർദ്ധം കൊണ്ടു ഹരിച്ച്[5] കോടിജ്യാ അപക്രമമായിട്ടു വരുത്താം ഫലഭേദമില്ല. ഇങ്ങനെ ആയനം വലനം.

7. അക്ഷവലനം

പിന്നെ സാക്ഷദേശത്തിങ്കൽ ഈ സ്വാഹോരാത്രവൃത്തവും കൂടിച്ചരിഞ്ഞി രിക്കയാൽ അച്ചരിവിനെ[1] അറിയാനായിക്കൊണ്ട് അവിടേക്കു നേരേ കിഴക്കു പടിഞ്ഞാറായിട്ട് ഒരു വൃത്തത്തെ കല്പിപ്പൂ. ഇതു സമമണ്ഡലത്തിങ്കേന്നു തല്ക്കാലച്ഛായാഭുജയോളം കേന്ദ്രവും നേമിയും എല്ലായവയവവും നീങ്ങി യിരുന്നൊന്ന്. ഘടികാമണ്ഡലത്തിന്. യാതൊരു പ്രകാരം സ്വാഹോരാത്രം, അവ്വണ്ണം ഇരുന്നൊന്നിത് സമമണ്ഡലത്തിന്. ഇതിന്നു "ഛായാകോടിവൃത്ത" മെന്നു പേർ. ഇവിടെ ഛായാകോടിവൃത്തത്തിനും സ്വാഹോരാത്രവൃത്ത ത്തിനും അപക്രമവൃത്തത്തിനും കൂടി യോഗമുണ്ട് ബിംബഘനമദ്ധ്യത്തി ങ്കൽ. നേമിയിങ്കൽ പിന്നെ മൂന്നും മൂന്നേടത്തു പുറപ്പെടും. അവിടെ ആദി ത്യബിംബത്തിന്റെ[2] നേരേ പടിഞ്ഞാറേപ്പുറത്തു പുറപ്പെടും ഛായാ കോടിവൃത്തം. ഇതിങ്കേന്നു തെക്കോട്ടു ചരിഞ്ഞുള്ളൂ സ്വാഹോരാത്രവൃത്തം[3]. ആകയാൽ ദക്ഷിണോത്തരവൃത്തത്തിന്റെ കിഴക്കേപ്പുറത്ത് ആദിത്യനെങ്കിൽ പടിഞ്ഞാറേപ്പുറത്ത് തെക്കു നീങ്ങി സ്വഹോരാത്രത്തിന്റെ പുറപ്പാട്. പിന്നെ ദക്ഷിണോത്തരവൃത്തത്തിന്റെ പടിഞ്ഞാറേപ്പുറത്ത് ഗ്രഹമെന്നിരിക്കിൽ വടക്കു നീങ്ങി സ്വാഹോരാത്രത്തിന്റെ പുറപ്പാട്. ഈ നീക്കത്തിന് "ആക്ഷം വലന"മെന്നു പേർ.

8. വലനദ്വയസംയോഗം

ഇങ്ങനെ ഉണ്ടായ വലനങ്ങൾ രണ്ടിനേയും ദിക്ക് ഒന്നെങ്കിൽ[1] കൂട്ടുകയും രണ്ടെങ്കിലന്തരിക്കയും[2] ചെയ്തത്, ഇച്ഛാകോടിവൃത്തത്തോട് അപക്രമ

6. 5. F. adds. ഫലം
7. 1. D. ഇച്ചരിവിനെ
2. B. സൂര്യബിംബത്തിന്റെ
3. C.E.F. സ്വാഹോരാത്രവൃത്തത്തിന്റെ
8. 1. F. ഒന്നാകിൽ
2. C. രണ്ടാകിൽ

വൃത്തത്തോടുള്ള അന്തരാളമുണ്ടാകും. ബിംബനേമിയിങ്കലേ അത് സാക്ഷദേശത്തിങ്കലേ വലനമാകുന്നത് വിക്ഷേപമില്ലാത്തപ്പോൾ. വിക്ഷേപമുള്ളപ്പോൾ പിന്നെ അതിങ്കേന്നു വിക്ഷേപദിക്കിങ്കൽ വിക്ഷേപത്തോളം[3] നീക്കമുണ്ട്. അവിടെ ത്രൈരാശികം കൊണ്ടു നടേ വരുന്ന[4] വിക്ഷേപം ബിംബാന്തരത്തിങ്കലേതായിട്ടിരിക്കും[5]. ആകയാൽ വിക്ഷേപത്തെ അർക്കബിംബാർദ്ധം കൊണ്ടു ഗുണിച്ച്, ബിംബാന്തരംകൊണ്ടു ഹരിപ്പൂ. ഫലം ആദിത്യബിംബനേമിയിങ്കലേ വിക്ഷേപവലനമാകുന്നത്. ഇങ്ങനെ, ഇതിനു തക്കവണ്ണവും കൂടി നീങ്ങും ബിംബനേമിയിങ്കൽ സ്പർശപ്രദേശവും മോക്ഷപ്രദേശവും. പിന്നെ ബിംബനേമിയിങ്കൽ കിഴക്കേപ്പുറത്ത് അതിനു തക്കവണ്ണം വലനത്തിന്റെ ദിക്കുകൾ വിപരീതമായിട്ടിരിക്കും എന്നേ വിശേഷമുള്ളൂ. ഇവിടെ ആദിത്യൻ ഗ്രഹിക്കപ്പെടുന്നതാകയാൽ ആദിത്യനെ ഗ്രാഹ്യഗ്രഹ[6]മെന്നു ചൊല്ലുന്നു. ഇവിടെ ഇങ്ങനെ വലനങ്ങളുടെ ത്രൈരാശികം ആയനത്തിങ്കലേതു ചൊല്ലീതായി. അതു തന്നെ ആക്ഷത്തിങ്കലേക്കും ന്യായം.

ഈവണ്ണം ഛായാകോടി-സ്വാഹോരാത്രവൃത്തങ്ങൾക്കു ബിംബഘനമധ്യത്തിങ്കൽ യോഗം, ദക്ഷിണോത്തരവൃത്തത്തിങ്കൽ പരമാന്തരാളം. അതു തല്ക്കാലനതോത്ക്രമജ്യാവിന്റെ അക്ഷാംശമായിട്ടിരിക്കും. ഗ്രഹത്തോടു ദക്ഷിണോത്തരവൃത്തത്തോടുള്ള അന്തരാളം സ്വാഹോരാത്രവൃത്തത്തിങ്കലേതു നതിയല്ലോ. എന്നിട്ട് ഇവിടെ നതജ്യാവു കോടിയാകുന്നത്[7].

അന്ത്യാപക്രമസ്ഥാനീയം അക്ഷമാകയാൽ നതജ്യാക്ഷജ്യാക്കൾ തങ്ങളിൽ ഗുണിച്ച് ത്രിജ്യകൊണ്ടു ഹരിച്ചത് ആക്ഷം വലനമാകുന്നത്. അക്ഷവലനത്തെ ഉണ്ടാക്കി ഗ്രാഹ്യബിംബത്തെ വരച്ച് അതിൽ പൂർവ്വാപരരേഖയും ദക്ഷിണോത്തരരേഖയും ഉണ്ടാക്കി പൂർവ്വാപരരേഖാഗ്രങ്ങളിൽ നിന്ന് ത്രിജ്യാവൃത്തത്തിങ്കലേതു, ഗ്രാഹ്യബിംബാർദ്ധത്തിങ്കലേക്ക് എത്ര എന്ന് ഗ്രാഹ്യബിംബനേമിയിങ്കലേ വലനം.

---

8. 3. F. വിക്ഷേപത്തോളവും
4. D. വരുത്തുന്ന
5. F. വരും
6. B. ഗ്രാഹിയെന്നു ചൊല്ലുന്നു
7. E. F. കോടിജ്യാവാകുന്നത്

9. ഗ്രഹണലേഖനം

ഇങ്ങനെ സ്പർശമോക്ഷേഷ്ടകാലങ്ങളിലേക്ക് വലനത്തെ ഉണ്ടാക്കി ഗ്രാഹ്യബിംബത്തെ വരച്ച്, അതിങ്കൽ പൂർവ്വാപരരേഖയും ദക്ഷിണോത്തരരേഖയും ഉണ്ടാക്കി, പിന്നെ പൂർവ്വാപരരേഖാഗ്രങ്ങളിൽ നിന്നും തല്ക്കാലവലനത്തോളം നീങ്ങി ഒരു ബിന്ദുവിനെ ഉണ്ടാക്കി, ആ ബിന്ദുവിങ്കലും ഗ്രാഹ്യബിംബഘനമദ്ധ്യത്തിങ്കലും കൂടി ഒരു വലനസൂത്രത്തെ ഉണ്ടാക്കി, ഈ സൂത്രത്തിങ്കൽ ഗ്രാഹ്യബിംബത്തിന്റെ കേന്ദ്രത്തിങ്കേന്നു തല്ക്കാലബിംബാന്തരത്തോളം അകന്നേടത്തു കേന്ദ്രമായിട്ടു ഗ്രാഹകബിംബത്തെ എഴുതൂ. അപ്പോൾ ഗ്രാഹകബിംബത്തിന്റെ പുറത്ത് അകപ്പെട്ട ഭാഗം ഗ്രാഹ്യബിംബത്തിങ്കൽ പ്രകാശമായിട്ടിരിക്കും. ഗ്രാഹകബിംബത്തിനകത്തകപ്പെട്ട ഗ്രാഹ്യബിംബഭാഗം മറഞ്ഞിരിക്കും. ഇങ്ങനെ ഗ്രഹണത്തിന്റെ സംസ്ഥാനത്തെ അറിയേണ്ടും[1] പ്രകാരം. ഇവിടെ ഗ്രാഹ്യബിംബത്തിങ്കലേക്കു വലനമുണ്ടാക്കേണമെന്നു നിയതമില്ല. ഇഷ്ടവ്യാസാർദ്ധവൃത്തത്തിങ്കലേക്ക് എങ്കിലുമാം ഉണ്ടാക്കുവാൻ. അപ്പോൾ ആ വൃത്തത്തിങ്കലേ ദിക്സൂത്രത്തിങ്കേന്നു വേണം വലനം നീക്കുവാൻ എന്നേ വിശേഷമുള്ളൂ. ഇങ്ങനെ[2] സൂര്യഗ്രഹണപ്രകാരം.

10. ചന്ദ്രഗ്രഹണത്തിൽ വിശേഷം

ചന്ദ്രഗ്രഹണത്തിങ്കൽ വിശേഷമാകുന്നതു പിന്നെ. ചന്ദ്രബിംബം ഗ്രാഹ്യമാകുന്നത്, ഭൂച്ഛായ ഗ്രാഹകമാകുന്നത്. അവിടെ ചന്ദ്രബിംബമാർഗ്ഗത്തിങ്കലേ ഭൂച്ഛായയുടെ വിസ്താരത്തെ "തമോബിംബം" എന്നു ചൊല്ലുന്നു. അവിടെ ഗ്രാഹ്യഗ്രാഹകബിംബങ്ങൾ രണ്ടിന്നും ദ്രഷ്ടാവിങ്കേന്ന് അകലമൊക്കുകയാൽ, നതിലംബനങ്ങൾ രണ്ടിനും തുല്യങ്ങളാകയാൽ, അവ രണ്ടിനേയും ഇവിടെ ഉപേക്ഷിക്കാം[1]. മറ്റുള്ള ന്യായങ്ങളെല്ലാമിവിടേയും തുല്യങ്ങൾ. ഇങ്ങനെ ഗ്രഹണപ്രകാരത്തെ ചൊല്ലീതായി.

---

9. 1. F. അറിയും
2. B. ഇതി സൂര്യഗ്രഹണപ്രകാരഃ
10. 1. F. ഉപേക്ഷിക്കുകയുമാം

ഇവിടെ പിന്നെ ചന്ദ്രാർക്കന്മാരുടെ കേന്ദ്രഭുജാഫലത്തിന് അഹർദളപരിധിസ്ഫുടമെന്നൊരു സംസ്കാരമുണ്ട്. അതു ഹേതുവായിട്ട് സ്ഫുടാന്തരമുണ്ടാകും. ആകയാൽ സമലിപ്താകാലത്തിനും നീക്കം വരും. ഇതിനേക്കൊണ്ടു ഗ്രഹണകാലത്തിനും നീക്കമുണ്ടാകുമെന്നൊരു പക്ഷം.

[ഗണിതയുക്തിഭാഷയിൽ
ഗ്രഹണം എന്ന
പന്ത്രണ്ടാമദ്ധ്യായം സമാപ്തം]

അദ്ധ്യായം പതിമൂന്ന്

വ്യതീപാതം

1. വ്യതീപാതലക്ഷണം

അനന്തരം വ്യതീപാതത്തെ ചൊല്ലുന്നു. അവിടെ ചന്ദ്രാർക്കന്മാരുടെ അപക്രമങ്ങളിൽ വച്ച് ഒന്നിന് ഓജപദമാകയാൽ വൃദ്ധിയും, മറ്റേതിനു യുഗ്മപദമാകയാൽ ക്ഷയവും വരുന്നേടത്തു യാതൊരിക്കൽ തങ്ങളിൽ സാമ്യമുണ്ടാകുന്നു, അന്നേരം വ്യതീപാതമാകുന്ന കാലം.

2. ഇഷ്ടക്രാന്ത്യാനയനം

ഇവിടെ ചന്ദ്രാർക്കന്മാരുടെ ഇഷ്ടാപക്രമത്തെ വരുത്തും[1] പ്രകാരം മുമ്പിൽചൊല്ലി. അനന്തരം ചന്ദ്രന്റെ തന്നെ ഇഷ്ടാപക്രമത്തെ പ്രകാരാന്തരേണ വരുത്തുമാറു ചൊല്ലുന്നു. ഇവിടെ പലവൃത്തമുള്ളേടത്ത് എല്ലാറ്റിനും വലിപ്പമൊക്കും എന്നും, ഒരു പ്രദേശത്തിങ്കൽ തന്നെ കേന്ദ്രമെന്നും നേമികൾ അകന്നിട്ടുമിരിക്കുന്നു എന്നും കല്പിക്കുമ്പോൾ എല്ലാ വൃത്തങ്ങളുടെ നേമിയും എല്ലാ വൃത്തങ്ങളുടെ നേമിയോടും രണ്ടിടത്തു സ്പർശിക്കും. രണ്ടിടത്ത് അകന്നിട്ടുമിരിക്കുമെന്നും നിയതം.

3. വിക്ഷേപം

ഇവിടെ അപക്രമവൃത്തവും ഘടികാവൃത്തവും ഇന്നേടത്തു യോഗം ഇത്ര പരമാന്തരാളമെന്നറിഞ്ഞ്, പിന്നെ അപക്രമവൃത്തവും വിക്ഷേപവൃത്തവും

---

2. 1. B. ഉണ്ടാക്കും

ഇന്നേടത്തു യോഗം. ഇത്ര പരമാന്തരാളം, ആ യോഗത്തിങ്കേന്നു വിക്ഷേപ വൃത്തത്തിങ്കൽ ഇത്ര ചെന്നേടത്തു ചന്ദ്രനെന്നും അറിഞ്ഞിരിക്കുമ്പോൾ ചന്ദ്ര ങ്കേന്നു ഘടികാവൃത്തം ഇത്ര അകലമുണ്ട് എന്ന് ആദിത്യന്റെ അപക്രമ ത്തെപ്പോലെ തന്നെയറിയാം.

ഇവിടെ ഘടികാവിക്ഷേപവൃത്തങ്ങൾക്ക് ഇന്നേടത്തു യോഗം, ഇത്ര പര മാന്തരാളം എന്നറിവാനുപായത്തെ[1] ചൊല്ലുന്നു. അവിടെ ഒരുനാൾ മീനമ ധ്യത്തിങ്കൽ അപക്രമഘടികാസംപാതം, ആ സംപാതത്തിങ്കേന്നു അപക്ര മവൃത്തം വടക്കോട്ട് അകലും അന്ന് കന്യാമധ്യത്തിങ്കേന്ന് തെക്കോട്ടകലു ന്നു. അകന്നുകൂടുമ്പോൾ 24 തീയതി അകലും അപക്രമവൃത്തത്തിങ്കൽ രാഹു നില്ക്കുന്നേടത്തു[2] വിക്ഷേപവൃത്തത്തിനു യോഗം, അവിടുന്നു വട ക്കോട്ട് അകലും, കേതു നില്ക്കുന്നേടത്തുന്ന് തെക്കോട്ടകലും. അപക്രമവൃ ത്തത്തിങ്കലേ[3] ഘടികാസംപാതത്തിങ്കലൂ രാഹുവെന്നും, അവിടം നിരക്ഷ ക്ഷിതിജത്തിങ്കൽ ഉദിക്ക ചെയ്യുന്നത് എന്നും കല്പിപ്പൂ. അപ്പോൾ ദക്ഷി ണോത്തരവൃത്തത്തിങ്കലു പരാമപക്രമവൃത്തവും പരമവിക്ഷേപവും. അവിടെ ഘടികാവൃത്തത്തിങ്കേന്ന് അപക്രമവൃത്തവും അതിങ്കേന്നു വിക്ഷേപവൃ ത്തവും ഒരു ദിക്കുനോക്കി അകലും. ആകയാൽ പരമാപക്രമവൃത്തവും പര മവിക്ഷേപവും കൂടിയോളമകലമുണ്ട് ഘടികാവൃത്തത്തിങ്കേന്നു വിക്ഷേപ വൃത്തത്തിന്റെ അയനാന്തപ്രദേശം. എന്നാൽ അന്ന് അതു ചന്ദ്രന്റെ പരമാ പക്രമമായിട്ടിരിക്കും. എന്നാൽ അന്ന് അത് പ്രമാണഫലമായിട്ടു വിഷുവദാ ദിചന്ദ്രന്റെ ഇഷ്ടാപക്രമത്തേയും വരുത്താം.

ഈവണ്ണമിരിക്കുമ്പോൾ ഉത്തരധ്രുവത്തിങ്കേന്നു പരമാപക്രമത്തോളം ഉയർന്നേടത്തു ദക്ഷിണോത്തരവൃത്തത്തിങ്കൽ ഉത്തരരാശികൂടം. ഇവിടുന്ന് പരമവിക്ഷേപത്തോളം ഉയർന്നേടത്തു ഉത്തരമാകുന്ന വിക്ഷേപപാർശ്വമാ കയാൽ പരമാപക്രമപരമവിക്ഷേപയോഗത്തോളം അകലമുണ്ട് ധ്രുവങ്കേന്ന് വിക്ഷേപപാർശ്വം ഘടികാവൃത്തത്തിങ്കേന്നു യാതൊരു പ്രകാരം ധ്രുവൻ, അപക്രമവൃത്തത്തിന് യാതൊരു പ്രകാരം, രാശികൂടവും അവ്വണ്ണമിരിപ്പൊ ന്ന്, വിക്ഷേപവൃത്തത്തിന് വിക്ഷേപപാർശ്വം. ആകയാൽ ധ്രുവനും വിക്ഷേ

3\. 1. C. അറിവാനുള്ള
2\. B. C.F. നിന്നേടത്ത്
3\. D. adds. വൃത്തത്തിങ്കേന്ന് മദ്ധ്യത്തിങ്കലേ

പപാർശ്വവും തങ്ങളിലുള്ള അകലത്തോട് ഒത്തിരിക്കും ഘടികാവിക്ഷേപ വൃത്തങ്ങളുടെ പരമാന്തരാളം. പിന്നെ ധ്രുവങ്കലും വിക്ഷേപപാർശ്വത്തിങ്കലും സ്പർശിച്ചിട്ട് ഒരു വൃത്തത്തെ കല്പിപ്പൂ. ഈ വൃത്തത്തിങ്കൽ തന്നെയായിരിക്കും ഘടികാവിക്ഷേപവൃത്തങ്ങളുടെ പരമാന്തരാളം.

എന്നാൽ ധ്രുവവിക്ഷേപപാർശ്വാന്തരാളമറിയേണ്ടുവത് പിന്നെ. അപക്രമസംസ്ഥാനത്തെ ഈവണ്ണം തന്നെ കല്പിക്കുമ്പോൾ [4]ചാപമധ്യത്തിങ്കലെ അയനാന്തത്തിലു രാഹു എന്നും ഇരിക്കുമ്പോൾ, വിഷുവത്തിങ്കൽ സ്പർശിച്ചിരിക്കുന്ന രാശികൂടവൃത്തത്തിങ്കൽ പൂർവ്വവിഷുവത്തിങ്കേന്നു പരമവിക്ഷേപത്തോളം വടക്കു നീങ്ങിയിരിക്കും വിക്ഷേപവൃത്തം. ആകയാൽ ഉത്തരരാശികൂടത്തിങ്കേന്ന് ഇത്ര പടിഞ്ഞാറു നീങ്ങി ഇരിക്കും വിക്ഷേപപാർശ്വം. ഇവിടെ പരമവിക്ഷേപവും പരമാപക്രമവും തങ്ങളിൽ ഭുജാകോടികളായിട്ടിരിക്കുമ്പോളേ കർണ്ണമായിട്ടിരിക്കും ധ്രുവവിക്ഷേപപാർശ്വാന്തരാളം. പിന്നെ വിക്ഷേപപാർശ്വധ്രുവങ്ങളിൽ സ്പർശിക്കുന്ന ഒരു വൃത്തത്തെ കല്പിപ്പൂ. ആ വൃത്തത്തിൽ വരുന്ന ഘടികാപക്രമങ്ങളുടെ പരമാന്തരാളം അവിടെ വിക്ഷേപവൃത്തായനാന്തമാകുന്നത്. ഈ വൃത്തത്തിന് 'വിക്ഷേപായനാന്തം' എന്നു പേർ. ഇതിന്നു ദക്ഷിണോത്തരവൃത്തത്തോടു യോഗം ധ്രുവങ്കൽ. ഇവിടുന്നു പരമാപക്രമത്തോളം ചെല്ലുമ്പോൾ പരമവിക്ഷേപത്തോളം പടിഞ്ഞാറു നീങ്ങും[5] വിക്ഷേപായനാന്തവൃത്തം. വൃത്തപാദം ചെല്ലുമ്പോൾ ദക്ഷിണോത്തരവൃത്തത്തിങ്കേന്ന് എത്ര പടിഞ്ഞാറു നീങ്ങുമെന്നു ഘടികാവൃത്തത്തിങ്കലുണ്ടാം പരമാന്തരാളം. ആകയാൽ ദക്ഷിണോത്തരവൃത്തത്തിങ്കേന്ന് അത്ര പടിഞ്ഞാറു നീങ്ങിയേടത്തു ഘടികാവൃത്തത്തിങ്കൽ വിക്ഷേപായനാന്തം. ആകയാൽ ക്ഷിതിജത്തിങ്കലെ പൂർവ്വവിഷുവത്തിങ്കേന്നു ഘടികാമണ്ഡലത്തിങ്കൽ[6] ഇത്ര മേല്പോട്ടു നീങ്ങിയേടത്തു വിക്ഷേപവിഷുവത്ത് എന്നും വരും, വൃത്തങ്ങളുടെ പരമാന്തരാളവും യോഗവും തങ്ങളിൽ വൃത്തപാദാന്തരിതമെല്ലോ എന്നിട്ട്. ഈ നീക്കത്തിന് 'വിക്ഷേപചലന'മെന്നു പേർ. എന്നാൽ അപക്രമവിഷുവദാദിയിങ്കൽ ഈ വിക്ഷേപചലനം സംസ്കരിച്ചാൽ വിക്ഷേപവിഷുവദാദിയാകും.

---

3. 4. C.D.E. om. ചാപ
5. C. F. നീങ്ങിയിരിക്കും
6. B. C. ഘടികാവൃത്തത്തിങ്കൽ

പിന്നെ യാതൊരിക്കൽ കന്യാമധ്യത്തിങ്കലേ വിഷുവത്തിങ്കൽ രാഹു നില്ക്കുന്നു, അപ്പോൾ ചാപമധ്യത്തിങ്കലെ അയനാന്തത്തിങ്കേന്ന് പരമവിക്ഷേപത്തോളം വടക്കു നീങ്ങും വിക്ഷേപവൃത്തം. ഉത്തരരാശികൂടത്തിങ്കേന്ന് അത്ര താണിരിക്കും വിക്ഷേപപാർശ്വം. അപ്പോൾ അവിടന്ന് ധ്രുവാന്തരാളം ചന്ദ്രന്റെ പരമാപക്രമമാകുന്നത്, അരകുറയ ഇരുപതു തീയതിയായിട്ടായിരിക്കും. ഈ വിക്ഷേപപാർശ്വധ്രുവാന്തരാളം ദക്ഷിണോത്തരവൃത്തത്തിങ്കലാകയാൽ ക്ഷേപായനാന്തം അപക്രമായനാന്തത്തിങ്കൽ തന്നെ. ക്ഷേപാപക്രമവിഷുവത്തുകളും ഒരിടത്തുതന്നെ. അപ്പോൾ വിക്ഷേപചലനമില്ല. പിന്നെ മിഥുനമധ്യത്തിൽ അയനാന്തത്തിങ്കൽ രാഹു നില്ക്കുമ്പോൾ കന്യാമധ്യത്തിങ്കൽ സ്പർശിച്ചിരിക്കുന്ന രാശികൂടവൃത്തത്തിന്മേൽ പരമവിക്ഷേപത്തോളം വടക്കുനീങ്ങിയേടത്തു സ്പർശിക്കും വിക്ഷേപവൃത്തം. ആകയാൽ ഉത്തരരാശികൂടത്തിങ്കേന്നു പരമവിക്ഷേപത്തോളം കിഴക്കു നീങ്ങിയിരിക്കും[7] ക്ഷേപപാർശ്വം. അവിടെയും കർണ്ണാകാരേണയിരിക്കും ക്ഷേപപാർശ്വധ്രുവാന്തരാളം. പിന്നേയും പൂർവ്വവിഷുവത്തിങ്കൽ രാഹുവാകുമ്പോൾ ഉദഗ്രാശികൂടത്തിങ്കേന്ന് മീത്തേ ഇരിക്കും വിക്ഷേപപാർശ്വം. ഇങ്ങനെ തന്നെ ദക്ഷിണരാശികൂടവൃത്തത്തിങ്കൽ ദക്ഷിണവിക്ഷേപപാർശ്വം. ഇങ്ങനെ രാശികൂടത്തിങ്കേന്നു പരമവിക്ഷേപാന്തരാളം അകന്നേടത്തു രാഹുവിന്റെ ഗതിക്കു തക്കവണ്ണം പരിഭ്രമിക്കുമ്പോൾ ക്ഷേപപാർശ്വം.

4. വിക്ഷേപചലനം

ഇവിടെ പരമവിക്ഷേപവ്യാസാർദ്ധമായിട്ട് ഒരു വൃത്തത്തെ കല്പിപ്പൂ. ഈ വൃത്തത്തിന് കേന്ദ്രം രാശികൂടത്തിങ്കേന്നു പരമവിക്ഷപശരത്തോളം ഭഗോളമധ്യം നോക്കി നീങ്ങിയേടത്തായിട്ടിരിക്കും. പിന്നെ ഇതിന്റെ കേന്ദ്രത്തിങ്കൽ നേമിയായിട്ട് അക്ഷദണ്ഡിങ്കൽ കേന്ദ്രമായിട്ട്, മറ്റൊരു വൃത്തത്തെ കല്പിപ്പൂ. അപ്പോൾ കക്ഷ്യാവൃത്തവും ഉച്ചനീചവൃത്തവും എന്ന പോലെ ഇരിക്കുമിവ രണ്ടും. ഇവിടെ അക്ഷദണ്ഡിങ്കേന്നു ക്ഷേപപാർശ്വോന്നതി ത്രിജ്യാസ്ഥാനീയം. പിന്നെ വിക്ഷേപപാർശ്വത്തിങ്കൽ യാതൊരിടത്തു ക്ഷേപപാർശ്വം

---

3. 7. C. വടക്കുനീങ്ങിയിരിക്കും

അവിടന്നു തന്റെ കേന്ദ്രത്തോടു മേൽകീഴുള്ളതു കോടിഫലസ്ഥാനീയം. കിഴക്കുപടിഞ്ഞാറായിട്ട് ദക്ഷിണവൃത്തത്തോടിട ഭുജാഫലസ്ഥാനീയം. ഇവിടെ പരിഭ്രമിക്കുന്ന വിക്ഷേപപാർശ്വം രാശികൂടത്തിങ്കേന്നു മിത്തേപ്പുറ ത്താകുമ്പോൾ ക്ഷേപപാർശ്വോന്നതിയിൽ കോടിഫലം കൂട്ടൂ, കീഴേപ്പുറ ത്താകുമ്പോൾ കളവൂ. അവിടെ മീനമധ്യത്തിങ്കൽ രാഹു നിൽക്കുന്നാൾ ധ്രുവങ്കേന്ന് എല്ലായിലും[1] ഉയരത്താകുന്നു, കന്യാമധ്യത്തിൽ രാഹു നില്ക്കു മ്പോൾ എല്ലായിലും കീഴാകുന്നു[2]. ആകയാൽ മേൽകീഴുള്ളതു കോടിഫല മെന്നും, മകരാദിയിൽ കൂട്ടൂ, കർക്ക്യാദിയിൽ കളവൂ എന്നും വന്നു.

അയനാന്തത്തിങ്കൽ ഭുജ തികയുന്നു. അവിടെ രാഹു നിൽക്കുമ്പോൾ ക്ഷേപപാർശ്വത്തിങ്കേന്നു കിഴക്കുപടിഞ്ഞാറു നീക്കമായാൽ കിഴക്കുപടിഞ്ഞാ റുള്ളതു ഭുജാഫലം. അവിടേയും തുലാദിയിൽ രാഹു നില്ക്കുമ്പോൾ[3] ദക്ഷി ണോത്തരവൃത്തത്തിങ്കേന്നു പടിഞ്ഞാറു ക്ഷേപപാർശ്വമാകയാൽ തുലാദി യിങ്കൽ വിക്ഷേപചലനം കൂട്ടുക വേണ്ടുവത്. മേഷാദിയിങ്കൽ ദക്ഷിണോ ത്തരവൃത്തത്തിന്റെ കിഴക്കേപ്പുറത്തു വിക്ഷേപപാർശ്വം എന്നിട്ടു[4] കളകവേ ണ്ടുവത്. ഇവിടെ വിഷുവദാദിരാഹുവിന്റെ ഭുജാകോടിജ്യാക്കളെ പരമവി ക്ഷേപംകൊണ്ടു ഗുണിച്ച് ത്രിജ്യകൊണ്ടു ഹരിപ്പൂ. എന്നാൽ ഭുജാകോടി ഫലങ്ങളുണ്ടാകും.

5. കർണ്ണാനയനം

പിന്നെ ഇവറ്റേക്കൊണ്ടു കർണ്ണം വരുത്തും പ്രകാരം. കർണ്ണമാകുന്നത് ധ്രുവവിക്ഷേപപാർശ്വാന്തരാളജ്യാവ്. ഈ ക്ഷേപായനാന്തവൃത്തം രാശികൂ ടത്തിങ്കൽ കൂടി സ്പർശിക്കുമ്പോൾ പരമാപക്രമവും പരമവിക്ഷേപവും തങ്ങ ളിൽ യോഗം താനന്തരം താൻ ചെയ്താൽ[1] ഇതരേതരകോടിഗുണനവും ത്രിജ്യാഹരണവും ചെയ്യേണം.

---

4. 1. F. adds. ഉയർന്നതാകുന്നു. മീനം മദ്ധ്യത്തിൽ രാഹു നിൽക്കുമ്പോൾ എല്ലായിലും കീഴാ കുന്നു.
2. D. താഴത്താകുന്നു
3. H. നില്ക്കുന്നനാൾ
4. F. adds. അത്

5. 1 B. ചെയ്താൽ

പിന്നെ പരമാപക്രമവും കോടിഫലവും തങ്ങളിൽ കൂട്ടുകതാനന്തരിക്കതാൻ ചെയ്യുന്നേടത്തും അന്ത്യക്ഷേപകോടിയും അന്ത്യാപക്രമകോടിയുമത്രേ ഗുണകാരമാകുന്നത്. ഇവിടെ ദക്ഷിണോത്തരവൃത്തൈകദേശത്തിങ്കലേ ജ്യാവായി ക്ഷേപപാർശ്വവൃത്തത്തിന്റെ കേന്ദ്രത്തിങ്കേന്നു തുടങ്ങി ഇതിന്റെ നേമിയോളം ചെല്ലുന്നതു പരമവിക്ഷേപമാകുന്നത്. ഇതിൽ ഒട്ടേടം ചെന്നതു കോടിഫലമാകുന്നത്. ഇത്രേ വിശേഷമുള്ളു, സംസ്ഥാനഭേദമില്ല, ആകയാൽ യോഗവിയോഗത്തിങ്കൽ ഗുണകാരഭേദമില്ല, ഗുണ്യഭേദമേയുള്ളൂ. ഇവിടെ പരമാപക്രമത്തെ പരമവിക്ഷേപകോടികൊണ്ടു ഗുണിച്ച് ത്രിജ്യകൊണ്ടു ഹരിച്ചഫലം ക്ഷേപപാർശ്വകേന്ദ്രത്തോട് അക്ഷദണ്ഡോടുള്ള അന്തരാളം. പിന്നെ കോടിഫലം ഇതിന്നു ക്ഷേപമായിരിക്കും[2]. പരമാപക്രമകോടികൊണ്ടു ഗുണിച്ച് ത്രിജ്യകൊണ്ടു ഹരിച്ചാൽ കോടിഫലാഗ്രത്തിങ്കേന്നു വിക്ഷേപ പാർശ്വാന്തരാളം ഭുജാഫലമാകുന്നത്. പിന്നെ ഇവറ്റേ തങ്ങളിൽ യോഗാന്തരങ്ങൾ[3] ചെയ്ത്, വർഗ്ഗിച്ച് ഭുജാഫലവർഗ്ഗം കൂട്ടി മൂലിച്ചതു ധ്രുവനോടു വിക്ഷേപപാർശ്വത്തോടിട അന്തരാളചാപഭാഗത്തിങ്കലേ ജ്യാവായിട്ടിരിക്കും.

6. വിക്ഷേപചലനം

എന്നാൽ ഘടികാവിക്ഷേപവൃത്തങ്ങളുടെ പരമാന്തരാളമാകുന്ന പരമാപക്രവും ഇതു തന്നെ. പിന്നെ ധ്രുവങ്കേന്നു ക്ഷേപായനാന്തവൃത്തത്തിങ്കൽ ക്ഷേപപാർശ്വത്തോളം ചെല്ലുമ്പോൾ ദക്ഷിണോത്തരാന്തരാളം[1], ഭുജാഫലത്തോളം വൃത്തപാദം[2] ചെല്ലുമ്പോളെത്ര എത്ര എന്നു ക്ഷേപായനാന്തവൃത്തത്തിങ്കേന്നു ദക്ഷിണോത്തരവൃത്തത്തിന്റെ പരമാന്തരാളം ഘടികാവൃത്തത്തിങ്കലുണ്ടാകും. ഇത് അയനാന്തങ്ങൾ രണ്ടും തങ്ങളിലുള്ള അന്തരാളമാകുന്നത്. ഇതു തന്നെ വിഷുവത്തുക്കളുടെ അന്തരാളമാകുന്നത്. എന്നിട്ട് ഇതിന് 'വിക്ഷേപചലന'മെന്നു പേർ. പിന്നെ സായനചന്ദ്രനിൽ വിക്ഷേപചലനം സംസ്കരിച്ചാൽ ഘടികാവിക്ഷേപവൃത്തസംപാതത്തിങ്കേന്നു ചന്ദ്രനോടുള്ള അന്തരാളം വിക്ഷേപവൃത്തത്തിങ്കലേത് ഉണ്ടാകും.

---

5. 2. A.E. ശേഷമായിരിക്കും
   3. B. യോഗംതാനന്തരം താൻ
6. 1. E. om. ദക്ഷിണോത്തരാന്തരാളം....to....വൃത്തപാദം
   2. B. om. വൃത്തപാദം

7. വ്യതീപാതകാലം

ഇങ്ങനെ വിക്ഷേപചലനവും അയനചലനവും സംസ്കരിച്ചിരിക്കുന്ന ചന്ദ്രനും അയനചലനം സംസ്കരിച്ചിരിക്കുന്ന ആദിത്യനും, ഇവ രണ്ടിലൊന്ന് ഓജപദത്തിങ്കലും മറ്റേതു യുഗ്മപദത്തിങ്കലുമെന്നിവണ്ണമിരിക്കുമ്പോൾ[1] അപക്രമസാമ്യം വരുന്നേടത്ത് വ്യതീപാതമാകുന്ന പുണ്യകാലം[2].

8. വ്യതീപാതാനയനം

ഈ അപക്രമങ്ങളുടെ സാമ്യകാലമറിയും പ്രകാരം പിന്നെ. ഓജയുഗ്മപദങ്ങളിൽ രണ്ടിൽ നിൽക്കുന്ന[1] ചന്ദ്രാർക്കന്മാരുടെ ഭുജാസാമ്യം വരുന്നു യാതൊരിക്കൽ എന്നിതിനെ ഊഹിച്ച് കല്പിച്ച് അന്നേരത്തെ ആദിത്യന്റെ[2] ഭുജാജ്യാവിനേക്കൊണ്ടു വരുന്ന ഇഷ്ടാപക്രമത്തോടു തുല്യമായിട്ട് ചന്ദ്രന്റെ അപക്രമമുണ്ടാവാൻ ഏതു ഭുജാജ്യാവു വേണ്ടുവത് എന്നിതിനെ ത്രൈരാശികം ചെയ്തു വരുത്തൂ. ഇവിടെ 'ദുഗ്ധലോകം' എന്നു പരമാപക്രമമാകുന്ന ആദിത്യന്[3] ഇതു ഭുജാജ്യാവാകുന്നത്. അപ്പോൾ തല്ക്കാലത്തിങ്കൽ വരുത്തിയത് അന്ത്യാപക്രമമാകുന്ന ചന്ദ്രന് ആദിത്യാപക്രമത്തോടു തുല്യമാവാൻ ഏതു ഭുജാജ്യാവാകുന്നത് ചന്ദ്രന്ന് എന്നു ത്രൈരാശികമാകുന്നത്. ഇവിടെ ആദിത്യന്റെ പരമാപക്രമം പ്രമാണം, ആദിത്യന്റെ ഭുജാജ്യാവു പ്രമാണഫലം. ചന്ദ്രന്റെ അന്ത്യാപക്രമം ഇച്ഛ. ചന്ദ്രന്റെ ഭുജാജ്യാവിച്ഛാഫലം. ഇവിടെ അന്ത്യാപക്രമം വലിയതിന്നു ഭുജാജ്യാവു ചെറുതായിട്ടിരിക്കും, ചെറിയതിന്ന് വലുതായിട്ടിരിക്കും. അന്നേരത്ത് അപക്രമസാമ്യം വരുന്നു. എന്നിട്ടു[3a] വിപരീതത്രൈരാശികം ഇവിടേയ്ക്കു വേണ്ടുവത്. ആകയാൽ ആദിത്യന്റെ

---

7. 1. B. എന്നിരിക്കുമ്പോൾ
   2. B. വ്യതീപാതകാലം
8. 1. B. രണ്ടു പദങ്ങളിൽ നില്ക്കുന്ന
   2. B. അർക്കന്റെ ; B. om. ആദിത്യന്റെ
   3. B. അർക്കന്ന്
   3a. B.adds, എന്നിട്ട് ഇവിടെ വ്യസ്തത്രൈരാശികം വേണ്ടുവതാകയാൽ C. ഇവിടെ

ഭുജാജ്യാവും അന്ത്യാപക്രമവും തങ്ങളിൽ ഗുണിച്ചതിങ്കേന്നു ചന്ദ്രന്റെ അന്ത്യാപക്രമത്തേക്കൊണ്ടു ഹരിച്ചഫലം ചന്ദ്രഭുജാജ്യാവ്. പിന്നെ ഇതിനെ ചാപിച്ചു പദവശാൽ അയനസന്ധിയിങ്കൽ താൻ[4] ഗോളസന്ധിയിങ്കൽ താൻ[5] സംസ്കരിച്ച് ചന്ദ്രനെ ഉണ്ടാക്കൂ.

പിന്നെ ആദിത്യങ്കേന്നുണ്ടാക്കിയ ചന്ദ്രനേയും തല്ക്കാലചന്ദ്രനേയും തങ്ങളിലന്തരിച്ചു രണ്ടേടത്തു വച്ച്, അർക്കചന്ദ്രന്മാരുടെ ഗതികൊണ്ടു ഗുണിച്ച്, ഗതിയോഗം[6] കൊണ്ടു ഹരിച്ചഫലം അതതിങ്കൽ സംസ്ക്കരിപ്പൂ. വ്യതീപാതം കഴിഞ്ഞുവെങ്കിൽ കളവൂ, മേൽവരുന്നൂവെങ്കിൽ[7] കൂട്ടൂ, പാതങ്കൽ വിപരീതമായിട്ട്, ഇങ്ങനെ അവിശേഷിപ്പൂ, ആദിത്യങ്കേന്നു[8] ഉണ്ടായ ചന്ദ്രഭുജാധനുസ്സും തല്ക്കാല ചന്ദ്രന്റെ ഭുജാധനുസ്സും സമമായിട്ടു വരുവോളം. അവിടെ ഓജപദത്തിങ്കലെ തല്ക്കാലചന്ദ്രന്റെ ഭുജാധനുസ്സ് വലുതാകിൽ കഴിഞ്ഞു. വ്യതീപാതം, ചെറുതാകിൽ മേലൂ. യുഗ്മപദത്തിങ്കൽ വിപരീതം. ഇവിടെ അർക്കചന്ദ്രൻമാർക്കുതാൻ ഭൂച്ഛായാ[9]ചന്ദ്രൻമാർക്കു താൻ സ്വാഹോരാത്രമൊന്നേയാകുമ്പോൾ വ്യതീപാതമുണ്ടാകുന്നു. പിന്നെ ബിംബൈകദേശത്തിനും സ്വാഹോരാത്രവ്യത്തൈക്യമില്ലാതാകുമ്പോൾ വ്യതീപാതമില്ല. ആകയാൽ മിക്കവാറും[10] നാലു നാഴിക വ്യതീപാതമായിട്ടിരിക്കും.

[ഗണിതയുക്തിഭാഷയിൽ
വ്യതീപാതമെന്ന
പതിമൂന്നാമദ്ധ്യായം സമാപ്തം]

8. 4. F. സന്ധിയിങ്കന്നു താൻ
5. F. സന്ധിയിങ്കന്നു താൻ
6. B. അതിന്റെ യോഗം
7. B. ഭാവിയെങ്കിൽ
8. B. അർക്കങ്കേന്നു
9. B. C. D. തല്ക്കാലചന്ദ്രച്ഛായാ
10. B. മിക്കതും

അദ്ധ്യായം പതിനാല്

മൌഢ്യവും ദർശനസംസ്കാരവും

1. ദർശനസംസ്കാരം

അനന്തരം ദർശനസംസ്കാരത്തെ ചൊല്ലുന്നു[1]. അതാകുന്നതു വിക്ഷേപിച്ചിരിക്കുന്ന ഗ്രഹം ക്ഷിതിജത്തിങ്കൽ ഉദിക്കുമ്പോൾ[2] അപക്രമമണ്ഡലത്തിന്റെ യാതൊരു പ്രദേശം ക്ഷിതിജത്തെ സ്പർശിക്കുന്നത് എന്നുള്ളത്. ഇവിടെ ഈവണ്ണം ക്ഷേത്രത്തെ കല്പിച്ചിട്ടു നിരൂപിപ്പൂ. ഉത്തരരാശികൂടം ഉയർന്നിരിക്കുമാറ് മേഷാദി മൂന്നു രാശികളിൽ എങ്ങാനും ഒരിടത്തിരിക്കുന്ന ഗ്രഹത്തിങ്കൽ സ്പർശിച്ചിരിക്കുന്ന രാശികൂടവൃത്തവും അപക്രമവൃത്തവും തങ്ങളിലുള്ള സംപാതം ക്ഷിതിജത്തിങ്കലുദിക്കുമാറ്. ഇവിടന്ന് ഉദഗ്രാശികൂടം നോക്കി വിക്ഷേപിച്ചിരിക്കുമാറ് ഗ്രഹമെന്നു കല്പിപ്പൂ. അപ്പോൾ ക്ഷിതിജത്തിങ്കൽ ഉയർന്നിരിക്കും ഗ്രഹം എന്നാൽ ഗ്രഹത്തിന്ന് അന്നേരത്ത് എത്ര ശങ്കുവെന്നതു നടേ ഉണ്ടാക്കുന്നത്. പിന്നെ ഈ ശങ്കു കോടിയായിട്ടിരിക്കുമ്പോൾ ഇതിന്നു കർണ്ണമായിട്ടു വിക്ഷേപകോടിവൃത്തത്തിങ്കലേ ഗ്രഹത്തിന് ക്ഷിതിജാന്തരാളമുണ്ടാകും[3]. ഇവിടെ ദൃക്ക്ഷേപവൃത്തത്തിങ്കൽ ഖമദ്ധ്യത്തിങ്കന്നു ദൃക്ക്ഷേപത്തോളം തെക്കു നീങ്ങിയേടത്ത് അപക്രമവൃത്തസംപാതം. ദൃക്ക്ഷേപവൃത്തത്തിങ്കൽ തന്നെ ക്ഷിതിജത്തിങ്കേന്ന് ദൃക്ക്ഷേപത്തോളം ഉയർന്നേടത്തു രാശികൂടം ഉത്തരം. ഉത്തരരാശികൂടം നോക്കി നീങ്ങുന്നത് ഉത്തരവിക്ഷേപം, ഗ്രഹസ്പൃഷ്ടരാശികൂടവും ക്ഷിതിജവും തങ്ങളിലുള്ള

---

1. 1. B. അഥ ദർശനസംസ്ക്കാരം
2. B. നില്ക്കുമ്പോൾ
3. D. F. ഗ്രഹക്ഷിതി;ജാന്തരാളമുണ്ടാകും

പരമാന്തരാളം ദൃക്ക്ഷേപം എന്നാൽ ക്ഷിതിജത്തിങ്കേന്നു വിക്ഷിപ്ത ഗ്രഹത്തോളം ചെല്ലുമ്പോൾ ക്ഷിതിജാന്തരാളം എത്രയെന്നു വിക്ഷിപ്തഗ്രഹത്തിന്റെ ശങ്കു ഉണ്ടാകും. പിന്നെ ദൃക്ക്ഷേപകോടിക്കു ത്രിജ്യാവു കർണ്ണം, ഈ ശങ്കുവിന്ന് എന്തു കർണ്ണമെന്നു ഗ്രഹക്ഷിതിജാന്തരാളത്തിങ്കലെ അപക്രമചാപഭാഗമുണ്ടാകുന്നതു പോലെ, ഈ വിക്ഷിപ്തഗ്രഹത്തോടു ക്ഷിതിജത്തോടുള്ള അന്തരാളത്തിങ്കലെ വിക്ഷേപകോടിവൃത്തഭാഗമുണ്ടാകും.

2. ഗ്രഹാസ്തോദയം – മൗഢ്യം

പിന്നെ വിക്ഷേപകോടിവൃത്തവും ക്ഷിതിജവുമുള്ള സംപാതത്തിങ്കൽ സ്പർശിച്ചിട്ട് ഒരു രാശികൂടവൃത്തത്തെ കല്പിപ്പൂ. ഇതിനും ഗ്രഹസ്പൃഷ്ടരാശികൂടവൃത്തത്തിനും രാശികൂടത്തിങ്കൽ യോഗം. ഈ യോഗത്തിങ്കേന്നു വിക്ഷേപകോടിയോളം ചെല്ലുന്നേടത്തു ഗ്രഹമിരിക്കുന്നു. അവിടെ ഈ രാശികുടവൃത്തങ്ങളുടെ അന്തരാളം വരുത്തിയ വിക്ഷേപഗ്രഹശങ്കുവിന്റെ കർണ്ണത്തോളം, അപ്പോളിവറ്റിന്റെ പരമാന്തരാളമെത്രയെന്ന് ഈ രാശികൂടവൃത്തങ്ങൾ രണ്ടിന്റേയും പരമാന്തരാളം അപക്രമവൃത്തത്തിങ്കലേത് ഉണ്ടാകും. ഇവിടെ ഗ്രഹസ്ഫുടം ലഗ്നമാകുമ്പോൾ ഇത്ര ഇലി ഉയർന്നിരിക്കുന്നു[1] ഗ്രഹം എന്നിട്ട് ഗ്രഹസ്ഫുടവും ഗ്രഹമുദിക്കുമ്പോളേ ലഗ്നവും തങ്ങളിലന്തരാളം ഈ രാശികൂടങ്ങളുടെ പരമാന്തരാളമായിട്ടിരിക്കും. ഇവിടെ ഇത്ര മുമ്പേ ഉദിക്കയാൽ ഗ്രഹസ്ഫുടത്തിങ്കേന്ന് ഈ അന്തരം കളഞ്ഞതു ഗ്രഹോദയത്തിങ്കലേ ലഗ്നമാകുന്നത്[2]. ഇങ്ങനെ ഉത്തരവിക്ഷേപത്തിങ്കൽ. ദക്ഷിണവിക്ഷേപത്തിങ്കൽ പിന്നെ ഇങ്ങനെ തന്നെ ക്ഷേത്രസംസ്ഥാനത്തെ കല്പിക്കുമ്പോൾ ക്ഷിതിജാപക്രമസമ്പാതത്തിങ്കേന്നു ഗ്രഹം ഗ്രഹസ്പൃഷ്ടരാശികൂടത്തിങ്കൽ[3] മേലേ തെക്കോട്ട് വിക്ഷേപിക്കയാൽ ക്ഷിതിജത്തിങ്കേന്ന് കീഴേപ്പുറത്ത് ഇരിക്കും ഗ്രഹം.

---

2. 1. D. E. ഉയർന്നിരിക്കും
2. B. ഗ്രഹോദയലഗ്നമാകുന്നത്
3. F. രാശികൂടവൃത്തത്തിങ്കൽ

ഈവണ്ണമിരിക്കുമ്പോൾ മുമ്പിൽ അധോമുഖശങ്കുവിനെക്കൊണ്ട് ഉദയാസ്തലഗ്നങ്ങളെ വരുത്തുവാൻ ചൊല്ലിയതു പോലെ വിക്ഷേപാഗ്രത്തിങ്കലിരിക്കുന്ന ഗ്രഹത്തിന്റെ ഉദയകാലത്തിങ്കലേ ലഗ്നവും ഗ്രഹസ്ഫുടവും തങ്ങളിലന്തരാളത്തിങ്കലേ കലകളുണ്ടാകും. അവിടെ ഗ്രഹം ഇത്ര പിന്നെ ഉദിപ്പൂ എന്നിട്ട്, ഈ അന്തരാളകലകളെ ഗ്രഹസ്ഫുടത്തിങ്കൽ കൂട്ടുകവേണ്ടുവതു ഗ്രഹോദയത്തിങ്കലേ ലഗ്നം വരുത്തുവാൻ.

ഈവണ്ണം ഗ്രഹാസ്തമയത്തിലേ അസ്തലഗ്നവും വരുത്തൂ. അവിടെ അധോമുഖശങ്കുവെങ്കിൽ ഗ്രഹം മുമ്പേ അസ്തമിക്കും, ഊർദ്ധ്വമുഖശങ്കുവെങ്കിൽ പിന്നെ അസ്തമിപ്പു ഗ്രഹസ്ഫുടാസ്തലഗ്നത്തിങ്കേന്ന്. എന്നിട്ട് ഋണധനങ്ങൾക്ക് പകർച്ചയുണ്ട്. അത്രേ വിശേഷമുള്ളു.

പിന്നെ ദക്ഷിണരാശികൂടം ക്ഷിതിജത്തിങ്കേന്ന് ഉയർന്നിരിക്കുന്നു എങ്കിൽ ദക്ഷിണവിക്ഷേപത്തിങ്കൽ ഗ്രഹം ഉയർന്നിരിപ്പൂ, ഉത്തരവിക്ഷേപത്തിങ്കൽ താണിരിപ്പൂ. ആകയാൽ അവിടെ ഉത്തരരാശികൂടോന്നതിയിങ്കൽ ചൊല്ലിയതിങ്കേന്നു വിപരീതമായിട്ടിരിക്കും ധനർണ്ണപ്രകാരം. ഇത്രേ വിശേഷമുള്ളൂ.

ഇവിടെ ദൃക്ക്ഷേപം ദക്ഷിണമാകുമ്പോൾ വടക്കേ രാശികൂടമുയർന്നിരിക്കും, ഉത്തരമാകുമ്പോൾ തെക്കേത്. ആകയാൽ വിക്ഷേപദൃക്ക്ഷേപങ്ങളുടെ ദിക്ക് ഒന്നേ എങ്കിൽ ഉദയത്തിങ്കൽ ദർശനസംസ്കാരഫലം ഗ്രഹത്തിങ്കൽ ധനം, ദിഗ്ഭേദമുണ്ടെങ്കിൽ ഋണം. അസ്തമയത്തിനു വിപരീതം.

3. ഗ്രഹങ്ങളുടെ ദർശനസംസ്കാരം

പിന്നെ ഈ ഗ്രഹോദയത്തിനും ആദിത്യനും കാലലഗ്നം വരുത്തി അന്തരിച്ചാൽ അന്തരം ഇത്ര തീയതി ഉണ്ടായിരിക്കുമ്പോൾ ഈ ഗ്രഹത്തെ കാണാം, ഇതിൽ കുറഞ്ഞാൽ കാണരുത്, എന്നുണ്ട്. അതിന്നു തക്കവണ്ണം പാടും, പിറപ്പും അറിയും പ്രകാരം പിന്നെ. വിക്ഷിപ്തഗ്രഹത്തിങ്കലേ മധ്യാഹ്നത്തിന്റെ മധ്യലഗ്നത്തെ വരുത്തുകയും, ഈവണ്ണം തന്നെ അവിടെ അക്ഷം കൂടാതെ വരുത്തിയ ദൃക്ക്ഷേപം കൊണ്ട് എന്നേ വിശേഷമുള്ളു.

ദക്ഷിണോത്തരവൃത്തം സാക്ഷത്തിങ്കലും നിരക്ഷത്തിങ്കലും ഒന്നേ അത്രേ. എന്നിട്ട് ഇങ്ങനെ ദർശനസംസ്കാരപ്രകാരം.

[ഗണിതയുക്തിഭാഷയിൽ
മൗഢ്യവും ദർശനസംസ്കാരവും
എന്ന പതിനാലാമദ്ധ്യായം സമാപ്തം]

അധ്യായം പതിനഞ്ച്

ചന്ദ്രശൃംഗോന്നതി

1. ചന്ദ്രസൂര്യന്മാരുടെ ദ്വിതീയസ്ഫുടകർണ്ണം

അനന്തരം ചന്ദ്രന്റെ ശൃംഗോന്നതിയെ ചൊല്ലുന്നു. അവിടെ നടേ ചന്ദ്രാർക്കന്മാരുടെ ദ്വിതീയസ്ഫുടകർണ്ണം വരുത്തൂ. ചന്ദ്രനു ദ്വിതീയസ്ഫുടസംസ്കാരവും ചെയ്യൂ. ഇവിടെ മുമ്പിൽ ചൊല്ലിയവണ്ണം[1] ഉച്ചനീചവ്യാസാർദ്ധത്തെ വരുത്തിയാൽ പിന്നെ അതിനൊരു സംസ്കാരം ചെയ്യണമെന്നു 'സിദ്ധാന്തശേഖര' പക്ഷം. അവിടെ ഈ വരുത്തിയ അന്ത്യഫലത്തെ ചന്ദ്രന്റെ മന്ദകർണ്ണം കൊണ്ടും അഞ്ചിലും ഗുണിച്ച് ത്രിജ്യകൊണ്ടു ഹരിക്കണമെന്നു 'മാനസ' പക്ഷം. എന്നാൽ ഇതിനെ വിചാരിക്കണം. അനന്തരം ദൃക്കർണ്ണമുണ്ടാക്കി ഭൂപൃഷ്ഠസംസ്കാരത്തെ ചെയ്തു നതിയേയും സംസ്കരിച്ച് ആദിത്യനും[2] നതിയെ ഉണ്ടാക്കി പിന്നെ ആദിത്യനും[3] ചന്ദ്രനും ലംബനം സംസ്കരിച്ച് അന്നേരത്തെ ആദിത്യന്റേയും ചന്ദ്രന്റേയും ബിംബഘനമധ്യങ്ങൾ തങ്ങളിൽ എത്ര അകലമുണ്ട് എന്നതിനേയും അറിയൂ.

2. സൂര്യ-ചന്ദ്ര-ബിംബാന്തരം

അവിടെ യാതൊരിക്കൽ നതിയും വിക്ഷേപവും ഇല്ലാഞ്ഞൂ[1] അന്നേരത്തെ സ്ഫുടാന്തരത്തിന്റെ ക്രമജ്യാവും ഉൽക്രമജ്യാവും വർഗ്ഗിച്ചു തങ്ങളിൽ കൂട്ടി മൂലിച്ചതു സമസ്തജ്യാവ്, ദ്രഷ്ടാവിങ്കൽ കേന്ദ്രമായി രണ്ടു ബിംബത്തി

---

1. 1. B. മുൻചൊല്ലിയവണ്ണം
   2. B. അർക്കനും
   3. അർക്കനും
2. 1. F. ഇല്ല

ങ്കലും സ്പർശിച്ചിരിക്കുന്ന വൃത്തത്തിങ്കലേത്. ഇവിടെ ബിംബാന്തരത്തെ അറിയും നേരത്ത് എളുപ്പത്തിനായിക്കൊണ്ട് അപക്രമമണ്ഡലത്തെ ദ്രഷ്ടാവിനേക്കുറിച്ച് സമമണ്ഡലം എന്ന പോലെ ഖമധ്യത്തെ സ്പർശിച്ച് നേരേ കിഴക്കുപടിഞ്ഞാറായി കല്പിപ്പൂ. ഖമധ്യത്തിങ്കലാദിത്യനേയും കല്പിപ്പൂ. ആദിത്യനെ സ്പർശിക്കുന്ന[2] രാശികൂടവൃത്തം ദക്ഷിണോത്തരമായിട്ടു കല്പിപ്പൂ. ഇവിടെ ഒട്ടകന്നിട്ടു ചന്ദ്രനേയും കല്പിപ്പൂ. ചന്ദ്രനെ സ്പർശിച്ചിട്ട് ഒരു രാശികൂടവൃത്തത്തേയും കല്പിപ്പൂ. പിന്നെ വൃത്തകേന്ദ്രത്തിങ്കേന്ന് ആദിത്യങ്കലും ചന്ദ്രങ്കലും സ്പർശിച്ചിട്ടു രണ്ടു സൂത്രങ്ങളേയും കല്പിപ്പൂ. അവിടെ നേരേ മേൽകീഴായിരിക്കും അർക്കസൂത്രം. അവിടന്ന് ഒട്ടു ചരിഞ്ഞിരിക്കും ചന്ദ്രസൂത്രം. അവിടെ ചന്ദ്രരാശികൂടാപക്രമസംപാതത്തിങ്കലഗ്രമായി ഊർദ്ധ്വസൂത്രത്തിങ്കൽ മൂലമായിരിക്കുന്നപ്പോന്ന്. രാശികൂടവൃത്തങ്ങൾ രണ്ടിന്റേയും അന്തരാളത്തിങ്കലേ അപക്രമചാപഭാഗത്തിങ്കലേ അർദ്ധജ്യാവു ഭുജാജ്യാവാകുന്നത്. ഇതിന്റെ മൂലത്തിങ്കേന്നു ഊർദ്ധ്വസൂത്രത്തിങ്കൽ ആദിത്യനോളമുള്ളതു ശരം. ഇവ രണ്ടിന്റേയും വർഗ്ഗയോഗമൂലം ബിംബാന്തരസമസ്തജ്യാവ്. ഇതിന്റെ അർദ്ധത്തിന്റെ ചാപത്തെ ഇരട്ടിച്ചതു ബിംബാന്തരചാപം.

നതിവിക്ഷേപമില്ലാത്തപ്പോൾ[3] യാതൊരിക്കൽ പിന്നെ ചന്ദ്രന്റെ രാശികൂടത്തിന്മേലേ വിക്ഷേപിക്കുന്നു, അപ്പോൾ വിക്ഷേപജ്യാവിന്റെ മൂലം ചന്ദ്രസൂത്രത്തിങ്കൽ ചന്ദ്രങ്കേന്നു വിക്ഷേശരത്തോളം കീഴുനീങ്ങി സ്പർശിക്കും. ചന്ദ്രസൂത്രാഗ്രവും ഊർദ്ധ്വസൂത്രവും തങ്ങളിലന്തരാളം ഭുജാജ്യാവ്. അപ്പോൾ വിക്ഷേപജ്യാമൂലത്തിങ്കേന്ന് ഊർദ്ധ്വസൂത്രാന്തരാളത്തെ ത്രൈരാശികം ചെയ്തുണ്ടാക്കണം. ചന്ദ്രസൂത്രാഗ്രത്തിങ്കേന്നു ഊർദ്ധ്വസൂത്രാന്തരാളം ദോർജ്യാവു, വിക്ഷേപശരത്തോളം കുറഞ്ഞേടത്തിന് എത്ര എന്നു ത്രൈരാശികമാകുന്നത്. വിക്ഷേപശരംകൊണ്ടു ത്രൈരാശികം ചെയ്ത് ദോർജ്യാവിങ്കേന്നു കളകിലുമാം. പിന്നെ വിക്ഷേപശരഫലവർഗ്ഗത്തെ ക്ഷേപശരവർഗ്ഗത്തിങ്കേന്നു കളഞ്ഞു മൂലിച്ചത് ചന്ദ്രസൂത്രാഗ്രത്തിങ്കൽ സ്പർശിക്കുന്ന[4] ദോർജ്യാവും ക്ഷേപമൂലത്തിങ്കൽ സ്പർശിക്കുന്ന[5] ദോർജ്യാവും തങ്ങളിൽ മേൽകീഴുള്ള അന്തരാളമാകുന്നത്. ഈ അന്തരാളത്തെ സ്ഫുടാന്തരോൽക്ര

2. 2. F. സ്പർശിച്ചിരിക്കുന്ന
3. F. വിക്ഷേപങ്ങളില്ലാത്തപ്പോൾ
4. C. സ്പർശിച്ചിരിക്കുന്ന
5. C. സ്പർശിച്ചിരിക്കുന്ന

മജ്യാവിങ്കൽ കൂട്ടൂ. എന്നാൽ അർക്കങ്കേന്നു വിക്ഷേപമൂലത്തിങ്കൽ സ്പർശിക്കുന്ന[6] ദോർജ്യാമൂലത്തോളമുണ്ടാകും. ഇപ്പോൾ ശരം കുറഞ്ഞൊന്നു നീളമേറും, ദോർജ്യാവോടു നീളം കുറയും. ഇവ രണ്ടിന്റേയും വർഗ്ഗയോഗമൂലം അർക്കങ്കേന്നു വിക്ഷേപജ്യാമൂലത്തോളമുള്ള സൂത്രമായിട്ടിരിക്കും.

ഇതിന്റെ വർഗ്ഗത്തിൽ വിക്ഷേപവർഗ്ഗം കൂട്ടി മൂലിച്ചാൽ ബിംബാന്തരസമസ്തജ്യാവായിട്ടു വരും. യാതൊരിക്കൽ പിന്നെ അർക്കന്നു നതി ഉള്ളു, അപ്പോൾ ഖമധ്യത്തിങ്കേന്നു ദക്ഷിണോത്തരത്തിങ്കലെ വിക്ഷേപിച്ചു എന്നു കല്പിപ്പൂ. അവിടെ സ്ഫുടാന്തരശരത്തിങ്കേന്ന് അർക്കന്റെ നതിശരത്തെ കളയേണം. ശേഷം ക്ഷേപശരം പോയ ദോർജ്യാമൂലത്തിങ്കേന്ന് അർക്കനതിജ്യാവിന്റെ മൂലത്തോളമുള്ള ഊർദ്ധ്വസൂത്രഖണ്ഡമുണ്ടാകും. ഇതു സ്ഫുടാന്തരശരത്തിങ്കേന്ന് അർക്കന്റെ നതിശരത്തെ കളഞ്ഞു ചന്ദ്രന്റെ ക്ഷേപശരത്തിന്റെ കോടിഫലത്തെ കൂട്ടീട്ടുമിരിപ്പോന്ന്. ഇത് ഒരു രാശിയാകുന്നത്, സ്ഫുടാന്തരദോർജ്യാവ് ഒരു രാശിയാകുന്നത്.

ആദിത്യനും ചന്ദ്രനും അപക്രമമണ്ഡലത്തിങ്കേന്ന് ഒരുപുറത്തേ നീക്കമെങ്കിൽ നത്യന്തരം, രണ്ടു പുറത്തെങ്കിൽ നതിയോഗം. ഇത് ഒരു രാശിയാകുന്നത്. ഇവിടെ നതി കുറഞ്ഞ ഗ്രഹത്തിങ്കേന്നു നേരേ[7] കല്പിക്കണം സ്ഫുടാന്തരദോർജ്യാശരങ്ങളെ എന്നേ വിശേഷമുള്ളു. ഇവ മൂന്നിന്റേയും വർഗ്ഗയോഗമൂലം കൊണ്ടു ബിംബാന്തരസമസ്തജ്യാവുണ്ടാകും. ഇങ്ങനെ സ്ഫുടാന്തരം മൂന്നു രാശിയിൽ കുറയുമ്പോൾ ഇതിൽ ഏറുന്നാളും ക്രാന്തിവൃത്തം ഇവ്വണ്ണം തന്നെ കല്പിപ്പൂ. ഖമദ്ധ്യത്തിങ്കേന്ന് ഇരുപുറവുമൊപ്പമകന്നിട്ട് ചന്ദ്രാർക്കന്മാരേയും കല്പിപ്പൂ. അപ്പോൾ രണ്ടിനും നതിയില്ലാതെ ഇരിക്കയാൽ സ്ഫുടാന്തരാർദ്ധത്തിന്റെ അർദ്ധജ്യാവിനെ ഇരട്ടിച്ചത് ബിംബാന്തരമാകുന്നത്. രണ്ടിനും നതിയുണ്ടാകുമ്പോൾ[8] സ്ഫുടാന്തരാർദ്ധത്തിന്റെ ജ്യാവുകൾ രാശികൂടവൃത്താപക്രമസംപാതത്തിങ്കേന്ന് ഊർദ്ധ്വസൂത്രത്തോടുള്ള അന്തരാളം. അവിടെ[9] അതതു നതിശരത്തിങ്കേന്ന് ഉണ്ടാക്കിയ ദോർജ്യാഫലത്തെ അതത് അർദ്ധത്തിങ്കേന്ന് കളവൂ. എന്നാൽ നതിജ്യാമൂലത്തിങ്കേന്ന് ഊർദ്ധ്വസൂത്രാന്തരാളമുണ്ടാകും. ഇവിടേയും പിന്നെ നതീടെ വലിപ്പത്തിനു

2. 6. F. സ്പർശിച്ചിരിക്കുന്ന
7. B. F. രേഖ
8. C. D.E. adds. പിന്നെ
9. D. ഇവിടെ

തക്കവണ്ണം ഊർദ്ധ്വസൂത്രത്തിങ്കൽ[10] മേൽകീഴായി സ്പർശിച്ചിരിക്കും.

പിന്നെ ഊർദ്ധ്വസൂത്രത്തിങ്കലെ ദോർജ്യാമൂലത്തിന്റെ അന്തരാളത്തെ ഉണ്ടാക്കൂ. അത് ഇവിടെ നതിശരത്തിന്റെ കോടിഫലമാകുന്നത് ഈ ശരത്തിന്റെ അഗ്രവും, മൂലവും തങ്ങളിൽ മേൽകീഴുള്ള അന്തരാളം[11]. എന്നാൽ രണ്ടു ശരത്തിന്റേയും കോടിഫലം തങ്ങളിലന്തരിച്ചത് ദോർജ്യാമൂലങ്ങളുടെ മേൽകീഴുള്ള അന്തരാളമാകുന്നത്[12]. ഇത് ഒരു രാശി. സ്ഫുടാന്തരാർദ്ധത്തിന്റെ ജ്യാക്കളിൽ നിന്നു തന്റെ തന്റെ നതിശരത്തിന്റെ ദോർജ്യാഫലത്തെ കളഞ്ഞതു സ്ഫുടാന്തരദോർജ്യാവ്. ഇവ രണ്ടിനേയും കൂട്ടിയതു രണ്ടാം രാശി. നത്യന്തരം താൻ നതിയോഗം താൻ മൂന്നാംരാശിയാകുന്നത് ഇവറ്റിന്റെ വർഗ്ഗയോഗമൂലം ബിംബാന്തരസമസ്തജ്യാവ്. നതിയോഗംതാനന്തരം താൻ ചന്ദ്രാർക്കന്മാരുടെ തെക്കുവടക്കുള്ള അന്തരം. നതിഫലം കളഞ്ഞിരിക്കുന്ന അന്തരാർദ്ധജ്യാക്കളുടെ യോഗം കിഴക്കുപടിഞ്ഞാറന്തരമാകുന്നത്. പിന്നെ നതിശരങ്ങളുടെ കോടിഫലാന്തരം മേൽകീഴുള്ളന്തരമാകുന്നത്. ഇങ്ങനെ മൂന്നിന്റേയും വർഗ്ഗയോഗമൂലം ബിംബാന്തരസമസ്തജ്യാവ്. ഇങ്ങനെ സ്ഫുടാന്തരം വൃത്തപാദത്തിങ്കലേറുമ്പോൾ ബിംബാന്തരാനയനപ്രകാരം. ഇതു ഗ്രഹണത്തിങ്കലെ ബിംബാന്തരം വരുത്തുന്നേടത്തും തുല്യന്യായം[13].

[ഗണിതയുക്തിഭാഷയിൽ ചന്ദ്രശൃംഗോന്നതി എന്ന പതിനഞ്ചാമദ്ധ്യായം സമാപ്തം]

ഗണിതയുക്തിഭാഷാ സമാപ്തം

---

2. 10 F. സുത്രാഗ്രത്തിങ്കന്ന്
11. B. C. E. അനന്തരം
12. C. F. അന്തരമാകുന്നത്
13. D. ഗ്രന്ഥാവസാനത്തിൽ ലേഖകന്റെ കുറിപ്പ്

"നൃലേഖി യുക്തിഭാഷാ വിപ്രണേ ബ്രഹ്മദത്തസംജ്ഞേന" |
'ഗോളപഥസ്ഥാഃ സ്യുഃ' കലിരഹിതാശ്ശോധയന്തസ്തേ ||
കരകൃതമപരാധം ക്ഷന്തുമർഹന്തി സന്തഃ |
ശ്രീഗുരുഭ്യോ നമഃ. ശ്രീ സാരസ്വത്യൈ നമഃ ||
വേദവ്യാസായ നമഃ. എന്റെ ചങ്ങണോങ്കുന്നത്തു ഭഗവതി ശരണമായിരിക്കണം.

അനുബന്ധം-I

സാങ്കേതികപദസൂചി : മലയാളം - ഇംഗ്ലീഷ്
(Glossary of Technical Terms : Malayalam - English)

ശ്രദ്ധിക്കുക-

1. സാങ്കേതികപദങ്ങൾ ആദ്യമായി അവതരിപ്പിക്കപ്പെട്ടിട്ടുള്ളതോ, നിർവ്വചിക്കപ്പെട്ടിട്ടുള്ളതോ ആയ സ്ഥാനങ്ങളിൽ, ആ സ്ഥാനങ്ങളുടെ അദ്ധ്യായം, വിഭാഗം, ഉപവിഭാഗം എന്നിവകളോടുകൂടി അവ രേഖപ്പെടുത്തപ്പെട്ടിരിക്കുന്നു.

2. സാങ്കേതികപദങ്ങൾ വീണ്ടും ഉപയോഗിക്കപ്പെട്ടിട്ടുള്ള സ്ഥാനങ്ങളെ ഇങ്ങിനെ രേഖപ്പെടുത്തിയിട്ടില്ല.

3. തന്റെ തന്നെ അർത്ഥത്തോടുകൂടി അല്പം സാങ്കേതികത്വവും ചേർത്ത് ഉപയോഗിച്ചിട്ടുള്ള സാധാരണപദങ്ങളെ (common words) അദ്ധ്യായ-വിഭാഗാദിപരാമർശങ്ങൾ ഇല്ലാതെ, 'c.w.' എന്ന ചുരുക്കപ്പേരിൽ രേഖപ്പെടുത്തിയിരിക്കുന്നു.

4. രണ്ടോ മൂന്നോ സാങ്കേതികപദങ്ങൾ ചേർന്നുണ്ടായ സാങ്കേതികപദങ്ങളെ (derived words) അദ്ധ്യായ-വിഭാഗാദിപരാമർശങ്ങൾ ഇല്ലാതെ 'd.w.' എന്ന ചുരുക്കപ്പേരിൽ രേഖപ്പെടുത്തിയിരിക്കുന്നു.

അംശം, I.6	1. Part; 2. Numerator; 3. Degree in angular measure
അംശക്ഷേത്രം, I.6. iii	Area segment
അംശഗുണനം, III.3	Multiplication of fractions
അംശഭാഗഹരണം, III.4	Division of fractions
അംഹസ്പതി, (c. w)	Intercalary month in which two sankrantis occur, considered inauspicious.
അക്ഷം, IX. 1; XI.2	1. Latitude; 2. Terrestrial latitude
അക്ഷക്ഷേത്രം, (d.w)	Latitudinal triangle
അക്ഷജ്യാ, (d.w)	Rsine terrestrial latitude
അക്ഷദണ്ഡം, IX.7	Axle of a wheel
അക്ഷദൃക് കർമ്മം, (d.w)	Reduction due to the latitude of the observer
അക്ഷവലനം, XII.5	1. Angle subtended at the body on the ecliptic by the arc joining the north point of the celestial horizon and the north pole of the equation; 2. Deflection due to the latitude of the observer.
അഗ്രം, VII.3	1. The extremity of a line or arc; 2. Remainder in division in Kuṭṭākāra
അഗ്രാ, XI. 14	Amplitude at rising, i.e., the north south distance of the rising point from the east-west line; the Rsine thereof,
അഗ്രാംഗുലം, X. 14	Agrā in terms of aṅgulas
അങ്ഗുലം, (c.w)	Linear measure, inch
അണുപരിമാണം, VIII.8	Infinitesimal
അതിദേശം, XI.19	Application or use a general rule

അധികം, (c.w)	Additive
അധികാബ്ദം, (d.w)	Additive lunar year
അധികശേഷം, V.1	The positive remainder after division
അധികാഗ്രഹാരം, (d.w)	The divisor in Sāgra- kuṭṭākāra which has numerically the greater remainder
അധിമാസം, V.1	Intercalary month
അധോമുഖശങ്കു	Downward gnomon
അന്തരചാപം, (d.w)	The intervening arc between two points in the circumference of the circle
അന്തരാളം, (d.w)	1. Difference; 2. The perpendicular distance from a point to a straight line or plane; 3. Divergence; 4. Intervening
അന്ത്യം, I.2	1. 10^{15} (Place and number); 2. The digit of highest denomination; 3. The last term in a series
അന്ത്യക്രാന്തി = പരമക്രാന്തി, (d.w)	Maximum declination, 24^0
അന്ത്യസ്ഥാനം, I.5.ii	1. The place of the digit of the highest denomination; 2. The ultimate place when arranged in a column.
അന്യോന്യഹരണം, (d.w)	Mutual continued division (as in finding G.C.M.)
അപക്രമം (അപക്രമധനുസ്, അപക്രാന്തി, അപമക്രാന്തി), IX.3	Declination of celestial body; obliquity of the ecliptic.
അപക്രമമണ്ഡലം (വൃത്തം, ക്രാന്തിവൃത്തം), VIII.16; IX.3, 12	Ecliptic, path of the Sun in the sky.
അപമണ്ഡലം, VIII.16; IX.3	Ecliptic

അപരപക്ഷം, (c.w)	The period from full moon to new moon
അപരവിഷുവത്ത്, IX.3	Point at which the Sun coursing along the Ecliptic crosses the Celestial sphere from the north to the south.
അപവർത്തനം, V.3	1. G.C.M; 2. Reducing a fraction or ratio to lowest terms; 3. Abrader
അപവർത്തനഹാരകം, V.3	Greatest Common Multiple (G.C.M.)
അബ്ജം, I.2	10^9 (number and place)
അമാവാസി, (c.w)	New Moon
അയനം, (c.w) XI. 3	1. Northward and southward motion of the Sun or other planets; 2. Declination
അയനചലനം, IX. 4	Precession of the equinoxes
അയനദൃക്കർമ്മം, (d.w)	Reduction for observation on the ecliptic
അയനവലനം, XII.5	1. Angle between the secondaries and the ecliptic of the place of the eclipsed body on the ecliptic; 2. Deflection due to declination
അയനസന്ധി, IX.3	Solstice
അയനാന്തം, IX.3	Solstice, vernal and autumnal
അയനാന്തവിപരീതവൃത്തം, IX.10	Reverse solsticial circle
അയനാന്തോന്നതി, (d.w)	Elevation of the Solstices
അയുതം, I.2	Number and place of 10,000
അർക്കാഗ്രാ, XI.13	1. Measure of the amplitude in the arc of the celestial horizon lying between the east point and point where the heavenly body concerned rises; 2. The distance from the extremity of the gnomonic shadow and the equinoctical shadow.

അർക്കാഗ്രാംഗുലം, XI.2	Measure of the arkāgrā in aṅgulas
അർദ്ധജ്യാ, (ജ്യാവ്, ജ്യാ) VII.3	Rsine
അല്പവൃത്തം, VIII.1	Smaller circles parallel to the Big circle, the ecliptic
അല്പശേഷം, V.3, 4	In Kuṭṭākāra the smaller of the last two remainders taken into consideration
അവമം (തിഥിക്ഷയം), V.1	Omitted lunar day.
അവർഗ്ഗസ്ഥാനം, (d.w)	Even place counting from the unit's place.
അവലംബകം, (c.w)	Plumb
അവാന്തരയുഗം, V. 3	A Unit of time. viz. 576 years or 210389 days adopted by ancient Hindu astronomers.
അവിശിഷ്ടം, V.3, 4	Obtained by successive approximation or iteration.
അവിശേഷം, V.3, 4	Successive approximation process of iteration
അവ്യക്തരാശി, (c.w)	An unknown quantity
അശ്രം, (c.w)	1. A side of a polygon; 2. An edge.
അഷ്ട്രാശ്രം, VI.2	Octogon
അസിതം, XII. 1, 2	Non-illuminated part of the moon in eclipse
അസു (പ്രാണൻ), (c.w)	Unit of time equal to 4 seconds
അസ്തമയം, IX.2	Setting, Diurnal or heliacal
അസ്തലഗ്നം, XI.3, 34	1. Lagna of time of planet's setting; 2. Setting or occident ecliptic.
അസ്ഫുടം, (d.w)	1. Rough; 2. Inexact
അഹർഗണം	Days elapsed from epoch

അഹർദലം, XII.7	Mid day
അഹോരാത്രവൃത്തം (ദ്യുവൃത്തം) (d.v)	Dirunal circle; Smaller circle, parallel to the Ghaṭikāmaṇḍala (celestial equator) along which stars rising north or south of the poles move.
ആകാശം, IX.1	1. Celestial Sphere; 2. Sky
ആകാശകക്ഷ്യാ (അംബരകക്ഷ്യാ, ഖകക്ഷ്യാ), IX.1	Boundary circle of the sky, having the linear distance which a planet travels in a yuga, equal to 124,74,72,05,76,000 yojanas, denoted by the expression ajñānitaamonamā sarāevpriyo nanu in Kaḷapayādi notation.
ആക്ഷം, (d.w)	Relating to Latitude
ആദി, VIII.1	1. Beginning; 2. Commencement; 3. Starting point
ആദിത്യമദ്ധ്യമം, (d.w)	The mean longitude of the Sun
ആദ്യകർണ്ണം, VIII.15	One of the diagonals of a quadrilateral taken for reference. The other is known as dvitīyakarṇa or itarakaṇa
ആദ്യസംകലിതം, VI.5.v	First integral or sum of an Arithmetic progression.
ആദ്യസ്ഥാനം, (d.w)	Unit's place
ആബാധ, VII.2	The two segments into which the base of a triangle is divided by the perpendicular from the vertex
ആയതചതുരശ്രം, (d.w)	Rectangle
ആയാമം, (c.w)	Length
ആയാമവിസ്താരം, (c.m)	Length and breadth
ആർക്ഷം, (നക്ഷത്രം) (c.m)	Sidereal

ആർത്തവത്സരം, (d.w)	1. Tropical year, from viṣuvat to viṣuvat; 2. Sāyanavatsara
ആശാഗ്രം, (d.w)	North-South distance of the rising point from the east-west line.
ആഹതി, (d.w)	Product
ഇച്ഛാ, IV.1	Requisition, being the third of the three quantities in the Rule of Three
ഇച്ഛാഫലം, IV.1	1. The desired consequent; 2.The fourth proportional.
ഇടം, (c.w)	Breadth
ഇതരകർണ്ണം, VII.15	The second diagonal in a quadrangle
ഇതരജ്യാവ്, (d.w)	The other co-ordinate
ഇതരേതരകോടി, (d.w)	The ordinate of the other Rsine
ഇന്ദുപാതം, (d.w)	Ascending node of the Moon
ഇന്ദൂച്ചം, (d.w)	Higher apsis of the Moon
ഇലി (ലിപ്ത, കല), VIII.1	1. Minute of angular measure; 1/360 of the circumference in angular measure
ഇഷ്ടം, (c.w)	Desired or given number
ഇഷ്ടകാലസ്വാഹോരാത്രം, XI.3	Day duration relating to the desired time.
ഇഷ്ടഗ്രഹണകാലം, XII.2	Moment of desired occultation
ഇഷ്ടജ്യാ, IX.1	Rsine at the desired point on the circumference of a circle
ഇഷ്ടദിക്ഛായാ, XI.20.iv	Shadow in desired direction
ഇഷ്ടദിഗ്വൃത്തം, XI. 20	Circle passing through the zenith and the planet
ഇഷ്ടദോഃകോടി ധനുസ്സ്, (d.w)	The complementary arc of any chosen arc

ഇഷ്ടപ്രദേശം, (d.w)	The desired piont
ഇഷ്ടഭുജാചാപഃ, VII.3	Arc of specified Rsine
ഇഷ്ടാപക്രമഃ, IX.9	Desired declination
ഇഷ്ടാപക്രമകോടിഃ, IX.9	Rcos. desired declination
ഇഷ്ടസംഖ്യ, I.6.ii	The desired number
ഉച്ചം, VIII.5	1. Higher apsis, especially pertaining to the epicycle of the equation of the centre; 2. Apogee of the Sun and the Moon; 3. Aphelion of the planets.
ഉച്ചനീചവൃത്തം, VIII.3	Epicycle
ഉച്ചനീചപരിധി, VIII.3	Epicycle
ഉച്ചനീചസൂത്രം, VIII.7, 8	See Ucca
ഉച്ചസൂത്രം, VIII.7,8	See Ucca
ഉജ്ജയിനി, IX.1	City in Central India, the meredian passing through which is taken as zero
ഉത്ക്രമജ്യാ, VII.4	Rversed sine
ഉത്തരവിഷുവത്ത്, IX.3	Autumnal equinox
ഉത്തരോത്തരസംകലിതൈക്യം, VI.14	Summation of Summation of progressive numbers
ഉദയം, IX.2, XI.3	1. Rising; 2. Heliacal rising; 3. Rising point of a star or constellation at the horizon
ഉദയകാലം, XI.3	Moment of the rising of a celestial body.
ഉദയജ്യാ, XI.3	1. Rsine of the amplitude of the rising point of the ecliptic; 2. Oriental sine; 3. Rsine of the amplitude of lagna in the east.
ഉദയലഗ്നം, XI.3	1. Rising sign; 2. Rising on orient ecliptic point
ഉദയാസ്തമയമാർഗ്ഗം, IX.2	Path of a Planet from rising to setting

ഉന്നതജ്യാ, XI.4, 26	Rsine of 90^0 less zenith distance
ഉന്നതപ്രാണൻ, XI.16	The Prāṇas in time yet to expire for a planet to set
ഉന്മമണ്ഡലം (ലങ്കാക്ഷിതിജം), IX. 7	1. Diurnal circle at Laṅkā; 2. East-West hour circle; Equinoctial colure; 3. Big circle passing through the North and South poles and the two East-West svastika; 4. Equitorial horizon.
ഉന്മീലനം, XII, 1,2	Emersion, in eclipse
ഉപപത്തി (യുക്തി), V.3, 8	Proof, Rationale
ഉപാധി, (c.w)	Assumption
ഉപാന്ത്യം, I.5.ii	1. Penultimate; 2. Penultimate term
ഊനശേഷം, (d.w)	The smallest number to be added to the dividend to make it exactly divisible by the given divisor
ഊനാഗ്രഹാരം, V	The divisor in Sāgra - Kuṭṭākāra which has numerically the smaller remainder
ഊനാധികധനുസ്സ്, (d.w)	The deficit or excess of an arc
ഊർദ്ധ്വം, (c.w)	The topmost; The earlier; Preceeding
ഊർദ്ധ്വാധോരേഖ, VIII.1, 3	Vertical
ഋക്ഷം, (നക്ഷത്രം) (c.w)	1. Asterism; 2. Star-group
ഋണം, (c.w)	1. Negative; 2. Subtractive quantity
ഏകം, I.2	1. Unit; 2. Unit's place; 3. One
ഏകദേശം, (d.w)	1. In the same straight line; 2. A part
ഏകദ്വിത്ര്യാദി, VI. 5.iv (ഏകാദിക്രമേണ)	1. Consecutive; 2. Numbers starting from unity

ഏകാദ്യേകോത്തരം, VI. 4 (ഏകാദ്യേകോത്തര സംകലിതം)	1+2+3+4 etc.
ഏകാദ്യേകോത്തരവർഗ്ഗസംകലിതം, VI. 4	$1^2+2^2+3^2+$-------------------
ഏകാദ്യേകോത്തരഘനസംകലിതം, VI.4.iii	$1^3+2^3+3^3+$-------------------
ഏകാദ്യേകോത്തരവർഗ്ഗവർഗ്ഗ സംകലിതം, VI.4.iii	$1^4+2^4+3^4+$-------------------
ഏകാദ്യേകോത്തര സമപഞ്ചഘാത സംകലിതം, V.4, iii, iv	$1^5+2^5+3^5+$---------------
ഏകാദ്യേകോത്തരസംകലിതം, VI.5.v	1+2+3+------------------
ഏകൈകോനം, (d.w)	Numbers descending by unity
ഏഷ്യചാപം, (d.w)	The arc to be traversed
ഓജം; - പദം, I.8.i; VII. 3	1. First and third quadrants of a circle; 2. Odd
കക്ഷ്യാ, VIII. 1, 2	Orbit
കക്ഷ്യാപ്രതിമണ്ഡലം, VIII. 2	Eccentric
കക്ഷ്യാമണ്ഡലം, VIII.7	1.Mean orbit; 2.Deferent; 3. Concentric
കക്ഷ്യാവൃത്തം, VIII.4	Orbital circle of a planet
കപാലം, (c.w)	Hemisphere
കരണം, (c.w)	Half-tithi period
കർക്കി, IX.3	Sign Kaṭaka, Cancer
കർക്ക്യാദി, IX.3	Commencing from the sign Karki or Cancer, the fourth zodiaed constellation
കർണ്ണം, VI.2, VII.3	1. The diagonal of a quadrilateral; 2. Hypotenuse of a right angled triangle; 3. Radias vector
കർണ്ണവൃത്തം, VIII. 7, 8	Hypotenuse circle

കർണ്ണവൃത്തജ്യാവ്, VIII. 7	Rsine in hypotenuse circle
കലാ, (അംശം, ലിപ്ത)	1. 1/21600 of the circumference of a circle; 2. Minute of arc
കലാഗതി, (d.w)	Daily motion of planets in terms of minutes of arc
കലാർദ്ധജ്യാ, (d.w)	The 24 Rsine differences in terms of minutes.
കലാവ്യാസം, (d.w)	Angular diameter in minutes
കലിദിനം, (കല്യഹർഗണം) (d.w)	Number of days elapsed since the Kali epoch
കലിയുഗം, (d.w)	The aeon which commenced on Feb,18th, 3102 B.C. at sunrise at Lanka
കല്യാദി, V.1	Commencing from Kali epoch
കല്യാദിധ്രുവം, (d.w)	Zero positions of Planets at the commencement of the Kali epoch
കാലകോടിജ്യാ, IX.11	Sine from the zenith with its tip at the point of contact of the Rāśikūṭa and Ghaṭikāvṛtta on the Ghaṭikāvṛtta
കാലകോട്യപക്രമം, IX.11	Declination of the Kālakoṭi on the Rāśikūṭavṛtta
കാലജ്യാ, (കാലദോർഗുണം), IX.12	Rsine of the angle between two points of time in degrees.
കാലഭാഗം, (കാലാംശം)	Degree of time at the rate of one hour equal to 15 degrees of time
കാലലഗ്നം, XI. 31, 32	Ecliptic point on the horizon at the desired time.
കുട്ടാകാരം, V. 3	Pulveriser, a type of indeterminate equation, called also Diophantine equation

കൃതി	Square
കൃഷ്ണപക്ഷം, (c.w)	Dark half of the lunar month
കേന്ദ്രം, VIII. 1,8; XI.1	1. Centre of a circle; 2. The particular point on the circumference from which the arc is measured; Anomaly 3. Mean anomaly or commutation; 4. Distance from Mandocca or Śīghrocca to mean planet
കേന്ദ്രഭ്രമണം, VIII.2	Movement of the Kendra
കോടി, VII.1	1. Abscissa; 2. Adjacent side of a rightangled triangle; 3. Corner rafters of kipped roof, 4.10^7 (number and place); 5. Complement of bhuja.
കോടിഖണ്ഡം, VII.2, 3	1. The difference between two successive abscissa; 2. The first differential of koṭijyā
കോടിചാപം, VII.5	Arc of R. cos
കോടിജ്യാ, VII.5	Rsine koṭi or Rcosine of bhujā
കോടിമൂലം, കോടഗ്രം, VII.2, 3	The point at which koṭi (R.cos) touches the circle at its statrting point and the other end is its end
കോടിവൃത്തം, VII.3	R cos circle
കോൺ, VI.1	1.Corner; 2.Direction; 3.Angle
കോണച്ഛായാ, XI. 20, iii	1. Shadow at the moment of passing the Karṇatta 2. Corner shadow
കോണവൃത്തം, (d.w)	Vertical circle extending from North-east to South-west or from North-west to South - east
കോണശങ്കു, XI. 20.iii	1.Šaṅku formed at the moment of passing the koṇavṛtta. 2.Corner Šaṅku

കോൽ, (c.w)	A unit of length equal to about 28 inches
ക്രമജ്യാ	Sum of the sine segments taken in order
ക്രമശങ്കു	Gnomon formed at the moment of passing the koṇavṛtta.
ക്രാന്തി, അപക്രമം, IX. 3	1. See Apakrama, 2.Declination
ക്രാന്തികോടി	Reverse declination
ക്രാന്തിജ്യാ, XI.21	Rsine declination
ക്രാന്തിമണ്ഡലം, (d.w) ക്രാന്തിവൃത്തം, (d.w)	1. Zodiacal circles; 2. Path of the Sun in the sky
ക്രിയാ, (c.w)	Sign Meṣa, Aries
ക്ഷിതിജം, IX.10	Terrestrial horizon passing through the four cardinal directions, where there is no latitude
ക്ഷിതിജ്യാവ്, XI. 14, 26	Sine on that part of the diurnal circle
ക്ഷേത്രം, I.5	Plane figure Geometrical figure
ക്ഷേത്രഫലം, I.5 v	Area of a plane or geometrical figure
ക്ഷേപം, V.1	1. Celestial latitude; 2. Additive quantity
ഖകക്ഷ്യാ, IX. 3	Ākāśakakṣyā
ഖഗോളം, VIII.1; IX. 3	Celestial sphere or globe
ഖണ്ഡം, I.8.ii	Part
ഖണ്ഡഗുണനം, (d.w.)	Multiplication by parts
ഖണ്ഡഗ്രഹണം, (d.w.)	Partial eclipse
ഖണ്ഡജ്യാ, VII.5	1. The difference between two successive ordinates; 2. The first differential of Bhujājyā (Rsine; Sine segment)
ഖണ്ഡജ്യാന്തരം, VI.7	The second differential of Jyā

ഖണ്ഡജ്യായോഗം, VI.8	Sum of sine segments
ഖമധ്യം, IX.3	1.Zenith; 2.Middle of the sky
ഖർവ്വം, I.2	10^{10} (Number and place)
ഗച്ഛം, VI.4	Number of terms in a Series
ഗച്ഛധനം, (d.w)	Sum of specified number terms in a Series
ഗണിതം, I.2.3	Mathematics
ഗതം, (c.w)	Elapsed portion of the days.
ഗതഗന്തവ്യപ്രാണൻ, XI.4	The prāṇas gone and to go
ഗതചാപം, (d.w)	The arc already traversed
ഗതി, VIII.1	1. Motion ; 2. Motion of celestial bodies
ഗതികലാ, (d.w)	Motion in terms of minutes of arc of a planet
ഗതിഭേദം, (d.w)	Difference in motion or rate of motion
ഗുണം, I.3	1.Multiplication; 2.Multiplier; 3.Rsine
ഗുണകം, I.5	Multiplier
ഗുണകാരം, I.5	Multiplier
ഗുണനം, I.5	Multiplication
ഗുണ്യം, I.5	Multiplicand
ഗുർവ്വക്ഷരം, (c.w)	20, One-sixtieth of a vinaḍi, 24/60/of a second in time measure
ഗോളം, VII.18, IX.7	1.Sphere; 2.Celestial sphere; 3. Globe
ഗോളകേന്ദ്രം, VIII.1.2	Centre of a sphere
ഗോളഘനം, VII.19	Volume of a sphere
ഗോളപൃഷ്ഠം, VII.18	Surface of a sphere
ഗോളപൃഷ്ഠഫലം, VII.18	Surface area of a sphere
ഗോളബന്ധം, IX.8	Construction of the armillary sphere

ഗോളമധ്യം, VIII.1	Centre of the sphere
ഗോളാദി, XI.5	The point of contact of the ghaṭikā and apakramavṛtta
ഗ്രഹം, VIII.1	Planet, including the Sun and the Moon, and the ucca or higher apsis, and pāta or ascending node
ഗ്രഹഗതി, VIII.1, 3, 4	Daily motion of a Planet
ഗ്രഹണം, XII.1-10	Eclipse
ഗ്രഹണകാലം, XII.1, 2	Duration of occulation during an eclipse
ഗ്രഹണപ്രദേശം, XII. 1	1. Portion of Sun or Moon eclipsed; 2. Magnitude of an eclipse
ഗ്രഹണമദ്ധ്യം, XII.2	Middle of eclipse
ഗ്രഹണലേഖനം, XII.9	Geometrical representation of the eclipse
ഗ്രഹണസംസ്ഥാനം, XII.3	State or situation of an eclipse at a particular time
ഗ്രഹഭുക്തി, VIII.1	Daily motion of a planet
ഗ്രഹഭ്രമണവൃത്തം, VIII, 3, 5	Circle of motion of a planet
ഗ്രഹയോഗം, (d.w)	Conjunction of two planets
ഗ്രഹവൃത്തകേന്ദ്രം, VIII. 1,2	Centre of a planet's orbit
ഗ്രഹസ്ഫുടം, VIII. 1	True longitude of a planet
ഗ്രഹാസ്തോദയം, XIV.2	Rising and setting of a planet
ഗ്രാസം, VII.22	The maximum width of the overlap of two intersecting circles or an eclipse and measure thereof.
ഗ്രാസോനവ്യാസം, (d.w)	The difference between the diameter and eclipsed portion in eclipse

ഗ്രാഹകം, XII.1	Eclipsing body in an eclipse
ഗ്രാഹകബിംബം, XII.1	Eclipsing body
ഗ്രാഹ്യം, XII.1	Eclipsed body in an eclipse
ഗ്രാഹ്യബിംബം, XII.1	Orb of the eclipsed body
ഘടികാ (നാഡിക), (c.w.)	Unit of time equal to 24 minutes
ഘടികാനതവൃത്തം, IX.10 (ഘടികാമണ്ഡലം)	Celestial Equator; path of the star rising exactly in the east and setting exactly in the west.
ഘനം, I.3	1.Cube of a number; 2. Solid body; 3. Sphere
ഘനക്ഷേത്രഫലം, (d.w)	Volume of a body
ഘനമദ്ധ്യം, (d.w)	Centre of a sphere
ഘനമൂലം, I.3	Cube root
ഘനസംകലിതം, VI. 5.iii,	Sum of a Series of cubes of natural numbers
ഘാതം, I.10	Product
ഘാതക്ഷേത്രം, I.5; v; I.8.ii	Rectangle
ചക്രം, (c.w.)	1. Circle; 2. Cycle
ചക്രകലാ (ചക്രലിപ്ത), (d.w.)	Minutes of arc contained in a circle being 21600
ചതുരശ്രം, (c.w.)	Quadrilateral
ചതുരശ്രഭൂമി, VII.18	The base of a quadrilateral. The opposite side is known as face (Mukham)
ചതുർയുഗം, V.1	A unit of time, viz. 4320000 years, adopted by ancient Hindu astronomers
ചന്ദ്രഗ്രഹണം, XII.2, 10	Lunar eclipse
ചന്ദ്രശൃംഗോന്നതി, XV. 1, 2	Measure of the Moon's phases

ചയം, VI. 4	The common difference in an Arithmetic progression
ചരം, VII.2	1. Arc of the celestial equator lying between the 6 o' clock circle and the hour circle of a heavenly body at rising; half the variation of a siderial day from 30 naḍikās; 2. Declinational ascensional difference
ചരകലാ, (d.w.)	Minutes of longitude corresponding to cara
ചരജ്യാ, (c.w)	Rsine caradala
ചരദളം (ചരാർദ്ധം) (d.w.)	Half ascensional difference
ചരപ്രാണം (ചരാസു) (d.w.)	Prāṇas or asus of ascensional difference
ചരാർദ്ധം, VII.I	Half ascensional difference
ചാന്ദ്രമാസം, V.I	1. Lunar month; 2. Period from one new moon to the next, equal to about 29.53 civil days
ചാപം, VII.I	1. Arc or segment of the circumference of a circle; 2. Constellation Dhanus
ചാപകോടി, (d.w.)	Complementary arc of Bhujacāpa
ചാപഖണ്ഡം, (d.w.)	Cāpa segment
ചാപഭുജാ, (d.w.)	An arc measured from Meṣādi and Tulādi in the anti-clock-wise direction in the first and third quadrants and in the clock-wise direction in the second and fourth quadrants
ചാപീകരണം, VI.6	Calculating the arc of a circle from its semichord
ചാരം, (c.w)	Motion
ഛാദകം (ഗ്രാഹകം), (d.w.)	Eclipsing body

ഛാദ്യം, (d.w)	Eclipsed body
ഛായ, X.I	1. Shadow; 2. Rsine of zenith distance, i.e., mahācchāyā
ഛായാകർണ്ണം, VII.17	Hypotenuse of a right angled triangle one of whose sides is the gnomon and the other is the shadow.
ഛായാകോടി, XI.3	R.Cos. shadow of a gnomon
ഛായാകോടിവൃത്തം, XII.7	Circle described by Rcos. shadow of gnomon
ഛായാഭുജ, XI.13	Rsine gnomonic shadow.
ഛായാലംബനം, XI.8, 37	Parallax of the gnomon
ഛേദം, III.2	Denominator
ഛേദകം, ഛേദ്യം, III.1	1.Figure; 2.Diagram; 3.Drawing
ജലധി, I.2	10^{14}, (number and place)
ജീവാ, ജ്യാ VI.19	Rsine
ജീവേപരസ്പരന്യായം, VII.8, 11	R sine (A plus or minus B)
ജൂകം, (c.w.)	Sign Tulā, Libra
ജ്യാ, ജ്യാവ്, (ജ്യാർദ്ധം), VII. 1	1.Semi-chord; 2.Ordinate of an arc; 3.Rsine line joining the two ends of an arc.
ജ്യാഖണ്ഡം, (d.w.)	1.Segment of arc; 2.Sine segment, 3.Sine difference.
ജ്യാചാപാന്തരം, (d.w.)	Difference between an arc and the corresponding semi-chord
ജ്യാപിണ്ഡം, (d.w.)	The semi-chords of one, two etc. parts of the arcs of a quadrant which is divided into any number of equal parts.

ജ്യാർദ്ധം, VII.I	See jya
ജ്യാവർഗ്ഗം, VII.7	Square of R sine
ജ്യാശരവർഗ്ഗയോഗമൂലം, (d.w.) VII.1	Root of the sum of the squares of R sine and R reversed sine
ജ്യാസംകലിതം, VII.5	The summation of semi-chords
ജ്യാ (സമസ്ത-), VII.1	Complete chord of the arc
ജ്യോതിർഗോളം, X.2, 3, 7	Celestial sphere
ജ്യോതിശ്ചക്രം, VIII.1, IX.1	Circle of asterisms
ഝഷം (മത്സ്യം), (c.w)	Figure of fish formed in a geometrical diagrams., like as in intersecting circles
തക്ഷണം, V.3	1.The method of abrasion; 2.The numbers by which the guṇakāra and phala are abraded
തമസ്സ്, (c.w)	1.Shadow cone of the earth at the Moon's distance; 2.Moon's ascending node.
തല്പര	One sixtieth of a vikalā or vili of angular measure
തഷ്ടം, V.3	Abraded
താഡനം	Multiplication
താരാഗ്രഹം, (d.w)	Star planets, viz., Mars, Mercury, Jupiter, Venus and Saturn
തിഥി, V.1	1.Lunar day, 2.Thirtieth part of the lunar or synodic month
തിഥിക്ഷയം, (d.w)	1.Omitted lunar day, 2. Subtractive day
തിഥ്യന്തം, (d.w)	End of the new moon tithi or the full moon tithi
തിർയ്യഗ്വൃത്തം, XI.20.i	Oblique or Transverse circle

തീയ്യതി (അംശം, ഭാഗം), IX.9	One degree of angular measure
തുംഗൻ, VIII.5	Apogee of the Moon
തുലാദി, IX.3	1.The six signs commencing from Tulā, 2.The other side of Meṣādi
തുല്യാകാരക്ഷേത്രം, (d.w)	Similar figures
തൃതീയകർണ്ണം, VII.10	In a cycle quadrilateral if any two sides are interchanged, a third diagonal is obtained which is called by this term
തൃതീയസംകലിതം, VI.5	Third integral
ത്രിജ്യാ, (ത്രിഭജ്യാ, ത്രിരാശിജ്യാ) (d.w)	1. Rsine 90^0; 2. The radius of length 3438 units, with the length of a minute of arc taken as unit and corresponding to unity in the tabular sines.
ത്രിഭജ്യാ, (ത്രിജ്യാ), (d.w)	Rsine 90 degrees
ത്രിരാശിജ്യാ (ത്രിജ്യാ), (d.w)	Rsine of 90^0, Rsine of three rāśis
ത്രിരാശ്യൂന കാലലഗ്നം, (d.w)	Kālalagna less 90^0
ത്രിശരാദി, (d.w)	Set of odd numbers (3, 5, 7, etc.)
ത്രൈരാശികം, IV.1.	1.Rule of Three, 2.Direct proportion
ത്രിഭുജം (ത്ര്യശ്രം), (d.w)	Triangle
ത്ര്യശ്രം വിഷമം, (d.w)	Scalene triangle with all three sides of a different lengths.
ദക്ഷിണോത്തരനതവൃത്തം, IX. 10	1. North-south big circle
ദക്ഷിണോത്തരമണ്ഡലം, IX.10	Meridian Circle
ദക്ഷിണോത്തരരേഖ, VIII.3	North-south line; Meridian; Solstical colure
ദക്ഷിണോത്തരവൃത്തം, IX.2	North-South Big circle passing through the zenith, round the celestial sphere
ദർശനസംസ്കാരം, XIV.1,3	Visibility correction of planets

ദളം, (C.W.)	Half
ദശം, 1.2	10 (number and place)
ദിക്ക്, (c.w)	Direction
ദിക്ചക്രം (ക്ഷിതിജം), (d.w)	Terrestrial horizon passing through the four cordinal directions at which that is no latitude.
ദിക്ജ്ഞാനം, IX.1, 15	Method of ascertaining the directions
ദിക്സാമ്യം, (d.w)	Same or parallel line or direction
ദിക്സൂത്രം, VI.1, 3	Straight lines indicating directions
ദിഗഗ്രാ, (d.w)	North-south distance of the rising point from the east-west line
ദിഗ്വൈപരീത്യം, VI.3	Perpendicularity
ദിനഭുക്തി, (d.w)	Motion per day
ദിവസം, (c.w)	Solar day
ദിവ്യദിനം, (d.w)	Divine day
ദിവ്യാബ്ദം, (d.w)	Divine year, equal to 360 years of men
ദൃക്കർണ്ണം, XI.17, 28	Hypotenuse with Dṛggolaśaṅku and Dṛggolacchāyā as sides
ദൃക്കർമ്മം, IX.6, 7	Reduction to observation
ദൃക്ഛായാ, XI.7	Parallax
ദൃക്ക്ഷേപം, XI.34	1. Ecliptic zenith distance; 2. Zenith distance of the non-agesimal or its Rsine
ദൃക്ക്ഷേപകോടി, XI.34	Rcos Dṛkkṣepa
ദൃക്ക്ഷേപജ്യാ, XI.34	Rsine Dṛkkṣepa
ദൃക്ക്ഷേപജ്യാകോടി, XI.34	Rcos Dṛkkṣepa
ദൃക്ക്ഷേപമണ്ഡലം (-വൃത്തം), XI.31	1. Vertical circle through the central ecliptic point. 2. Secondary to the ecliptic passing through the zenith.

ദൃക്ക്ഷേപലഗ്നം, (d.w)	Nonagesimal; point on the ecliptic 90^0 less from the lagna or rising point of the ecliptic
ദൃക്ക്ഷേപവൃത്തം, XI.31	1. Vertical circle through the central ecliptic point. 2. Secondary to the ecliptic passing through the zenith
ദൃക്ക്ഷേപശങ്കു, XI.34	Gnomon re- ecliptic zenith distance
ദൃഗ്ഗതി, (d.w)	Arc of the ecliptic measured from the central ecliptic point or its Rsine; Rsine altitude of the nongesimal
ദൃഗ്ഗതിജ്യാ, (d.w)	Rsine of the attitude of the nonagesimal points of the ecliptic
ദൃഗ്ഗോളം, X.7	1. Visible celestial sphere; 2. Khagola and Bhagola together
ദൃഗ്ഗോളച്ഛായ, XI.7	Shadow relating to Dṛggola
ദൃഗ്ഗോളശങ്കു, XI.7	Gnomon relating to Dṛggola
ദൃഗ്ജ്യാ, (d.w)	Rsine of the zenith distance
ദൃഗ്വൃത്തം (ദൃങ്മണ്ഡലം)[1], XI.6, 20	Vertical circle passing through the zenith of the observer and the planet
ദൃങ്മണ്ഡലം, XI. 6, 20	Vertical circle in the Dṛggola
ദൃങ്മധ്യം, (d.w)	Centre of the eye-level of the seer on the surface of the earth
ദൃഢം, (c.w)	Reduced by the G.C.M., i.e. converted into primes of each other in indeterminate equations
ദൃഢക്ഷേപം (ശുദ്ധി), (d.w)	Additive and subtractive divided by the G.C. M of dividend and divisor in Kuṭṭākāra
ദൃഢഭാജകം, V.3	Reduced divisor (by the G.C.M)
ദൃഢഭാജ്യം, V.3	Reduced dividend by G.C.M

ദേശാന്തരം, (d.w)	1. Longitude; 2. Difference in terrestrial longitude; 3. Correction for terrestrial longitude
ദേശാന്തരകാലം, (d.w)	Time difference due to terrestrial longitude
ദേശാന്തരസംസ്കാരം, (d.w)	Correction for local longitude
ദോസ് (ഭുജ), (d.w)	1. Side of a triangle, 2. Ordinate of an arc, 3. Opposite side of a right angled triangle
ദ്യുഗണം (കലിദിനം), (d.w)	Number of days from Kali epoch
ദ്യുജ്യാ, (d.w)	Day - radius
ദ്യുവൃത്തം (അഹോരാത്രവൃത്തം), IX.9	Diurnal circle. Smaller circles parallel to the Ghaṭikāmaṇḍala (celestial equator), along which stars rise north or south of the poles
ദ്വാത്രിംശദശ്രം, (d.w)	A polygon of 32 sides
ദ്വാദശാംഗുലശങ്കു, XI.2	A gnomon 12 digits long used by the ancient Hindu mathematicians in the measurement of shadows
ദ്വാദശാംഗുലശങ്കുച്ഛായ, XI.2,10	Shadow of a 12 digit gnomon
ദ്വിതീയകർണ്ണം, VII.10	The seemed hypotenuse in a poygon
ദ്വിതീയസംസ്കാരഹാരകം, (d.w)	The divisor used to calculate a second correction after a first correction
ദ്വിതീയസങ്കലിതം, (d.w)	Sum of the series of second integrals
ധനം, (c.w)	1. Positive, 2. Additive
ധനുസ്സ്, (c.w)	Arc of a circle
ധ്രുവം, IX.1	1. Celestial pole, pole-star, north or south; 2. Zero positions of planets at epoch
ധ്രുവവൃത്തം (ധ്രുവകവൃത്തം), IX.7	Meridian circle
ധ്രുവനക്ഷത്രം, IX.1	Pole star
ധ്രുവോന്നതി, IX.7,8	Elevation of the celestial pole

നക്ഷത്രം, VIII.1,2	Star; Asterism; Constellation
നക്ഷത്രകക്ഷ്യാ, (ഭകക്ഷ്യാ) (c.w)	Orbit of the asterisms, equal to 17,32,60,008 yojanas, denoted by the expression janā nu nītiraṅgasarpa being 50 times the orbit of the sun.
നക്ഷത്രഗോളം, IX.1,2	The starry sphere
നതം, IX.10	Meridian zenith distance; Hour angle; Interval between mid-day and time taken
നതകോടിജ്യാവ്, IX.12	Rcos. of the hour angle
നതജ്യാവ്, IX, 20.v	Rsine of zenith distance or hour angle
നതദൃക്ക്ഷേപവൃത്തം, XI.21i	Circle touching the zenith and Natsamamandala
നതജ്യ, XI.20.v	Rsine hour angle
നതനാഡി, IX.10	Interval in nādis between midday and time taken
നതപ്രാണം, (d.w)	Prānas of zenith distance
നതഭാഗം, (നതാംശം) (d.w)	Degree of zenith distance
നതവൃത്തം, IX.10	A Big Circle which passes through the sides (pārśva) of another Big Circle around the sphere
നതസമമണ്ഡലം, XI.21	Prione vertical at the meridian
നതി, XI.2, 35	Parallax in celestial latitude
നതികലാ, (d.w)	Nati in minutes
നതിയോഗം, XV.2	Sum of two parallaxes in celestial latitude
നതിലംബനലിപ്താ, XI.35	R.cos. Parallax in celestial longitudes in terms of minutes of arc
നത്യന്തരം, XV.2	Difference beteween parallaxes in celestial latitude

നാക്ഷത്രവർഷം, നാക്ഷത്രസംവത്സരം (d.w)	1. Sidereal year; 2. Equivalant to meṣādi to meṣādi; Nirayana year; Solar Year
നാഡീവൃത്തം, (നാഡീവലയം) (d.w)	....
നാഭികേന്ദ്രം	Centre of a circle
നാഭ്യുച്ഛ്രയം (നാഭ്യുത്സേധം), (d.w)	Elavation of nābhi (Centre)
നാഴിക, (c.w)	Measure of time equal to 1/60th of a solar day, i.e; 24 minutes
നിഖർവം, I.2	10^{11} (number and place)
നിമീലനം	Immersion, in eclipse
നിരക്ഷം, IX.2	Region of zero latitude, i.e. terrestrial equator
നിരക്ഷക്ഷിതിജം, IX.7	Equatorial horizon
നിരക്ഷപ്രദേശം, (-ദേശം) IX.1	Equatorial region
നിരക്ഷരേഖ, (d.w)	Equator
നിരന്തരസംഖ്യ, I.4	Consecutive numbers
നീചം, V.III.1	Perigee or perihelion
നിചോച്ചമണ്ഡലം, VIII.1	Epicycle
നേമി, VIII.3; XI.1	Circumference of a Circle
പക്ഷം, (c.w)	Light or dark half of the lunar month
പങ്ക്തി, (c.w)	Column; Ten, (Number and place)
പഞ്ചരാശികം, (d.w)	Compound proportion involving five terms
പഠിതജ്യാ (മഹാജ്യാ), VII.3,4	The 24 specified Rsines
പദം, VII.2,3	1. Square root; 2. Terms of a series; 3. Quadrant of a circle
പരക്രാന്തി, പരമക്രാന്ത്രി, (d.w)	Maximum declination, 24^0
പരമഗ്രാസം, (d.w)	Maximum eclipse or obscuration

പരമസ്വാഹോരാത്രം, IX.9	Longest day in the year
പരമ്പര, (c.w)	A series
പരല്പേര്	1. Word and letter numerals; 2.Numbers formed through letters, words and phrases
പരമാന്തരാളം	Maximum distance between two things
പരമാന്തരാളം, IX.5	A big circle which passes through the two sides (parsua) of the other Big Circle around a sphere
പരമാപക്രമം, IX.9	Maximum declination of a celestial body from the Ecliptic to its orbit
പരമാപക്രമജീവാ, IX.9	Rsine of the greatest declination
പരശങ്കു, പരമശങ്കു, (d.w)	Rsine of greatest altitude, i.e, Rsine of meridian altitude
പരാർദ്ധം, I.2	10^{17} (Number and place)
പരികർമ്മം, I.2	Arithmetical processes or manipulations
പരിധി (നേമി)	Cicumference
പരിഭ്രമണം, VI (c.w)	A complete revolution of a planet along the zodiac with reference to a fixed star
പരിലേഖം, (പരിലേഖനം) XII.9	Graphical or diagrammatic representation
പര്യയം, (ഭഗണം) V.1, 3; VIII.3	1.Revolution; 2. Number of revolutions of a planet in a yuga
പർവ്വാന്തം, (d.w)	The time when moon is in conjunction with or opposition to the sun; End point of the new or full moon
പലജ്യാ,	Sine latitude
പലഭാ,	Equinoctical shadow

പാട് (മൌഢ്യം), XIV.3	Invisibility of a planet due to its light or retrograde motion opposite to the disc of the Sun
പാതൻ, VIII.16	Mode, Generally ascending node
പാർശ്വം, VI.2	Side, Surface
പിതൃദിനം, (c.w)	Day of the manes
പിറപ്പ്, XIV.3	Rising or reappearance of a planet after pāṭu (Mauḍhya) which see.
പൂർവ്വവിഷുവത്ത്, IX.3	1. Point at which the sun coarising along the Ecliptic crosses the celestial equator from the south to the north; 2. Vernal equinox
പൂർവ്വാപരരേഖ, VIII.3	East-west line or direction; Prime vertical
പൂർവ്വാപരബിന്ദു, XI.1	East and west points
പൂർവ്വാപരവൃത്തം, IX.3	East-west Big Circle passing throug the zenith round the celestial globe
പൃഷ്ഠം, (c.w)	Surface
പ്രതത്പര	One sixtieth of a tatpara in angular measure
പ്രതിപത് (പ്രതിപദം), (c.w)	The first day of a lunar fortnight
പ്രതിഭുജം, (d.w)	Opposite side
പ്രതിമണ്ഡലം, VIII.3	Eccentric circle with its centre on the circumference of a planet's orbit of a circle
പ്രതിമണ്ഡലകർണ്ണം, VIII.7	Distance of the planet on the eccentric
പ്രതിമണ്ഡലസ്ഫുടം, VIII. I,3	True longitude of a planet in the eccentric circle
പ്രതിമന്ദോച്ചം, (d.w)	Perigee as opposed to apogee
പ്രത്യക്കപാലം, (d.w)	The hemisphere other than the one that is being considered in a sphere

പ്രഭാഗജാതി, III.1	Fractions of fractions
പ്രമാണം, IV.1	Antecedant; First term of a proportion, i.e argument in a Rule of Three.
പ്രമാണഫലം, IV.1	1. The consequent; 2. Second term in a proportion
പ്രയുതം, I.2	10^{16}, (number and place)
പ്രവഹഭ്രമണം, IX.3, XI.4	Revolution of the planets due to the provector wind
പ്രവഹമാരുതം, പ്രവാഹവായു, IX.3; XI.4	Provector wind
പ്രസ്താരം, (c.w)	Number of combinations
പ്രാക്കപാലം, (d.w)	The eastern hemisphere
പ്രാഗ്ലഗ്നം, (d.w)	Orient rising of the ecliptic
പ്രാണം, XI.4	Unit of time equal to one-sixth of a vināḍi or four sidereal seconds
ഫലം, (c.w)	1. Fruit, in the Rule of Three; 2. Result; 3. Bhūja
ബഡവാമുഖം IX.1	1. Terrestrial south pole. 2. The place in the South of the earth from where the south polar star is right above.
ബാഹ്യ, XI.1	Lateral side of a rt. angled triangle; Semi-chord; Rsine
ബിംബം, XII.4	Disc of Planet
ബിംബഘനമധ്യാന്തരം, XII.2	Sum of the semi-diameters of a Planet less the eclipsed part
ബിംബമാനം, XII. 3, 4	Measure of the discs of Planets
ബിംബാന്തരം, XII.3	Sum of the semi-diameters of two planets minus the eclipsed part.

ഭം (നക്ഷത്രം) (c.w)	Asterism : Star
ഭകക്ഷ്യാ, (d.w)	Path of the asterisms
ഭകൂടം, (രാശികൂടം) VIII.16	The two apexes of the circles cutting the ecliptic at rt. angles.
ഭഗണം (പര്യയം), V.1, V.3, VIII.1	1. Revolution of a planet along the Ecliptic; 2. Number of revolutions of a planet during a certain period. 3. 12 rasis or 360 degrees
ഭഗോളം, VIII.2, IX.3	1. Sphere of asterisms; 2. Zodiacal sphere, with its centre at the Earth's centre.
ഭഗോളമധ്യം, VIII.2	Centre of the zodiacal shpere
ഭഗോളശങ്കു, XI.5	Gnomon with reference to the surface of the bhagola
ഭചക്രം, (ഭമണ്ഡലം) (d.w)	Circle of asterisms
ഭപഞ്ജരം, (d.w)	Circle of asterisms
ഭാഗം, (അംശം, തീയതി)	1. $\frac{1}{360}$ of a circle, 2. Degree of angular measure
ഭാഗജാതി, III.1	Fraction
ഭാഗഹരണം, III.3	Division
ഭാഗാനുബന്ധം,	Associated fraction
ഭാഗാപവാഹം,	Dissociated fraction
ഭാജകം, V.3	Divisor (General and in Kuṭṭākāra)
ഭാജ്യം, V.3	Dividend; The multiplicand in Kuṭṭākāra
ഭിന്നമൂലം, III.5	Square root of fractions
ഭിന്നവർഗ്ഗം, III.5	Square of fractions
ഭിന്നസംഖ്യ, 1.III	Fraction

ഭുക്തി (= ഗതി), VIII.1	Motion; daily motion
ഭുജ, VI.2; VII.9	1. Lateral side of a rt. angled triangle; 2. of the angle, the degrees gone in the odd quadrants and to go in the even quadrants.
ഭുജാഖണ്ഡം, (d.w)	The difference between two successive ordinates
ഭുജാജ്യാ, (d.w)	Rsine of an angle
ഭുജാന്തരഫലം, (d.w)	Correction for the equation of time due to the eccentricity of the ecliptic
ഭുജാഫലം, VIII.9	Equation of the centre.
ഭൂമി, VI.2; IX.7	One side of a triangle or quadrilateral taken for reference, generally the trase; Earth
ഭൂഗോളം, IX.1	Earth-sphere
ഭൂച്ഛായാ, XII.4	Earth's shadow
ഭോഗം, (ഭുക്തി), VIII.1	1. Motion; 2. Daily motion
ഭൂതാരാഗ്രഹവിവരം, (d.w)	Angular distance between the Earth and a Planet.
ഭൂദിനം, V.1	1.Terrestrial day, 2.Civil day; 3. Sunrise to sunrise; 4.The number of terrestrial days in a yuga or kalpa
ഭൂപരിധി, (d.w)	Circumference of the Earth, 3350 Yojanas.
ഭൂപാർശ്വം, IX.7	Side of the Earth
ഭൂഭ്രമണം, VIII.1	Earth's rotation
ഭൂമധ്യം, VIII.1	Centre of the Earth
ഭൂമധ്യരേഖ, (d.w)	Terrestrial equator
ഭൂവ്യാസാർദ്ധം, (d.w)	Radius of the Earth
മകരാദി, (d.w)	The six signs commencing from Makara (Capricorn)

മണ്ഡപം, VI.3	A square with a pyramidal roof usually found in Hindu temples
മണ്ഡലം, (c.w)	1. Circle; 2. Orb
മതി,	Small tentative multiplier in Kuṭṭākāra got by guessing correctly according to the conditions given
മതിഫലം, (d.w)	The result corresponding to a given mati
മത്സ്യം (ഝഷം), (c.w)	The overlapping portion of two intersecting circles, taking the form of a fish.
മധ്യം, I.2	10^{16} (number and place); Middle point; Mean (Planet etc.)
മധ്യകാലം, (d.w)	Mean time.
മധ്യഗതി, (d.w)	Mean motion of Planets; Mean daily motion
മധ്യഗ്രഹം, (d.w)	Mean Planet
മധ്യഗ്രഹണം, XII.1	Mid-eclipse
മധ്യച്ഛായ, (d.w)	Mid-day shadow
മധ്യജ്യ, (d.w)	Meridian sine, i.e. Rsine of the zenith distance of the meridian ecliptic point
മധ്യന്ദിനച്ഛായ, XI.12	Mid-day shadow
മധ്യഭുക്തി, (d.w)	Mean daily motion
മധ്യമം, VIII.7	1.Mean; 2.Mean longitude of a Planet
മധ്യമഗതി, (മദ്ധ്യഗതി), VIII.3	Mean motion of a Planet
മധ്യലഗ്നം, XI.32,33	Meridian ecliptic piont
മധ്യസ്ഫുടം, VIII.7	Mean Planet
മധ്യാഹ്നം, (c.w)	Mid-day
മധ്യാഹ്നച്ഛായ, XI. 12	Mid-day shadow

മധ്യാഹ്നാഗ്രാംഗുലം, (d.w)	Measure of amplitude at noon in terms of aṅgula
മന്ദം, VIII.13	1. Slow; mandocca, 2. Apogee of slow motion; See also mandocca
മന്ദ,(നീചോച്ച) വൃത്തം VIII.13	1. Manda epicycle; 2. Epicycle of the equation of the centre
മന്ദകർണ്ണം, VIII.8,13	Hypotenuse associated with mandocca; radius vector
മന്ദകർണ്ണവൃത്തം, VIII.14	Circle extended by Mandakarṇa
മന്ദകർമ്മം, VIII.1, 2, 13	1. Manda operation in planetary computation
മന്ദകേന്ദ്രം, VIII.13	Manda anomaly
മന്ദകേന്ദ്രഫലം, VIII.13	1. Manda correction; 2. Equation of the centure
മന്ദപരിധി,	Epicycle of the equation of the centre
മന്ദസ്ഫുടം, VIII.13	True longitude of Planet at the aper of the slowest motion
മന്ദോച്ചം, (തുംഗൻ) VIII.3	1. Apogee or aphelion; 2. Higher apsis relating to the epicycle of the equation of the centre
മന്ദോച്ചനീചവൃത്തം, (മന്ദവൃത്തം), VIII.3	Manda-nīca epicycle
മരുത്, IX.3	Proveetor wind, supposed to make the planets revolve
മഹാച്ഛായ, XI.5	1. Great shadow; the distance from the foot of the Mahāśaṅku to the centre of the Earth; Rsine zenith distance; 2.The gnomonic shadow subtended on the horizon by the sun on the diurnal circle.
മഹാജ്യാ, (പഠിതജ്യാ), VII.3,4	The 24 Rsines used for computation
മഹാപത്മം, I.2	10^{12}, (number and place)

മഹാമേരു, (മേരു) IX.1	Mount Meru, taken to mark the Terrestrial pole in the north
മഹാവൃത്തം, IX.9	The Big Circle around a sphere, touching its two opposite sides with radius being that of the sphere
മഹാശങ്കു, XI.5	1. Great gnomon; 2.The perpendicular dropped from the Sun to the earth-line; 3. Rsine altitude
മഹാശേഷം,	In Kuṭṭākākara, the greater of the last two remainders taken into consideration.
മാനം, (c.w)	1. Measure; 2. An arbitrary unit of measurement.
മൂലം, I.9; VII.3	1. The starting point of a line or arc; 2. Square root, cube root etc.
മൂലസംകലിതം, VI.5, i; VI.5.v	Sum of a Series of natural numebrs
മൃഗം, (c.w)	Sign Makara or Capricorn
മേരു, (മഹാമേരു) IX.1	1. Terrestrial North pole; 2.The place in the north of the earth from where the North polar star is right above; 3. Situated 90 degrees north of Laṅkā.
മേഷാദി, VIII.3, VIII.1	1. First point of Aries; 2. Commencing point of the ecliptic.
മോക്ഷം,	1.Emergence, in eclipse; 2.Last point of contact.
മൗഢ്യം,(ക്രമം, വക്രം) XIV.1,2	Invisibility of a Planet due to its right or retrograde motion opposite the disc of the Sun
യവകോടി, IX.1	An astronomically postulated city on the Terrestrial Equator, 90 degrees east of Laṅkā.

യാമ്യം, (d.w)	Southern.
യാമ്യഗോളം, (d.w)	1. Celestial sphere as viewed from the south; 2. Southern celestial sphere.
യാമ്യോത്തരരേഖ (ദക്ഷിണോത്തരരേഖ), VIII.3	South-north line, meridian
യുക്തി, (c.w)	Proof, Rationals
യുഗം, (c.w)	Aeon
യുഗഭഗണം, V.3; VIII.1	Number of revolutions of a planet during a yuga (aeon)
യുഗ്മം, (-പദം) I.8, i; VIII.3	1. Even; 2. Second or fourth quardeant in a circle.
യുഗ്മസ്ഥാനം, (d.w.)	Even place cunting from unit's place
യോഗം, I.3	1. Conjuction of two planets; 2. Sum, 3. Daily yoga, nityayoga, twentyseven in number and named viṣkambha, Prīti, Āyuṣmān, etc. being Sun plus Moon; cf. candro yogo 'rkayuktaḥ; 3. addition
യോഗചാപം, (d.w)	Arc whose semi-chord is equal to the sum of two given semi-chords.
യോജന, VIII.1	Unit of linear measure, equal to about seven miles.
യോജനഗതി, VIII.1	Daily motion of Planets in yojanas
യോജനവ്യാസം, (d.w)	Diameter in yojanas
രാശി, VII.1	1. A number; 2. One sign in the zodiac equal to 30 degrees in angular measure. 3. $\frac{1}{12}$ of the circumference in angular measure.

രാശികൂടം, VIII.16	1. The two apexes of circles cutting the ecliptic at rt. angles. 2. The two points on the celestial sphere 90^0 degrees north and south of the ecliptic from where the rāśi-s (signs) are counted.
രാശികുടവൃത്തം, VIII.16;IX.10	The circle commencing from the Rāśikūṭas and cutting the ecliptic at internvals of one rāśi (30 degrees) each.
രാശികൂടശങ്കു, XI.31	Gnomon at the rāśikūṭas
രാശികൂടോന്നതി,	Altitude of the rāśikūṭas
രാശിഗോളം, IX.3	Zodiacal sphere. See also Bhagola.
രാശിചക്രം,	Ecliptic.
രാശിപ്രമാണം, VII.1	Measure of the rāśi
രാശ്യുദയം, (d.w)	Rising of the signs.
രാഹു, (=പാതൻ) (c.w)	Node of Moon, esp. the ascending node
രൂപം, I.4; III.1	1. Unity; 2.One; 3. Form
രൂപവിഭാഗം, (d.w)	Division by magnitude.
രോമകവിഷയം IX.1	Astronomically postulated city in the Terrestrial Equator, 90 degrees east of Laṅkā.
ലക്ഷം, I.2	10^5 (number and place), Lakh.
ലഗ്നം, XI.31	1. Ecliptic point on the horizon; 2. Rising point of the ecliptic
ലഗ്നസമമണ്ഡലം, XI.31	Prime vertical as the Orient ecliptic point
ലഘുവൃത്തം, VIII.1	Smaller circle parallel to the Mahāvṛtta (Big circle) in a sphere
ലങ്ക, IX.1	Laṅkā., a city postulated astronomically on the Earth's equator at zero longitude.

ലങ്കാക്ഷിതിജം, IX.2, (ഉദ്വൃത്തം, ഉന്മണ്ഡലം)	Diurnal circle as Laṅkā; East-West hour circle; Equinoctial colure. Big circle passing through the North and South poles and the two East-West Svastika
ലങ്കോദയം, IX.1	Time of the rising of the signs at Lanka, i.e, right ascensins of the signs.
ലങ്കോദയജ്യാ, IX.II	Sine right ascension
ലംബം, VI.2; VII.1,9	1. Altitude; 2. Co-latitude; Perpendicular; Vertical
ലംബകം, XI.2	Plumb
ലംബജ്യാ, (d.w)	Rsine co-latitude, i.e, Rcos latitude
ലംബനം, X.2	Rcos latitude; Parallax in longitude, or difference between the parallaxes in longitude of the Sun and the Moon in terms of time.
ലംബനനാഴിക, (d.w)	Parallax in longitude in terms of nāḍikās
ലംബനയോജനം, (d.w)	Parallax in terms of yojanas
ലാടം, XIII.2	A type of vyatīpāta, which occurs when Sun plus Moon is equal to 180^0 degrees.
ലിപ്ത (ഇലി), കല	Minute of arc in angular measure.
ലിപ്താവ്യാസം, (d.w)	Angular diameter in minutes
വക്രം,	Retrograde.
വക്രഗതി, (d.w)	Retrograde motion of a planet.
വണ്ണമൊപ്പിക്കുക,	Convert fractions to the same denomination
വർഗ്ഗം, I.3,8.i	Square.
വർഗ്ഗക്ഷേത്രം, I.8.i.ii	Square area, place, space.
വർഗ്ഗമൂലം I.9	Square root.

വർഗ്ഗവർഗ്ഗം V.7	Square of squares.
വർഗ്ഗവർഗ്ഗസംകലിതം, VI.4	Summation of squares of squares
വർഗ്ഗസംങ്കലിതം, VI.4,5.ii	Sum of a series of squares of natural numbers
വർഗ്ഗസ്ഥാനം, I.9; XII.6,7	The odd place counting from the unit's place
വലനം, XII. 6,7	Deflection of a planet due to akṣa, or ayana
വലനദ്വയസംയോഗം, XII.8	Sum of akṣa and ayana valanas
വല്ല്യുപസംഹാരം, V.4	A particular kind of operation in Kuṭṭākāra
വല്ലി, V.3	1. Series of results in Kuṭṭaka, i.e, Kuṭṭākāra operation; 2. Column of numbers
വായു (പ്രവഹവായു), IX.3	Provector wind supposed to make the planets revolve
വായുഗോളം, IX.3	Atmopheric spheres
വികല (വിലി, വിലിപ്ത)	1. One sixtieth of a minute of angular measure, 2. One second.
വിക്ഷിപ്തം, VIII.16	Having celestial latitude, deviated from the ecliptic
വിക്ഷിപ്തഗ്രഹക്രാന്തി, IX.11	Declination of a planet in its polar latitude
വിക്ഷേപം, VIII.16	1. Celestial latitude; 2. Polar latitudes. Latitude of the Moon or a planet
വിക്ഷേപകോടിവൃത്തം, VIII.16	Circle on which Rcos celestial latitude is measured.
വിക്ഷേപചലനം, XIII.6	Precession of the equinoxes
വിക്ഷേപമണ്ഡലം (വിമണ്ഡലം), VIII.16	Orbit of a Planet
വിക്ഷേപലഗ്നം, XIII.3	Celestial latitude at the Orient ecliptic point
വിനാഴിക (വിനാഡി, വിഘടികാ), (c.w)	One-sixtieth of a nāḍikā; 24 seconds.
വിപരീതകർണ്ണം, VIII. 10,11,12	Reverse hypotenuse.

വിപരീതച്ഛായ, XI. 11	Reverse computation from gnomonic shadow
വിപരീതദൃഗ്വൃത്തം, XI.20. i	Reverse computation from Dṛgvṛtta
വിപരീതവൃത്തം, IX.10; XI.20. i	Circle computed reversely
വിമർദാർദ്ധം	Half total obscuration in an eclipse.
വിയോഗം, I.iii; III.1	Subtraction
വിലി, വിലിപ്ത (വികല)	Second of arc in angular measure
വിവരം, (c.w)	Difference
വിശേഷം, (c.w)	Difference
വിശ്ലേഷം, (c.w)	Difference
വിഷമം, (c.w)	1. Odd; 2. Odd number.
വിഷുവത്ത്, IX.3	1. Equinox 2. Point of intersection of the ecliptic and (krāntivṛtta or Apakrama-vṛtta) and the celestial equator (Ghaṭikāṇḍala) 3. Vernal: March 21; Autumnal Sept.23
വിഷുവച്ഛായ, XI.2	Equinoctial shadow at midday
വിഷുവജ്ജീവ, (-ജ്യാ), XI.3	Rsine of latitude at equinox.
വിഷുവത്കർണ്ണം, IX.3	Hypotenuse of equinoctial shadow.
വിഷുവദ്ഭാ (വിഷുവച്ഛായ), IX.3	Equinoctial shadow, i.e Midday shadow of a 12-digit gnomon when the Sun is at the equinox
വിഷുവദ്വിപരീതനതവൃത്തം, IX.9,10	The circle cutting the Celestial Equator.
വിഷുവന്മണ്ഡലം (ഘടികാമണ്ഡലം, ഘടികാവൃത്തം), IX.3	1. Celestial Equator. 2. Path of a star rising exactly in the east and setting exactly in the west
വിഷ്കംഭം I.3	1. Diameter; 2. The first of 27 daily yogas, being Sun plus Moon

വിസ്താരം, (c.w)	Breadth
വൃത്തം (സമവൃത്തം), VIII.1	1. Circle; 2. Perfect Circle
വൃത്തകേന്ദ്രം, VIII.3, IX.1	Centre of a circle.
വൃത്തനേമി, VIII.3; IX.1	Circumference of a circle
വൃത്തപരിധി, VI.9	Circumference of a circle
വൃത്തപാതം, VIII.1	The two points at which two Big circles around a shpere intersect.
വൃത്തപാദം, VII.2;	1. Quadrant; 2. Quarter of a Circle; 90 degrees
വൃത്തപാർശ്വം, VII.2	The two ends of the axis around which a sphere is made to rotate; Two directly opposite sides of a sphere on the line of its diameter
വൃത്താന്തർഗ്ഗത ചതുരശ്രം, VII.10	A cyclic quadrilateral
വൃന്ദം, I.2	10^9 (number and place)
വൈധൃതം, XIII.2	The type of Vyatīpāta which occurs at a time when the sum of the longitudes of the Sun and the Moon amounts to 12 signs or 360 degrees
വ്യക്തി, (c.w)	Unity.
വ്യതീപാതം, XIII.2	The time when Sun Plus moon equals six signs i.e, 180^0
വ്യതീപാതകാലം, XIII.2	Duration of Vyatīpāta
വ്യവകലിതം, I.4	Subtraction
വ്യസ്തകുട്ടാകാരം, (d.w)	Inverse process in Kuṭṭākāra
വ്യസ്തത്രൈരാശികം, IV.2	Inverse proportion
വ്യാപ്തിഗ്രഹണം, (d.w)	Generalisation

വ്യാസം, (c.w)	Diameter of a circle or sphere
വ്യാസാർദ്ധം, (d.w)	Semi - diameter, radius
ശങ്കു, I.2; IX.1, 21	1. Gnomon; 2. 12-digit gnomon; 3. Mahāśaṅku or great gnomon, the perpendicular dropped from the Sun to the earth-line, or the Rsine altitude; 4. The number 10^{13}.
ശങ്കുകോടി, (d.w)	Complement of altitude or zenith distance
ശംകഗ്രം, XI.13	North-south distance of the rising or setting point from the tip of the shadow, i.e. agrā. 2. Natijyā; 3. Distance of the planet's projection on the plane of the horizon from the rising-setting line.
ശതം, I.2	10^2 (number and place); Hundred
ശരം, VII.2	1. Arrow, 2. Rversed sine 3. Sag or height of an arc
ശരഖണ്ഡം, VII.16	Parts of the height of an arc
ശരോനവ്യാസം, VII.16	Diameter less śara
ശിഷ്ടം, (c.w)	Remainder in an operation
ശിഷ്ടചാപം, VII.4	The difference between the given cāpa and the nearest Mahājyācāpa
ശീഘ്രം, VIII.1,2,19	Higher apsis of the equation of the epicycle in the equation of conjunction
ശ്രീഘ്രകർണ്ണം, VIII.8-12	1. Hypotenuse associated with śīghrocca; 2. Geocentric radius vector
ശീഘ്രകർമ്മം, VIII.1,2, 14	Śīghra operation in planetary computation
ശീഘ്രകേന്ദ്രം, VIII.10, 11	Centre of the śīghra epicycle
ശീഘ്രപരിധി, VIII.16	Epicycle of the equation of conjunction

ശീഘ്രവൃത്തം, VIII.6	śīghra epicycle.
ശീഘ്രസ്ഫുടം, VIII.14	True longitude of a planet at śīghra position, i.e, apex of its swiftest motion
ശീഘ്രോച്ചം, VIII.6	1. Higher apsis of the epicycle related to the equation of conjunction. 2. Apex of the fastest motion of a planet
ശാഘ്രോച്ചനീചവൃത്തം, VIII.16	Śīghra epicycle
ശുദ്ധി, (c.w)	Subtraction
ശൂന്യം, (c.w)	Zero
ശൃംഗോന്നതി, XV.1,2	Elevation of the lunar horns
ശേഷം, (ശിഷ്ടം) (c.w)	Remainder in an operation
ശോധ്യഫലം (d.w)	Correction to be applied to a result
ശ്രുതി(കർണ്ണം), (c.w)	Hypotenuse
ശ്രേഢി, I.8.v	Series
ശ്രേഢീക്ഷേത്രം, I.8.v	A figure representing a series graphically
ഷഡശ്രം, VII.1	1. Hexagon; 2. Regular hexagon
ഷോഡശാശ്രം, VI.2	Polygon of 16 sides.
സംവത്സരം, V.1 (സൌരസംവത്സരം)	1. Siderial year; 2. Time taken by the Sun starting from the vernal equinox (Pūrva-visuvat) to return again to the Equinox
സംവർഗ്ഗം	Product
സംസർപം	The lunar month preceding a lunar month called Aṃhaspati which latter does not contain a saṅkrānti
സംസ്കാരം, (c.w)	Correction by addition or subtraction
സങ്കലനം, I.4	Addition

സങ്കലിതം, I.4. VII.5	1. Sum of a Sseries of natural numbers; 2. Addition
സങ്കലിതസംകലിതം, VI.5.ii	Integral of an integral
സങ്കലിതൈക്യം VI.4	Sum of the integrals
സംക്രാന്തി	1. The moment a planet enters into a sign of the zodiac; 2. Entry from one sign to the next
സംഖ്യാസ്വരൂപം, I.2	Nature of numbers
സദൃശം, (c.w)	1. of the same denomination or kind; 2. Similar
സമം, (c.w)	Level, Equal
സമഘാതം,	Product of like terms
സമച്ഛായ, XI.17	Prime vertical shadow
സമച്ഛേദം, III.2	Same denominator
സമത്ര്യശ്രം, VII.1	Equilateral triangle
സമനിലം, (ഭൂമി) XI.1	1.Plane ground; 2.Level space; 3.Horizontal
സമപ്രോതം, (സമപ്രോതവൃത്തം), (d.w)	Secondary to the prime vertical
സമമണ്ഡലം, IX.7	Prime vertical
സമരേഖ, IX.1	Prime vertical
സമലംബചതുരശ്രം, (d.w)	Trapezium
സമവിതാനം, III.1; VII.1	Level
സമശങ്കു, (സമമണ്ഡല ശങ്കു) XI.16	Rsine of altitude of a celestial body when upon the prime vertical
സമസംഖ്യ, (d.w)	Even number
സമസ്തഗ്രഹണം (പൂർണ്ണഗ്രഹണം), XII.5	Total eclipse
സമസ്തജ്യാ, VII.1	Rsine of a full arc

സമാന്തരരേഖ, (d.w)	Parallel straight line
സമ്പർക്കാർദ്ധം, (d.w)	Half the sum of the eclipsed and eclipsing bodies
സമ്പാതജീവാ	Common chord of the same denomination or nature
സർവദോർയ്യുതിദളം, VII.15	Semi-perimeter.
സർവസാധാരണത്വം, (c.w)	Universality.
സവർണ്ണം, (c.w)	Of the same denomination or nature
സഹസ്രം, I.2	1. 10^3 (number and place); 2. One thousand.
സാഗ്രം	1.With remainder; 2.A kind of Kuṭṭākākāra
സാധനം, (c.w)	Given data.
സാർപമസ്തകം	Vyatīpāta when the Sun plus Moon is equal to 7 degrees 16 minutes
സാവനദിനം, V.i	1. Civil day; 2. Duration from sunrise to sunrise; 3. Solar day
സിതം	Illuminated part of the Moon, Phase of the Moon
സിദ്ധപുരം, IX.1	An astronomically postulated city on the Terrestrial Equator; 180 degrees into opposite to Laṅkā.
സൂത്രം, (c.w.)	1. Line; 2. Direction; 3. formula
സൂര്യഗ്രഹണം, XII.2	Solar eclipse.
സൂര്യസ്ഫുടം, VIII.7	True longitude of the Sun
സൌമ്യം, (c.w)	Northern
സൌമ്യഗോളം, (c.w)	The Northern hemisphere
സൌരം, V.1	Solar.
സൌരാബ്ദം, (d.w)	Solar year

സ്ഥാനവിഭാഗം, (d.w)	Division according to place
സ്ഥിത്യർദ്ധം	Half duration of an eclipse
സ്ഥൌല്യം, (d.w)	1. Difference from the correct value, 2. Error
സ്പർശം, (d.w)	1. First contact in an eclipse; 2. Touch
സ്ഫുടം, VIII.7	True longitude of a planet
സ്ഫുടം, (ഗ്രഹം), VIII, 13	True position of a planet
സ്ഫുടക്രിയ, VIII.1	Computation of true longitude of a planet
സ്ഫുടമധ്യാന്തരാളം, VIII.7	Difference between the true and mean longitudes of a planet
സ്ഫുടമധ്യാന്തരാളചാപം, VIII.7	Arc of the longitude between the true and mean of a planet
സ്ഫുടവിക്ഷേപം, (d.w.)	Celestial latitude as corrected for parallax
സ്ഫുടാന്തരം, (d.w.)	Difference between true longitudes
സ്വം, (d.w.)	1. Addition, 2. Additive quantity
സ്വദേശക്ഷിതിജം, IX. 7	Horizon at one's place or the place of observation
സ്വദേശനതം, (d.w.), XI.21.i	Meridian zenith distance, at one's place or the place of observation
സ്വദേശനതകോടി, (d.w.), XI.21.i	R.cos of Śvadesanata
സ്ഫുടഗതി, VIII.1, 8	True daily motion of a planet
സ്ഫുടഗ്രഹം, VIII.1, 8	True longitude of a planet
സ്ഫുടന്യായം VIII.2	Rationate or method for exactitude
സ്വാഹോരാത്രവൃത്തം, (മണ്ഡലം), ദ്യുജ്യാവൃത്തം	Diurnal circle
സ്വോർദ്ധ്വം	The number above the penultimate in Kuṭṭākākāram

ഹനനം, (c.w)	Multiplication
ഹരണം, I.7	Division
ഹരണഫലം	Quotient
ഹാരകം	Divisor
ഹാര്യം, I.7	Dividend
ഹൃതശേഷം, I.7	Remainder after division

അനുബന്ധം-II

INDEX OF QUOTATIONS

ഉദ്ധൃതശ്ലോകങ്ങളുടെ സൂചി

ശ്ലോകം	ആധാരഗ്രന്ഥം	നിർദേശം
അത്രേശകോണഗാരിഷ്ടഃ	ലീലാവതി	VII. 15
അന്തരയോഗേ കാര്യ്യേ		VII. 15
അന്തേ സമസംഖ്യാദള		VI. 10
അന്ത്യക്രാന്തീഷ്ടതത്കോട്യാ	സിദ്ധാന്തദർപ്പണം, 28, 29	IX. 12
അന്ത്യദ്യുജ്യേഷ്ടഭക്രാന്ത്യോഃ	സിദ്ധാന്തദർപ്പണം	IX. 12
അന്യോന്യഹാരാഭിഹതൗ	ലീലാവതി , 30	VI. 8
അവ്യക്തവർഗ്ഗഘനവഗ്ഗ		VI. 8
ഇഷ്ടജ്യാത്രിജ്യയോർഘാതാൽ	തന്ത്രസംഗ്രഹവ്യാഖ്യാ, II 206	VI. 6
ഇഷ്ടദോഃകോടിധനുഷോഃ	തന്ത്രസംഗ്രഹം, II 10 B	VII.4
ഇഷ്ടോനയുക്തേന	ലീലാവതി , 16	XI. 20.ii
ഇഷ്ടോനയുഗ്രാശിവധഃ കൃതിഃ	ലീലാവതി , 20	VII. 15
ഋണമൃണധനയോർഘാതോ	ബ്രഹ്മസ്ഫുടസിദ്ധാന്തം, 183	VI. 8
ഏകദശശതസഹസ്രായുത	ലീലാവതി , 10	I. 2
ഏകവിംശതിയുതം ശതദ്വയം	ലീലാവതി , 247	V. 4
ഏവം തദൈവാത്ര യദാ	ലീലാവതി , 246	V. 4
ഓജാനാം സംയുതേസ്ത്യക്ത്വാ	തന്ത്രസംഗ്രഹവ്യാഖ്യാ, II.208	VI.6
ഗ്രാസോനേ ദ്വേ വൃത്തേ		VII.16
ഛായയോഃ കർണ്ണയോരന്തരേ	ലീലാവതി , 232	VII.17
തദാദിതസ്ത്രിസംഖ്യാപ്തം	തന്ത്രസംഗ്രഹവ്യാഖ്യാ , II.210	VI. 6
തസ്യാ ഊർദ്ധ്വഗതായാഃ		VI. 8

ത്രിശരാദിവിഷമസംഖ്യാ		VI. 3
ദ്വ്യാദിയുജാം വാ കൃതയോ		VI. 9
ദ്വ്യാദേശ്ചതുരാദേർവ്വാ		VI. 9
പഞ്ചാശദേകസഹിതാ	ലീലാവതി, 97	VII. 15
പരസ്പരം ഭാജിതയോഃ	ലീലാവതി, 243	V. 4
പ്രതിഭുജദളകൃതി		VII.15
പ്രഥമാദിഫലേഭ്യോ/ഥ	തന്ത്രസംഗ്രഹവ്യാഖ്യാ, II.207	VI.6
ഭാജ്യോ ഹാരഃ ക്ഷേപകഃ	ലീലാവതി, 242	V. 4
മിഥോ ഭജേത്തൌ ദൃഢൌ	ലീലാവതി, 244	V.4
ലബ്ധീനാമവസാനം സ്യാത്	തന്ത്രസംഗ്രഹവ്യാഖ്യാ, II.209	VI.6
ലംബഗുണം ഭൂമ്യർദ്ധം		VII.15
വർഗ്ഗയോഗോ ദ്വയോ രാശ്യോഃ		VII.15
വിഷമാണാം യുതേസ്ത്യക്ത്വാ	തന്ത്രസംഗ്രഹവ്യാഖ്യാ, II.211	VI.6
വൃത്തേ ശരവർഗ്ഗോർദ്ധ	ആര്യഭടീയം, ഗണിതപാദം, 17	VII.19
വ്യസ്തത്രൈരാശികഫലമിച്ഛാഭക്തം	ബ്രഹ്മസ്ഫുടസിദ്ധാന്തം, ഗണിത, II	XII. 3
വ്യാസാച്ഛരോനാച്ഛര	ലീലാവതി, 204	VI. 16
വ്യാസാദ് വാരിധിനിഹതാത്		VI. 9
സമപഞ്ചാഹതയോയാ	തന്ത്രസംഗ്രഹവ്യാഖ്യാ, II.287	VI. 9
സമയുതിഫലമപഹായ		VI. 11
സർവ്വദോർയുതിദളം ചതുഃസ്ഥിതം	ലീലാവതി, 167	VII. 15
സ്വോർദ്ധ്വേ ഹത്യേന്ത്യേന	ലീലാവതി, 245	V. 4

Index